LINEAR MODEL METHODOLOGY

LINEAR MODEL METHODOLOGY

André I. Khuri

CRC Press
Taylor & Francis Group
Boca Raton London New York

CRC Press is an imprint of the
Taylor & Francis Group an **informa** business
A CHAPMAN & HALL BOOK

Chapman & Hall/CRC
Taylor & Francis Group
6000 Broken Sound Parkway NW, Suite 300
Boca Raton, FL 33487-2742

Printed in the United States of America on acid-free paper
10 9 8 7 6 5 4 3 2 1

International Standard Book Number: 978-1-58488-481-1 (Hardback)

Library of Congress Cataloging-in-Publication Data

Khuri, André I., 1940-
Linear model methodology / André I. Khuri.
p. cm.
Includes bibliographical references and index.
ISBN 978-1-58488-481-1 (hardcover : alk. paper)
1. Linear models (Statistics)--Textbooks. I. Title.

QA279.K47 2010
519.5--dc22 2009027269

Visit the Taylor & Francis Web site at
http://www.taylorandfrancis.com

and the CRC Press Web site at
http://www.crcpress.com

To my wife, Ronnie

and

our grandchildren, George Nicholas and Gabriella Nicole Louh

May Peace Triumph on Earth.
It Is Humanity's Only Hope.

Contents

Preface

The purpose of this book is to provide a comprehensive coverage of the basic concepts and methodologies used in the area of *linear models*. Given the importance of this subject in both statistical theory and experimental research, a good understanding of its fundamental principles and theory is essential. Special emphasis has therefore been placed on the clarity of the presentation combined with a rigorous development of the theory underlying linear models. This undertaking is supported by a large number of examples, numerical or otherwise, in order to illustrate the applicability of the various methods presented in this book. Furthermore, all chapters, except for Chapter 1, are equipped with numerous exercises, some of which are designed to give the reader added insight into the subject area discussed in a given chapter. In addition, an extensive bibliography is provided for the benefit of the interested reader who can use it for more in-depth study of linear models and related areas.

This book covers a wide variety of topics in linear models that incorporate both the classical approach as well as the more recent trends and modeling techniques that have been developed in the last 30 years. Coverage of the material is done in a manner that reflects contemporary advances made in linear models. However, it does not include topics on regression analysis, such as model selection, multicollinearity, or regression diagnostics. These topics are discussed in detail in numerous regression textbooks and are better taught in methods courses. The focus of this book is more on the theory of linear models.

This book is intended for graduate students who need to take a course or two in linear models. In fact, a sizable portion of the book evolved from material I used to teach a couple of courses in linear models at the University of Florida in the last 20 years. In this respect, Chapters 1 through 8 can be taught as a one-semester course followed by coverage of Chapters 9 through 13 as a second course on linear models. Chapters 11 and 12 can be particularly helpful to graduate students looking for dissertation topics. This book can also be useful for practicing statisticians and researchers who have an interest in linear models, but did not have sufficient exposure to this area during their educational training. This book is self-contained, but a course in introductory statistics and some knowledge of matrix algebra and calculus would be helpful.

This book contains 13 chapters. Chapter 1 gives some historical perspectives on the evolution of certain methods and techniques used in linear models. Chapter 2 reviews some fundamental concepts concerning vector spaces and linear transformations. This chapter provides the groundwork for Chapter 3, which deals with the basic concepts and results in matrix algebra that are relevant to the study of linear models. The latter chapter is not intended to provide detailed proofs for all the stated theorems and results. Doing so is beyond the scope of this book and can detract from its focus. Instead, Chapter 3 intends to make these theorems accessible to the reader since they are utilized in the development of the methodology in the remaining chapters. The references in the bibliography given at the end of the book can be consulted for more detailed coverage of matrix algebra. It is important here to recognize that matrices have played a central role in the development of the modern approach in linear models. A working knowledge of matrices and their properties is therefore crucial to the understanding of the theory of linear models.

Chapter 4 discusses the multivariate normal distribution and some related distributions. Chapter 5 presents a study of quadratic forms and their distributional properties under the normality assumption. Quadratic forms play an important role in the formulation of analysis of variance. Chapter 6 deals with the analysis of full-rank linear models. These models encompass regression and response surface models whose model matrices have full column ranks. The analysis of linear models that are not of full rank is the subject of Chapter 7. Such models are typically encountered in analysis of variance situations. Chapter 8 develops general rules for the analysis of balanced data. The methodology presented in this chapter provides a systematic approach for setting up a complete analysis of the data that includes hypothesis testing and interval estimation concerning certain unknown parameters of a given linear model.

Chapters 4 through 8 make up the core material in the study of classical linear models. They also include more recent techniques for solving some well-known problems, such as those that pertain to the distribution and independence of quadratic forms in Chapter 5, the analysis of estimable linear functions and contrasts in Chapter 7, and the general treatment of balanced random and mixed-effects models in Chapter 8.

Chapters 9 through 13 cover more contemporary topics in linear models and can therefore be regarded as forming the second part of this book, whereas Chapters 1 through 8 make up the first part. In particular, Chapter 9 addresses the adequacy of Satterthwaite's approximation, a popular and frequently used technique in analysis of variance. Chapter 10 discusses the analysis of unbalanced data for linear models with all fixed effects. Chapter 11 also deals with unbalanced data, but it considers linear models whose effects are either all random (random-effects models), or possibly include a combination of fixed and random effects (mixed-effects models). This chapter discusses estimation of variance components and estimable linear functions of the fixed effects in a given mixed-effects model. It also provides detailed coverage of approximate and exact tests concerning certain random and mixed-effects models.

Chapter 12 discusses several more recent topics in linear models. These include heteroscedastic linear models, response surface models with random effects, and linear multiresponse models. Finally, Chapter 13 presents an introduction to generalized linear models. These models represent an extension of classical linear models and provide a unified approach for the modeling of discrete as well as continuous response data.

I would like to thank all those who reviewed and commented on a preliminary outline of the book manuscript. I am also grateful to my wife, Ronnie, for her support and patience during the five years it has taken me to complete the writing of this book.

André I. Khuri
Gainesville, Florida

Author

André I. Khuri, PhD, is a professor emeritus in the Department of Statistics at the University of Florida in Gainesville. He is the author of *Advanced Calculus with Applications in Statistics* (Wiley); the coauthor (with John Cornell) of *Response Surfaces* (Dekker); the coauthor (with Thomas Mathew and Bimal K. Sinha) of *Statistical Tests for Mixed Linear Models* (Wiley); and the editor of *Response Surface Methodology and Related Topics* (World Scientific).

1

Linear Models: Some Historical Perspectives

Quite often, experimental research work requires the empirical identification of the relationship between an observable response variable, Y, and a set of associated variables, or factors, believed to have an effect on Y. In general, such a relationship, if it exists, is unknown, but is usually assumed to be of a particular form, provided that it can adequately describe the dependence of Y on the associated variables (or factors). This results in the establishment of the so-called *postulated model* which contains a number of unknown parameters, in addition to a random experimental error term. The role of this error term is to account for the extra variation in Y that cannot be explained by the postulated model. In particular, if the unknown parameters appear linearly in such a model, then it is called a *linear model*.

In this book, we consider two types of linear models depending on the nature of the factors that affect the response variable Y. If the factors are quantitative (that is, they can be measured on a continuous scale, such as the temperature and pressure of a certain chemical reaction), then the model is called a *regression model*. For example, we may have a regression model of the form,

$$Y = \beta_0 + \sum_{i=1}^{k} \beta_i x_i + \epsilon, \tag{1.1}$$

where

$x_1, x_2, \ldots, x_k$ are mathematical variables that represent the levels of the associated factors

$\beta_0, \beta_1, \ldots, \beta_k$ are unknown parameters

ϵ is a random experimental error term

It is common to refer to $x_1, x_2, \ldots, x_k$ as *control*, *input*, or *explanatory* variables. A more general expression for a regression model is one of the form

$$Y = \boldsymbol{f}'(\boldsymbol{x})\boldsymbol{\beta} + \epsilon, \tag{1.2}$$

where

$\boldsymbol{x} = (x_1, x_2, \ldots, x_k)'$

$\boldsymbol{f}'(\boldsymbol{x})$ is a known vector function whose elements are powers and cross products of powers of $x_1, x_2, \ldots, x_k$ up to a certain degree

$\boldsymbol{\beta}$ is a vector of unknown parameters

The model in (1.2) is called a *polynomial model* in $x_1, x_2, \ldots, x_k$. For example, a second-degree polynomial model in just x_1, x_2 is of the form

$$Y = \beta_0 + \beta_1 x_1 + \beta_2 x_2 + \beta_{12} x_1 x_2 + \beta_{11} x_1^2 + \beta_{22} x_2^2 + \epsilon.$$

Estimates of the elements of $\boldsymbol{\beta}$ in (1.2) can be obtained by running a series of experiments in which the response Y is measured (or observed) for particular settings of $x_1, x_2, \ldots, x_k$. The resulting values can then be used as input data in an appropriate estimation method.

If the factors affecting the response are qualitative (that is, their levels are not necessarily measurable, but can be described, such as machines and operators in an industrial experiment), then the model is called an *analysis of variance* (ANOVA) model. For example, we may have the models

$$Y_{ij} = \mu + \alpha_i + \epsilon_{ij}, \tag{1.3}$$

$$Y_{ijk} = \mu + \alpha_i + \beta_j + (\alpha\beta)_{ij} + \epsilon_{ijk}, \tag{1.4}$$

$$Y_{ijk} = \mu + \alpha_i + \beta_{ij} + \epsilon_{ijk}, \tag{1.5}$$

which will be described and discussed later on in this book. Model (1.3) is the one-way classification model, the one in (1.4) is the two-way crossed classification with interaction model, and in (1.5) we have the two-fold nested classification model. The parameters that appear in these models represent the various effects that influence the response. For example, in (1.3), α_i represents the effect of level i of a given factor. In the second model in (1.4), α_i and β_j represent the effects of levels i and j, respectively, of two given factors, and $(\alpha\beta)_{ij}$ is their interaction effect. In (1.5), β_{ij} represents the effect of the jth level of a factor which is nested within the ith level of the factor represented by α_i.

Given a set of observations on Y, both regression and ANOVA models can be expressed in matrix form as

$$\boldsymbol{Y} = \boldsymbol{X}\boldsymbol{\beta} + \boldsymbol{\epsilon}, \tag{1.6}$$

where

$\boldsymbol{Y}$ is the vector of observations on Y
$\boldsymbol{X}$ is a known matrix called the model matrix
$\boldsymbol{\beta}$ is the vector of unknown parameters
$\boldsymbol{\epsilon}$ is the vector of random experimental errors

In the case of a regression model, as in (1.2), the rows of $\boldsymbol{X}$ are values of $\boldsymbol{f}'(\boldsymbol{x})$ at various settings of $\boldsymbol{x}$ inside the region of experimentation. If, however, (1.6) represents an ANOVA model, then the elements of $\boldsymbol{X}$ consist of zeros and ones.

Typically, in an ANOVA model, the interest is in estimating means associated with the levels of the factors under consideration, in addition to testing certain hypotheses concerning these means. Such hypotheses are set up for the purpose of assessing the significance of the associated factors. On the other hand, in a regression model, in addition to testing significance of its control variables, estimates of the model's unknown parameters can be obtained for the purpose of estimating the mean response (that is, the mean of Y) as well as predicting "future" response values within a certain region of interest.

Linear models have played an important role in many aspects of statistical experimental research for the past 75 years. Furthermore, the theory of linear models has been instrumental in the development of several areas in statistics, such as regression analysis, analysis of variance, experimental design, response surface methodology, multivariate analysis, time series analysis, and growth curve analysis, to name just a few.

In the remainder of this chapter, we provide some brief history concerning certain key concepts and techniques used in the early development of linear models.

1.1 The Invention of Least Squares

The origin of linear models can be traced back to the early nineteenth century. Undoubtedly, the tool that has made it possible to develop the theory of linear models is the *method of least squares*. This method, which evolved shortly after 1800, is used to estimate the unknown parameters in a given linear model. It was initially developed in response to the needs of scientists in the fields of astronomy and geodesy. From the historical point of view, there has been some dispute as to who was the first to introduce this method (see Stigler, 1981; 1986, Chapter 1). The method was first published in 1805 by Adrien Marie Legendre (1752–1833) as an appendix entitled "Sur la méthode des moindres quarrés" (on the method of least squares), which appeared in Legendre's book, *Nouvelles Méthodes Pour la Détermination des Orbites des Comètes* (*New Methods for the Determination of the Orbits of the Comets*). Four years later, Carl Friedrich Gauss (1777–1855) published the method in 1809 in Volume 2 of his work (written in Latin) on celestial mechanics entitled *Theoria Motus Corporum Coelestium in Sectionibus Conicis Solem Ambientium* (*The Theory of the Motion of Heavenly Bodies Moving Around the Sun in Conic Sections*). Gauss, however, claimed that he had been using the method since 1795. His claim was the source of the aforementioned controversy. Plackett (1972) presented an account of the circumstances in which the discovery of the method took place and the course of the ensuing controversy. He also included interesting translations of letters exchanged between Legendre and Gauss and between Gauss and other mathematicians of his time. Stigler (1981) stated that "It is argued (though not conclusively) that Gauss probably possessed the method well before Legendre, but that he was unsuccessful in communicating it to his contemporaries." It should be mentioned here, however, that Gauss went far beyond Legendre in linking the method to probability and providing algorithms for the computation of estimates (see Stigler, 1981, p. 472). In fact, the first discussion of the model in (1.1) in which the probability distribution of the error term was explicitly considered was in Gauss's 1809 work (see Seal, 1967, Section 1.3).

1.2 The Gauss–Markov Theorem

The *Gauss–Markov theorem* is an important theorem associated with least-squares estimation. It represents a turning point in the early development of the theory of linear models. The theorem was first proved by Gauss during the period 1821–1823. It states that among all the unbiased estimates of a linear function of the parameters which are expressible as linear combinations of the observations (elements of the response vector $\boldsymbol{Y}$), the one produced by the least-squares procedure has minimum variance. Such an estimate became known as the *best linear unbiased estimate* (BLUE). Gauss's result has therefore provided a strong impetus to the use of least squares as a method of parameter estimation due to this optimal property.

Another version of Gauss's proof was given by Andrey Markov (1856–1922) in 1912. His proof was described by Neyman (1934) as being "elegant." Neyman believed that Markov's contribution, which was written in Russian, had been overlooked in the West. As a compromise, the name *Gauss–Markov theorem* was adopted.

It should be noted that Gauss's proof assumed a linear model with uncorrelated errors having zero means and equal variances. An extension of this proof to the case of correlated errors with a known variance–covariance matrix was given by Aitken (1935). It is interesting here to remark that Aitken gave the first formulation of the theorem in terms of matrices.

1.3 Estimability

Estimability is an important property, particularly for ANOVA models where the matrix $\boldsymbol{X}$ in (1.6) is not of full column rank (see the treatment of such models in Chapter 7). In this case, the least-squares equations (or normal equations) do not yield a unique solution for estimating the parameter vector $\boldsymbol{\beta}$. However, for some particular linear functions of $\boldsymbol{\beta}$, namely $\boldsymbol{\lambda}'\boldsymbol{\beta}$, where $\boldsymbol{\lambda}'$ is a linear combination of the rows of $\boldsymbol{X}$, the corresponding estimate, $\boldsymbol{\lambda}'\hat{\boldsymbol{\beta}}$, is unique. Such linear functions are said to be estimable. Here, the elements of $\hat{\boldsymbol{\beta}}$ are obtained by using any solution to the normal equations. Thus, even though $\hat{\boldsymbol{\beta}}$ is not unique, the value of $\boldsymbol{\lambda}'\hat{\boldsymbol{\beta}}$ is unique as it remains invariant to the choice of $\hat{\boldsymbol{\beta}}$. Furthermore, $\boldsymbol{\lambda}'\hat{\boldsymbol{\beta}}$ is the BLUE of $\boldsymbol{\lambda}'\boldsymbol{\beta}$ (see Theorem 7.6 which is an extension of the Gauss–Markov theorem to less-than-full-rank models). By contrast, for regression models where the matrix $\boldsymbol{X}$ is of full column rank, $\hat{\boldsymbol{\beta}}$ is unique and is the BLUE of $\boldsymbol{\beta}$. We conclude that whenever $\boldsymbol{\lambda}'\boldsymbol{\beta}$ is estimable (in the case of an ANOVA model), the properties of $\boldsymbol{\lambda}'\hat{\boldsymbol{\beta}}$ are the same as those under a regression model.

The notion of estimability in linear models was first introduced by Bose (1944). It has since become an important consideration in the analysis of ANOVA models. Seely (1977) pointed out that in some linear model textbooks, little justification is given to the requirement that the elements of $\boldsymbol{A\beta}$ be estimable when testing a null hypothesis concerning $\boldsymbol{A\beta}$, where $\boldsymbol{A}$ is a known matrix of full row rank. Such justification, however, is given in the book by Searle (1971) where it is shown that unless the elements of $\boldsymbol{A\beta}$ are estimable, the numerator sum of squares in the associated F-ratio is not well defined (see Section 7.4.2 regarding testing a hypothesis concerning $\boldsymbol{A\beta}$ when the elements of $\boldsymbol{A\beta}$ are estimable).

1.4 Maximum Likelihood Estimation

Suppose that we have a random variable, Y, whose distribution depends on some unknown parameters denoted by $\theta_1, \theta_2, \ldots, \theta_p$. Let $g(y, \boldsymbol{\theta})$ denote the density function, or probability mass function, of Y depending on whether Y is a continuous or a discrete random variable, respectively, where $\boldsymbol{\theta} = (\theta_1, \theta_2, \ldots, \theta_p)'$ and y is a value of Y. Let us also suppose that we have a sample of n independent observations on Y denoted by $Y_1, Y_2, \ldots, Y_n$. Then, the density function (or probability mass function) of $\boldsymbol{Y} = (Y_1, Y_2, \ldots, Y_n)'$ is given by

$$h(\boldsymbol{y}, \boldsymbol{\theta}) = \prod_{i=1}^{n} g(y_i, \boldsymbol{\theta}), \tag{1.7}$$

where

$\boldsymbol{y} = (y_1, y_2, \ldots, y_n)'$
y_i is a given value of Y_i $(i = 1, 2, \ldots, n)$

By definition, the *likelihood function*, $\mathcal{L}(\boldsymbol{\theta}, \boldsymbol{Y})$, for the sample, $Y_1, Y_2, \ldots, Y_n$, is a function of $\boldsymbol{\theta}$, which for a given value, $\boldsymbol{y}$, of $\boldsymbol{Y}$ is equal to the density function in (1.7), that is,

$$\mathcal{L}(\boldsymbol{\theta}, \boldsymbol{y}) = h(\boldsymbol{y}, \boldsymbol{\theta}). \tag{1.8}$$

Note that in (1.8) the likelihood function is obtained by reversing the roles of $\boldsymbol{y}$ and $\boldsymbol{\theta}$ so that $\mathcal{L}(\boldsymbol{\theta}, \boldsymbol{y})$ is viewed as a function of $\boldsymbol{\theta}$ for the given value, $\boldsymbol{y}$, of $\boldsymbol{Y}$.

The *method of maximum likelihood* estimates θ by finding the value, $\hat{\theta}$, of θ that maximizes $\mathcal{L}(\theta, y)$ over a certain parameter space of θ for each y in some set S. The resulting value is called the *maximum likelihood estimate* of θ. Note that in this method, we try to find the values of the parameters that would have most likely produced the data obtained from the observed sample. It should also be noted that $\hat{\theta}$ may not be unique since the likelihood function can possibly attain its maximum value at several locations inside the parameter space.

The method of maximum likelihood is generally attributed to R. A. Fisher (1890–1962) who propounded it as a means of parameter estimation in his two renowned papers, Fisher (1922, 1925). Even though other methods similar to this method have been in existence prior to the work of Fisher, the definition of likelihood itself appears to be entirely his own (see Edwards, 1974). The introduction of this method by Fisher has led to the establishment of a whole new branch of statistical theory. Aldrich (1997) stated that "the making of maximum likelihood was one of the most important developments in 20th century statistics." The method has since become a very important tool in the repertoire of linear models as well as generalized linear models (see Chapter 13). Interesting accounts concerning the history of this method and the work of R. A. Fisher can be found in Norden (1972), Edwards (1974), and Aldrich (1997), among others.

1.5 Analysis of Variance

Analysis of variance (ANOVA) is a statistical technique developed by R. A. Fisher in the 1920s in order to facilitate the analysis and interpretation of data from field trials and laboratory experiments. Fisher's (1918) paper on population genetics introduced the terms "variance" and "analysis of variance." However, it was after the publication of Fisher's (1925) book that ANOVA became widely used as an important tool in experimental research. The ANOVA table devised by Fisher provides a convenient tabulation of sums of squares that measure the amounts of variation associated with the various effects in a given model. Using ratios of mean squares (sums of squares divided by their corresponding degrees of freedom), it is possible to derive test statistics for certain hypotheses concerning the effects under consideration. These statistics have, under certain assumptions which include normality, F-distributions (the symbol F was introduced in honor of R. A. Fisher by G. W. Snedecor (1934)).

The initial development of ANOVA was designed for what are now called fixed-effects models (or just fixed models). By definition, a *fixed-effects model* is one in which all of its effects, except for the error term, are represented by fixed unknown parameters. The F-statistics that can be derived from the corresponding ANOVA table test the hypotheses that the effects of the levels of each factor are all equal. A fixed-effects model is also called *Model I*. On the other hand, if all the effects in the model, except for the general (or grand) mean, are random, then the model is called a *random-effects model* (or just random model), a *variance components model*, or *Model II*. The terminology of "Model I" and "Model II" is due to Eisenhart (1947), and the use of "variance components" refers to the variances of the random effects in the model. We recall that Fisher (1918) introduced the term "variance" in the literature and he implicitly employed variance components models. His (1918) paper was a milestone to variance components theory. The third model type is the so-called *mixed-effects model* (or just mixed model) which contains random effects (besides the error term) as well as fixed effects (besides the general mean). This model is also referred to as *Model III*. Whenever a model contains a random effect, the interest is in estimating its variance component in addition to testing its significance. The determination of which effects are to be considered as fixed and which as random depends on the nature of the levels of the associated factors. If the levels of a factor are of particular interest, as in the consideration of particular varieties of corn in an agricultural experiment, then the levels are said to have *fixed effects*. However, if the levels are selected at random from a large population, as in the random sampling of machine parts from a large warehouse in an industrial experiment, then the corresponding factor is said to have a *random effect*.

Variance components models were used by astronomers, long before they were known to statisticians. These models can be traced back to the works of the astronomers Airy (1861) and Chauvenet (1863) or even earlier. For example, Airy made an explicit use of a variance components model for the one-way model. Fisher's (1925) book further advanced variance components theory by initiating what has come to be known as the *ANOVA method of estimation*. In this method, the so-called *ANOVA estimates* of variance components in a given random or mixed model are obtained by equating the mean squares of the random effects in the corresponding ANOVA table to their expected values. The resulting equations, which are linear, are then solved for the variance components to be estimated. Maximum likelihood estimation, which was developed by R. A. Fisher, as was previously mentioned, can also be used to estimate variance components. This was first attempted by Crump (1947, 1951). Other methods of estimation of variance components were later developed by other authors (see the survey article by Khuri and Sahai, 1985; also Robinson, 1987; Searle, 1995).

1.5.1 Balanced and Unbalanced Data

One important consideration before undertaking any analysis of a given data set is whether the data are balanced or unbalanced. A data set is said to be *balanced* if the numbers of observations in the subclasses of the data are all equal. When these numbers are not equal, including perhaps situations where no observations exist in some subclasses, the data set is said to be *unbalanced*. While the analysis of balanced data is quite developed with well-defined methods for estimation and hypothesis testing for the associated models, the same cannot be said about unbalanced data (see Chapters 8, 10, and 11). One of the difficulties with unbalanced data is the lack of a unique ANOVA table, as is the case with balanced data. This makes it difficult to come up with a unified approach for the analysis of unbalanced data. Furthermore, unlike the case of balanced data, there are no exact tests that can be applied to random or mixed models, except in a small number of special cases (see Chapter 11).

Some of the early papers dealing with the analysis of unbalanced data are those by Yates (1934), who developed the methods of *unweighted* and *weighted squares of means* for data with no empty cells, and Wald (1940, 1941). Henderson's (1953) landmark paper presented three methods for estimating variance components, the last of which became the standard procedure for obtaining ANOVA estimates of the variance components for both random and mixed models. A coverage of the history of unbalanced data analysis from Yates's (1934) paper to the beginning of the computational revolution in the 1960s was given by Herr (1986).

1.6 Quadratic Forms and Craig's Theorem

Let $\boldsymbol{A}$ be a known symmetric matrix of order $n \times n$, and let $\boldsymbol{Y}$ be a random vector of n elements. By definition, $\boldsymbol{Y}'\boldsymbol{A}\boldsymbol{Y}$ is called a *quadratic form* in $\boldsymbol{Y}$ with respect to the matrix $\boldsymbol{A}$. Quadratic forms play an important role in analysis of variance. More specifically, if $\boldsymbol{Y}$ is a data vector associated with a given linear model, then every sum of squares in the corresponding ANOVA table can be expressed as a quadratic form in $\boldsymbol{Y}$. In addition, if $\boldsymbol{Y}$ has the multivariate normal distribution with a mean $\boldsymbol{\mu}$ and a nonsingular variance–covariance matrix, $\boldsymbol{\Sigma}$, that is, $\boldsymbol{Y} \sim N(\boldsymbol{\mu}, \boldsymbol{\Sigma})$, then it would be of interest to know the distributions of the sums of squares (or quadratic forms) in the ANOVA table. This is important since these sums of squares can be used to develop test statistics concerning the various effects in the associated linear model.

There are two important theorems that pertain to quadratic forms in normally distributed random vectors. The first theorem concerns the distribution of $\boldsymbol{Y}'\boldsymbol{A}\boldsymbol{Y}$ where $\boldsymbol{Y} \sim N(\boldsymbol{\mu}, \boldsymbol{\Sigma})$. This theorem states that a necessary and sufficient condition for $\boldsymbol{Y}'\boldsymbol{A}\boldsymbol{Y}$ to have a chi-squared distribution with r degrees of freedom is that $\boldsymbol{A}\boldsymbol{\Sigma}$ should be idempotent of rank r (see Section 3.9 for a definition of an idempotent matrix). The second theorem concerns the stochastic independence of two quadratic forms in $\boldsymbol{Y}$, $\boldsymbol{Y}'\boldsymbol{A}\boldsymbol{Y}$ and $\boldsymbol{Y}'\boldsymbol{B}\boldsymbol{Y}$, where $\boldsymbol{A}$ and $\boldsymbol{B}$ are symmetric matrices and $\boldsymbol{Y} \sim N(\boldsymbol{\mu}, \boldsymbol{\Sigma})$. A necessary and sufficient condition for these quadratic forms to be stochastically independent is that $\boldsymbol{A}\boldsymbol{\Sigma}\boldsymbol{B} = \boldsymbol{0}$. The proofs of these two theorems are given in Chapter 5.

The independence theorem was first considered by Craig (1943) and is therefore referred to as *Craig's theorem*. Craig, however, assumed that $\boldsymbol{Y}$ has the standard multivariate normal distribution, $N(\boldsymbol{0}, \boldsymbol{I}_n)$. Several authors have since worked on extending Craig's result to the general case where $\boldsymbol{Y} \sim N(\boldsymbol{\mu}, \boldsymbol{\Sigma})$. While the sufficiency part of this theorem is easy to prove (that is, showing that $\boldsymbol{A\Sigma B} = \boldsymbol{0}$ implies independence of the quadratic forms), the necessity part (showing that independence implies $\boldsymbol{A\Sigma B} = \boldsymbol{0}$) in the noncentral case (that is, $\boldsymbol{\mu} \neq \boldsymbol{0}$) turned out to be quite difficult to prove. Driscoll and Gundberg (1986) gave a history of the development of Craig's theorem in the noncentral case. They pointed out that authors of earlier papers and textbooks had given "incorrect or incomplete coverage of Craig's theorem and its proof." They did indicate, however, that a correct proof was finally given by Laha (1956). Laha's approach was based on a difficult mathematical result that he did not actually prove. Reid and Driscoll (1988) discovered that Ogawa (1950) had apparently been the first person to give a correct and complete proof of Craig's theorem for the noncentral case. They then proceeded to give "an accessible" proof of Craig's theorem (in the noncentral case) which only required the use of linear algebra and calculus.

The theorem concerning the distribution of $\boldsymbol{Y}'\boldsymbol{AY}$ also has a similar history. As before, the sufficiency part of this theorem (showing that idempotency of $\boldsymbol{A\Sigma}$ implies chi-squaredness of $\boldsymbol{Y}'\boldsymbol{AY}$) is easy to prove. However, the necessity part (showing that chi-squaredness of $\boldsymbol{Y}'\boldsymbol{AY}$ implies idempotency of $\boldsymbol{A\Sigma}$) is not easy to prove, particularly when $\boldsymbol{\mu} \neq \boldsymbol{0}$ (the noncentral case). Proofs given in the statistical literature of the latter part either assume that $\boldsymbol{\mu} = \boldsymbol{0}$, which is a relatively easy case, or invoke the same result by Laha (1956) that was mentioned earlier. An alternative proof of the necessity part was given more recently in Khuri (1999) using only simple calculus tools and basic matrix results.

A more general theorem dealing with quadratic forms in normally distributed random vectors is the one due to Cochran (1934). This theorem, which is discussed in Chapter 5 (see Theorem 5.7), is a remarkable one since it has useful applications in analysis of variance, particularly in the case of fixed-effects models as will be shown in Chapter 5.

1.7 The Role of Matrix Algebra

Matrices were introduced in mathematics around the middle of the nineteenth century. Their use in statistics, however, did not begin until the 1930s with the publication of Turnbull and Aitken's (1932) book. Matrices have since become a very important tool in many areas of statistics, particularly in linear models and multivariate analysis. In fact, it can be easily said that the modern development of the theory of linear models is due in large part to matrix algebra. The use of matrices has made the derivation of vital results in linear models much easier, faster, and more direct than what was practiced earlier. Matrix algebra replaced the tedious use of the summation notation that was commonplace in statistics before 1930.

Craig (1943) used determinants in his proof of the independence theorem of quadratic forms. Cramér (1946) had a whole chapter on matrices and determinants. Books by Kempthorne (1952) and Rao (1952) seem to be the first to have made considerable use of matrix algebra. C. R. Rao is credited with having introduced into statistics the concept of generalized inverses of singular matrices in the 1960s (see Rao, 1962, 1966) using the work of Moore (1920) and Penrose (1955) (see the definition of a generalized inverse of a matrix in Section 3.7.1). Generalized inverses became very useful in finding a solution to the normal equations in the case of a linear model whose model matrix is not of full column rank. They were also instrumental in understanding estimability of linear functions of the model's unknown parameters. This has led to the development of a unified theory of least squares (see Rao, 1973b).

Without any doubt, the availability of modern computers and computer software has made the actual execution of matrix computations in linear models and statistics, in general, a very simple task. Matrix operations, that nowadays take only few seconds to execute, used to take hours if not days before the advent of modern computers. Searle (1999) reported that "During graduate student days in a small computing group at Cornell, there was great excitement when in 1959 we inverted a 10-by-10 matrix in seven minutes. After all, only a year or two earlier, a friend had inverted a 40-by-40 matrix, by hand, using electric (Marchant or Monroe) calculators. That took six weeks!"

1.8 The Geometric Approach

Another approach to linear models, other than the well-known algebraic one, is the *geometric* or *coordinate-free approach*. This approach, although interesting, has not attracted a great deal of attention. Actually, there are certain concepts in linear models, such as least squares, that naturally elicit the use of the geometric approach.

R. A. Fisher used to think geometrically from time to time (see Fisher, 1915). W. H. Kruskal (1919–2005) was one of the early advocates of the coordinate-free approach to linear models. His (1961) paper described the geometric version of the Gauss–Markov theorem. Kruskal (1968) addressed the question of equality of ordinary least squares and best linear unbiased estimates using a coordinate-free approach. In Kruskal (1975), an analytic geometric approach was used in dealing with generalized inverses of matrices. Herr (1980) reviewed several papers that used the geometric approach to linear models, starting with Fisher's (1915) paper and ending with Kruskal's (1975) paper. L. Fisher (1973) presented a proof of Cochran's theorem that emphasized the geometric approach. Schey (1985) used geometrical arguments to explain certain features of orthogonal contrasts in the context of the one-way ANOVA model (see Section 7.7). Other papers that considered the geometric approach to linear models include Eaton (1970) and Haberman (1975). More recently, Wichura (2006) devoted an entire book on the geometric approach to the theory of linear models. His book discussed optimal properties of various methods of estimating and testing hypotheses concerning the unknown parameters in linear models.

2

Basic Elements of Linear Algebra

2.1 Introduction

This chapter provides the groundwork for Chapter 3, which gives a basic introduction to matrix algebra. Both chapters are vital for the understanding of the theory of linear models.

The present chapter reviews some fundamental concepts concerning vector spaces and subspaces, linear dependence and independence of elements in a vector space, direct sums, bases, and dimensions of vector spaces, in addition to linear transformations. The main purpose of this chapter is to familiarize the reader with these concepts, but without delving deeply into the theory of linear algebra. The references at the end of the book can be consulted for a more detailed study of the subject area.

In this chapter, as well as in the remainder of the book, the set of all real numbers is denoted by R, and its elements are referred to as scalars. The set of all n-tuples of real numbers will be denoted by R^n $(n \geq 1)$.

2.2 Vector Spaces

A *vector space* over R is a set V of elements, which can be added or multiplied by scalars, in such a way that the sum of two elements of V is an element of V, and the product of an element of V by a scalar is an element of V. More specifically, the following properties must be satisfied:

(1) For every $\boldsymbol{u}, \boldsymbol{v}$ in V, $\boldsymbol{u} + \boldsymbol{v}$ is a uniquely defined element of V.

(2) For every $\boldsymbol{u}$ in V and any scalar α, $\alpha\boldsymbol{u}$ is an element in V.

(3) $\boldsymbol{u} + \boldsymbol{v} = \boldsymbol{v} + \boldsymbol{u}$ for all $\boldsymbol{u}, \boldsymbol{v}$ in V.

(4) $\boldsymbol{u} + (\boldsymbol{v} + \boldsymbol{w}) = (\boldsymbol{u} + \boldsymbol{v}) + \boldsymbol{w}$ for all $\boldsymbol{u}, \boldsymbol{v}, \boldsymbol{w}$ in V.

(5) There exists an element in V, called the *zero element* and is denoted by $\boldsymbol{0}$, such that $\boldsymbol{0} + \boldsymbol{u} = \boldsymbol{u} + \boldsymbol{0} = \boldsymbol{u}$ for every $\boldsymbol{u}$ in V.

(6) For each $\boldsymbol{u}$ in V, there exists a unique element $-\boldsymbol{u}$ in V such that $\boldsymbol{u} + (-\boldsymbol{u}) = (-\boldsymbol{u}) + \boldsymbol{u} = \boldsymbol{0}$.

(7) For every $\boldsymbol{u}$ and $\boldsymbol{v}$ in V and any scalar α, $\alpha(\boldsymbol{u} + \boldsymbol{v}) = \alpha\boldsymbol{u} + \alpha\boldsymbol{v}$.

(8) $(\alpha + \beta)\boldsymbol{u} = \alpha\boldsymbol{u} + \beta\boldsymbol{u}$ for any scalars α and β and any $\boldsymbol{u}$ in V.

(9) $\alpha(\beta\boldsymbol{u}) = (\alpha\beta)\boldsymbol{u}$ for any scalars α and β and any $\boldsymbol{u}$ in V.

(10) For every $\boldsymbol{u}$ in V, $1\boldsymbol{u} = \boldsymbol{u}$, where 1 is the number one, and $0\boldsymbol{u} = \boldsymbol{0}$, where 0 is the number zero.

Example 2.1 Let $V = R^n$ be the set of all n-tuples of elements in R, $n \geq 1$. Let $(u_1, u_2, \ldots, u_n)$ and $(v_1, v_2, \ldots, v_n)$ be two elements in V. Their sum is defined as $(u_1 + v_1, u_2 + v_2, \ldots, u_n + v_n)$. Also, for any scalar α, $\alpha(u_1, u_2, \ldots, u_n)$ is defined as $(\alpha u_1, \alpha u_2, \ldots, \alpha u_n)$. It is easy to verify that properties (1) through (10) are satisfied. The zero element is the n-tuple $(0, 0, \ldots, 0)$.

Example 2.2 Let $V = P(x)$ be the set of all polynomials in x over R. Any two polynomials can be added to produce a third polynomial, and the product of a polynomial with a scalar is also a polynomial. Thus V is a vector space.

Example 2.3 In $P(x)$ of Example 2.2, let $P_k(x)$ be the set of all polynomials of degree k or less. Then $P_k(x)$ is a vector space, and any element in $P_k(x)$ can be written as $\sum_{n=0}^{k} a_n x^n$, where $a_0, a_1, \ldots, a_k$ are scalars.

Example 2.4 The set of all positive numbers is not a vector space, since multiplying any positive number by a negative scalar produces a negative number.

2.3 Vector Subspaces

Let V be a vector space over R, and let W be a subset of V. Then W is said to be a *vector subspace* of V if it satisfies the following conditions:

(1) If $\boldsymbol{u}$ and $\boldsymbol{v}$ are any two elements in W, then their sum $\boldsymbol{u} + \boldsymbol{v}$ is an element of W.

(2) If $\boldsymbol{u}$ is any element in W, and if α is any scalar, then $\alpha\boldsymbol{u}$ is an element in W.

(3) The zero element, $\boldsymbol{0}$, of V is also an element of W.

It follows that in order for W to be a vector subspace of V, it must itself be a vector space, that is, properties (1) through (10) in Section 2.2, which are satisfied for all elements of V, must also be satisfied for all elements of W.

Example 2.5 The vector space defined in Example 2.3 is a subspace of the vector space defined in Example 2.2.

Example 2.6 The set V of all functions defined on the closed interval $[-1,1]$ is a vector space. The set W of all continuous functions defined on $[-1,1]$ is a subspace of V.

Example 2.7 The set W of all pairs (x_1, x_2) in R^2 such that $x_2 - ax_1 = 0$, where a is a scalar, is a vector subspace of R^2. However, if this equation is replaced with $x_2 - ax_1 = b$, where $b \neq 0$, then W is no longer a vector subspace since the zero element (0, 0) in R^2 does not belong to W.

Example 2.8 Let W be the set of all elements $(u_1, u_2, \ldots, u_n)$ in R^n such that $u_n = 0$. Then W is a vector subspace of R^n.

It should be noted that if W_1 and W_2 are subspaces of V, then their intersection $W_1 \cap W_2$ is clearly a subspace of V. However, the union $W_1 \cup W_2$ of W_1 and W_2 is not necessarily a subspace of V. For example, if W_i is the set of all ordered pairs (x_1, x_2) in R^2 such that $x_2 - a_i x_1 = 0$, where a_i is a scalar $(i = 1, 2)$, then W_1 and W_2 are vector subspaces of R^2 (see Example 2.7). However, $W_1 \cup W_2$ is not a vector subspace of R^2 since for the pairs $(1, a_1)$ and $(1, a_2)$, which belong to W_1 and W_2, respectively, and hence belong to $W_1 \cup W_2$, the sum $(1, a_1) + (1, a_2) = (2, a_1 + a_2)$ belongs to neither W_1 nor W_2, and therefore does not belong to $W_1 \cup W_2$, if $a_1 \neq a_2$.

Definition 2.1 Let $W_1, W_2, \ldots, W_n$ be vector subspaces of the vector space V. The *direct sum* of $W_1, W_2, \ldots, W_n$, denoted by $\oplus_{i=1}^{n} W_i$, consists of all elements $\boldsymbol{u}$ in V that can be uniquely expressed as $\boldsymbol{u} = \sum_{i=1}^{n} \boldsymbol{u}_i$, where $\boldsymbol{u}_i \in W_i$, $i = 1, 2, \ldots, n$.

From this definition, it follows that two elements, $\boldsymbol{u}_1 = \sum_{i=1}^{n} \boldsymbol{u}_{1i}$ and $\boldsymbol{u}_2 = \sum_{i=1}^{n} \boldsymbol{u}_{2i}$, in $\oplus_{i=1}^{n} W_i$ are equal if and only if for each i, $\boldsymbol{u}_{1i} = \boldsymbol{u}_{2i}$ $(i = 1, 2, \ldots, n)$. The addition of two such elements is defined to be $\boldsymbol{u}_1 + \boldsymbol{u}_2 = \sum_{i=1}^{n} (\boldsymbol{u}_{1i} + \boldsymbol{u}_{2i})$. Furthermore, if α is a scalar and $\boldsymbol{u} = \sum_{i=1}^{n} \boldsymbol{u}_i$ is an element in $\oplus_{i=1}^{n} W_i$, then $\alpha \boldsymbol{u}$ is defined as $\sum_{i=1}^{n} (\alpha \boldsymbol{u}_i)$.

It is easy to verify that if $W_1, W_2, \ldots, W_n$ are vector subspaces of V, then $\oplus_{i=1}^{n} W_i$ is also a vector subspace of V. In addition, $\cap_{i=1}^{n} W_i$ consists of just the zero element $\boldsymbol{0}$ of V. To prove this last assertion, let $\boldsymbol{u}$ be an element of $\cap_{i=1}^{n} W_i$. Then $\boldsymbol{u}$ belongs to W_i for all i $(i = 1, 2, \ldots, n)$. It follows that sums of the form $\sum_{i=1}^{n} \boldsymbol{u}_i$, where only one $\boldsymbol{u}_i$ is equal to $\boldsymbol{u}$ and the remaining ones are equal to the zero element in the corresponding vector subspaces $(i = 1, 2, \ldots, n)$, must belong to $\oplus_{i=1}^{n} W_i$. But, all such sums are equal to $\boldsymbol{u}$, and $\boldsymbol{u}_i = \boldsymbol{0}$ for all i since the representation of $\boldsymbol{u}$ as $\sum_{i=1}^{n} \boldsymbol{u}_i$ is unique. Consequently, $\boldsymbol{u} = \boldsymbol{0}$.

2.4 Bases and Dimensions of Vector Spaces

The elements of a vector space V can be represented as linear combinations of a set of elements of V that form a basis of V. To understand what a basis is, let us consider the following definitions:

Definition 2.2 Let V be a vector space over R, and let $\boldsymbol{u}_1, \boldsymbol{u}_2, \ldots, \boldsymbol{u}_n$ be elements of V. Then, $\boldsymbol{u}_1, \boldsymbol{u}_2, \ldots, \boldsymbol{u}_n$ are *linearly dependent* if there exist scalars, $\alpha_1, \alpha_2, \ldots, \alpha_n$, not all equal to zero, such that $\sum_{i=1}^{n} \alpha_i \boldsymbol{u}_i = \boldsymbol{0}$. If, however, $\sum_{i=1}^{n} \alpha_i \boldsymbol{u}_i = \boldsymbol{0}$ is true only when all the α_i's are zero, then $\boldsymbol{u}_1, \boldsymbol{u}_2, \ldots, \boldsymbol{u}_n$ are said to be *linearly independent*.

It should be noted that if $\boldsymbol{u}_1, \boldsymbol{u}_2, \ldots, \boldsymbol{u}_n$ are linearly independent, then none of them can be zero. To see this, suppose, for example, that $\boldsymbol{u}_1 = \boldsymbol{0}$. Then $\alpha_1 \boldsymbol{u}_1 + 0\boldsymbol{u}_2 + \cdots + 0\boldsymbol{u}_n = \boldsymbol{0}$ for any $\alpha_1 \neq 0$, which implies that $\boldsymbol{u}_1, \boldsymbol{u}_2, \ldots, \boldsymbol{u}_n$ are linearly dependent, a contradiction. From this we can conclude that any set of elements of V that contains the zero element $\boldsymbol{0}$ must be linearly dependent. Also, if $\boldsymbol{u}_1, \boldsymbol{u}_2, \ldots, \boldsymbol{u}_n$ are linearly dependent, then at least one of them can be expressed as a linear combination of the remaining elements. This easily follows from the fact that in $\sum_{i=1}^{n} \alpha_i \boldsymbol{u}_i = \boldsymbol{0}$, at least one α_i is not equal to zero when $\boldsymbol{u}_1, \boldsymbol{u}_2, \ldots, \boldsymbol{u}_n$ are linearly dependent.

Definition 2.3 Let $\boldsymbol{u}_1, \boldsymbol{u}_2, \ldots, \boldsymbol{u}_n$ be n elements in a vector space V. The collection of all linear combinations of the form $\sum_{i=1}^{n} \alpha_i \boldsymbol{u}_i$, where $\alpha_1, \alpha_2, \ldots, \alpha_n$ are scalars, is called the *linear span* of $\boldsymbol{u}_1, \boldsymbol{u}_2, \ldots, \boldsymbol{u}_n$ and is denoted by $L(\boldsymbol{u}_1, \boldsymbol{u}_2, \ldots, \boldsymbol{u}_n)$.

It is easy to see that $L(\boldsymbol{u}_1, \boldsymbol{u}_2, \ldots, \boldsymbol{u}_n)$ is a vector subspace of V. Such a subspace is said to be *spanned* (that is, generated) by $\boldsymbol{u}_1, \boldsymbol{u}_2, \ldots, \boldsymbol{u}_n$. For example, if $\boldsymbol{u}_1 = (a_1, b_1)$ and $\boldsymbol{u}_2 = (a_2, b_2)$ are elements in the vector space $V = R^2$, then the set of elements

$$\alpha_1 \boldsymbol{u}_1 + \alpha_2 \boldsymbol{u}_2 = (\alpha_1 a_1 + \alpha_2 a_2, \alpha_1 b_1 + \alpha_2 b_2)$$

forms a linear span of $\boldsymbol{u}_1, \boldsymbol{u}_2$.

Definition 2.4 Let V be a vector space. If there exist linearly independent elements $\boldsymbol{u}_1, \boldsymbol{u}_2, \ldots, \boldsymbol{u}_n$ in V such that $V = L(\boldsymbol{u}_1, \boldsymbol{u}_2, \ldots, \boldsymbol{u}_n)$, then $\boldsymbol{u}_1, \boldsymbol{u}_2, \ldots, \boldsymbol{u}_n$ form a *basis* of V. The number, n, of elements in this basis is called the *dimension* of V, and is denoted by $dim(V)$. In case V consists of just the zero element, its dimension is set to zero.

For example, consider the functions x, e^x, which are defined on the interval [0,1]. These functions are linearly independent. To show this, suppose that there exist scalars, α_1 and α_2, such that

$$\alpha_1 x + \alpha_2 e^x = 0$$

for all x in [0,1]. Differentiating the two sides of this relation with respect to x gives

$$\alpha_1 + \alpha_2 e^x = 0.$$

By subtracting the first relation from the second, we obtain $\alpha_1(1 - x) = 0$ for all x in [0,1], and hence $\alpha_1 = 0$. Using the first relation, it follows that $\alpha_2 = 0$. Hence, x and e^x are linearly independent. If $W = L(x, e^x)$ is the linear span of x and e^x, then x and e^x form a basis of W, which is of dimension 2.

It should be noted that a basis of a vector space is not unique. However, its dimension, $dim(V)$, is unique. Furthermore, if $\boldsymbol{u}_1, \boldsymbol{u}_2, \ldots, \boldsymbol{u}_n$ form a basis of V, and if $\boldsymbol{u}$ is a given element in V, then there exists a unique set of scalars, $\alpha_1, \alpha_2, \ldots, \alpha_n$, such that $\boldsymbol{u} = \sum_{i=1}^n \alpha_i \boldsymbol{u}_i$. To show this, suppose that there exists another set of scalars, $\beta_1, \beta_2, \ldots, \beta_n$, such that $\boldsymbol{u} = \sum_{i=1}^n \beta_i \boldsymbol{u}_i$. Then $\sum_{i=1}^n (\alpha_i - \beta_i)\boldsymbol{u}_i = \boldsymbol{0}$, which implies that $\alpha_i = \beta_i$ for all i since the $\boldsymbol{u}_i$'s are linearly independent.

Theorem 2.1 Let $W_1, W_2, \ldots, W_n$ be vector subspaces of the vector space V. Then

$$dim\left[\oplus_{i=1}^n W_i\right] = \sum_{i=1}^n dim(W_i).$$

Proof. The proof is left to the reader as an exercise.

2.5 Linear Transformations

Let U and V be two vector spaces over R. Suppose that T is a function that maps U into V. This fact is written symbolically as $T : U \to V$. The function T is said to be a *linear transformation* (or a *linear map*) on U into V if

$$T(\alpha_1 \boldsymbol{u}_1 + \alpha_2 \boldsymbol{u}_2) = \alpha_1 T(\boldsymbol{u}_1) + \alpha_2 T(\boldsymbol{u}_2)$$

for all $\boldsymbol{u}_1$, $\boldsymbol{u}_2$ in U and any scalars α_1 and α_2.

For example, let $U = V = P(x)$, where $P(x)$ is the vector space of all polynomials in x over R (see Example 2.2). Define $T : U \to V$ such that $T\left[f(x)\right] = f'(x)$, where $f(x)$ is a polynomial in $P(x)$ and $f'(x)$ denotes its derivative. Then T is a linear transformation.

In general, if $T : U \to V$, then $T(U)$ is the image of U under T, or the range of T. It consists of all elements in V of the form $T(\boldsymbol{u})$ for $\boldsymbol{u}$ in U. If T is a linear transformation, then $T(U)$ is a vector subspace of V. This is true because if $\boldsymbol{v}_1$ and $\boldsymbol{v}_2$ are in $T(U)$, then there exist $\boldsymbol{u}_1$ and $\boldsymbol{u}_2$ in U such that $\boldsymbol{v}_1 = T(\boldsymbol{u}_1)$ and $\boldsymbol{v}_2 = T(\boldsymbol{u}_2)$. Hence, $\boldsymbol{v}_1 + \boldsymbol{v}_2 = T(\boldsymbol{u}_1) + T(\boldsymbol{u}_2) = T(\boldsymbol{u}_1 + \boldsymbol{u}_2)$, which belongs to $T(U)$. Also, if α is a scalar and $\boldsymbol{v}$ is an element in $T(U)$ such that $\boldsymbol{v} = T(\boldsymbol{u})$ for some $\boldsymbol{u}$ in U, then $\alpha\boldsymbol{v} = \alpha T(\boldsymbol{u}) = T(\alpha\boldsymbol{u})$, which belongs to $T(U)$. Thus, $T(U)$ is a vector subspace of V.

Definition 2.5 Let $T : U \to V$ be a linear transformation. The *kernel* of T, or the *null space* of T, is the set of all elements $\boldsymbol{u}$ in U such that $T(\boldsymbol{u}) = \boldsymbol{0}$, where $\boldsymbol{0}$ is the zero element in V. Such a set is denoted by *ker* T.

Example 2.9 Let $T : R^3 \to R^2$ be defined such that $T(x_1, x_2, x_3) = (x_1, x_2)$ for any x_1, x_2, x_3 in R. Then, T is a linear transformation whose kernel consists of all elements in R^3 whose first two coordinates are equal to zero, that is, the elements of the form $(0, 0, x_3)$ with arbitrary values for x_3.

Example 2.10 Let $T : R^3 \to R$ be defined such that $T(x_1, x_2, x_3) = 2x_1 + x_2 - 3x_3$. The kernel of T consists of all elements (x_1, x_2, x_3) in R^3 such that $2x_1 + x_2 - 3x_3 = 0$. This represents a plane in R^3 passing through the origin.

Theorem 2.2 Let $T : U \to V$ be a linear transformation. Then, *ker* T is a vector subspace of U.

Proof. Let $\boldsymbol{u}_1$ and $\boldsymbol{u}_2$ be two elements in *ker* T. Then $T(\boldsymbol{u}_1) = T(\boldsymbol{u}_2) = \boldsymbol{0}$, and $\boldsymbol{u}_1 + \boldsymbol{u}_2$ must therefore belong to *ker* T since $T(\boldsymbol{u}_1 + \boldsymbol{u}_2) = T(\boldsymbol{u}_1) + T(\boldsymbol{u}_2) = \boldsymbol{0}$. Also, if α is a scalar and $\boldsymbol{u}$ is an element in *ker* T, then $\alpha\boldsymbol{u}$ is an element in *ker* T since $T(\alpha\boldsymbol{u}) = \alpha T(\boldsymbol{u}) = \boldsymbol{0}$. Hence, *ker* T is a vector subspace of U.

Theorem 2.3 Let $T : U \to V$ be a linear transformation. Then

$$dim(U) = dim(ker\ T) + dim[T(U)].$$

Proof. Let $dim(U) = n$, $dim(ker\ T) = p$, and $dim[T(U)] = q$. Let $\boldsymbol{u}_1, \boldsymbol{u}_2, \ldots, \boldsymbol{u}_p$ be a basis of *ker* T, and $\boldsymbol{v}_1, \boldsymbol{v}_2, \ldots, \boldsymbol{v}_q$ be a basis of $T(U)$. Then there exist elements $\boldsymbol{w}_1, \boldsymbol{w}_2, \ldots, \boldsymbol{w}_q$ in U such that $T(\boldsymbol{w}_i) = \boldsymbol{v}_i$, $i = 1, 2, \ldots, q$. The objective here is to show that $\boldsymbol{u}_1, \boldsymbol{u}_2, \ldots, \boldsymbol{u}_p; \boldsymbol{w}_1, \boldsymbol{w}_2, \ldots, \boldsymbol{w}_q$ form a basis of U, that is, they are linearly independent and span U.

Suppose that there exist scalars $\alpha_1, \alpha_2, \ldots, \alpha_p;\ \beta_1, \beta_2, \ldots, \beta_q$ such that

$$\sum_{i=1}^{p} \alpha_i \boldsymbol{u}_i + \sum_{i=1}^{q} \beta_i \boldsymbol{w}_i = \boldsymbol{0}_{\mathbf{u}}, \tag{2.1}$$

where $\mathbf{0_u}$ is the zero element in U. Then,

$$\begin{aligned}\mathbf{0_v} &= T\left(\sum_{i=1}^{p} \alpha_i \boldsymbol{u}_i + \sum_{i=1}^{q} \beta_i \boldsymbol{w}_i\right)\\ &= \sum_{i=1}^{p} \alpha_i T(\boldsymbol{u}_i) + \sum_{i=1}^{q} \beta_i T(\boldsymbol{w}_i)\\ &= \sum_{i=1}^{q} \beta_i T(\boldsymbol{w}_i)\\ &= \sum_{i=1}^{q} \beta_i \boldsymbol{v}_i,\end{aligned}$$

where $\mathbf{0_v}$ is the zero element in V. Since the $\boldsymbol{v}_i$'s are linearly independent, $\beta_i = 0$ for $i = 1, 2, \ldots, q$. Furthermore, from (2.1) it can be concluded that $\alpha_i = 0$ for $i = 1, 2, \ldots, p$ since the $\boldsymbol{u}_i$'s are linearly independent. It follows from (2.1) that $\boldsymbol{u}_1, \boldsymbol{u}_2, \ldots, \boldsymbol{u}_p; \boldsymbol{w}_1, \boldsymbol{w}_2, \ldots, \boldsymbol{w}_q$ are linearly independent.

Let us now show that $\boldsymbol{u}_1, \boldsymbol{u}_2, \ldots, \boldsymbol{u}_p; \boldsymbol{w}_1, \boldsymbol{w}_2, \ldots, \boldsymbol{w}_q$ span U, that is, if $\boldsymbol{u}$ is any element in U, then it belongs to the linear span $L(\boldsymbol{u}_1, \boldsymbol{u}_2, \ldots, \boldsymbol{u}_p; \boldsymbol{w}_1, \boldsymbol{w}_2, \ldots, \boldsymbol{w}_q)$: Let $\boldsymbol{v} = T(\boldsymbol{u})$. Then there exist scalars $a_1, a_2, \ldots, a_q$ such that $\boldsymbol{v} = \sum_{i=1}^{q} a_i \boldsymbol{v}_i$. Hence,

$$\begin{aligned}T(\boldsymbol{u}) &= \sum_{i=1}^{q} a_i T(\boldsymbol{w}_i)\\ &= T\left(\sum_{i=1}^{q} a_i \boldsymbol{w}_i\right).\end{aligned}$$

Thus,

$$T\left(\boldsymbol{u} - \sum_{i=1}^{q} a_i \boldsymbol{w}_i\right) = \mathbf{0_v}.$$

This indicates that $\boldsymbol{u} - \sum_{i=1}^{q} a_i \boldsymbol{w}_i$ is an element in *ker* T. Therefore,

$$\boldsymbol{u} - \sum_{i=1}^{q} a_i \boldsymbol{w}_i = \sum_{i=1}^{p} b_i \boldsymbol{u}_i \tag{2.2}$$

for some scalars $b_1, b_2, \ldots, b_p$. From (2.2) it follows that

$$\boldsymbol{u} = \sum_{i=1}^{p} b_i \boldsymbol{u}_i + \sum_{i=1}^{q} a_i \boldsymbol{w}_i.$$

This shows that $\boldsymbol{u}$ belongs to the linear span of $\boldsymbol{u}_1, \boldsymbol{u}_2, \ldots, \boldsymbol{u}_p; \boldsymbol{w}_1, \boldsymbol{w}_2, \ldots, \boldsymbol{w}_q$. Consequently, these elements form a basis of U. Hence, $n = p + q$. □

Corollary 2.1 $T : U \to V$ is a one-to-one linear transformation if and only if $dim(ker\ T) = 0$, that is, $ker\ T$ consists of just the zero element.

Proof. By definition, $T : U \to V$ is a *one-to-one transformation* if whenever $T(u_1) = T(u_2)$ for u_1, u_2 in U, then $u_1 = u_2$. Thus, if T is one-to-one, then for any u in $ker\ T$, $T(u) = T(\mathbf{0_u}) = \mathbf{0_v}$, which indicates that $u = \mathbf{0_u}$. Hence, $ker\ T$ consists of just the zero element, and $dim(ker\ T) = 0$. Vice versa, suppose that $ker\ T$ consists of just the zero element. If u_1 and u_2 are elements in U such that $T(u_1) = T(u_2)$, then $T(u_1 - u_2) = \mathbf{0_v}$, which implies that $u_1 - u_2$ belongs to $ker\ T$. Thus $u_1 - u_2 = \mathbf{0_u}$, which indicates that T is a one-to-one linear transformation. □

Exercises

2.1 Let U and V be two vector spaces over R. The *Cartesian product* $U \times V$ is defined as the set of all ordered pairs (u, v), where u and v are elements in U and V, respectively. The sum of two elements, (u_1, v_1) and (u_2, v_2) in $U \times V$, is defined as $(u_1 + u_2, v_1 + v_2)$, and if α is a scalar, then $\alpha(u, v)$ is defined as $(\alpha u, \alpha v)$, where (u, v) is an element in $U \times V$. Show that $U \times V$ is a vector space.

2.2 Show that if W_1 and W_2 are two subspaces of a vector space V, then their intersection $W_1 \cap W_2$ is a subspace of V.

2.3 Let V be the vector space consisting of all continuous functions on $[-1, 1]$. Let W be the set of all continuous functions in V that have first and second derivatives on $[-1, 1]$. Show that W is a vector subspace of V.

2.4 Let u_1, u_2, u_3, u_4 be four elements in a vector space V. Show that $L(u_1, u_2)$ is a subspace of $L(u_1, u_2, u_3, u_4)$ (see Definition 2.3).

2.5 Suppose that a vector space V has dimension n. Show that if $u_1, u_2, \ldots, u_m$ are linearly independent elements in V, then $m \leq n$.

2.6 Let $u_1, u_2, \ldots, u_n$ be linearly independent elements in a vector space V, and let W be their linear span $L(u_1, u_2, \ldots, u_n)$. If v is any element in V that is not in W, then show that $u_1, u_2, \ldots, u_n, v$ are linearly independent.

2.7 Let V be a vector space, and suppose that $u_1, u_2, \ldots, u_n$ are linearly independent elements in V. Show that there exists a basis of V that contains these n elements.

2.8 Prove Theorem 2.1.

2.9 Let $T : U \to V$ be a linear transformation. Show that T is one-to-one if and only if whenever $\boldsymbol{u}_1, \boldsymbol{u}_2, \ldots, \boldsymbol{u}_n$ are linearly independent in U, then $T(\boldsymbol{u}_1), T(\boldsymbol{u}_2), \ldots, T(\boldsymbol{u}_n)$ are linearly independent in V.

2.10 Let $T : U \to V$ be a linear transformation. Suppose that U has a basis consisting of the two elements $\boldsymbol{u}_1$ and $\boldsymbol{u}_2$. Show that either $T(\boldsymbol{u}_1)$ and $T(\boldsymbol{u}_2)$ are linearly independent, or $T(U)$ has dimension 1, or $T(U)$ consists of just the zero element in V.

2.11 Let V be the vector space consisting of functions having derivatives of all orders in some neighborhood of a point x_0 in R. Let $T : V \to V$ be a linear transformation such that for any element f in V, $T(f) = f'$, the derivative of f.

(a) What is the kernel of T?

(b) Let $T^* : V \to V$ be defined such that $T^*(f) = T(f) - 2f$ for any f in V. What is the kernel of T^*?

2.12 Let $T : R^3 \to R$ be a linear transformation such that

$$T(x_1, x_2, x_3) = x_1 - 2x_2 + 4x_3.$$

What is the dimension of the kernel of T?

2.13 Let W be a subspace of the vector space V. If $dim(W) = dim(V)$, then $W = V$.

2.14 Let $T : U \to V$ be a linear transformation. Suppose that $dim(U) = dim(V)$. If $dim(ker\ T) = 0$, or if $T(U) = V$, then T is one-to-one and onto (T is a transformation from U *onto* V if $T(U) = V$).

2.15 Let U and V be two vector spaces over R, and T be the function $T : U \to V$. Then, T is said to be an *isomorphism* of U onto V if it satisfies the following conditions:

(i) T is linear

(ii) T is one-to-one

(iii) T is onto

In this case, U and V are said to be *isomorphic*.

Suppose now that $U = W_1 \oplus W_2$, the direct sum, and $V = W_1 \times W_2$, the Cartesian product of W_1 and W_2 (see Exercise 2.1), where W_1 and W_2 are vector spaces over R. Let $T : U \to V$ be defined such that for any $\boldsymbol{u} \in U$, $T(\boldsymbol{u}) = (\boldsymbol{u}_1, \boldsymbol{u}_2)$, where $\boldsymbol{u}_1$ and $\boldsymbol{u}_2$ provide a unique representation of $\boldsymbol{u}$ as $\boldsymbol{u} = \boldsymbol{u}_1 + \boldsymbol{u}_2$ with $\boldsymbol{u}_1 \in W_1$, $\boldsymbol{u}_2 \in W_2$. Show that T is an isomorphism of U onto V.

3

Basic Concepts in Matrix Algebra

Matrix algebra plays a vital role in the development of the theory of linear models. There is hardly any aspect of linear models that does not utilize some matrix notation or methodology. In fact, knowledge of matrix algebra is nowadays considered to be quite indispensable to the understanding of the fundamentals of linear models.

The purpose of this chapter is to provide an exposition of the basic concepts and results in matrix algebra, particularly those that have widespread applications in linear models. Given the expository nature of this chapter, theorems will, for the most part, be stated without proofs since the emphasis will be on using these theorems rather than proving them. The references given at the end of this book can be consulted for a complete account of the proofs.

3.1 Introduction and Notation

According to Eves (1969, p. 366), the English mathematician Arthur Cayley (1821–1895) and other algebraists of his time were instrumental in the modern development of matrix algebra. The term *matrix* was first applied in 1850. Cayley was motivated by the need to have a contracted notation to represent a set of linear equations of the form

$$\sum_{j=1}^{n} a_{ij}x_j = y_i, \quad i = 1, 2, \ldots, m,$$

where the a_{ij}'s are scalars that were detached from the variables x_j to produce the single matrix equation

$$\begin{bmatrix} a_{11} & a_{12} & \ldots & a_{1n} \\ a_{21} & a_{22} & \ldots & a_{2n} \\ \ldots & \ldots & \ldots & \ldots \\ a_{m1} & a_{m2} & \ldots & a_{mn} \end{bmatrix} \begin{bmatrix} x_1 \\ x_2 \\ \ldots \\ x_n \end{bmatrix} = \begin{bmatrix} y_1 \\ y_2 \\ \ldots \\ y_m \end{bmatrix}. \tag{3.1}$$

Cayley regarded such a scheme as an operator acting upon the variables $x_1, x_2, \ldots, x_n$. Farebrother (1997) and Grattan–Guiness (1994, p. 67) also give credit to other mathematicians in the eighteenth and nineteenth centuries who may well have made greater, although indirect, contributions than did Cayley. An interesting article concerning the origin of matrices and their introduction into statistics was written by Searle (1999). He reported that the year 1930 was "a good starting point for the entry of matrices into statistics. That was the year of Volume 1 of the *Annals of Mathematical Statistics* with its very first paper, Wicksell (1930), being *Remarks on Regression*." This was followed by Turnbull and Aitken's (1932) book with several applications of matrices to statistics.

3.1.1 Notation

The rectangular array consisting of the m rows and n columns, as shown on the left-hand side of equation (3.1), is called a *matrix* of order $m \times n$. The matrix as a whole is denoted by a boldface capital letter, for example, $\boldsymbol{A}$, and the scalar a_{ij} is called its (i, j)th element $(i = 1, 2, \ldots, m; j = 1, 2, \ldots, n)$. In some cases, it is more convenient to represent $\boldsymbol{A}$ using the notation $\boldsymbol{A} = (a_{ij})$.

It should be noted that equation (3.1) represents a linear transformation, $T : R^n \to R^m$. If $\boldsymbol{u}_1, \boldsymbol{u}_2, \ldots, \boldsymbol{u}_n$ form a basis for R^n, and if $\boldsymbol{v}_1, \boldsymbol{v}_2, \ldots, \boldsymbol{v}_m$ form a basis for R^m, then each $T(\boldsymbol{u}_j)$, $j = 1, 2, \ldots, n$, has a unique representation in R^m as a linear combination of $\boldsymbol{v}_1, \boldsymbol{v}_2, \ldots, \boldsymbol{v}_m$ of the form

$$T(\boldsymbol{u}_j) = \sum_{i=1}^{m} a_{ij}\boldsymbol{v}_i, \quad j = 1, 2, \ldots, n.$$

Once the bases in R^n and R^m have been selected, the linear transformation T is completely determined by the elements of $\boldsymbol{A}$: if $\boldsymbol{u}$ is any element in R^n, then $\boldsymbol{u} = \sum_{j=1}^{n} b_j\boldsymbol{u}_j$ for some scalars, $b_1, b_2, \ldots, b_n$. Hence,

$$\begin{aligned} T(\boldsymbol{u}) &= \sum_{j=1}^{n} b_j T(\boldsymbol{u}_j) \\ &= \sum_{i=1}^{m}\sum_{j=1}^{n} a_{ij} b_j \boldsymbol{v}_i. \end{aligned}$$

The matrix $\boldsymbol{A}$ is then said to be the *matrix representation* of T with respect to the aforementioned bases.

3.2 Some Particular Types of Matrices

(a) *Square matrix*. If the number of rows of the matrix $\boldsymbol{A}$ in (3.1) is equal to the number of columns, that is, $m = n$, then $\boldsymbol{A}$ is called a square matrix of order $n \times n$.

(b) *Diagonal matrix*. If the off-diagonal elements of a square matrix A of order $n \times n$ are all equal to zero, then A is called a diagonal matrix and is written as

$$A = \text{diag}(a_{11}, a_{22}, \ldots, a_{nn}).$$

(c) *Identity matrix*. If the diagonal elements of the matrix A in (b) are all equal to 1, then A is called an identity matrix of order $n \times n$, and is denoted by $\boldsymbol{I}_n$.

(d) *Matrix of ones*. If the elements of a matrix of order $m \times n$ are all equal to 1, then it is called a matrix of ones of order $m \times n$, and is denoted by $\boldsymbol{J}_{m\times n}$. If $m = n$, then it is denoted by $\boldsymbol{J}_n$.

(e) *Zero matrix*. If the elements of a matrix of order $m \times n$ are all equal to 0, then it is called a zero matrix of order $m \times n$, and is denoted by $\mathbf{0}_{m\times n}$, or just $\mathbf{0}$.

(f) *Triangular matrix*. If the elements of a square matrix that are below its diagonal are all equal to 0, then the matrix is called upper triangular. If, however, the elements that are above the diagonal are equal to 0, then the matrix is called lower triangular. A triangular matrix is a square matrix that is either upper triangular or lower triangular.

(g) *Row vector*. A matrix of order $1 \times n$ is called a row vector.

(h) *Column vector*. A matrix of order $n \times 1$ is called a column vector.

3.3 Basic Matrix Operations

(a) **Equality of matrices**. Let $\boldsymbol{A} = (a_{ij})$ and $\boldsymbol{B} = (b_{ij})$ be two matrices of the same order, $m \times n$. Then, $\boldsymbol{A} = \boldsymbol{B}$ if and only if $a_{ij} = b_{ij}$ for all $i = 1, 2, \ldots, m$; $j = 1, 2, \ldots, n$.

(b) **Addition of matrices**. Let $\boldsymbol{A} = (a_{ij})$ and $\boldsymbol{B} = (b_{ij})$ be two matrices of the same order, $m \times n$. Then, $\boldsymbol{A} + \boldsymbol{B}$ is a matrix $\boldsymbol{C} = (c_{ij})$ of order $m \times n$ such that $c_{ij} = a_{ij} + b_{ij}$ for $i = 1, 2, \ldots, m$; $j = 1, 2, \ldots, n$.

(c) **Scalar multiplication**. Let α be a scalar and let $\boldsymbol{A} = (a_{ij})$ be a matrix of order $m \times n$. Then $\alpha A = (\alpha a_{ij})$.

(d) **Product of matrices**. Let $\boldsymbol{A} = (a_{ij})$ and $\boldsymbol{B} = (b_{ij})$ be matrices of orders $m \times n$ and $n \times p$, respectively. The product $\boldsymbol{AB}$ is a matrix $\boldsymbol{C} = (c_{ij})$ of order $m \times p$ such that $c_{ij} = \sum_{k=1}^{n} a_{ik}b_{kj}$ for $i = 1, 2, \ldots, m$; $j = 1, 2, \ldots, p$. Note that this product requires that the number of columns of $\boldsymbol{A}$ be equal to the number of rows of $\boldsymbol{B}$.

(e) **The transpose of a matrix**. Let $\boldsymbol{A} = (a_{ij})$ be a matrix of order $m \times n$. The transpose of $\boldsymbol{A}$, denoted by $\boldsymbol{A}'$, is a matrix of order $n \times m$ whose rows are the columns of $\boldsymbol{A}$. Thus, the (i,j)th element of $\boldsymbol{A}'$ is equal to the (j,i)th element of $\boldsymbol{A}$, $i = 1, 2, \ldots, n$; $j = 1, 2, \ldots, m$. For example, if

$$\boldsymbol{A} = \begin{bmatrix} 1 & 5 & -1 \\ 2 & 1 & 7 \end{bmatrix},$$

then

$$\boldsymbol{A}' = \begin{bmatrix} 1 & 2 \\ 5 & 1 \\ -1 & 7 \end{bmatrix}.$$

Note that $(\boldsymbol{A}')' = \boldsymbol{A}$.

Let us now consider the following special cases:

(i) If $\boldsymbol{A}' = \boldsymbol{A}$, then $\boldsymbol{A}$ is said to be a *symmetric matrix*. Such a matrix must necessarily be square.

(ii) If $\boldsymbol{A}' = -\boldsymbol{A}$, then $\boldsymbol{A}$ is said to be a *skew-symmetric matrix*. This matrix must also be square. Furthermore, its diagonal elements must all be equal to zero.

(iii) If $\boldsymbol{a}$ is a column vector, then $\boldsymbol{a}'$ is a row vector.

(f) **The trace of a matrix**. Let $\boldsymbol{A} = (a_{ij})$ be a square matrix of order $n \times n$. The trace of $\boldsymbol{A}$, denoted by $tr(\boldsymbol{A})$, is the sum of its diagonal elements, that is,

$$tr(\boldsymbol{A}) = \sum_{i=1}^{n} a_{ii}.$$

It is easy to show that

(i) $tr(\boldsymbol{A}) = tr(\boldsymbol{A}')$.

(ii) If $\boldsymbol{A}$ is of order $m \times n$, and $\boldsymbol{B}$ is of order $n \times m$, then $tr(\boldsymbol{AB}) = tr(\boldsymbol{BA})$.

(iii) If $\boldsymbol{A}, \boldsymbol{B}, \boldsymbol{C}$ are matrices of orders $m \times n$, $n \times p$, and $p \times m$, respectively, then

$$tr(\boldsymbol{ABC}) = tr(\boldsymbol{BCA}) = tr(\boldsymbol{CAB}).$$

Properties (ii) and (iii) indicate that the trace of a product of matrices is invariant under a cyclic permutation of the matrices.

(iv) $tr(\alpha \boldsymbol{A}) = \alpha\, tr(\boldsymbol{A})$, where α is a scalar and $\boldsymbol{A}$ is a square matrix.

(v) $tr(\boldsymbol{A} + \boldsymbol{B}) = tr(\boldsymbol{A}) + tr(\boldsymbol{B})$, where $\boldsymbol{A}$ and $\boldsymbol{B}$ are square matrices of order $n \times n$.

(vi) $tr(\boldsymbol{A}'\boldsymbol{A}) = 0$ if and only if $\boldsymbol{A} = \boldsymbol{0}$.

3.4 Partitioned Matrices

Let $A = (a_{ij})$ be a matrix of order $m \times n$. A submatrix B of A is a matrix which can be obtained from A by deleting a certain number of rows and columns. In particular, if A is a square matrix of order $n \times n$, and if rows $i_1, i_2, \ldots, i_p$ and columns $i_1, i_2, \ldots, i_p$ $(p < n)$ are deleted from A, then the resulting submatrix is called a *principal submatrix* of A. If the deleted rows and columns are the last p rows and the last p columns, respectively, then the resulting submatrix is called a *leading principal submatrix* of order $(n-p) \times (n-p)$.

Definition 3.1 A partitioned matrix is a matrix that consists of several submatrices obtained by drawing horizontal lines between its rows and/or vertical lines between its columns.

If an $m \times n$ matrix A is partitioned into rc submatrices, A_{ij}, $i = 1, 2, \ldots, r$; $j = 1, 2, \ldots, c$, by drawing lines between certain rows and columns, then A can be expressed as

$$A = \begin{bmatrix} A_{11} & A_{12} & \ldots & A_{1c} \\ A_{21} & A_{22} & \ldots & A_{2c} \\ \ldots & \ldots & \ldots & \ldots \\ A_{r1} & A_{r2} & \ldots & A_{rc} \end{bmatrix}.$$

The submatrix A_{ij} is of order $m_i \times n_j$, where the m_i's and n_j's are positive integers such that $\sum_{i=1}^{r} m_i = m$ and $\sum_{j=1}^{c} n_j = n$. The matrix A can also be written as $A = (A_{ij})$, where it is understood that $i = 1, 2, \ldots, r$ and $j = 1, 2, \ldots, c$.

In particular, if $r = c$ and $A_{ij} = \mathbf{0}$ for $i \neq j$, then A is called a *block-diagonal matrix*, and is written as

$$A = \text{diag}(A_{11}, A_{22}, \ldots, A_{rr}).$$

Definition 3.2 Let $A = (a_{ij})$ and $B = (b_{ij})$ be matrices of orders $m_1 \times n_1$ and $m_2 \times n_2$, respectively. The direct (Kronecker) product of A and B, denoted by $A \otimes B$, is a matrix of order $m_1 m_2 \times n_1 n_2$ defined as a partitioned matrix of the form

$$A \otimes B = \begin{bmatrix} a_{11}B & a_{12}B & \ldots & a_{1n_1}B \\ a_{21}B & a_{22}B & \ldots & a_{2n_1}B \\ \ldots & \ldots & \ldots & \ldots \\ a_{m_1 1}B & a_{m_1 2}B & \ldots & a_{m_1 n_1}B \end{bmatrix},$$

which can be written as $A \otimes B = (a_{ij}B)$.

Properties of the direct product can be found in several matrix algebra books and articles, for example, Searle (1982, Section 10.7), Graybill (1983, Section 8.8), Magnus and Neudecker (1988, Chapter 2), Harville (1997, Chapter 16), Henderson and Searle (1981), and Khuri (1982). The following are some of these properties:

(a) $(\boldsymbol{A} \otimes \boldsymbol{B})' = \boldsymbol{A}' \otimes \boldsymbol{B}'$.

(b) $\boldsymbol{A} \otimes (\boldsymbol{B} \otimes \boldsymbol{C}) = (\boldsymbol{A} \otimes \boldsymbol{B}) \otimes \boldsymbol{C}$.

(c) $(\boldsymbol{A} \otimes \boldsymbol{B})(\boldsymbol{C} \otimes \boldsymbol{D}) = \boldsymbol{AC} \otimes \boldsymbol{BD}$, if $\boldsymbol{AC}$ and $\boldsymbol{BD}$ are defined.

(d) $tr(\boldsymbol{A} \otimes \boldsymbol{B}) = tr(\boldsymbol{A})tr(\boldsymbol{B})$, if $\boldsymbol{A}$ and $\boldsymbol{B}$ are square matrices.

The article by Henderson, Pukelsheim, and Searle (1983) gives a detailed account of the history of the direct product.

Definition 3.3 Let $\boldsymbol{A}_1, \boldsymbol{A}_2, \ldots, \boldsymbol{A}_k$ be matrices of orders $m_i \times n_i (i = 1, 2, \ldots, k)$. The direct sum of these matrices, denoted by $\oplus_{i=1}^{k} \boldsymbol{A}_i$, is a block-diagonal matrix of order $m \times n$, where $m = \sum_{i=1}^{k} m_i$ and $n = \sum_{i=1}^{k} n_i$, of the form

$$\oplus_{i=1}^{k} \mathbf{A}_i = \text{diag}(\mathbf{A}_1, \mathbf{A}_2, \ldots, \mathbf{A}_k).$$

Direct sums are discussed in some matrix algebra books, for example, Searle (1982, Section 10.6) and Graybill (1983, Section 8.8). Some properties of direct sums are

(a) $\oplus_{i=1}^{k} \boldsymbol{A}_i + \oplus_{i=1}^{k} \boldsymbol{B}_i = \oplus_{i=1}^{k} (\boldsymbol{A}_i + \boldsymbol{B}_i)$, if $\boldsymbol{A}_i$ and $\boldsymbol{B}_i$ are of the same order for $i = 1, 2, \ldots, k$.

(b) $\left[\oplus_{i=1}^{k} \boldsymbol{A}_i\right]\left[\oplus_{i=1}^{k} \boldsymbol{B}_i\right] = \oplus_{i=1}^{k} \boldsymbol{A}_i \boldsymbol{B}_i$, if $\boldsymbol{A}_i \boldsymbol{B}_i$ is defined for $i = 1, 2, \ldots, k$.

(c) $\left[\oplus_{i=1}^{k} \boldsymbol{A}_i\right]' = \oplus_{i=1}^{k} \boldsymbol{A}_i'$.

(d) $tr\left[\oplus_{i=1}^{k} \boldsymbol{A}_i\right] = \sum_{i=1}^{k} tr(\boldsymbol{A}_i)$, if $\boldsymbol{A}_i$ is a square matrix for $i = 1, 2, \ldots, k$.

3.5 Determinants

Historically, *determinants* were considered before matrices. According to Smith (1958, p. 273), the Chinese had some knowledge of determinants as early as about AD 1300. In the West, the theory of determinants is believed to have originated with the German mathematician Gottfried Leibniz (1646–1716) in 1693. However, the actual development of the theory of determinants did not begin until the publication of a book by Gabriel Cramer (1704–1752) (see Price, 1947, p. 85) in 1750.

Originally, a determinant was defined in terms of a system of linear equations. The word "determinant" originated with reference to the fact that a determinant "determines" whether the system has a unique solution, which occurs when the determinant is not equal to zero. Alexandre Vandermonde (1735–1796) was the first to recognize determinants independently from a system of linear equations. Arthur Cayley (1821–1895) is credited with having been the first to introduce the common present-day notation (of a determinant) of vertical bars enclosing a square matrix. For more interesting facts about the history of determinants, see the article by Price (1947).

Let $A = (a_{ij})$ be a square matrix of order $n \times n$. The determinant of A, denoted by $det(A)$, is a scalar that can be computed iteratively as

$$det(\mathbf{A}) = \sum_{j=1}^{n}(-1)^{i+j}\, a_{ij}\, det(\mathbf{A}_{ij}), \quad i = 1, 2, \ldots, n, \tag{3.2}$$

where $\mathbf{A}_{ij}$ is a submatrix of $\mathbf{A}$ obtained by deleting row i and column j. The determinant of $\mathbf{A}_{ij}$ is obtained in terms of determinants of submatrices of order $(n-2) \times (n-2)$ using a formula similar to the one for $det(\mathbf{A})$. This process is repeated several times until the submatrices on the right-hand side of formula (3.2) become of order 2×2. By definition, the determinant of the 2×2 matrix,

$$\mathbf{B} = \begin{bmatrix} b_{11} & b_{12} \\ b_{21} & b_{22} \end{bmatrix},$$

is given by $det(\mathbf{B}) = b_{11}b_{22} - b_{12}b_{21}$. Thus, by an iterative application of the formula in (3.2), the value of $det(\mathbf{A})$ can be determined. For example, if $\mathbf{A}$ is the matrix

$$\mathbf{A} = \begin{bmatrix} 2 & -1 & 1 \\ 3 & 0 & 2 \\ 2 & 1 & 5, \end{bmatrix}, \tag{3.3}$$

then, by expanding $det(\mathbf{A})$ according to the elements of the first row of $\mathbf{A}$, we get

$$\begin{aligned} det(\mathbf{A}) &= 2\,det\left(\begin{bmatrix} 0 & 2 \\ 1 & 5 \end{bmatrix}\right) + det\left(\begin{bmatrix} 3 & 2 \\ 2 & 5 \end{bmatrix}\right) + det\left(\begin{bmatrix} 3 & 0 \\ 2 & 1 \end{bmatrix}\right) \\ &= 2(-2) + 11 + 3 \\ &= 10. \end{aligned}$$

The determinant of $\mathbf{A}_{ij}$ in formula (3.2) is called a *minor* of $\mathbf{A}$ of order $n-1$. The quantity $(-1)^{i+j}\, det(\mathbf{A}_{ij})$ is called a *cofactor* of the corresponding element, a_{ij}, of $\mathbf{A}$, and is denoted by a_{ij}^{c}. The determinant of a principal submatrix of a square matrix $\mathbf{A}$ is called a *principal minor*, and the determinant of a leading principal submatrix is called a *leading principal minor*. Minors can also be defined for a general matrix A of order $m \times n$: if we remove all but p rows and the same number of columns from A, where $p \leq \min(m, n)$, then the determinant of the resulting submatrix is called a minor of A of order p.

Note that in formula (3.2), the expansion of $det(A)$ is carried out by the elements of any row of A, for example, row i ($i = 1, 2, \ldots, n$). It can also be carried out by the elements of any column of A. In the latter case, $det(A)$ is given by the equivalent formula

$$det(A) = \sum_{i=1}^{n} (-1)^{i+j} a_{ij}\, det(A_{ij}), \quad j = 1, 2, \ldots, n.$$

For example, expanding the determinant of the matrix A in (3.3) by the elements of the first column, we obtain

$$\begin{aligned} det(A) &= 2\,det\left(\begin{bmatrix} 0 & 2 \\ 1 & 5 \end{bmatrix}\right) - 3\,det\left(\begin{bmatrix} -1 & 1 \\ 1 & 5 \end{bmatrix}\right) + 2\,det\left(\begin{bmatrix} -1 & 1 \\ 0 & 2 \end{bmatrix}\right) \\ &= 2(-2) - 3(-6) + 2(-2) \\ &= 10. \end{aligned}$$

The following are some basic properties of determinants (see, for example, Aitken, 1958, Chapter 2; Searle, 1982, Section 4.3; Harville, 1997, Section 13.2):

(a) $det(AB) = det(A)det(B)$, if A and B are $n \times n$ matrices.

(b) $det(A') = det(A)$.

(c) $det(\alpha A) = \alpha^n det(A)$, if α is a scalar, and A is a matrix of order $n \times n$.

(d) If any two rows (or columns) of A are identical, then $det(A) = 0$.

(e) If any two rows (or columns) of A are interchanged, then $det(A)$ is multiplied by -1.

(f) If $det(A) = 0$, then A is called a *singular matrix*; otherwise, if $det(A) \neq 0$, then A is called a *nonsingular matrix*.

(g) If A and B are matrices of orders $m \times m$ and $n \times n$, respectively, then

$$det(A \otimes B) = [det(A)]^n \, [det(B)]^m .$$

(h) $det(A \oplus B) = det(A)det(B)$.

(i) If an $n \times n$ matrix A is upper (or lower) triangular, then $det(A) = \prod_{i=1}^{n} a_{ii}$, where a_{ii} is the ith diagonal element of A ($i = 1, 2, \ldots, n$).

(j) If A is partitioned as

$$A = \begin{bmatrix} A_{11} & A_{12} \\ A_{21} & A_{22} \end{bmatrix},$$

where A_{ij} is of order $n_i \times n_j$ ($i, j = 1, 2$), then

$$det(\boldsymbol{A}) = \begin{cases} det(\boldsymbol{A}_{11})det(\boldsymbol{A}_{22} - \boldsymbol{A}_{21}\boldsymbol{A}_{11}^{-1}\boldsymbol{A}_{12}), & \text{if } \boldsymbol{A}_{11} \text{ is nonsingular,} \\ det(\boldsymbol{A}_{22})det(\boldsymbol{A}_{11} - \boldsymbol{A}_{12}\boldsymbol{A}_{22}^{-1}\boldsymbol{A}_{21}), & \text{if } \boldsymbol{A}_{22} \text{ is nonsingular.} \end{cases}$$

(k) If $\boldsymbol{A}$ is a block-diagonal matrix, $\boldsymbol{A} = \text{diag}(\boldsymbol{A}_1, \boldsymbol{A}_2, \ldots, \boldsymbol{A}_n)$, where $\boldsymbol{A}_i$ is a square matrix, then $det(\boldsymbol{A}) = \prod_{i=1}^{n} det(\boldsymbol{A}_i)$.

3.6 The Rank of a Matrix

The rank of a matrix $\boldsymbol{A}$ is the number of linearly independent rows (or columns), and is denoted by $rank(\boldsymbol{A})$.

Suppose that $\boldsymbol{A}$ is of order $m \times n$. Let $\boldsymbol{u}'_1, \boldsymbol{u}'_2, \ldots, \boldsymbol{u}'_m$ denote the row vectors of $\boldsymbol{A}$, and let $\boldsymbol{v}_1, \boldsymbol{v}_2, \ldots, \boldsymbol{v}_n$ denote its column vectors. The linear spans associated with the rows and columns of $\boldsymbol{A}$ are $V_1 = L(\boldsymbol{u}'_1, \boldsymbol{u}'_2, \ldots, \boldsymbol{u}'_m)$, $V_2 = L(\boldsymbol{v}_1, \boldsymbol{v}_2, \ldots, \boldsymbol{v}_n)$, respectively. The dimensions of these two vector spaces are equal and equal to $rank(\boldsymbol{A})$.

Another equivalent definition for the rank of $\boldsymbol{A}$, which does not use the notion of vector spaces, is the following: if all the minors of $\boldsymbol{A}$ of order $r + 1$ and higher (if they exist) are zero while at least one minor of $\boldsymbol{A}$ of order r is not zero, then $\boldsymbol{A}$ has rank equal to r (see, for example, Aitken, 1958, p. 60). Note that there need not be minors of order $r + 1$ if the rank of $\boldsymbol{A}$ is r. For example, the matrix

$$\boldsymbol{A} = \begin{bmatrix} 1 & 0 & 2 \\ 3 & 4 & 7 \end{bmatrix}$$

has rank 2 since it has at least one nonzero minor or order 2, namely, the determinant of the submatrix

$$\boldsymbol{B} = \begin{bmatrix} 1 & 0 \\ 3 & 4 \end{bmatrix},$$

but there are no minors of order 3 in $\boldsymbol{A}$ since it has only two rows. Thus the rank of $\boldsymbol{A}$ can be defined as the largest order of a nonzero minor of $\boldsymbol{A}$.

Some properties associated with the rank of a matrix are (see, for example, Graybill, 1983, Chapter 1; Harville, 1997, Section 4.4; Marsaglia and Styan, 1974):

(a) $rank(\boldsymbol{A}) = rank(\boldsymbol{A}')$.

(b) The rank of $\boldsymbol{A}$ is invariant under multiplication by a nonsingular matrix. Thus, if $\boldsymbol{A}$ is an $m \times n$ matrix, and $\boldsymbol{P}$ and $\boldsymbol{Q}$ are nonsingular matrices of orders $m \times m$ and $n \times n$, respectively, then $rank(\boldsymbol{A}) = rank(\boldsymbol{PA}) = rank(\boldsymbol{AQ})$.

(c) $rank(\boldsymbol{A}) = rank(\boldsymbol{AA}') = rank(\boldsymbol{A}'\boldsymbol{A})$.

(d) For any matrices $\boldsymbol{A}_1, \boldsymbol{A}_2, \ldots, \boldsymbol{A}_k$ having the same number of rows,

$$rank(\boldsymbol{A}_1 : \boldsymbol{A}_2 : \ldots : \boldsymbol{A}_k) \leq \sum_{i=1}^{k} rank(\boldsymbol{A}_i).$$

In particular, if these matrices have also the same number of columns, then

$$rank\left(\sum_{i=1}^{k} \boldsymbol{A}_i\right) \leq rank(\boldsymbol{A}_1 : \boldsymbol{A}_2 : \ldots : \boldsymbol{A}_k).$$

We conclude that for any matrices, $\boldsymbol{A}_1, \boldsymbol{A}_2, \ldots, \boldsymbol{A}_k$, having the same number of rows and columns,

$$rank\left(\sum_{i=1}^{k} \boldsymbol{A}_i\right) \leq \sum_{i=1}^{k} rank(\boldsymbol{A}_i).$$

Equality is achieved if and only if there are nonsingular matrices, $\boldsymbol{F}$ and $\boldsymbol{G}$, such that $\boldsymbol{A}_i = \boldsymbol{F}\boldsymbol{D}_i\boldsymbol{G}$, where $\boldsymbol{D}_i$ is a diagonal matrix with diagonal elements equal to zeros and ones ($i = 1, 2, \ldots, k$) such that $\boldsymbol{D}_i\boldsymbol{D}_j = \boldsymbol{0}$, if $i \neq j$. This result can be found in Marsaglia and Styan (1974, Theorem 12, p. 283).

(e) If $\boldsymbol{A}$ and $\boldsymbol{B}$ are matrices of orders $m \times n$ and $n \times q$, respectively, then

$$rank(\boldsymbol{A}) + rank(\boldsymbol{B}) - n \leq rank(\boldsymbol{AB}) \leq \min\{rank(\boldsymbol{A}), rank(\boldsymbol{B})\}$$

This is known as *Sylvester's law*.

(f) $rank(\boldsymbol{A} \otimes \boldsymbol{B}) = rank(\boldsymbol{A})rank(\boldsymbol{B})$.

(g) $rank(\boldsymbol{A} \oplus \boldsymbol{B}) = rank(\boldsymbol{A}) + rank(\boldsymbol{B})$.

Definition 3.4 Let $\boldsymbol{A}$ be a matrix of order $m \times n$ and *rank* r. Then,

(a) $\boldsymbol{A}$ is of full row rank if $r = m < n$.

(b) $\boldsymbol{A}$ is of full column rank if $r = n < m$.

(c) $\boldsymbol{A}$ is of full rank if $r = m = n$. In this case, the determinant of $\boldsymbol{A}$ is not equal to zero, that is, $\boldsymbol{A}$ is a nonsingular matrix.

3.7 The Inverse of a Matrix

Let A be a nonsingular matrix of order $n \times n$. The *inverse* of A, denoted by A^{-1}, is an $n \times n$ matrix that satisfies the condition $AA^{-1} = A^{-1}A = I_n$. Such a matrix is unique.

The inverse of A can be computed as follows: let a^c_{ij} denote the cofactor of a_{ij} (see Section 3.5). Define the matrix C as $C = (a^c_{ij})$. The inverse of A is given by the formula,

$$A^{-1} = \frac{C'}{det(A)},$$

where C' is the transpose of C. The matrix C' is called the *adjoint* of A, and is denoted by *adj* A. For example, if A is the matrix

$$A = \begin{bmatrix} 1 & 0 \\ 2 & 5 \end{bmatrix},$$

then

$$adj\, A = \begin{bmatrix} 5 & 0 \\ -2 & 1 \end{bmatrix},$$

and

$$A^{-1} = \frac{1}{5}\begin{bmatrix} 5 & 0 \\ -2 & 1 \end{bmatrix}.$$

The following are some properties associated with the inverse operation (see, for example, Searle, 1982, Chapter 5; Harville, 1997, Chapter 8):

(a) $(AB)^{-1} = B^{-1}A^{-1}$.

(b) $(A')^{-1} = (A^{-1})'$.

(c) $det(A^{-1}) = \frac{1}{det(A)}$, if A is nonsingular.

(d) $(A^{-1})^{-1} = A$.

(e) $(A \otimes B)^{-1} = A^{-1} \otimes B^{-1}$.

(f) $(A \oplus B)^{-1} = A^{-1} \oplus B^{-1}$.

(g) If A is partitioned as

$$A = \begin{bmatrix} A_{11} & A_{12} \\ A_{21} & A_{22} \end{bmatrix},$$

where A_{ij} is of order $n_i \times n_j$ $(i, j = 1, 2)$, then the inverse of A is given by

$$A^{-1} = \begin{bmatrix} B_{11} & B_{12} \\ B_{21} & B_{22} \end{bmatrix},$$

where

$$B_{11} = (A_{11} - A_{12}A_{22}^{-1}A_{21})^{-1},$$
$$B_{12} = -B_{11}A_{12}A_{22}^{-1},$$
$$B_{21} = -A_{22}^{-1}A_{21}B_{11},$$
$$B_{22} = A_{22}^{-1} + A_{22}^{-1}A_{21}B_{11}A_{12}A_{22}^{-1}.$$

3.7.1 Generalized Inverse of a Matrix

A *generalized inverse* (or g-inverse) of an $m \times n$ matrix A is any $n \times m$ matrix B such that $ABA = A$. We denote such a matrix by A^-. Thus,

$$AA^-A = A. \tag{3.4}$$

Note that A^- is defined even if A is not a square matrix. If A is a square matrix, it does not have to be nonsingular. A generalized inverse is not unique. In fact, condition (3.4) can be satisfied by infinitely many matrices (see, for example, Searle, 1982, Chapter 8). Algorithms for computing generalized inverses can be found in, for example, Searle (1982, Chapter 8) and Harville (1997, Chapter 9).

If A is nonsingular, then condition (3.4) is satisfied by only the inverse A^{-1}. Thus, A^{-1} is a special case of A^- when A is nonsingular. The following are some properties concerning generalized inverses (see Searle, 1982, Section 8.6; Harville, 1997, Sections 9.3 and 9.4):

(a) If A is symmetric, then A^- can be chosen to be symmetric.

(b) $A(A'A)^-A'$ is invariant to the choice of a generalized inverse of $A'A$.

(c) $A(A'A)^-A'A = A$ for any matrix A.

(d) $rank(A^-) \geq rank(A)$ for any matrix A.

(e) $rank(A^-A) = rank(AA^-) = rank(A)$ for any matrix A.

3.8 Eigenvalues and Eigenvectors

Let A be a square matrix of order $n \times n$. A scalar λ is an *eigenvalue* (characteristic root or latent root) of A if there exits a nonzero vector, x, such that

$$Ax = \lambda x. \tag{3.5}$$

Such a vector is called an *eigenvector* (characteristic vector or latent vector) of A corresponding to λ. Equation (3.5) can be written as

$$(A - \lambda I_n)x = 0, \quad x \neq 0. \tag{3.6}$$

This equation indicates that x belongs to the *kernel* (null space) of $A - \lambda I_n$. Equation (3.6) also indicates that the columns of $A - \lambda I_n$ are linearly dependent. Hence, the rank of $A - \lambda I_n$ must be less than n, which implies that $A - \lambda I_n$ is a singular matrix and its determinant is therefore equal to zero. We thus have

$$det(A - \lambda I_n) = 0.$$

This is called the *characteristic equation* of A. Note that the left-hand side of the equation is a polynomial in λ of degree n called the *characteristic polynomial*, and the set of all of its distinct roots forms the so-called *spectrum* of A. If a particular root has multiplicity equal to m (≥ 1), then it is called an eigenvalue of A of multiplicity m.

The following are some properties associated with eigenvalues and eigenvectors (see, for example, Searle, 1982, Chapter 11; Marcus and Minc, 1964, Chapter 2; Graybill, 1983, Chapter 3; Magnus and Neudecker, 1988, Chapter 1):

(a) The eigenvalues of a symmetric matrix are real.

(b) If A is a symmetric matrix, then its rank is equal to the number of its nonzero eigenvalues. Thus, if $\lambda_1, \lambda_2, \ldots, \lambda_k$ are the distinct nonzero eigenvalues of A, then $rank(A) = \sum_{i=1}^{k} m_i$, where m_i is the multiplicity of λ_i ($i = 1, 2, \ldots, k$).

(c) Suppose that $\lambda_1, \lambda_2, \ldots, \lambda_k$ are the distinct nonzero eigenvalues of A. If $v_1, v_2, \ldots, v_k$ are the eigenvectors of A corresponding to $\lambda_1, \lambda_2, \ldots, \lambda_k$, respectively, then $v_1, v_2, \ldots, v_k$ are linearly independent. In particular, if A is symmetric, then $v_1, v_2, \ldots, v_k$ are orthogonal to one another, that is, $v_i' v_j = 0$ for $i \neq j$ ($i, j = 1, 2, \ldots, k$).

(d) Let $\lambda_1, \lambda_2, \ldots, \lambda_n$ be the eigenvalues of A, then

   (i) $tr(A) = \sum_{i=1}^{n} \lambda_i$.

   (ii) $det(A) = \prod_{i=1}^{n} \lambda_i$.

(e) Let A and B be two matrices of orders $m \times m$ and $n \times n$, respectively. Let $\lambda_1, \lambda_2, \ldots, \lambda_m$ be the eigenvalues of A, and $\tau_1, \tau_2, \ldots, \tau_n$ be the eigenvalues of B. Then,

   (i) The eigenvalues of $A \otimes B$ are of the form $\lambda_i \tau_j$ ($i = 1, 2, \ldots, m$; $j = 1, 2, \ldots, n$).

   (ii) The eigenvalues of $A \oplus B$ are $\lambda_1, \lambda_2, \ldots, \lambda_m; \tau_1, \tau_2, \ldots, \tau_n$.

(f) Let A and B be two matrices of orders $m \times n$ and $n \times m (n \geq m)$, respectively. The nonzero eigenvalues of AB are the same as the nonzero eigenvalues of BA. In particular, if $m = n$, then all the eigenvalues of AB (not just the nonzero ones) are the same as those of BA (see Magnus and Neudecker, 1988, Theorem 9, p. 14).

(g) The eigenvalues of a triangular matrix A (lower or upper) are equal to the diagonal elements of A.

3.9 Idempotent and Orthogonal Matrices

There are two particular types of matrices that play an important role in the theory of linear models. They are called idempotent and orthogonal matrices.

A square matrix, A, is *idempotent* if $A^2 = A$. For example, the matrix $A = I_n - \frac{1}{n}J_n$, where J_n is the matrix of ones of order $n \times n$, is idempotent.

Idempotent matrices are used frequently in linear models, particularly in connection with the distribution of quadratic forms in normal random variables, as will be seen later in Chapter 5.

The following are some properties of idempotent matrices (see Graybill, 1983, Section 12.3; Harville, 1997, Chapter 10):

If A is idempotent, then

(a) The eigenvalues of A are equal to zeros and ones.

(b) $rank(A) = tr(A)$.

(c) For any integer k greater than or equal to one, $A^k = A$.

In addition to the above properties, it can be easily shown, using property (c) in Section 3.7.1, that for any matrix A, the matrix $A(A'A)^{-}A'$ is idempotent of rank equal to the rank of A.

A square matrix, A, of order $n \times n$ is *orthogonal* if $A'A = I_n$. Thus, if A is orthogonal, then the absolute value of its determinant is equal to 1. Furthermore, its inverse A^{-1} is equal to its transpose A'.

3.9.1 Parameterization of Orthogonal Matrices

Since $A'A = I_n$ for an orthogonal matrix A of order $n \times n$, the n^2 elements of A are subject to $\frac{n(n+1)}{2}$ equality constraints. These elements can therefore be represented by $n^2 - \frac{n(n+1)}{2} = \frac{n(n-1)}{2}$ independent parameters. There are several methods to parameterize an orthogonal matrix (for a review of such methods, see Khuri and Good, 1989). Two of these methods are described here.

(I) **Exponential representation.** If A is an orthogonal matrix with determinant equal to 1, then it can be represented as

$$A = \exp(T), \tag{3.7}$$

where T is a skew-symmetric matrix of order $n \times n$ (see Gantmacher, 1959, p. 288). The elements of T above its main diagonal can be used to parameterize A. The exponential function in (3.7) is defined as the sum of the infinite series of matrices,

$$\exp(T) = \sum_{i=0}^{\infty} \frac{1}{i!} T^i, \tag{3.8}$$

where $T^0 = I_n$.

If A is given, to find T, we first find the eigenvalues of A. These are of the form $e^{\pm i\phi_1}, e^{\pm i\phi_2}, \ldots, e^{\pm i\phi_q}$, 1, where the eigenvalue 1 is of multiplicity $n - 2q$, and i is the complex imaginary number $\sqrt{-1}$. Note that $e^{\pm i\phi_j} = \cos\phi_j \pm i \sin\phi_j$ $(j = 1, 2, \ldots, q)$. If we denote the matrix,

$$\begin{bmatrix} a & b \\ -b & a \end{bmatrix}$$

by $[a + bi]$, then A can be written (Gantmacher, 1959, p. 288) as the product of three real matrices,

$$A = Q\,\mathrm{diag}\{[e^{i\phi_1}], \ldots, [e^{i\phi_q}], 1, \ldots, 1\}\, Q',$$

where Q is an orthogonal matrix of the form

$$Q = [x_1 : y_1 : x_2 : y_2 : \ldots : x_q : y_q : x_{2q+1} : \ldots : x_n]$$

such that $x_j + i y_j$ is an eigenvector of A with eigenvalue $e^{i\phi_j}$ $(j = 1, 2, \ldots, q)$ and x_k is an eigenvector with eigenvalue 1 $(k = 2q+1, \ldots, n)$. The skew-symmetric matrix T can then be defined by the formula

$$T = Q\,\mathrm{diag}\{[i\phi_1], \ldots, [i\phi_q], 0, \ldots, 0\}\, Q',$$

which gives $A = \exp(T)$ since $e^{[i\phi_j]} = [e^{i\phi_j}]$ $(j = 1, 2, \ldots, q)$.

Example 3.1 Consider the orthogonal matrix

$$A = \frac{1}{\sqrt{6}} \begin{bmatrix} \sqrt{2} & \sqrt{2} & \sqrt{2} \\ \sqrt{3} & -\sqrt{3} & 0 \\ 1 & 1 & -2 \end{bmatrix}$$

whose determinant is $det(A) = 1$. The eigenvalues of A are $e^{i\phi_1}$, $e^{-i\phi_1}$, 1, where $\phi_1 = 0.9265\pi$. Let $x_1 + iy_1$ be an eigenvector of A with the eigenvalue $e^{i\phi_1}$ (x_1 and y_1 have each a length equal to 1). Then, we have the equation

$$A(x_1 + iy_1) = (\cos\phi_1 + i\sin\phi_1)(x_1 + iy_1),$$

which can be represented in the form

$$\begin{aligned} Ax_1 &= \cos\phi_1\, x_1 - \sin\phi_1\, y_1, \\ Ay_1 &= \sin\phi_1\, x_1 + \cos\phi_1\, y_1. \end{aligned}$$

These equations imply that $x_1' y_1 = 0$, as a result of A being orthogonal, hence, $(Ax_1)'(Ax_1) = x_1' x_1$, $(Ay_1)'(Ay_1) = y_1' y_1$. There is no unique solution to these equations. One solution is given by

$$x_1 = \begin{bmatrix} -0.358933 \\ 0.161094 \\ 0.919356 \end{bmatrix}, \quad y_1 = \begin{bmatrix} -0.292181 \\ 0.916094 \\ -0.274595 \end{bmatrix}.$$

Note that $x_1' y_1 = 0$, and the length of each vector is chosen to be equal to 1.

Let us now find an eigenvector of A corresponding to the eigenvalue 1. This vector satisfies the equation,

$$(A - I_3)\, x_3 = \mathbf{0},$$

which also does not have a unique solution. A solution is given by

$$x_3 = \begin{bmatrix} 0.886452 \\ 0.367180 \\ 0.281747 \end{bmatrix}.$$

Note that x_3 is orthogonal to x_1 and y_1, and is chosen to have a length equal to 1. The matrix A can therefore be expressed as

$$A = Q \begin{bmatrix} \cos\phi_1 & \sin\phi_1 & 0 \\ -\sin\phi_1 & \cos\phi_1 & 0 \\ 0 & 0 & 1 \end{bmatrix} Q',$$

where $Q = [x_1 : y_1 : x_3]$. The skew-symmetric matrix T can then be written as

$$\begin{aligned} T &= Q \begin{bmatrix} 0 & \phi_1 & 0 \\ -\phi_1 & 0 & 0 \\ 0 & 0 & 0 \end{bmatrix} Q' \\ &= Q \begin{bmatrix} 0 & 0.9265\pi & 0 \\ -0.9265\pi & 0 & 0 \\ 0 & 0 & 0 \end{bmatrix} Q', \end{aligned}$$

which satisfies equation (3.7).

(II) **Cayley's representation** (Arthur Cayley, 1821–1895). If A is an orthogonal matrix of order $n \times n$ that does not have the eigenvalue -1, then it can be written in Cayley's form (see Gantmacher, 1959, p. 289), namely,

$$A = (I_n - U)(I_n + U)^{-1},$$

where U is a skew-symmetric matrix of order $n \times n$.

The need to parameterize orthogonal matrices arises in several situations. For example, in some simulation experiments, parameterization is used to generate random orthogonal matrices. Heiberger, Velleman, and Ypelaar (1983) used this technique to construct test data with special properties for multivariate linear models. See also Anderson, Olkin, and Underhill (1987), and Olkin (1990).

3.10 Quadratic Forms

Let $A = (a_{ij})$ be a symmetric matrix of order $n \times n$, and let $x = (x_1, x_2, \ldots, x_n)'$ be a column vector of order $n \times 1$. The scalar function

$$\begin{aligned} Q(x) &= x'Ax \\ &= \sum_{i=1}^{n}\sum_{j=1}^{n} a_{ij}x_i x_j \end{aligned}$$

is called a *quadratic form* in x.

Quadratic forms play an important role in linear models. For example, in a typical analysis of variance (ANOVA) table associated with a given model and a given data set, any sum of squares can be represented as a quadratic form in the data vector.

A quadratic form is said to be

(a) *Positive definite* if $x'Ax > 0$ for all $x \neq 0$, and is zero only when $x = 0$.

(b) *Positive semidefinite* if $x'Ax \geq 0$ for all x, and $x'Ax = 0$ for some $x \neq 0$.

(c) *Nonnegative definite* if $x'Ax \geq 0$ for all x.

(d) *Negative definite* if $x'(-A)x$ is positive definite.

(e) *Negative semidefinite* if $x'(-A)x$ is positive semidefinite.

The above definitions also apply to the matrix A of the quadratic form. Thus, A is said to be positive definite, positive semidefinite, nonnegative definite, negative definite, or negative semidefinite if $x'Ax$ is positive definite, positive semidefinite, nonnegative definite, negative definite, or negative semidefinite, respectively.

Theorem 3.1 Let A be a symmetric matrix. Then, A is positive definite if and only if either of the following two conditions is satisfied:

(a) The eigenvalues of A are all positive.

(b) The leading principal minors of A are all positive.

Proof. See Lancaster (1969, Theorem 2.14.4).

Theorem 3.2 Let A be a symmetric matrix. Then, A is positive semidefinite if and only if its eigenvalues are nonnegative with at least one equal to zero.

Proof. See Basilevsky (1983, Theorem 5.10).

Theorem 3.3 Let A be a matrix of order $m \times n$ and rank r. Then,

(a) AA' and $A'A$ are both nonnegative definite.

(b) $A'A$ is positive semidefinite if $r < n$.

(c) $A'A$ is positive definite if $r = n$.

Proof. See Graybill (1983, Corollary 12.2.2).

3.11 Decomposition Theorems

The following theorems show how certain matrices can be reduced to particular forms. These theorems have many useful applications in the theory of linear models.

Theorem 3.4 (The Spectral Decomposition Theorem) Let A be a symmetric matrix of order $n \times n$. There exits an orthogonal matrix P such that

$$A = P\Lambda P', \tag{3.9}$$

where $\Lambda = \text{diag}(\lambda_1, \lambda_2, \ldots, \lambda_n)$ is a diagonal matrix with diagonal elements equal to the eigenvalues of A. The columns of P are eigenvectors of A corresponding to the λ_i's. Thus, if P is partitioned as

$$P = [p_1 : p_2 : \ldots : p_n],$$

where p_i is the ith column of P, then p_i is an eigenvector of A corresponding to the ith eigenvalue λ_i $(i = 1, 2, \ldots, n)$. Formula (3.9) can then be written as

$$A = \sum_{i=1}^{n} \lambda_i p_i p_i'. \tag{3.10}$$

Proof. See Basilevsky (1983, Theorem 5.8, p. 200).

Corollary 3.1 Let A be a symmetric matrix of order $n \times n$ and rank r. Then, A can be written as $A = BB'$, where B is of order $n \times r$ and rank r (that is, B is a full-column-rank matrix). The matrix B is real only when A is nonnegative definite.

Proof. This follows directly from the Spectral Decomposition Theorem and writing Λ as

$$\Lambda = \text{diag}(\Lambda_1, \mathbf{0}),$$

where

Λ_1 is a diagonal matrix of order $r \times r$ whose diagonal elements are the nonzero eigenvalues of A

$\mathbf{0}$ is a zero matrix of order $(n - r) \times (n - r)$

Let us now decompose Λ as

$$\Lambda = \begin{bmatrix} \Lambda_1^{1/2} \\ \mathbf{0} \end{bmatrix} \begin{bmatrix} \Lambda_1^{1/2} & \mathbf{0} \end{bmatrix}.$$

Here, $\Lambda_1^{1/2}$ is a diagonal matrix whose diagonal elements are the square roots of those of Λ_1. Note that the diagonal elements of $\Lambda_1^{1/2}$ are not necessarily real numbers. The matrix B can then be chosen as

$$B = P \begin{bmatrix} \Lambda_1^{1/2} \\ \mathbf{0} \end{bmatrix},$$

which is of order $n \times r$ and rank r (see Searle, 1971, Lemma 7, p. 37).

Theorem 3.5 (The Singular-Value Decomposition) Let A be a matrix of order $m \times n$ $(m \leq n)$ and rank r. There exist orthogonal matrices P and Q such that

$$A = P[D \quad \mathbf{0}]Q',$$

where

$D = \text{diag}(\lambda_1, \lambda_2, \ldots, \lambda_m)$ is a diagonal matrix with nonnegative diagonal elements

$\mathbf{0}$ is a zero matrix of order $m \times (n - m)$

The positive diagonal elements of D are the square roots of the positive eigenvalues of AA' (or, equivalently, of $A'A$), and are called the *singular values* of A.

Proof. See, for example, Searle (1982, pp. 316–317), Harville (1997, Section 21.12).

Theorem 3.6 (The Cholesky Decomposition) Let A be a symmetric matrix of order $n \times n$.

(a) If A is positive definite, then there exits a unique upper triangular matrix T with positive diagonal elements such that $A = T'T$.

(b) If A is nonnegative definite with rank equal to r, then there exits a unique upper triangular matrix U with r positive diagonal elements and with $n - r$ zero rows such that $A = U'U$.

Proof. See Harville (1997, Theorem 14.5.16, p. 231).

The following three theorems are useful for the simultaneous diagonalization of matrices.

Theorem 3.7 Let A and B be symmetric matrices of order $n \times n$. If A is positive definite, then there exits a nonsingular matrix Q such that

$$Q'AQ = I_n \quad \text{and} \quad Q'BQ = D,$$

where D is a diagonal matrix whose diagonal elements are the roots of the polynomial equation $det(B - \lambda A) = 0$.

Proof. See Graybill (1983, Theorem 12.2.13).

Theorem 3.8 Let A and B be nonnegative definite matrices (neither one has to be positive definite). Then, there exits a nonsingular matrix Q such that $Q'AQ$ and $Q'BQ$ are each diagonal.

Proof. See Graybill (1983, Theorem 12.2.13) and Newcomb (1960) for a detailed proof.

Theorems 3.7 and 3.8 show the existence of a nonsingular matrix that diagonalizes two particular matrices. In general, if $A_1, A_2, \ldots, A_k$ are $n \times n$ matrices, then they are said to be *simultaneously diagonalizable* if there exists a nonsingular matrix Q such that

$$Q^{-1}A_1Q = D_1, \quad Q^{-1}A_2Q = D_2, \ldots, \quad Q^{-1}A_kQ = D_k,$$

where $D_1, D_2, \ldots, D_k$ are diagonal matrices.

The next theorem gives the condition for the existence of such a matrix when $A_1, A_2, \ldots, A_k$ are symmetric.

Theorem 3.9 Let $A_1, A_2, \ldots, A_k$ be symmetric matrices of order $n \times n$. Then, there exits an orthogonal matrix P such that

$$A_i = P\Lambda_i P', \quad i = 1, 2, \ldots, k,$$

where Λ_i is a diagonal matrix, if and only if

$$A_iA_j = A_jA_i, \quad \text{for all } i \neq j \; (i, j = 1, 2, \ldots, k),$$

that is, the matrices commute in pairs.

Proof. See Harville (1997, Theorem 21.13.1).

3.12 Some Matrix Inequalities

Several well-known inequalities, such as the Cauchy–Schwarz, Hölder's, Minkowski's, and Jensen's inequalities, have been effectively used in mathematical statistics. In this section, additional inequalities that pertain to matrices are presented. These inequalities are useful in many aspects of linear models ranging from computational considerations to theoretical developments.

Let A be a symmetric matrix of order $n \times n$. Suppose that its eigenvalues are arranged in a descending order of magnitude. Let $e_i(A)$ denote the ith eigenvalue of A so that $e_1(A) \geq e_2(A) \geq \cdots \geq e_n(A)$. Thus, $e_1(A)$ and $e_n(A)$ are, respectively, the largest and the smallest of the eigenvalues. They are so designated by writing $e_{\max}(A) = e_1(A)$, $e_{\min}(A) = e_n(A)$.

Theorem 3.10 Consider the ratio $\frac{x'Ax}{x'x}$, which is called *Rayleigh's quotient* for A. Then,

$$e_{\min}(A) \leq \frac{x'Ax}{x'x} \leq e_{\max}(A).$$

The lower and upper bounds can be achieved by choosing x to be an eigenvector corresponding to $e_{\min}(A)$ and $e_{\max}(A)$, respectively. Thus,

$$\inf_{x \neq 0} \left[\frac{x'Ax}{x'x}\right] = e_{\min}(A),$$

$$\sup_{x \neq 0} \left[\frac{x'Ax}{x'x}\right] = e_{\max}(A).$$

Proof. This follows directly from applying the Spectral Decomposition Theorem (Theorem 3.4).

Theorem 3.11 If A is a symmetric matrix and B is a positive definite matrix, both of order $n \times n$, then

$$e_{\min}(\mathbf{B}^{-1}A) \leq \frac{x'Ax}{x'Bx} \leq e_{\max}(B^{-1}A).$$

Note that the eigenvalues of $B^{-1}A$ are real since they are the same as those of the matrix $C = B^{-\frac{1}{2}}AB^{-\frac{1}{2}}$, which is symmetric. This follows from applying property (f) in Section 3.8. Here, $B^{-\frac{1}{2}}$ is a matrix defined as follows: by the Spectral Decomposition Theorem, B can be written as $B = P\Lambda P'$, where $\Lambda = \text{diag}(\lambda_1, \lambda_2, \ldots, \lambda_n)$ is a diagonal matrix whose diagonal elements are the eigenvalues of B and P is an orthogonal matrix whose columns are the corresponding eigenvectors of B. Then, $B^{-\frac{1}{2}}$ is defined as

$$B^{-\frac{1}{2}} = P\Lambda^{-\frac{1}{2}}P',$$

where

$$\Lambda^{-\frac{1}{2}} = \text{diag}\left(\lambda_1^{-\frac{1}{2}}, \lambda_2^{-\frac{1}{2}}, \ldots, \lambda_n^{-\frac{1}{2}}\right).$$

Theorem 3.12 Let A and B be matrices of the same order. Then,

$$[tr(A'B)]^2 \leq [tr(A'A)][tr(B'B)].$$

Equality holds if and only if one of these two matrices is a scalar multiple of the other. This is the matrix analogue of the *Cauchy–Schwarz* inequality (see Magnus and Neudecker, 1988, Theorem 2, p. 201).

Theorem 3.13

(a) Let $A = (a_{ij})$ be positive definite of order $n \times n$. Then,

$$det(A) \leq \prod_{i=1}^{n} a_{ii}.$$

Equality holds if and only if A is diagonal.

(b) For any matrix $A = (a_{ij})$ of order $n \times n$,

$$[det(A)]^2 \leq \prod_{i=1}^{n}\left(\sum_{j=1}^{n} a_{ij}^2\right).$$

Equality holds if and only if AA' is a diagonal matrix or A has a zero row.

Proof. See Magnus and Neudecker (1988, Theorems 28 and 18, pp. 23 and 214, respectively). The inequality in part (b) is called *Hadamard's inequality*.

Theorem 3.14 For any two positive semidefinite matrices, A and B, of order $n \times n$ such that $A \neq 0$ and $B \neq 0$,

$$[det(A + B)]^{\frac{1}{n}} \geq [det(A)]^{\frac{1}{n}} + [det(B)]^{\frac{1}{n}}.$$

Equality holds if and only if $det(A + B) = 0$, or $A = \alpha B$ for some $\alpha > 0$.

This is called *Minkowski's determinant inequality* (see Magnus and Neudecker, 1988, Theorem 28, p. 227).

Theorem 3.15 If A is a positive semidefinite matrix and B is a positive definite matrix, both of order $n \times n$, then for any i ($i = 1, 2, \ldots, n$),

$$e_i(A)e_{\min}(B) \leq e_i(AB) \leq e_i(A)e_{\max}(B).$$

In particular,

$$e_{\min}(A)e_{\min}(B) \leq e_i(AB) \leq e_{\max}(A)e_{\max}(B).$$

Furthermore, if A is positive definite, then for any i $(i = 1, 2, \ldots, n)$,

$$\frac{e_i^2(\boldsymbol{AB})}{e_{\max}(\boldsymbol{A})e_{\max}(\boldsymbol{B})} \le e_i(\boldsymbol{A})e_i(\boldsymbol{B}) \le \frac{e_i^2(\boldsymbol{AB})}{e_{\min}(\boldsymbol{A})e_{\min}(\boldsymbol{B})}.$$

Proof. See Anderson and Gupta (1963, Corollary 2.2.1).

Theorem 3.16 Let A and $\boldsymbol{B}$ be symmetric matrices of order $n \times n$. Then,

(a) $e_i(A) \le e_i(A + \boldsymbol{B})$, $\quad i = 1, 2, \ldots, n$, if $\boldsymbol{B}$ is nonnegative definite.

(b) $e_i(A) < e_i(A + \boldsymbol{B})$, $\quad i = 1, 2, \ldots, n$, if $\boldsymbol{B}$ is positive definite.

Proof. See Bellman (1997, Theorem 3, p. 117).

Corollary 3.2 Let A be a positive definite matrix and $\boldsymbol{B}$ be a positive semidefinite matrix, both of order $n \times n$. Then,

$$det(\boldsymbol{A}) \le det(\boldsymbol{A} + \boldsymbol{B}).$$

Equality holds if and only if $\boldsymbol{B} = \boldsymbol{0}$.

Proof. See Magnus and Neudecker (1988, Theorem 22, p. 21).

Theorem 3.17 (Schur's Theorem) Let $\boldsymbol{A} = (a_{ij})$ be a symmetric matrix of order $n \times n$, and let $\| \boldsymbol{A} \|_2$ denote its Euclidean norm, that is,

$$\| \boldsymbol{A} \|_2 = \left(\sum_{i=1}^{n} \sum_{j=1}^{n} a_{ij}^2 \right)^{\frac{1}{2}}.$$

Then,

$$\sum_{i=1}^{n} e_i^2(\boldsymbol{A}) = \| \boldsymbol{A} \|_2^2 .$$

Proof. See Lancaster (1969, Theorem 7.3.1).

Since $\| \boldsymbol{A} \|_2 \le n\, [\max_{i,j} | a_{ij} |]$, then from Theorem 3.17 we conclude that

$$| e_{\max}(\boldsymbol{A}) | \le n \left[\max_{i,j} | a_{ij} | \right].$$

Theorem 3.18 Let A be a symmetric matrix of order $n \times n$, and let m and s be scalars defined as

$$m = \frac{tr(\boldsymbol{A})}{n}, \quad s = \left[\frac{tr(\boldsymbol{A}^2)}{n} - m^2 \right]^{\frac{1}{2}}.$$

Then,

(a) $m - s(n-1)^{\frac{1}{2}} \le e_{\min}(A) \le m - \frac{s}{(n-1)^{\frac{1}{2}}}$.

(b) $m + \frac{s}{(n-1)^{\frac{1}{2}}} \le e_{\max}(A) \le m + s(n-1)^{\frac{1}{2}}$.

(c) $e_{\max}(A) - e_{\min}(A) \le s(2n)^{\frac{1}{2}}$.

Proof. See Wolkowicz and Styan (1980, Theorems 2.1 and 2.5).

3.13 Function of Matrices

Consider the function $f(x)$ defined on R, the set of all real numbers. If in the formula for $f(x)$, x is replaced with a matrix $\boldsymbol{A}$, then $f(\boldsymbol{A})$ is said to be a *matrix function*. We have already seen examples of such a function. For example, the function $f(\boldsymbol{A}) = \boldsymbol{A}^{-\frac{1}{2}}$ defined in Section 3.12 is for a positive definite matrix $\boldsymbol{A}$. Another example is the exponential function $\exp(\boldsymbol{A})$, where $\boldsymbol{A}$ is an $n \times n$ matrix, and, if we recall from Section 3.9.1, $\exp(\boldsymbol{A})$ is expressible as the sum of the power series,

$$\exp(\boldsymbol{A}) = \boldsymbol{I}_n + \sum_{i=1}^{\infty} \frac{1}{i!}\boldsymbol{A}^i. \tag{3.11}$$

Obviously, this representation is meaningful provided that the infinite series in (3.11) is convergent. In order to understand such convergence, the following definitions are needed:

Definition 3.5 Let $\boldsymbol{A}$ be a matrix of order $m \times n$. A *norm* of $\boldsymbol{A}$, denoted by $\| \boldsymbol{A} \|$, is a scalar function with the following properties:

(a) $\| \boldsymbol{A} \| \ge 0$, and $\| \boldsymbol{A} \| = 0$ if and only if $\boldsymbol{A} = \boldsymbol{0}$.

(b) $\| c\boldsymbol{A} \| = | c | \, \| \boldsymbol{A} \|$, where c is a scalar.

(c) $\| \boldsymbol{A} + \boldsymbol{B} \| \le \| \boldsymbol{A} \| + \| \boldsymbol{B} \|$, where $\boldsymbol{B}$ is any matrix of order $m \times n$.

(d) $\| \boldsymbol{AC} \| \le \| \boldsymbol{A} \| \, \| \boldsymbol{C} \|$, where $\boldsymbol{C}$ is any matrix for which the product $\boldsymbol{AC}$ is defined.

An example of a matrix norm is the *Euclidean norm* $\| \boldsymbol{A} \|_2$ defined as

$$\| \boldsymbol{A} \|_2 = \left(\sum_{i=1}^{m} \sum_{j=1}^{n} a_{ij}^2 \right)^{\frac{1}{2}} \tag{3.12}$$

for a matrix $\boldsymbol{A} = (a_{ij})$ of order $m \times n$. Another example of a matrix norm is the *spectral norm* $\parallel \boldsymbol{A} \parallel_s = \left[e_{\max}(\boldsymbol{A}'\boldsymbol{A})\right]^{\frac{1}{2}}$, where, if we recall, $e_{\max}(\boldsymbol{A}'\boldsymbol{A})$ is the largest eigenvalue of $\boldsymbol{A}'\boldsymbol{A}$.

Definition 3.6 Let $\{\boldsymbol{A}_k\}_{k=1}^{\infty}$ be an infinite sequence of matrices of order $m \times n$. The infinite series $\sum_{k=1}^{\infty} \boldsymbol{A}_k$ is said to converge to the $m \times n$ matrix $\boldsymbol{S} = (s_{ij})$ if the series $\sum_{k=1}^{\infty} a_{ijk}$ converges for all $i = 1, 2, \ldots, m$; $j = 1, 2, \ldots, n$, where a_{ijk} is the (i, j)th element of $\boldsymbol{A}_k$, and

$$\sum_{k=1}^{\infty} a_{ijk} = s_{ij}, \quad i = 1, 2, \ldots, m; \quad j = 1, 2, \ldots, n. \tag{3.13}$$

The series $\sum_{k=1}^{\infty} \boldsymbol{A}_k$ is divergent if at least one of the series in (3.13) is divergent.

Theorem 3.19 Let $\boldsymbol{A}$ be a symmetric matrix of order $n \times n$ such that $\parallel \boldsymbol{A} \parallel < 1$, where $\parallel \boldsymbol{A} \parallel$ is any matrix norm of $\boldsymbol{A}$. Then, $\sum_{k=0}^{\infty} \boldsymbol{A}^k$ converges to $(\boldsymbol{I}_n - \boldsymbol{A})^{-1}$, where $\boldsymbol{A}^0 = \boldsymbol{I}_n$.

Proof. See Khuri (2003, Corollary 5.5.1, p. 181).

Let us now consider a general method for determining convergence of a power series in a square matrix $\boldsymbol{A}$.

Theorem 3.20 Suppose that the $n \times n$ matrix $\boldsymbol{A}$ is diagonalizable, that is, there exits a nonsingular matrix $\boldsymbol{Q}$ such that

$$\boldsymbol{Q}^{-1}\boldsymbol{A}\boldsymbol{Q} = \text{diag}(\lambda_1, \lambda_2, \ldots, \lambda_n),$$

where $\lambda_1, \lambda_2, \ldots, \lambda_n$ are the eigenvalues of $\boldsymbol{A}$, which are not necessarily real valued. Let $f(z)$ be an analytic function defined on an open set S containing all the eigenvalues of $\boldsymbol{A}$. Then,

(a) The function $f(\boldsymbol{A})$ is defined as

$$f(\boldsymbol{A}) = \boldsymbol{Q}\,\text{diag}\left[f(\lambda_1), f(\lambda_2), \ldots, f(\lambda_n)\right]\boldsymbol{Q}^{-1}.$$

(b) The function $f(\boldsymbol{A})$ can be represented as the sum of a convergent power series of the form

$$f(\boldsymbol{A}) = \sum_{k=0}^{\infty} c_k \boldsymbol{A}^k,$$

if the power series $\sum_{k=0}^{\infty} c_k \lambda_1^k$, $\sum_{k=0}^{\infty} c_k \lambda_2^k, \ldots, \sum_{k=0}^{\infty} c_k \lambda_n^k$ are all convergent and represent $f(\lambda_1), f(\lambda_2), \ldots, f(\lambda_n)$, respectively.

Proof. See Golub and Van Loan (1983) [part (a) is Corollary 11.1.2, p. 382 and part (b) is Theorem 11.2.3, p. 390].

For example, for the matrix functions $\exp(A)$, $\sin(A)$, $\cos(A)$ we have the following series representations:

$$\exp(A) = \sum_{k=0}^{\infty} \frac{1}{k!} A^k,$$

$$\sin(A) = \sum_{k=0}^{\infty} (-1)^k \frac{1}{(2k+1)!} A^{2k+1},$$

$$\cos(A) = \sum_{k=0}^{\infty} (-1)^k \frac{1}{(2k)!} A^{2k}.$$

Corollary 3.3 Let A be a symmetric matrix of order $n \times n$ such that $|\lambda_i| < 1$ for $i = 1, 2, \ldots, n$, where λ_i is the ith eigenvalue of A. Then, the series $\sum_{k=0}^{\infty} A^k$ converges to $(I_n - A)^{-1}$.

Proof. The matrix A is diagonalizable by the Spectral Decomposition Theorem (Theorem 3.4). Furthermore, since $|\lambda_i| < 1$, the power series $\sum_{k=0}^{\infty} \lambda_i^k$ is absolutely convergent, hence convergent, and

$$\sum_{k=0}^{\infty} \lambda_i^k = \frac{1}{1 - \lambda_i}, \quad i = 1, 2, \ldots, n.$$

It follows from Theorem 3.20 that

$$\sum_{k=0}^{\infty} A^k = (I_n - A)^{-1}.$$

□

Another series representation of a well-known matrix function is given by

$$\log(I_n - A) = \sum_{k=1}^{\infty} \frac{1}{k} A^k,$$

where A is a symmetric matrix whose eigenvalues fall inside the open interval $(-1, 1)$.

3.14 Matrix Differentiation

In some cases, there may be a need to take the derivative of a matrix function. The use of this derivative can greatly simplify certain computations such as the finding of the optimum of a particular matrix function, as is the case with *maximum likelihood* and *least-squares* estimation techniques. Several theorems on matrix differentiation are now stated. Further details can be found in, for example, Dwyer (1967), Neudecker (1969), Nel (1980), Rogers (1980), Graham (1981, Chapters 4 through 6), Searle (1982, Chapter 12), Graybill (1983, Chapter 10), and Harville (1997, Chapter 15).

Definition 3.7 Derivative of a scalar function of $\boldsymbol{X}$ with respect to $\boldsymbol{X}$.

Let $\boldsymbol{X} = (x_{ij})$ be an $m \times n$ matrix whose elements are mathematically independent real variables. Let $f(\boldsymbol{X})$ be a real-valued matrix function that depends on $\boldsymbol{X}$. Then, the derivative of $f(\boldsymbol{X})$ with respect to $\boldsymbol{X}$, denoted by $\frac{\partial f(\boldsymbol{X})}{\partial \boldsymbol{X}}$, is an $m \times n$ matrix whose (i,j)th element is the partial derivative $\frac{\partial f}{\partial x_{ij}}$, $i = 1,2,\ldots,m$; $j = 1,2,\ldots,n$. In particular, if $\boldsymbol{X}$ is a column vector $\boldsymbol{x} = (x_1, x_2, \ldots, x_m)'$ of m elements, then $\frac{\partial f(\boldsymbol{x})}{\partial \boldsymbol{x}}$ is a column vector whose ith element is $\frac{\partial f(\boldsymbol{x})}{\partial x_i}$, $i = 1,2,\ldots,m$. Similarly, if $\boldsymbol{X}$ is a row vector $\boldsymbol{x}' = (x_1, x_2, \ldots, x_n)$ of n elements, then $\frac{\partial f(\boldsymbol{x})}{\partial \boldsymbol{x}'}$ is a row vector whose ith element is $\frac{\partial f(\boldsymbol{x})}{\partial x_i}$, $i = 1,2,\ldots,n$.

Definition 3.8 Derivative of a vector function of $\boldsymbol{x}$ with respect to $\boldsymbol{x}$.

Let $\boldsymbol{y}(\boldsymbol{x}) = [y_1(\boldsymbol{x}), y_2(\boldsymbol{x}), \ldots, y_n(\boldsymbol{x})]'$ be a vector function whose elements are scalar functions of $\boldsymbol{x} = (x_1, x_2, \ldots, x_m)'$. Then, the derivative of $\boldsymbol{y}'$ with respect to $\boldsymbol{x}$ is the $m \times n$ matrix

$$\frac{\partial \boldsymbol{y}'(\boldsymbol{x})}{\partial \boldsymbol{x}} = \left[\frac{\partial y_1(\boldsymbol{x})}{\partial \boldsymbol{x}} : \frac{\partial y_2(\boldsymbol{x})}{\partial \boldsymbol{x}} : \cdots : \frac{\partial y_n(\boldsymbol{x})}{\partial \boldsymbol{x}}\right].$$

The transpose of this matrix is denoted by $\frac{\partial \boldsymbol{y}(\boldsymbol{x})}{\partial \boldsymbol{x}'}$.

If $\boldsymbol{y}(\boldsymbol{x})$ is a vector-valued function of $\boldsymbol{x}$ and $\boldsymbol{x}$ is a vector-valued function of $\mathbf{z}$ with q elements, then

$$\frac{\partial \boldsymbol{y}(\boldsymbol{x})}{\partial \mathbf{z}'} = \frac{\partial \boldsymbol{y}(\boldsymbol{x})}{\partial \boldsymbol{x}'} \frac{\partial \boldsymbol{x}}{\partial \mathbf{z}'}.$$

Here, $\frac{\partial \boldsymbol{y}(\boldsymbol{x})}{\partial \boldsymbol{x}'}$ is an $n \times m$ matrix and $\frac{\partial \boldsymbol{x}}{\partial \mathbf{z}'}$ is an $m \times q$ matrix resulting in $\frac{\partial \boldsymbol{y}(\boldsymbol{x})}{\partial \mathbf{z}'}$ being an $n \times q$ matrix. This formula gives the so-called *vector chain rule*.

Definition 3.9 Derivative of a matrix with respect to a scalar.

Let $\boldsymbol{Y} = (y_{ij})$ be an $m \times n$ matrix whose elements depend on p mathematically independent real variables, $u_1, u_2, \ldots, u_p$. Then, the partial derivative of $\boldsymbol{Y}$ with respect to u_k ($k = 1,2,\ldots,p$) is the matrix $\frac{\partial \boldsymbol{Y}}{\partial u_k}$ whose (i,j)th element is $\frac{\partial y_{ij}}{\partial u_k}$.

Definition 3.10 Derivative of a matrix with respect to another matrix.

Let $Y = (y_{ij})$ be matrix of order $m \times n$ whose elements are functions of the elements of a matrix $X = (x_{ij})$, which is of order $p \times q$. Then, the derivative of Y with respect to X, denoted by $\frac{\partial Y}{\partial X}$, is given by the partitioned matrix

$$\frac{\partial Y}{\partial X} = \begin{bmatrix} \frac{\partial Y}{\partial x_{11}} & \frac{\partial Y}{\partial x_{12}} & \cdots & \frac{\partial Y}{\partial x_{1q}} \\ \frac{\partial Y}{\partial x_{21}} & \frac{\partial Y}{\partial x_{22}} & \cdots & \frac{\partial Y}{\partial x_{2q}} \\ \cdots & \cdots & \cdots & \cdots \\ \frac{\partial Y}{\partial x_{p1}} & \frac{\partial Y}{\partial x_{p2}} & \cdots & \frac{\partial Y}{\partial x_{pq}} \end{bmatrix},$$

where, as in Definition 3.9, $\frac{\partial Y}{\partial x_{ij}}$ is the derivative of Y with respect to x_{ij}. Since $\frac{\partial Y}{\partial x_{ij}}$ is of order $m \times n$, the matrix $\frac{\partial Y}{\partial X}$ is of order $mp \times nq$.

For example, if

$$X = \begin{bmatrix} x_{11} & x_{12} \\ x_{21} & x_{22} \end{bmatrix},$$

and

$$Y = \begin{bmatrix} x_{11}x_{12} & x_{21}^2 \\ \sin(x_{11} + x_{21}) & e^{x_{22}} \end{bmatrix},$$

then,

$$\frac{\partial Y}{\partial x_{11}} = \begin{bmatrix} x_{12} & 0 \\ \cos(x_{11} + x_{21}) & 0 \end{bmatrix},$$

$$\frac{\partial Y}{\partial x_{12}} = \begin{bmatrix} x_{11} & 0 \\ 0 & 0 \end{bmatrix},$$

$$\frac{\partial Y}{\partial x_{21}} = \begin{bmatrix} 0 & 2x_{21} \\ \cos(x_{11} + x_{21}) & 0 \end{bmatrix},$$

$$\frac{\partial Y}{\partial x_{22}} = \begin{bmatrix} 0 & 0 \\ 0 & e^{x_{22}} \end{bmatrix}.$$

Hence,

$$\frac{\partial Y}{\partial X} = \begin{bmatrix} x_{12} & 0 & x_{11} & 0 \\ \cos(x_{11} + x_{21}) & 0 & 0 & 0 \\ 0 & 2x_{21} & 0 & 0 \\ \cos(x_{11} + x_{21}) & 0 & 0 & e^{x_{22}} \end{bmatrix}.$$

Theorem 3.21 Let $f(x) = x'a$, where a is a constant vector. Then,

$$\frac{\partial(x'\mathbf{a})}{\partial x} = \mathbf{a}.$$

Corollary 3.4 Let A be a constant matrix of order $m \times n$. Then,

$$\frac{\partial(x'A)}{\partial x} = A.$$

Theorem 3.22 Let $f(x) = x'Ax$, where A is a constant symmetric matrix. Then,

$$\frac{\partial f(x)}{\partial x} = 2Ax.$$

Theorem 3.23 Let $X = (x_{ij})$ be an $n \times n$ matrix of real variables.

(a) If the elements of X are mathematically independent, then

$$\frac{\partial[det(X)]}{\partial X} = (x_{ij}^c),$$

where x_{ij}^c is the cofactor of x_{ij} $(i, j = 1, 2, \ldots, n)$.

(b) If X is symmetric, and its elements are mathematically independent, except for $x_{ij} = x_{ji}$, then

$$\frac{\partial[det(X)]}{\partial X} = 2\left(x_{ij}^c\right) - \text{diag}(x_{11}^c, x_{22}^c, \ldots, x_{nn}^c).$$

Corollary 3.5 Let $X = (x_{ij})$ be an $n \times n$ nonsingular matrix such that $det(X) > 0$.

(a) If the elements of X are mathematically independent, then

$$\frac{\partial \left\{\log[det(X)]\right\}}{\partial X} = (X^{-1})'.$$

(b) If X is symmetric and its elements are mathematically independent, except for $x_{ij} = x_{ji}$, $i \neq j$, then

$$\frac{\partial \left\{\log[det(X)]\right\}}{\partial X} = 2X^{-1} - \text{diag}(x^{11}, x^{22}, \ldots, x^{nn}),$$

where x^{ii} is the ith diagonal element of X^{-1} $(i = 1, 2, \ldots, n)$.

Theorem 3.24 Let X be an $n \times n$ nonsingular matrix whose elements are functions of a scalar t. Then,

$$\frac{\partial X^{-1}}{\partial t} = -X^{-1}\frac{\partial X}{\partial t}X^{-1}.$$

Corollary 3.6 Let $X = (x_{ij})$ be an $n \times n$ nonsingular matrix.

(a) If the elements of X are mathematically independent, then

$$\frac{\partial X^{-1}}{\partial x_{ij}} = -X^{-1}\Delta_{ij}X^{-1},$$

where Δ_{ij} is an $n \times n$ matrix whose (i,j)th element is equal to 1 and the remaining elements are equal to 0.

(b) If X is symmetric and its elements are mathematically independent, except for $x_{ij} = x_{ji}$, $i \neq j$, then

$$\frac{\partial X^{-1}}{\partial x_{ij}} = -X^{-1}\Delta^{*}_{ij}X^{-1},$$

where Δ^{*}_{ij} is an $n \times n$ matrix whose (i,j)th and (j,i)th elements are equal to 1 and the remaining elements are equal to 0.

Exercises

3.1 Let X be an $n \times p$ matrix of rank p $(n \geq p)$. What is the rank of $XX'X$?

3.2 Let A and B be $n \times n$ symmetric matrices, and let $C = AB - BA$.

(a) Show that $tr(CC') = 2\,tr(A^2B^2) - 2\,tr[(AB)^2]$.

(b) Deduce from (a) that

$$tr[(AB)^2] \leq tr(A^2B^2).$$

(c) Under what condition is the equality in (b) attained?

3.3 Let A be a positive definite matrix of order $n \times n$. Show that

$$det(A) \leq \prod_{i=1}^{n} a_{ii},$$

where a_{ii} is the ith diagonal element of A $(i = 1, 2, \ldots, n)$.

[Hint: Prove this inequality by mathematical induction.]

3.4 Let A be an $n \times n$ matrix that satisfies the equation

$$A^2 + 2A + I_n = \mathbf{0}.$$

(a) Show that $\boldsymbol{A}$ is nonsingular.

(b) How would you evaluate $\boldsymbol{A}^{-1}$?

3.5 Let $\boldsymbol{A} = (a_{ij})$ be an $n \times n$ matrix and let $adj\, \boldsymbol{A}$ be its adjoint. Show that

$$\begin{aligned} \boldsymbol{A}(adj\, \boldsymbol{A}) &= (adj\, \boldsymbol{A})\boldsymbol{A} \\ &= det(\boldsymbol{A})\boldsymbol{I}_n. \end{aligned}$$

3.6 Consider the characteristic equation for an $n \times n$ matrix, which can be written as

$$\begin{aligned} det(\boldsymbol{A} - \lambda \boldsymbol{I}_n) &= a_0 + a_1\lambda + a_2\lambda^2 + \cdots + a_n\lambda^n \\ &= 0, \end{aligned}$$

where $a_0, a_1, \ldots, a_n$ are known coefficients that depend on the elements of $\boldsymbol{A}$.

(a) Show that $\boldsymbol{A}$ satisfies its characteristic equation, that is,

$$a_0\boldsymbol{I}_n + a_1\boldsymbol{A} + a_2\boldsymbol{A}^2 + \cdots + a_n\boldsymbol{A}^n = \boldsymbol{0}.$$

This result is known as the *Cayley–Hamilton Theorem*.
[Hint: Let $\boldsymbol{B} = adj\ (\boldsymbol{A} - \lambda \boldsymbol{I}_n)$. The elements of $\boldsymbol{B}$ are polynomials of degree $n-1$ or less in λ. We can then express $\boldsymbol{B}$ as

$$\boldsymbol{B} = \boldsymbol{B}_0 + \lambda \boldsymbol{B}_1 + \lambda^2 \boldsymbol{B}_2 + \cdots + \lambda^{n-1}\boldsymbol{B}_{n-1},$$

where $\boldsymbol{B}_0, \boldsymbol{B}_1, \ldots, \boldsymbol{B}_{n-1}$ are matrices of order $n \times n$ that do not depend on λ. Applying now the result in Exercise 3.5 to $\boldsymbol{B}$, we get

$$\begin{aligned} (\boldsymbol{A} - \lambda \boldsymbol{I}_n)\boldsymbol{B} &= det(\boldsymbol{A} - \lambda \boldsymbol{I}_n)\boldsymbol{I}_n \\ &= (a_0 + a_1\lambda + \cdots + a_n\lambda^n)\boldsymbol{I}_n. \end{aligned}$$

By comparing the coefficients of the powers of λ on both sides of this equation, we conclude that

$$a_0\boldsymbol{I}_n + a_1\boldsymbol{A} + a_2\boldsymbol{A}^2 + \cdots + a_n\boldsymbol{A}^n = \boldsymbol{0}].$$

(b) Deduce from (a) that if $\lambda_1, \lambda_2, \ldots, \lambda_n$ are the eigenvalues of $\boldsymbol{A}$, then

$$(\boldsymbol{A} - \lambda_1 \boldsymbol{I}_n)(\boldsymbol{A} - \lambda_2 \boldsymbol{I}_n) \ldots (\boldsymbol{A} - \lambda_n \boldsymbol{I}_n) = \boldsymbol{0}.$$

(c) Show how to obtain the inverse of $\boldsymbol{A}$, if $\boldsymbol{A}$ is nonsingular, using the Cayley–Hamilton Theorem.

3.7 Verify the Cayley–Hamilton Theorem for the 3×3 matrix

$$\boldsymbol{A} = \begin{bmatrix} 2 & 2 & 1 \\ 1 & 3 & 1 \\ 1 & 2 & 2 \end{bmatrix}.$$

3.8 Two $n \times n$ matrices, $\boldsymbol{A}$ and $\boldsymbol{B}$, are said to be similar if there is a nonsingular $n \times n$ matrix $\boldsymbol{P}$ such that $\boldsymbol{A} = \boldsymbol{PBP}^{-1}$. Show that an $n \times n$ matrix $\boldsymbol{A}$ is similar to a diagonal matrix if and only if $\boldsymbol{A}$ has n linearly independent eigenvectors.

3.9 Let $\boldsymbol{A}$ be an $n \times n$ matrix. Show that if the eigenvalues of $\boldsymbol{A}$ are distinct, then $\boldsymbol{A}$ is similar to a diagonal matrix.

3.10 Let $\boldsymbol{A}$ and $\boldsymbol{B}$ be matrices of order $n \times n$; $\boldsymbol{B}$ is similar to $\boldsymbol{A}$ and $\boldsymbol{A} = \boldsymbol{CD}$, where $\boldsymbol{C}$ and $\boldsymbol{D}$ are symmetric. Show that $\boldsymbol{B}$ can be written as the product of two symmetric matrices.

3.11 Let $\boldsymbol{A}$ be a nonsingular matrix of order $n \times n$ and $\boldsymbol{a}$ be an $n \times 1$ vector. Show that the matrix $\boldsymbol{B}$ is nonsingular, where

$$\boldsymbol{B} = \boldsymbol{A} - \frac{1}{c}\boldsymbol{a}\boldsymbol{a}',$$

where c is a nonzero scalar such that $c \neq \boldsymbol{a}'\boldsymbol{A}^{-1}\boldsymbol{a}$.

3.12 Let $\boldsymbol{A}$ be the 2×2 matrix

$$\boldsymbol{A} = \begin{bmatrix} 0 & -2 \\ 1 & 3 \end{bmatrix}.$$

(a) Show that $\boldsymbol{A}$ can be written as

$$\boldsymbol{A} = \boldsymbol{Q}\begin{bmatrix} 1 & 0 \\ 0 & 2 \end{bmatrix}\boldsymbol{Q}^{-1},$$

where

$$\boldsymbol{Q} = \begin{bmatrix} 2 & 1 \\ -1 & -1 \end{bmatrix}$$

(b) Show that

$$\exp(\boldsymbol{A}) = \begin{bmatrix} 2e - e^2 & 2e - 2e^2 \\ e^2 - e & 2e^2 - e \end{bmatrix},$$

where $e = \exp(1)$.

3.13 Suppose that $\boldsymbol{A}$ is similar to a diagonal matrix (see Exercise 3.8). Show that

$$det[\exp(\boldsymbol{A})] = \exp[tr(\boldsymbol{A})].$$

3.14 Let $\boldsymbol{A}$ be a positive semidefinite matrix of order $n \times n$. Show that

$$[det(\boldsymbol{A})]^{1/n} \leq \frac{1}{n}tr(\boldsymbol{A}).$$

Under what condition does equality hold?

3.15 Let A be a positive definite matrix matrix of order $n \times n$. Show that

$$tr(A)tr(A^{-1}) \geq n^2.$$

Under what condition does equality hold?

3.16 Let $A = (a_{ij})$ be a symmetric matrix of order $n \times n$. Show that

$$e_{\min}(A) \leq a_{ii} \leq e_{\max}(A), \quad i = 1, 2, \ldots, n.$$

3.17 Suppose that A is a skew-symmetric matrix such that $A^2 = -I$. Show that A must be an orthogonal matrix.

3.18 Let A and B be symmetric $n \times n$ matrices. Show that

$$tr(AB) \leq \frac{1}{2} tr(A^2 + B^2).$$

Under what condition does equality hold?

3.19 Consider the orthogonal matrix

$$A = \frac{1}{3} \begin{bmatrix} 1 & 2 & 2 \\ 2 & 1 & -2 \\ -2 & 2 & -1 \end{bmatrix}.$$

Find a skew-symmetric matrix, T, such that $A = \exp(T)$.

3.20 Let A and B be matrices of order $m \times n$ $(m \geq n)$.

(a) Show that if $det(A'B) \neq 0$, then both A and B have full column ranks.

(b) Show that if the condition in (a) is valid, then

$$det(B'B) \geq det[B'A(A'A)^{-1}A'B].$$

(c) Show that whether the condition in (a) is satisfied or not,

$$[det(A'B)]^2 \leq det(A'A)det(B'B).$$

3.21 Let A and B be two symmetric matrices of order $n \times n$. Show that for any scalar α, $0 \leq \alpha \leq 1$,

$$e_{\min}[\alpha A + (1-\alpha)B] \geq \alpha\, e_{\min}(A) + (1-\alpha)\, e_{\min}(B)$$
$$e_{\max}[\alpha A + (1-\alpha)B] \leq \alpha\, e_{\max}(A) + (1-\alpha)\, e_{\max}(B)$$

The first and second inequalities indicate that the smallest eigenvalue and the largest eigenvalue functions are concave and convex, respectively, on the space of symmetric matrices.

3.22 Let A be an $n \times n$ matrix. Show that

$$\exp(A \otimes I_m) = \exp(A) \otimes I_m.$$

3.23 Let A be a positive definite matrix of order $n \times n$ with eigenvalues $\lambda_1 \geq \lambda_2 \geq \cdots \geq \lambda_n > 0$. Show that

$$1 \leq (x'Ax)(x'A^{-1}x) \leq \frac{1}{4}\left[\left(\frac{\lambda_1}{\lambda_n}\right)^{1/2} + \left(\frac{\lambda_n}{\lambda_1}\right)^{1/2}\right]^2,$$

for any vector x of n elements and unit length. This inequality is known as the *Kantorovich inequality* (see Marcus and Minc, 1964, p. 117).

3.24 Let $a_1, a_2, \ldots, a_n$ and $b_1, b_2, \ldots, b_n$ be two sets of vectors of m elements each. Prove the following identity:

$$\left(\sum_{i=1}^{n} a_i a_i'\right)\left(\sum_{j=1}^{n} b_j b_j'\right) = \left(\sum_{i=1}^{n} a_i b_i'\right)^2 + \sum_{i<j}(a_i' b_j - a_j' b_i)(a_i b_j' - a_j b_i').$$

This identity can also be expressed as

$$AA'BB' = AB'AB' + \sum_{i<j}(a_i' b_j - a_j' b_i)(a_i b_j' - a_j b_i'),$$

where $A = [a_1 : a_2 : \cdots : a_n]$, $B = [b_1 : b_2 : \cdots : b_n]$.

Note: This identity is a generalization of the so-called *Lagrange identity* for real numbers, namely,

$$\left(\sum_{i=1}^{n} \alpha_i \beta_i\right)^2 = \left(\sum_{i=1}^{n} \alpha_i^2\right)\left(\sum_{j=1}^{n} \beta_j^2\right) - \sum_{i<j}(\alpha_i \beta_j - \alpha_j \beta_i)^2,$$

where $\alpha_1, \alpha_2, \ldots, \alpha_n$ and $\beta_1, \beta_2, \ldots, \beta_n$ are two sets of real numbers. Such an identity can be viewed as a generalization of the well-known *Cauchy–Schwarz inequality* since it implies that

$$\left(\sum_{i=1}^{n} \alpha_i \beta_i\right)^2 \leq \left(\sum_{i=1}^{n} \alpha_i^2\right)\left(\sum_{j=1}^{n} \beta_j^2\right).$$

For more details, see Trenkler (2004).

3.25 Let A and B be two symmetric idempotent matrices of order $n \times n$. Show that the following statements are equivalent:

(a) $AB = B$.

(b) $A - B$ is idempotent.

(c) The column space of B is a vector subspace of the column space of A.

3.26 Let A and B be two matrices of order $n \times n$. Show that the following conditions are equivalent:

(a) $AA'BB' = AB'AB'$.

(b) $tr(AA'BB') = tr(AB'AB')$.

(c) $A'B = B'A$.

3.27 Let A and B be $n \times n$ orthogonal matrices. Show that the following conditions are equivalent:

(a) $(AB)^2 = I_n$.

(b) $tr[(AB)^2] = n$.

(c) $BA = (BA)'$.

3.28 (a) Show that $\frac{\partial[tr(AX)]}{\partial X} = A'$, where A is of order $n \times m$ and X is of order $m \times n$.

(b) Show that $\frac{\partial[tr(X'AX)]}{\partial X} = (A' + A)X$, where A is of order $m \times m$ and X is of order $m \times n$.

(c) Show that $\frac{\partial[tr(XAX')]}{\partial X} = X(A + A')$, where A is of order $n \times n$ and X is of order $m \times n$.

3.29 Let X be a matrix of order $m \times n$ and let A be a symmetric matrix of order $n \times n$. Show that $\frac{\partial det(C)}{\partial X} X' = 2\, det(C) I_m$, where $C = XAX'$. (See Wolkowicz, 1994, p. 658.)

3.30 Consider Definition 3.10.

(a) Let X and Y be matrices of orders $p \times q$ and $m \times n$, respectively. Show that $\frac{\partial Y}{\partial X}$ can be expressed as

$$\frac{\partial Y}{\partial X} = \sum_{i=1}^{p} \sum_{j=1}^{q} \mathrm{E}_{ij} \otimes \frac{\partial Y}{\partial x_{ij}},$$

where E_{ij} is a matrix of order $p \times q$ whose elements are equal to 0, except for the (i, j)th element, which is equal to 1.

(b) If X, Y, Z are matrices of orders $p \times q$, $q \times r$, and $s \times t$, respectively, then show that

$$\frac{\partial (XY)}{\partial Z} = \frac{\partial X}{\partial Z}(I_t \otimes Y) + (I_s \otimes X)\frac{\partial Y}{\partial Z}.$$

4

The Multivariate Normal Distribution

This chapter provides an exposition concerning the normal distribution, its properties and characterizing functions, in addition to other distributions that are derived from it. The importance of this distribution stems in part from its being the cornerstone upon which many theorems in classical linear model methodology are based.

The reader is expected to be acquainted with the concepts of discrete and continuous random variables, their independence, and probability distributions. In particular, a continuous random variable, X, is said to be *absolutely continuous* if its cumulative distribution function, namely, $F(x) = P(X \leq x)$, is differentiable for all x in a set $\mathcal{A}$. In this case, the derivative of $F(x)$ is called the *density function* of X and is denoted by $f(x)$. The function $F(x)$ can therefore be expressed as an integral of the form

$$F(x) = \int_{-\infty}^{x} f(t)dt, \quad x \in \mathcal{A}.$$

In general, a continuous random variable is not necessarily absolutely continuous. For more information concerning random variables and their associated distributions, the reader is referred to, for example, Casella and Berger (2002), Hogg and Craig (1978), Lindgren (1976), and Mood, Graybill, and Boes (1973).

4.1 History of the Normal Distribution

The normal distribution was first introduced by Abraham de Moivre (1667–1754) in an article in 1733 in the context of approximating the binomial distribution. His article was not discovered until 1924 by Karl Pearson (1857–1936). Carl Friedrich Gauss (1777–1855) justified the method of least squares rigorously in 1809 by assuming a normal distribution in connection with the analysis of measurement errors. He also used it to analyze astronomical data in 1809. The name "normal distribution" was coined independently by Charles Peirce (1839–1914), Francis Galton (1822–1911), and Wilhelm Lexis (1837–1914) around 1875 (see Stigler, 1986).

4.2 The Univariate Normal Distribution

The normal distribution belongs to a family of absolutely continuous distributions. Its density function, $f(x)$, is given by

$$f(x) = \frac{1}{\sqrt{2\pi\sigma^2}} \exp\left[-\frac{1}{2\sigma^2}(x-\mu)^2\right], \quad -\infty < x < \infty. \tag{4.1}$$

A random variable, X, having this density function is said to be *normally distributed*, and this fact is denoted by writing $X \sim N(\mu, \sigma^2)$. In formula (4.1), $\mu = E(X)$ is the mean, or expected value, of X, and $\sigma^2 = \text{Var}(X)$ is the variance of X. In particular, if $\mu = 0$ and $\sigma^2 = 1$, then X is said to have the *standard normal distribution*, which is usually denoted by Z, that is, $Z \sim N(0, 1)$. Using (4.1), Z has the density function,

$$g(z) = \frac{1}{\sqrt{2\pi}} \exp\left(-\frac{z^2}{2}\right), \quad -\infty < z < \infty. \tag{4.2}$$

The moment generating function (abbreviated m.g.f.) of $X \sim N(\mu, \sigma^2)$ is of the form

$$\begin{aligned}\phi_X(t) &= E(e^{tX}) \\ &= \int_{-\infty}^{\infty} e^{tx} \frac{1}{\sqrt{2\pi\sigma^2}} \exp\left[-\frac{1}{2\sigma^2}(x-\mu)^2\right] dx \\ &= \exp\left(\mu t + \frac{1}{2}\sigma^2 t^2\right), \end{aligned} \tag{4.3}$$

where t is a mathematical variable. This function is defined for all values of t in R, the set of all real numbers. The m.g.f. provides a characterization of the normal distribution. It can also be used to derive all the noncentral moments of X, that is, values of $\mu_n' = E(X^n)$ for $n = 1, 2, \ldots$.

To evaluate μ_n', the nth derivative of $\phi_X(t)$ is evaluated at $t = 0$, that is,

$$\mu_n' = \left.\frac{d^n[\phi_X(t)]}{dt^n}\right|_{t=0}, \quad n = 1, 2, \ldots .$$

Note that $\phi_X(t)$ in (4.3) has derivatives of all orders in R. Equivalently, values of μ_n' can be obtained by expanding the function $\phi_X(t)$ around $t = 0$ using Maclaurin's series. In this case, the coefficient of $\frac{t^n}{n!}$ in such an expression is equal to $\mu_n' (n = 1, 2, \ldots)$.

An alternative characterization of the normal distribution is given by the cumulant generating function defined by the formula

$$\psi_X(t) = \log[\phi_X(t)], \tag{4.4}$$

where log(.) is the natural logarithmic function. The coefficient of $\frac{t^r}{r!}$ in Maclaurin's series expansion of $\psi_X(t)$ around $t = 0$ is called the rth *cumulant* and is denoted by $\kappa_r (r = 1, 2, \ldots)$. Hence, from (4.3) we have

$$\psi_X(t) = \mu t + \frac{1}{2}\sigma^2 t^2. \tag{4.5}$$

It follows that all the cumulants of the normal distribution are zero, except for the first two, namely, $\kappa_1 = \mu$, $\kappa_2 = \sigma^2$.

4.3 The Multivariate Normal Distribution

Let $\mathbf{Z} = (Z_1, Z_2, \ldots, Z_n)'$ be a vector of n mutually independent and identically distributed random variables having the standard normal distribution $N(0, 1)$. This vector is said to have the *multivariate standard normal distribution* with a mean vector $\mathbf{0}$ and a variance–covariance matrix $\boldsymbol{I}_n$, and is represented symbolically by writing $\mathbf{Z} \sim N(\mathbf{0}, \boldsymbol{I}_n)$. Since the Z_i's are independent, their joint density function, or just the density function of $\mathbf{Z}$, denoted by $g(\boldsymbol{z})$, is the product of their marginal density functions. These are of the form given in (4.2). Hence,

$$\begin{aligned} g(\boldsymbol{z}) &= \prod_{i=1}^{n}\left[\frac{1}{\sqrt{2\pi}}\exp\left(-\frac{z_i^2}{2}\right)\right] \\ &= (2\pi)^{\frac{-n}{2}}\exp\left(-\frac{1}{2}\boldsymbol{z}'\boldsymbol{z}\right), \end{aligned} \tag{4.6}$$

where $\boldsymbol{z} \in R^n$, the n-dimensional Euclidean space.

In general, let $\boldsymbol{X} = (X_1, X_2, \ldots, X_n)'$ be a random vector such that $E(X_i) = \mu_i$ and $\text{Var}(X_i) = \sigma_{ii} (i = 1, 2, \ldots, n)$, and $\text{Cov}(X_i, X_j) = \sigma_{ij}$ is the covariance of X_i and X_j, $i \neq j$. The mean vector and variance–covariance matrix of $\boldsymbol{X}$ are

$$E(\boldsymbol{X}) = \boldsymbol{\mu} \tag{4.7}$$

$$\text{Var}(\boldsymbol{X}) = \boldsymbol{\Sigma}, \tag{4.8}$$

respectively, where $\boldsymbol{\mu} = (\mu_1, \mu_2, \ldots, \mu_n)'$, and $\boldsymbol{\Sigma} = (\sigma_{ij})$ is given by

$$\begin{aligned} \boldsymbol{\Sigma} &= E[(\boldsymbol{X} - \boldsymbol{\mu})(\boldsymbol{X} - \boldsymbol{\mu})'] \\ &= E(\boldsymbol{X}\boldsymbol{X}') - \boldsymbol{\mu}\boldsymbol{\mu}'. \end{aligned} \tag{4.9}$$

In addition, the matrix

$$\boldsymbol{R} = \boldsymbol{D}^{-1}\boldsymbol{\Sigma}\boldsymbol{D}^{-1},$$

where D a diagonal matrix whose diagonal elements are the square roots of $\sigma_{11}, \sigma_{22}, \ldots, \sigma_{nn}$, is called the *correlation matrix*. Its (i,j)th element is usually denoted by ρ_{ij}, where $\rho_{ij} = \frac{\sigma_{ij}}{\sqrt{\sigma_{ii}\sigma_{jj}}}$ is the correlation of X_i and X_j, $i, j = 1, 2, \ldots, n$.

In general, the matrix $\boldsymbol{\Sigma}$ is nonnegative definite. Formula (4.9) is a special case of a more general one giving the covariance matrix of two random vectors, $\boldsymbol{X}_1$ and $\boldsymbol{X}_2$, namely,

$$\begin{aligned}\boldsymbol{\Sigma}_{12} &= E[(\boldsymbol{X}_1 - \boldsymbol{\mu}_1)(\boldsymbol{X}_2 - \boldsymbol{\mu}_2)'] \\ &= E(\boldsymbol{X}_1\boldsymbol{X}_2') - \boldsymbol{\mu}_1\boldsymbol{\mu}_2', \end{aligned} \tag{4.10}$$

where $\boldsymbol{\mu}_1$ and $\boldsymbol{\mu}_2$ are the mean vectors of $\boldsymbol{X}_1$ and $\boldsymbol{X}_2$, respectively. If $\boldsymbol{X}_1$ has n_1 elements and $\boldsymbol{X}_2$ has n_2 elements, then $\boldsymbol{\Sigma}_{12}$ is a matrix of order $n_1 \times n_2$.

If $\boldsymbol{A}$ is a constant matrix, then it is easy to show that the mean vector and variance–covariance matrix of $\boldsymbol{Y} = \boldsymbol{A}\boldsymbol{X}$, assuming that this product is well defined, are of the form

$$E(\boldsymbol{Y}) = \boldsymbol{A}\boldsymbol{\mu} \tag{4.11}$$

$$\begin{aligned}\text{Var}(\boldsymbol{Y}) &= E[\boldsymbol{A}(\boldsymbol{X} - \boldsymbol{\mu})(\boldsymbol{X} - \boldsymbol{\mu})'\boldsymbol{A}'] \\ &= \boldsymbol{A}\boldsymbol{\Sigma}\boldsymbol{A}'. \end{aligned} \tag{4.12}$$

More generally, if $\boldsymbol{X}_1$ and $\boldsymbol{X}_2$ are two random vectors with mean vectors $\boldsymbol{\mu}_1$ and $\boldsymbol{\mu}_2$, respectively, and if $\boldsymbol{A}$ and $\boldsymbol{B}$ are constant matrices, then the covariance matrix of $\boldsymbol{A}\boldsymbol{X}_1$ and $\boldsymbol{B}\boldsymbol{X}_2$ is given by

$$\begin{aligned}\text{Cov}(\boldsymbol{A}\boldsymbol{X}_1, \boldsymbol{B}\boldsymbol{X}_2) &= E[\boldsymbol{A}(\boldsymbol{X}_1 - \boldsymbol{\mu}_1)(\boldsymbol{X}_2 - \boldsymbol{\mu}_2)'\boldsymbol{B}'] \\ &= \boldsymbol{A}\boldsymbol{\Sigma}_{12}\boldsymbol{B}', \end{aligned} \tag{4.13}$$

where $\boldsymbol{\Sigma}_{12}$ is the covariance matrix of $\boldsymbol{X}_1$ and $\boldsymbol{X}_2$.

In particular, if $\boldsymbol{Z} \sim N(\boldsymbol{0}, \boldsymbol{I}_n)$, $\boldsymbol{\mu}$ is a vector of n elements, and $\boldsymbol{\Sigma}$ is a positive definite matrix of order $n \times n$, then the random vector

$$\boldsymbol{X} = \boldsymbol{\mu} + \boldsymbol{\Sigma}^{\frac{1}{2}}\boldsymbol{Z} \tag{4.14}$$

has the multivariate normal distribution with a mean vector $\boldsymbol{\mu}$ and a variance–covariance matrix $\boldsymbol{\Sigma}$. This is denoted by writing $\boldsymbol{X} \sim N(\boldsymbol{\mu}, \boldsymbol{\Sigma})$. The matrix $\boldsymbol{\Sigma}^{\frac{1}{2}}$ is obtained as in Section 3.12 by first applying the Spectral Decomposition Theorem (Theorem 3.4) to $\boldsymbol{\Sigma}$, which gives

$$\boldsymbol{\Sigma} = \boldsymbol{P}\boldsymbol{\Lambda}\boldsymbol{P}',$$

where

$\boldsymbol{\Lambda}$ is a diagonal matrix whose diagonal elements are the eigenvalues of $\boldsymbol{\Sigma}$, which are positive

$\boldsymbol{P}$ is an orthogonal matrix whose columns are the corresponding eigenvectors of $\boldsymbol{\Sigma}$

The matrix $\boldsymbol{\Sigma}^{\frac{1}{2}}$ is then given by

$$\boldsymbol{\Sigma}^{\frac{1}{2}} = \boldsymbol{P}\boldsymbol{\Lambda}^{\frac{1}{2}}\boldsymbol{P}',$$

where $\boldsymbol{\Lambda}^{\frac{1}{2}}$ is a diagonal matrix whose diagonal elements are the square roots of the corresponding diagonal elements of $\boldsymbol{\Lambda}$. Note that $\boldsymbol{X}$ in (4.14) is a linear function of $\boldsymbol{Z}$. Hence, its density functions is given by

$$f(\boldsymbol{x}) = g(\boldsymbol{z}) \mid det\ (\boldsymbol{J})|, \tag{4.15}$$

where $g(\boldsymbol{z})$ is the density functions of $\boldsymbol{Z}$ as in (4.6), and $\boldsymbol{J}$ is the *Jacobian matrix*

$$\begin{aligned} \boldsymbol{J} &= \frac{\partial \boldsymbol{z}'}{\partial \boldsymbol{x}} \\ &= \frac{\partial}{\partial \boldsymbol{x}}\left[(\boldsymbol{x} - \boldsymbol{\mu})'\,\boldsymbol{\Sigma}^{-\frac{1}{2}}\right] \\ &= \boldsymbol{\Sigma}^{-\frac{1}{2}}. \end{aligned}$$

The justification for (4.15) can be found in Khuri (2003, Theorem 7.11.1). Thus,

$$|det(\boldsymbol{J})| = [det(\boldsymbol{\Sigma})]^{-\frac{1}{2}}, \tag{4.16}$$

since $\boldsymbol{\Sigma}$ is positive definite. Making now the proper substitution in (4.15) gives the density function for a multivariate normal distribution with a mean vector $\boldsymbol{\mu}$ and a variance–covariance matrix $\boldsymbol{\Sigma}$, namely,

$$f(\boldsymbol{x}) = \frac{1}{(2\pi)^{\frac{n}{2}}\,[det(\boldsymbol{\Sigma})]^{\frac{1}{2}}} \exp\left[-\frac{1}{2}(\boldsymbol{x} - \boldsymbol{\mu})'\boldsymbol{\Sigma}^{-1}(\boldsymbol{x} - \boldsymbol{\mu})\right], \tag{4.17}$$

where $\boldsymbol{x} \in R^n$.

4.4 The Moment Generating Function

4.4.1 The General Case

In general, the *moment generating function* (m.g.f.) of a random vector $\boldsymbol{X}$ with a density function $h(\boldsymbol{x})$, where $\boldsymbol{x} \in R^n$, is defined as

$$\phi_{\boldsymbol{X}}(\boldsymbol{t}) = E\left(e^{\boldsymbol{t}'\boldsymbol{X}}\right), \tag{4.18}$$

where $\boldsymbol{t} = (t_1, t_2, \ldots, t_n)'$ is a vector in R^n whose elements are mathematical variables. The domain of $\phi_{\boldsymbol{X}}(\boldsymbol{t})$ is assumed to contain the point $\boldsymbol{t} = \boldsymbol{0}$. Thus,

$$\phi_{\boldsymbol{X}}(\boldsymbol{t}) = \int_{R^n} e^{\boldsymbol{t}'\boldsymbol{x}} h(\boldsymbol{x})d\boldsymbol{x}, \tag{4.19}$$

where the integral in (4.19) is n-dimensional over R^n, $\boldsymbol{x} = (x_1, x_2, \ldots, x_n)'$, and $d\boldsymbol{x}$ denotes $dx_1 dx_2 \ldots dx_n$.

The m.g.f. of $\boldsymbol{X}$ provides a characterization of the distribution of $\boldsymbol{X}$. It can also be used to derive the noncentral multivariate moments of $\boldsymbol{X}$, which are of the form

$$\mu'_{r_1 r_2 \cdots r_n} = \int_{R^n} x_1^{r_1} x_2^{r_2} \ldots x_n^{r_n} h(\boldsymbol{x}) d\boldsymbol{x}, \tag{4.20}$$

where $r_1, r_2, \ldots, r_n$ are nonnegative integers. If $\phi_{\boldsymbol{X}}(\boldsymbol{t})$ has partial derivatives of all orders with respect to $t_1, t_2, \ldots, t_n$ in an open subset of R^n that contains the point $\boldsymbol{t} = \boldsymbol{0}$, then

$$\mu'_{r_1 r_2 \cdots r_n} = \left. \frac{\partial^r [\phi_{\boldsymbol{X}}(\boldsymbol{t})]}{\partial t_1^{r_1} \partial t_2^{r_1} \ldots \partial t_n^{r_n}} \right|_{\boldsymbol{t}=\boldsymbol{0}}, \tag{4.21}$$

where $r = \sum_{i=1}^{n} r_i$.

Another advantage of the m.g.f. is its use in providing a check on the independence of two random vectors, $\boldsymbol{X}_1$ and $\boldsymbol{X}_2$. More specifically, let $\phi_{\boldsymbol{X}}(\boldsymbol{t})$ be the m.g.f. of $\boldsymbol{X} = (\boldsymbol{X}_1', \boldsymbol{X}_2')'$. Then, $\boldsymbol{X}_1$ and $\boldsymbol{X}_2$ are independent if and only if

$$\phi_{\boldsymbol{X}}(\boldsymbol{t}) = \phi_{\boldsymbol{X}_1}(\boldsymbol{t}_1) \phi_{\boldsymbol{X}_2}(\boldsymbol{t}_2) \tag{4.22}$$

for all values of $\boldsymbol{t}$ in an open subset of R^n that contains the point $\boldsymbol{t} = \boldsymbol{0}$ (see, for example, Arnold, 1981, Lemma 3.4; Graybill, 1976, Section 2.4). In (4.22), $\boldsymbol{t}_1$ and $\boldsymbol{t}_2$ partition $\boldsymbol{t}$ in a manner similar to that of $\boldsymbol{X}$, and $\phi_{\boldsymbol{X}_1}(\boldsymbol{t}_1)$, $\phi_{\boldsymbol{X}_2}(\boldsymbol{t}_2)$ are the corresponding moment generating functions of $\boldsymbol{X}_1$ and $\boldsymbol{X}_2$, respectively. More generally, if $\boldsymbol{X}$ and $\boldsymbol{t}$ are partitioned as $\boldsymbol{X} = (\boldsymbol{X}_1' : \boldsymbol{X}_2' : \ldots : \boldsymbol{X}_r')'$, $\boldsymbol{t} = (\boldsymbol{t}_1' : \boldsymbol{t}_2' : \ldots : \boldsymbol{t}_r')'$, $r \geq 2$, then $\boldsymbol{X}_1, \boldsymbol{X}_2, \ldots, \boldsymbol{X}_r$ are said to be *mutually independent* if and only if

$$\phi_{\boldsymbol{X}}(\boldsymbol{t}) = \prod_{i=1}^{r} \phi_{\boldsymbol{X}_i}(\boldsymbol{t}_i) \tag{4.23}$$

for all values of $\boldsymbol{t}$ in an open subset of R^n that includes the point $\boldsymbol{t} = \boldsymbol{0}$, where $\phi_{\boldsymbol{X}_i}(\boldsymbol{t}_i)$ is the m.g.f. of $\boldsymbol{X}_i (i = 1, 2, \ldots, r)$. Unless otherwise stated, mutually independent random vectors (or variables) are usually referred to as just independent.

An alternative characterization of the distribution of $\boldsymbol{X}$ is through its *cumulant generating function* (c.g.f.), which is defined as

$$\psi_{\boldsymbol{X}}(\boldsymbol{t}) = \log[\phi_{\boldsymbol{X}}(\boldsymbol{t})], \tag{4.24}$$

where $\log(\cdot)$ is the natural logarithmic function. Cumulants of a multivariate distribution can be defined as a generalization of the univariate case. For simplicity, we shall consider a bivariate situation involving two random variables X_1 and X_2. Generalizations to more than two variables can be carried out without any difficulty.

Let $\psi_{X_1,X_2}(t_1, t_2)$ denote the c.g.f. of X_1 and X_2. Maclaurin's series expansion of $\psi_{X_1,X_2}(t_1, t_2)$ around $(t_1, t_2) = (0, 0)$ is of the form

$$\psi_{X_1,X_1}(t_1, t_2) = \sum_{n=1}^{\infty} \frac{1}{n!} \sum_{r_1+r_2=n} \frac{n!}{r_1!r_2!} t_1^{r_1} t_2^{r_2} \frac{\partial^n \psi_{X_1,X_2}(0,0)}{\partial t_1^{r_1} \partial t_2^{r_2}}$$

$$= \sum_{n=1}^{\infty} \sum_{r_1+r_2=n} \frac{1}{r_1!r_2!} t_1^{r_1} t_2^{r_1} \frac{\partial^n \psi_{X_1,X_2}(0,0)}{\partial t_1^{r_1} \partial t_2^{r_2}},$$

where $\frac{\partial^n \psi_{X_1,X_2}(0,0)}{\partial t_1^{r_1} \partial t_2^{r_2}}$ is the value of the nth partial derivative $\frac{\partial^n \psi_{X_1,X_2}(t_1,t_2)}{\partial t_1^{r_1} \partial t_2^{r_2}}$ at $(t_1, t_2) = (0, 0)$. The coefficient of $\frac{t_1^{r_1} t_2^{r_2}}{r_1!r_2!}$ in this expansion is denoted by $\kappa_{r_1 r_2}$ and is called the *cumulant of the bivariate distribution* of X_1 and X_2 of order (r_1, r_2), or just the (r_1, r_2)th cumulant of X_1 and X_2.

Using condition (4.22), it can be stated that X_1 and X_2 are independent if and only if

$$\psi_{X_1,X_2}(t_1, t_2) = \psi_{X_1}(t_1) + \psi_{X_2}(t_2)$$

for all values of (t_1, t_2) in an open subset of R^2 that contains the point (0,0), where $\psi_{X_1}(t_1)$ and $\psi_{X_2}(t_2)$ are the marginal cumulant generating functions of X_1 and X_2, respectively.

4.4.2 The Case of the Multivariate Normal

The following theorem gives the m.g.f. of a multivariate normal distribution:

Theorem 4.1 The moment generating function of $\mathbf{X} \sim N(\boldsymbol{\mu}, \boldsymbol{\Sigma})$ is given by

$$\phi_{\mathbf{X}}(\boldsymbol{t}) = \exp\left(\boldsymbol{t}'\boldsymbol{\mu} + \frac{1}{2}\boldsymbol{t}'\boldsymbol{\Sigma}\boldsymbol{t}\right). \tag{4.25}$$

Proof. Substituting the normal density function in (4.17) into formula (4.19), we get

$$\phi_{\mathbf{X}}(\boldsymbol{t}) = \frac{1}{(2\pi)^{n/2}[det(\boldsymbol{\Sigma})]^{\frac{1}{2}}} \int_{R^n} \exp\left[\boldsymbol{t}'\boldsymbol{x} - \frac{1}{2}(\boldsymbol{x} - \boldsymbol{\mu})'\boldsymbol{\Sigma}^{-1}(\boldsymbol{x} - \boldsymbol{\mu})\right] d\boldsymbol{x}. \tag{4.26}$$

Let $\boldsymbol{y} = \boldsymbol{x} - \boldsymbol{\mu} - \boldsymbol{\Sigma}\boldsymbol{t}$. The Jacobian matrix of this transformation is $\frac{\partial \boldsymbol{x}'}{\partial \boldsymbol{y}} = \boldsymbol{I}_n$. Formula (4.26) can then be written as (see Khuri, 2003, Section 7.9.4)

$$\phi_{\mathbf{X}}(\boldsymbol{t}) = \frac{1}{(2\pi)^{n/2}[det(\boldsymbol{\Sigma})]^{\frac{1}{2}}} \int_{R^n} \exp\left(\boldsymbol{t}'\boldsymbol{\mu} + \frac{1}{2}\boldsymbol{t}'\boldsymbol{\Sigma}\boldsymbol{t} - \frac{1}{2}\boldsymbol{y}'\boldsymbol{\Sigma}^{-1}\boldsymbol{y}\right) d\boldsymbol{y}$$

$$= \frac{1}{(2\pi)^{n/2}[det(\boldsymbol{\Sigma})]^{\frac{1}{2}}} \exp\left(\boldsymbol{t}'\boldsymbol{\mu} + \frac{1}{2}\boldsymbol{t}'\boldsymbol{\Sigma}\boldsymbol{t}\right) \int_{R^n} \exp\left(-\frac{1}{2}\boldsymbol{y}'\boldsymbol{\Sigma}^{-1}\boldsymbol{y}\right) d\boldsymbol{y} \tag{4.27}$$

Now, let $z = \Sigma^{-\frac{1}{2}} y$. The integral in (4.27) can be written as (see Khuri, 2003, formula 7.67, p. 300)

$$\begin{aligned}\int_{R^n} \exp\left(-\frac{1}{2} y' \Sigma^{-1} y\right) dy &= \int_{R^n} \exp\left(-\frac{1}{2} z' z\right) \left| det\left(\frac{\partial y'}{\partial z}\right)\right| dz \\ &= \int_{R^n} \exp\left(-\frac{1}{2} z' z\right) \left| det\left(\Sigma^{1/2}\right)\right| dz \\ &= [det(\Sigma)]^{1/2} \prod_{i=1}^{n} \left(\int_{-\infty}^{\infty} e^{-\frac{1}{2} z_i^2} dz_i\right),\end{aligned}$$

since Σ is positive definite. But,

$$\int_{-\infty}^{\infty} \exp\left(-\frac{1}{2} z_i^2\right) dz_i = (2\pi)^{1/2}, \quad i = 1, 2, \ldots, n.$$

Hence,

$$\int_{R^n} \exp\left(-\frac{1}{2} y' \Sigma^{-1} y\right) dy = (2\pi)^{n/2} [det(\Sigma)]^{1/2}. \tag{4.28}$$

Substituting the term on the right-hand side of (4.28) into formula (4.27) produces the desired result, namely,

$$\phi_X(t) = \exp\left(t' \mu + \frac{1}{2} t' \Sigma t\right). \qquad \square$$

Corollary 4.1 Suppose that $X \sim N(\mu, \Sigma)$. Let A be a constant matrix of order $m \times n$ and rank $m(\leq n)$, where n is the number of elements of X. Then,

$$AX \sim N(A\mu, A\Sigma A').$$

Proof. Let $Y = AX$. The m.g.f. of Y is

$$\begin{aligned}\phi_Y(t) &= E(e^{t'AX}) \\ &= \phi_X(t'A) \\ &= \exp\left(t'A\mu + \frac{1}{2} t' A \Sigma A' t\right).\end{aligned}$$

By comparing this m.g.f. with the one in (4.25), it can be seen that Y is normally distributed with a mean vector $A\mu$ and a variance–covariance matrix $A\Sigma A'$. $\square$

From Corollary 4.1 we conclude that if X is normally distributed, then any portion of it is also normally distributed. For example, if X is partitioned as $X = (X_1' : X_2')'$, then both X_1 and X_2 are normally distributed since they can be written as $X_1 = A_1 X$, $X_2 = A_2 X$, where $A_1 = [I_{n_1} : \mathbf{0}_{n_1 \times n_2}]$, $A_2 = [\mathbf{0}_{n_2 \times n_1} : I_{n_2}]$, with n_i being the number of elements of $X_i (i = 1, 2)$. If μ and Σ are partitional accordingly as $\mu = (\mu_1' : \mu_2')'$, and

$$\Sigma = \begin{bmatrix} \Sigma_{11} & \Sigma_{12} \\ \Sigma_{12}' & \Sigma_{22} \end{bmatrix}, \tag{4.29}$$

then $X_1 \sim N(\mu_1, \Sigma_{11})$, $X_2 \sim N(\mu_2, \Sigma_{22})$.

A slightly more general result gives the distribution of the random vector $Y = AX + b$, where b is a constant vector. In this case, if $X \sim N(\mu, \Sigma)$, then $Y \sim N(A\mu + b, A\Sigma A')$.

Corollary 4.2 Suppose that $X \sim N(\mu, \Sigma)$ is partitioned as $X = (X_1' : X_2')'$. Then, X_1 and X_2 are independent if and only if their covariance matrix Σ_{12} is zero.

Proof. Let the mean vector μ be partitioned as $\mu = (\mu_1' : \mu_2')'$, and Σ be partitional as in (4.29). If X_1 and X_2 are independent, then it is obvious that $\Sigma_{12} = \mathbf{0}$. Vice versa, if $\Sigma_{12} = \mathbf{0}$, then by Theorem 4.1, the m.g.f. of X is

$$\phi_X(t) = \exp\left(t_1'\mu_1 + t_2'\mu_2 + \frac{1}{2}t_1'\Sigma_{11}t_1 + \frac{1}{2}t_2'\Sigma_{22}t_2\right), \tag{4.30}$$

where t_1 and t_2 are the corresponding portions of t. Formula (4.30) results from noting that

$$\begin{aligned} t'\Sigma t &= (t_1' : t_2') \begin{bmatrix} \Sigma_{11} & \mathbf{0} \\ \mathbf{0}' & \Sigma_{22} \end{bmatrix} \begin{pmatrix} t_1 \\ t_2 \end{pmatrix} \\ &= t_1'\Sigma_{11}t_1 + t_{22}'\Sigma_{22}t_2. \end{aligned}$$

Hence, $\phi_X(t)$ can be written as

$$\phi_X(t) = \phi_{X_1}(t_1)\phi_{X_2}(t_2), \tag{4.31}$$

where $\phi_{X_1}(t_1)$ and $\phi_{X_2}(t_2)$ are the moment generating functions of X_1 and X_2, respectively, which are normally distributed by Corollary 4.1. Formula (4.31) indicates that X_1 and X_2 are independent. □

Corollary 4.2 can be easily extended to include a partitioning of X of the form $X = (X_1' : X_2' : \ldots : X_r')'$, $r \geq 2$. In this case, if X is normally distributed, then $X_1, X_2, \ldots, X_r$ are mutually independent if and only if $\Sigma_{ij} = \mathbf{0}$ for all $i \neq j$, where Σ_{ij} is the covariance matrix of X_i and $X_j (i, j = 1, 2, \ldots, r)$.

4.5 Conditional Distribution

Let $X \sim N(\mu, \Sigma)$ be partitioned as $(X_1' : X_2')'$. The corresponding partitioning of μ is $(\mu_1' : \mu_2')'$, and Σ is partitioned as in (4.29). If $\Sigma_{12} \neq \mathbf{0}$, then X_1 and X_2 are not independent by Corollary 4.2. By definition, the *conditional distribution* of X_1 given X_2, written symbolically as $X_1|X_2$, is the distribution of X_1 when X_2 is held constant. Obviously, if X_1 and X_2 are dependent, such a distribution is different from that of X_1.

The dependence of X_1 on X_2 is now to be displayed. For this purpose, let Y be defined as $Y = X_1 - AX_2$, where A is a constant matrix. The vector Y is normally distributed by Corollary 4.1 since it is a linear function of X. The matrix A is chosen so that Y and X_2 are independent. Note that the joint distribution of Y and X_2 is a multivariate normal because it can be easily shown that $(Y' : X_2')'$ is a nonsingular linear transformation of X. Hence, by Corollary 4.2, Y and X_2 are independent if and only if their covariance matrix is zero. Thus, the matrix A can be determined by equating $\text{Cov}(Y, X_2)$ to zero, where

$$\text{Cov}(Y, X_2) = \Sigma_{12} - A\Sigma_{22}.$$

It follows that Y and X_2 are independent if $A = \Sigma_{12}\Sigma_{22}^{-1}$. The vector Y can therefore be written as

$$Y = X_1 - \Sigma_{12}\Sigma_{22}^{-1}X_2. \tag{4.32}$$

Hence,

$$X_1 = Y + \Sigma_{12}\Sigma_{22}^{-1}X_2.$$

This shows that X_1 is the sum of two component vectors: Y, which is independent of X_2, and a linear function of X_2. Consequently, $X_1|X_2$ is written as

$$X_1|X_2 = Y + \Sigma_{12}\Sigma_{22}^{-1}X_2, \tag{4.33}$$

where $\Sigma_{12}\Sigma_{22}^{-1}X_2$ is treated as a constant vector. From (4.32), the mean of Y is $\mu_1 - \Sigma_{12}\Sigma_{22}^{-1}\mu_2$, and its variance–covariance matrix is given by

$$\begin{aligned}
\text{Var}(Y) &= \text{Var}(X_1) - 2\,\text{Cov}(X_1, \Sigma_{12}\Sigma_{22}^{-1}X_2) \\
&\quad + \text{Var}(\Sigma_{12}\Sigma_{22}^{-1}X_2) \\
&= \Sigma_{11} - 2\Sigma_{12}\Sigma_{22}^{-1}\Sigma_{21} \\
&\quad + \Sigma_{12}\Sigma_{22}^{-1}\Sigma_{22}\Sigma_{22}^{-1}\Sigma_{21} \\
&= \Sigma_{11} - \Sigma_{12}\Sigma_{22}^{-1}\Sigma_{21}.
\end{aligned}$$

Using (4.33), we finally conclude that

$$X_1|X_2 \sim N[\mu_1 + \Sigma_{12}\Sigma_{22}^{-1}(X_2 - \mu_2), W],$$

where

$$W = \Sigma_{11} - \Sigma_{12}\Sigma_{22}^{-1}\Sigma_{21}.$$

4.6 The Singular Multivariate Normal Distribution

We recall from Corollary 4.1 that when $X \sim N(\mu, \Sigma)$, $AX \sim N(A\mu,\ A\Sigma A')$ provided that A is of full row rank. This condition is needed in order for $A\Sigma A'$ to be nonsingular. Note that $A\Sigma A'$ is actually positive definite if A is of full row rank, since Σ is positive definite.

Let us now consider the distribution of AX, where $X \sim N(\mu, \Sigma)$ and A is not of full row rank. Let A be of order $m \times n$ and rank $r(<m)$. The elements of AX are therefore linearly dependent. As a result, the variance–covariance matrix of AX, namely $A\Sigma A'$, is singular. Without any loss of generality, the matrix A can be partitioned as $A = [A_1' : A_2']'$, where A_1 is of order $r \times n$ and rank r, and the remaining $(m - r)$ rows of A that make up the matrix A_2 are linearly dependent on those of A_1. Hence, A_2 can be written as $A_2 = BA_1$ for some matrix B. We then have

$$\begin{aligned} AX &= \begin{bmatrix} A_1X \\ BA_1X \end{bmatrix} \\ &= \begin{bmatrix} I_r \\ B \end{bmatrix} A_1X. \end{aligned} \tag{4.34}$$

Note that $A_1X \sim N(A_1\mu,\ A_1\Sigma A_1')$ since A_1 is of full row rank. This shows that AX is a linear function of a multivariate normal random variable, namely A_1X, which is of a lower dimension than that of X. In this case, AX is said to have the *singular multivariate normal distribution.*

For example, suppose that $X = (X_1,\ X_2,\ X_3)'$, where $Y = (X_1,\ X_2)'$ has the bivariate normal distribution and $X_3 = 2X_1 - 3X_2$. Then X has the singular multivariate normal distribution and is written as

$$X = \begin{bmatrix} I_2 \\ b' \end{bmatrix} Y,$$

where $b = (2, -3)'$.

Situations that involve the use of the singular normal distribution arise in connection with linear models that are less than full rank, as will be seen in Chapter 7.

4.7 Related Distributions

Several well-known distributions in statistics can be derived from the multivariate normal distribution. These distributions are frequently used in the study of linear models.

4.7.1 The Central Chi-Squared Distribution

Let $\boldsymbol{X} \sim N(\boldsymbol{0}, \boldsymbol{I}_n)$. Then $U = \boldsymbol{X}'\boldsymbol{X}$ has the *central chi-squared distribution* with n degrees of freedom, written symbolically as $U \sim \chi^2_n$. It is customary to drop the word "central" when referring to this distribution.

The chi-squared distribution is a special case of the *gamma distribution*. By definition, a random variable W has the gamma distribution with parameters α and β, written symbolically as $W \sim G(\alpha, \beta)$, if its density function is of the form

$$f(w) = \frac{1}{\Gamma(\alpha)\beta^{\alpha}} w^{\alpha-1} e^{-w/\beta}, \quad 0 < w < \infty, \tag{4.35}$$

where α and β are positive constants and $\Gamma(\alpha)$ is the so-called *gamma function*, which is given by

$$\Gamma(\alpha) = \int_0^{\infty} e^{-x} x^{\alpha-1} dx.$$

The gamma distribution is absolutely continuous; its mean is $E(W) = \alpha\beta$ and its variance is $\mathrm{Var}(W) = \alpha\beta^2$. It can be shown that its moment generating function is

$$\phi_W(t) = (1 - \beta t)^{-\alpha}, \quad t < \frac{1}{\beta}. \tag{4.36}$$

In particular, if $\alpha = \frac{n}{2}$ and $\beta = 2$, where n is a positive integer, then W has the chi-squared distribution with n degrees of freedom. Thus, if $U \sim \chi^2_n$, then from (4.35), its density function is

$$f(u) = \frac{1}{\Gamma(\frac{n}{2})2^{n/2}} u^{n/2-1} e^{-u/2}, \quad 0 < u < \infty, \tag{4.37}$$

and the corresponding mean, variance, and moment generating function of U are, respectively, $E(U) = n$, $\mathrm{Var}(U) = 2n$,

$$\phi_U(t) = (1 - 2t)^{-n/2}, \quad t < \frac{1}{2}. \tag{4.38}$$

4.7.2 The Noncentral Chi-Squared Distribution

If the mean of the normal random vector $\boldsymbol{X}$ in Section 4.7.1 is nonzero, that is, $\boldsymbol{X} \sim N(\boldsymbol{\mu}, \boldsymbol{I}_n)$ with $\boldsymbol{\mu} \neq \boldsymbol{0}$, then $U = \boldsymbol{X}'\boldsymbol{X}$ has the *noncentral chi-squared distribution* with n degrees of freedom and a *noncentrality parameter* $\lambda = \boldsymbol{\mu}'\boldsymbol{\mu}$. This distribution is written symbolically as $\chi^2_n(\lambda)$. Note that when $\lambda = 0$, the distribution is reduced to the central chi-squared distribution χ^2_n.

The density function of $U \sim \chi^2_n(\lambda)$ is (see Johnson and Kotz, 1970, Chapter 28)

$$g(u) = \frac{e^{-\frac{1}{2}(\lambda+u)}}{2^{\frac{n}{2}}} \sum_{i=0}^{\infty} \frac{(\frac{\lambda}{2})^i}{i!} \frac{u^{\frac{n}{2}+i-1}}{\Gamma(\frac{n}{2}+i)2^i}, \quad 0 < u < \infty. \tag{4.39}$$

It can be seen that $g(u)$ is expressible as the sum of an infinite series whose ith term is of the form

$$h_i(u,\lambda) = p_i(\lambda) f_i(u), \quad i = 0, 1, \ldots, \tag{4.40}$$

where

$$p_i(\lambda) = \frac{e^{-\lambda/2}(\lambda/2)^i}{i!}$$

is the probability that a Poisson random variable with a mean $\frac{\lambda}{2}$ attains the value i, and $f_i(u)$ is the density function of a central chi-squared random variable with $(n + 2i)$ degrees of freedom $(i = 0, 1, 2, \ldots)$.

The infinite series in (4.39) is a power series in u. Its convergence depends on the value of u. We now show that this series is actually uniformly convergent with respect to u on $(0, \infty)$. For a definition of uniform convergence of a power series, see, for example, Khuri (2003, Chapter 5).

Theorem 4.2 The infinite series in (4.39) converges uniformly with respect to u on $(0, \infty)$.

Proof. It is sufficient to show uniform convergence of the series

$$S(\lambda) = \sum_{i=0}^{\infty} \frac{\left(\frac{\lambda}{2}\right)^i}{i!} \frac{u^i e^{-\frac{u}{2}}}{2^i \Gamma\left(\frac{n}{2}+i\right)},$$

since the series in (4.39) is a constant multiple of $S(\lambda)$. We first note that for $i = 0, 1, 2, \ldots,$ and for $u > 0$,

$$e^{\frac{u}{2}} > \frac{(\frac{u}{2})^i}{i!}.$$

This follows from the fact that the right-hand side of the above inequality is the ith term of Maclaurin's series expansion of $e^{\frac{u}{2}}$. Hence, for all $u > 0$,

$$\frac{(\frac{\lambda}{2})^i}{i!} \frac{u^i e^{-\frac{u}{2}}}{2^i \Gamma(\frac{n}{2}+i)} < \frac{(\frac{\lambda}{2})^i}{\Gamma(\frac{n}{2}+i)}, \quad i = 0, 1, \ldots$$

But, the right-hand side of this inequality is the ith term of a convergent infinite series of constant terms. This can be easily shown by applying the ratio test of convergence of infinite series of positive terms (see, for example, Khuri, 2003, p. 148):

$$\frac{(\frac{\lambda}{2})^{i+1}}{\Gamma(\frac{n}{2}+i+1)} \frac{\Gamma(\frac{n}{2}+i)}{(\frac{\lambda}{2})^i} = \frac{\lambda}{n+2i},$$

since $\Gamma(\frac{n}{2}+i+1) = (\frac{n}{2}+i)\Gamma(\frac{n}{2}+i)$. It can be seen that as i goes to infinity, the limit of this ratio is zero, which is less than 1. This proves convergence of $\sum_{i=0}^{\infty} \frac{(\frac{\lambda}{2})^i}{\Gamma(\frac{n}{2}+i)}$. It follows that the series $S(\lambda)$ is uniformly convergent on $(0, \infty)$ by the Weierstrass M-test (see, for example, Khuri, 2003, Theorem 5.3.2). □

Theorem 4.2 is needed for the derivation of the moment generating function of a noncentral chi-squared distribution, as will be shown in the next theorem.

Theorem 4.3 Let $U \sim \chi_n^2(\lambda)$. The moment generating function (m.g.f.) of U, denoted by $\phi_U(t, \lambda)$, is given by

$$\phi_U(t,\lambda) = (1-2t)^{-\frac{n}{2}} \exp[\lambda t(1-2t)^{-1}], \quad t < \frac{1}{2}. \tag{4.41}$$

Proof. We have that

$$\phi_U(t,\lambda) = E(e^{tU}).$$

Using the density function of U given in (4.39), we get

$$\phi_U(t,\lambda) = \int_0^{\infty} e^{tu} \frac{e^{-\frac{1}{2}(\lambda+u)}}{2^{\frac{n}{2}}} \sum_{i=0}^{\infty} \frac{(\frac{\lambda}{2})^i}{i!} \frac{u^{\frac{n}{2}+i-1}}{2^i\Gamma(\frac{n}{2}+i)} du. \tag{4.42}$$

Since the infinite series in (4.42) is uniformly convergent, the integration and summation operations can be interchanged without affecting the value of the integral (see, for example, Fulks, 1978, Corollary 14.3f, p. 515). We then have

$$\phi_U(t,\lambda) = \sum_{i=0}^{\infty} \frac{e^{-\frac{\lambda}{2}}(\frac{\lambda}{2})^i}{i!} \int_0^{\infty} e^{tu} \frac{u^{\frac{n}{2}+i-1}e^{-\frac{u}{2}}}{2^{\frac{n}{2}+i}\Gamma(\frac{n}{2}+i)} du. \tag{4.43}$$

We note that the integral in (4.43) is the m.g.f. of a central chi-squared distribution with $(n+2i)$ degrees of freedom. By formula (4.38), the latter m.g.f. is equal to $(1-2t)^{-(n+2i)/2}$. It follows that

$$\begin{aligned}\phi_U(t,\lambda) &= \sum_{i=0}^{\infty}\frac{e^{-\frac{\lambda}{2}}(\frac{\lambda}{2})^i}{i!}(1-2t)^{-(n+2i)/2}\\ &= e^{-\frac{\lambda}{2}}(1-2t)^{-\frac{n}{2}}\sum_{i=0}^{\infty}\frac{1}{i!}\left(\frac{\lambda/2}{1-2t}\right)^i\\ &= e^{-\frac{\lambda}{2}}(1-2t)^{-\frac{n}{2}}\exp\left(\frac{\lambda/2}{1-2t}\right)\\ &= (1-2t)^{-\frac{n}{2}}\exp[\lambda t(1-2t)^{-1}].\end{aligned}$$

□

Using the m.g.f. from Theorem 4.3, it can be shown that the mean and variance of $U \sim \chi^2_n(\lambda)$ are $E(U) = n + \lambda$ and $\text{Var}(U) = 2n + 4\lambda$.

Theorem 4.4 Let U_1 and U_2 be independently distributed as $\chi^2_{n_1}(\lambda_1)$ and $X^2_{n_2}(\lambda_2)$, respectively. Then,

$$U = U_1 + U_2 \sim \chi^2_{n_1+n_2}(\lambda_1 + \lambda_2).$$

Proof. From Theorem 4.3, the moment generating functions of U_1 and U_2 are, respectively,

$$\begin{aligned}\phi_{U_1}(t,\lambda_1) &= (1-2t)^{-\frac{n_1}{2}}\exp[\lambda_1 t(1-2t)^{-1}]\\ \phi_{U_2}(t,\lambda_2) &= (1-2t)^{-\frac{n_2}{2}}\exp[\lambda_2 t(1-2t)^{-1}]\end{aligned}$$

Hence, the m.g.f. of $U = U_1 + U_2$ is

$$\begin{aligned}E(e^{tU}) &= E(e^{tU_1})E(e^{tU_2})\\ &= \phi_{U_1}(t,\lambda_1)\phi_{U_2}(t,\lambda_2),\end{aligned}$$

since U_1 and U_1 are independent. It follows that

$$E(e^{tU}) = (1-2t)^{-(n_1+n_2)/2}\exp[(\lambda_1+\lambda_2)t(1-2t)^{-1}].$$

Thus, $U \sim \chi^2_{n_1+n_2}(\lambda_1 + \lambda_2)$. □

Theorem 4.4 can be generalized so that if $U_1, U_2, \ldots, U_k$ are mutually independent such that $U_i \sim \chi^2_{n_i}(\lambda_i)$, $i = 1, 2, \ldots, k$, then $\sum_{i=1}^k U_i \sim \chi^2_{\sum_{i=1}^k n_i}(\sum_{i=1}^k \lambda_i)$.

4.7.3 The *t*-Distribution

Suppose that $Z \sim N(0,1)$, $U \sim \chi^2_n$, and Z and U are independent. Then,

$$V = \frac{Z}{(U/n)^{1/2}}$$

has the *central t-distribution* (or just the t-distribution) with n degrees of freedom, and is denoted symbolically by t_n. The mean of V is zero and its variance is equal to $n/(n-2)$, $n > 2$. The corresponding density function is given by

$$f(v) = \frac{\Gamma(\frac{n+1}{2})}{(n\pi)^{\frac{1}{2}}\Gamma(\frac{n}{2})}\left(1 + \frac{v^2}{n}\right)^{-\frac{n+1}{2}}, \quad -\infty < v < \infty.$$

If the mean of Z is not equal to zero, that is, $Z \sim N(\mu, 1)$, then V has the so-called *noncentral t-distribution* with n degrees of freedom and a noncentrality parameter μ. This distribution is denoted symbolically by $t_n(\mu)$. More details concerning the noncentral t-distribution can be found in Johnson and Kotz (1970, Chapter 31). See also Evans, Hastings, and Peacock (2000, Chapter 39).

4.7.4 The *F*-Distribution

Let $U_1 \sim \chi^2_{n_1}$, $U_2 \sim \chi^2_{n_2}$ be independently distributed chi-squared variates, then,

$$Y = \frac{U_1/n_1}{U_2/n_2}$$

has the *central F-distribution* (or just the F-distribution) with n_1 and n_2 degrees of freedom. This is written symbolically as $Y \sim F_{n_1,n_2}$. The mean of Y is $E(Y) = n_2/(n_2-2)$, $n_2 > 2$, and its variance is

$$\text{Var}(Y) = \frac{2n_2^2(n_1+n_2-2)}{n_1(n_2-2)^2(n_2-4)}, \quad n_2 > 4.$$

The density function of Y is given by

$$f(y) = \frac{\Gamma(\frac{n_1+n_2}{2})}{\Gamma(\frac{n_1}{2})\Gamma(\frac{n_2}{2})}\left(\frac{n_1}{n_2}\right)^{\frac{n_1}{2}}\left(1 + \frac{n_1}{n_2}y\right)^{-\frac{n_1+n_2}{2}} y^{\frac{n_1}{2}-1}, \quad 0 < y < \infty.$$

If, however, U_1 has the noncentral chi-squared distribution $\chi^2_{n_1}(\lambda_1)$ with n_1 degrees of freedom and a noncentrality parameter λ_1, and $U_2 \sim \chi^2_{n_2}$ independently of U_1, then Y has the so-called *noncentral F-distribution* with n_1 and n_2 degrees of freedom and a noncentrality parameter λ_1. This is written symbolically as $Y \sim F_{n_1,n_2}(\lambda_1)$. Furthermore, if $U_1 \sim \chi^2_{n_1}(\lambda_1)$, $U_2 \sim \chi^2_{n_2}(\lambda_2)$, and U_1 is independent of U_2, then Y has the *doubly noncentral F-distribution* with n_1 and n_2 degrees of freedom and noncentrality parameters λ_1 and λ_2. This distribution is denoted by $F_{n_1,n_2}(\lambda_1, \lambda_2)$. More details concerning the F-distribution can be found in Johnson and Kotz (1970, Chapters 26 and 30). See also Evans, Hastings, and Peacock (2000, Chapters 17 and 18).

4.7.5 The Wishart Distribution

Let $X_1, X_2, \ldots, X_n$ be mutually independent and identically distributed as $N(\mathbf{0}, \boldsymbol{\Sigma})$ with each having m elements. Then, the $m \times m$ random matrix

$$A = \sum_{i=1}^{n} X_i X_i'$$

has the (*central*) *Wishart distribution* with n degrees of freedom and a variance–covariance matrix $\boldsymbol{\Sigma}$, written symbolically as $A \sim W_m(n, \boldsymbol{\Sigma})$. If $n \geq m$, then it can be shown that A is positive definite with probability 1 (see Seber, 1984, Section 2.3). This distribution represents a generalization of the chi-squared distribution. If $m = 1$ and $\boldsymbol{\Sigma} = 1$, then A reduces to χ^2_n. It can also be shown (see Exercise 4.7) that if c is a nonzero constant vector with m elements, then $c'Ac \sim \sigma^2_c \chi^2_n$, where $\sigma^2_c = c'\boldsymbol{\Sigma}c$. In particular, if the elements of c are all equal to zero, except for the ith element, which is equal to 1, then $c'Ac$ gives the ith diagonal element, a_{ii}, of A and therefore $a_{ii} \sim \sigma_{ii}\chi^2_n$, where σ_{ii} is the ith diagonal element of $\boldsymbol{\Sigma}$.

If, in the aforementioned definition of the Wishart distribution, the mean of X_i is μ_i rather than $\mathbf{0}$ ($i = 1, 2, \ldots, n$), then A is said to have the *noncentral Wishart distribution* with n degrees of freedom, a variance–covariance matrix $\boldsymbol{\Sigma}$, and a *noncentratily parameter matrix* defined to be

$$\boldsymbol{\Omega} = \boldsymbol{\Sigma}^{-\frac{1}{2}} \left(\sum_{i=1}^{n} \mu_i \mu_i' \right) \boldsymbol{\Sigma}^{-\frac{1}{2}}.$$

In this case, we write $A \sim W_m(n,\ \boldsymbol{\Sigma},\ \boldsymbol{\Omega})$. This distribution generalizes the noncentral chi-squared distribution. If $m = 1$, $\boldsymbol{\Sigma} = 1$, then A has the noncentral chi-squared distribution with n degrees of freedom and a noncentrality parameter $\lambda = \sum_{i=1}^{n} \mu_i^2$. Additional properties concerning the Wishart distribution can be found in standard multivariate textbooks such as Muirhead (1982) and Seber (1984).

4.8 Examples and Additional Results

Let $X_1, X_2, \ldots, X_n$ be mutually independent and normally distributed random variables with means μ and variances σ^2. These random variables form a random sample of size n from the normal distribution $N(\mu, \sigma^2)$. Let $\bar{X} = \frac{1}{n}\sum_{i=1}^{n} X_i$ be the sample mean and s^2 be the sample variance given by

$$s^2 = \frac{1}{n-1}\sum_{i=1}^{n}(X_i - \bar{X})^2.$$

Under these conditions, $\bar{X}$ and s^2 have the following known properties (see, for example, Casella and Berger, 2002, Section 5.3, p. 218):

(1) $\bar{X}$ and s^2 are independent random variables

(2) $\bar{X} \sim N(\mu, \frac{\sigma^2}{n})$

(3) $\frac{(n-1)s^2}{\sigma^2} \sim \chi^2_{n-1}$

For a proof of properties (1) and (3), see Exercise 4.18. The proof of property (2) is straightforward on the basis of Corollary 4.1.

On the basis of these properties, it can be deduced that the random variable

$$V = \frac{\bar{X} - \mu}{\frac{s}{\sqrt{n}}} \tag{4.44}$$

has the t-distribution with $(n-1)$ degrees of freedom. This follows from the fact that V is obtained by dividing the standard normal variate, $Z = \frac{\bar{X}-\mu}{\sigma/\sqrt{n}}$, by the square root of $\frac{U}{n-1}$, where $U = \frac{(n-1)s^2}{\sigma^2}$, which is independent of $\bar{X}$ and is distributed as χ^2_{n-1}. Hence, by the definition of the t-distribution in Section 4.7.3, $V \sim t_{n-1}$.

Properties (1) through (3) can be extended to samples from a multivariate normal distribution. Their multivariate analogs are given by the following theorem (see, for example, Seber, 1984, Section 3.2.2, p. 63).

Theorem 4.5 Let $\boldsymbol{X}_1, \boldsymbol{X}_2, \ldots, \boldsymbol{X}_n$ be mutually independent and normally distributed as $N(\boldsymbol{\mu}, \boldsymbol{\Sigma})$ with each having m elements. Let $\bar{\boldsymbol{X}} = \frac{1}{n}\sum_{i=1}^{n} \boldsymbol{X}_i$ and $\boldsymbol{S}$ be the so-called *sample variance–covariance matrix* given by

$$\boldsymbol{S} = \frac{1}{n-1}\sum_{i=1}^{n}(\boldsymbol{X}_i - \bar{\boldsymbol{X}})(\boldsymbol{X}_i - \bar{\boldsymbol{X}})'$$

If $n - 1 \geq m$, then

(1) $\bar{\boldsymbol{X}}$ and $\boldsymbol{S}$ are independent

(2) $\bar{\boldsymbol{X}} \sim N(\boldsymbol{\mu}, \frac{1}{n}\boldsymbol{\Sigma})$

(3) $(n-1)\boldsymbol{S}$ has the central Wishart distribution $W_m(n-1, \boldsymbol{\Sigma})$ with $(n-1)$ degrees of freedom

(4) $T^2 = n(\bar{\boldsymbol{X}} - \boldsymbol{\mu})'\boldsymbol{S}^{-1}(\bar{\boldsymbol{X}} - \boldsymbol{\mu})$ has the so-called *Hotelling's T^2-distribution* with m and $n-1$ degrees of freedom

For a proof of parts (1) and (3), see Exercise 4.19. The proof of part (2) is straightforward. Note that T^2 is the multivariate analog of the square of the t-random variable given in (4.44). Properties of its distribution can be found in standard multivariate textbooks (see, for example, Seber, 1984, Section 2.4, p. 28).

4.8.1 Some Misconceptions about the Normal Distribution

Some properties, associated with the normal distribution can be misconceived leading to incorrect statements about this distribution. Here are some examples.

Example 4.1 We may recall from Corollary 4.1 that if $\boldsymbol{X} = (X_1, X_2, \ldots, X_n)'$ is normally distributed as $N(\boldsymbol{\mu}, \boldsymbol{\Sigma})$, then any portion of $\boldsymbol{X}$ is also normally distributed, including the individual elements of $\boldsymbol{X}$ (marginal distributions of $\boldsymbol{X}$). The reverse, however, is not necessarily true, that is, if the elements of a random vector $\boldsymbol{X}$ are each normally distributed, $\boldsymbol{X}$ itself may not be normally distributed. Kowalski (1973) provided several examples of bivariate (that is, $n = 2$) nonnormal distributions with normal marginals. The following is a description of one of his examples:

Let $\boldsymbol{X} = (X_1, X_2)'$ have the density function

$$g(x_1, x_2) = \omega_1 f_1(x_1, x_2) + \omega_2 f_2(x_1, x_2), \tag{4.45}$$

where $\omega_1 + \omega_2 = 1, \omega_i \geq 0,\ i = 1, 2$, and f_i is the density function for the bivariate normal $N(\boldsymbol{0}, \boldsymbol{\Sigma}_i)$, where

$$\boldsymbol{\Sigma}_i = \begin{bmatrix} 1 & \rho_i \\ \rho_i & 1 \end{bmatrix}, \quad i = 1, 2.$$

The moment generating function (m.g.f.) corresponding to the distribution in (4.45) is

$$\begin{aligned} \phi(\boldsymbol{t}) &= E[\exp(t_1 X_1 + t_2 X_2)] \\ &= \int_{-\infty}^{\infty}\int_{-\infty}^{\infty} \exp(t_1 x_1 + t_2 x_2) g(x_1, x_2) dx_1 dx_2 \\ &= \omega_1 \phi_1(\boldsymbol{t}) + \omega_2 \phi_2(\boldsymbol{t}), \end{aligned} \tag{4.46}$$

where $\boldsymbol{t} = (t_1, t_2)'$, and $\phi_1(\boldsymbol{t})$ and $\phi_2(\boldsymbol{t})$ are the m.g.f.'s corresponding to f_1 and f_2, respectively, that is,

$$\phi_i(\boldsymbol{t}) = \exp\left(\frac{1}{2}\boldsymbol{t}'\boldsymbol{\Sigma}_i\boldsymbol{t}\right), \quad i = 1, 2.$$

The m.g.f.'s for the marginal distributions of X_1 and X_2 are obtained from $\phi(\boldsymbol{t})$ in (4.46) by replacing $\boldsymbol{t}$ with $(t_1, 0)'$ and $(0, t_2)'$, respectively. Doing so yields $e^{\frac{1}{2}t_1^2}$ and $e^{\frac{1}{2}t_2^2}$ as the m.g.f.'s for X_1, X_2, respectively. This implies that the marginal distributions of $\boldsymbol{X}$ are normally distributed as $N(0, 1)$. The vector $\boldsymbol{X}$, however, is not normally distributed unless $\rho_1 = \rho_2$. This can be seen from the form of the m.g.f. $\phi(\boldsymbol{t})$, which, when $\rho_1 = \rho_2$, is equal to $\phi_1(\boldsymbol{t})$ [or $\phi_2(\boldsymbol{t})$]. Hence, if $\rho_1 \neq \rho_2$, $\boldsymbol{X}$ is not normally distributed, even though its marginal distributions are normal.

Example 4.2 On the basis of Corollary 4.2, if $\boldsymbol{X} = (X_1, X_2)'$ has the bivariate normal distribution, then X_1 and X_2 are independent if and only if $\text{Cov}(X_1, X_2) = 0$. It is not true, however, that any two normally distributed random variables, X_1 and X_2, that have a zero covariance must also be independent. In other words, if $\boldsymbol{X}$ does not have the bivariate normal distribution, but X_1 and X_2 are normally distributed and $\text{Cov}(X_1, X_2) = 0$, then X_1 and X_2 need not be independent. As a counterexample, let us again consider Example 4.1. The normally distributed random variables X_1 and X_2 are not independent since the m.g.f. $\phi(\boldsymbol{t})$ in (4.46) is not the product of their marginal m.g.f.'s. But, the covariance of X_1 and X_2 can be zero if $\rho_1 \neq \rho_2$ since

$$\text{Cov}(X_1, X_2) = \omega_1\rho_1 + \omega_2\rho_2. \tag{4.47}$$

Formula (4.47) follows from the fact that

$$\begin{aligned}
\text{Cov}(X_1, X_2) &= E(X_1X_2) - E(X_1)E(X_2) \\
&= E(X_1X_2) \\
&= \left.\frac{\partial^2\phi(\boldsymbol{t})}{\partial t_1 \partial t_2}\right|_{\boldsymbol{t}=\boldsymbol{0}} \\
&= \omega_1\rho_1 + \omega_2\rho_2,
\end{aligned}$$

since $E(X_1) = E(X_2) = 0$.

Melnick and Tenenbein (1982) discussed several incorrect statements about the normal distribution and provided corresponding counter examples.

4.8.2 Characterization Results

There are several properties that characterize the normal distribution. By this we mean properties that are true if and only if the distribution is normal. Here are some examples of such properties.

Result 4.1 Independence of $\bar{X}$ and s^2.

We recall from a property stated earlier in this section that the sample mean $\bar{X}$ and the sample variance s^2 of a random sample from a normal distribution $N(\mu, \sigma^2)$ are independent. The converse is also true, that is, independence of $\bar{X}$ and s^2 implies normality of the parent population. This is given by the following theorem:

Theorem 4.6 Let $X_1, X_2, \ldots, X_n$ be a random sample from a parent distribution ($n \geq 2$), and let $\bar{X}$ and s^2 be their sample mean and sample variance, respectively. A necessary and sufficient condition for the independence of $\bar{X}$ and s^2 is that the parent distribution be normal.

This result was first shown by Geary (1936) and later by Lukacs (1942) who gave a simpler proof. See also Geisser (1956) and Kagan, Linnik, and Rao (1973, Theorem 4.2.1, p. 103).

Result 4.2 Normality of a Linear Combination of Random Variables

Let $\boldsymbol{X} = (X_1, X_2, \ldots, X_n)'$ be normally distributed as $N(\boldsymbol{\mu}, \boldsymbol{\Sigma})$. From Corollary 4.1, any linear combination, $\sum_{i=1}^{n} a_i X_i$, of the elements of $\boldsymbol{X}$ is normally distributed, where the a_i's are constant coefficients. This property provides another characterization of the normal distribution, which is based on the following theorem:

Theorem 4.7 Let $X_1, X_2, \ldots, X_n$ be random variables. If the distribution of $\sum_{i=1}^{n} a_i X_i$ is normal for any set of real numbers, $a_1, a_2, \ldots, a_n$, not all zero, then the joint distribution of $X_1, X_2, \ldots, X_n$ must be a multivariate normal.

Proof. Let $\boldsymbol{X} = (X_1, X_2, \ldots, X_n)'$, $\boldsymbol{a} = (a_1, a_2, \ldots, a_n)'$. Suppose that the mean of $\boldsymbol{X}$ is $\boldsymbol{\mu}$ and the variance–covariance matrix is $\boldsymbol{\Sigma}$. The mean and variance of $\boldsymbol{a}'\boldsymbol{X}(= \sum_{i=1}^{n} a_i X_i)$ are $\boldsymbol{a}'\boldsymbol{\mu}$ and $\boldsymbol{a}'\boldsymbol{\Sigma}\boldsymbol{a}$, respectively (see formulas 4.11 and 4.12). If $\boldsymbol{a}'\boldsymbol{X}$ is normally distributed, then its moment generating function is

$$E(e^{t\,\boldsymbol{a}'\boldsymbol{X}}) = \exp\left[(\boldsymbol{a}'\boldsymbol{\mu})t + \frac{1}{2}(\boldsymbol{a}'\boldsymbol{\Sigma}\boldsymbol{a})t^2\right].$$

Let $\boldsymbol{u}' = t\boldsymbol{a}'$. Then,

$$E(e^{\boldsymbol{u}'\boldsymbol{X}}) = \exp\left(\boldsymbol{u}'\boldsymbol{\mu} + \frac{1}{2}\boldsymbol{u}'\boldsymbol{\Sigma}\boldsymbol{u}\right),$$

which is the moment generating function of a multivariate normal with a mean vector $\boldsymbol{\mu}$ and a variance–covariance matrix $\boldsymbol{\Sigma}$. Thus, $\boldsymbol{X} \sim N(\boldsymbol{\mu}, \boldsymbol{\Sigma})$. □

Result 4.3 Independence of Two Linear Combinations of Random Variables.

Let $X_1, X_2, \ldots, X_n$ be mutually independent random variables, and let $V_1 = \sum_{i=1}^{n} a_i X_i$ and $V_2 = \sum_{i=1}^{n} b_i X_i$ be two linear combinations of the X_i's. If V_1 and V_2 are independent, then X_i must be normally distributed if $a_i b_i \neq 0$, $i = 1, 2, \ldots, n$.

This result is known as the *Darmois–Skitovic Theorem* (see Kagan, Linnik, and Rao, 1973, p. 89). The same result also holds if $X_1, X_2, \ldots, X_n$ are replaced by $\boldsymbol{X}_1, \boldsymbol{X}_2, \ldots, \boldsymbol{X}_n$, which are mutually independent $m \times 1$ random vectors (see Rao, 1973a, p. 525). This is given in the next result.

Result 4.4 Independence of Two Linear Combinations of Random Vectors.

If $\boldsymbol{X}_1, \boldsymbol{X}_2, \ldots, \boldsymbol{X}_n$ are n mutually independent $m \times 1$ random vectors such that $\boldsymbol{V}_1 = \sum_{i=1}^{n} a_i \boldsymbol{X}_i$ and $\boldsymbol{V}_2 = \sum_{i=1}^{n} b_i \boldsymbol{X}_i$ are independent, then $\boldsymbol{X}_i$ is normally distributed for any i such that $a_i b_i \neq 0$.

Another generalization of the Darmois–Skitovic Theorem is given by the following result:

Result 4.5 Independence of Two Sums of Linear Transforms of Random Vectors.

Let $\boldsymbol{X}_1$, $\boldsymbol{X}_2, \ldots, \boldsymbol{X}_n$ be mutually independent $m \times 1$ random vectors, and let $\boldsymbol{A}_1$, $\boldsymbol{A}_2, \ldots, \boldsymbol{A}_n$; $\boldsymbol{B}_1, \boldsymbol{B}_2, \ldots, \boldsymbol{B}_n$ be nonsingular $m \times m$ matrices. If $\sum_{i=1}^{n} \boldsymbol{A}_i \boldsymbol{X}_i$ is independent of $\sum_{i=1}^{n} \boldsymbol{B}_i \boldsymbol{X}_i$, then the $\boldsymbol{X}_i$'s are normally distributed.

This result was proved by Ghurye and Olkin (1962). See also Johnson and Kotz (1972, p. 59).

Result 4.6 Normality of a Sum of Independent Random Variables.

If $\mathbf{X}$ and $\mathbf{Y}$ are two independent $m \times 1$ random vectors such that $\mathbf{X}+\mathbf{Y}$ is normally distributed, then both $\mathbf{X}$ and $\mathbf{Y}$ are normally distributed.

This is a well-known result. Its proof can be found in several textbooks (see, for example, Rao, 1973a, p. 525; Muirhead, 1982, p. 14).

Exercises

4.1 Suppose that $\mathbf{X} \sim N(\boldsymbol{\mu}, \boldsymbol{\Sigma})$, where $\boldsymbol{\mu} = (1, 0, -1)'$ and $\boldsymbol{\Sigma}$ is given by

$$\boldsymbol{\Sigma} = \frac{1}{13}\begin{bmatrix} 8 & -2 & 1 \\ -2 & 7 & 3 \\ 1 & 3 & 5 \end{bmatrix}.$$

Let X_i be the ith element of $\mathbf{X}(i = 1, 2, 3)$.

(a) Find the distribution of $X_1 + 2X_2$.

(b) Find the joint density function of X_1 and X_2.

(c) Find the conditional distribution of X_1, given X_2 and X_3.

(d) Find the joint density function of $Y_1 = X_1+X_2+X_3$, $Y_2 = X_1-2X_3$.

(e) Find the moment generating function of $\mathbf{Y} = (Y_1, Y_2)'$, where Y_1 and Y_2 are the same as in part (d).

4.2 The moment generating function of a random vector $\mathbf{X} = (X_1, X_2, X_3)'$ is given by

$$\phi_{\mathbf{X}}(\mathbf{t}) = \exp\left(t_1 + t_2 + 2t_3 + 2t_1^2 + t_2^2 + \frac{3}{2}t_3^2 + t_1t_2 + t_2t_3\right).$$

(a) Find the moment generating function of X_1.

(b) Find the moment generating function of $(X_1, X_2)'$.

(c) Find the value of $P(X_1 + X_2 > X_3)$.

4.3 Suppose that $\mathbf{X} \sim N(\boldsymbol{\mu}, \boldsymbol{\Sigma})$, where $\boldsymbol{\mu} = (5.6, 5.1)'$ and $\boldsymbol{\Sigma}$ is given by

$$\boldsymbol{\Sigma} = \begin{bmatrix} 0.04 & 0.024 \\ 0.024 & 0.04 \end{bmatrix}.$$

Compute the value of the following probability:

$$P(5.20 < X_2 < 5.85 | X_1 = 6.1),$$

where X_1 and X_2 are the elements of $\mathbf{X}$.

4.4 Show that the density function, $f(x)$, for the normal distribution $N(\mu, \sigma^2)$ satisfies the differential equation,

$$\frac{df(x)}{dx} = \frac{(\mu - x)}{\sigma^2} f(x).$$

[Note: This shows that the normal distribution is a member of a general family of distributions known as *Pearson distributions* defined by the differential equation,

$$\frac{df(x)}{dx} = \frac{(x - a)f(x)}{b_0 + b_1 x + b_2 x^2},$$

where a, b_0, b_1, and b_2 are constant scalars (see Kendall and Stuart, 1963, Chapter 6, p. 148).]

4.5 Suppose that $\boldsymbol{X} = (X_1, X_2)'$ has the bivariate normal distribution $N(\boldsymbol{0}, \boldsymbol{\Sigma})$, where

$$\boldsymbol{\Sigma} = \begin{bmatrix} \sigma_{11} & \sigma_{12} \\ \sigma_{12} & \sigma_{22} \end{bmatrix}.$$

Show that the conditional distribution of $X_1|X_2$ has mean $= \rho X_2 \sqrt{\frac{\sigma_{11}}{\sigma_{22}}}$ and variance $= \sigma_{11}(1-\rho^2)$, where ρ is the correlation coefficient between X_1 and X_2 given by

$$\rho = \frac{\sigma_{12}}{(\sigma_{11}\sigma_{22})^{1/2}}$$

4.6 Suppose that $\boldsymbol{X} = (X_1, X_2, X_3)'$ has the multivariate normal distribution such that $\mu_1 = \mu_2 = -1, \mu_3 = 0, \sigma_{11} = 4, \sigma_{22} = 8, \sigma_{33} = 12$, where μ_i is the mean of X_i and σ_{ii} is the variance of $X_i (i = 1, 2, 3)$. Suppose also that X_1 is independent of $X_2 - X_1$, X_1 is independent of $X_3 - X_2$, and X_2 is independent of $X_3 - X_2$.

(a) Find the variance–covariance matrix of $\boldsymbol{X}$.

(b) Find $P(X_1 + X_2 + 2X_3 < 2)$.

(c) Find the conditional distribution of X_1 given X_2.

(d) Find the conditional distribution of X_1 given X_2 and X_3

4.7 Let $\boldsymbol{X}_1, \boldsymbol{X}_2, \ldots, \boldsymbol{X}_n$ be mutually independent and identically distributed as $N(\boldsymbol{0}, \boldsymbol{\Sigma})$. Let $\boldsymbol{A} = \sum_{i=1}^{n} \boldsymbol{X}_i \boldsymbol{X}_i'$. Show that $\boldsymbol{c}'\boldsymbol{A}\boldsymbol{c} \sim \sigma_c^2 \chi_n^2$ where $\boldsymbol{c}$ is a constant vector and $\sigma_c^2 = \boldsymbol{c}'\boldsymbol{\Sigma}\boldsymbol{c}$.

4.8 Suppose that $\boldsymbol{X} = (X_1, X_2)'$ has the bivariate normal distribution with a mean vector $\boldsymbol{\mu} = \lambda \boldsymbol{1}_2$, and a variance–covariance matrix $\boldsymbol{\Sigma} = (\sigma_{ij})$, where λ is the common mean of X_1 and X_2. Let $\rho = \frac{\sigma_{12}}{(\sigma_{11}\sigma_{22})^{1/2}}$ be the correlation coefficient between X_1 and X_2, and let r and θ be the polar coordinates corresponding to $X_1 - \lambda$ and $X_2 - \lambda$, that is, $X_1 - \lambda = r\cos\theta$, $X_2 - \lambda = r\sin\theta$.

(a) Find the joint density function of r and θ.

(b) Consider the probability value

$$p = P(|X_1 - \lambda| < |X_2 - \lambda|).$$

Use part (a) to find an expression for p as a double integral, and show that it can be written as

$$p = 2\int_{\frac{\pi}{4}}^{\frac{3\pi}{4}}\left[\int_0^{\infty} g(r,\theta)dr\right]d\theta,$$

where $g(r,\theta)$ is the joint density function of r and θ.

(c) Show that the value of p in part (b) is equal to

$$p = 1 - \frac{1}{\pi}\arctan\left\{\frac{2[\sigma_{11}\sigma_{22}(1-\rho^2)]^{\frac{1}{2}}}{\sigma_{22}-\sigma_{11}}\right\},$$

if $\sigma_{22} > \sigma_{11}$, where arctan(.) is the inverse of the tangent function and has values inside the interval $[-\frac{\pi}{2}, \frac{\pi}{2}]$.

[Note: A large value of p indicates that X_1 is closer to λ than X_2, which means that X_1 is a better estimator of λ than X_2.]

[Hint: See the article by Lowerre (1983).]

4.9 Suppose that $f(x)$ is a density function of the form

$$f(x) = \exp[\tau\mu x + s(x) + q(\mu)], \quad -\infty < x < \infty,$$

where

μ is the mean of the corresponding distribution,
$s(x)$ and $q(\mu)$ are functions of x and μ, respectively
τ is a positive constant

(a) Show that $\mu = -\frac{1}{\tau}\frac{dq(\mu)}{d\mu}$.

(b) Find the moment generating function for this distribution.

(c) Deduce from part (b) that this distribution must be normal with mean μ and variance $= \frac{1}{\tau}$.

[Note: This exercise is based on Theorem 1 by Anderson (1971)]

4.10 Let $\mathbf{X} = (X_1, X_2, \ldots, X_n)'$ be a random vector of n elements ($n \geq 3$) with the density function,

$$f(x) = \frac{1}{(2\pi)^{n/2}}e^{-\frac{1}{2}\sum_{i=1}^{n}x_i^2}\left[1 + \prod_{i=1}^{n}(x_i e^{-\frac{1}{2}x_i^2})\right].$$

It can be seen that $\mathbf{X}$ is not normally distributed.

(a) Show that the marginal distributions of the elements of $\mathbf{X}$ are normally distributed as $N(0, 1)$.

(b) Let $\mathbf{X}^{(j)}$ be a vector of $(n-1)$ elements obtained from $\mathbf{X}$ by just deleting $X_j (j = 1, 2, \ldots, n)$. Show that $\mathbf{X}^{(j)}$ has the multivariate normal distribution, and that its elements are mutually independent $N(0, 1)$ random variables.

(c) Show that the elements of $\mathbf{X}$ are pairwise independent, that is, any two elements of $\mathbf{X}$ are independent, but are not mutually independent.

(d) Deduce that any proper subset of $\{X_1, X_2, \ldots, X_n\}$ consists of random variables that are jointly normally distributed and mutually independent.

[Note: This example demonstrates that pairwise independence does not necessarily imply mutual independence, and that a proper subset of a random vector $\mathbf{X}$ may be jointly normally distributed, yet $\mathbf{X}$ itself does not have the multivariate normal distribution. For more details, see Pierce and Dykstra (1969)].

4.11 Let $X_1, X_2, \ldots, X_n$ be a random sample from the $N(\mu, \sigma^2)$ distribution. Using properties (1) through (3) in Section 4.8, we can write

$$P\left(-a < \frac{\bar{X} - \mu}{\frac{\sigma}{\sqrt{n}}} < a\right) = 1 - \alpha_1,$$

and

$$P\left(b < \frac{1}{\sigma^2}\sum_{i=1}^{n}(X_i - \bar{X})^2 < c\right) = 1 - \alpha_2,$$

where a is the upper $(\alpha_1/2)100$th percentile of the standard normal distribution, b and c are the lower and upper $(\alpha_2/2)100$th percentiles of the χ^2_{n-1} distribution, and $\bar{X} = \frac{1}{n}\sum_{i=1}^{n} X_i$.

(a) Show that

$$P\left[\bar{X} - a\frac{\sigma}{\sqrt{n}} < \mu < \bar{X} + a\frac{\sigma}{\sqrt{n}},\ \frac{1}{c}\sum_{i=1}^{n}(X_i - \bar{X})^2 < \sigma^2 < \frac{1}{b}\sum_{i=1}^{n}(X_i - \bar{X})^2\right]$$
$$= (1 - \alpha_1)(1 - \alpha_2).$$

This gives a so-called $(1 - \alpha_1)(1 - \alpha_2)100\%$ confidence region for (μ, σ^2).

(b) Draw the region in part (a).

(c) Show that by projecting this confidence region onto the vertical σ^2-*axis*, we obtain a conservative $(1-\alpha_1)(1-\alpha_2)100\%$ confidence interval on σ^2, that is,

$$P(d_1 < \sigma^2 < d_2) \geq (1-\alpha_1)(1-\alpha_2),$$

where $[d_1, d_2]$ is the projection of the confidence region on the σ^2-*axis*.

(d) Show that

$$\frac{n}{\sigma^2}(\bar{X}-\mu)^2 + \frac{1}{\sigma^2}\sum_{i=1}^{n}(X_i-\bar{X})^2 \sim \chi_n^2$$

(e) Use the result in part (d) to obtain another confidence region on (μ, σ^2).

[Note: For additional details concerning these results and others, see Arnold and Shavelle (1998).]

4.12 It is known that if $f(x_1, x_2)$ is the joint density function of X_1 and X_2, where $-\infty < X_1 < \infty, -\infty < X_2 < \infty$, then the marginal density function of X_1 is given by $\int_{-\infty}^{\infty} f(x_1, x_2)dx_2$. Use this method to show that if $(X_1, X_2)' \sim N(\mathbf{0}, \boldsymbol{I}_2)$, then $Y = X_1 + X_2$ is distributed as $N(0, 2)$.

[Hint: Use a particular change of variables.]

4.13 Let X_1 and X_2 be two independent normal variables with means, μ_1, μ_2, and variances, σ_1^2, σ_2^2, respectively. Let $Y = \min(X_1, X_2)$.

(a) Show that the density function of Y is given by

$$f(y) = f_1(y) + f_2(y), \quad -\infty < y < \infty,$$

where

$$f_1(y) = \frac{1}{\sigma_1}\phi\left(\frac{y-\mu_1}{\sigma_1}\right)\Phi\left(-\frac{y-\mu_2}{\sigma_2}\right),$$
$$f_2(y) = \frac{1}{\sigma_2}\phi\left(\frac{y-\mu_2}{\sigma_2}\right)\Phi\left(-\frac{y-\mu_1}{\sigma_1}\right),$$

and where $\phi(x) = \frac{1}{\sqrt{2\pi}}e^{-\frac{x^2}{2}}$, and $\Phi(x) = \int_{-\infty}^{x}\phi(u)du$.

(b) Find the moment generating function of Y and show that it is equal to

$$\phi_Y(t) = \phi_1(t) + \phi_2(t),$$

where

$$\phi_1(t) = \exp\left(\mu_1 t + \frac{1}{2}\sigma_1^2 t^2\right)\Phi\left[\frac{\mu_2 - \mu_1 - \sigma_1^2 t}{(\sigma_1^2 + \sigma_2^2)^{1/2}}\right]$$

$$\phi_2(t) = \exp\left(\mu_2 t + \frac{1}{2}\sigma_2^2 t^2\right)\Phi\left[\frac{\mu_1 - \mu_2 - \sigma_2^2 t}{(\sigma_1^2 + \sigma_2^2)^{1/2}}\right].$$

(c) Make use of the moment generating function in part (b) to find the mean and variance of Y.

[Note: For more details, see Cain (1994) and Tong (1990, p. 147).]

4.14 Let X_1 and X_2 be distributed independently as $N(0, 1)$.

(a) Find the joint density function of

$$Y_1 = X_1^2 + X_2^2, \quad Y_2 = \frac{X_1}{X_2},$$

and show that it is equal to

$$f(y_1, y_2) = \frac{1}{2\pi}e^{-\frac{1}{2}y_1}(1 + y_2^2)^{-1}.$$

(b) Find the marginal density functions of Y_1 and Y_2 and show that Y_2 has the Canchy distribution with the density function

$$f_2(y_2) = \frac{1}{\pi(1 + y_2^2)}.$$

(c) Deduce from part (b) that Y_1 and Y_2 are independent.

4.15 (a) Show that for every $z > 0$,

$$\left(\frac{1}{z} - \frac{1}{z^3}\right)\phi(z) < 1 - \Phi(z) < \frac{\phi(z)}{z},$$

where $\phi(z) = \frac{1}{\sqrt{2\pi}}e^{-\frac{1}{2}z^2}$, $\Phi(z) = \int_{-\infty}^{z}\phi(x)dx$.

(b) Deduce from part (a) that for a large z, $1 - \Phi(z)$ is approximately equal to $\frac{\phi(z)}{z}$.

(c) Deduce from part (a) that if $X \sim N(\mu, \sigma^2)$, then

$$\sigma\left(\frac{1}{z} - \frac{1}{z^3}\right) < \frac{P(X > x)}{f(x)} < \frac{\sigma}{z},$$

where $z = \frac{x-\mu}{\sigma} > 0$, and $f(x)$ is the density function of X.

[Note: The ratio $\frac{1}{f(x)}P(X > x)$ is known as *Mill's ratio*.]

[Hint: For a proof of part (a), see Lemma 2 in Feller (1957, p. 166).]

4.16 Show that Mill's ratio, described in Exercise 4.15, satisfies the following inequality:

$$\frac{1-\Phi(z)}{\phi(z)} \geq \frac{(4+z^2)^{1/2}-z}{2}$$

for $z > 0$, where $\phi(z)$ and $\Phi(z)$ are the same as in Exercise 4.15.

[Hint: See the note by Birnbaum (1942).]

4.17 Let X_1 and X_2 be two independent random variables distributed as $N(\mu, \sigma^2)$. Let s_2^2 be the corresponding sample variance

$$s_2^2 = \sum_{i=1}^{2}(X_i - \bar{X}_2)^2,$$

where $\bar{X}_2 = \frac{1}{2}(X_1 + X_2)$.

(a) Show that $s_2^2 = \frac{1}{2}(X_1 - X_2)^2$.

(b) Show that $\bar{X}_2$ and s_2^2 are independent.

(c) Show that $\frac{s_2^2}{\sigma^2} \sim \chi_1^2$.

[Note: Parts (b) and (c) provide a verification of properties (1) and (3), respectively, in Section 4.8 when $n = 2$.]

4.18 Let $X_1, X_2, \ldots, X_n$ be mutually independent and identically distributed as $N(\mu, \sigma^2)$. Let $\bar{X}_n$ and s_n^2 be the corresponding sample mean and sample variance, respectively. The objective here is to prove properties (1) and (3) in Section 4.8 by mathematical induction on the sample size n. Exercise 4.17 shows that these properties are true when $n = 2$. To show now that the same properties hold for a sample of size $n + 1$, if it is assumed that they hold for a sample of size n. For this purpose, the validity of the following parts must be established:

(a) Show that $\bar{X}_{n+1} = \frac{1}{n+1}(n\bar{X}_n + X_{n+1})$, and

$$ns_{n+1}^2 = (n-1)s_n^2 + \frac{n}{n+1}(X_{n+1} - \bar{X}_n)^2.$$

(b) Show that $X_{n+1} - \bar{X}_n$ is normally distributed with mean zero and variance $\sigma^2(1 + \frac{1}{n})$.

(c) Deduce from parts (a) and (b) that if $\bar{X}_n$ and s_n^2 are independent and $\frac{(n-1)s_n^2}{\sigma^2} \sim \chi_{n-1}^2$, then $\frac{ns_{n+1}^2}{\sigma^2} \sim \chi_n^2$.

(d) Show that $n\bar{X}_n + X_{n+1}$ is independent of $X_{n+1} - \bar{X}_n$.

(e) Deduce from parts (a) and (d) that if $\bar{X}_n$ and s_n^2 are independent, then so are $\bar{X}_{n+1}$ and s_{n+1}^2.

[Note: Parts (c) and (e) prove the validity of properties (1) and (3) for a sample of size $n+1$. Hence, these properties should be valid for all $n \geq 2$. More details concerning this proof using the mathematical induction argument can be found in Stigler (1984).]

4.19 Let $\mathbf{X}_1, \mathbf{X}_2, \ldots, \mathbf{X}_n (n \geq 2)$ be mutually independent and identically distributed as $N(\boldsymbol{\mu}, \boldsymbol{\Sigma})$, each having m elements $(m \leq n-1)$. Let $\bar{\mathbf{X}}_n = \frac{1}{n}\sum_{i=1}^{n} \mathbf{X}_i$ be the sample mean vector and

$$\mathbf{S}_n = \frac{1}{n-1}\sum_{i=1}^{n}(\mathbf{X}_i - \bar{\mathbf{X}}_n)(\mathbf{X}_i - \bar{\mathbf{X}}_n)'$$

be the sample variance–covariance matrix.

(a) Show that $\bar{\mathbf{X}}_n$ and $\mathbf{S}_n$ are independently distributed for all n.

(b) Use the method of mathematical induction on the sample size n to show that $(n-1)\mathbf{S}_n$ has the Wishart distribution $W_m(n-1, \boldsymbol{\Sigma})$ with $n-1$ degrees of freedom.

[Hint: To prove part (a), show first that

$$\text{Cov}(\mathbf{X}_i - \bar{\mathbf{X}}_n, \bar{\mathbf{X}}_n) = 0, \quad i = 1, 2, \ldots, n.$$

Then, $[(\mathbf{X}_1 - \bar{\mathbf{X}}_n)', (\mathbf{X}_2 - \bar{\mathbf{X}}_n)', \ldots, (\mathbf{X}_n - \bar{\mathbf{X}}_n)']'$ is uncorrelated with, and hence independent of, $\bar{\mathbf{X}}_n$. To prove part (b), show that

(1) (b) is true for $n = 2$, and

(2) if (b) is true for a sample of size n, then it must be true for a sample of size $n+1$.

To show (1), use the fact that

$$\frac{1}{\sqrt{2}}(\mathbf{X}_1 - \mathbf{X}_2) \sim N(\mathbf{0}, \boldsymbol{\Sigma}),$$

and that

$$\mathbf{S}_2 = \frac{1}{2}(\mathbf{X}_1 - \mathbf{X}_2)(\mathbf{X}_1 - \mathbf{X}_2)'.$$

To prove (2), establish first the following identities:

$$\bar{\mathbf{X}}_{n+1} = \frac{1}{n+1}(n\bar{\mathbf{X}}_n + \mathbf{X}_{n+1})$$

$$n\mathbf{S}_{n+1} = (n-1)\mathbf{S}_n + \frac{n}{n+1}(\mathbf{X}_{n+1} - \bar{\mathbf{X}}_n)(\mathbf{X}_{n+1} - \bar{\mathbf{X}}_n)',$$

then use the fact that $(n-1)\mathbf{S}_n \sim W_m(n-1, \mathbf{\Sigma})$ by the induction assumption, $\bar{\mathbf{X}}_n$ and $\mathbf{S}_n$ are independent, and that

$$\mathbf{X}_{n+1} - \bar{\mathbf{X}}_n \sim N\left[\mathbf{0}, \left(1 + \frac{1}{n}\right)\mathbf{\Sigma}\right]$$

to show that $n\mathbf{S}_{n+1} \sim W_m(n, \mathbf{\Sigma})$. Hence, part (b) is true for all n. For more details concerning these results, see Ghosh (1996).]

5

Quadratic Forms in Normal Variables

The subject of quadratic forms in normal random variables has received a great deal of attention since the early development of the theory of linear models. This is mainly due to the important role quadratic forms play in the statistical analysis of linear models. The sums of squares in a given *ANOVA* (analysis of variance) table, for example, can each be represented as a quadratic form in the corresponding data vector. These sums of squares are used to develop appropriate test statistics.

The purpose of this chapter is to provide a detailed coverage of the basic results pertaining to the distribution of quadratic forms in normally distributed random vectors. The coverage also includes results concerning the independence of a linear form from a quadratic form as well as the independence of two or more quadratic forms.

5.1 The Moment Generating Function

Let $X'AX$ be a quadratic form in X, where A is a symmetric matrix of order $n \times n$ and X is distributed as $N(\mu, \Sigma)$. The *moment generating function* of $X'AX$ is given by

$$\begin{aligned}\phi(t) &= E[\exp(t X'AX)] \\ &= \int_{R^n} \exp(t x'Ax) f(x)\, dx,\end{aligned}$$

where $f(x)$ is the density function of X as shown in formula (4.17). Hence,

$$\begin{aligned}\phi(t) &= \frac{1}{(2\pi)^{n/2}[det(\Sigma)]^{1/2}} \int_{R^n} \exp\left[t x'Ax - \frac{1}{2}(x-\mu)'\Sigma^{-1}(x-\mu)\right] dx \\ &= \frac{\exp(-\frac{1}{2}\mu'\Sigma^{-1}\mu)}{(2\pi)^{n/2}[det(\Sigma)]^{1/2}} \int_{R^n} \exp\left(-\frac{1}{2}x'W_t^{-1}x + x'\Sigma^{-1}\mu\right) dx, \qquad (5.1)\end{aligned}$$

where W_t is a positive definite matrix whose inverse is

$$W_t^{-1} = \Sigma^{-1/2}(I_n - 2t\,\Sigma^{1/2}A\Sigma^{1/2})\Sigma^{-1/2}, \qquad (5.2)$$

and $\Sigma^{-1/2}$ is the inverse of $\Sigma^{1/2}$, the square root of Σ, which was defined in Section 4.3. In Appendix 5.A, it is shown that there exists a positive number, t_0, such that W_t^{-1}, and hence W_t, is positive definite if $| t |< t_0$, where t_0 is given by

$$t_0 = \frac{1}{2 \max_i | \lambda_i |}, \tag{5.3}$$

and λ_i $(i = 1, 2, \ldots, n)$ is the ith eigenvalue of the matrix $\Sigma^{1/2} A \Sigma^{1/2}$. Making now the following change of variables:

$$x = y + W_t \Sigma^{-1} \mu,$$

formula (5.1) can be written as

$$\phi(t) = \frac{\exp\left(-\frac{1}{2}\mu'\Sigma^{-1}\mu\right)}{(2\pi)^{n/2}[det(\Sigma)]^{1/2}} \exp\left(\frac{1}{2}\mu'\Sigma^{-1}W_t\Sigma^{-1}\mu\right) \int_{R^n} \exp\left(-\frac{1}{2}y'W_t^{-1}y\right) dy. \tag{5.4}$$

Note that

$$\int_{R^n} \exp\left(-\frac{1}{2}y'W_t^{-1}y\right) dy = (2\pi)^{n/2}[det(W_t)]^{1/2}. \tag{5.5}$$

The proof of (5.5) is the same as the one used to derive formula (4.28). From (5.4) and (5.5) we then have

$$\phi(t) = \frac{[det(W_t)]^{1/2}}{[det(\Sigma)]^{1/2}} \exp\left\{-\frac{1}{2}\mu'\Sigma^{-1/2}\left[I_n - \left(I_n - 2t\,\Sigma^{1/2}A\Sigma^{1/2}\right)^{-1}\right]\Sigma^{-1/2}\mu\right\},$$

or equivalently,

$$\phi(t) = \frac{\exp\left\{-\frac{1}{2}\mu'\Sigma^{-1/2}\left[I_n - \left(I_n - 2t\,\Sigma^{1/2}A\Sigma^{1/2}\right)^{-1}\right]\Sigma^{-1/2}\mu\right\}}{\left[det\left(I_n - 2t\,\Sigma^{1/2}A\Sigma^{1/2}\right)\right]^{1/2}}. \tag{5.6}$$

This function is defined for $| t |< t_0$, where t_0 is given by (5.3).

Theorem 5.1 Let $X \sim N(\mu, \Sigma)$. Then, the rth *cumulant* of $X'AX$ is

$$\kappa_r(X'AX) = 2^{r-1}(r-1)!\{tr[(A\Sigma)^r] + r\,\mu'(A\Sigma)^{r-1}A\mu\}, \quad r = 1, 2, \ldots \tag{5.7}$$

Proof. We may recall from Section 4.2 that the rth cumulant of $X'AX$ is the coefficient of $\frac{t^r}{r!}$ $(r = 1, 2, \ldots)$ in Mclaurin's series expansion of $\psi(t)$, the *cumulant generating function* of $\boldsymbol{X}'\boldsymbol{A}\boldsymbol{X}$, which is the natural logarithm of $\phi(t)$ in (5.6), that is,

$$\begin{aligned}\psi(t) = &-\frac{1}{2}\log\left[det\left(\boldsymbol{I}_n - 2t\,\boldsymbol{\Sigma}^{1/2}\boldsymbol{A}\boldsymbol{\Sigma}^{1/2}\right)\right] \\ &-\frac{1}{2}\boldsymbol{\mu}'\boldsymbol{\Sigma}^{-1/2}\left[\boldsymbol{I}_n - \left(\boldsymbol{I}_n - 2t\,\boldsymbol{\Sigma}^{1/2}\boldsymbol{A}\boldsymbol{\Sigma}^{1/2}\right)^{-1}\right]\boldsymbol{\Sigma}^{-1/2}\boldsymbol{\mu}. \qquad (5.8)\end{aligned}$$

Note that

$$det\left(\boldsymbol{I}_n - 2t\,\boldsymbol{\Sigma}^{1/2}\boldsymbol{A}\boldsymbol{\Sigma}^{1/2}\right) = \prod_{i=1}^{n}(1 - 2t\lambda_i), \qquad (5.9)$$

where λ_i is the ith eigenvalue of $\boldsymbol{\Sigma}^{1/2}\boldsymbol{A}\boldsymbol{\Sigma}^{1/2}$ $(i = 1, 2, \ldots, n)$. Taking the logarithm of both sides of (5.9), we get

$$\log\left[det\left(\boldsymbol{I}_n - 2t\,\boldsymbol{\Sigma}^{1/2}\boldsymbol{A}\boldsymbol{\Sigma}^{1/2}\right)\right] = \sum_{i=1}^{n}\log(1 - 2t\lambda_i).$$

If $| 2t\lambda_i | < 1$, then $\log(1 - 2t\lambda_i)$ can be represented by the power series,

$$\log(1 - 2t\lambda_i) = -\sum_{r=1}^{\infty}\frac{(2t\lambda_i)^r}{r}, \quad i = 1, 2, \ldots, n,$$

which is absolutely convergent for all t such that $| 2t\lambda_i | < 1$, $i = 1, 2, \ldots, n$. This condition is satisfied if $| t | < t_0$, where t_0 is given by (5.3). In this case,

$$\begin{aligned}-\frac{1}{2}\log\left[det\left(\boldsymbol{I}_n - 2t\,\boldsymbol{\Sigma}^{1/2}\boldsymbol{A}\boldsymbol{\Sigma}^{1/2}\right)\right] &= \frac{1}{2}\sum_{i=1}^{n}\sum_{r=1}^{\infty}\frac{(2t\lambda_i)^r}{r} \\ &= \sum_{r=1}^{\infty}\frac{(2t)^r}{2r}\sum_{i=1}^{n}\lambda_i^r \\ &= \sum_{r=1}^{\infty}\frac{(2t)^r}{2r}\,tr\left[\left(\boldsymbol{\Sigma}^{1/2}\boldsymbol{A}\boldsymbol{\Sigma}^{1/2}\right)^r\right] \\ &= \sum_{r=1}^{\infty}\frac{t^r}{r!}\left\{2^{r-1}(r-1)!\,tr\left[\left(\boldsymbol{\Sigma}^{1/2}\boldsymbol{A}\boldsymbol{\Sigma}^{1/2}\right)^r\right]\right\}.\end{aligned}$$

(5.10)

Furthermore,

$$\begin{aligned}\left(\boldsymbol{I}_n - 2t\,\boldsymbol{\Sigma}^{1/2}\boldsymbol{A}\boldsymbol{\Sigma}^{1/2}\right)^{-1} &= \sum_{r=0}^{\infty}\left(2t\,\boldsymbol{\Sigma}^{1/2}\boldsymbol{A}\boldsymbol{\Sigma}^{1/2}\right)^r \\ &= \boldsymbol{I}_n + \sum_{r=1}^{\infty}(2t)^r\left(\boldsymbol{\Sigma}^{1/2}\boldsymbol{A}\boldsymbol{\Sigma}^{1/2}\right)^r. \qquad (5.11)\end{aligned}$$

The power series in the symmetric matrix $2t\boldsymbol{\Sigma}^{1/2}\boldsymbol{A}\boldsymbol{\Sigma}^{1/2}$ on the right-hand side of (5.11) converges to the matrix on the left-hand side if $|\ 2t\lambda_i\ | < 1$ for $i = 1, 2, \ldots, n$ (see Theorem 5.5.2 in Khuri, 2003, p. 180). As was seen earlier, this condition is satisfied if $|\ t\ | < t_0$. Hence,

$$-\frac{1}{2}\boldsymbol{\mu}'\boldsymbol{\Sigma}^{-1/2}\left[\boldsymbol{I}_n - \left(\boldsymbol{I}_n - 2t\boldsymbol{\Sigma}^{1/2}\boldsymbol{A}\boldsymbol{\Sigma}^{1/2}\right)^{-1}\right]\boldsymbol{\Sigma}^{-1/2}\boldsymbol{\mu}$$
$$= \frac{1}{2}\sum_{r=1}^{\infty}(2t)^r\boldsymbol{\mu}'\boldsymbol{\Sigma}^{-1/2}\left(\boldsymbol{\Sigma}^{1/2}\boldsymbol{A}\boldsymbol{\Sigma}^{1/2}\right)^r\boldsymbol{\Sigma}^{-1/2}\boldsymbol{\mu}$$
$$= \sum_{r=1}^{\infty}\frac{t^r}{r!}\left(2^{r-1}r!\right)\boldsymbol{\mu}'\boldsymbol{\Sigma}^{-1/2}\left(\boldsymbol{\Sigma}^{1/2}\boldsymbol{A}\boldsymbol{\Sigma}^{1/2}\right)^r\boldsymbol{\Sigma}^{-1/2}\boldsymbol{\mu}. \tag{5.12}$$

Substituting the right-hand sides of (5.10) and (5.12) in (5.8) yields a Maclaurin's series expansion of $\psi(t)$, which is convergent if $|\ t\ | < t_0$, where t_0 is given by (5.3). The coefficient of $\frac{t^r}{r!}$ in this expansion gives $\kappa_r(\boldsymbol{X}'\boldsymbol{A}\boldsymbol{X})$, which is equal to

$$\kappa_r\left(\boldsymbol{X}'\boldsymbol{A}\boldsymbol{X}\right) = 2^{r-1}(r-1)!\,tr\left[\left(\boldsymbol{\Sigma}^{1/2}\boldsymbol{A}\boldsymbol{\Sigma}^{1/2}\right)^r\right]$$
$$+ 2^{r-1}r!\,\boldsymbol{\mu}'\boldsymbol{\Sigma}^{-1/2}\left(\boldsymbol{\Sigma}^{1/2}\boldsymbol{A}\boldsymbol{\Sigma}^{1/2}\right)^r\boldsymbol{\Sigma}^{-1/2}\boldsymbol{\mu}, \quad r = 1, 2, \ldots \tag{5.13}$$

Note that

$$tr\left[\left(\boldsymbol{\Sigma}^{1/2}\boldsymbol{A}\boldsymbol{\Sigma}^{1/2}\right)^r\right] = tr\left[(\boldsymbol{A}\boldsymbol{\Sigma})^r\right],$$

and

$$\boldsymbol{\mu}'\boldsymbol{\Sigma}^{-1/2}\left(\boldsymbol{\Sigma}^{1/2}\boldsymbol{A}\boldsymbol{\Sigma}^{1/2}\right)^r\boldsymbol{\Sigma}^{-1/2}\boldsymbol{\mu} = \boldsymbol{\mu}'(\boldsymbol{A}\boldsymbol{\Sigma})^{r-1}\boldsymbol{A}\boldsymbol{\mu}.$$

Making the proper substitution in (5.13) results in formula (5.7). □

Corollary 5.1 If $\boldsymbol{X} \sim N(\boldsymbol{\mu}, \boldsymbol{\Sigma})$, then the mean and variance of $\boldsymbol{X}'\boldsymbol{A}\boldsymbol{X}$ are

$$E\left(\boldsymbol{X}'\boldsymbol{A}\boldsymbol{X}\right) = \boldsymbol{\mu}'\boldsymbol{A}\boldsymbol{\mu} + tr(\boldsymbol{A}\boldsymbol{\Sigma}),$$
$$\text{Var}\left(\boldsymbol{X}'\boldsymbol{A}\boldsymbol{X}\right) = 4\,\boldsymbol{\mu}'\boldsymbol{A}\boldsymbol{\Sigma}\boldsymbol{A}\boldsymbol{\mu} + 2\,tr\left[(\boldsymbol{A}\boldsymbol{\Sigma})^2\right].$$

Proof. The mean and variance of $\boldsymbol{X}'\boldsymbol{A}\boldsymbol{X}$ are, respectively, the first and second cumulants of $\boldsymbol{X}'\boldsymbol{A}\boldsymbol{X}$. Hence, from formula (5.7) we obtain

$$E(\boldsymbol{X}'\boldsymbol{A}\boldsymbol{X}) = \kappa_1(\boldsymbol{X}'\boldsymbol{A}\boldsymbol{X})$$
$$= \boldsymbol{\mu}'\boldsymbol{A}\boldsymbol{\mu} + tr(\boldsymbol{A}\boldsymbol{\Sigma}),$$
$$\text{Var}(\boldsymbol{X}'\boldsymbol{A}\boldsymbol{X}) = \kappa_2(\boldsymbol{X}'\boldsymbol{A}\boldsymbol{X})$$
$$= 4\,\boldsymbol{\mu}'\boldsymbol{A}\boldsymbol{\Sigma}\boldsymbol{A}\boldsymbol{\mu} + 2\,tr[(\boldsymbol{A}\boldsymbol{\Sigma})^2].$$

□

It should be noted that the expression for $E(X'AX)$ is valid even if X is not normally distributed. This is shown in the next theorem.

Theorem 5.2 If X is any random vector with a mean μ and a variance–covariance matrix Σ, then

$$E\left(X'AX\right) = \mu' A\mu + tr(A\Sigma).$$

Proof. Since $X'AX$ is a scalar, it is equal to its trace. Hence,

$$\begin{aligned} E\left(X'AX\right) &= E\left[tr\left(X'AX\right)\right] \\ &= E\left[tr\left(AXX'\right)\right] \\ &= tr\left[E\left(AXX'\right)\right] \\ &= tr\left[AE\left(XX'\right)\right] \\ &= tr\left[A\left(\Sigma + \mu\mu'\right)\right], \text{ by (4.9)} \\ &= tr\left(A\mu\mu'\right) + tr\left(A\Sigma\right) \\ &= \mu' A\mu + tr(A\Sigma). \end{aligned}$$

□

The next theorem describes the covariance of $X'AX$ and a linear form BX, where B is a constant matrix and X is normally distributed.

Theorem 5.3 Let $X'AX$ be a quadratic form and BX be a linear form, where B is a constant matrix of order $m \times n$. If $X \sim N(\mu, \Sigma)$, then

$$\text{Cov}\left(BX, X'AX\right) = 2\,B\Sigma A\mu. \tag{5.14}$$

Proof. We have that

$$\begin{aligned} \text{Cov}\left(BX, X'AX\right) &= E\left\{(BX - B\mu)\left[X'AX - E(X'AX)\right]\right\} \\ &= E\left\{(BX - B\mu)\left[X'AX - \mu' A\mu - tr\left(A\Sigma\right)\right]\right\} \\ &= E\left[(BX - B\mu)\left(X'AX - \mu' A\mu\right)\right], \end{aligned} \tag{5.15}$$

since the expected value of $BX - B\mu$ is equal to the zero vector. By centering X around its mean, we can write

$$(BX - B\mu)\left(X'AX - \mu' A\mu\right) = B(X - \mu)[(X - \mu)' A(X - \mu) + 2\,(X - \mu)' A\mu].$$

Let $X - \mu = Y$. Then, $Y \sim N(\mathbf{0},\ \Sigma)$, and

$$\text{Cov}\left(BX, X'AX\right) = E\left[BY\left(Y'AY + 2\,Y'A\mu\right)\right]. \tag{5.16}$$

Now,

$$\begin{aligned} E\left[(BY)\,(2\,Y'A\mu)\right] &= 2\,BE\left(YY'\right)A\mu \\ &= 2\,B\Sigma A\mu, \end{aligned} \tag{5.17}$$

and

$$E\left[(BY)\left(Y'AY\right)\right] = BE\left[(Y)\left(Y'AY\right)\right]. \tag{5.18}$$

Note that $(Y)(Y'AY)$ is a scalar multiple of the random vector Y. Its ith element is equal to

$$Y_i(Y'AY) = Y_i \sum_{j=1}^{n}\sum_{k=1}^{n} a_{jk} Y_j Y_k, \quad i = 1, 2, \ldots, n,$$

where Y_i is the ith element of Y and a_{jk} is the (j,k)th element of A $(j,k = 1,2,\ldots,n)$. But, for any values of i,j,k $(= 1,2,\ldots,n)$,

$$E(Y_iY_jY_k) = 0,$$

since by (4.17),

$$E(Y_iY_jY_k) = \frac{1}{(2\pi)^{n/2}[det(\Sigma)]^{1/2}} \int_{R^n} y_i y_j y_k \exp\left(-\frac{1}{2}y'\Sigma^{-1}y\right) dy,$$

and the integrand is an odd function over R^n. It follows that $E[Y_i(Y'\,AY)] = 0$ for $i = 1,2,\ldots,n$. Hence, $E[(Y)(Y'\,AY)] = \mathbf{0}_1$, where $\mathbf{0}_1$ is a zero vector of order $n \times 1$. Substituting this value in (5.18), we get

$$E[(BY)(Y'AY)] = \mathbf{0}_2, \tag{5.19}$$

where $\mathbf{0}_2$ is a zero vector of order $m \times 1$. From (5.16), (5.17), and (5.19), we conclude (5.14). □

5.2 Distribution of Quadratic Forms

In this section, we begin our study of the distribution of quadratic forms in normal random vectors.

Lemma 5.1 Let A be a symmetric matrix of order $n \times n$ and X be normally distributed as $N(\mu, \Sigma)$. Then, $X'AX$ can be expressed as

$$X'AX = \sum_{i=1}^{k} \gamma_i W_i, \tag{5.20}$$

where $\gamma_1, \gamma_2, \ldots, \gamma_k$ are the distinct nonzero eigenvalues of $\Sigma^{1/2}A\Sigma^{1/2}$ (or, equivalently, the matrix $A\Sigma$) with multiplicities $\nu_1, \nu_2, \ldots, \nu_k$, respectively, and the $W_i's$ are mutually independent such that $W_i \sim \chi^2_{\nu_i}(\theta_i)$, where

$$\theta_i = \mu'\Sigma^{-1/2}P_iP_i'\Sigma^{-1/2}\mu, \tag{5.21}$$

and $\boldsymbol{P}_i$ is a matrix of order $n\times\nu_i$ whose columns are orthonormal eigenvectors of $\boldsymbol{\Sigma}^{1/2}\boldsymbol{A}\boldsymbol{\Sigma}^{1/2}$ corresponding to ν_i $(i=1,2,\ldots,k)$.

Proof. Let $\boldsymbol{X}^*=\boldsymbol{\Sigma}^{-1/2}\boldsymbol{X}$. Then, $\boldsymbol{X}^*\sim N(\boldsymbol{\Sigma}^{-1/2}\boldsymbol{\mu},\boldsymbol{I}_n)$, and $\boldsymbol{X}'\boldsymbol{A}\boldsymbol{X}=(\boldsymbol{X}^*)'\boldsymbol{\Sigma}^{1/2}\boldsymbol{A}\boldsymbol{\Sigma}^{1/2}\boldsymbol{X}^*$. Let $\gamma_1,\gamma_2,\ldots,\gamma_k$ be the distinct nonzero eigenvalues of $\boldsymbol{\Sigma}^{1/2}\boldsymbol{A}\boldsymbol{\Sigma}^{1/2}$ with multiplicities $\nu_1,\nu_2,\ldots,\nu_k$, respectively. By applying the Spectral Decomposition Theorem (Theorem 3.4) to the symmetric matrix $\boldsymbol{\Sigma}^{1/2}\boldsymbol{A}\boldsymbol{\Sigma}^{1/2}$, we obtain

$$\boldsymbol{\Sigma}^{1/2}\boldsymbol{A}\boldsymbol{\Sigma}^{1/2}=\boldsymbol{P}\boldsymbol{\Gamma}\boldsymbol{P}', \tag{5.22}$$

where
- $\boldsymbol{\Gamma}$ is a diagonal matrix of eigenvalues of $\boldsymbol{\Sigma}^{1/2}\boldsymbol{A}\boldsymbol{\Sigma}^{1/2}$
- $\boldsymbol{P}$ is an orthogonal matrix whose columns are the corresponding eigenvectors

The matrix $\boldsymbol{\Gamma}$ can be written as

$$\boldsymbol{\Gamma}=\mathrm{diag}(\gamma_1\boldsymbol{I}_{\nu_1},\gamma_2\boldsymbol{I}_{\nu_2},\ldots,\gamma_k\boldsymbol{I}_{\nu_k},\boldsymbol{0}), \tag{5.23}$$

where $\boldsymbol{0}$ is a diagonal matrix whose diagonal elements are the $n-r$ zero eigenvalues of $\boldsymbol{\Sigma}^{1/2}\boldsymbol{A}\boldsymbol{\Sigma}^{1/2}$ and $r=\sum_{i=1}^{k}\nu_i$, which is the rank of $\boldsymbol{A}$.

Let now $\boldsymbol{Z}=\boldsymbol{P}'\boldsymbol{X}^*$. Then, $\boldsymbol{Z}\sim N(\boldsymbol{P}'\boldsymbol{\Sigma}^{-1/2}\boldsymbol{\mu},\boldsymbol{I}_n)$. Let us also partition $\boldsymbol{P}$ as $\boldsymbol{P}=[\boldsymbol{P}_1:\boldsymbol{P}_2:\ldots:\boldsymbol{P}_k:\boldsymbol{P}_{k+1}]$, where $\boldsymbol{P}_i$ is a matrix of order $n\times\nu_i$ whose columns are orthonormal eigenvectors of $\boldsymbol{\Sigma}^{1/2}\boldsymbol{A}\boldsymbol{\Sigma}^{1/2}$ corresponding to γ_i $(i=1,2,\ldots,k)$, and the columns of $\boldsymbol{P}_{k+1}$ are orthonormal eigenvectors corresponding to the zero eigenvalue of multiplicity $n-r$. Using (5.22) we can then write

$$\begin{aligned}\boldsymbol{X}'\boldsymbol{A}\boldsymbol{X}&=(\boldsymbol{X}^*)'\boldsymbol{P}\boldsymbol{\Gamma}\boldsymbol{P}'\boldsymbol{X}^*\\&=\boldsymbol{Z}'\boldsymbol{\Gamma}\boldsymbol{Z}\\&=\sum_{i=1}^{k}\gamma_i\boldsymbol{Z}_i'\boldsymbol{Z}_i\\&=\sum_{i=1}^{k}\gamma_iW_i,\end{aligned}$$

where
- $\boldsymbol{Z}_i=\boldsymbol{P}_i'\boldsymbol{X}^*$
- $W_i=\boldsymbol{Z}_i'\boldsymbol{Z}_i$ $(i=1,2,\ldots,k)$

Since $\boldsymbol{Z}_i\sim N(\boldsymbol{P}_i'\boldsymbol{\Sigma}^{-1/2}\boldsymbol{\mu},\boldsymbol{I}_{\nu_i})$, $W_i\sim\chi^2_{\nu_i}(\theta_i)$, where $\theta_i=\boldsymbol{\mu}'\boldsymbol{\Sigma}^{-1/2}\boldsymbol{P}_i\boldsymbol{P}_i'\boldsymbol{\Sigma}^{-1/2}\boldsymbol{\mu}$, $i=1,2,\ldots,k$ (see Section 4.7.2). Furthermore, $W_1,W_2,\ldots,W_k$ are mutually independent by the fact that $\boldsymbol{Z}_1,\boldsymbol{Z}_2,\ldots,\boldsymbol{Z}_k$ are mutually independent. The latter assertion follows since $\boldsymbol{Z}$ is normally distributed and $\mathrm{Cov}(\boldsymbol{Z}_i,\boldsymbol{Z}_j)=\boldsymbol{P}_i'\boldsymbol{P}_j=\boldsymbol{0}$ for $i\neq j$, which implies mutual independence of the $\boldsymbol{Z}_i$'s by Corollary 4.2. This completes the proof of the lemma. □

Lemma 5.1 gives a representation of the distribution of $X'AX$ as a linear combination of mutually independent chi-squared variates when X has the multivariate normal distribution. This lemma wil be instrumental in proving the following important result, which gives the condition for $X'AX$ to have the chi-squared distribution.

Theorem 5.4 Let A be a symmetric matrix of order $n \times n$ and X be normally distributed as $N(\mu, \Sigma)$. A necessary and sufficient condition for $X'AX$ to have the noncentral chi-squared distribution $\chi_r^2(\theta)$, where $\theta = \mu' A \mu$, is that $A\Sigma$ be idempotent of rank r.

Proof. We shall first prove the sufficiency part, then prove the necessity part, which is quite more involved.

Sufficiency. Suppose that $A\Sigma$ is idempotent of rank r. To show that $X'AX$ is distributed as $\chi_r^2(\theta)$, where $\theta = \mu' A \mu$. It should first be noted that $A\Sigma$ is idempotent if and only if the symmetric matrix $\Sigma^{1/2} A \Sigma^{1/2}$ is idempotent. The proof of this assertion is given in Appendix 5.B.

The proof of sufficiency can be easily established by invoking Lemma 5.1. If $A\Sigma$ is idempotent of rank r, then so is the matrix $\Sigma^{1/2} A \Sigma^{1/2}$. Hence, $\Sigma^{1/2} A \Sigma^{1/2}$ must have r eigenvalues equal to 1 and $n - r$ eigenvalues equal to zero (see property (a) in Section 3.9 concerning the eigenvalues of an idempotent matrix). Using the representation (5.20) in Lemma 5.1, it can be concluded that $X'AX = W_1 \sim \chi_r^2(\theta_1)$ since $k = 1$, $\gamma_1 = 1$, and $\nu_1 = r$. The noncentrality parameter θ_1 in (5.21) is given by

$$\theta_1 = \mu' \Sigma^{-1/2} P_1 P_1' \Sigma^{-1/2} \mu, \tag{5.24}$$

where P_1 is such that $P = [P_1 : P_2]$ is the orthogonal matrix of orthonormal eigenvectors of $\Sigma^{1/2} A \Sigma^{1/2}$ with P_1 and P_2 being the portions of P corresponding to the eigenvalues 1 and 0, respectively. Hence, by formula (5.22),

$$\Sigma^{1/2} A \Sigma^{1/2} = P_1 P_1',$$

since from (5.23), $\Gamma = \mathrm{diag}(I_r, 0)$. By making the substitution in (5.24) we conclude that $\theta_1 = \mu' A \mu$. We shall drop subscript 1 and use just $\theta = \mu' A \mu$ since there is only one noncentrality parameter.

An alternative proof of sufficiency is based on showing that the moment generating function of $X'AX$ is the same as the one for $\chi_r^2(\theta)$ when $A\Sigma$ is idempotent of rank r (see Exercise 5.12).

Necessity. Suppose that $X'AX \sim \chi_r^2(\theta)$. To show that $A\Sigma$ is idempotent of rank r. The proof was given by Khuri (1999) and is reproduced here.

Let us begin by considering the representation (5.20) of Lemma 5.1. If $X' AX \sim \chi_r^2(\theta)$, then by Theorem 4.3, its moment generating function (m.g.f.), denoted here by $\phi(t, \theta)$, is

$$\phi(t,\theta) = (1-2t)^{-r/2}\exp\left[\theta t(1-2t)^{-1}\right]$$
$$= (1-2t)^{-r/2}\exp\left[-\frac{\theta}{2}\left(1-\frac{1}{1-2t}\right)\right], \quad t < \frac{1}{2}. \tag{5.25}$$

Similarly, since $W_i \sim \chi^2_{\nu_i}(\theta_i)$, $i = 1, 2, \ldots, k$, its m.g.f., which is denoted by $\phi_i(t, \theta_i)$, is

$$\phi_i(t,\theta_i) = (1-2t)^{-\nu_i/2}\exp\left[-\frac{\theta_i}{2}\left(1-\frac{1}{1-2t}\right)\right], \quad t < \frac{1}{2}, i = 1, 2, \ldots, k.$$

Furthermore, since the W_i's in (5.20) are mutually independent, then the m.g.f. of the right-hand side of (5.20), denoted by $\phi_s(t, \theta_1, \theta_2, \ldots, \theta_k)$, is of the form

$$\phi_s(t,\theta_1,\theta_2,\ldots,\theta_k)$$
$$= \prod_{i=1}^{k}(1-2\gamma_i t)^{-\nu_i/2}\exp\left[-\frac{\theta_i}{2}\left(1-\frac{1}{1-2\gamma_i t}\right)\right], \quad |\, t \,| < \frac{1}{2\max_i |\, \gamma_i \,|}. \tag{5.26}$$

Note that in (5.26), $\gamma_i t$ must be less than $\frac{1}{2}$ for $i = 1, 2, \ldots, k$. This is satisfied if $|\, t \,| < \frac{1}{2\max_i|\gamma_i|}$. Equating the right-hand sides of (5.25) and (5.26), we get

$$(1-2t)^{-r/2}\exp\left[-\frac{\theta}{2}\left(1-\frac{1}{1-2t}\right)\right]$$
$$= \prod_{i=1}^{k}(1-2\gamma_i t)^{-\nu_i/2}\exp\left[-\frac{\theta_i}{2}\left(1-\frac{1}{1-2\gamma_i t}\right)\right], \quad |\, t \,| < t^*, \tag{5.27}$$

where

$$t^* = \min\left(\frac{1}{2}, \frac{1}{2\max_i |\, \gamma_i \,|}\right).$$

Taking the natural logarithms of both sides of (5.27), we obtain

$$-\frac{r}{2}\log(1-2t) - \frac{\theta}{2}\left(1-\frac{1}{1-2t}\right)$$
$$= \sum_{i=1}^{k}\left[-\frac{\nu_i}{2}\log(1-2\gamma_i t)] - \frac{\theta_i}{2}\left(1-\frac{1}{1-2\gamma_i t}\right)\right], \quad |\, t \,| < t^*. \tag{5.28}$$

Now, for $\mid u \mid< 1$, Maclaurin's series expansions of the functions $\log(1-u)$ and $1-\frac{1}{1-u}$ are given by

$$\log(1-u) = -\sum_{j=1}^{\infty} \frac{u^j}{j}, \tag{5.29}$$

$$1-\frac{1}{1-u} = -\sum_{j=1}^{\infty} u^j, \tag{5.30}$$

respectively. Applying these expressions to the appropriate terms in (5.28), we get

$$\frac{r}{2}\sum_{j=1}^{\infty}\frac{(2\,t)^j}{j} + \frac{\theta}{2}\sum_{j=1}^{\infty}(2\,t)^j = \sum_{i=1}^{k}\left[\frac{\nu_i}{2}\sum_{j=1}^{\infty}\frac{(2\,\gamma_i\,t)^j}{j} + \frac{\theta_i}{2}\sum_{j=1}^{\infty}(2\,\gamma_i\,t)^j\right], \quad \mid t \mid < t^*. \tag{5.31}$$

Let us now equate the coefficients of $(2\,t)^j$ on both sides of (5.31). We obtain

$$\frac{r}{2j} + \frac{\theta}{2} = \sum_{i=1}^{k}\left(\frac{\nu_i}{2j} + \frac{\theta_i}{2}\right)\gamma_i^j, \quad j = 1, 2, \ldots \tag{5.32}$$

Formula (5.32) implies that $\mid \gamma_i \mid\leq 1$ for $i = 1, 2, \ldots, k$, otherwise, if $\mid \gamma_i \mid> 1$, then by letting j go to infinity, the right-hand side of (5.32) will tend to infinity. This leads to a contradiction since, as $j \to \infty$, the limit of the left-hand side of (5.32) is equal to $\frac{\theta}{2} < \infty$.

The next step is to show that $\mid \gamma_i \mid= 1$ for $i = 1, 2, \ldots, k$. If $\mid \gamma_i \mid< 1$, then by taking the sum over j of the terms on both sides of (5.32), we obtain

$$\sum_{j=1}^{\infty}\left(\frac{r}{2j} + \frac{\theta}{2}\right) = \sum_{i=1}^{k}\frac{\nu_i}{2}\left(\sum_{j=1}^{\infty}\frac{\gamma_i^j}{j}\right) + \frac{1}{2}\sum_{i=1}^{k}\theta_i\sum_{j=1}^{\infty}\gamma_i^j. \tag{5.33}$$

Using formulas (5.29) and (5.30) in (5.33), we get

$$\sum_{j=1}^{\infty}\left(\frac{r}{2j} + \frac{\theta}{2}\right) = -\sum_{i=1}^{k}\frac{\nu_i}{2}\log(1-\gamma_i) - \frac{1}{2}\sum_{i=1}^{k}\theta_i\left(1 - \frac{1}{1-\gamma_i}\right). \tag{5.34}$$

The equality in (5.34) is not possible since the left-hand side is infinite (both $\sum_{j=1}^{\infty}\frac{1}{j}$ and $\sum_{j=1}^{\infty}\theta$ are divergent series), whereas the right-hand side is finite. This contradiction leads to the conclusion that $\mid \gamma_i \mid= 1$ for $i = 1, 2, \ldots, k$.

We recall from Lemma 5.1 that k is the number of distinct nonzero eigenvalues of $\boldsymbol{\Sigma}^{1/2}\boldsymbol{A}\boldsymbol{\Sigma}^{1/2}$ (or, equivalently, of the matrix $\boldsymbol{A\Sigma}$). The next objective is to show that $k = 1$ and $\gamma_1 = 1$.

Since $|\gamma_i| = 1$ for $i = 1, 2, \ldots, k$, there can only be three cases to consider, namely,

(a) $k = 2,\ \gamma_1 = 1,\ \gamma_2 = -1.$

(b) $k = 1,\ \gamma_1 = -1.$

(c) $k = 1,\ \gamma_1 = 1.$

We now show that Case (c) is the only valid one.

Case (a). Choosing j to be an even integer in (5.32), we obtain

$$\frac{r}{2j} + \frac{\theta}{2} = \sum_{i=1}^{2}\left(\frac{\nu_i}{2j} + \frac{\theta_i}{2}\right), \quad j = 2, 4, \ldots \tag{5.35}$$

Letting j go to infinity (as an even integer) in (5.35), we get $\theta = \theta_1 + \theta_2$. Hence, (5.35) shows that $r = \nu_1 + \nu_2 = rank(\boldsymbol{\Sigma}^{1/2}\boldsymbol{A}\boldsymbol{\Sigma}^{1/2}) = rank(\boldsymbol{A}\boldsymbol{\Sigma})$. Choosing now j to be an odd integer in (5.32) yields the following result:

$$\frac{r}{2j} + \frac{\theta}{2} = \frac{\nu_1}{2j} + \frac{\theta_1}{2} - \frac{\nu_2}{2j} - \frac{\theta_2}{2}, \quad j = 1, 3, \ldots \tag{5.36}$$

Substituting $\theta = \theta_1 + \theta_2,\ r = \nu_1 + \nu_2$ in (5.36), we obtain

$$\frac{\nu_2}{2j} + \frac{\theta_2}{2} = -\frac{\nu_2}{2j} - \frac{\theta_2}{2}, \quad j = 1, 3, \ldots$$

This is not possible since $\nu_i > 0$ and $\theta_i \geq 0$ for $i = 1, 2$. Case (a) is therefore invalid.

Case (b). This case is also invalid. Choosing here j to be an odd integer in (5.32), we get

$$\frac{r}{2j} + \frac{\theta}{2} = -\frac{\nu_1}{2j} - \frac{\theta_1}{2}, \quad j = 1, 3, \ldots$$

This equality is not possible since the left-hand side is positive whereas the right-hand side is negative.

Having determined that Cases (a) and (b) are not valid, Case (c) must therefore be the only valid one. In this case, (5.32) takes the form

$$\frac{r}{2j} + \frac{\theta}{2} = \frac{\nu_1}{2j} + \frac{\theta_1}{2}, \quad j = 1, 2, \ldots$$

Letting j go to infinity, we find that $\theta_1 = \theta$, and hence $\nu_1 = r$. It follows that the matrix $\boldsymbol{\Sigma}^{1/2}\boldsymbol{A}\boldsymbol{\Sigma}^{1/2}$ has only one nonzero eigenvalue equal to 1 of multiplicity $\nu_1 = r$, and is therefore of rank r. Since this matrix is symmetric, it is easy to show, using the Spectral Decomposition Theorem (Theorem 3.4), that it must also be idempotent. This implies that $\boldsymbol{A}\boldsymbol{\Sigma}$ is idempotent and of rank r. □

Corollary 5.2 (Good, 1969) Let A and X be the same as in Theorem 5.4. A necessary and sufficient condition for $X'AX$ to have the $\chi^2_r(\theta)$ distribution is that $tr\,(A\Sigma) = tr\left[(A\Sigma)^2\right] = r$, and $rank\,(A\Sigma) = r$.

Proof.

Necessity. If $X'AX \sim \chi^2_r(\theta)$, then from the proof of the necessity part of Theorem 5.4, the matrix $\Sigma^{1/2}A\Sigma^{1/2}$ must be idempotent of rank r. Hence, $rank(A\Sigma) = rank(\Sigma^{1/2}A\Sigma^{1/2}) = r$, and

$$\begin{aligned} tr\left[\left(\Sigma^{1/2}A\Sigma^{1/2}\right)^2\right] &= tr\left(\Sigma^{1/2}A\Sigma^{1/2}\right) \\ &= rank\left(\Sigma^{1/2}A\Sigma^{1/2}\right) \\ &= r. \end{aligned}$$

But, $tr(\Sigma^{1/2}A\Sigma^{1/2}) = tr(A\Sigma)$, and $tr[(\Sigma^{1/2}A\Sigma^{1/2})^2] = tr[(A\Sigma)^2]$. Thus, $tr(A\Sigma) = tr[(A\Sigma)^2] = r$.

Sufficiency. Suppose that the given condition is true. Then, $tr(\Sigma^{1/2}A\Sigma^{1/2}) = tr[(\Sigma^{1/2}A\Sigma^{1/2})^2] = r$ and $rank(\Sigma^{1/2}A\Sigma^{1/2}) = r$. This implies that $\Sigma^{1/2}A\Sigma^{1/2}$ has r nonzero eigenvalues $\lambda_1, \lambda_2, \ldots, \lambda_r$, and $\sum_{i=1}^r \lambda_i = \sum_{i=1}^r \lambda_i^2 = r$ (Note: $\lambda_1^2, \lambda_2^2, \ldots, \lambda_r^2$ are the nonzero eigenvalues of the square of $\Sigma^{1/2}A\Sigma^{1/2}$). Therefore,

$$\begin{aligned} \sum_{i=1}^r (\lambda_i - 1)^2 &= \sum_{i=1}^r \lambda_i^2 - 2\sum_{i=1}^r \lambda_i + r \\ &= \sum_{i=1}^r \lambda_i - 2\sum_{i=1}^r \lambda_i + r \\ &= 0. \end{aligned}$$

Thus, $\Sigma^{1/2}A\Sigma^{1/2}$ has r nonzero eigenvalues equal to 1. It must therefore be idempotent of rank r. Hence, $A\Sigma$ is idempotent of rank r. Consequently, $X'AX \sim \chi^2_r(\theta)$ by Theorem 5.4. □

The following corollaries can be easily proved.

Corollary 5.3 If $X \sim N(\mathbf{0}, \Sigma)$, then $X'AX \sim \chi^2_r$ if and only if $A\Sigma$ is idempotent of rank r.

Corollary 5.4 If $X \sim N(\mu, \Sigma)$, then $X'\Sigma^{-1}X$ has the noncentral chi-squared distribution with n degrees of freedom (n is the number of elements of X) and a noncentrality parameter $\mu'\Sigma^{-1}\mu$.

Corollary 5.5 If $X \sim N(\mu, \sigma^2 I_n)$, then $\frac{1}{\sigma^2}X'X$ has the noncentral chi-squared distribution with n degrees of freedom and a noncentrality parameter $\frac{\mu'\mu}{\sigma^2}$.

Example 5.1 Let $X_1, X_2, \ldots, X_n$ be mutually independent and normally distributed random variables such that $X_i \sim N(\mu, \sigma^2)$, $i = 1, 2, \ldots, n$. In Section 4.8, it was stated that $\frac{(n-1)s^2}{\sigma^2} \sim \chi^2_{n-1}$, where s^2 is the sample variance given by

$$s^2 = \frac{1}{n-1}\sum_{i=1}^{n}(X_i - \bar{X})^2,$$

and $\bar{X} = \frac{1}{n}\sum_{i=1}^{n} X_i$. This result can now be easily shown on the basis of Theorem 5.4.

Let $\boldsymbol{X} = (X_1, X_2, \ldots, X_n)'$. Then, $\boldsymbol{X} \sim N(\mu\boldsymbol{1}_n, \sigma^2\boldsymbol{I}_n)$, and s^2 can be written as

$$s^2 = \frac{1}{n-1}\boldsymbol{X}'(\boldsymbol{I}_n - \frac{1}{n}\boldsymbol{J}_n)\boldsymbol{X},$$

where $\boldsymbol{J}_n$ is the matrix of ones of order $n \times n$. Then, by Theorem 5.4,

$$\begin{aligned}\frac{(n-1)s^2}{\sigma^2} &= \frac{1}{\sigma^2}\boldsymbol{X}'(\boldsymbol{I}_n - \frac{1}{n}\boldsymbol{J}_n)\boldsymbol{X}\\ &\sim \chi^2_{n-1},\end{aligned}$$

since

$$\begin{aligned}\boldsymbol{A\Sigma} &= \left[\frac{1}{\sigma^2}\left(\boldsymbol{I}_n - \frac{1}{n}\boldsymbol{J}_n\right)\right]\left(\sigma^2\boldsymbol{I}_n\right)\\ &= \boldsymbol{I}_n - \frac{1}{n}\boldsymbol{J}_n,\end{aligned}$$

which is idempotent of rank $n-1$. Furthermore, the noncentrality parameter is zero since

$$\begin{aligned}\theta &= \boldsymbol{\mu}'\boldsymbol{A}\boldsymbol{\mu}\\ &= \frac{1}{\sigma^2}\mu^2\boldsymbol{1}_n'(\boldsymbol{I}_n - \frac{1}{n}\boldsymbol{J}_n)\boldsymbol{1}_n\\ &= 0.\end{aligned}$$

Example 5.2 Consider the balanced fixed-effects one-way model,

$$Y_{ij} = \mu + \alpha_i + \epsilon_{ij}, \quad i = 1, 2, \ldots, a; \quad j = 1, 2, \ldots, m, \tag{5.37}$$

where α_i represents the effect of level i of a factor denoted by A $(i = 1, 2, \ldots, a)$ and ϵ_{ij} is a random experimental error. It is assumed that μ and α_i are fixed unknown parameters and that the ϵ_{ij}'s are independently distributed as $N(0, \sigma^2_\epsilon)$, $i = 1, 2, \ldots, a$; $j = 1, 2, \ldots, m$. Model (5.37) can be written in vector form as

$$\boldsymbol{Y} = \mu(\boldsymbol{1}_a \otimes \boldsymbol{1}_m) + (\boldsymbol{I}_a \otimes \boldsymbol{1}_m)\boldsymbol{\alpha} + \boldsymbol{\epsilon}, \tag{5.38}$$

where $\boldsymbol{\alpha} = (\alpha_1, \alpha_2, \ldots, \alpha_a)'$, $\boldsymbol{Y} = (Y_{11}, Y_{12}, \ldots, Y_{1m}, \ldots, Y_{a1}, Y_{a2}, \ldots, Y_{am})'$, and $\boldsymbol{\epsilon} = (\epsilon_{11}, \epsilon_{12}, \ldots, \epsilon_{1m}, \ldots, \epsilon_{a1}, \epsilon_{a2}, \ldots, \epsilon_{am})'$. Under the aforementioned assumptions, $\boldsymbol{Y} \sim N(\boldsymbol{\mu}, \boldsymbol{\Sigma})$, where

$$\boldsymbol{\mu} = E(\boldsymbol{Y}) = \mu(\boldsymbol{1}_a \otimes \boldsymbol{1}_m) + (\boldsymbol{I}_a \otimes \boldsymbol{1}_m)\boldsymbol{\alpha} \tag{5.39}$$

and

$$\boldsymbol{\Sigma} = \text{Var}(\boldsymbol{Y}) = \sigma_\epsilon^2(\boldsymbol{I}_a \otimes \boldsymbol{I}_m) = \sigma_\epsilon^2 \boldsymbol{I}_{am}. \tag{5.40}$$

The sums of squares associated with factor A and the error term are

$$SS_A = \frac{1}{m}\sum_{i=1}^{a} Y_{i.}^2 - \frac{1}{am}Y_{..}^2,$$
$$SS_E = \sum_{i=1}^{a}\sum_{j=1}^{m} Y_{ij}^2 - \frac{1}{m}\sum_{i=1}^{a} Y_{i.}^2,$$

respectively, where $Y_{i.} = \sum_{j=1}^{m} Y_{ij}$, $Y_{..} = \sum_{i=1}^{a}\sum_{j=1}^{m} Y_{ij}$. These sums of squares can be expressed as quadratic forms, namely,

$$SS_A = \boldsymbol{Y}'\left[\frac{1}{m}(\boldsymbol{I}_a \otimes \boldsymbol{J}_m) - \frac{1}{am}(\boldsymbol{J}_a \otimes \boldsymbol{J}_m)\right]\boldsymbol{Y}, \tag{5.41}$$

$$SS_E = \boldsymbol{Y}'\left[\boldsymbol{I}_a \otimes \boldsymbol{I}_m - \frac{1}{m}(\boldsymbol{I}_a \otimes \boldsymbol{J}_m)\right]\boldsymbol{Y}. \tag{5.42}$$

Then, by Theorem 5.4,

$$\frac{SS_A}{\sigma_\epsilon^2} \sim \chi_{a-1}^2(\theta), \tag{5.43}$$

$$\frac{SS_E}{\sigma_\epsilon^2} \sim \chi_{a(m-1)}^2. \tag{5.44}$$

The distribution in (5.43) is a noncentral chi-squared since

$$\frac{1}{\sigma_\epsilon^2}\left[\frac{1}{m}(\boldsymbol{I}_a \otimes \boldsymbol{J}_m) - \frac{1}{am}(\boldsymbol{J}_a \otimes \boldsymbol{J}_m)\right]\boldsymbol{\Sigma} = \frac{1}{m}(\boldsymbol{I}_a \otimes \boldsymbol{J}_m) - \frac{1}{am}(\boldsymbol{J}_a \otimes \boldsymbol{J}_m),$$

which is idempotent of rank $a-1$. The corresponding noncentrality parameter is

$$\begin{aligned}\theta &= \frac{1}{\sigma_\epsilon^2}[\mu(\boldsymbol{1}_a' \otimes \boldsymbol{1}_m') + \boldsymbol{\alpha}'(\boldsymbol{I}_a \otimes \boldsymbol{1}_m')]\left[\frac{1}{m}(\boldsymbol{I}_a \otimes \boldsymbol{J}_m) - \frac{1}{am}(\boldsymbol{J}_a \otimes \boldsymbol{J}_m)\right] \\ &\quad \times [\mu(\boldsymbol{1}_a \otimes \boldsymbol{1}_m) + (\boldsymbol{I}_a \otimes \boldsymbol{1}_m)\boldsymbol{\alpha}] \\ &= \frac{1}{\sigma_\epsilon^2}\boldsymbol{\alpha}'(\boldsymbol{I}_a \otimes \boldsymbol{1}_m')\left[\frac{1}{m}(\boldsymbol{I}_a \otimes \boldsymbol{J}_m) - \frac{1}{am}(\boldsymbol{J}_a \otimes \boldsymbol{J}_m)\right](\boldsymbol{I}_a \otimes \boldsymbol{1}_m)\boldsymbol{\alpha}\end{aligned}$$

$$= \frac{m}{\sigma_\epsilon^2}\alpha'(I_a - \frac{1}{a}J_a)\alpha$$

$$= \frac{m}{\sigma_\epsilon^2}\sum_{i=1}^{a}(\alpha_i - \bar{\alpha}_.)^2,$$

where $\bar{\alpha}_. = \frac{1}{a}\sum_{i=1}^{a}\alpha_i$. Furthermore, the distribution in (5.44) is a central chi-squared since

$$\frac{1}{\sigma_\epsilon^2}\left[I_a \otimes I_m - \frac{1}{m}\left(I_a \otimes J_m\right)\right]\Sigma = I_a \otimes I_m - \frac{1}{m}\left(I_a \otimes J_m\right),$$

which is idempotent of rank $a(m-1)$. The noncentrality parameter is zero since it can be easily verified that

$$\left[\mu\left(\mathbf{1}_a' \otimes \mathbf{1}_m'\right) + \alpha'\left(I_a \otimes \mathbf{1}_m'\right)\right]\left[I_a \otimes I_m - \frac{1}{m}\left(I_a \otimes J_m\right)\right]$$
$$\times\left[\mu\left(\mathbf{1}_a \otimes \mathbf{1}_m\right) + \left(I_a \otimes \mathbf{1}_m\right)\alpha\right] = 0.$$

5.3 Independence of Quadratic Forms

This section covers another important area in the distribution theory of quadratic forms, namely, that of independence of two quadratic forms. The development of this area has had a long history starting with the work of Craig (1943), Aitken (1950), Ogawa (1950), and Laha (1956). Several other authors have subsequently contributed to that development. Craig (1943) established a theorem giving the necessary and sufficient condition for the independence of two quadratic forms in normal variates. A historical account concerning the development of Craig's theorem was given in Section 1.6. See also Driscoll and Gundberg (1986).

Theorem 5.5 Let A and B be symmetric matrices of order $n \times n$, and let X be normally distributed as $N(\mu, \Sigma)$. A necessary and sufficient condition for $X'AX$ and $X'BX$ to be independent is that

$$A\Sigma B = \mathbf{0}. \tag{5.45}$$

Proof. As in Theorem 5.4, the sufficiency part of the present theorem is relatively easy, but the necessity part is more involved.

Sufficiency. Suppose that condition (5.45) is true. To show that $X'AX$ and $X'BX$ are independent.

Condition (5.45) is equivalent to

$$(\Sigma^{1/2}A\Sigma^{1/2})(\Sigma^{1/2}B\Sigma^{1/2}) = \mathbf{0}, \tag{5.46}$$

which implies that

$$\left(\Sigma^{1/2}A\Sigma^{1/2}\right)\left(\Sigma^{1/2}B\Sigma^{1/2}\right) = \left(\Sigma^{1/2}B\Sigma^{1/2}\right)\left(\Sigma^{1/2}A\Sigma^{1/2}\right).$$

This indicates that the two symmetric matrices, $\Sigma^{1/2}A\Sigma^{1/2}$ and $\Sigma^{1/2}B\Sigma^{1/2}$, commute. By Theorem 3.9, there exists an orthogonal matrix P such that

$$\Sigma^{1/2}A\Sigma^{1/2} = P\Lambda_1 P',$$
$$\Sigma^{1/2}B\Sigma^{1/2} = P\Lambda_2 P',$$

where Λ_1 and Λ_2 are diagonal matrices. From (5.46) it follows that

$$\Lambda_1\Lambda_2 = 0. \tag{5.47}$$

The columns of P can be arranged so that, in light of (5.47), Λ_1 and Λ_2 are written as

$$\Lambda_1 = \begin{bmatrix} D_1 & 0 \\ 0' & 0 \end{bmatrix},$$
$$\Lambda_2 = \begin{bmatrix} 0 & 0 \\ 0' & D_2 \end{bmatrix},$$

where D_1 and D_2 are diagonal matrices whose diagonal elements are not all equal to zero.

Let $Z = P'\Sigma^{-1/2}X$. Then, $Z \sim N(P'\Sigma^{-1/2}\mu, I_n)$. Note that the elements of Z are mutually independent. Furthermore,

$$X'AX = Z'\begin{bmatrix} D_1 & 0 \\ 0' & 0 \end{bmatrix}Z,$$
$$X'BX = Z'\begin{bmatrix} 0 & 0 \\ 0' & D_2 \end{bmatrix}Z.$$

If we partition Z as $[Z_1' : Z_2']'$, where the number of rows of Z_i is the same as the number of columns of D_i $(i = 1, 2)$, we get

$$X'AX = Z_1'D_1Z_1,$$
$$X'BX = Z_2'D_2Z_2.$$

We conclude that $X'AX$ and $X'BX$ must be independent because they depend on Z_1 and Z_2, respectively, which are independent since the elements of Z are mutually independent. □

Necessity. This proof is adapted from an article by Reid and Driscoll (1988).

Suppose that $X'AX$ and $X'BX$ are independently distributed. To show that condition (5.45) is true.

Let us consider the joint *cumulant generating function* (c.g.f.) of $X'AX$ and $X'BX$, which is given by

$$\begin{aligned}\psi_{A,B}(s,t) &= \log E\left[\exp\left(s\,X'AX + t\,X'BX\right)\right] \\ &= \log E\left\{\exp\left[X'(s\,A + t\,B)X\right]\right\}.\end{aligned}$$

Note that $E\{\exp[X'(s\,A + t\,B)X]\}$ can be derived using formula (5.6) after replacing $t\,A$ with $s\,A + t\,B$. We then have

$$\begin{aligned}\psi_{A,B}(s,t) = &-\frac{1}{2}\mu'\Sigma^{-1/2}\left\{I_n - \left[I_n - \Sigma^{1/2}(2\,s\,A + 2\,t\,B)\Sigma^{1/2}\right]^{-1}\right\}\Sigma^{-1/2}\mu \\ &-\frac{1}{2}\log\left\{det\left[I_n - \Sigma^{1/2}(2\,s\,A + 2\,t\,B)\Sigma^{1/2}\right]\right\}. \qquad (5.48)\end{aligned}$$

Using formulas (5.10) and (5.12) in (5.48), we obtain

$$\begin{aligned}\psi_{A,B}(s,t) = \sum_{r=1}^{\infty}\Bigg[&\frac{2^{r-1}\,r!}{r!}\mu'\Sigma^{-1/2}\left[\Sigma^{1/2}(s\,A + t\,B)\Sigma^{1/2}\right]^r\Sigma^{-1/2}\mu \\ &+\frac{2^{r-1}\,(r-1)!}{r!}tr\left\{\left[\Sigma^{1/2}(s\,A + t\,B)\Sigma^{1/2}\right]^r\right\}\Bigg]. \qquad (5.49)\end{aligned}$$

The infinite series in (5.49) is convergent provided that the point (s,t) falls within a small region $\Re$ around the origin $(0, 0)$.

Now, $X'AX$ and $X'BX$ are independent if and only if $\psi_{A,B}(s,t)$ is the sum of the marginal c.g.f.'s of $X'AX$ and $X'BX$. The latter two are denoted by $\psi_A(s), \psi_B(t)$, respectively, and are obtained by putting $t = 0$ then $s = 0$, respectively, in $\psi_{A,B}(s,t)$. We then have

$$\begin{aligned}\psi_{A,B}(s,t) &= \psi_A(s) + \psi_B(t) \\ &= \sum_{r=1}^{\infty}\left\{\frac{2^{r-1}\,r!}{r!}\mu'\Sigma^{-1/2}\left(s\,\Sigma^{1/2}A\Sigma^{1/2}\right)^r\Sigma^{-1/2}\mu\right. \\ &\qquad\left.+\frac{2^{r-1}\,(r-1)!}{r!}tr\left[\left(s\,\Sigma^{1/2}A\Sigma^{1/2}\right)^r\right]\right\} \\ &\quad+\sum_{r=1}^{\infty}\left\{\frac{2^{r-1}\,r!}{r!}\mu'\Sigma^{-1/2}\left(t\,\Sigma^{1/2}B\Sigma^{1/2}\right)^r\Sigma^{-1/2}\mu\right. \\ &\qquad\left.+\frac{2^{r-1}\,(r-1)!}{r!}tr\left[\left(t\Sigma^{1/2}B\Sigma^{1/2}\right)^r\right]\right\}. \qquad (5.50)\end{aligned}$$

Since s and t are arbitrary within the region $\Re$, the rth terms of the infinite series on the right-hand sides of (5.49) and (5.50) must be equal for $r = 1, 2, \ldots$.

Dividing these terms by $\frac{2^{r-1}(r-1)!}{r!}$, we get

$$r\,\mu'\Sigma^{-1/2}\left[\Sigma^{1/2}(sA+tB)\,\Sigma^{1/2}\right]^r\Sigma^{-1/2}\mu + tr\left\{\left[\Sigma^{1/2}(sA+tB)\Sigma^{1/2}\right]^r\right\}$$
$$= r\,\mu'\Sigma^{-1/2}\left(s\,\Sigma^{1/2}A\Sigma^{1/2}\right)^r\Sigma^{-1/2}\mu + tr\left[\left(s\,\Sigma^{1/2}A\Sigma^{1/2}\right)^r\right]$$
$$+ r\,\mu'\Sigma^{-1/2}\left(t\,\Sigma^{1/2}B\Sigma^{1/2}\right)^r\Sigma^{-1/2}\mu + tr\left[\left(t\,\Sigma^{1/2}B\Sigma^{1/2}\right)^r\right], \quad r=1,2,\ldots \tag{5.51}$$

For any fixed, but arbitrary, values of $(s,\,t)$ within the region $\Re$, let us consider the distinct nonzero eigenvalues of $s\,\Sigma^{1/2}A\Sigma^{1/2}$, $t\,\Sigma^{1/2}B\Sigma^{1/2}$, and those of $\Sigma^{1/2}(sA+tB)\Sigma^{1/2}$. Let $\tau_1,\tau_2,\ldots,\tau_k$ denote the totality of all of these eigenvalues. If, for example, τ_i is an eigenvalue of $s\,\Sigma^{1/2}A\Sigma^{1/2}$, let $\nu_i(s\,\Sigma^{1/2}A\Sigma^{1/2})$ denote its multiplicity, and let $P_i(s\,\Sigma^{1/2}A\Sigma^{1/2})$ be the matrix whose columns are the corresponding orthonormal eigenvectors of $s\,\Sigma^{1/2}A\Sigma^{1/2}$ in the spectral decomposition of $s\,\Sigma^{1/2}A\Sigma^{1/2}$. If τ_i is not an eigenvalue of $s\,\Sigma^{1/2}A\Sigma^{1/2}$, $\nu_i(s\,\Sigma^{1/2}A\Sigma^{1/2})$ is set equal to zero, and $P_i(s\,\Sigma^{1/2}A\Sigma^{1/2})$ is taken to be a zero matrix. We can then write

$$\mu'\Sigma^{-1/2}\left(s\,\Sigma^{1/2}A\Sigma^{1/2}\right)^r\Sigma^{-1/2}\mu = \sum_{i=1}^{k}\tau_i^r\mu_i\left(s\,\Sigma^{1/2}A\Sigma^{1/2}\right),$$

where

$$\mu_i\left(s\,\Sigma^{1/2}A\Sigma^{1/2}\right) = \mu'\Sigma^{-1/2}P_i\left(s\,\Sigma^{1/2}A\Sigma^{1/2}\right)P_i'\left(s\,\Sigma^{1/2}A\Sigma^{1/2}\right)\Sigma^{-1/2}\mu,$$

and

$$tr\left[\left(s\,\Sigma^{1/2}A\Sigma^{1/2}\right)^r\right] = \sum_{i=1}^{k}\tau_i^r\nu_i\left(s\,\Sigma^{1/2}A\Sigma^{1/2}\right).$$

Similar definitions of ν_i and P_i can be made with regard to $t\,\Sigma^{1/2}B\Sigma^{1/2}$ and $\Sigma^{1/2}(sA+tB)\Sigma^{1/2}$. Formula (5.51) can then be expressed as

$$\sum_{i=1}^{k}\tau_i^r\left\{\nu_i\left[\Sigma^{1/2}(sA+tB)\,\Sigma^{1/2}\right] - \nu_i\left(s\,\Sigma^{1/2}A\Sigma^{1/2}\right) - \nu_i\left(t\,\Sigma^{1/2}B\Sigma^{1/2}\right)\right\}$$
$$+\sum_{i=1}^{k}r\,\tau_i^r\left\{\mu_i\left[\Sigma^{1/2}(sA+tB)\,\Sigma^{1/2}\right] - \mu_i\left(s\,\Sigma^{1/2}A\Sigma^{1/2}\right)\right.$$
$$\left.-\mu_i\left(t\,\Sigma^{1/2}B\Sigma^{1/2}\right)\right\} = 0, \quad r=1,2,\ldots \tag{5.52}$$

The first $2k$ of these equations can be written in matrix form as

$$M\delta = \mathbf{0}, \tag{5.53}$$

where $M = (m_{i,j})$ is a matrix of order $2k \times 2k$ whose (i,j)th element is such that $m_{i,j} = \tau_j^i$ and $m_{i,k+j} = i\,\tau_j^i$ for $i = 1, 2, \ldots, 2k$ and $j = 1, 2, \ldots, k$, and where $\boldsymbol{\delta}$ is a vector of order $2k \times 1$ with the following elements

$$\delta_i = \nu_i\left[\boldsymbol{\Sigma}^{1/2}(s\boldsymbol{A} + t\boldsymbol{B})\boldsymbol{\Sigma}^{1/2}\right] - \nu_i\left(s\,\boldsymbol{\Sigma}^{1/2}\boldsymbol{A}\boldsymbol{\Sigma}^{1/2}\right) - \nu_i\left(t\,\boldsymbol{\Sigma}^{1/2}\boldsymbol{B}\boldsymbol{\Sigma}^{1/2}\right), \quad i = 1, 2, \ldots, k,$$

and

$$\delta_{k+i} = \mu_i\left[\boldsymbol{\Sigma}^{1/2}(s\boldsymbol{A} + t\boldsymbol{B})\boldsymbol{\Sigma}^{1/2}\right] - \mu_i\left(s\,\boldsymbol{\Sigma}^{1/2}\boldsymbol{A}\boldsymbol{\Sigma}^{1/2}\right) - \mu_i\left(t\,\boldsymbol{\Sigma}^{1/2}\boldsymbol{B}\boldsymbol{\Sigma}^{1/2}\right), \quad i = 1, 2, \ldots, k.$$

The matrix $\boldsymbol{M}$ in (5.53) can be shown to be nonsingular (for details, see Reid and Driscoll, 1988). From (5.53) we can then conclude that

$$\nu_i\left[\boldsymbol{\Sigma}^{1/2}(s\boldsymbol{A} + t\boldsymbol{B})\boldsymbol{\Sigma}^{1/2}\right] = \nu_i\left(s\,\boldsymbol{\Sigma}^{1/2}\boldsymbol{A}\boldsymbol{\Sigma}^{1/2}\right) + \nu_i\left(t\,\boldsymbol{\Sigma}^{1/2}\boldsymbol{B}\boldsymbol{\Sigma}^{1/2}\right), \quad i = 1, 2, \ldots, k, \tag{5.54}$$

$$\mu_i\left[\boldsymbol{\Sigma}^{1/2}(s\boldsymbol{A} + t\boldsymbol{B})\boldsymbol{\Sigma}^{1/2}\right] = \mu_i\left(s\,\boldsymbol{\Sigma}^{1/2}\boldsymbol{A}\boldsymbol{\Sigma}^{1/2}\right) + \mu_i\left(t\,\boldsymbol{\Sigma}^{1/2}\boldsymbol{B}\boldsymbol{\Sigma}^{1/2}\right), \quad i = 1, 2, \ldots, k. \tag{5.55}$$

Multiplying the ith equation in (5.54) by τ_i^r $(r = 1, 2, \ldots)$ and summing the resulting k equations over i, we obtain

$$tr\left\{\left[\boldsymbol{\Sigma}^{1/2}(s\boldsymbol{A} + t\boldsymbol{B})\boldsymbol{\Sigma}^{1/2}\right]^r\right\} = tr\left[\left(s\,\boldsymbol{\Sigma}^{1/2}\boldsymbol{A}\boldsymbol{\Sigma}^{1/2}\right)^r\right] + tr\left[\left(t\,\boldsymbol{\Sigma}^{1/2}\boldsymbol{B}\boldsymbol{\Sigma}^{1/2}\right)^r\right], \quad r = 1, 2, \ldots$$

In particular, for $r = 4$, we get

$$tr\left\{\left[\boldsymbol{\Sigma}^{1/2}(s\boldsymbol{A} + t\boldsymbol{B})\boldsymbol{\Sigma}^{1/2}\right]^4\right\} = s^4\,tr\left[\left(\boldsymbol{\Sigma}^{1/2}\boldsymbol{A}\boldsymbol{\Sigma}^{1/2}\right)^4\right] + t^4\,tr\left[\left(\boldsymbol{\Sigma}^{1/2}\boldsymbol{B}\boldsymbol{\Sigma}^{1/2}\right)^4\right], \tag{5.56}$$

for all (s, t) in $\Re$. Differentiating the two sides of (5.56) four times, twice with respect to s and twice with respect to t, we finally obtain

$$tr\left\{\left[\left(\boldsymbol{\Sigma}^{1/2}\boldsymbol{A}\boldsymbol{\Sigma}^{1/2}\right)\left(\boldsymbol{\Sigma}^{1/2}\boldsymbol{B}\boldsymbol{\Sigma}^{1/2}\right) + \left(\boldsymbol{\Sigma}^{1/2}\boldsymbol{B}\boldsymbol{\Sigma}^{1/2}\right)\left(\boldsymbol{\Sigma}^{1/2}\boldsymbol{A}\boldsymbol{\Sigma}^{1/2}\right)\right]^2\right\} + 2\,tr\left[\left(\boldsymbol{\Sigma}^{1/2}\boldsymbol{A}\boldsymbol{\Sigma}^{1/2}\right)\left(\boldsymbol{\Sigma}^{1/2}\boldsymbol{B}\boldsymbol{\Sigma}^{1/2}\right)^2\left(\boldsymbol{\Sigma}^{1/2}\boldsymbol{A}\boldsymbol{\Sigma}^{1/2}\right)\right] = 0. \tag{5.57}$$

Each of the trace terms in (5.57) is of the form $tr(\boldsymbol{C}'\boldsymbol{C})$, and is therefore non-negative. We can then conclude from (5.57) that each term must be zero. In particular,

$$tr\left[\left(\boldsymbol{\Sigma}^{1/2}\boldsymbol{A}\boldsymbol{\Sigma}^{1/2}\right)\left(\boldsymbol{\Sigma}^{1/2}\boldsymbol{B}\boldsymbol{\Sigma}^{1/2}\right)^2\left(\boldsymbol{\Sigma}^{1/2}\boldsymbol{A}\boldsymbol{\Sigma}^{1/2}\right)\right] = 0. \tag{5.58}$$

Formula (5.58) implies that the matrix $\boldsymbol{C} = (\boldsymbol{\Sigma}^{1/2}\,\boldsymbol{A}\boldsymbol{\Sigma}^{1/2})(\boldsymbol{\Sigma}^{1/2}\,\boldsymbol{B}\boldsymbol{\Sigma}^{1/2})$ must be zero. Hence, $\boldsymbol{A}\boldsymbol{\Sigma}\boldsymbol{B} = \boldsymbol{0}$. □

Example 5.3 Consider again Example 5.2, and in particular the sums of squares, SS_A and SS_E, given in (5.41) and (5.42), respectively. We recall that $\boldsymbol{Y} \sim N(\boldsymbol{\mu}, \boldsymbol{\Sigma})$, where $\boldsymbol{\mu}$ and $\boldsymbol{\Sigma}$ are described in (5.39) and (5.40), respectively. The quadratic forms representing SS_A and SS_E are independent since

$$\left[\frac{1}{m}(\boldsymbol{I}_a \otimes \boldsymbol{J}_m) - \frac{1}{a\,m}(\boldsymbol{J}_a \otimes \boldsymbol{J}_m)\right]\left(\sigma_\epsilon^2 \boldsymbol{I}_{a\,m}\right)\left[\boldsymbol{I}_a \otimes \boldsymbol{I}_m - \frac{1}{m}(\boldsymbol{I}_a \otimes \boldsymbol{J}_m)\right]$$
$$= \sigma_\epsilon^2\left[\frac{1}{m}(\boldsymbol{I}_a \otimes \boldsymbol{J}_m) - \frac{1}{m}(\boldsymbol{I}_a \otimes \boldsymbol{J}_m) - \frac{1}{a\,m}(\boldsymbol{J}_a \otimes \boldsymbol{J}_m) + \frac{1}{a\,m}(\boldsymbol{J}_a \otimes \boldsymbol{J}_m)\right] = \boldsymbol{0}.$$

Example 5.4 This is a counterexample to demonstrate that if $\boldsymbol{X}$ is not normally distributed, then independence of $\boldsymbol{X}'\boldsymbol{A}\boldsymbol{X}$ and $\boldsymbol{X}'\boldsymbol{B}\boldsymbol{X}$ does not necessarily imply that $\boldsymbol{A}\boldsymbol{\Sigma}\boldsymbol{B} = \boldsymbol{0}$.

Let $\boldsymbol{X} = (X_1, X_2, \ldots, X_n)'$, where the X_i's are mutually independent and identically distributed discrete random variables such that $X_i = -1, 1$ with probabilities equal to $\frac{1}{2}(i = 1, 2, \ldots, n)$. Thus $E(\boldsymbol{X}) = \boldsymbol{0}$ and $\text{Var}(\boldsymbol{X}) = \boldsymbol{\Sigma} = \boldsymbol{I}_n$. Let $\boldsymbol{A} = \boldsymbol{B} = \boldsymbol{I}_n$. Then, $\boldsymbol{X}'\boldsymbol{A}\boldsymbol{X} = \boldsymbol{X}'\boldsymbol{B}\boldsymbol{X} = \sum_{i=1}^n X_i^2 = n$ with probability 1. It follows that $\boldsymbol{X}'\boldsymbol{A}\boldsymbol{X}$ and $\boldsymbol{X}'\boldsymbol{B}\boldsymbol{X}$ are independent (the joint probability mass function of $\boldsymbol{X}'\boldsymbol{A}\boldsymbol{X}$ and $\boldsymbol{X}'\boldsymbol{B}\boldsymbol{X}$ is the product of their marginal probability mass functions since, in this example, each quadratic form assumes one single value with probability 1). But, $\boldsymbol{A}\boldsymbol{\Sigma}\boldsymbol{B} = \boldsymbol{I}_n \neq \boldsymbol{0}$.

5.4 Independence of Linear and Quadratic Forms

In this section, we consider the condition under which a quadratic form, $\boldsymbol{X}'\boldsymbol{A}\boldsymbol{X}$, is independent of a linear form, $\boldsymbol{B}\boldsymbol{X}$, when $\boldsymbol{X}$ is normally distributed.

Theorem 5.6 Let $\boldsymbol{A}$ be a symmetric matrix of order $n \times n$, $\boldsymbol{B}$ be a matrix of order $m \times n$, and let $\boldsymbol{X} \sim N(\boldsymbol{\mu}, \boldsymbol{\Sigma})$. A necessary and sufficient condition for $\boldsymbol{X}'\boldsymbol{A}\boldsymbol{X}$ and $\boldsymbol{B}\boldsymbol{X}$ to be independent is that

$$\boldsymbol{B}\boldsymbol{\Sigma}\boldsymbol{A} = \boldsymbol{0}. \tag{5.59}$$

Proof.

Sufficiency. Suppose that condition (5.59) is true. To show that $\boldsymbol{X}'\boldsymbol{A}\boldsymbol{X}$ and $\boldsymbol{B}\boldsymbol{X}$ are independent.

Condition (5.59) is equivalent to

$$\boldsymbol{B}\boldsymbol{\Sigma}^{1/2}\boldsymbol{\Sigma}^{1/2}\boldsymbol{A}\boldsymbol{\Sigma}^{1/2} = \boldsymbol{0}.$$

Using formula (5.22) in this equation, we obtain

$$\boldsymbol{B}\boldsymbol{\Sigma}^{1/2}\boldsymbol{P}\boldsymbol{\Gamma}\boldsymbol{P}' = \boldsymbol{0},$$

where $\boldsymbol{\Gamma}$ is described in (5.23). Hence,

$$\boldsymbol{B}\boldsymbol{\Sigma}^{1/2}\sum_{j=1}^{k}\gamma_j\boldsymbol{P}_j\boldsymbol{P}_j' = \boldsymbol{0}, \tag{5.60}$$

where, if we recall, $\gamma_1, \gamma_2, \ldots, \gamma_k$ are the distinct nonzero eigenvalues of $\boldsymbol{\Sigma}^{1/2}\boldsymbol{A}\boldsymbol{\Sigma}^{1/2}$, and $\boldsymbol{P}_i$ is the column portion of $\boldsymbol{P}$ corresponding to γ_i ($i = 1, 2, \ldots, k$). Multiplying the two sides of (5.60) on the right by $\boldsymbol{P}_i$ ($i = 1, 2, \ldots, k$) and recalling that $\boldsymbol{P}_i'\boldsymbol{P}_i = \boldsymbol{I}_{\nu_i}, \boldsymbol{P}_j'\boldsymbol{P}_i = \boldsymbol{0}$, $i \neq j$, we conclude that

$$\boldsymbol{B}\boldsymbol{\Sigma}^{1/2}\boldsymbol{P}_i = \boldsymbol{0}, \quad i = 1, 2, \ldots, k. \tag{5.61}$$

Now, from Lemma 5.1,

$$\boldsymbol{X}'\boldsymbol{A}\boldsymbol{X} = \sum_{i=1}^{k}\gamma_i\boldsymbol{Z}_i'\boldsymbol{Z}_i, \tag{5.62}$$

where $\boldsymbol{Z}_i = \boldsymbol{P}_i'\boldsymbol{\Sigma}^{-1/2}\boldsymbol{X}$, $i = 1, 2, \ldots, k$. The covariance matrix of $\boldsymbol{B}\boldsymbol{X}$ and $\boldsymbol{Z}_i$ is zero since

$$\begin{aligned}\text{Cov}(\boldsymbol{B}\boldsymbol{X}, \boldsymbol{P}_i'\boldsymbol{\Sigma}^{-1/2}\boldsymbol{X}) &= \boldsymbol{B}\boldsymbol{\Sigma}\boldsymbol{\Sigma}^{-1/2}\boldsymbol{P}_i \\ &= \boldsymbol{B}\boldsymbol{\Sigma}^{1/2}\boldsymbol{P}_i \\ &= \boldsymbol{0}, \quad i = 1, 2, \ldots, k.\end{aligned} \tag{5.63}$$

If $\boldsymbol{B}$ is of full row rank, then each of $\boldsymbol{B}\boldsymbol{X}$ and $\boldsymbol{Z}_i$ is a full row-rank linear transformation of $\boldsymbol{X}$, which is normally distributed. Furthermore, because of (5.61), the rows of $\boldsymbol{B}$ are linearly independent of those of $\boldsymbol{P}_i'\boldsymbol{\Sigma}^{-1/2}$ (see Exercise 5.21). Hence, the random vector

$$\begin{bmatrix}\boldsymbol{B}\boldsymbol{X} \\ \boldsymbol{Z}_i\end{bmatrix} = \begin{bmatrix}\boldsymbol{B} \\ \boldsymbol{P}_i'\boldsymbol{\Sigma}^{-1/2}\end{bmatrix}\boldsymbol{X}, \quad i = 1, 2, \ldots, k,$$

is normally distributed. Since the covariance matrix of $\boldsymbol{B}\boldsymbol{X}$ and $\boldsymbol{Z}_i$ is zero, $\boldsymbol{B}\boldsymbol{X}$ and $\boldsymbol{Z}_i$ must be independent by Corollary 4.2 ($i = 1, 2, \ldots, k$). It follows that

$\boldsymbol{BX}$ is independent of $\boldsymbol{X'AX}$ in light of (5.62). If, however, $\boldsymbol{B}$ is not of full row rank, then $\boldsymbol{B}$ can be expressed as $\boldsymbol{B} = [\boldsymbol{B}_1' : \boldsymbol{B}_2']'$, where $\boldsymbol{B}_1$ is a matrix of order $r_1 \times n$ and rank r_1 (r_1 is the rank of $\boldsymbol{B}$), and $\boldsymbol{B}_2$ is a matrix of order $(m - r_1) \times n$ such that $\boldsymbol{B}_2 = \boldsymbol{C}_2\boldsymbol{B}_1$ for some matrix $\boldsymbol{C}_2$. Then, from (5.63) we conclude that $\boldsymbol{B}_1\boldsymbol{X}$ and $\boldsymbol{Z}_i$ have a zero covariance matrix. This implies independence of $\boldsymbol{B}_1\boldsymbol{X}$ and $\boldsymbol{Z}_i$, and hence of $\boldsymbol{BX}$ and $\boldsymbol{Z}_i (i = 1, 2, \ldots, k)$ since

$$\boldsymbol{BX} = \begin{bmatrix} \boldsymbol{I}_{r_1} \\ \boldsymbol{C}_2 \end{bmatrix} \boldsymbol{B}_1\boldsymbol{X}.$$

Consequently, $\boldsymbol{BX}$ is independent of $\boldsymbol{X'AX}$. □

Necessity. Suppose that $\boldsymbol{BX}$ and $\boldsymbol{X'AX}$ are independent. To show that condition (5.59) is true. We shall consider two cases depending on whether or not $\boldsymbol{B}$ is of full row rank.

If $\boldsymbol{B}$ is of full row rank, then independence of $\boldsymbol{BX}$ and $\boldsymbol{X'AX}$ implies that $\boldsymbol{X'B'BX}$ and $\boldsymbol{X'AX}$ are also independent. Hence, by Theorem 5.5,

$$\boldsymbol{B'B\Sigma A} = \boldsymbol{0}. \tag{5.64}$$

Multiplying the two sides of (5.64) on the left by $\boldsymbol{B}$, we get

$$(\boldsymbol{BB'})\boldsymbol{B\Sigma A} = \boldsymbol{0}. \tag{5.65}$$

Since $\boldsymbol{BB'}$ is nonsingular, we conclude from (5.65) that $\boldsymbol{B\Sigma A} = \boldsymbol{0}$.

If $\boldsymbol{B}$ is not of full row rank, then, as before in the proof of sufficiency, we can write $\boldsymbol{B} = [\boldsymbol{B}_1' : \boldsymbol{B}_2']'$, where $\boldsymbol{B}_1$ is of full row rank and $\boldsymbol{B}_2 = \boldsymbol{C}_2\boldsymbol{B}_1$. Let us now express $\boldsymbol{B}_1\boldsymbol{X}$ as

$$\boldsymbol{B}_1\boldsymbol{X} = [\boldsymbol{I}_{r_1} : \boldsymbol{0}_1]\boldsymbol{BX},$$

where $\boldsymbol{0}_1$ is a zero matrix of order $r_1 \times (m - r_1)$. Independence of $\boldsymbol{BX}$ and $\boldsymbol{X'AX}$ implies the same for $\boldsymbol{B}_1\boldsymbol{X}$ and $\boldsymbol{X'AX}$. Hence, from the previous case, we conclude that $\boldsymbol{B}_1\boldsymbol{\Sigma A} = \boldsymbol{0}$. Furthermore, $\boldsymbol{B}_2\boldsymbol{\Sigma A} = \boldsymbol{C}_2\boldsymbol{B}_1\boldsymbol{\Sigma A} = \boldsymbol{0}$. Consequently,

$$\begin{aligned} \boldsymbol{B\Sigma A} &= \begin{bmatrix} \boldsymbol{B}_1 \\ \boldsymbol{B}_2 \end{bmatrix} \boldsymbol{\Sigma A} \\ &= \boldsymbol{0}. \end{aligned}$$

□

Example 5.5 Consider again Example 5.1. In Section 4.8, it was stated that $\bar{X}$ and s^2 are independent. We are now in a position to prove this very easily on the basis of Theorem 5.6. We have that

$$\begin{aligned} \bar{X} &= \frac{1}{n} \sum_{i=1}^{n} X_i \\ &= \frac{1}{n} \boldsymbol{1}_n' \boldsymbol{X}, \\ s^2 &= \frac{1}{n-1} \boldsymbol{X}'(\boldsymbol{I}_n - \frac{1}{n}\boldsymbol{J}_n)\boldsymbol{X}, \end{aligned}$$

where $\boldsymbol{X} \sim N(\mu \mathbf{1}_n, \sigma^2 \boldsymbol{I}_n)$. Letting $\boldsymbol{B} = \frac{1}{n}\mathbf{1}'_n$ and $\boldsymbol{A} = \frac{1}{n-1}(\boldsymbol{I}_n - \frac{1}{n}\boldsymbol{J}_n)$, and noting that

$$\begin{aligned} \boldsymbol{B}\boldsymbol{\Sigma}\boldsymbol{A} &= \frac{1}{n(n-1)}\mathbf{1}'_n(\sigma^2 \boldsymbol{I}_n)(\boldsymbol{I}_n - \frac{1}{n}\boldsymbol{J}_n) \\ &= \mathbf{0}, \end{aligned}$$

we conclude that $\bar{X}$ and s^2 are independent. □

5.5 Independence and Chi-Squaredness of Several Quadratic Forms

Let $\boldsymbol{X} \sim N(\mu, \sigma^2 \boldsymbol{I}_n)$. Suppose that $\boldsymbol{X}'\boldsymbol{X}$ can be partitioned as

$$\boldsymbol{X}'\boldsymbol{X} = \sum_{i=1}^{p} \boldsymbol{X}'\boldsymbol{A}_i\boldsymbol{X}, \tag{5.66}$$

that is,

$$\boldsymbol{I}_n = \sum_{i=1}^{p} \boldsymbol{A}_i, \tag{5.67}$$

where $\boldsymbol{A}_i$ is a symmetric matrix of order $n \times n$ and rank k_i $(i = 1, 2, \ldots, p)$. The following theorem, which is due to Cochran (1934), has useful applications in the analysis of variance for fixed-effects models, as will be seen later.

Theorem 5.7 Let $\boldsymbol{X} \sim N(\mu, \sigma^2 \boldsymbol{I}_n)$ and $\boldsymbol{A}_i$ be a symmetric matrix of order $n \times n$ and rank k_i $(i = 1, 2, \ldots, p)$ such that $\boldsymbol{I}_n = \sum_{i=1}^{p} \boldsymbol{A}_i$. Then, any one of the following three conditions implies the other two:

(a) $n = \sum_{i=1}^{p} k_i$, that is, the ranks of $\boldsymbol{A}_1, \boldsymbol{A}_2, \ldots, \boldsymbol{A}_p$ sum to the rank of $\boldsymbol{I}_n$.

(b) $\frac{1}{\sigma^2}\boldsymbol{X}'\boldsymbol{A}_i\boldsymbol{X} \sim \chi^2_{k_i}(\theta_i)$, where $\theta_i = \frac{1}{\sigma^2}\mu'\boldsymbol{A}_i\mu$, $i = 1, 2, \ldots, p$.

(c) $\boldsymbol{X}'\boldsymbol{A}_1\boldsymbol{X}$, $\boldsymbol{X}'\boldsymbol{A}_2\boldsymbol{X}$, $\ldots$, $\boldsymbol{X}'\boldsymbol{A}_p\boldsymbol{X}$ are mutually independent.

Proof. The proof consists of the following parts:

I. (a) implies (b). If (a) holds, then by applying the result in property (d) of Section 3.6, we can write $\boldsymbol{A}_i = \boldsymbol{F}\boldsymbol{D}_i\boldsymbol{G}$, where $\boldsymbol{F}$ and $\boldsymbol{G}$ are nonsingular matrices and $\boldsymbol{D}_i$ is a diagonal matrix with diagonal elements equal to zeros and ones such that $\boldsymbol{D}_i\boldsymbol{D}_j = \mathbf{0}$, $i \neq j$ $(i, j = 1, 2, \ldots, p)$. It follows that $\sum_{i=1}^{p} \boldsymbol{D}_i$ is a diagonal matrix of order $n \times n$ whose diagonal elements are also equal to zeros and ones. But, because of (5.67), $\sum_{i=1}^{p} \boldsymbol{D}_i$ must be of rank n and therefore cannot have zero diagonal elements. Hence, $\sum_{i=1}^{p} \boldsymbol{D}_i = \boldsymbol{I}_n$. Using (5.67) we conclude that $\boldsymbol{FG} = \boldsymbol{I}_n$, that is, $\boldsymbol{G} = \boldsymbol{F}^{-1}$. We therefore have

$$\boldsymbol{A}_i = \boldsymbol{F}\boldsymbol{D}_i\boldsymbol{F}^{-1}, \quad i = 1, 2, \ldots, p. \tag{5.68}$$

Thus, $\boldsymbol{A}_i^2 = \boldsymbol{F}\boldsymbol{D}_i^2\boldsymbol{F}^{-1} = \boldsymbol{F}\boldsymbol{D}_i\boldsymbol{F}^{-1} = \boldsymbol{A}_i$. This implies that $\boldsymbol{A}_i$ is an idempotent matrix. By Theorem 5.4, $\frac{1}{\sigma^2}\boldsymbol{X}'\boldsymbol{A}_i\boldsymbol{X} \sim \chi^2_{k_i}(\theta_i)$ since $(\frac{\boldsymbol{A}_i}{\sigma^2})(\sigma^2\boldsymbol{I}_n) = \boldsymbol{A}_i$ is idempotent of rank k_i, where $\theta_i = \frac{1}{\sigma^2}\boldsymbol{\mu}'\boldsymbol{A}_i\boldsymbol{\mu}$, $i = 1, 2, \ldots, p$.

II. (b) implies (c). If (b) is true, then $\boldsymbol{A}_i$ is idempotent by Theorem 5.4 $(i = 1, 2, \ldots, p)$. By squaring the two sides of (5.67), we get

$$\begin{aligned}\boldsymbol{I}_n &= \sum_{i=1}^{p} \boldsymbol{A}_i^2 + \sum_{i \neq j} \boldsymbol{A}_i\boldsymbol{A}_j \\ &= \sum_{i=1}^{p} \boldsymbol{A}_i + \sum_{i \neq j} \boldsymbol{A}_i\boldsymbol{A}_j \\ &= \boldsymbol{I}_n + \sum_{i \neq j} \boldsymbol{A}_i\boldsymbol{A}_j.\end{aligned}$$

Hence,

$$\sum_{i \neq j} \boldsymbol{A}_i\boldsymbol{A}_j = \boldsymbol{0},$$

or,

$$\sum_{i \neq j} \boldsymbol{A}_i^2\boldsymbol{A}_j^2 = \boldsymbol{0}.$$

Taking the traces on both sides, we get

$$\sum_{i \neq j} tr(\boldsymbol{A}_i^2\boldsymbol{A}_j^2) = 0,$$

or,

$$\sum_{i \neq j} tr[(\boldsymbol{A}_j\boldsymbol{A}_i)(\boldsymbol{A}_i\boldsymbol{A}_j)] = 0. \tag{5.69}$$

But, $tr[(\boldsymbol{A}_j\boldsymbol{A}_i)(\boldsymbol{A}_i\boldsymbol{A}_j)]$ is of the form $tr(\boldsymbol{C}'\boldsymbol{C})$, where $\boldsymbol{C} = \boldsymbol{A}_i\boldsymbol{A}_j$, and $tr(\boldsymbol{C}'\boldsymbol{C}) \geq 0$. From (5.69) it follows that $tr[(\boldsymbol{A}_j\boldsymbol{A}_i)(\boldsymbol{A}_i\boldsymbol{A}_j)] = 0$ and hence $\boldsymbol{A}_j\boldsymbol{A}_i = \boldsymbol{0}$, $i \neq j$. This establishes pairwise independence of the quadratic forms by Theorem 5.5. We now proceed to show that the quadratic forms are in fact mutually independent. This can be achieved by using the moment generating function approach.

Since $A_iA_j = \mathbf{0}$, we have that $A_iA_j = A_jA_i$, $i \neq j$. By Theorem 3.9, there exists an orthogonal matrix P such that

$$A_i = P\Lambda_i P', \quad i = 1, 2, \ldots, p, \tag{5.70}$$

where Λ_i is a diagonal matrix; its diagonal elements are equal to zeros and ones since A_i is idempotent. Furthermore, $\Lambda_i\Lambda_j = \mathbf{0}$ for $i \neq j$. Thus, by rearranging the columns of P, it is possible to express $\Lambda_1, \Lambda_2, \ldots, \Lambda_p$ as

$$\begin{aligned}
\Lambda_1 &= \mathrm{diag}(I_{k_1}, \mathbf{0}, \ldots, \mathbf{0})\\
\Lambda_2 &= \mathrm{diag}(\mathbf{0}, I_{k_2}, \mathbf{0}, \ldots, \mathbf{0})\\
&\;\;\vdots\\
\Lambda_p &= \mathrm{diag}(\mathbf{0}, \mathbf{0}, \ldots, I_{k_p}),
\end{aligned}$$

where, if we recall, $k_i = rank(A_i)$, $i = 1, 2, \ldots, p$.

Let us now consider the joint moment generating function of $X'A_1^*X, X'A_2^*X, \ldots, X'A_p^*X$, where $A_i^* = \frac{1}{\sigma^2}A_i$, $i = 1, 2, \ldots, p$, namely,

$$\begin{aligned}
\phi_*(t) &= E[\exp(t_1X'A_1^*X + t_2X'A_2^*X + \cdots + t_pX'A_p^*X)]\\
&= E\left\{\exp\left[X'\left(\sum_{i=1}^{p} t_iA_i^*\right)X\right]\right\},
\end{aligned}$$

where $t = (t_1, t_2, \ldots, t_p)'$. Using (5.6) after replacing tA with $\sum_{i=1}^{p} t_iA_i^*$ and Σ with $\sigma^2 I_n$, $\phi_*(t)$ can be written as

$$\phi_*(t) = \frac{\exp\left\{-\frac{1}{2\sigma^2}\mu'[I_n - (I_n - 2\sigma^2\sum_{i=1}^{p} t_iA_i^*)^{-1}]\mu\right\}}{[\,det(I_n - 2\sigma^2\sum_{i=1}^{p} t_iA_i^*)]^{1/2}}. \tag{5.71}$$

Note that

$$\begin{aligned}
I_n - \left(I_n - 2\sigma^2\sum_{i=1}^{p} t_iA_i^*\right)^{-1} &= I_n - \left[I_n - 2P\left(\sum_{i=1}^{p} t_i\Lambda_i\right)P'\right]^{-1}\\
&= I_n - \left[I_n - 2P\,\mathrm{diag}\left(t_1I_{k_1}, t_2I_{k_2}, \ldots, t_pI_{k_p}\right)P'\right]^{-1}\\
&= I_n - P\,\mathrm{diag}\Big[(1-2t_1)^{-1}I_{k_1}, (1-2t_2)^{-1}I_{k_2}, \ldots,\\
&\quad (1-2t_p)^{-1}I_{k_p}\Big]P'\\
&= -P\,\mathrm{diag}\left(\frac{2t_1}{1-2t_1}I_{k_1}, \frac{2t_2}{1-2t_2}I_{k_2}, \ldots, \frac{2t_p}{1-2t_p}I_{k_p}\right)P'\\
&= -\sum_{i=1}^{p} P_i\left(\frac{2t_i}{1-2t_i}\right)P_i',
\end{aligned}$$

where $P_1, P_2, \ldots, P_p$ are the portions of P corresponding to $I_{k_1}, I_{k_2}, \ldots, I_{k_p}$, respectively, such that $P = [P_1 : P_2 : \ldots : P_p]$. Hence, the numerator of (5.71) can be written as

$$\exp\left\{-\frac{1}{2\sigma^2}\mu'\left[I_n - \left(I_n - 2\sigma^2\sum_{i=1}^{p} t_i A_i^*\right)^{-1}\right]\mu\right\}$$

$$= \exp\left\{\frac{1}{2\sigma^2}\sum_{i=1}^{p}\mu' P_i\left(\frac{2t_i}{1-2t_i}\right)P_i'\mu\right\} = \exp\left[\sum_{i=1}^{p}\theta_i t_i(1-2t_i)^{-1}\right], \quad (5.72)$$

where $\theta_i = \mu' P_i P_i' \mu/\sigma^2$, $i = 1, 2, \ldots, p$. Furthermore, the denominator of (5.71) can be expressed as

$$\left[det\left(I_n - 2\sigma^2\sum_{i=1}^{p} t_i A_i^*\right)\right]^{1/2}$$

$$= \left[det\left\{P\,\text{diag}\left[(1-2t_1)I_{k_1}, (1-2t_2)I_{k_2}, \ldots, (1-2t_p)I_{k_p}\right]P'\right\}\right]^{1/2}$$

$$= \prod_{i=1}^{p}(1-2t_i)^{k_i/2}. \quad (5.73)$$

From (5.71) through (5.73) we finally conclude that

$$\phi_*(t) = \prod_{i=1}^{p}\phi_i(t_i, \theta_i), \quad (5.74)$$

where

$$\phi_i(t_i, \theta_i) = (1-2t_i)^{-k_i/2}\exp\left[\theta_i t_i(1-2t_i)^{-1}\right], \quad i = 1, 2, \ldots, p.$$

Note that $\phi_i(t_i, \theta_i)$ is the moment generating function of $X'A_i^*X$, which, according to (b), is distributed as $\chi^2_{k_i}(\theta_i)$ (see formula (4.41) for the moment generating function of a noncentral chi-squared distribution). From (5.74) it follows that $X'A_1^*X$, $X'A_2^*X$, $\ldots$, $X'A_p^*X$, and hence, $X'A_1X$, $X'A_2X$, $\ldots$, $X'A_pX$, are mutually independent.

III. (b) implies (a). If (b) holds, then A_i is idempotent by Theorem 5.4 ($i = 1, 2, \ldots, p$). Taking the trace of both sides of (5.67) and recalling that $tr(A_i) = rank(A_i) = k_i$, we get $n = \sum_{i=1}^{p} k_i$, which establishes (a).

IV. (c) implies (b). If (c) is true, then by Theorem 5.5, $A_iA_j = 0$, $i \neq j$. Using the same argument as before in II, A_i ($i = 1, 2, \ldots, p$) can be written as in (5.70), where $\Lambda_1 = \text{diag}(\Delta_1, 0, \ldots, 0), \Lambda_2 = \text{diag}(0, \Delta_2, \ldots, 0), \ldots, \Lambda_p = \text{diag}(0, 0, \ldots, \Delta_p)$, and Δ_i is a diagonal matrix of order $k_i \times k_i$ whose diagonal elements are the nonzero eigenvalues of A_i ($i = 1, 2, \ldots, p$). From (5.67) we conclude that

$$\begin{aligned} I_n &= \sum_{i=1}^{p} \Lambda_i \\ &= \text{diag}(\Delta_1, \Delta_2, \ldots, \Delta_p). \end{aligned}$$

This indicates that $\Delta_i = I_{k_i}$, $i = 1, 2, \ldots, p$. It follows that A_i must be idempotent. Hence, $\frac{1}{\sigma^2} X' A_i X \sim \chi^2_{k_i}(\theta_i)$, $i = 1, 2, \ldots, p$.

Having shown that (a) implies (b) and conversely in I and III, and that (b) implies (c) and conversely in II and IV, we conclude that (a), (b), and (c) are equivalent. □

Example 5.6 Consider the following ANOVA table (Table 5.1) from a fixed-effects linear model,

$$Y = X\beta + \epsilon, \tag{5.75}$$

where Y is a response vector of order $n \times 1$ with a mean vector $\mu = X\beta$ and a variance–covariance matrix $\Sigma = \sigma^2 I_n$, X is a known matrix, not necessarily of full column rank, β is a vector of fixed unknown parameters, and ϵ is a random experimental error vector such that $\epsilon \sim N(\mathbf{0}, \sigma^2 I_n)$. In Table 5.1, q is the number of effects in the model, T_0 represents the *null effect* associated with the grand mean in the model (or intercept for a regression model) with $k_0 = 1$, and T_q represents the experimental error (or residual) effect. Each sum of squares can be written as a quadratic form, namely, $SS_i = Y' A_i Y$, $i = 0, 1, \ldots, q$, and SS_{Total} is the uncorrected (uncorrected for the mean) sum of squares, $Y'Y$. Thus,

$$\begin{aligned} Y'Y &= \sum_{i=0}^{q} SS_i \\ &= \sum_{i=0}^{q} Y' A_i Y, \end{aligned}$$

TABLE 5.1
ANOVA Table for a Fixed-Effects Model

Source	DF	SS	MS	E(MS)	F
T_0	k_0	SS_0	MS_0	$\frac{\sigma^2}{k_0}\theta_0 + \sigma^2$	
T_1	k_1	SS_1	MS_1	$\frac{\sigma^2}{k_1}\theta_1 + \sigma^2$	$\frac{MS_1}{MS_E}$
.	.	.	.	.	.
.	.	.	.	.	.
.	.	.	.	.	.
T_{q-1}	k_{q-1}	SS_{q-1}	MS_{q-1}	$\frac{\sigma^2}{k_{q-1}}\theta_{q-1} + \sigma^2$	$\frac{MS_{q-1}}{MS_E}$
T_q(Error)	k_q	$SS_q(SS_E)$	$MS_q(MS_E)$	σ^2	
Total	n	SS_{Total}			

that is,

$$I_n = \sum_{i=0}^{q} A_i,$$

and

$$n = \sum_{i=0}^{q} k_i. \tag{5.76}$$

Note that SS_q is the residual (or error) sum of squares, which is denoted by SS_E. In addition, the number of degrees of freedom for the ith source (effect), namely k_i, is the same as the rank of A_i ($i = 0, 1, 2, \ldots, q$). This follows from the fact that k_i actually represents the number of linearly independent elements of $A_i^{1/2}Y$ ($A_i^{1/2}$ is well defined here since A_i is nonnegative definite because $SS_i = Y'A_iY$ is a sum of squares, which is also equal to the square of the Euclidean norm of $A_i^{1/2}Y$, that is, $SS_i = \| A_i^{1/2}Y \|_2^2$, $i = 0, 1, \ldots, q$). Furthermore, the rank of $A_i^{1/2}$ is the same as the rank of A_i. Hence, from condition (5.76) and Theorem 5.7 we conclude that $SS_0, SS_1, \ldots, SS_q$ are mutually independent and $\frac{1}{\sigma^2}SS_i \sim \chi^2_{k_i}(\theta_i)$, where $\theta_i = \frac{1}{\sigma^2}\beta'X'A_iX\beta$, $i = 0, 1, \ldots, q$. Note that for $i = q$, $\theta_i = 0$, and therefore $\frac{1}{\sigma^2}SS_q \sim \chi^2_{k_q}$ since (see Chapter 7)

$$\begin{aligned} SS_q &= SS_E \\ &= Y'[I_n - X(X'X)^{-}X']Y. \end{aligned}$$

Hence,

$$\begin{aligned} \theta_q &= \frac{1}{\sigma^2}\beta'X'[I_n - X(X'X)^{-}X']X\beta \\ &= 0. \end{aligned}$$

The expected value of the ith mean square, $MS_i = \frac{1}{k_i}SS_i$, $i = 0, 1, \ldots, q$, is therefore given by

$$\begin{aligned} E(MS_i) &= \frac{\sigma^2}{k_i}E\left(\frac{1}{\sigma^2}SS_i\right) \\ &= \frac{\sigma^2}{k_i}E[\chi^2_{k_i}(\theta_i)] \\ &= \frac{\sigma^2}{k_i}(k_i + \theta_i) \\ &= \sigma^2 + \frac{\sigma^2}{k_i}\theta_i, \quad i = 0, 1, \ldots, q-1, \end{aligned}$$

$$\begin{aligned} E(MS_q) &= E(MS_E) \\ &= \frac{\sigma^2}{k_q} E\left(\frac{1}{\sigma^2} SS_E\right) \\ &= \frac{\sigma^2}{k_q} E(\chi^2_{k_q}) \\ &= \sigma^2. \end{aligned}$$

It follows that $F_i = \frac{MS_i}{MS_E}$ is a test statistic for testing the null hypothesis, $H_{0i} : \theta_i = 0$, against the alternative hypothesis, $H_{ai} : \theta_i \neq 0$, $i = 1, 2, \ldots, q-1$. Under H_{0i}, F_i has the F-distribution with k_i and k_q degrees of freedom $(i = 1, 2, \ldots, q-1)$. Given the form of $E(MS_i)$, H_{0i} can be rejected at the α-level of significance if $F_i > F_{\alpha, k_i, k_q}$, where F_{α, k_i, k_q} denotes the upper α-quantile of the F-distribution with k_i and k_q degrees of freedom $(i = 1, 2, \ldots, q-1)$.

A particular case of model (5.75) is the balanced fixed-effects one-way model described in Example 5.2. In this case, $q = 2$, $n = am$, and the corresponding ANOVA table is displayed as Table 5.2, where $SS_0 = \frac{1}{am} \boldsymbol{Y}'(\boldsymbol{J}_a \otimes \boldsymbol{J}_m)\boldsymbol{Y}$, SS_A and SS_E are given in (5.41) and (5.42), respectively, and

$$\begin{aligned} \theta_0 &= \frac{1}{\sigma^2} \left[\mu(\mathbf{1}'_a \otimes \mathbf{1}'_m) + \boldsymbol{\alpha}' \left(\boldsymbol{I}_a \otimes \mathbf{1}'_m\right)\right] \\ &\quad \times \left[\frac{1}{am} \left(\boldsymbol{J}_a \otimes \boldsymbol{J}_m\right)\right] \left[\mu \left(\mathbf{1}_a \otimes \mathbf{1}_m\right) + \left(\boldsymbol{I}_a \otimes \mathbf{1}_m\right) \boldsymbol{\alpha}\right] \\ &= \frac{1}{\sigma^2} \left[a\, m\, \mu^2 + 2\, m\, n \left(\sum_{i=1}^{a} \alpha_i\right) + \frac{m}{a} \left(\sum_{i=1}^{a} \alpha_i\right)^2\right], \\ \theta_1 &= \frac{m}{\sigma^2} \sum_{i=1}^{a} (\alpha_i - \bar{\alpha}_.)^2, \end{aligned}$$

where $\bar{\alpha}_. = \frac{1}{a} \sum_{i=1}^{a} \alpha_i$, as was shown in Example 5.2. All sums of squares are mutually independent with $\frac{1}{\sigma^2} SS_0 \sim \chi^2_1(\theta_0)$, $\frac{1}{\sigma^2} SS_A \sim \chi^2_{a-1}(\theta_1)$, and $\frac{1}{\sigma^2} SS_E \sim \chi^2_{a(m-1)}$. This confirms the results established in Examples 5.2 and 5.3

TABLE 5.2
ANOVA Table for the One-Way Model

Source	DF	*SS*	*MS*	*E(MS)*	*F*
Null effect	1	SS_0	MS_0	$\sigma^2\,\theta_0 + \sigma^2$	
A	$a-1$	SS_A	MS_A	$\frac{\sigma^2}{a-1}\,\theta_1 + \sigma^2$	$\frac{MS_A}{MS_E}$
Error	$a(m-1)$	SS_E	MS_E	σ^2	
Total	am	SS_{Total}			

with regard to the distribution and independence of SS_A and SS_E. Note that the null hypothesis $H_{01} : \theta_1 = 0$ is equivalent to $H_{01} : \alpha_1 = \alpha_2 = \cdots = \alpha_a$, or that the means of the a levels of factor A are equal.

5.6 Computing the Distribution of Quadratic Forms

Quite often, it may be of interest to evaluate the cumulative distribution function of $\boldsymbol{X}'\boldsymbol{A}\boldsymbol{X}$, where $\boldsymbol{A}$ is a symmetric matrix of order $n \times n$ and $\boldsymbol{X} \sim N(\boldsymbol{\mu}, \boldsymbol{\Sigma})$, that is,

$$F(u) = P(\boldsymbol{X}'\boldsymbol{A}\boldsymbol{X} \leq u), \tag{5.77}$$

for a given value of u. Alternatively, one may compute the value of u corresponding to a given value of p, where $p = F(u)$. Such a value of u is called the pth *quantile* of $\boldsymbol{X}'\boldsymbol{A}\boldsymbol{X}$, which we denote by u_p.

To evaluate $F(u)$ in (5.77), it is convenient to use the representation (5.20) expressing $\boldsymbol{X}'\boldsymbol{A}\boldsymbol{X}$ as a linear combination of mutually independent chi-squared variates. Formula (5.77) can then be written as

$$F(u) = P\left(\sum_{i=1}^{k} \gamma_i W_i \leq u\right), \tag{5.78}$$

where, if we recall, $\gamma_1, \gamma_2, \ldots, \gamma_k$ are the distinct nonzero eigenvalues of $\boldsymbol{\Sigma}^{1/2}\boldsymbol{A}\boldsymbol{\Sigma}^{1/2}$ (or, equivalently, the matrix $\boldsymbol{A}\boldsymbol{\Sigma}$) with multiplicities $\nu_1, \nu_2, \ldots, \nu_k$, and the W_i's are mutually independent such that $W_i \sim \chi^2_{\nu_i}(\theta_i)$, $i = 1, 2, \ldots, k$. In this case, the value of $F(u)$ can be easily calculated using a computer algorithm given by Davies (1980), which is based on a method proposed by Davies (1973). This algorithm is described in Davies (1980) as *Algorithm AS 155*, and can be easily accessed through *STATLIB*, which is an e-mail and file transfer protocol (FTP)-based retrieval system for statistical software.

For example, suppose that $\boldsymbol{X} \sim N(\boldsymbol{0}, \boldsymbol{\Sigma})$, where

$$\boldsymbol{\Sigma} = \begin{bmatrix} 3 & 5 & 1 \\ 5 & 13 & 0 \\ 1 & 0 & 1 \end{bmatrix}. \tag{5.79}$$

Let $\boldsymbol{A}$ be the symmetric matrix

$$\boldsymbol{A} = \begin{bmatrix} 1 & 1 & 2 \\ 1 & 2 & 3 \\ 2 & 3 & 1 \end{bmatrix}.$$

The eigenvalues of $\boldsymbol{\Sigma}^{1/2}\boldsymbol{A}\boldsymbol{\Sigma}^{1/2}$ (or $\boldsymbol{A}\boldsymbol{\Sigma}$) are -2.5473, 0.0338, 46.5135. Formula (5.78) takes the form,

$$F(u) = P(-2.5473\, W_1 + 0.0338\, W_2 + 46.5135\, W_3 \leq u),$$

TABLE 5.3
Quantiles of $-2.5473\, W_1 + 0.0338\, W_2 + 46.5135\, W_3$

p	pth Quantile (u_p)
0.25	2.680 (first quartile)
0.50	18.835 (median)
0.75	59.155 (third quartile)
0.90	123.490 (90th percentile)
0.95	176.198 (95th percentile)

where W_1, W_2, and W_3 are mutually independent variates, each distributed as χ_1^2. Using Davies' (1980) algorithm, quantiles of $-2.5473\, W_1 + 0.0338\, W_2 + 46.5135\, W_3$ can be obtained. Some of these quantiles are displayed in Table 5.3.

5.6.1 Distribution of a Ratio of Quadratic Forms

One interesting application of Davies' algorithm is in providing a tabulation of the values of the cumulative distribution function of a ratio of two quadratic forms, namely,

$$h(\boldsymbol{X}) = \frac{\boldsymbol{X}'\boldsymbol{A}_1\boldsymbol{X}}{\boldsymbol{X}'\boldsymbol{A}_2\boldsymbol{X}}, \tag{5.80}$$

where $\boldsymbol{X} \sim N(\boldsymbol{\mu}, \boldsymbol{\Sigma})$, and $\boldsymbol{A}_1$ and $\boldsymbol{A}_2$ are symmetric matrices with $\boldsymbol{A}_2$ assumed to be positive semidefinite. Ratios such as $h(\boldsymbol{X})$ are frequently encountered in statistics, particularly in *analysis of variance* as well as in econometrics. The exact distribution of $h(\boldsymbol{X})$ is known only in some special cases, but is mathematically intractable, in general. Several methods were proposed to approximate this distribution. Gurland (1955) approximated the density function of $h(\boldsymbol{X})$ using an infinite series of *Laguerre polynomials*. Lugannani and Rice (1984) used numerical techniques to evaluate this density function. More recently, Lieberman (1994) used the *saddlepoint method*, which was introduced by Daniels (1954), to approximate the distribution of $h(\boldsymbol{X})$.

Let $G(u)$ denote the cumulative distribution function of $h(\boldsymbol{X})$. Then,

$$G(u) = P\left(\frac{\boldsymbol{X}'\boldsymbol{A}_1\boldsymbol{X}}{\boldsymbol{X}'\boldsymbol{A}_2\boldsymbol{X}} \leq u\right), \tag{5.81}$$

which can be written as

$$G(u) = P(\boldsymbol{X}'\boldsymbol{A}_u\boldsymbol{X} \leq 0), \tag{5.82}$$

where $\boldsymbol{A}_u = \boldsymbol{A}_1 - u\,\boldsymbol{A}_2$. Expressing $\boldsymbol{X}'\boldsymbol{A}_u\boldsymbol{X}$ as a linear combination of mutually independent chi-squared variates, as was done earlier in (5.78), we get

$$G(u) = P\left(\sum_{i=1}^{l} \gamma_{ui}\, W_{ui} \leq 0\right), \tag{5.83}$$

TABLE 5.4
Values of $G(u)$ Using Formula (5.83)

u	$G(u)$
1.0	0.0
1.25	0.01516
1.50	0.01848
2.0	0.69573
2.75	0.99414
3.0	1.0

where $\gamma_{u1}, \gamma_{u2}, \ldots, \gamma_{ul}$ are the distinct nonzero eigenvalues of $\boldsymbol{\Sigma}^{1/2}\boldsymbol{A}_u\boldsymbol{\Sigma}^{1/2}$ with multiplicities $\nu_{u1}, \nu_{u2}, \ldots, \nu_{ul}$, and the W_{ui}'s are mutually independent such that $W_{ui} \sim \chi^2_{\nu_{ui}}(\theta_{ui})$, $i = 1, 2, \ldots, l$. Thus, for a given value of u, $G(u)$ can be easily computed on the basis of Davies' algorithm.

As an example, let us consider the distribution of

$$h(\boldsymbol{X}) = \frac{X_1^2 + 2X_2^2 + 3X_3^2}{X_1^2 + X_2^2 + X_3^2},$$

where $\boldsymbol{X} = (X_1, X_2, X_3)' \sim N(\boldsymbol{0}, \boldsymbol{\Sigma})$ and $\boldsymbol{\Sigma}$ is the same as in (5.79). In this case, $\boldsymbol{A}_1 = \text{diag}(1, 2, 3)$ and $\boldsymbol{A}_2 = \boldsymbol{I}_3$. Hence, $\boldsymbol{A}_u = \boldsymbol{A}_1 - u\boldsymbol{A}_2 = \text{diag}(1-u, 2-u, 3-u)$. For a given value of u, the eigenvalues of $\boldsymbol{\Sigma}^{1/2}\boldsymbol{A}_u\boldsymbol{\Sigma}^{1/2}$ are obtained and used in formula (5.83). The corresponding values of $G(u)$ can be computed using Davies' algorithm. Some of these values are shown in Table 5.4.

Appendix 5.A: Positive Definiteness of the Matrix $\boldsymbol{W}_t^{-1}$ in (5.2)

To show that there exists a positive number t_0 such that $\boldsymbol{W}_t^{-1}$ is positive definite if $|t| < t_0$, where

$$t_0 = \frac{1}{2\,\max_i \mid \lambda_i \mid}, \tag{5.A.1}$$

and λ_i is the ith eigenvalue of the matrix $\boldsymbol{\Sigma}^{1/2}\boldsymbol{A}\boldsymbol{\Sigma}^{1/2}$ $(i = 1, 2, \ldots, n)$.

It is clear that $\boldsymbol{W}_t^{-1}$ in (5.2) is positive definite if and only if the matrix $(\boldsymbol{I}_n - 2t\,\boldsymbol{\Sigma}^{1/2}\boldsymbol{A}\boldsymbol{\Sigma}^{1/2})$ is positive definite. The latter matrix is positive definite if and only if its eigenvalues, namely, $1 - 2t\lambda_i$, $i = 1, 2, \ldots, n$, are positive. Note that

$$2t\lambda_i \le 2 \mid t \mid \max_i \mid \lambda_i \mid, \quad i = 1, 2, \ldots, n.$$

Hence, in order for $1 - 2t\lambda_i$ to be positive, it is sufficient to choose t such that

$$2 \mid t \mid \max_i \mid \lambda_i \mid < 1.$$

Thus, choosing $|t| < t_0$, where t_0 is as shown in (5.A.1), guarantees that $1 - 2t\lambda_i > 0$ for $i = 1, 2, \ldots, n$. Note that the denominator in (5.A.1) cannot be zero since if $\max_i |\lambda_i| = 0$, then $\lambda_i = 0$ for all i, which implies that $A = \mathbf{0}$. □

Appendix 5.B: $A\boldsymbol{\Sigma}$ Is Idempotent If and Only If $\boldsymbol{\Sigma}^{1/2}A\boldsymbol{\Sigma}^{1/2}$ Is Idempotent

Suppose that $A\boldsymbol{\Sigma}$ is idempotent. Then,

$$A\boldsymbol{\Sigma}A\boldsymbol{\Sigma} = A\boldsymbol{\Sigma}. \tag{5.B.1}$$

Multiplying (5.B.1) on the left by $\boldsymbol{\Sigma}^{1/2}$ and on the right by $\boldsymbol{\Sigma}^{-1/2}$, we get

$$\boldsymbol{\Sigma}^{1/2}A\boldsymbol{\Sigma}^{1/2}\boldsymbol{\Sigma}^{1/2}A\boldsymbol{\Sigma}^{1/2} = \boldsymbol{\Sigma}^{1/2}A\boldsymbol{\Sigma}^{1/2}. \tag{5.B.2}$$

This shows that $\boldsymbol{\Sigma}^{1/2}A\boldsymbol{\Sigma}^{1/2}$ is idempotent.

Vice versa, if $\boldsymbol{\Sigma}^{1/2}A\boldsymbol{\Sigma}^{1/2}$ is idempotent, then by multiplying (5.B.2) on the left by $\boldsymbol{\Sigma}^{-1/2}$ and on the right by $\boldsymbol{\Sigma}^{1/2}$, we obtain (5.B.1). Therefore, $A\boldsymbol{\Sigma}$ is idempotent. □

Exercises

5.1 Suppose that $\boldsymbol{X} = (X_1, X_2, X_3)'$ is distributed as $N(\boldsymbol{\mu}, \boldsymbol{\Sigma})$ such that

$$\begin{aligned} Q &= (\boldsymbol{X} - \boldsymbol{\mu})'\boldsymbol{\Sigma}^{-1}(\boldsymbol{X} - \boldsymbol{\mu}) \\ &= 2X_1^2 + 3X_2^2 + 4X_3^2 + 2X_1X_2 - 2X_1X_3 - 4X_2X_3 \\ &\quad - 6X_1 - 6X_2 + 10X_3 + 8. \end{aligned}$$

(a) Find $\boldsymbol{\mu}$ and $\boldsymbol{\Sigma}$.

(b) Find the moment generating function of Q.

(c) Find the moment generating function for the conditional distribution of X_1 given X_2 and X_3.

5.2 Let $\boldsymbol{X} = (\boldsymbol{X}_1' : \boldsymbol{X}_2')'$ be normally distributed. The corresponding mean vector $\boldsymbol{\mu}$ and variance–covariance matrix $\boldsymbol{\Sigma}$ are partitioned as $\boldsymbol{\mu} = (\boldsymbol{\mu}_1' : \boldsymbol{\mu}_2')'$, and

$$\boldsymbol{\Sigma} = \begin{bmatrix} \boldsymbol{\Sigma}_{11} & \boldsymbol{\Sigma}_{12} \\ \boldsymbol{\Sigma}_{12}' & \boldsymbol{\Sigma}_{22} \end{bmatrix}.$$

The numbers of elements in $\boldsymbol{X}_1$ and $\boldsymbol{X}_2$ are n_1 and n_2, respectively. Let $\boldsymbol{A}$ be a constant matrix of order $n_1 \times n_2$.

(a) Show that

$$E(\boldsymbol{X}_1'\boldsymbol{A}\boldsymbol{X}_2) = \boldsymbol{\mu}_1'\boldsymbol{A}\boldsymbol{\mu}_2 + tr(\boldsymbol{A}\boldsymbol{\Sigma}_{12}').$$

(b) Find $\mathrm{Var}(\boldsymbol{X}_1'\boldsymbol{A}\boldsymbol{X}_2)$.

[Note: The expression $\boldsymbol{X}_1'\boldsymbol{A}\boldsymbol{X}_2$ is called a *bilinear form*.]

5.3 Consider the quadratic forms $\boldsymbol{X}'\boldsymbol{A}\boldsymbol{X}$, $\boldsymbol{X}'\boldsymbol{B}\boldsymbol{X}$, where $\boldsymbol{X} \sim N(\boldsymbol{\mu}, \boldsymbol{\Sigma})$. Show that

$$\mathrm{Cov}(\boldsymbol{X}'\boldsymbol{A}\boldsymbol{X}, \boldsymbol{X}'\boldsymbol{B}\boldsymbol{X}) = 2\,tr(\boldsymbol{A}\boldsymbol{\Sigma}\boldsymbol{B}\boldsymbol{\Sigma}) + 4\,\boldsymbol{\mu}'\boldsymbol{A}\boldsymbol{\Sigma}\boldsymbol{B}\boldsymbol{\mu}.$$

5.4 Let $\boldsymbol{X} = (X_1, X_2)' \sim N(\boldsymbol{0}, \boldsymbol{\Sigma})$, where

$$\boldsymbol{\Sigma} = \begin{bmatrix} \sigma_{11} & \sigma_{12} \\ \sigma_{12} & \sigma_{22} \end{bmatrix}.$$

Find $\mathrm{Cov}(X_1^2 + 3\,X_1 - 2,\; 2\,X_2^2 + 5\,X_2 + 3)$.

5.5 Let $Q_1 = \boldsymbol{X}'\boldsymbol{A}_1\boldsymbol{X}$ and $Q_2 = \boldsymbol{X}'\boldsymbol{A}_2\boldsymbol{X}$ be two quadratic forms, where $\boldsymbol{X} \sim N(\boldsymbol{\mu}, \boldsymbol{\Sigma})$ and $\boldsymbol{A}_1$ and $\boldsymbol{A}_2$ are nonnegative definite matrices. Show that a quadratic form, $\boldsymbol{X}'\boldsymbol{B}\boldsymbol{X}$, is distributed independently of $Q_1 + Q_2$ if and only if it is distributed independently of Q_1 and of Q_2.

[Hint: See Bhat (1962)].

5.6 Consider Exercise 5.5 again. Show that a linear form, $\boldsymbol{C}\boldsymbol{X}$, where $\boldsymbol{C}$ is a matrix of full row rank, is distributed independently of $Q_1 + Q_2$ if and only if it is distributed independently of Q_1 and of Q_2.

5.7 Let $Q_i = \boldsymbol{X}'\boldsymbol{A}_i\boldsymbol{X}$, where $\boldsymbol{X} \sim N(\boldsymbol{\mu}, \boldsymbol{\Sigma})$, and $\boldsymbol{A}_i$ is nonnegative definite, $i = 1, 2, \ldots, n$. Show that a linear form, $\boldsymbol{C}\boldsymbol{X}$, is distributed independently of $\sum_{i=1}^{n} Q_i$ if and only if it is distributed independently of Q_i for all $i = 1, 2, \ldots, n$.

[Note: This is an extension of the result in Exercise 5.6, and can be proved by mathematical induction.]

5.8 Let $\mathbf{X} \sim N(\boldsymbol{\mu}, \boldsymbol{\Sigma})$, and let X_i be the ith element of $\mathbf{X}$ ($i = 1, 2, \ldots, n$) and σ_{ij} be the (i, j)th element of $\boldsymbol{\Sigma}$ ($i, j = 1, 2, \ldots, n$). Show that $\bar{X} = \frac{1}{n}\sum_{i=1}^{n} X_i$ is distributed independently of $\sum_{i=1}^{n}(X_i - \bar{X})^2$ if and only if $\bar{\sigma}_{i.} = \bar{\sigma}_{..}$ for all $i = 1, 2, \ldots, n$, where $\bar{\sigma}_{i.} = \frac{1}{n}\sum_{j=1}^{n} \sigma_{ij}$ ($i = 1, 2, \ldots, n$), and $\bar{\sigma}_{..} = \frac{1}{n}\sum_{i=1}^{n} \bar{\sigma}_{i.}$.

[Hint: Make use of the result in Exercise 5.7.]

5.9 Consider Exercise 5.8 again. Show that if $\bar{X}$ and $\sum_{i=1}^{n}(X_i - \bar{X})^2$ are independent, then $\bar{X}$ is independent of the sample range, that is, $\max_{i \neq i'}(X_i - X_{i'})$.

5.10 Let $\mathbf{X} \sim N(\boldsymbol{\mu}, \boldsymbol{\Sigma})$. Show that the quadratic form $\mathbf{X}'\mathbf{A}\mathbf{X}$ has the chi-squared distribution with n degrees of freedom if and only if $\mathbf{A} = \boldsymbol{\Sigma}^{-1}$.

5.11 Let $\mathbf{X} = (X_1, X_2, \ldots, X_n)'$ be distributed as $N(\boldsymbol{\mu}, \boldsymbol{\Sigma})$. Let $\bar{X} = \frac{1}{n}\sum_{i=1}^{n} X_i$. Show that $n\bar{X}^2$ and $\sum_{i=1}^{n}(X_i - \bar{X})^2$ are distributed independently as $c_1\chi_1^2(\lambda_1)$ and $c_2\chi_{n-1}^2(\lambda_2)$, respectively, where c_1 and c_2 are nonnegative constants, if and only if

$$\boldsymbol{\Sigma} = c_2\mathbf{I}_n + \frac{c_1 - c_2}{n}\mathbf{J}_n,$$

where $\mathbf{I}_n$ is the identity matrix and $\mathbf{J}_n$ is the matrix of ones, both of order $n \times n$.

[Note: This result implies that, for a correlated data set from a normal population, the square of the sample mean and the sample variance are distributed independently as $c\chi^2$ variates if and only if the X_i's have a common variance and a common covariance.]

5.12 Prove the sufficiency part of Theorem 5.4 by showing that if $\mathbf{A}\boldsymbol{\Sigma}$ is idempotent of rank r, then the moment generating function of $\mathbf{X}'\mathbf{A}\mathbf{X}$ is the same as the one for $\chi_r^2(\theta)$.

5.13 Consider Example 5.2. Suppose here that $\alpha_i \sim N(0, \sigma_\alpha^2)$ and that the α_i's and ϵ_{ij}'s are independently distributed with $\epsilon_{ij} \sim N(0, \sigma_\epsilon^2)$, $i = 1, 2, \ldots, a$; $j = 1, 2, \ldots, m$. Let SS_A and SS_E be the same sum of squares associated with factor A and the error term, respectively.

(a) Show that $\frac{SS_A}{m\,\sigma_\alpha^2 + \sigma_\epsilon^2} \sim \chi_{a-1}^2$ and $\frac{SS_E}{\sigma_\epsilon^2} \sim \chi_{a(m-1)}^2$.

(b) Show that SS_A and SS_E are independent.

5.14 Consider again Exercise 5.13. Let $\hat{\sigma}_\alpha^2 = \frac{1}{m}\left[\frac{SS_A}{a-1} - \frac{SS_E}{a(m-1)}\right]$ be the *analysis of variance* (ANOVA) estimator of σ_α^2.

(a) Find $\text{Var}(\hat{\sigma}^2_\alpha)$.

(b) Find the probability of a negative $\hat{\sigma}^2_\alpha$. Can the values of a and m be chosen so that the value of this probability is reduced?

5.15 Consider the so-called *unbalanced random one-way model*,

$$Y_{ij} = \mu + \alpha_i + \epsilon_{ij}, \quad i = 1, 2, \ldots, a; j = 1, 2, \ldots, n_i,$$

where $\alpha_i \sim N(0, \sigma^2_\alpha)$, $\epsilon_{ij} \sim N(0, \sigma^2_\epsilon)$, and the α_i's and ϵ_{ij}'s are independently distributed. Note that $n_1, n_2, \ldots, n_a$ are not necessarily equal. Let SS_A and SS_E be the sums of squares associated with α_i and ϵ_{ij}, respectively, namely,

$$SS_A = \sum_{i=1}^{a} \frac{Y_{i.}^2}{n_i} - \frac{Y_{..}^2}{n_.},$$

$$SS_E = \sum_{i=1}^{a} \sum_{j=1}^{n_i} Y_{ij}^2 - \sum_{i=1}^{a} \frac{Y_{i.}^2}{n_i},$$

where $Y_{i.} = \sum_{j=1}^{n_i} Y_{ij}$, $Y_{..} = \sum_{i=1}^{a} \sum_{j=1}^{n_i} Y_{ij}$, $n_. = \sum_{i=1}^{a} n_i$.

(a) Show that SS_A and SS_E are independent.

(b) Show that $\frac{SS_E}{\sigma^2_\epsilon} \sim \chi^2_{n_.-a}$.

(c) What distribution does SS_A have?

(d) Show that if $\sigma^2_\alpha = 0$, then $\frac{SS_A}{\sigma^2_\epsilon} \sim \chi^2_{a-1}$.

5.16 Let $\hat{\sigma}^2_\alpha$ be the ANOVA estimator of σ^2_α in Exercise 5.15, which is given by

$$\hat{\sigma}^2_\alpha = \frac{1}{d}\left[\frac{SS_A}{a-1} - \frac{SS_E}{n_. - a}\right],$$

where

$$d = \frac{1}{a-1}\left(n_. - \frac{1}{n_.}\sum_{i=1}^{a} n_i^2\right).$$

(a) Show that $E(SS_A) = d(a-1)\sigma^2_\alpha + (a-1)\sigma^2_\epsilon$, and verify that $E(\hat{\sigma}^2_\alpha) = \sigma^2_\alpha$.

(b) Find $\text{Var}(\hat{\sigma}^2_\alpha)$.

(c) Show that for a fixed value of $n_.$, $\text{Var}(\hat{\sigma}^2_\alpha)$ attains a minimum for all σ^2_α, σ^2_ϵ if and only if the data set is balanced, that is, $n_1 = n_2 = \ldots = n_a$.

5.17 Consider again Exercise 5.16. Show how to compute the exact probability of a negative $\hat{\sigma}^2_\alpha$.

[Hint: Express SS_A as a linear combination of independent central chi-squared variates, then apply the methodology described in Section 5.6.]

5.18 Give an expression for the moment generating function of $\hat{\sigma}^2_\alpha$ in Exercise 5.16.

[Hint: Use the same hint given in Exercise 5.17 with regard to SS_A.]

5.19 Consider the sums of squares, SS_A and SS_E, described in Exercise 5.15. Show that

$$E\left\{\frac{1}{d}\left[\left(\frac{n_. - a - 2}{a - 1}\right)\frac{SS_A}{SS_E} - 1\right]\right\} = \frac{\sigma^2_\alpha}{\sigma^2_\epsilon},$$

where d is the same as in Exercise 5.16.

5.20 Consider the quadratic forms, $Q_1 = \boldsymbol{X}'\boldsymbol{A}\boldsymbol{X}$, $Q_2 = \boldsymbol{X}'\boldsymbol{B}\boldsymbol{X}$, where $\boldsymbol{X} \sim N(\boldsymbol{0}, \boldsymbol{\Sigma})$ and $\boldsymbol{A}$ and $\boldsymbol{B}$ are nonnegative definite matrices of order $n \times n$.

(a) Show that if Q_1 and Q_2 are uncorrelated, then they are also independent.

(b) Let $\boldsymbol{\Sigma} = \boldsymbol{I}_n$. Show that Q_1 and Q_2 are independent if and only if $tr(\boldsymbol{A}\boldsymbol{B}) = 0$.

[Note: Part (a) was proved by Matérn (1949).]

5.21 Consider the proof of sufficiency for Theorem 5.6. Show that condition (5.61) implies that the rows of $\boldsymbol{B}$ are linearly independent of the rows of $\boldsymbol{P}_i'\boldsymbol{\Sigma}^{-1/2}$, $i = 1, 2, \ldots, k$.

5.22 Let $\boldsymbol{X} \sim N(\boldsymbol{0}, \boldsymbol{I}_n)$. Show that $\boldsymbol{X}'\boldsymbol{A}\boldsymbol{X}$ and $\boldsymbol{X}'\boldsymbol{B}\boldsymbol{X}$ are independent if and only if $det(\boldsymbol{I}_n - s\,\boldsymbol{A} - t\,\boldsymbol{B}) = [det(\boldsymbol{I}_n - s\,\boldsymbol{A})][det(\boldsymbol{I}_n - t\,\boldsymbol{B})]$ for all values of s and t.

5.23 Consider Exercise 5.22. Show that if $\boldsymbol{X}'\boldsymbol{A}\boldsymbol{X}$ and $\boldsymbol{X}'\boldsymbol{B}\boldsymbol{X}$ are independent, then the rank of $\boldsymbol{A} + \boldsymbol{B}$ is the sum of the ranks of $\boldsymbol{A}$ and $\boldsymbol{B}$.

[Hint: Choose $s = t$ in Exercise 5.22.]

5.24 (Kawada, 1950) Let $Q_1 = \boldsymbol{X}'\boldsymbol{A}\boldsymbol{X}$, $Q_2 = \boldsymbol{X}'\boldsymbol{B}\boldsymbol{X}$, where $\boldsymbol{X} \sim N(\boldsymbol{0}, \boldsymbol{I}_n)$. Let T_{ij} be defined as

$$T_{ij} = E(Q_1^i\,Q_2^j) - E(Q_1^i)\,E(Q_2^j), \quad i, j = 1, 2.$$

Show that

(a) $T_{11} = 2\,tr(\boldsymbol{A}\boldsymbol{B})$.

(b) $T_{12} = 8\,tr(\boldsymbol{A}\boldsymbol{B}^2) + 4\,tr(\boldsymbol{A}\boldsymbol{B})\,tr(\boldsymbol{B})$.

(c) $T_{21} = 8\,tr(\mathbf{A}^2\mathbf{B}) + 4\,tr(\mathbf{AB})\,tr(\mathbf{A})$.

(d) $T_{22} = 32\,tr(\mathbf{A}^2\mathbf{B}^2) + 16\,tr[(\mathbf{AB})^2] + 16\,tr(\mathbf{AB}^2)\,tr(\mathbf{A}) + 16\,tr(\mathbf{A}^2\mathbf{B})\,tr(\mathbf{B}) + 8\,tr(\mathbf{AB})\,tr(\mathbf{A})\,tr(\mathbf{B}) + 8\,tr[(\mathbf{AB})^2]$.

5.25 (Kawada, 1950) Deduce from Exercise 5.24 that if $T_{ij} = 0$ for $i, j = 1, 2$, then Q_1 and Q_2 are independent.

5.26 Let $\mathbf{X} = (X_1, X_2)' \sim N(\boldsymbol{\mu}, \boldsymbol{\Sigma})$, where $\boldsymbol{\mu} = (1.25, 1.75)'$, and $\boldsymbol{\Sigma}$ is given by

$$\boldsymbol{\Sigma} = \begin{bmatrix} 1 & 0.5 \\ 0.5 & 1 \end{bmatrix}.$$

Let $G(u)$ be defined as

$$G(u) = P\left(\frac{2\,X_1X_2}{2\,X_1^2 + X_2^2} \leq u\right).$$

Find the values of $G(u)$ at $u = \frac{1}{2}, \frac{1}{3}, \frac{1}{4}$ using Davies' (1980) algorithm.

5.27 Consider the one-way model in Exercise 5.15 under the following assumptions:

(i) The α_i's are independently distributed as $N(0, \sigma^2_\alpha)$.

(ii) The ϵ_{ij}'s are independently distributed as $N(0, \sigma^2_i), i = 1, 2, \ldots, a;$ $j = 1, 2, \ldots, n_i$.

(iii) The α_i's and ϵ_{ij}'s are independent.

(a) Under assumptions (i), (ii), and (iii), what distributions do SS_A and SS_E have, where SS_A and SS_E are the same sums of squares as in Exercise 5.15?

(b) Are SS_A and SS_E independent under assumptions (i), (ii), and (iii)?

(c) If $\hat{\sigma}^2_\alpha$ is the ANOVA estimator of σ^2_α in Exercise 5.16, show that

$$E(\hat{\sigma}^2_\alpha) = \sigma^2_\alpha + \frac{1}{d}\left[\frac{1}{a-1}\sum_{i=1}^{a}\left(1 - \frac{n_i}{n_.}\right)\sigma^2_i - \frac{1}{n_. - a}\sum_{i=1}^{a}(n_i - 1)\sigma^2_i\right],$$

and hence $\hat{\sigma}^2_\alpha$ is a biased estimator of σ^2_α under assumptions (i), (ii), and (iii).

5.28 Consider again the same model and the same assumptions as in Exercise 5.15.

(a) Show that $F = \frac{n_.-a}{a-1}\frac{SS_A}{SS_E}$ is a test statistic for testing the null hypothesis $H_0 : \sigma^2_\alpha = 0$.

(b) Find the power of the test in (a) at the 5% level of significance, given that $\frac{\sigma^2_\alpha}{\sigma^2_\epsilon} = 1.5$, and that $a = 4,\ n_1 = 8,\ n_2 = 10,\ n_3 = 6,\ n_4 = 12$. [Hint: Use Davies' (1980) algorithm.]

6

Full-Rank Linear Models

One of the objectives of an experimental investigation is the empirical determination of the functional relationship that may exist between a response variable, Y, and a set of control (or input) variables denoted by $x_1, x_2, \ldots, x_k$. The response Y is assumed to have a continuous distribution, and the x_i's are nonstochastic variables whose settings can be controlled, or determined, by the experimenter. These settings are measured on a continuous scale. For example, the yield, Y, of peanuts, in pounds per acre, is influenced by two control variables, x_1 and x_2, representing the amounts of two different fertilizers.

In general, the relationship between Y and $x_1, x_2, \ldots, x_k$ is unknown. It is therefore customary to start out the experimental investigation by postulating a simple relationship of the form

$$Y = \beta_0 + \sum_{i=1}^{k} \beta_i x_i + \epsilon, \tag{6.1}$$

where ϵ is an experimental error term associated with the measured, or observed, response at a point $\boldsymbol{x} = (x_1, x_2, \ldots, x_k)'$ in a region of interest, $\mathfrak{R}$, and $\beta_0, \beta_1, \ldots, \beta_k$ are fixed unknown parameters. Model (6.1) is called a *multiple linear regression model*. In particular, when $k = 1$, it is called a *simple linear regression model*. A more general model than the one in (6.1) is given by

$$Y = \boldsymbol{f}'(\boldsymbol{x})\,\boldsymbol{\beta} + \epsilon, \tag{6.2}$$

where $\boldsymbol{\beta} = (\beta_1, \beta_2, \ldots, \beta_p)'$ is a vector of p unknown parameters and $\boldsymbol{f}(\boldsymbol{x})$ is a $p \times 1$ vector whose first element is equal to one and its remaining $p - 1$ elements are polynomial functions of $x_1, x_2, \ldots, x_k$. These functions are in the form of powers and cross products of powers of the x_i's up to degree d (≥ 1). Thus (6.2) represents a complete polynomial model of degree d. For example, the model in (6.1) is of degree 1 with $\boldsymbol{f}(\boldsymbol{x}) = (1, x_1, x_2, \ldots, x_k)'$, and the model

$$Y = \beta_0 + \sum_{i=1}^{k} \beta_i x_i + \sum_{i<j} \beta_{ij} x_i x_j + \sum_{i=1}^{k} \beta_{ii} x_i^2 + \epsilon$$

is of degree 2 with

$$\boldsymbol{f}(\boldsymbol{x}) = (1, x_1, x_2, \ldots, x_k, x_1 x_2, x_1 x_3, \ldots, x_{k-1} x_k, x_1^2, x_2^2, \ldots, x_k^2)'.$$

Model (6.2) is referred to as a *linear model* due to the fact that the elements of β appear linearly in the model. All linear regression models can be represented by such a model. The error term ϵ in (6.2) is assumed to have a continuous distribution whose mean, or expected value, is $E(\epsilon) = 0$. Since the elements of x are nonstochastic, the mean of Y at x, denoted by $\mu(x)$ and called the *mean response* at x, is

$$\mu(x) = f'(x)\,\beta. \tag{6.3}$$

Model (6.2) is said to be inadequate, or to suffer from *lack of fit*, if the true mean response $\mu(x)$ is not equal to the expression on the right-hand side of (6.3), that is, when $E(\epsilon)$ in (6.2) is not equal to zero. This occurs when $\mu(x)$ depends on some unknown function of $x_1, x_2, \ldots, x_k$ besides $f'(x)\beta$, or on other variables not accounted for by the model.

6.1 Least-Squares Estimation

In order to estimate β in model (6.2), a series of n experiments ($n > p$) are carried out, in each of which the response Y is observed at different settings of the control variables, $x_1, x_2, \ldots, x_k$. Let Y_u denote the observed response value at x_u, where $x_u = (x_{u1}, x_{u2}, \ldots, x_{uk})'$ with x_{ui} denoting the uth setting of x_i at the uth experimental run ($i = 1, 2, \ldots, k$; $u = 1, 2, \ldots, n$). From (6.2) we then have

$$Y_u = f'(x_u)\,\beta + \epsilon_u, \quad u = 1, 2, \ldots, n, \tag{6.4}$$

where ϵ_u is the experimental error associated with Y_u ($u = 1, 2, \ldots, n$). Model (6.4) can be expressed in matrix form as

$$Y = X\beta + \epsilon, \tag{6.5}$$

where $Y = (Y_1, Y_2, \ldots, Y_n)'$, X is an $n \times p$ matrix whose uth row is $f'(x_u)$ and $\epsilon = (\epsilon_1, \epsilon_2, \ldots, \epsilon_n)'$. The matrix X is assumed to be of full column rank, that is, *rank*(X) $= p$. In this case, model (6.5) is said to be of full rank. In addition, it is assumed that $E(\epsilon) = \mathbf{0}$, and $\mathrm{Var}(\epsilon) = \sigma^2 I_n$, where σ^2 is unknown and I_n is the identity matrix of order $n \times n$. This implies that the response values, $Y_1, Y_2, \ldots, Y_n$, are uncorrelated and have variances equal to σ^2. Thus the expected value of Y is $E(Y) = X\beta$ and its variance–covariance matrix is $\mathrm{Var}(Y) = \sigma^2 I_n$. We refer to $E(Y)$ as the *mean response vector* and is denoted by μ.

Under the above assumptions, estimation of β in model (6.5) can be achieved by using the *method of ordinary least squares* (OLS). By definition, the OLS estimator of β, denoted by $\hat{\beta}$, is the vector that minimizes the square of the Euclidean norm of $\boldsymbol{Y} - \boldsymbol{X}\boldsymbol{\beta}$, that is,

$$\begin{aligned} S(\boldsymbol{\beta}) &= \| \boldsymbol{Y} - \boldsymbol{X}\boldsymbol{\beta} \|^2 \\ &= (\boldsymbol{Y} - \boldsymbol{X}\boldsymbol{\beta})' (\boldsymbol{Y} - \boldsymbol{X}\boldsymbol{\beta}) \\ &= \boldsymbol{Y}' \boldsymbol{Y} - 2\,\boldsymbol{\beta}'\boldsymbol{X}'\boldsymbol{Y} + \boldsymbol{\beta}'\boldsymbol{X}'\boldsymbol{X}\boldsymbol{\beta}, \end{aligned} \tag{6.6}$$

with respect to $\boldsymbol{\beta}$. Since $S(\boldsymbol{\beta})$ has first-order partial derivatives with respect to the elements of $\boldsymbol{\beta}$, a necessary condition for $S(\boldsymbol{\beta})$ to have a minimum at $\boldsymbol{\beta} = \hat{\boldsymbol{\beta}}$ is that $\frac{\partial [S(\boldsymbol{\beta})]}{\partial \boldsymbol{\beta}} = \boldsymbol{0}$ at $\boldsymbol{\beta} = \hat{\boldsymbol{\beta}}$, that is,

$$\left[\frac{\partial}{\partial \boldsymbol{\beta}} \left(\boldsymbol{Y}'\boldsymbol{Y} - 2\,\boldsymbol{\beta}'\boldsymbol{X}'\boldsymbol{Y} + \boldsymbol{\beta}'\boldsymbol{X}'\boldsymbol{X}\boldsymbol{\beta}\right)\right]_{\boldsymbol{\beta}=\hat{\boldsymbol{\beta}}} = \boldsymbol{0}. \tag{6.7}$$

Applying Theorems 3.21 and 3.22, we can write

$$\begin{aligned} \frac{\partial}{\partial \boldsymbol{\beta}}(\boldsymbol{\beta}'\boldsymbol{X}'\boldsymbol{Y}) &= \boldsymbol{X}'\boldsymbol{Y} \\ \frac{\partial}{\partial \boldsymbol{\beta}}(\boldsymbol{\beta}'\boldsymbol{X}'\boldsymbol{X}\boldsymbol{\beta}) &= 2\,\boldsymbol{X}'\boldsymbol{X}\boldsymbol{\beta}. \end{aligned}$$

Making the substitution in (6.7), we obtain

$$-2\,\boldsymbol{X}'\boldsymbol{Y} + 2\,\boldsymbol{X}'\boldsymbol{X}\hat{\boldsymbol{\beta}} = \boldsymbol{0}. \tag{6.8}$$

Solving (6.8) for $\hat{\boldsymbol{\beta}}$, after noting that $\boldsymbol{X}'\,\boldsymbol{X}$ is a nonsingular matrix by the fact that $\boldsymbol{X}$ is of full column rank (see property (c) in Section 3.6), we get

$$\hat{\boldsymbol{\beta}} = (\boldsymbol{X}'\boldsymbol{X})^{-1}\,\boldsymbol{X}'\boldsymbol{Y}. \tag{6.9}$$

Note that $S(\boldsymbol{\beta})$ achieves its absolute minimum over the parameter space of $\boldsymbol{\beta}$ at $\hat{\boldsymbol{\beta}}$ since (6.8) has a unique solution given by $\hat{\boldsymbol{\beta}}$, and the *Hessian matrix* of second-order partial derivatives of $S(\boldsymbol{\beta})$ with respect to the elements of $\boldsymbol{\beta}$, namely the matrix,

$$\begin{aligned} \frac{\partial}{\partial \boldsymbol{\beta}'}\left[\frac{\partial}{\partial \boldsymbol{\beta}} S(\boldsymbol{\beta})\right] &= \frac{\partial}{\partial \boldsymbol{\beta}'}\left[-2\,\boldsymbol{X}'\boldsymbol{Y} + 2\,\boldsymbol{X}'\boldsymbol{X}\boldsymbol{\beta}\right] \\ &= 2\,\boldsymbol{X}'\boldsymbol{X}, \end{aligned}$$

is positive definite [see Theorem 3.3(c) in Chapter 3 and Corollary 7.7.1 in Khuri (2003)]. Alternatively, $S(\boldsymbol{\beta})$ can be shown to have an absolute minimum at $\hat{\boldsymbol{\beta}}$ by simply writing $S(\boldsymbol{\beta})$ in (6.6) as

$$S(\boldsymbol{\beta}) = \| \boldsymbol{Y} - \boldsymbol{X}\hat{\boldsymbol{\beta}} \|^2 + \| \boldsymbol{X}\hat{\boldsymbol{\beta}} - \boldsymbol{X}\boldsymbol{\beta} \|^2 . \tag{6.10}$$

Equality (6.10) follows from writing $Y - X\beta$ as $Y - X\hat{\beta} + X\hat{\beta} - X\beta$ and noting that

$$\begin{aligned}
&\left(Y - X\hat{\beta} + X\hat{\beta} - X\beta\right)'\left(Y - X\hat{\beta} + X\hat{\beta} - X\beta\right) \\
&\quad = \left(Y - X\hat{\beta}\right)'\left(Y - X\hat{\beta}\right) + 2\left(Y - X\hat{\beta}\right)'\left(X\hat{\beta} - X\beta\right) \\
&\quad + \left(X\hat{\beta} - X\beta\right)'\left(X\hat{\beta} - X\beta\right).
\end{aligned} \tag{6.11}$$

The middle term on the right-hand side of (6.11) is zero because $\hat{\beta}$ satisfies Equation (6.8). From (6.10) we conclude that for all β in the parameter space,

$$\| Y - X\beta \|^2 \geq \| Y - X\hat{\beta} \|^2 .$$

Equality is achieved if and only if $\beta = \hat{\beta}$. This is true because

$$\| X\hat{\beta} - X\beta \|^2 = \left(\hat{\beta} - \beta\right)' X'X \left(\hat{\beta} - \beta\right),$$

and the right-hand side is zero if and only if $\hat{\beta} - \beta = \mathbf{0}$ since $X'X$ is positive definite, as was pointed out earlier. It follows that the absolute minimum of $S(\beta)$ is

$$\begin{aligned}
S\left(\hat{\beta}\right) &= \| Y - X\hat{\beta} \|^2 \\
&= \left(Y - X\hat{\beta}\right)' \left(Y - X\hat{\beta}\right) \\
&= \left[Y - X\left(X'X\right)^{-1} X'Y\right]' \left[Y - X\left(X'X\right)^{-1} X'Y\right] \\
&= Y'\left[I_n - X\left(X'X\right)^{-1} X'\right] Y,
\end{aligned} \tag{6.12}$$

since $I_n - X(X'X)^{-1}X'$ is an idempotent matrix. The right-hand side of (6.12) is called the *error (or residual) sum of squares*, and is denoted by SS_E. We thus have

$$SS_E = Y'\left[I_n - X\left(X'X\right)^{-1} X'\right] Y, \tag{6.13}$$

which has $n - p$ degrees of freedom since $I_n - X(X'X)^{-1}X'$ is of rank $n - p$.

6.1.1 Estimation of the Mean Response

An estimator of the mean response, $\mu(x)$, in (6.3) is given by

$$\hat{\mu}(x) = f'(x)\hat{\beta}.$$

This is also called the *predicted response* at x, which is denoted by $\hat{Y}(x)$. Hence,

$$\begin{aligned}
\hat{Y}(x) &= f'(x)\,\hat{\beta} \\
&= f'(x)\left(X'X\right)^{-1} X'Y.
\end{aligned} \tag{6.14}$$

The $n \times 1$ vector $\hat{Y}$ whose uth element is $\hat{Y}(x_u)$, where x_u is the setting of x at the uth experimental run ($u = 1, 2, \ldots, n$), is the vector of predicted responses. From (6.14) we then have

$$\begin{aligned}\hat{Y} &= X\hat{\beta} \\ &= X\left(X'X\right)^{-1}X'Y. \end{aligned} \tag{6.15}$$

The vector

$$Y - \hat{Y} = \left[I_n - X\left(X'X\right)^{-1}X'\right]Y \tag{6.16}$$

is called the *residual vector*. Thus Y can be written as

$$Y = \hat{Y} + \left(Y - \hat{Y}\right).$$

Note that $\hat{Y}$ is orthogonal to $Y - \hat{Y}$ since

$$\begin{aligned}\hat{Y}'\left(Y - \hat{Y}\right) &= Y'X\left(X'X\right)^{-1}X'\left[I_n - X\left(X'X\right)^{-1}X'\right]Y \\ &= \mathbf{0}.\end{aligned}$$

We also note from (6.15) that $\hat{Y}$ belongs to the column space of X, which is the linear span of the columns of X (see Definition 2.3). Let us denote this linear span by $\mathcal{C}(X)$. Furthermore, the mean response vector, namely $\mu = X\beta$, also belongs to $\mathcal{C}(X)$. It is easy to see that $Y - \hat{Y}$ is orthogonal to all vectors in $\mathcal{C}(X)$. The vector $\hat{Y}$ can therefore be regarded as the orthogonal projection of Y on $\mathcal{C}(X)$ through the matrix $X(X'X)^{-1}X'$ in (6.15).

The square of the Euclidean norm of $\hat{Y}$, namely $\|X\hat{\beta}\|^2$, is called the *regression sum of squares* and is denoted by SS_{Reg}. We thus have

$$\begin{aligned}SS_{Reg} &= \| X\hat{\beta} \|^2 \\ &= \hat{\beta}'X'X\hat{\beta} \\ &= Y'X\left(X'X\right)^{-1}X'Y. \end{aligned} \tag{6.17}$$

This sum of squares has p degrees of freedom since the matrix $X(X'X)^{-1}X'$, which is idempotent, is of rank p. Note that SS_{Reg} and SS_E [see (6.13)] provide a partitioning of $\| Y \|^2 = Y'Y$ since

$$Y'Y = Y'X\left(X'X\right)^{-1}X'Y + Y'\left[I_n - X\left(X'X\right)^{-1}X'\right]Y. \tag{6.18}$$

Such a partitioning is usually displayed in an *analysis of variance* (ANOVA) table of the form (Table 6.1).

A historical note

The method of least squares was first published in 1805 by the French mathematician Adrien-Marie Legendre (1752–1833). However, in 1809, Carl Friedrich Gauss (1777–1855) claimed that he had been using the method since 1795, but ended up publishing it in 1809. More details concerning the origin of this method were given in Section 1.1.

TABLE 6.1
An ANOVA Table for a Regression Model

Source	DF	SS	MS
Regression	p	SS_{Reg}	SS_{Reg}/p
Error	$n-p$	SS_E	$SS_E/(n-p)$
Total	n	$\boldsymbol{Y}'\boldsymbol{Y}$	

6.2 Properties of Ordinary Least-Squares Estimation

Let us again consider model (6.5) under the assumption that $E(\boldsymbol{\epsilon}) = \boldsymbol{0}$ and $\text{Var}(\boldsymbol{\epsilon}) = \sigma^2 \boldsymbol{I}_n$. Thus, $E(\boldsymbol{Y}) = \boldsymbol{X\beta}$ and $\text{Var}(\boldsymbol{Y}) = \sigma^2 \boldsymbol{I}_n$. The least-squares estimator of $\boldsymbol{\beta}$ is given by $\hat{\boldsymbol{\beta}}$ as in (6.9).

A number of results and properties associated with $\hat{\boldsymbol{\beta}}$ will be discussed in this section. Some of these results are easy to show, others are derived in more detail.

6.2.1 Distributional Properties

(a) $E(\hat{\boldsymbol{\beta}}) = \boldsymbol{\beta}$, that is, $\hat{\boldsymbol{\beta}}$ is an unbiased estimator of $\boldsymbol{\beta}$.

This is true since

$$\begin{aligned} E(\hat{\boldsymbol{\beta}}) &= (\boldsymbol{X}'\boldsymbol{X})^{-1}\boldsymbol{X}'E(\boldsymbol{Y}) \\ &= (\boldsymbol{X}'\boldsymbol{X})^{-1}\boldsymbol{X}'\boldsymbol{X\beta} \\ &= \boldsymbol{\beta}. \end{aligned}$$

(b) $\text{Var}(\hat{\boldsymbol{\beta}}) = \sigma^2\,(\boldsymbol{X}'\boldsymbol{X})^{-1}$.

This follows from the fact that

$$\begin{aligned} \text{Var}\left(\hat{\boldsymbol{\beta}}\right) &= \left(\boldsymbol{X}'\boldsymbol{X}\right)^{-1}\boldsymbol{X}'\text{Var}(\boldsymbol{Y})\boldsymbol{X}(\boldsymbol{X}'\boldsymbol{X})^{-1} \\ &= \left(\boldsymbol{X}'\boldsymbol{X}\right)^{-1}\boldsymbol{X}'(\sigma^2\,\boldsymbol{I}_n)\boldsymbol{X}(\boldsymbol{X}'\boldsymbol{X})^{-1} \\ &= \sigma^2\left(\boldsymbol{X}'\boldsymbol{X}\right)^{-1}. \end{aligned}$$

(c) $E(MS_E) = \sigma^2$, where MS_E is the error (residual) mean square defined by

$$MS_E = \frac{SS_E}{n-p}, \tag{6.19}$$

and SS_E is the error (residual) sum of squares in (6.13), which has $n-p$ degrees of freedom. This result follows directly from applying

Theorem 5.2 to SS_E:

$$\begin{aligned}
E(MS_E) &= \frac{1}{n-p}\,E(SS_E) \\
&= \frac{1}{n-p}\left\{\beta'X'\left[I_n - X\left(X'X\right)^{-1}X'\right]X\beta \right.\\
&\quad \left. +\sigma^2\, tr\left[I_n - X\left(X'X\right)^{-1}X'\right]\right\} \\
&= \frac{1}{n-p}\left[\sigma^2\,(n-p)\right] \\
&= \sigma^2.
\end{aligned}$$

Hence, MS_E is an unbiased estimator of σ^2. We denote such an estimator by $\hat{\sigma}^2$.

(d) $E(MS_{Reg}) = \frac{1}{p}\beta' X' X\beta + \sigma^2$, where MS_{Reg} is the regression mean square defined by

$$MS_{Reg} = \frac{1}{p}\,SS_{Reg}, \tag{6.20}$$

and SS_{Reg} is the regression sum of squares given in (6.17), which has p degrees of freedom. This result also follows from applying Theorem 5.2 to SS_{Reg}:

$$\begin{aligned}
E(MS_{Reg}) &= \frac{1}{p}\,E(SS_{Reg}) \\
&= \frac{1}{p}\left\{\beta'X'\left[X\left(X'X\right)^{-1}X'\right]X\beta + \sigma^2\, tr\left[X\left(X'X\right)^{-1}X'\right]\right\} \\
&= \frac{1}{p}\left(\beta'X'X\beta + p\,\sigma^2\right) \\
&= \frac{1}{p}\,\beta'X'X\beta + \sigma^2.
\end{aligned}$$

6.2.1.1 *Properties under the Normality Assumption*

In addition to the aforementioned assumptions concerning ϵ, let ϵ be now assumed to have the normal distribution $N(\mathbf{0}, \sigma^2 I_n)$. This results in a number of added properties concerning $\hat{\beta}$. In particular, we have

(e) $\hat{\beta} \sim N[\beta, \sigma^2(X'X)^{-1}]$.

This follows from applying Corollary 4.1 to $\hat{\beta} = (X'X)^{-1}X'Y$, which is a linear function of Y, and Y is distributed as $N(X\beta, \sigma^2 I_n)$. Hence, by using properties (a) and (b) in Section 6.2.1, we conclude that $\hat{\beta} \sim N[\beta, \sigma^2(X'X)^{-1}]$.

The next four results give additional properties concerning the distributions of SS_E and SS_{Reg}, their independence, and the independence of $\hat{\beta}$ from MS_E.

(f) $\frac{1}{\sigma^2} SS_E \sim \chi^2_{n-p}$.

This can be shown by applying Theorem 5.4 to SS_E in (6.13) and noting that $\frac{1}{\sigma^2}[I_n - X(X'X)^{-1}X'](\sigma^2 I_n)$ is idempotent of rank $n-p$. Furthermore, the corresponding noncentrality parameter is zero since

$$\beta' X' \left[I_n - X\left(X'X\right)^{-1} X'\right] X\beta = 0.$$

(g) $\frac{1}{\sigma^2} SS_{Reg} \sim \chi^2_p(\theta)$, where $\theta = \frac{1}{\sigma^2}\beta' X' X\beta$.

Here, application of Theorem 5.4 to SS_{Reg} in (6.17) yields the desired result since $\frac{1}{\sigma^2}[X(X'X)^{-1}X'](\sigma^2 I_n)$ is idempotent of rank p, and the noncentrality parameter is

$$\begin{aligned}\theta &= \beta' X' \left[\frac{1}{\sigma^2} X\left(X'X\right)^{-1} X'\right] X\beta \\ &= \frac{1}{\sigma^2}\beta' X' X\beta.\end{aligned}$$

(h) SS_{Reg} and SS_E are independent.

This results from applying Theorem 5.5 to SS_{Reg} and SS_E. More specifically, using condition (5.45), we get

$$\left[X\left(X'X\right)^{-1} X'\right]\left(\sigma^2 I_n\right)\left[I_n - X\left(X'X\right)^{-1} X'\right] = \mathbf{0}.$$

(i) $\hat{\beta}$ and MS_E are independent.

The proof of this result is based on an application of Theorem 5.6 to the linear form, $\hat{\beta} = (X'X)^{-1}X'Y$, and the quadratic form, $MS_E = \frac{1}{n-p} Y'[I_n - X(X'X)^{-1}X']Y$. In this case, using condition (5.59), we obtain

$$\left(X'X\right)^{-1} X'\left(\sigma^2 I_n\right)\left\{\frac{1}{n-p}\left[I_n - X\left(X'X\right)^{-1} X'\right]\right\} = \mathbf{0}.$$

6.2.2 The Gauss–Markov Theorem

This well-known theorem gives an optimal property concerning least-squares estimation.

Theorem 6.1 Let $c'\beta$ be a linear function of β, where c is a given nonzero constant vector. If $E(\epsilon) = \mathbf{0}$ and $\text{Var}(\epsilon) = \sigma^2 I_n$, where ϵ is the experimental error vector in (6.5), then $c'\hat{\beta} = c'(X'X)^{-1}X'Y$ is the *best linear unbiased estimator* (BLUE) of $c'\beta$. By best, it is meant that $c'\hat{\beta}$ has the smallest variance among all linear unbiased estimators of $c'\beta$.

Proof. It is clear that $c'\hat{\beta}$ is a linear function of Y and that it is unbiased for $c'\beta$ [see property 6.2.1(a)]. Let us now show that $c'\hat{\beta}$ has the smallest variance among all linear unbiased estimators of $c'\beta$.

Let $\lambda'Y$ be any linear unbiased estimator of $c'\beta$, that is, $E(\lambda'Y) = c'\beta$. This implies that

$$\lambda'X\beta = c'\beta \tag{6.21}$$

for all β in R^p, the p-dimensional Euclidean space. It follows that

$$\lambda'X = c'. \tag{6.22}$$

Now, the variance of $\lambda'Y$ is

$$\text{Var}(\lambda'Y) = \lambda'\lambda\,\sigma^2. \tag{6.23}$$

Furthermore, from property 6.2.1(b), we have

$$\text{Var}\left(c'\hat{\beta}\right) = \sigma^2 c'\,(X'X)^{-1}\,c. \tag{6.24}$$

Using (6.22) in (6.24), we get

$$\text{Var}\left(c'\hat{\beta}\right) = \sigma^2\,\lambda'X\,(X'X)^{-1}\,X'\lambda. \tag{6.25}$$

From (6.23) and (6.25) we can then write

$$\begin{aligned}\text{Var}(\lambda'Y) - \text{Var}\left(c'\hat{\beta}\right) &= \sigma^2\lambda'\left[I_n - X\,(X'X)^{-1}\,X'\right]\lambda \\ &\geq 0. \end{aligned} \tag{6.26}$$

The inequality in (6.26) follows from the fact that the matrix $I_n - X(X'X)^{-1}X'$ is idempotent, hence positive semidefinite. This shows that

$$\text{Var}\left(c'\hat{\beta}\right) \leq \text{Var}(\lambda'Y). \tag{6.27}$$

Equality in (6.27) is achieved if and only if $c'\hat{\beta} = \lambda'Y$. This follows from (6.26) and noting that $\text{Var}(\lambda'Y) = \text{Var}(c'\hat{\beta})$ if and only if

$$\left[I_n - X\,(X'X)^{-1}\,X'\right]\lambda = \mathbf{0},$$

or, equivalently, if and only if

$$\lambda' = \lambda'X\,(X'X)^{-1}\,X'.$$

Hence,

$$\begin{aligned}\lambda'Y &= \lambda'X\,(X'X)^{-1}\,X'Y \\ &= c'\hat{\beta}. \end{aligned}$$

□

Theorem 6.1 can be generalized so that it applies to a vector linear function of $\boldsymbol{\beta}$, $\boldsymbol{C\beta}$, where $\boldsymbol{C}$ is a given constant matrix of order $q \times p$ and rank q ($\leq p$). This is given by the following corollary:

Corollary 6.1 Let $\boldsymbol{C}$ be a given matrix of order $q \times p$ and rank q ($\leq p$). Then, $\boldsymbol{C}\hat{\boldsymbol{\beta}}$ is the best linear unbiased estimator (BLUE) of $\boldsymbol{C\beta}$. By best, it is meant that the matrix,

$$\boldsymbol{B} = \text{Var}(\boldsymbol{\Lambda Y}) - \text{Var}\left(\boldsymbol{C}\hat{\boldsymbol{\beta}}\right)$$

is positive semidefinite, where $\boldsymbol{\Lambda Y}$ is any vector linear function of $\boldsymbol{Y}$ that is unbiased for $\boldsymbol{C\beta}$.

Proof. The unbiasedness of $\boldsymbol{\Lambda Y}$ for $\boldsymbol{C\beta}$ implies that $\boldsymbol{\Lambda X} = \boldsymbol{C}$. The $q \times n$ matrix $\boldsymbol{\Lambda}$ must therefore be of rank q since $\boldsymbol{C}$ is of rank q. Now, the matrix $\boldsymbol{B}$ is positive semidefinite if and only if

$$\boldsymbol{t}'\boldsymbol{Bt} \geq 0, \tag{6.28}$$

for all $\boldsymbol{t} \in R^q$ with $\boldsymbol{t}'\boldsymbol{Bt} = 0$ for some $\boldsymbol{t} \neq \boldsymbol{0}$. But,

$$\begin{aligned} \boldsymbol{t}'\boldsymbol{Bt} &= \text{Var}\left(\boldsymbol{t}'\boldsymbol{\Lambda Y}\right) - \text{Var}\left(\boldsymbol{t}'\boldsymbol{C}\hat{\boldsymbol{\beta}}\right) \\ &= \text{Var}\left(\boldsymbol{\lambda}_t'\boldsymbol{Y}\right) - \text{Var}\left(\boldsymbol{c}_t'\hat{\boldsymbol{\beta}}\right), \end{aligned} \tag{6.29}$$

where $\boldsymbol{\lambda}_t' = \boldsymbol{t}'\boldsymbol{\Lambda}$ and $\boldsymbol{c}_t' = \boldsymbol{t}'\boldsymbol{C}$. Note that $\boldsymbol{\lambda}_t'\boldsymbol{Y}$ is a linear function of $\boldsymbol{Y}$, which is unbiased for $\boldsymbol{c}_t'\boldsymbol{\beta}$. The latter assertion is true because

$$\begin{aligned} E(\boldsymbol{\lambda}_t'\boldsymbol{Y}) &= \boldsymbol{\lambda}_t'\boldsymbol{X\beta} \\ &= \boldsymbol{t}'\boldsymbol{\Lambda X\beta} \\ &= \boldsymbol{t}'\boldsymbol{C\beta} \\ &= \boldsymbol{c}_t'\boldsymbol{\beta}. \end{aligned} \tag{6.30}$$

From Theorem 6.1 we conclude that $\text{Var}(\boldsymbol{\lambda}_t'\boldsymbol{Y}) \geq \text{Var}(\boldsymbol{c}_t'\hat{\boldsymbol{\beta}})$ for all $\boldsymbol{t} \in R^q$, which implies (6.28). It remains to show that $\boldsymbol{t}'\boldsymbol{Bt} = 0$ for some $\boldsymbol{t} \neq \boldsymbol{0}$.

Suppose that $\boldsymbol{t}'\boldsymbol{Bt} = 0$. Then, from (6.29) we have

$$\sigma^2\left[\boldsymbol{\lambda}_t'\boldsymbol{\lambda}_t - \boldsymbol{c}_t'\left(\boldsymbol{X}'\boldsymbol{X}\right)^{-1}\boldsymbol{c}_t\right] = 0. \tag{6.31}$$

Since $\boldsymbol{\lambda}_t' = \boldsymbol{t}'\boldsymbol{\Lambda}$, $\boldsymbol{c}_t' = \boldsymbol{t}'\boldsymbol{C} = \boldsymbol{t}'\boldsymbol{\Lambda X}$, (6.31) can be written as

$$\sigma^2\boldsymbol{t}'\boldsymbol{\Lambda}\left[\boldsymbol{I}_n - \boldsymbol{X}\left(\boldsymbol{X}'\boldsymbol{X}\right)^{-1}\boldsymbol{X}'\right]\boldsymbol{\Lambda}'\boldsymbol{t} = 0. \tag{6.32}$$

From (6.32) we conclude that there is some nonzero vector $\boldsymbol{u} = \boldsymbol{\Lambda}'\boldsymbol{t}$ for which the equality in (6.32) is true, since the matrix $\boldsymbol{I}_n - \boldsymbol{X}(\boldsymbol{X}'\boldsymbol{X})^{-1}\boldsymbol{X}'$ is positive semidefinite. For such a vector, $\boldsymbol{t} = (\boldsymbol{\Lambda\Lambda}')^{-1}\boldsymbol{\Lambda u} \neq \boldsymbol{0}$, which implies that $\boldsymbol{t}'\boldsymbol{Bt} = 0$ for some nonzero $\boldsymbol{t}$. □

A special case of Corollary 6.1 when $\boldsymbol{C} = \boldsymbol{I}_p$ shows that $\hat{\boldsymbol{\beta}}$ is the BLUE of $\boldsymbol{\beta}$.

6.3 Generalized Least-Squares Estimation

In this section, we discuss again the estimation of β in model (6.5), but under a more general setup concerning the variance–covariance matrix of ϵ. Here, we consider that $\text{Var}(\epsilon) = \sigma^2 V$, where V is a known positive definite matrix. Estimation of β in this case can be easily reduced to the case discussed in Section 6.1. Multiplying both sides of (6.5) on the left by $V^{-1/2}$, we get

$$Y_v = X_v\beta + \epsilon_v, \tag{6.33}$$

where $Y_v = V^{-1/2}Y$, $X_v = V^{-1/2}X$, and $\epsilon_v = V^{-1/2}\epsilon$. Note that X_v is of full column rank and that $E(\epsilon_v) = \mathbf{0}$, $\text{Var}(\epsilon_v) = V^{-1/2}(\sigma^2 V)V^{-1/2} = \sigma^2 I_n$. The OLS estimator of β in model (6.33) is therefore given by

$$\begin{aligned}\hat{\beta}_v &= \left(X_v'X_v\right)^{-1} X_v'Y_v \\ &= \left(X'V^{-1}X\right)^{-1} X'V^{-1}Y. \end{aligned} \tag{6.34}$$

We call $\hat{\beta}_v$ the *generalized least-squares estimator* (GLSE) of β for model (6.5). This estimator is unbiased for β and its variance–covariance matrix is

$$\begin{aligned}\text{Var}\left(\hat{\beta}_v\right) &= \left(X'V^{-1}X\right)^{-1} X'V^{-1}\left(\sigma^2 V\right)V^{-1}X\left(X'V^{-1}X\right)^{-1} \\ &= \sigma^2\left(X'V^{-1}X\right)^{-1}. \end{aligned} \tag{6.35}$$

Applying the Gauss–Markov Theorem (Theorem 6.1) to model (6.33) we conclude that $c'\hat{\beta}_v = c'(X'V^{-1}X)^{-1}X'V^{-1}Y$ is the BLUE of $c'\beta$, where c is a given nonzero constant vector. In addition, using Corollary 6.1, if $C\beta$ is a vector linear function of β, where C is a given matrix of order $q \times p$ and rank q $(\leq p)$, then $C\hat{\beta}_v$ is the BLUE of $C\beta$. In particular, $\hat{\beta}_v$ is the BLUE of β.

6.4 Least-Squares Estimation under Linear Restrictions on β

The parameter vector β in model (6.5) may be subject to some linear restrictions of the form

$$A\beta = m, \tag{6.36}$$

where

A is a known matrix of order $r \times p$ and rank r $(\leq p)$

m is a known vector

For example, in a simple linear regression model, $Y = \beta_0 + \beta_1 x + \epsilon$, estimation of the slope β_1 may be needed when $\beta_0 = 0$, that is, for a model with a zero Y-intercept. Also, in a multiple linear regression model, such as $Y = \beta_0 + \beta_1 x_1 + \beta_2 x_2 + \epsilon$, the mean response, $\mu(x) = \beta_0 + \beta_1 x_1 + \beta_2 x_2$, may be set equal to zero when $x_1 = 1,\ x_2 = 3$. In this case, $\beta_0,\ \beta_1,\ \beta_2$ are subject to the linear restriction, $\beta_0 + \beta_1 + 3\beta_2 = 0$.

In this section, we consider least-squares estimation of $\boldsymbol{\beta}$ when $\boldsymbol{\beta}$ is subject to linear restrictions of the form given in (6.36). We make the same assumptions on $\boldsymbol{\epsilon}$ as in Section 6.1, namely, $E(\boldsymbol{\epsilon}) = \mathbf{0}$ and $\text{Var}(\boldsymbol{\epsilon}) = \sigma^2 \boldsymbol{I}_n$.

This particular type of estimation can be derived by minimizing $S(\boldsymbol{\beta})$ in (6.6) under the equality restrictions (6.36). A convenient way to do this is to use the method of *Lagrange multipliers* (see, for example, Khuri, 2003, Section 7.8). Consider the function,

$$T(\boldsymbol{\beta}, \boldsymbol{\kappa}) = S(\boldsymbol{\beta}) + \boldsymbol{\kappa}'(\boldsymbol{A}\boldsymbol{\beta} - \boldsymbol{m}), \tag{6.37}$$

where $\boldsymbol{\kappa}$ is a vector of r Lagrange multipliers. Differentiating $T(\boldsymbol{\beta}, \boldsymbol{\kappa})$ with respect to $\boldsymbol{\beta}$ and equating the derivative to zero, we obtain

$$-2\boldsymbol{X}'\boldsymbol{Y} + 2\boldsymbol{X}'\boldsymbol{X}\boldsymbol{\beta} + \boldsymbol{A}'\boldsymbol{\kappa} = \mathbf{0}. \tag{6.38}$$

Solving (6.38) with respect to $\boldsymbol{\beta}$ and denoting this solution by $\hat{\boldsymbol{\beta}}_r$, we get

$$\hat{\boldsymbol{\beta}}_r = (\boldsymbol{X}'\boldsymbol{X})^{-1}\left[\boldsymbol{X}'\boldsymbol{Y} - \frac{1}{2}\boldsymbol{A}'\boldsymbol{\kappa}\right]. \tag{6.39}$$

Substituting $\hat{\boldsymbol{\beta}}_r$ for $\boldsymbol{\beta}$ in (6.36), we obtain

$$\boldsymbol{A}\,(\boldsymbol{X}'\boldsymbol{X})^{-1}\left[\boldsymbol{X}'\boldsymbol{Y} - \frac{1}{2}\boldsymbol{A}'\boldsymbol{\kappa}\right] = \boldsymbol{m}.$$

Solving this equation for $\boldsymbol{\kappa}$, we get

$$\boldsymbol{\kappa} = 2\left[\boldsymbol{A}\,(\boldsymbol{X}'\boldsymbol{X})^{-1}\boldsymbol{A}'\right]^{-1}\left[\boldsymbol{A}\,(\boldsymbol{X}'\boldsymbol{X})^{-1}\boldsymbol{X}'\boldsymbol{Y} - \boldsymbol{m}\right]. \tag{6.40}$$

From (6.39) and (6.40) we then have the solution

$$\begin{aligned}\hat{\boldsymbol{\beta}}_r &= (\boldsymbol{X}'\boldsymbol{X})^{-1}\boldsymbol{X}'\boldsymbol{Y} - (\boldsymbol{X}'\boldsymbol{X})^{-1}\boldsymbol{A}'\left[\boldsymbol{A}\,(\boldsymbol{X}'\boldsymbol{X})^{-1}\boldsymbol{A}'\right]^{-1}\left[\boldsymbol{A}\,(\boldsymbol{X}'\boldsymbol{X})^{-1}\boldsymbol{X}'\boldsymbol{Y} - \boldsymbol{m}\right]\\ &= \hat{\boldsymbol{\beta}} - (\boldsymbol{X}'\boldsymbol{X})^{-1}\boldsymbol{A}'\left[\boldsymbol{A}\,(\boldsymbol{X}'\boldsymbol{X})^{-1}\boldsymbol{A}'\right]^{-1}\left(\boldsymbol{A}\hat{\boldsymbol{\beta}} - \boldsymbol{m}\right),\end{aligned} \tag{6.41}$$

where $\hat{\boldsymbol{\beta}}$ is the OLS estimator given in (6.9). This solution is called the *restricted least-squares estimator* of $\boldsymbol{\beta}$. It is easy to see that $\hat{\boldsymbol{\beta}}_r$ satisfies the equality restrictions in (6.36). Furthermore, $S(\beta)$ attains its minimum value, $S(\hat{\beta}_r)$, over the parameter space constrained by (6.36) when $\beta = \hat{\beta}_r$. It can also be verified that

$$\begin{aligned} S\left(\hat{\beta}_r\right) &= \left(Y - X\hat{\beta}\right)'\left(Y - X\hat{\beta}\right) + \left(\hat{\beta} - \hat{\beta}_r\right)' X'X \left(\hat{\beta} - \hat{\beta}_r\right) \\ &= SS_E + \left(\hat{\beta} - \hat{\beta}_r\right)' X'X \left(\hat{\beta} - \hat{\beta}_r\right). \end{aligned} \tag{6.42}$$

Hence, $S(\hat{\beta}_r) > SS_E$ since equality is attained if and only if $\hat{\beta} = \hat{\beta}_r$, which is not possible. Using (6.41), formula (6.42) can also be written as

$$S\left(\hat{\beta}_r\right) = SS_E + \left(A\hat{\beta} - m\right)'\left[A\left(X'X\right)^{-1}A'\right]^{-1}\left(A\hat{\beta} - m\right).$$

The assertion that $S(\beta)$ attains its minimum value at $\hat{\beta}_r$ in the constrained parameter space can be verified as follows: Let β be any vector in the constrained parameter space. Then, $A\beta = m = A\hat{\beta}_r$. Hence, $A(\hat{\beta}_r - \beta) = \mathbf{0}$. Therefore,

$$\begin{aligned} S(\beta) &= (Y - X\beta)'(Y - X\beta) \\ &= \left(Y - X\hat{\beta}_r + X\hat{\beta}_r - X\beta\right)'\left(Y - X\hat{\beta}_r + X\hat{\beta}_r - X\beta\right) \\ &= \left(Y - X\hat{\beta}_r\right)'\left(Y - X\hat{\beta}_r\right) + 2\left(Y - X\hat{\beta}_r\right)'\left(X\hat{\beta}_r - X\beta\right) \\ &\quad + \left(X\hat{\beta}_r - X\beta\right)'\left(X\hat{\beta}_r - X\beta\right). \end{aligned}$$

But, from (6.41) we have

$$\begin{aligned} \left(Y - X\hat{\beta}_r\right)'\left(X\hat{\beta}_r - X\beta\right) &= \left\{Y - X\hat{\beta} + X\left(X'X\right)^{-1}A'\left[A\left(X'X\right)^{-1}A'\right]^{-1}\right. \\ &\quad \times \left.\left(A\hat{\beta} - m\right)\right\}' X\left(\hat{\beta}_r - \beta\right) \\ &= \left(A\hat{\beta} - m\right)'\left[A\left(X'X\right)^{-1}A'\right]^{-1} A\left(X'X\right)^{-1}X'X \\ &\quad \times \left(\hat{\beta}_r - \beta\right) \\ &= \mathbf{0}, \quad \text{since } A\left(\hat{\beta}_r - \beta\right) = \mathbf{0}. \end{aligned}$$

It follows that

$$S(\beta) = S\left(\hat{\beta}_r\right) + \left(\hat{\beta}_r - \beta\right)' X'X \left(\hat{\beta}_r - \beta\right).$$

Hence, $S(\beta) \geq S(\hat{\beta}_r)$, and equality is attained if and only if $\beta = \hat{\beta}_r$.

6.5 Maximum Likelihood Estimation

This type of estimation requires that particular assumptions be made concerning the distribution of the error term in model (6.5). In this section, we assume that $\boldsymbol{\epsilon}$ is normally distributed as $N(\mathbf{0}, \sigma^2 \boldsymbol{I}_n)$. Hence, $\boldsymbol{Y} \sim N(\boldsymbol{X}\boldsymbol{\beta}, \sigma^2 \boldsymbol{I}_n)$. Then, the corresponding likelihood function, which, for a given value, $\boldsymbol{y}$, of $\boldsymbol{Y}$, is the same as the density function of $\boldsymbol{Y}$, but is treated as a function of $\boldsymbol{\beta}$ and σ^2 and is therefore given by

$$\mathcal{L}(\boldsymbol{\beta}, \sigma^2, \boldsymbol{y}) = \frac{1}{(2\pi\sigma^2)^{n/2}} \exp\left[-\frac{1}{2\sigma^2}(\boldsymbol{y} - \boldsymbol{X}\boldsymbol{\beta})'(\boldsymbol{y} - \boldsymbol{X}\boldsymbol{\beta})\right]. \tag{6.43}$$

By definition, the *maximum likelihood estimates* (MLE) of $\boldsymbol{\beta}$ and σ^2 are those values of $\boldsymbol{\beta}$ and σ^2 that maximize the likelihood function, or equivalently, the natural logarithm of $\mathcal{L}$, namely

$$l\left(\boldsymbol{\beta}, \sigma^2, \boldsymbol{y}\right) = -\frac{n}{2}\log(2\pi) - \frac{n}{2}\log\left(\sigma^2\right) - \frac{1}{2\sigma^2}(\boldsymbol{y} - \boldsymbol{X}\boldsymbol{\beta})'(\boldsymbol{y} - \boldsymbol{X}\boldsymbol{\beta}).$$

To find the MLE of $\boldsymbol{\beta}$ and σ^2, we proceed as follows: We first find the stationary values of $\boldsymbol{\beta}$ and σ^2 for which the partial derivatives of $l(\boldsymbol{\beta}, \sigma^2, \boldsymbol{y})$ with respect to $\boldsymbol{\beta}$ and σ^2 are equal to zero. The next step is to verify that these values maximize $l(\boldsymbol{\beta}, \sigma^2, \boldsymbol{y})$. Setting the partial derivatives of $l(\boldsymbol{\beta}, \sigma^2, \boldsymbol{y})$ with respect to $\boldsymbol{\beta}$ and σ^2 to zero, we get

$$\begin{aligned}\frac{\partial l\left(\boldsymbol{\beta}, \sigma^2, \boldsymbol{y}\right)}{\partial \boldsymbol{\beta}} &= -\frac{1}{2\sigma^2}\left(-2\boldsymbol{X}'\boldsymbol{y} + 2\boldsymbol{X}'\boldsymbol{X}\boldsymbol{\beta}\right) \\ &= \mathbf{0} \end{aligned} \tag{6.44}$$

$$\begin{aligned}\frac{\partial l\left(\boldsymbol{\beta}, \sigma^2, \boldsymbol{y}\right)}{\partial \sigma^2} &= -\frac{n}{2\sigma^2} + \frac{1}{2\sigma^4}(\boldsymbol{y} - \boldsymbol{X}\boldsymbol{\beta})'(\boldsymbol{y} - \boldsymbol{X}\boldsymbol{\beta}) \\ &= 0. \end{aligned} \tag{6.45}$$

Let $\tilde{\boldsymbol{\beta}}$ and $\tilde{\sigma}^2$ denote the solution of equations 6.44 and 6.45 for $\boldsymbol{\beta}$ and σ^2, respectively. From (6.44) we find that

$$\tilde{\boldsymbol{\beta}} = (\boldsymbol{X}'\boldsymbol{X})^{-1}\boldsymbol{X}'\boldsymbol{Y}, \tag{6.46}$$

which is the same as $\hat{\boldsymbol{\beta}}$, the OLS estimator of $\boldsymbol{\beta}$ in (6.9). Note that $\boldsymbol{Y}$ was used in place of $\boldsymbol{y}$ in (6.46) since the latter originated from the likelihood function in (6.43) where it was treated as a mathematical variable. In formula (6.46), however, $\boldsymbol{Y}$ is treated as a random vector since it is data dependent. From (6.45) we get

$$\tilde{\sigma}^2 = \frac{1}{n}\left(\boldsymbol{Y} - \boldsymbol{X}\tilde{\boldsymbol{\beta}}\right)'\left(\boldsymbol{Y} - \boldsymbol{X}\tilde{\boldsymbol{\beta}}\right). \tag{6.47}$$

We recall that $\hat{\sigma}^2 = MS_E$ is an unbiased estimator of σ^2 [see property 6.2.1(c)]. The estimator $\tilde{\sigma}^2$, however, is not unbiased for σ^2.

Let us now verify that $\tilde{\beta}$ and $\tilde{\sigma}^2$ are indeed maximal values for $l(\beta, \sigma^2, y)$ with respect to β and σ^2. For this purpose, we consider the matrix of second-order partial derivatives of $l(\beta, \sigma^2, y)$ with respect to β and σ^2. This is the *Hessian matrix* and is given by

$$M = \begin{bmatrix} \frac{\partial}{\partial \beta'}\left[\frac{\partial l(\beta, \sigma^2, y)}{\partial \beta}\right] & \frac{\partial}{\partial \sigma^2}\left[\frac{\partial l(\beta, \sigma^2, y)}{\partial \beta}\right] \\ \frac{\partial}{\partial \beta'}\left[\frac{\partial l(\beta, \sigma^2, y)}{\partial \sigma^2}\right] & \frac{\partial}{\partial \sigma^2}\left[\frac{\partial l(\beta, \sigma^2, y)}{\partial \sigma^2}\right] \end{bmatrix}.$$

Evaluating the matrix M at $\beta = \tilde{\beta}$, and $\sigma^2 = \tilde{\sigma}^2$, we get

$$M = \begin{bmatrix} -\frac{1}{\tilde{\sigma}^2} X'X & \frac{1}{2\tilde{\sigma}^4}\left(-2X'y + 2X'X\tilde{\beta}\right) \\ \frac{1}{2\tilde{\sigma}^4}\left(-2y'X + 2\tilde{\beta}'X'X\right) & \frac{n}{2\tilde{\sigma}^4} - \frac{1}{\tilde{\sigma}^6}\left(y - X\tilde{\beta}\right)'\left(y - X\tilde{\beta}\right) \end{bmatrix}. \quad (6.48)$$

Making use of (6.44) and (6.47) in (6.48), we obtain

$$M = \begin{bmatrix} -\frac{1}{\tilde{\sigma}^2} X'X & \mathbf{0} \\ \mathbf{0}' & -\frac{n}{2\tilde{\sigma}^4} \end{bmatrix}, \quad (6.49)$$

which is clearly negative definite. Hence, $l(\beta, \sigma^2, y)$ has a local maximum at $\beta = \tilde{\beta}$, $\sigma^2 = \tilde{\sigma}^2$ (see, for example, Khuri, 2003, Corollary 7.7.1). Since this is the only local maximum, it must also be the absolute maximum. Hence, $\tilde{\beta}$ and $\tilde{\sigma}^2$ are the MLE of β and σ^2, respectively. The maximum value of $\mathcal{L}(\beta, \sigma^2, y)$ in (6.43) is

$$\max_{\beta, \sigma^2} \mathcal{L}\left(\beta, \sigma^2, y\right) = \frac{1}{\left(2\pi\tilde{\sigma}^2\right)^{n/2}} e^{-n/2}. \quad (6.50)$$

6.5.1 Properties of Maximum Likelihood Estimators

We have that $\tilde{\beta} = \hat{\beta}$ and $\tilde{\sigma}^2 = \frac{SS_E}{n} = \frac{n-p}{n} MS_E$. On the basis of the properties given in Section 6.2.1.1, $\tilde{\beta} \sim N(\beta, \sigma^2(X'X)^{-1})$, $\tilde{\sigma}^2 \sim \frac{\sigma^2}{n}\chi^2_{n-p}$, and $\tilde{\beta}$ and $\tilde{\sigma}^2$ are independent. In addition, $\tilde{\beta}$ and $\tilde{\sigma}^2$ have two more properties, namely, *sufficiency* and *completeness*. In order to understand these properties, the following definitions are needed:

Definition 6.1 Let Y be a random vector whose distribution depends on a parameter vector, $\phi \in \Omega$, where Ω is some parameter space. A statistic $U(Y)$ is a sufficient statistic for ϕ if the conditional distribution of Y given the value of $U(Y)$ does not depend on ϕ.

In practice, the determination of sufficiency is more easily accomplished by using the following well-known theorem in statistical inference, namely, the *Factorization Theorem* (see, for example, Casella and Berger, 2002, Theorem 6.2.6).

Theorem 6.2 (Factorization Theorem) Let $g(\boldsymbol{y}, \boldsymbol{\phi})$ denote the density function of the random vector $\boldsymbol{Y}$. A statistic $\boldsymbol{U}(\boldsymbol{Y})$ is a sufficient statistic for $\boldsymbol{\phi} \in \Omega$ if and only if $g(\boldsymbol{y}, \boldsymbol{\phi})$ can be written as

$$g(\boldsymbol{y}, \boldsymbol{\phi}) = g_1(\boldsymbol{y})\, g_2(\boldsymbol{u}(\boldsymbol{y}), \boldsymbol{\phi})$$

for all $\boldsymbol{\phi} \in \Omega$, where g_1 is a function of $\boldsymbol{y}$ only and g_2 is a function of $\boldsymbol{u}(\boldsymbol{y})$ and $\boldsymbol{\phi}$.

Definition 6.2 Let $\boldsymbol{Y}$ be a random vector with the density function $g(\boldsymbol{y}, \boldsymbol{\phi})$, which depends on a parameter vector, $\boldsymbol{\phi} \in \Omega$. Let $\mathcal{F}$ denote the family of distributions $\{g(\boldsymbol{y}, \boldsymbol{\phi}), \boldsymbol{\phi} \in \Omega\}$. This family is said to be *complete* if for every function $h(\boldsymbol{Y})$ for which

$$E[h(\boldsymbol{Y})] = 0, \quad \text{for all } \boldsymbol{\phi} \in \Omega,$$

then $h(\boldsymbol{Y}) = 0$ with probability equal to 1 for all $\boldsymbol{\phi} \in \Omega$.

Note that completeness is a property of a family of distributions, not of a particular distribution. For example, let $Y \sim N(\mu, 1)$, then

$$g(y, \mu) = \frac{1}{\sqrt{2\pi}} \exp\left[-\frac{1}{2}(y - \mu)^2\right], \quad -\infty < y < \infty, \tag{6.51}$$

where $-\infty < \mu < \infty$. Let $h(Y)$ be a function such that $E[h(Y)] = 0$ for all μ. Then,

$$\frac{1}{\sqrt{2\pi}} \int_{-\infty}^{\infty} h(y) \exp\left[-\frac{1}{2}(y - \mu)^2\right] dy = 0, \quad -\infty < \mu < \infty,$$

which implies that

$$\int_{-\infty}^{\infty} h(y) e^{-y^2/2} e^{y\mu} dy = 0, \; -\infty < \mu < \infty. \tag{6.52}$$

The left-hand side of (6.52) is the two-sided *Laplace transformation* of the function $h(y)e^{-y^2/2}$ [see Chapter III in Zemanian (1987)]. Since the two-sided Laplace transformation of the zero function is also equal to zero, we conclude that

$$h(y)e^{-y^2/2} = 0, \tag{6.53}$$

and hence $h(y) = 0$ almost everywhere (with respect to *Lebesgue measure*). This assertion is based on the *uniqueness property* of the two-sided Laplace transformation [see Theorem 3.5.2 in Zemanian (1987, p. 69)]. We thus have $P[h(Y) = 0] = 1$, which indicates that the family of normal distributions, $N(\mu, 1)$, is complete.

We can clearly see that we would not have been able to conclude from (6.52) that $h(y) = 0$ if μ had just a fixed value. This explains our earlier remark that completeness is a property of a family of distributions, but not of a particular distribution. For example, if $h(Y) = Y$ and $Y \sim N(0, 1)$, then having $E[h(Y)] = E(Y) = 0$ does not imply that $h(Y) = 0$.

Definition 6.3 Let $\boldsymbol{U}(\boldsymbol{Y})$ be a statistic whose distribution belongs to a family of distributions that is complete. Then, $\boldsymbol{U}(\boldsymbol{Y})$ is said to be a complete statistic.

For example, if $\bar{Y}$ is the sample mean of a sample of size n from a normal distribution $N(\mu, 1)$, $-\infty < \mu < \infty$, then $\bar{Y} \sim N(\mu, \frac{1}{n})$. Since this family of distributions is complete, as was seen earlier, we conclude that $\bar{Y}$ is a complete statistic.

The completeness of the family of normal distributions $N(\mu, 1)$ can be derived as a special case using a more general family of distributions called the *exponential family*.

Definition 6.4 Let $\mathcal{F} = \{g(y, \phi), \phi \in \Omega\}$ be a family of density functions (or probability mass functions) such that

$$g(\boldsymbol{y}, \boldsymbol{\phi}) = \varphi(\boldsymbol{y})\, c(\boldsymbol{\phi}) \exp\left[\sum_{i=1}^{k} \omega_i(\boldsymbol{\phi}) t_i(\boldsymbol{y})\right], \tag{6.54}$$

where $\varphi(\boldsymbol{y}) \geq 0$ and $t_1(\boldsymbol{y}), t_2(\boldsymbol{y}), \ldots, t_k(\boldsymbol{y})$ are real-valued functions of $\boldsymbol{y}$ only, and $c(\boldsymbol{\phi}) \geq 0$ and $\omega_1(\boldsymbol{\phi}), \omega_2(\boldsymbol{\phi}), \ldots, \omega_k(\boldsymbol{\phi})$ are real-valued functions of $\boldsymbol{\phi}$ only. Then, $\mathcal{F}$ is called an exponential family.

Several well-known distributions belong to the exponential family. These include the normal, gamma, and beta distributions, among the continuous distributions; and the binomial, Poisson, and negative binomial, among the discrete distributions. For example, for the family of normal distributions, $N(\mu, \sigma^2)$, we have

$$\begin{aligned} g(y, \boldsymbol{\phi}) &= \frac{1}{\sqrt{2\pi\sigma^2}} \exp\left[-\frac{1}{2\sigma^2}(y-\mu)^2\right], \quad -\infty < \mu < \infty, \sigma > 0, \quad (6.55) \\ &= \frac{1}{\sqrt{2\pi\sigma^2}} \exp\left(-\frac{\mu^2}{2\sigma^2}\right) \exp\left(-\frac{y^2}{2\sigma^2} + \frac{\mu y}{\sigma^2}\right). \end{aligned}$$

Comparing this with the expression in (6.54), we see that $\boldsymbol{\phi} = (\mu, \sigma^2)'$, $\varphi(y) = 1$, $c(\boldsymbol{\phi}) = \frac{1}{\sqrt{2\pi\sigma^2}} \exp\left(-\frac{\mu^2}{2\sigma^2}\right)$, $\omega_1(\boldsymbol{\phi}) = -\frac{1}{2\sigma^2}$, $t_1(y) = y^2$, $\omega_2(\boldsymbol{\phi}) = \frac{\mu}{\sigma^2}$, and $t_2(y) = y$.

The exponential family has several nice properties. One of these properties is given by the following well-known theorem (see, for example, Arnold, 1981, Theorem 1.2, p. 2; Graybill, 1976, Theorem 2.7.8, p. 79; Wasan, 1970, Theorem 2, p. 64):

Theorem 6.3 Consider the exponential family defined in formula (6.54). Let $\boldsymbol{\omega}(\boldsymbol{\phi}) = [\omega_1(\boldsymbol{\phi}), \omega_2(\boldsymbol{\phi}), \ldots, \omega_k(\boldsymbol{\phi})]'$ and $\boldsymbol{t}(\boldsymbol{y}) = [t_1(\boldsymbol{y}), t_2(\boldsymbol{y}), \ldots, t_k(\boldsymbol{y})]'$. Then, $\boldsymbol{t}(\boldsymbol{Y})$ is a complete and sufficient statistic provided that the set $\{\boldsymbol{\omega}(\boldsymbol{\phi}), \boldsymbol{\phi} \in \Omega\}$ contains a nondegenerate k-dimensional rectangle (i.e., has a nonempty interior).

After this series of definitions and theorems, we are now ready to show that the maximum likelihood estimators, $\tilde{\boldsymbol{\beta}}$ and $\tilde{\sigma}^2$, have the properties of sufficiency and completeness.

Theorem 6.4 Let $\boldsymbol{Y} \sim N(\boldsymbol{X}\boldsymbol{\beta}, \sigma^2\boldsymbol{I}_n)$. Then, the maximum likelihood estimators of $\boldsymbol{\beta}$ and σ^2, namely, $\tilde{\boldsymbol{\beta}} = (\boldsymbol{X}'\boldsymbol{X})^{-1}\boldsymbol{X}'\boldsymbol{Y}$, and

$$\tilde{\sigma}^2 = \frac{1}{n}\boldsymbol{Y}'\left[\boldsymbol{I}_n - \boldsymbol{X}\left(\boldsymbol{X}'\boldsymbol{X}\right)^{-1}\boldsymbol{X}'\right]\boldsymbol{Y}, \tag{6.56}$$

are complete and sufficient statistics for $\boldsymbol{\beta}$ and σ^2.

Proof. The density function, $g(\boldsymbol{y}, \boldsymbol{\phi})$, of $\boldsymbol{Y}$ is the same as the likelihood function in (6.43), where $\boldsymbol{\phi} = (\boldsymbol{\beta}', \sigma^2)'$. We can then write

$$\begin{aligned} g(\boldsymbol{y}, \boldsymbol{\phi}) &= \frac{1}{(2\pi\sigma^2)^{n/2}} \exp\left[-\frac{1}{2\sigma^2}(\boldsymbol{y} - \boldsymbol{X}\boldsymbol{\beta})'(\boldsymbol{y} - \boldsymbol{X}\boldsymbol{\beta})\right] \\ &= \frac{1}{(2\pi\sigma^2)^{n/2}} \exp\left\{-\frac{1}{2\sigma^2}\left[\left(\boldsymbol{y} - \boldsymbol{X}\tilde{\boldsymbol{\beta}}\right)'\left(\boldsymbol{y} - \boldsymbol{X}\tilde{\boldsymbol{\beta}}\right)\right.\right. \\ &\quad \left.\left. + \left(\tilde{\boldsymbol{\beta}} - \boldsymbol{\beta}\right)'\boldsymbol{X}'\boldsymbol{X}\left(\tilde{\boldsymbol{\beta}} - \boldsymbol{\beta}\right)\right]\right\} \\ &= \frac{1}{(2\pi\sigma^2)^{n/2}} \exp\left\{-\frac{1}{2\sigma^2}\left[n\tilde{\sigma}^2 + \left(\tilde{\boldsymbol{\beta}} - \boldsymbol{\beta}\right)'\boldsymbol{X}'\boldsymbol{X}\left(\tilde{\boldsymbol{\beta}} - \boldsymbol{\beta}\right)\right]\right\}. \end{aligned} \tag{6.57}$$

We note that the right-hand side of (6.57) is a function of $\tilde{\sigma}^2$, $\tilde{\boldsymbol{\beta}}$, and the elements of $\boldsymbol{\phi}$. Hence, by the Factorization Theorem (Theorem 6.2), the statistic $(\tilde{\boldsymbol{\beta}}', \tilde{\sigma}^2)'$ is sufficient for $\boldsymbol{\phi}$ [the function g_1 in Theorem 6.2, in this case, is identically equal to one, and the function g_2 is equal to the right-hand side of (6.57)].

Now, to show completeness, let us rewrite (6.57) as

$$\begin{aligned} g(\boldsymbol{y}, \boldsymbol{\phi}) &= \frac{1}{(2\pi\sigma^2)^{n/2}} \exp\left(-\frac{1}{2\sigma^2}\boldsymbol{\beta}'\boldsymbol{X}'\boldsymbol{X}\boldsymbol{\beta}\right) \\ &\quad \times \exp\left\{-\frac{1}{2\sigma^2}\left[n\tilde{\sigma}^2 + \tilde{\boldsymbol{\beta}}'\boldsymbol{X}'\boldsymbol{X}\tilde{\boldsymbol{\beta}}\right] + \frac{1}{\sigma^2}\boldsymbol{\beta}'\boldsymbol{X}'\boldsymbol{X}\tilde{\boldsymbol{\beta}}\right\}. \end{aligned} \tag{6.58}$$

By comparing (6.58) with (6.54) we find that $g(\boldsymbol{y}, \boldsymbol{\phi})$ belongs to the exponential family with $k = p + 1$,

$$\begin{aligned}
\varphi(\boldsymbol{y}) &= 1, \\
c(\boldsymbol{\phi}) &= \frac{1}{(2\pi\sigma^2)^{n/2}} \exp\left(-\frac{1}{2\sigma^2}\boldsymbol{\beta}'\boldsymbol{X}'\boldsymbol{X}\boldsymbol{\beta}\right), \\
\omega_1(\boldsymbol{\phi}) &= -\frac{1}{2\sigma^2}, \\
t_1(\boldsymbol{y}) &= n\tilde{\sigma}^2 + \tilde{\boldsymbol{\beta}}'\boldsymbol{X}'\boldsymbol{X}\tilde{\boldsymbol{\beta}}, \qquad (6.59) \\
\boldsymbol{\omega}_2'(\boldsymbol{\phi}) &= \frac{1}{\sigma^2}\boldsymbol{\beta}', \\
\boldsymbol{t}_2(\boldsymbol{y}) &= \boldsymbol{X}'\boldsymbol{X}\tilde{\boldsymbol{\beta}}. \qquad (6.60)
\end{aligned}$$

Furthermore, the set

$$\begin{aligned}
\boldsymbol{\omega}(\boldsymbol{\phi}) &= \left[\omega_1(\boldsymbol{\phi}), \boldsymbol{\omega}_2'(\boldsymbol{\phi})\right]' \\
&= \left(-\frac{1}{2\sigma^2}, \frac{1}{\sigma^2}\boldsymbol{\beta}'\right)',
\end{aligned}$$

is a subset of a $(p+1)$-dimensional Euclidean space with a negative first coordinate, and this subset has a nonempty interior. Hence, by Theorem 6.3, $\boldsymbol{t}(\boldsymbol{Y}) = [t_1(\boldsymbol{Y}), \boldsymbol{t}_2'(\boldsymbol{Y})]'$ is a complete statistic. But, from (6.59) and (6.60) we can solve for $\tilde{\boldsymbol{\beta}}$ and $\tilde{\sigma}^2$ in terms of $t_1(\boldsymbol{Y})$ and $\boldsymbol{t}_2(\boldsymbol{Y})$, and we obtain,

$$\begin{aligned}
\tilde{\boldsymbol{\beta}} &= (\boldsymbol{X}'\boldsymbol{X})^{-1}\, \boldsymbol{t}_2(\boldsymbol{Y}), \\
\tilde{\sigma}^2 &= \frac{1}{n}\left[t_1(\boldsymbol{Y}) - \boldsymbol{t}_2'(\boldsymbol{Y})\,(\boldsymbol{X}'\boldsymbol{X})^{-1}\,(\boldsymbol{X}'\boldsymbol{X})\,(\boldsymbol{X}'\boldsymbol{X})^{-1}\,\boldsymbol{t}_2(\boldsymbol{Y})\right] \\
&= \frac{1}{n}\left[t_1(\boldsymbol{Y}) - \boldsymbol{t}_2'(\boldsymbol{Y})\,(\boldsymbol{X}'\boldsymbol{X})^{-1}\,\boldsymbol{t}_2(\boldsymbol{Y})\right].
\end{aligned}$$

It follows that $(\tilde{\boldsymbol{\beta}}', \tilde{\sigma}^2)'$ is a complete statistic (any invertible function of a statistic with a complete family has a complete family; see Arnold, 1981, Lemma 1.3, p. 3). We finally conclude that $(\tilde{\boldsymbol{\beta}}', \tilde{\sigma}^2)'$ is a complete and sufficient statistic for $(\boldsymbol{\beta}', \sigma^2)'$. □

Corollary 6.2 Let $\boldsymbol{Y} \sim N(\boldsymbol{X\beta}, \sigma^2\boldsymbol{I}_n)$, where $\boldsymbol{X}$ is of order $n \times p$ and rank p $(< n)$. Then, $\hat{\boldsymbol{\beta}} = (\boldsymbol{X}'\boldsymbol{X})^{-1}\boldsymbol{X}'\boldsymbol{Y}$, and

$$MS_E = \frac{1}{n-p}\boldsymbol{Y}'\left[\boldsymbol{I}_n - \boldsymbol{X}\,(\boldsymbol{X}'\boldsymbol{X})^{-1}\,\boldsymbol{X}'\right]\boldsymbol{Y},$$

are complete and sufficient statistics for $\boldsymbol{\beta}$ and σ^2.

The completeness and sufficiency of $\hat{\boldsymbol{\beta}}$ and MS_E in Corollary 6.2, combined with their being unbiased for $\boldsymbol{\beta}$ and σ^2, give these estimators a certain optimal property, which is described in the next theorem.

Theorem 6.5 Let $Y = X\beta + \epsilon$, where $\epsilon \sim N(\mathbf{0}, \sigma^2 I_n)$. Then, $\hat{\beta}$ and MS_E are the unique unbiased estimators of β and σ^2 with the smallest variance in the class of all unbiased estimators of β and σ^2, that is, for all β and σ^2,

$$\text{Var}\left(\hat{\beta}\right) \leq \text{Var}\left(\beta^*\right) \tag{6.61}$$

$$\text{Var}(MS_E) \leq \text{Var}\left(\sigma^{*2}\right), \tag{6.62}$$

where β^* and σ^{*2} are estimators that belong to the class of all unbiased estimators of β and σ^2, respectively, and the inequality in (6.61) means that the matrix $\text{Var}(\beta^*) - \text{Var}(\hat{\beta})$ is positive semidefinite.

This theorem is based on the so-called *Lehmann–Scheffé Theorem* (see, for example, Casella and Berger, 2002, p. 369). It gives $\hat{\beta}$ and MS_E the distinction of being *uniformly minimum variance unbiased estimators* (UMVUE) of β and σ^2 (see also Graybill, 1976, Theorem 6.2.2, p. 176).

It should be noted that the earlier optimal property concerning $\hat{\beta}$, given by the Gauss–Markov Theorem in Section 6.2.2, states that $\hat{\beta}$ has the smallest variance among all linear unbiased estimators, that is, $\hat{\beta}$ is the best estimator in the class of all linear unbiased estimators of β. Theorem 6.5, however, gives a stronger result, namely, that $\hat{\beta}$ is the best estimator in the class of all unbiased estimators, including those estimators that are linear. This stronger result is due to the normality assumption made earlier concerning ϵ in Theorem 6.5. The Gauss–Markov Theorem does not require such an assumption.

6.6 Inference Concerning β

Consider model (6.5) where it is assumed that $\epsilon \sim N(\mathbf{0}, \sigma^2 I_n)$. In this section, a test statistic is derived concerning the general linear hypothesis,

$$H_0 : A\beta = m, \tag{6.63}$$

against the alternative hypothesis,

$$H_a : A\beta \neq m, \tag{6.64}$$

where A is a known matrix of order $r \times p$ and rank r $(\leq p)$, and m is a known vector of r elements.

It is easy to see that under H_0,

$$A\hat{\beta} \sim N\left[m, \sigma^2 A\left(X'X\right)^{-1} A'\right],$$

where $\hat{\beta}$ is the OLS estimator of β given in (6.9). Hence, under H_0,

$$\frac{1}{\sigma^2}\left(A\hat{\beta} - m\right)'\left[A\left(X'X\right)^{-1} A'\right]^{-1}\left(A\hat{\beta} - m\right) \sim \chi^2_r.$$

Furthermore, $\hat{\beta}$ is independent of MS_E according to property (i) in Section 6.2.1.1, and $\frac{n-p}{\sigma^2}MS_E \sim \chi^2_{n-p}$ by property (f) in Section 6.2.1.1. It follows that under H_0, the statistic,

$$F = \frac{\left(A\hat{\beta} - m\right)'\left[A\left(X'X\right)^{-1}A'\right]^{-1}\left(A\hat{\beta} - m\right)}{rMS_E}, \tag{6.65}$$

has the F-distribution with r and $n - p$ degrees of freedom. Under the alternative hypothesis H_a,

$$\begin{aligned} E&\left\{\left(A\hat{\beta} - m\right)'\left[A\left(X'X\right)^{-1}A'\right]^{-1}\left(A\hat{\beta} - m\right)\right\} \\ &= (m_a - m)'\left[A\left(X'X\right)^{-1}A'\right]^{-1}(m_a - m) \\ &\quad + tr\left\{\left[A\left(X'X\right)^{-1}A'\right]^{-1}\left[A\left(X'X\right)^{-1}A'\right]\sigma^2\right\} \\ &= (m_a - m)'\left[A\left(X'X\right)^{-1}A'\right]^{-1}(m_a - m) + r\sigma^2, \end{aligned} \tag{6.66}$$

as can be seen from applying Theorem 5.2, where m_a is an alternative value of $A\beta$ under H_a ($m_a \neq m$). Note that

$$(m_a - m)'\left[A\left(X'X\right)^{-1}A'\right]^{-1}(m_a - m) > 0,$$

since $A(X'X)^{-1}A'$ is a positive definite matrix. Consequently, a large value of the test statistic F in (6.65) leads to a rejection of H_0. Thus, H_0 is rejected at the α-level if $F \geq F_{\alpha,r,n-p}$.

Under the alternative value, $A\beta = m_a$, the test statistic F in (6.65) has the noncentral F-distribution with r and $n - p$ degrees of freedom and a noncentrality parameter given by

$$\theta = \frac{1}{\sigma^2}(m_a - m)'\left[A\left(X'X\right)^{-1}A'\right]^{-1}(m_a - m).$$

Hence, the corresponding power of the test is

$$\text{Power} = P\left[F > F_{\alpha,r,n-p} \mid F \sim F_{r,n-p}(\theta)\right].$$

A special case of the hypothesis in (6.63) is

$$H_0 : \beta = \beta_0,$$

where $A = I_p$ and $m = \beta_0$. In this case, the statistic F in (6.65) reduces to

$$F = \frac{\left(\hat{\beta} - \beta_0\right)' X'X \left(\hat{\beta} - \beta_0\right)}{pMS_E}, \tag{6.67}$$

which has the F-distribution with p and $n-p$ degrees of freedom under H_0.

6.6.1 Confidence Regions and Confidence Intervals

Given that $\epsilon \sim N(\mathbf{0}, \sigma^2 I_n)$, it is easy to derive a confidence region on a vector linear function of β such as $\lambda = A\beta$, where A is the same matrix as in (6.63). A $(1-\alpha)100\%$ confidence region on λ is given by

$$\left(A\hat{\beta} - \lambda\right)' \left[A\left(X'X\right)^{-1} A'\right]^{-1} \left(A\hat{\beta} - \lambda\right) \le r\, MS_E\, F_{\alpha,r,n-p}. \tag{6.68}$$

In the event $\lambda = a'\beta$, where a' is a known vector of p elements, it would be more convenient to use the t-distribution to derive a $(1-\alpha)100\%$ confidence interval on $a'\beta$ of the form

$$a'\hat{\beta} \mp t_{\frac{\alpha}{2},n-p} \left[a'\left(X'X\right)^{-1} a\, MS_E\right]^{1/2}. \tag{6.69}$$

This is based on the fact that

$$\frac{a'\hat{\beta} - a'\beta}{\left[a'\left(X'X\right)^{-1} a\, MS_E\right]^{1/2}}$$

has the t-distribution with $n-p$ degrees of freedom.

6.6.1.1 *Simultaneous Confidence Intervals*

The confidence interval in (6.69) provides a coverage probability of $1-\alpha$ for a particular linear function, $a'\beta$, of β. In some cases, it may be desired to have confidence intervals on all linear functions of β of the form $l'\beta$, where $l \in R^p$, the p- dimensional Euclidean space, with a joint coverage probability of $1-\alpha$. More specifically, if C_l denotes such a confidence interval on $l'\beta$, then

$$P\{\bigcap_{l \in R^p} [l'\beta \in C_l]\} = 1 - \alpha,$$

or, equivalently,

$$P\left[l'\beta \in C_l\, , \forall\, l \in R^p\right] = 1 - \alpha.$$

Such intervals are referred to as *simultaneous confidence intervals*. To construct these intervals, the following lemma is needed.

Lemma 6.1 The inequality, $\mid b'x \mid \le c(b'b)^{1/2}$, holds for all b if and only if $x'x \le c^2$ $(c > 0)$.

Proof. If $x'x \leq c^2$, then by the Cauchy–Schwarz inequality,

$$| b'x | \leq c \left(b'b\right)^{1/2}. \tag{6.70}$$

Vice versa, if (6.70) is true for all b, then, in particular for $b = x$, we have

$$| x'x | \leq c \left(x'x\right)^{1/2}. \tag{6.71}$$

Whenever $x \neq 0$, we conclude from (6.71) that $x'x \leq c^2$. If $x = 0$, then $x'x \leq c^2$ is already satisfied. □

Theorem 6.6 $(1 - \alpha)100\%$ simultaneous confidence intervals on all linear functions of β of the form $l'\beta$, $l \in R^p$, are given by

$$l'\hat{\beta} \mp \left[p\, MS_E\, F_{\alpha,p,n-p}\, l'\left(X'X\right)^{-1} l\right]^{1/2}.$$

Proof. Using formula (6.68), a $(1 - \alpha)100\%$ confidence region on β can be written as

$$\left(\hat{\beta} - \beta\right)' X'X \left(\hat{\beta} - \beta\right) \leq p\, MS_E\, F_{\alpha,p,n-p}. \tag{6.72}$$

Let $x = (X'X)^{1/2}(\hat{\beta} - \beta)$, $c^2 = p\, MS_E\, F_{\alpha,p,n-p}$. Using Lemma 6.1, inequality (6.72) is equivalent to

$$| b'\left(X'X\right)^{1/2}\left(\hat{\beta} - \beta\right) | \leq \left(b'b\right)^{1/2} \left(p\, MS_E\, F_{\alpha,p,n-p}\right)^{1/2}, \quad \forall\, b \in R^p. \tag{6.73}$$

Let now $l' = b'(X'X)^{1/2}$. Thus, $b' = l'(X'X)^{-1/2}$. Substituting in (6.73), we get

$$| l'\left(\hat{\beta} - \beta\right) | \leq \left[l'\left(X'X\right)^{-1} l\right]^{1/2} \left(p\, MS_E\, F_{\alpha,p,n-p}\right)^{1/2}, \tag{6.74}$$

for all $l \in R^p$ with a joint probability of $1 - \alpha$. From (6.74) we conclude that

$$P\left\{l'\beta \in l'\hat{\beta} \mp \left[p\, MS_E\, F_{\alpha,p,n-p}\, l'\left(X'X\right)^{-1} l\right]^{1/2}, \quad \forall\, l \in R^p\right\} = 1 - \alpha. \tag{6.75}$$

The intervals in (6.75) are called *Scheffé's simultaneous confidence intervals* on all linear functions of β. □

6.6.2 The Likelihood Ratio Approach to Hypothesis Testing

An alternative method for deriving the test statistic for the null hypothesis H_0 in (6.63) is based on the *likelihood ratio principle* (see, for example, Casella and Berger, 2002, Section 8.2.1). This is related to the *maximum likelihood estimation* discussed in Section 6.5. By definition, the *likelihood ratio test statistic*, λ, for testing H_0 is given by

$$\lambda = \frac{\max_{H_0} \mathcal{L}\left(\beta, \sigma^2, y\right)}{\max_{\beta,\, \sigma^2} \mathcal{L}\left(\beta, \sigma^2, y\right)}, \tag{6.76}$$

where $\max_{\beta,\sigma^2} \mathcal{L}(\beta, \sigma^2, y)$ is the maximum of the likelihood function in (6.43) maximized over the entire parameter space of $\beta \in R^p$ and σ^2 $(0 < \sigma^2 < \infty)$, and $\max_{H_0} \mathcal{L}(\beta, \sigma^2, y)$ denotes the maximum value of the likelihood function maximized over a restricted parameter space of β defined by $A\beta = m$. Note that $\lambda \leq 1$ and small values of λ lead to the rejection of H_0.

Recall that $\max_{\beta,\sigma^2} \mathcal{L}(\beta, \sigma^2, y)$ is given by the right-hand side of (6.50). It can also be shown (see Exercise 6.7) that

$$\max_{H_0} \mathcal{L}\left(\beta, \sigma^2, y\right) = \left[\frac{2\pi}{n}\left(y - X\hat{\beta}_r\right)'\left(y - X\hat{\beta}_r\right)\right]^{-n/2} e^{-n/2}, \tag{6.77}$$

where $\hat{\beta}_r$ is the restricted least-squares estimator of β,

$$\hat{\beta}_r = \hat{\beta} - (X'X)^{-1} A'\left[A\,(X'X)^{-1} A'\right]^{-1}\left(A\hat{\beta} - m\right), \tag{6.78}$$

as was seen earlier in Section 6.4. Using formulas (6.50) and (6.77), the expression in (6.76) can be written as

$$\lambda = \left[\frac{\left(y - X\hat{\beta}\right)'\left(y - X\hat{\beta}\right)}{\left(y - X\hat{\beta}_r\right)'\left(y - X\hat{\beta}_r\right)}\right]^{n/2}. \tag{6.79}$$

But,

$$\left(y - X\hat{\beta}\right)'\left(y - X\hat{\beta}\right) = SS_E,$$

as can be seen from (6.13) for a realized value, y, of Y. In addition, from (6.41) and (6.42), we have

$$\begin{aligned}\left(y - X\hat{\beta}_r\right)'\left(y - X\hat{\beta}_r\right) &= SS_E + \left(\hat{\beta} - \hat{\beta}_r\right)' X'X\left(\hat{\beta} - \hat{\beta}_r\right)\\ &= SS_E + \left(A\hat{\beta} - m\right)'\left[A\,(X'X)^{-1} A'\right]^{-1}\left(A\hat{\beta} - m\right).\end{aligned}$$

We then have,

$$\lambda = \left[\frac{SS_E}{SS_E + Q}\right]^{n/2}, \tag{6.80}$$

where

$$Q = \left(A\hat{\beta} - m\right)'\left[A\,(X'X)^{-1} A'\right]^{-1}\left(A\hat{\beta} - m\right).$$

Thus, the ratio λ is a monotone decreasing function of $\frac{Q}{SS_E}$. Since a small value of λ leads to the rejection of H_0, a large value of $\frac{Q}{SS_E}$, or equivalently, of $\frac{Q}{r\,MS_E}$, where $MS_E = \frac{SS_E}{n-p}$, will have the same effect. But, $\frac{Q}{r\,MS_E}$ is equal to the test statistic F in (6.65). We conclude that the likelihood ratio principle results in the same test statistic as the one based on the distributional properties associated with $\hat{\beta}$ under the assumption of normality.

6.7 Examples and Applications

In this section, several applications of the methodology described in the present chapter will be discussed. In addition, a number of numerical examples will be presented to illustrate the implementation of this methodology.

6.7.1 Confidence Region for the Location of the Optimum

One of the objectives of *response surface methodology* is the adequate estimation of the optimum mean response within a certain region of interest. The precision of the estimated optimum is often indicated via a confidence region on the optimum. Such an optimum is usually considered within a certain constrained region.

Let $\mu(\boldsymbol{x})$ denote the mean response at a point $\boldsymbol{x} = (x_1, x_2, \ldots, x_k)'$ in a region of interest, $\mathfrak{R}$. Suppose that $\mu(\boldsymbol{x})$ is represented by a second-degree model of the form

$$\mu(\boldsymbol{x}) = \beta_0 + \boldsymbol{x}'\boldsymbol{\beta}^* + \boldsymbol{x}'\boldsymbol{B}\boldsymbol{x}, \tag{6.81}$$

where $\boldsymbol{\beta}^* = (\beta_1, \beta_2, \ldots, \beta_k)'$, $\boldsymbol{B}$ is a symmetric matrix of order $k \times k$ whose ith diagonal element is β_{ii} and (i, j)th off-diagonal element is $\frac{1}{2}\beta_{ij}$ $(i \neq j)$; β_0 and the elements of $\boldsymbol{\beta}^*$ and $\boldsymbol{B}$ are fixed unknown parameters. Suppose that the region $\mathfrak{R}$ is constrained by the functions $g_1(\boldsymbol{x}), g_2(\boldsymbol{x}), \ldots, g_m(\boldsymbol{x})$ such that

$$g_i(\boldsymbol{x}) = c_i, \quad i = 1, 2, \ldots, m, \tag{6.82}$$

where $c_1, c_2, \ldots, c_m$ are given constants and $g_i(\boldsymbol{x})$ is a quadratic function in $\boldsymbol{x}$ of the form

$$g_i(\boldsymbol{x}) = \gamma_{0i} + \boldsymbol{x}'\boldsymbol{\gamma}_i + \boldsymbol{x}'\boldsymbol{\Gamma}_i\boldsymbol{x}, \quad i = 1, 2, \ldots, m, \tag{6.83}$$

where γ_{0i} and the elements of $\boldsymbol{\gamma}_i$ and $\boldsymbol{\Gamma}_i$ are known parameters. We need to optimize $\mu(\boldsymbol{x})$ subject to the constraints given in (6.82). This can be accomplished by applying the *method of Lagrange multipliers* to the function

$$M(\boldsymbol{x}) = \mu(\boldsymbol{x}) - \sum_{i=1}^{m} \lambda_i \left[g_i(\boldsymbol{x}) - c_i\right], \tag{6.84}$$

where $\lambda_1, \lambda_2, \ldots, \lambda_m$ are Lagrange multipliers. Taking the partial derivative of $M(\boldsymbol{x})$ with respect to $\boldsymbol{x}$ for fixed λ_i, we get

$$\frac{\partial M(\boldsymbol{x})}{\partial \boldsymbol{x}} = 2\left(\boldsymbol{B} - \sum_{i=1}^{m} \lambda_i \boldsymbol{\Gamma}_i\right)\boldsymbol{x} + \boldsymbol{\beta}^* - \sum_{i=1}^{m} \lambda_i \boldsymbol{\gamma}_i. \tag{6.85}$$

Let $\boldsymbol{\xi}$ be the location of the true constrained optimum. If the model in (6.81) is correct, then $\boldsymbol{\xi}$ must satisfy the equation

$$2\left(\boldsymbol{B} - \sum_{i=1}^{m} \lambda_i \boldsymbol{\Gamma}_i\right)\boldsymbol{\xi} + \left(\boldsymbol{\beta}^* - \sum_{i=1}^{m} \lambda_i \boldsymbol{\gamma}_i\right) = \boldsymbol{0}. \tag{6.86}$$

Let $\hat{\delta}$ denote the estimator of the quantity on the left-hand side of (6.86), which is formed by replacing the unknown parameters of the model in (6.81) with their unbiased least-squares estimators, that is,

$$\hat{\delta} = 2\left(\hat{B} - \sum_{i=1}^{m} \lambda_i \Gamma_i\right)\xi + \hat{\beta}^* - \sum_{i=1}^{m} \lambda_i \gamma_i. \tag{6.87}$$

Assuming that the response variable $Y(x)$ at $x \in \mathfrak{R}$ is normally distributed with mean $\mu(x)$ and variance σ^2, and the response data are uncorrelated, the least-squares estimators in (6.87) are normally distributed. Hence, $\hat{\delta} \sim N(\mathbf{0}, V)$, where the elements of V are obtained from appropriate functions of the elements of the variance–covariance matrix of the parameter estimators in (6.87). Consequently, a $(1-\alpha)100\%$ confidence region on ξ for fixed λ_i ($i = 1, 2, \ldots, m$) is defined by the inequality

$$\hat{\delta}' V^{-1} \hat{\delta} \leq \chi^2_{\alpha,k}, \tag{6.88}$$

where $\chi^2_{\alpha,k}$ is the upper α-quantile of the chi-squared distribution with k degrees of freedom.

In a typical response surface investigation, the unconstrained optimum may fall outside the region where the data are collected. Such an optimum is undesirable because it can only be extrapolated and is therefore unreliable. In this case, certain constraints are considered in order to restrict optimization within the experimental region to a fixed distance from the region's center.

Example 6.1 In some biomedical studies, the fitted model may not be linear as in (6.81), but can be expressed as an exponential function of $\mu(x) = \beta_0 + x'\beta^* + x'Bx$, which is monotone increasing. Hence, any constrained optimization of this function is equivalent to a constrained optimization of $\mu(x)$. Stablein, Carter, and Wampler (1983) used such a model in their determination of the optimum combination and its confidence bounds in a murine cancer chemotherapy experiment. Different combinations of the levels of two drugs, namely, 5-Fluorouracil (5FU) and Teniposide (VM26) were used. In this experiment, leukemia cells were injected intraperitoneally into each of 127 mice on Day 0 and the treatment, consisting of an injection of the combination of the two drugs, was administered on Day 7. The data set consisting of the treatment combination levels and resulting survival times, in days, is given in Stablein, Carter, and Wampler (1983, Table 1). The same data set is reproduced in Table 6.2. A proportional-hazards analysis was performed using the model

$$L^*(t) = L_0(t)\exp\left(x'\beta^* + x'Bx\right), \tag{6.89}$$

TABLE 6.2
Data from the Murine Cancer Chemotherapy Experiment

Treatment Levels		
5FU (mg/kg)	**VM26 (mg/kg)**	**Days of Survival**
0.00	0.00	8, 9(2), 10(5)
0.00	9.71	10, 13(5), 14(2)
0.00	19.40	8, 10, 13, 14(4), 15
0.00	25.90	9, 14(4), 15(3)
35.60	9.71	13, 14(3), 15(3), 17
48.50	4.85	9, 13(2), 14(3), 15(2)
48.50	19.40	14(2), 15(2), 16(4)
97.10	0.00	8(2), 10, 11, 12(2), 14, 16
97.10	3.56	8, 9(2), 11(2), 13(2), 16
97.10	9.71	8, 10, 11, 16(2), 17(2), 18
97.10	25.90	16(3), 17, 18(3), 19
194.00	0.00	10, 13(6), 14
194.00	4.85	11(2), 14(3), 16, 17
194.00	19.40	8, 14, 16, 20(4), 21
259.00	0.00	9, 11, 12(3), 13(3)
259.00	9.71	16(2), 17, 18(2), 19(2), 20

Source: Reprinted from Stablein, D.M. et al., *Biometrics*, 39, 759, 1983. With permission.

Note: The number in parentheses indicates the number of animals failing on the day in question.

where $\frac{L^*(t)}{L_0(t)}$ is the hazard at time t of the treatment group relative to that of the untreated control group, and $\boldsymbol{x} = (x_1, x_2)'$ is the vector of coded dosage levels for the two drugs given by

$$x_1 = \frac{\text{5FU dose (mg/kg)} - 130}{130}$$

$$x_2 = \frac{\text{VM26 dose (mg/kg)} - 13}{13}.$$

Of interest here is the minimization of the natural logarithm of the relative hazard function, that is, minimizing $\boldsymbol{x}'\boldsymbol{\beta}^* + \boldsymbol{x}'\boldsymbol{B}\boldsymbol{x}$ within a circle of radius r centered at (0,0).

Estimates of the parameters in model (6.89) were obtained on the basis of maximum likelihood using the data in Table 6.2. These estimates are shown in Table 6.3. Note that the asymptotic properties of the maximum likelihood estimates, including the estimated asymptotic variance–covariance matrix of the estimates, were used here.

A constrained minimum within a distance, r, of the experimental region's center can be determined by requiring $\boldsymbol{x}'\boldsymbol{x} \leq r^2$ and minimizing the function $\boldsymbol{x}'\boldsymbol{\beta}^* + \boldsymbol{x}'\boldsymbol{B}\boldsymbol{x} - \lambda(\boldsymbol{x}'\boldsymbol{x} - r^2)$. To determine the treatment with minimum value of the logarithm of the relative hazard on a circle of radius r, the value of λ must

TABLE 6.3
Parameter Estimates for Model (6.89)

Parameter	Estimate	p-Value
β_1	−1.2312	<0.001
β_2	−1.5084	<0.001
β_{11}	0.5467	0.046
β_{22}	0.8850	<0.001
β_{12}	0.7186	0.026

Source: Reprinted from Stablein, D.M. et al., *Biometrics*, 39, 759, 1983. With permission.

be chosen smaller than the smallest eigenvalue of $\boldsymbol{B}$ (see Khuri and Cornell, 1996, p. 192), where from Table 6.3, $\boldsymbol{B}$ is the 2×2 matrix

$$\boldsymbol{B} = \begin{bmatrix} 0.5467 & 0.3593 \\ 0.3593 & 0.8850 \end{bmatrix}.$$

Its eigenvalues are 1.1130 and 0.3187. Hence, λ should be chosen less than 0.3187 in order to achieve a minimum. For example, for $\lambda = -0.6$, the constrained minimum is estimated to be on a circle of radius 1 with dosages of 5FU = 227.9 mg/kg, and of VM26 = 22.0 mg/kg (see Stablein, Carter, and Wampler, 1983, p. 762). An estimated asymptotic confidence region was placed around the location of the true minimum by using asymptotically unbiased maximum likelihood estimates of the parameters and setting $\lambda = -0.6$. This region was obtained by identifying all the points in the two-dimensional experimental region that satisfy the inequality in (6.88) (see Figure 6.1). The experimental region excluded dosage values that were believed from clinical knowledge to be toxic. This includes the location of the unconstrained minimum. More details about this can be found in Stablein, Carter, and Wampler (1983).

6.7.2 Confidence Interval on the True Optimum

In this section, a confidence interval on the value of the true optimum of the mean response is developed. For simplicity reasons, an unconstrained optimum is considered here.

Consider again model (6.81). The stationary point of $\mu(\boldsymbol{x})$, that is, the point at which $\frac{\partial \mu(\boldsymbol{x})}{\partial \boldsymbol{x}}$ is equal to the zero vector, is given by

$$\boldsymbol{x}_0 = -\frac{1}{2}\boldsymbol{B}^{-1}\boldsymbol{\beta}^*. \tag{6.90}$$

Then, at $\boldsymbol{x}_0$, $\mu(\boldsymbol{x})$ has a minimum value if $\boldsymbol{B}$ is positive definite, and a maximum value if $\boldsymbol{B}$ is negative definite. If $\boldsymbol{B}$ is neither positive definite nor negative definite, then $\boldsymbol{x}_0$ is a saddle point.

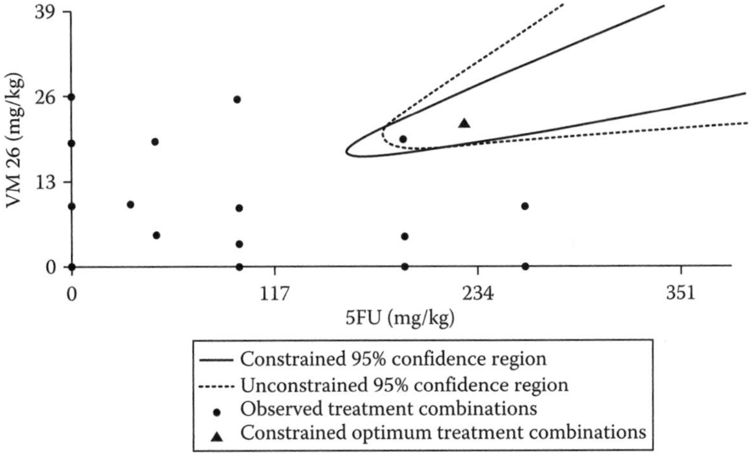

FIGURE 6.1
Constrained and unconstrained confidence regions for the 5FU-VM26 experiment. (Reprinted from Stablein, D.M. et al., *Biometrics*, 39, 759, 1983. With permission.)

Let $\boldsymbol{\beta}$ denote the vector consisting of all the unknown parameters in model (6.81). If $\boldsymbol{Y}$ is the vector of observed response values at n design settings of $\boldsymbol{x}$, then $\boldsymbol{Y}$ is represented by model (6.5). Assuming that the random error vector $\boldsymbol{\epsilon}$ in this model is distributed as $N(\mathbf{0}, \sigma^2\boldsymbol{I}_n)$, the $(1-\alpha)100\%$ confidence region on $\boldsymbol{\beta}$ is given by

$$C = \left\{\boldsymbol{\gamma} : \left(\hat{\boldsymbol{\beta}} - \boldsymbol{\gamma}\right)' \boldsymbol{X}'\boldsymbol{X}\left(\hat{\boldsymbol{\beta}} - \boldsymbol{\gamma}\right) \leq p\, MS_E\, F_{\alpha,p,n-p}\right\}, \tag{6.91}$$

where $\hat{\boldsymbol{\beta}} = (\boldsymbol{X}'\boldsymbol{X})^{-1}\boldsymbol{X}'\boldsymbol{Y}$ and MS_E is the error mean square. The objective here is to use the confidence region in (6.91) in order to obtain a confidence interval on the mean response at $\boldsymbol{x}_0$ in (6.90).

It is known that if $g(\boldsymbol{\beta})$ is any continuous function of $\boldsymbol{\beta}$, then

$$P\left\{\min_{\boldsymbol{\gamma}\in C}\left[g(\boldsymbol{\gamma})\right] \leq g(\boldsymbol{\beta}) \leq \max_{\boldsymbol{\gamma}\in C}\left[g(\boldsymbol{\gamma})\right]\right\} \geq 1-\alpha. \tag{6.92}$$

This inequality follows from the fact that if $\boldsymbol{\beta} \in C$, then $g(\boldsymbol{\beta}) \in g(C)$ and therefore

$$\begin{aligned} P\left[g(\boldsymbol{\beta}) \in g(C)\right] &\geq P[\boldsymbol{\beta} \in C] \\ &= 1-\alpha. \end{aligned}$$

Note that because of the continuity of the function g,

$$g(C) = \left\{\min_{\boldsymbol{\gamma}\in C}\left[g\left(\boldsymbol{\gamma}\right)\right], \max_{\boldsymbol{\gamma}\in C}\left[g(\boldsymbol{\gamma})\right]\right\}. \tag{6.93}$$

Thus the interval $\{\min_{\gamma\in C}[g(\gamma)],\ \max_{\gamma\in C}[g(\gamma)]\}$ represents a conservative confidence interval on $g(\beta)$ with a confidence coefficient greater than or equal to $1-\alpha$.

From (6.81), the value of $\mu(\boldsymbol{x})$ at $\boldsymbol{x}=\boldsymbol{x}_0$, as given by (6.90), is

$$\mu(\boldsymbol{x}_0)=\beta_0-\frac{1}{4}\boldsymbol{\beta}^{*'}\boldsymbol{B}^{-1}\boldsymbol{\beta}^{*}. \tag{6.94}$$

The right-hand side of (6.94) represents a particular function of $\boldsymbol{\beta}$. We can then write $\mu(\boldsymbol{x}_0)=g(\boldsymbol{\beta})$ and apply (6.92) to obtain a conservative confidence interval on the true optimum of the mean response (assuming that $\boldsymbol{B}$ is either positive definite or negative definite). Such an interval is therefore of the form

$$\left[\min_{\boldsymbol{\beta}\in C}\left\{\beta_0-\frac{1}{4}\boldsymbol{\beta}^{*'}\boldsymbol{B}^{-1}\boldsymbol{\beta}^{*}\right\},\ \max_{\boldsymbol{\beta}\in C}\left\{\beta_0-\frac{1}{4}\boldsymbol{\beta}^{*'}\boldsymbol{B}^{-1}\boldsymbol{\beta}^{*}\right\}\right], \tag{6.95}$$

which has a coverage probability greater than or equal to $1-\alpha$. This confidence interval was used by Carter et al. (1984) in their analysis of survival data from a preclinical cancer chemotherapy experiment involving the combination of two drugs.

The computation of the bounds of the confidence interval in (6.95) requires first the identification of points in the confidence region C in (6.91). To accomplish this, Carter et al. (1984) used the following linear transformation in order to reduce C to a hypersphere: Let $\boldsymbol{P}$ be an orthogonal matrix such that $\boldsymbol{X}'\boldsymbol{X}=\boldsymbol{P}\boldsymbol{\Lambda}\boldsymbol{P}'$, where $\boldsymbol{\Lambda}$ is a diagonal matrix whose diagonal elements are the eigenvalues of $\boldsymbol{X}'\boldsymbol{X}$, which are positive. Then,

$$\left(\hat{\boldsymbol{\beta}}-\boldsymbol{\gamma}\right)'\boldsymbol{X}'\boldsymbol{X}\left(\hat{\boldsymbol{\beta}}-\boldsymbol{\gamma}\right)=\left(\hat{\boldsymbol{\beta}}-\boldsymbol{\gamma}\right)'\boldsymbol{P}\boldsymbol{\Lambda}\boldsymbol{P}'\left(\hat{\boldsymbol{\beta}}-\boldsymbol{\gamma}\right).$$

Let $\boldsymbol{z}=\boldsymbol{\Lambda}^{1/2}\boldsymbol{P}'(\hat{\boldsymbol{\beta}}-\boldsymbol{\gamma})$. The inequality in (6.91) can be written as

$$\boldsymbol{z}'\boldsymbol{z}\leq p\,MS_E\,F_{\alpha,p,n-p},$$

which represents a hypersphere centered at the origin of radius $=(p\,MS_E\,F_{\alpha,p,n-p})^{1/2}$. Polar coordinates can then be easily used to select points $\boldsymbol{z}$ in this hypersphere and hence points $\boldsymbol{\gamma}$ in C of the form

$$\boldsymbol{\gamma}=\hat{\boldsymbol{\beta}}-\boldsymbol{P}\boldsymbol{\Lambda}^{-1/2}\boldsymbol{z}.$$

On this basis, a large number of points can be chosen in C, and the corresponding values of $\beta_0-\frac{1}{4}\boldsymbol{\beta}^{*'}\boldsymbol{B}^{-1}\boldsymbol{\beta}^{*}$ can be computed and used to determine the bounds in (6.95).

Carter et al. (1984) used the proportional-hazards model (6.89) in their analysis of survival data from a cancer chemotherapy experiment, as was mentioned earlier. In this case,

$$\left[\min_{\boldsymbol{\beta}\in C}\left(-\frac{1}{4}\boldsymbol{\beta}^{*'}\boldsymbol{B}^{-1}\boldsymbol{\beta}^{*}\right),\ \max_{\boldsymbol{\beta}\in C}\left(-\frac{1}{4}\boldsymbol{\beta}^{*'}\boldsymbol{B}^{-1}\boldsymbol{\beta}^{*}\right)\right]$$

gives a conservative confidence interval on $\log\left[\frac{L^*(t)}{L_0(t)}\right]$ at the location of the optimum. Note that the confidence region C used by Carter et al. (1984) was in fact not the same as in (6.91), but was rather constructed using the asymptotic properties of the maximum likelihood estimates of the model parameters in (6.89). For more details, see Carter et al. (1984, p. 1128).

6.7.3 Confidence Interval for a Ratio

In some situations, it may be of interest to obtain a confidence interval on a ratio of linear combinations of the parameters in model (6.5), where it is assumed that $\boldsymbol{\epsilon} \sim N(\mathbf{0}, \sigma^2 \boldsymbol{I}_n)$. Let us therefore consider finding confidence limits for the ratio $\psi = \frac{\boldsymbol{a}_1'\boldsymbol{\beta}}{\boldsymbol{a}_2'\boldsymbol{\beta}}$, where $\boldsymbol{a}_1$ and $\boldsymbol{a}_2$ are vectors of known constants. Let Δ be defined as $\Delta = \boldsymbol{a}_1'\boldsymbol{\beta} - \psi\, \boldsymbol{a}_2'\boldsymbol{\beta}$. An unbiased estimate of Δ is $\hat{\Delta} = \boldsymbol{a}_1'\hat{\boldsymbol{\beta}} - \psi\, \boldsymbol{a}_2'\hat{\boldsymbol{\beta}}$. Hence,

$$
\begin{aligned}
E(\hat{\Delta}) &= 0, \\
\mathrm{Var}(\hat{\Delta}) &= \boldsymbol{a}_1'\left(\boldsymbol{X}'\boldsymbol{X}\right)^{-1}\boldsymbol{a}_1\sigma^2 - 2\,\psi\, \boldsymbol{a}_1'\left(\boldsymbol{X}'\boldsymbol{X}\right)^{-1}\boldsymbol{a}_2\sigma^2 + \psi^2 \boldsymbol{a}_2'\left(\boldsymbol{X}'\boldsymbol{X}\right)^{-1}\boldsymbol{a}_2\sigma^2 \\
&= \left(d_{11} - 2\,\psi\, d_{12} + \psi^2 d_{22}\right)\sigma^2,
\end{aligned}
$$

where $d_{11} = \boldsymbol{a}_1'(\boldsymbol{X}'\boldsymbol{X})^{-1}\boldsymbol{a}_1$, $d_{12} = \boldsymbol{a}_1'(\boldsymbol{X}'\boldsymbol{X})^{-1}\boldsymbol{a}_2$, and $d_{22} = \boldsymbol{a}_2'(\boldsymbol{X}'\boldsymbol{X})^{-1}\boldsymbol{a}_2$. It follows that

$$
\frac{\hat{\Delta}^2}{\left(d_{11} - 2\,\psi d_{12} + \psi^2 d_{22}\right) MS_E} \sim F_{1,n-p}.
$$

Consequently,

$$
P\left[\left(\boldsymbol{a}_1'\hat{\boldsymbol{\beta}} - \psi\boldsymbol{a}_2'\hat{\boldsymbol{\beta}}\right)^2 - \left(d_{11} - 2\,\psi d_{12} + \psi^2 d_{22}\right) MS_E F_{\alpha,1,n-p} \leq 0\right] = 1 - \alpha. \tag{6.96}
$$

The probability statement in (6.96) can be rewritten as

$$
P\left[A\psi^2 - 2\,B\psi + C \leq 0\right] = 1 - \alpha, \tag{6.97}
$$

where,

$$
\begin{aligned}
A &= \left(\boldsymbol{a}_2'\hat{\boldsymbol{\beta}}\right)^2 - d_{22} MS_E F_{\alpha,1,n-p} \\
B &= \left(\boldsymbol{a}_1'\hat{\boldsymbol{\beta}}\right)\left(\boldsymbol{a}_2'\hat{\boldsymbol{\beta}}\right) - d_{12} MS_E F_{\alpha,1,n-p} \\
C &= \left(\boldsymbol{a}_1'\hat{\boldsymbol{\beta}}\right)^2 - d_{11} MS_E F_{\alpha,1,n-p}.
\end{aligned}
$$

In order to obtain confidence limits on ψ from (6.97), the equation, $A\psi^2 - 2\,B\psi + C = 0$, must have two distinct real roots, ψ_1 and ψ_2 ($\psi_1 < \psi_2$), and the inequality $A\psi^2 - 2B\psi + C \leq 0$ must be satisfied by all those values of ψ such that $\psi_1 \leq \psi \leq \psi_2$. This occurs whenever $B^2 - AC > 0$ and $A > 0$. Under these conditions, the probability statement in (6.97) is equivalent to

$$P[\psi_1 \leq \psi \leq \psi_2] = 1 - \alpha, \tag{6.98}$$

where

$$\psi_1 = \frac{B - (B^2 - AC)^{1/2}}{A} \tag{6.99}$$

$$\psi_2 = \frac{B + (B^2 - AC)^{1/2}}{A}. \tag{6.100}$$

Hence, a confidence interval $[\psi_1, \psi_2]$ on ψ exists provided that $B^2 - AC > 0$ and $A > 0$. If these conditions are not satisfied, then the probability statement in (6.97) can only provide a so-called *confidence set* on ψ, which consists of all values of ψ that satisfy $A\psi^2 - 2B\psi + C \leq 0$.

Example 6.2 An experiment was conducted in order to determine if there is a relationship between arterial oxygen tension, x (millimeters of mercury) and cerebral blood flow, Y, in human beings. Fifteen patients were used in the study and the resulting data are given in Table 6.4.

The data set in Table 6.4 was used to fit the quadratic model

$$Y = \beta_0 + \beta_1 x + \beta_2 x^2 + \epsilon. \tag{6.101}$$

TABLE 6.4
Data for Example 6.2

Arterial Oxygen Tension, x	Cerebral Blood Flow, Y
603.4	80.33
582.5	79.80
556.2	77.20
594.6	79.21
558.9	77.44
575.2	78.01
580.1	79.53
451.2	74.46
404.0	75.22
484.0	74.58
452.4	75.90
448.4	75.80
334.8	80.67
320.3	82.60
350.3	78.20

Note: The background information concerning this data set is described in Walpole and Myers (1985, p. 364).

TABLE 6.5
Parameter Estimates for Model (6.101)

Parameter	Estimate	Standard Error	t-Value	p-Value
β_0	144.3982	5.7380	25.17	<0.0001
β_1	−0.2971	0.0254	−11.70	<0.0001
β_2	0.00032	0.000027	11.72	<0.0001

TABLE 6.6
ANOVA Table for Model (6.101)

Source	DF	SS	MS	F	p-Value
Regression	2	76.2360	38.1180	68.75	<0.0001
Error	12	6.6534	0.5544		
Total	14	82.8894			

The least-squares estimates of the model parameters are shown in Table 6.5, and the corresponding ANOVA table is given in Table 6.6.

We note that model (6.101) provides a good fit to the data, and all the t-statistic values for the parameters are highly significant.

Let x_0 denote the location of the stationary point of the mean response $\mu(x)$. If the quadratic model in (6.101) is correct, then x_0 must satisfy the equation $\beta_1 + 2\beta_2 x_0 = 0$. Hence, $x_0 = -\frac{\beta_1}{2\beta_2}$, which can be written as

$$x_0 = \frac{\boldsymbol{a}_1'\boldsymbol{\beta}}{\boldsymbol{a}_2'\boldsymbol{\beta}}, \tag{6.102}$$

where $\boldsymbol{\beta} = (\beta_0, \beta_1, \beta_2)'$, $\boldsymbol{a}_1' = (0, -1, 0)$, $\boldsymbol{a}_2' = (0, 0, 2)$. In this case, $A = 3.89 \times 10^{-7} > 0$ and $B^2 - AC = 7.25 \times 10^{-12} > 0$. We have an absolute (unconstrained) minimum at x_0 if $\beta_2 > 0$. Replacing β_1 and β_2 with their least-squares estimates from Table 6.5, we get the following estimate of x_0:

$$\begin{aligned}\hat{x}_0 &= -\frac{\hat{\beta}_1}{2\,\hat{\beta}_2} \\ &= 464.219,\end{aligned}$$

which falls within the experimental region. Using formulas (6.99) and (6.100), the 95% confidence bounds on $\psi = x_0 = -\frac{\beta_1}{2\beta_2}$ are $\psi_1 = 461.372$, $\psi_2 = 475.234$. Thus, with a 0.95 probability, the interval [461.372, 475.234] contains the arterial oxygen tension value that minimizes the mean cerebral blood flow.

6.7.4 Demonstrating the Gauss–Markov Theorem

The Gauss–Markov Theorem (Theorem 6.1) guarantees optimality of the least-squares estimator, $\hat{\boldsymbol{\beta}} = (\boldsymbol{X}'\boldsymbol{X})^{-1}\boldsymbol{X}'\boldsymbol{Y}$, for model (6.5) under the conditions $E(\boldsymbol{\epsilon}) = \boldsymbol{0}$, $\text{Var}(\boldsymbol{\epsilon}) = \sigma^2\boldsymbol{I}_n$. In particular, any element of $\hat{\boldsymbol{\beta}}$ has a variance smaller than or equal to that of any other linear unbiased estimator of the corresponding element of β. In this section, the Gauss–Markov Theorem is demonstrated by presenting an example of such alternative unbiased linear estimators and verifying that they are less efficient than the corresponding least-squares estimators. The following example was described in Jeske (1994):

Consider the simple linear regression model,

$$Y_u = \beta_0 + \beta_1 x_u + \epsilon_u, \quad u = 1, 2, \ldots, n, \tag{6.103}$$

where the ϵ_u's are mutually independent with zero mean and variance σ^2. The best linear unbiased estimators (BLUE) of β_0 and β_1 are

$$\hat{\beta}_1 = \frac{S_{xY}}{S_{xx}} \tag{6.104}$$

$$\hat{\beta}_0 = \bar{Y} - \hat{\beta}_1 \bar{x}, \tag{6.105}$$

where $\bar{Y} = \frac{1}{n}\sum_{u=1}^{n} Y_u$, $\bar{x} = \frac{1}{n}\sum_{u=1}^{n} x_u$, $S_{xY} = \sum_{u=1}^{n}(x_u - \bar{x})(Y_u - \bar{Y})$, and $S_{xx} = \sum_{u=1}^{n}(x_u - \bar{x})^2$. The variances of $\hat{\beta}_0$ and $\hat{\beta}_1$ are

$$\text{Var}(\hat{\beta}_1) = \frac{\sigma^2}{S_{xx}} \tag{6.106}$$

$$\text{Var}(\hat{\beta}_0) = \frac{\sigma^2 \sum_{u=1}^{n} x_u^2}{n\, S_{xx}}. \tag{6.107}$$

Jeske (1994) introduced the following estimators of β_0 and β_1:

$$\tilde{\beta}_1 = \frac{S_{x^{-1}Y}}{S_{xx^{-1}}} \tag{6.108}$$

$$\tilde{\beta}_0 = \bar{Y}_\omega - \tilde{\beta}_1 \bar{x}_h, \tag{6.109}$$

where $\bar{x}_h = \left(\frac{1}{n}\sum_{u=1}^{n}\frac{1}{x_u}\right)^{-1}$ is the harmonic mean of $x_1, x_2, \ldots, x_n$ (assuming that $x_u > 0$ for $u = 1, 2, \ldots, n$), $\bar{Y}_\omega = \left(\sum_{u=1}^{n}\frac{1}{x_u}\right)^{-1}\sum_{u=1}^{n}\frac{Y_u}{x_u}$ is the weighted sample mean of $Y_1, Y_2, \ldots, Y_n$ with $\frac{1}{x_u}$ $(u = 1, 2, \ldots, n)$ as weights, $S_{x^{-1}Y} = \sum_{u=1}^{n}\left(\frac{1}{x_u} - \frac{1}{\bar{x}_h}\right)(Y_u - \bar{Y})$, $S_{xx^{-1}} = \sum_{u=1}^{n}(x_u - \bar{x})\left(\frac{1}{x_u} - \frac{1}{\bar{x}_h}\right)$. It is clear that $\tilde{\beta}_0$ and $\tilde{\beta}_1$ are linear estimators that are also unbiased for β_0 and β_1, respectively, since

$$\begin{aligned} E\left(\tilde{\beta}_1\right) &= \frac{1}{S_{xx^{-1}}}\sum_{u=1}^{n}\left(\frac{1}{x_u} - \frac{1}{\bar{x}_h}\right)[\beta_1 (x_u - \bar{x})] \\ &= \beta_1, \\ E(\tilde{\beta}_0) &= \left(\sum_{u=1}^{n}\frac{1}{x_u}\right)^{-1}\sum_{u=1}^{n}\frac{1}{x_u}(\beta_0 + \beta_1 x_u) - \beta_1 \bar{x}_h \\ &= \beta_0 + \beta_1 \bar{x}_h - \beta_1 \bar{x}_h \\ &= \beta_0. \end{aligned}$$

Furthermore, the variance of $\tilde{\beta}_1$ is given by

$$\begin{aligned}\text{Var}\left(\tilde{\beta}_1\right) &= \frac{1}{S^2_{xx^{-1}}}\text{Var}\left[\sum_{u=1}^{n}\left(\frac{1}{x_u}-\frac{1}{\bar{x}_h}\right)Y_u\right]\\ &= \sigma^2\frac{S_{x^{-1}x^{-1}}}{S^2_{xx^{-1}}},\end{aligned} \tag{6.110}$$

where $S_{x^{-1}x^{-1}} = \sum_{u=1}^{n}\left(\frac{1}{x_u}-\frac{1}{\bar{x}_h}\right)^2$. Formula (6.110) can be written as

$$\text{Var}\left(\tilde{\beta}_1\right) = \frac{\text{Var}\left(\hat{\beta}_1\right)}{r^2_{xx^{-1}}}, \tag{6.111}$$

where

$$r^2_{xx^{-1}} = \frac{S^2_{xx^{-1}}}{S_{xx}\,S_{x^{-1}x^{-1}}} \tag{6.112}$$

is the square of the sample correlation coefficient between x_u and $\frac{1}{x_u}$, $u = 1, 2, \ldots, n$. Since $0 < r^2_{xx^{-1}} < 1$, it is clear from (6.111) that $\text{Var}(\tilde{\beta}_1) > \text{Var}(\hat{\beta}_1)$, and hence $\tilde{\beta}_1$ is less efficient than $\hat{\beta}_1$. In addition, using (6.109), the variance of $\tilde{\beta}_0$ is obtained as follows:

$$\text{Var}\left(\tilde{\beta}_0\right) = \text{Var}\left(\bar{Y}_\omega\right) - 2\,\bar{x}_h\,\text{Cov}\left(\bar{Y}_\omega, \tilde{\beta}_1\right) + \bar{x}_h^2\,\text{Var}\left(\tilde{\beta}_1\right). \tag{6.113}$$

Note that

$$\begin{aligned}\text{Var}(\bar{Y}_\omega) &= \frac{\bar{x}_h^2}{n^2}\left(\sum_{u=1}^{n}\frac{1}{x_u^2}\right)\sigma^2\\ &= \frac{\bar{x}_h^2}{n^2}\left(S_{x^{-1}x^{-1}} + \frac{n}{\bar{x}_h^2}\right)\sigma^2\\ &= \frac{\sigma^2}{n} + \frac{\bar{x}_h^2\,S_{x^{-1}x^{-1}}}{n^2}\,\sigma^2,\end{aligned} \tag{6.114}$$

and

$$\begin{aligned}\text{Cov}(\bar{Y}_\omega\,,\,\tilde{\beta}_1) &= \text{Cov}\left[\left(\sum_{u=1}^{n}\frac{1}{x_u}\right)^{-1}\sum_{u=1}^{n}\frac{Y_u}{x_u},\ \frac{1}{S_{xx^{-1}}}\sum_{u=1}^{n}\left(\frac{1}{x_u}-\frac{1}{\bar{x}_h}\right)Y_u\right]\\ &= \frac{\bar{x}_h}{n\,S_{xx^{-1}}}\left[\sum_{u=1}^{n}\frac{1}{x_u}\left(\frac{1}{x_u}-\frac{1}{\bar{x}_h}\right)\right]\sigma^2\\ &= \frac{\bar{x}_h}{n}\,\frac{S_{x^{-1}x^{-1}}}{S_{xx^{-1}}}\,\sigma^2.\end{aligned} \tag{6.115}$$

Using (6.110), (6.114), and (6.115) in (6.113), we get

$$\begin{aligned}
\mathrm{Var}(\tilde{\beta}_0) &= \left(\frac{\sigma^2}{n} + \frac{\sigma^2}{n^2}\,\bar{x}_h^2\, S_{x^{-1}x^{-1}}\right) \\
&\quad - 2\,\bar{x}_h \left(\frac{\bar{x}_h\, S_{x^{-1}x^{-1}}}{n\, S_{xx^{-1}}}\,\sigma^2\right) + \bar{x}_h^2 \left(\frac{S_{x^{-1}x^{-1}}}{S^2_{xx^{-1}}}\,\sigma^2\right) \\
&= \frac{\sigma^2}{n} + \sigma^2\,\bar{x}_h^2 S_{x^{-1}x^{-1}} \left[\frac{1}{n^2} - \frac{2}{n\, S_{xx^{-1}}} + \frac{1}{S^2_{xx^{-1}}}\right] \\
&= \frac{\sigma^2}{n} + \sigma^2\,\bar{x}_h^2 S_{x^{-1}x^{-1}} \left(\frac{1}{n} - \frac{1}{S_{xx^{-1}}}\right)^2 \\
&= \frac{\sigma^2}{n} + \sigma^2\,\bar{x}_h^2 S_{x^{-1}x^{-1}} \left(-\frac{\bar{x}}{\bar{x}_h S_{xx^{-1}}}\right)^2 \\
&= \frac{\sigma^2}{n} + \frac{\bar{x}^2 S_{x^{-1}x^{-1}}}{S^2_{xx^{-1}}}\,\sigma^2. \qquad (6.116)
\end{aligned}$$

It can be verified that formula (6.116) can be expressed as (see Exercise 6.10)

$$\mathrm{Var}\left(\tilde{\beta}_0\right) = \frac{\mathrm{Var}\left(\hat{\beta}_0\right)}{r^2_{xx^{-1}}} - \frac{1 - r^2_{xx^{-1}}}{n r^2_{xx^{-1}}}\,\sigma^2. \qquad (6.117)$$

Note that

$$\frac{\mathrm{Var}\left(\hat{\beta}_0\right)}{r^2_{xx^{-1}}} - \frac{1 - r^2_{xx^{-1}}}{n r^2_{xx^{-1}}}\,\sigma^2 > \mathrm{Var}\left(\hat{\beta}_0\right),$$

since

$$\mathrm{Var}\left(\hat{\beta}_0\right)\left[\frac{1 - r^2_{xx^{-1}}}{r^2_{xx^{-1}}}\right] - \frac{1 - r^2_{xx^{-1}}}{n\, r^2_{xx^{-1}}}\,\sigma^2 > 0. \qquad (6.118)$$

Inequality (6.118) is true because $\mathrm{Var}(\hat{\beta}_0) > \frac{\sigma^2}{n}$, which results from using formula (6.107) and the fact that $\sum_{u=1}^{n} x_u^2 > S_{xx}$. It follows that $\tilde{\beta}_0$ is less efficient than $\hat{\beta}_0$. It can also be seen from (6.118) that the smaller $r^2_{xx^{-1}}$ is, the larger the term on the left-hand side of (6.118). Thus, the loss of efficiency in using $\tilde{\beta}_0$, as measured by the size of the difference $\mathrm{Var}(\tilde{\beta}_0) - \mathrm{Var}(\hat{\beta}_0)$, is a strictly decreasing function of $r^2_{xx^{-1}}$. The same remark can be made regarding using $\tilde{\beta}_1$ [see formula (6.111)].

6.7.5 Comparison of Two Linear Models

In this section, a comparison is made of the parameter vectors for two linear models. The need for such a comparison arises in situations where the modeling of a response of interest is carried out under two different experimental conditions. In this case, it would be of interest to determine whether or not the parameters of the associated models are different.

Consider the two models

$$Y_1 = X_1\beta_1 + \epsilon_1 \tag{6.119}$$
$$Y_2 = X_2\beta_2 + \epsilon_2, \tag{6.120}$$

where

X_1 and X_2 are of orders $n_1 \times p$ and $n_2 \times p$, both of rank p

ϵ_1 and ϵ_2 are normally distributed as $N(\mathbf{0}, \sigma_1^2 I_{n_1})$ and $N(\mathbf{0}, \sigma_2^2 I_{n_2})$, respectively

Suppose that it is of interest to test the hypothesis

$$H_0 : \beta_1 = \beta_2 \tag{6.121}$$

against the alternative hypothesis

$$H_a : \beta_1 \neq \beta_2.$$

The hypothesis (6.121) is called the *hypothesis of concurrence*. Other hypotheses can also be considered, for example, equalities involving only portions of β_1 and β_2.

Two cases will be considered, depending on whether or not ϵ_1 and ϵ_2 are independent.

Case 1. ϵ_1 and ϵ_2 are independently distributed.

Models (6.119) and (6.120) can be combined into a single linear model of the form

$$Y = X\beta + \epsilon, \tag{6.122}$$

where $Y = (Y_1' : Y_2')'$, $X = \text{diag}(X_1, X_2)$, $\beta = (\beta_1' : \beta_2')'$, and $\epsilon = (\epsilon_1' : \epsilon_2')'$. Since ϵ_1 and ϵ_2 are independent and normally distributed, ϵ is also normally distributed as $N(\mathbf{0}, \Delta)$, where $\Delta = \text{diag}(\sigma_1^2 I_{n_1}, \sigma_2^2 I_{n_2})$. The hypothesis (6.121) can then be written as

$$H_0 : A\beta = \mathbf{0}, \tag{6.123}$$

where $A = [I_p : -I_p]$. Hence, under H_0,

$$\left(A\hat{\beta}\right)' \left[A\left(X'\Delta^{-1}X\right)^{-1} A'\right]^{-1} A\hat{\beta} \sim \chi_p^2, \tag{6.124}$$

where

$$\begin{aligned}\hat{\beta} &= \left(X'\Delta^{-1}X\right)^{-1} X'\Delta^{-1}Y \\ &= \left[\hat{\beta}_1' : \hat{\beta}_2'\right]',\end{aligned}$$

and $\hat{\beta}_i = (X_i'X_i)^{-1}X_i'Y_i$ ($i = 1, 2$). The chi-squared random variable in (6.124) can be written as

$$\left(\hat{\beta}_1 - \hat{\beta}_2\right)'\left[\sigma_1^2\left(X_1'X_1\right)^{-1} + \sigma_2^2\left(X_2'X_2\right)^{-1}\right]^{-1}\left(\hat{\beta}_1 - \hat{\beta}_2\right) \sim \chi_p^2. \qquad (6.125)$$

Furthermore, $\frac{(n_i-1)MS_{E_i}}{\sigma_i^2} \sim \chi^2_{n_i-p}$ independently for $i = 1, 2$, where

$$MS_{E_i} = \frac{1}{n_i - p} Y_i'[I_n - X_i(X_i'X_i)^{-1}X_i']Y_i, \quad i = 1, 2,$$

is the residual mean square for the ith model ($i = 1, 2$). Since MS_{E_i} ($i = 1, 2$) is independent of both $\hat{\beta}_1$ and $\hat{\beta}_2$, and hence of the chi-squared distribution in (6.125), the F-ratio,

$$F = \left(\frac{n_1 + n_2 - 2p}{p}\right)\frac{\left(\hat{\beta}_1 - \hat{\beta}_2\right)'\left[\sigma_1^2\left(X_1'X_1\right)^{-1} + \sigma_2^2\left(X_2'X_2\right)^{-1}\right]^{-1}\left(\hat{\beta}_1 - \hat{\beta}_2\right)}{(n_1 - p)\,MS_{E_1}/\sigma_1^2 + (n_2 - p)\,MS_{E_2}/\sigma_2^2}$$

has the F-distribution with p and $n_1 + n_2 - 2p$ degrees of freedom under H_0. It can be noted that this F-ratio depends on the unknown value of $\frac{\sigma_1^2}{\sigma_2^2}$ and cannot therefore be used as a test statistic for testing H_0. If, however, $\frac{\sigma_1^2}{\sigma_2^2} = c$, where c is a known positive constant, then

$$F = \left(\frac{n_1 + n_2 - 2p}{p}\right)\frac{\left(\hat{\beta}_1 - \hat{\beta}_2\right)'\left[c\left(X_1'X_1\right)^{-1} + \left(X_2'X_2\right)^{-1}\right]^{-1}\left(\hat{\beta}_1 - \hat{\beta}_2\right)}{(n_1 - p)\,MS_{E_1}/c + (n_2 - p)\,MS_{E_2}}$$

is the test statistic, which, under H_0, has the F-distribution with p and $n_1 + n_2 - 2p$ degrees of freedom. The hypothesis H_0 can be rejected at the α-level if $F \geq F_{\alpha, p, n_1+n_2-2p}$. In general, when σ_1^2 and σ_2^2 are unknown and the ratio $\frac{\sigma_1^2}{\sigma_2^2}$ is also unknown, σ_1^2 and σ_2^2 in the formula for F can be replaced by MS_{E_1} and MS_{E_2}, respectively. In this case, F_{p, n_1+n_2-2p} is considered an approximation to the distribution of the resulting F-statistic under H_0 (see Ali and Silver, 1985).

Case 2. ϵ_1 and ϵ_2 are correlated.

This case arises when, for example, the data used to fit the two models are obtained from the same experimental units at two different treatment conditions.

Consider again the models in (6.119) and (6.120). Assume that X_1 and X_2 have the same number of rows, n. The models can be written as

$$Y_1 = \gamma_{10}\mathbf{1}_n + Z_1\gamma_1 + \epsilon_1 \qquad (6.126)$$

$$Y_2 = \gamma_{20}\mathbf{1}_n + Z_2\gamma_2 + \epsilon_2, \qquad (6.127)$$

where Z_i is a matrix of order $n \times (p-1)$ such that $[\mathbf{1}_n : Z_i] = X_i$, and $(\gamma_{i0} : \boldsymbol{\gamma}_i') = \boldsymbol{\beta}_i'$, $i = 1, 2$. Suppose that $\text{Cov}(\boldsymbol{\epsilon}_1, \boldsymbol{\epsilon}_2) = \rho\sigma_1\sigma_2 \boldsymbol{I}_n$, where ρ is an unknown correlation coefficient. Two hypotheses can be tested, namely,

$$H_{0p} : \boldsymbol{\gamma}_1 = \boldsymbol{\gamma}_2, \tag{6.128}$$

which is called the *hypothesis of parallelism*, and

$$H_{0c} : \gamma_{10} = \gamma_{20}; \ \boldsymbol{\gamma}_1 = \boldsymbol{\gamma}_2, \tag{6.129}$$

which is called the *hypothesis of concurrence*. If we were to use the same approach as in Case 1, which utilized weighted least squares, we would obtain an F-ratio that depends on the variance–covariance matrix $\boldsymbol{\Gamma}$ for the combined error vector $\boldsymbol{\epsilon}$, that is,

$$\boldsymbol{\Gamma} = \begin{bmatrix} \sigma_1^2 \boldsymbol{I}_n & \rho\sigma_1\sigma_2 \boldsymbol{I}_n \\ \rho\sigma_1\sigma_2 \boldsymbol{I}_n & \sigma_2^2 \boldsymbol{I}_n \end{bmatrix}.$$

Since $\boldsymbol{\Gamma}$ is unknown, it would be necessary to estimate it and use the estimate in the F-ratio. This, however, would result in an approximate asymptotic test (see Zellner, 1962). The approach considered in this section uses instead an exact test which was developed by Smith and Choi (1982). This test does not require estimating $\boldsymbol{\Gamma}$. The following is a description of this approach.

Let $\boldsymbol{d} = \boldsymbol{Y}_1 - \boldsymbol{Y}_2$. Then, from models (6.126) and (6.127), we get

$$\boldsymbol{d} = \boldsymbol{C}_1 \boldsymbol{\delta}_1 + \boldsymbol{\eta}, \tag{6.130}$$

where $\boldsymbol{C}_1 = [\mathbf{1}_n : \boldsymbol{Z}_1 : -\boldsymbol{Z}_2]$, $\boldsymbol{\delta}_1 = (\gamma_{10} - \gamma_{20} : \boldsymbol{\gamma}_1' : \boldsymbol{\gamma}_2')'$, and $\boldsymbol{\eta} = \boldsymbol{\epsilon}_1 - \boldsymbol{\epsilon}_2$. Hence, $\boldsymbol{\eta} \sim N(\mathbf{0}, \sigma_d^2 \boldsymbol{I}_n)$, where $\sigma_d^2 = \sigma_1^2 + \sigma_2^2 - 2\rho\sigma_1\sigma_2$.

The tests for parallelism and concurrence consider 3 cases according to the number, q, of the $p-1$ covariates in the models that have been measured at the same levels for both treatment conditions. These cases are (i) $q = 0$, (ii) $0 < q < p-1$, (iii) $q = p-1$, where p is the number of parameters in each model, including the intercept.

Case (i). $q = 0$.

If the columns of $\boldsymbol{C}_1$ in (6.130) are linearly independent, then $\boldsymbol{C}_1$ is of full column rank, which is equal to $2p-1$. The hypothesis of parallelism in (6.128) can be written as

$$H_{0p} : \boldsymbol{A}_p \boldsymbol{\delta}_1 = \mathbf{0},$$

where $\boldsymbol{A}_p = [\mathbf{0} : \boldsymbol{I}_{p-1} : -\boldsymbol{I}_{p-1}]$ is of order $(p-1) \times (2p-1)$ and rank $p-1$. The corresponding test statistic is

$$F_p = \frac{\left(\boldsymbol{A}_p \hat{\boldsymbol{\delta}}_1\right)' \left[\boldsymbol{A}_p \left(\boldsymbol{C}_1' \boldsymbol{C}_1\right)^{-1} \boldsymbol{A}_p'\right]^{-1} \left(\boldsymbol{A}_p \hat{\boldsymbol{\delta}}_1\right)}{(p-1) MS_E^{(1)}}, \tag{6.131}$$

where $\hat{\boldsymbol{\delta}}_1 = (\boldsymbol{C}_1'\boldsymbol{C}_1)^{-1}\boldsymbol{C}_1'\boldsymbol{d}$ and

$$MS_E^{(1)} = \frac{1}{n-2p+1}\,\boldsymbol{d}'\left[\boldsymbol{I}_n - \boldsymbol{C}_1\left(\boldsymbol{C}_1'\boldsymbol{C}_1\right)^{-1}\boldsymbol{C}_1'\right]\boldsymbol{d}.$$

Under H_{0p}, F_p has the F-distribution with $p-1$ and $n-2p+1$ degrees of freedom.

The hypothesis of concurrence, H_{0c}, in (6.129) can be expressed as

$$H_{0c} : \boldsymbol{A}_c\boldsymbol{\delta}_1 = \boldsymbol{0},$$

where

$$\boldsymbol{A}_c = \begin{bmatrix} 1 & 0 & 0 \\ \boldsymbol{0} & \boldsymbol{I}_{p-1} & -\boldsymbol{I}_{p-1} \end{bmatrix}$$

is a matrix of order $p \times (2p-1)$ and rank p. The corresponding test statistic is

$$F_c = \frac{\left(\boldsymbol{A}_c\hat{\boldsymbol{\delta}}_1\right)'\left[\boldsymbol{A}_c\left(\boldsymbol{C}_1'\boldsymbol{C}_1\right)^{-1}\boldsymbol{A}_c'\right]^{-1}\left(\boldsymbol{A}_c\hat{\boldsymbol{\delta}}_1\right)}{pMS_E^{(1)}}.$$

Under H_{0c}, F_c has the F-distribution with p and $n-2p+1$ degrees of freedom.

Case (ii). $0 < q < p-1$.

In this case, the matrices $\boldsymbol{Z}_1$ and $\boldsymbol{Z}_2$ in (6.126) and (6.127) can be written as

$$\boldsymbol{Z}_1 = [\boldsymbol{Z}_0 : \boldsymbol{Z}_{11}]$$
$$\boldsymbol{Z}_2 = [\boldsymbol{Z}_0 : \boldsymbol{Z}_{22}],$$

where $\boldsymbol{Z}_0$ is of order $n \times q$ whose rows represent the settings of the covariates in both models that have been measured at the same levels for both treatment conditions (assuming that these covariates were written first in the models). The matrices, $\boldsymbol{Z}_{11}$ and $\boldsymbol{Z}_{22}$ in $\boldsymbol{Z}_1$ and $\boldsymbol{Z}_2$ are different. In this case, the model for $\boldsymbol{d} = \boldsymbol{Y}_1 - \boldsymbol{Y}_2$ is

$$\boldsymbol{d} = \boldsymbol{C}_2\boldsymbol{\delta}_2 + \boldsymbol{\eta},$$

where

$$\boldsymbol{C}_2 = [\boldsymbol{1}_n : \boldsymbol{Z}_0 : \boldsymbol{Z}_{11} : -\boldsymbol{Z}_{22}]$$
$$\boldsymbol{\delta}_2 = \left[\gamma_{10} - \gamma_{20} : \boldsymbol{\gamma}_1^{(0)'} - \boldsymbol{\gamma}_2^{(0)'} : \boldsymbol{\gamma}_1^{(1)'} : \boldsymbol{\gamma}_2^{(1)'}\right]',$$

where $\boldsymbol{\gamma}_i^{(0)}$, $\boldsymbol{\gamma}_i^{(1)}$ are the portions of $\boldsymbol{\gamma}_i$ that correspond to $\boldsymbol{Z}_0$ and $\boldsymbol{Z}_{ii}$, respectively $(i = 1, 2)$. In this case, $\boldsymbol{C}_2$ is of full column rank, which is equal to $2p-q-1$.

The hypothesis of parallelism takes the form

$$H_{0p} : \boldsymbol{B}_p\boldsymbol{\delta}_2 = \boldsymbol{0},$$

where

$$B_p = \text{diag}(M_1, M_2)$$

is a matrix of order $(p-1) \times (2p-q-1)$ and rank $p-1$, $M_1 = [\mathbf{0} : I_q]$ is a $q \times (q+1)$ matrix, and $M_2 = [I_{p-1-q} : -I_{p-1-q}]$ is a $(p-1-q) \times (2p-2-2q)$ matrix. The corresponding test statistic is

$$F_p = \frac{\left(B_p \hat{\delta}_2\right)' \left[B_p \left(C_2' C_2\right)^{-1} B_p'\right]^{-1} \left(B_p \hat{\delta}_2\right)}{(p-1) MS_E^{(2)}},$$

where $\hat{\delta}_2 = (C_2' C_2)^{-1} C_2 d$ and

$$MS_E^{(2)} = \frac{1}{n-2p+q+1} d' \left[I_n - C_2 \left(C_2' C_2\right)^{-1} C_2'\right] d.$$

Under H_{0p}, F_p has the F-distribution with $p-1$ and $n-2p+q+1$ degrees of freedom.

As for the hypothesis of concurrence, it can be written as

$$H_{0c} : B_c \delta_2 = \mathbf{0},$$

where

$$B_c = \begin{bmatrix} 1 & \mathbf{0}' \\ \mathbf{0} & B_p \end{bmatrix}$$

is a $p \times (2p-q)$ matrix of rank p. The relevant test statistic is given by

$$F_c = \frac{\left(B_c \hat{\delta}_2\right)' \left[B_c \left(C_2' C_2\right)^{-1} B_c'\right]^{-1} \left(B_c \hat{\delta}_2\right)}{p MS_E^{(2)}},$$

which has the F-distribution with p and $n-2p+q+1$ degrees of freedom under H_{0c}.

Case (iii). $q = p-1$.

In this case, $Z_1 = Z_2$, that is, all $p-1$ covariates are measured at the same levels for both treatment conditions. The model for $d = Y_1 - Y_2$ can then be written as

$$d = C_3 \delta_3 + \eta,$$

where

$$C_3 = [\mathbf{1}_n : Z_1]$$
$$\delta_3 = \left[\gamma_{10} - \gamma_{20} : \gamma_1' - \gamma_2'\right]'.$$

The matrix C_3 is of full column rank, which is equal to p.

The hypothesis of parallelism is

$$H_{0p} : \boldsymbol{D}_p \boldsymbol{\delta}_3 = \boldsymbol{0},$$

where $\boldsymbol{D}_p = [\boldsymbol{0} : \boldsymbol{I}_{p-1}]$ is of order $(p-1) \times p$ and rank $p-1$. The relevant test statistic is

$$F_p = \frac{\left(\boldsymbol{D}_p \hat{\boldsymbol{\delta}}_3\right)' \left[\boldsymbol{D}_p \left(\boldsymbol{C}_3' \boldsymbol{C}_3\right)^{-1} \boldsymbol{D}_p'\right]^{-1} \left(\boldsymbol{D}_p \hat{\boldsymbol{\delta}}_3\right)}{(p-1) MS_E^{(3)}},$$

where $\hat{\boldsymbol{\delta}}_3 = (\boldsymbol{C}_3' \boldsymbol{C}_3)^{-1} \boldsymbol{C}_3' \boldsymbol{d}$ and

$$MS_E^{(3)} = \frac{1}{n-p} \boldsymbol{d}' \left[\boldsymbol{I}_n - \boldsymbol{C}_3 \left(\boldsymbol{C}_3' \boldsymbol{C}_3\right)^{-1} \boldsymbol{C}_3'\right] \boldsymbol{d}.$$

Under H_{0p}, F_p has the F-distribution with $p-1$ and $n-p$ degrees of freedom.

Finally, for the hypothesis of concurrence, we have

$$H_{0c} : \boldsymbol{\delta}_3 = \boldsymbol{0},$$

and the corresponding test statistic is

$$F_c = \frac{\hat{\boldsymbol{\delta}}_3' \boldsymbol{C}_3' \boldsymbol{C}_3 \hat{\boldsymbol{\delta}}_3}{p MS_E^{(3)}},$$

which has the F-distribution with p and $n-p$ degrees of freedom under H_{0c}.

Example 6.3 This example, which was given by Smith and Choi (1982), is concerned with the effect of body weight, x, on glucose tolerance, Y, for an individual. Twenty six healthy males ingested a standard glucose solution and the plasma glucose level of each individual was measured at 1 and 3 h after ingestion. The resulting data are given in Table 6.7.

First-degree models were fitted to the glucose values at 1 and 3 h, respectively. These were found to be

$$\hat{Y}_1 = -9.07 + 0.72\,x$$
$$\hat{Y}_2 = 103.29 - 0.18\,x.$$

In this case, $p = 2$ and $q = 1$. The hypothesis of parallelism has the test statistic value $F_p = 12.8164$ with 1 and 24 degrees of freedom. The corresponding p-value is 0.0015, which gives little support for the hypothesis. Thus, the rates of change of glucose level with respect to weight at 1 and 3 h are not the same.

TABLE 6.7
Weight (in lb) and Glucose Values (mg/100 ml)

	Glucose Value			Glucose Value	
Weight	**1 h**	**3 h**	**Weight**	**1 h**	**3 h**
175	103	76	185	110	50
140	108	50	160	118	91
165	98	76	150	70	66
187	123	50	185	90	116
202	158	48	165	98	108
140	93	106	177	116	64
212	150	61	140	108	68
206	139	44	128	98	48
169	110	62	165	92	89
155	76	96	180	134	86
178	148	79	215	157	85
185	135	49	225	136	66
205	141	45	181	134	102

Source: Reprinted from Smith, P.J. and Choi, S.C., *Technometrics*, 24, 123, 1982. With permission.

Exercises

6.1 Let $X_1, X_2, \ldots, X_n$ be random variables with mean μ and variance σ^2, but are not necessarily mutually independent. Let s^2 be the sample variance.

(a) Show that

$$E(s^2) \leq \frac{n\sigma^2}{n-1}.$$

(b) Can the upper bound in (a) be attained?

6.2 Let $\boldsymbol{Y} = \boldsymbol{X}\boldsymbol{\beta} + \boldsymbol{\epsilon}$, where $\boldsymbol{X}$ is of order $n \times p$ and rank $p(<n)$. Suppose that $E(\boldsymbol{\epsilon}) = \boldsymbol{0}$ and $\text{Var}(\boldsymbol{\epsilon}) = \boldsymbol{\Sigma} = (\sigma_{ij})$. Let

$$MS_E = \frac{1}{n-p}\boldsymbol{Y}'[\boldsymbol{I}_n - \boldsymbol{X}(\boldsymbol{X}'\boldsymbol{X})^{-1}\boldsymbol{X}']\boldsymbol{Y}.$$

(a) Show that

$$E(MS_E) \leq \frac{1}{n-p}\sum_{i=1}^{n}\sigma_{ii}.$$

(b) Can the upper bound in (a) be attained?

(c) Deduce from (a) that if $\sigma_{ii} = \sigma^2$ for $i = 1, 2, \ldots, n$, then

$$| E(MS_E) - \sigma^2 | \leq \sigma^2 \max\left\{1, \frac{p}{n-p}\right\}.$$

The quantity $E(MS_E) - \sigma^2$ represents the bias of MS_E.

[Note: For more details about this exercise, see Dufour (1986)].

6.3 Consider the linear model,

$$\boldsymbol{Y} = \boldsymbol{X\beta} + \boldsymbol{\epsilon},$$

where $\boldsymbol{X}$ is $n \times p$ of rank p $(< n)$, and $\boldsymbol{\epsilon} \sim N(\boldsymbol{0}, \sigma^2 \boldsymbol{I}_n)$.

(a) Show that

$$E\left(\frac{1}{MS_E}\right) = \frac{n-p}{\sigma^2(n-p-2)},$$

provided that $n > p + 2$, where MS_E is the error mean square.

(b) Find the uniformly minimum variance unbiased estimator of $\frac{\boldsymbol{a}'\boldsymbol{\beta}}{\sigma^2}$, where $\boldsymbol{a}$ is a constant vector.

6.4 Consider the simple linear regression model,

$$Y_u = \beta_0 + \beta_1 x_u + \epsilon_u, \quad u = 1, 2, \ldots, n,$$

where $\epsilon_1, \epsilon_2, \ldots, \epsilon_n$ are mutually independent and distributed as $N(0, \sigma^2)$ $(n > 2)$.

(a) Show that $\frac{n-4}{(n-2)MS_E}$ is an unbiased estimator of $\frac{1}{\sigma^2}$, where MS_E is the error mean square.

(b) Find the uniformly minimum variance unbiased estimator (UMVUE) of $2\,\sigma^2 + 5\,\beta_1$.

(c) Find the UMVUE of $\frac{\beta_0}{\sigma^2}$.

(d) Show that $\text{Var}(\hat{\beta}_0)$ achieves its minimum value if $x_1, x_2, \ldots, x_n$ are chosen such that $\sum_{u=1}^{n} x_u = 0$, where $\hat{\beta}_0$ is the least-squares estimator of β_0.

6.5 Consider again the simple linear regression model in Exercise 6.4. Let $\mu(x) = \beta_0 + \beta_1 x$ denote the mean response at a point x in the experimental region. Let x_0 be the value of x at which $\mu = 0$. Under what conditions is it possible to obtain a 95% confidence interval on x_0?

6.6 Consider the same linear model as in Exercise 6.3. Make use of the confidence region on $\boldsymbol{\beta}$, given by the inequality in (6.72), to obtain simultaneous confidence intervals on the elements of $\boldsymbol{\beta}$ with a joint confidence coefficient greater than or equal to $1 - \alpha$.

6.7 Prove formula (6.77).

6.8 Consider the full-rank model

$$Y = X\beta + \epsilon,$$

where X is an $n \times p$ matrix of rank p. Suppose that the mean and variance–covariance matrix of Y are given by

$$E(Y) = X\beta + Z\gamma,$$
$$\text{Var}(Y) = \sigma^2 I_n,$$

where Z is an $n \times q$ matrix such that $W = [X : Z]$ is of full column rank, and γ is a vector of unknown parameters.

(a) Show that $\hat{\beta} = (X'X)^{-1}X'Y$ is a biased estimator of β.

(b) Show that, in general, MS_E, is a biased estimator of σ^2, where

$$MS_E = \frac{1}{n-p} Y'[I_n - X(X'X)^{-1}X']Y,$$

and that $E(MS_E) \geq \sigma^2$.

(c) Under what conditions can MS_E be unbiased for σ^2?

(d) Show that $\widetilde{SS_E} \leq SS_E$, where $SS_E = (n-p)MS_E$ and $\widetilde{SS_E} = Y'[I_n - W(W'W)^{-1}W']Y$.

6.9 Let $e = Y - \hat{Y}$ be the vector of residuals given in formula (6.16). Show that $\text{Var}(e_i) < \sigma^2$, $i = 1, 2, \ldots, n$, where e_i is the ith element of e ($i = 1, 2, \ldots, n$), and σ^2 is the error variance.

6.10 Prove formula (6.117).

6.11 Consider the full-rank model

$$Y = X\beta + \epsilon,$$

where $E(\epsilon) = \mathbf{0}$ and $\text{Var}(\epsilon) = \Sigma$. Show that the BLUE of β, namely, $(X'\Sigma^{-1}X)^{-1}X'\Sigma^{-1}Y$, is equal to the ordinary least-squares estimator, $(X'X)^{-1}X'Y$, if and only if there exists a nonsingular matrix, F, such that $\Sigma X = XF$.

[Hint: See Theorem 6.8.1 in Graybill (1976)].

6.12 Consider again the same model as in Exercise 6.11. Show that the ordinary least-squares estimator of β is BLUE if and only if

$$X'\Sigma^{-1}(I_n - W) = \mathbf{0},$$

where $W = X(X'X)^{-1}X'$.

[Hint: See Milliken and Albohali (1984)].

6.13 Consider fitting the second-degree model,

$$Y = \beta_0 + \sum_{i=1}^{k} \beta_i x_i + \sum_{i<j}^{k} \beta_{ij} x_i x_j + \epsilon \tag{6.132}$$

to a given data set. Let $z_1, z_2, \ldots, z_k$ denote the coded variables corresponding to $x_1, x_2, \ldots, x_k$, respectively, such that

$$z_i = \frac{x_i - a_i}{b_i}, \quad i = 1, 2, \ldots, k,$$

where a_i and b_i are known constants. Applying this transformation to model (6.132), we obtain

$$Y = \gamma_0 + \sum_{i=1}^{k} \gamma_i z_i + \sum_{i<j}^{k} \gamma_{ij} z_i z_j + \epsilon, \tag{6.133}$$

where the γ_i's and γ_{ij}'s are unknown parameters. Using the given data, models (6.132) and (6.133) can be expressed as

$$\boldsymbol{Y} = \boldsymbol{X}\boldsymbol{\beta} + \boldsymbol{\epsilon} \tag{6.134}$$

$$\boldsymbol{Y} = \boldsymbol{Z}\boldsymbol{\gamma} + \boldsymbol{\epsilon}. \tag{6.135}$$

(a) Show that the column spaces of $\boldsymbol{X}$ and $\boldsymbol{Z}$ are identical [in this case, models (6.134) and (6.135) are said to be *equivalent*].

(b) Show that $\boldsymbol{X}\hat{\boldsymbol{\beta}} = \boldsymbol{Z}\hat{\boldsymbol{\gamma}}$, where $\hat{\boldsymbol{\beta}} = (\boldsymbol{X}'\boldsymbol{X})^{-1}\boldsymbol{X}'\boldsymbol{Y}, \hat{\boldsymbol{\gamma}} = (\boldsymbol{Z}'\boldsymbol{Z})^{-1}\boldsymbol{Z}'\boldsymbol{Y}$.

(c) Show that the regression sum of squares and the error sum of squares are the same for models (6.134) and (6.135).

6.14 Let $\boldsymbol{Y} = \boldsymbol{X}\boldsymbol{\beta} + \boldsymbol{\epsilon}$ be partitioned as $\boldsymbol{Y} = \boldsymbol{X}_1\boldsymbol{\beta}_1 + \boldsymbol{X}_2\boldsymbol{\beta}_2 + \boldsymbol{\epsilon}$, where $\boldsymbol{X}$ is $n \times p$ of rank p and $\boldsymbol{\epsilon} \sim N(\boldsymbol{0}, \sigma^2 \boldsymbol{I}_n)$. Let $R(\boldsymbol{\beta}_2|\boldsymbol{\beta}_1)$ be defined as

$$R(\boldsymbol{\beta}_2|\boldsymbol{\beta}_1) = \boldsymbol{Y}'\boldsymbol{X}\left(\boldsymbol{X}'\boldsymbol{X}\right)^{-1}\boldsymbol{X}'\boldsymbol{Y} - \boldsymbol{Y}'\boldsymbol{X}_1\left(\boldsymbol{X}_1'\boldsymbol{X}_1\right)^{-1}\boldsymbol{X}_1'\boldsymbol{Y}.$$

This represents the increase in the regression sum of squares due to the addition of $\boldsymbol{X}_2\boldsymbol{\beta}_2$ to a model that contains only $\boldsymbol{X}_1\boldsymbol{\beta}_1$.

(a) Show that $\frac{1}{\sigma^2}R(\boldsymbol{\beta}_2|\boldsymbol{\beta}_1)$ has the noncentral chi-squared distribution.

(b) Show that $R(\boldsymbol{\beta}_2|\boldsymbol{\beta}_1)$ is independent of both $R(\boldsymbol{\beta}_1)$ and SS_E, where $R(\boldsymbol{\beta}_1) = \boldsymbol{Y}'\boldsymbol{X}_1(\boldsymbol{X}_1'\boldsymbol{X}_1)^{-1}\boldsymbol{X}_1'\boldsymbol{Y}$ and SS_E is the error sum of squares for the full model, that is, $\boldsymbol{Y} = \boldsymbol{X}\boldsymbol{\beta} + \boldsymbol{\epsilon}$.

(c) Find the expected value of $R(\boldsymbol{\beta}_2|\boldsymbol{\beta}_1)$.

(d) Deduce from (b) and (c) the hypothesis that can be tested by the F-ratio,

$$F = \frac{R(\beta_2|\beta_1)}{p_2 MS_E},$$

where p_2 is the number of columns of $\mathbf{X}_2$ and $MS_E = \frac{SS_E}{n-p}$.

6.15 Consider the model,

$$Y_u = \beta_0 + \sum_{i=1}^{k} \beta_i x_{ui} + \epsilon_u, \quad u = 1, 2, \ldots, n,$$

which can be written as $\boldsymbol{Y} = \boldsymbol{X}\boldsymbol{\beta} + \boldsymbol{\epsilon}$. The matrix $\boldsymbol{X}$ is partitioned as $\boldsymbol{X} = [\mathbf{1}_n : \boldsymbol{D}]$, where $\boldsymbol{D}$ is $n \times k$ of rank k whose uth row consists of the settings $x_{u1}, x_{u2}, \ldots, x_{uk}$, $u = 1, 2, \ldots, n$.

(a) Show that $d^{ii} \geq \frac{1}{d_{ii}}$, where d_{ii} is the ith diagonal element of $\boldsymbol{D}'\boldsymbol{D}$ and d^{ii} is the ith diagonal element of $(\boldsymbol{D}'\boldsymbol{D})^{-1}$.

(b) Under what conditions can the equality in (a) be attained?

(c) What is the advantage of having a design (i.e., the matrix $\boldsymbol{D}$) that satisfies the conditions in (b)?

6.16 Consider the model,

$$Y(\boldsymbol{x}) = \boldsymbol{f}'(\boldsymbol{x})\boldsymbol{\beta} + \epsilon,$$

where $\epsilon \sim N(0, \sigma^2)$, $\boldsymbol{f}'(\boldsymbol{x})\boldsymbol{\beta}$ represents a polynomial of degree d in the elements of $\boldsymbol{x}$. The response Y is observed at n settings of $\boldsymbol{x}$, namely, $\boldsymbol{x}_1, \boldsymbol{x}_2, \ldots, \boldsymbol{x}_n$. The resulting data set is used to fit the model and obtain $\hat{\boldsymbol{\beta}}$, the ordinary least-squares estimate of $\boldsymbol{\beta}$. The usual assumption of independence of the error terms is considered valid.

Suppose that Y is to be predicted at k "new" points, $\boldsymbol{x}_{n+1}, \boldsymbol{x}_{n+2}, \ldots, \boldsymbol{x}_{n+k}$. Let $\boldsymbol{Y}^* = [Y(\boldsymbol{x}_{n+1}), Y(\boldsymbol{x}_{n+2}), \ldots, Y(\boldsymbol{x}_{n+k})]'$, and let $\boldsymbol{X}^*$ be the corresponding matrix of order $k \times p$ whose uth row is $\boldsymbol{f}'(\boldsymbol{x}_u)$, $u = n+1, n+2, \ldots, n+k$, where p is the number of elements of $\boldsymbol{\beta}$.

(a) Find the variance–covariance matrix of $\boldsymbol{Y}^* - \boldsymbol{X}^*\hat{\boldsymbol{\beta}}$.

(b) Use (a) to obtain a region that contains $\boldsymbol{Y}^*$ with probability $1 - \alpha$.

[Note: Such a region is called a *prediction region* on $\boldsymbol{Y}^*$. In particular, if $k = 1$, we get the well-known prediction interval on a "new" response value.]

(c) Use (b) to obtain simultaneous prediction intervals on the elements of $\boldsymbol{Y}^*$ with a joint coverage probability greater than or equal to $1 - \alpha$.

6.17 Consider models (6.119) and (6.120) and the null hypothesis $H_0 : \boldsymbol{\beta}_1 = \boldsymbol{\beta}_2$ (see Section 6.7.5). Suppose that $\boldsymbol{\epsilon}_1$ and $\boldsymbol{\epsilon}_2$ are independently distributed. Using the likelihood ratio test, show that a test statistic for testing H_0, when the error variances are equal, is given by the ratio

$$T = \frac{(SS_E - SS_{E_1} - SS_{E_2})/p}{(SS_{E_1} + SS_{E_2})/(n_1 + n_2 - 2p)},$$

where $SS_{E_i} = \boldsymbol{Y}_i'[\boldsymbol{I}_n - \boldsymbol{X}_i(\boldsymbol{X}_i'\boldsymbol{X}_i)^{-1}\boldsymbol{X}_i']\boldsymbol{Y}_i$, $i = 1, 2$, and $SS_E = \boldsymbol{Y}'[\boldsymbol{I}_n - \boldsymbol{X}_0(\boldsymbol{X}_0'\boldsymbol{X}_0)^{-1}\boldsymbol{X}_0']\boldsymbol{Y}$, where $\boldsymbol{Y} = (\boldsymbol{Y}_1' : \boldsymbol{Y}_2')'$ and $\boldsymbol{X}_0 = [\boldsymbol{X}_1' : \boldsymbol{X}_2']'$.

6.18 Suppose that the full-rank model, $\boldsymbol{Y} = \boldsymbol{X}\boldsymbol{\beta} + \boldsymbol{\epsilon}$, is partitioned as $\boldsymbol{Y} = \boldsymbol{X}_1\boldsymbol{\beta}_1 + \boldsymbol{X}_2\boldsymbol{\beta}_2 + \boldsymbol{\epsilon}$, where $\boldsymbol{X}_1$ is $n \times p_1$, $\boldsymbol{X}_2$ is $n \times p_2$, and $\boldsymbol{\epsilon} \sim N(\boldsymbol{0}, \sigma^2\boldsymbol{I}_n)$.

(a) Show that the least-squares estimate of $\boldsymbol{\beta}_2$ can be written as $\hat{\boldsymbol{\beta}}_2 = \boldsymbol{C}'\boldsymbol{Y}$, where $\boldsymbol{C} = \boldsymbol{X}_1\boldsymbol{C}_{12} + \boldsymbol{X}_2\boldsymbol{C}_{22}$, $\boldsymbol{C}_{12}$ and $\boldsymbol{C}_{22}$ are matrices of orders $p_1 \times p_2$ and $p_2 \times p_2$, respectively, that depend on $\boldsymbol{X}_1$ and $\boldsymbol{X}_2$, and $\boldsymbol{C}_{22}$ is positive definite.

(b) Let SS_{E_1} be the residual sum of squares for the regression of $\boldsymbol{Y}$ on $\boldsymbol{X}_1$ alone. Show that SS_{E_1} and $\hat{\boldsymbol{\beta}}_2$ are not independent.

(c) Show that $[\boldsymbol{I}_n - \boldsymbol{X}_1(\boldsymbol{X}_1'\boldsymbol{X}_1)^{-1}\boldsymbol{X}_1']\boldsymbol{C}$ is of rank p_2.

6.19 An experiment was conducted to investigate the effect of five control variables on the selective H_2SO_4 hydrolysis of waxy maize starch granules. These variables were x_1 = temperature, x_2 = acid concentration, x_3 = starch concentration, x_4 = hydrolysis duration (time), and x_5 = stirring speed. The measured response, Y = hydrolysis yield (wt%), was calculated as the ratio between the weight of freeze-dried hydrolyzed particles and the initial weight of native granules for an aliquot of 50 mL taken in the 250 mL of hydrolyzed suspensions. The original and coded settings of x_1, x_2, x_3, x_4, x_5 are given in the following table:

Variable	Unit	Low Level ($x_i = -1$)	Medium Level ($x_i = 0$)	High Level ($x_i = 1$)
x_1	°C	35	37.5	40
x_2	mol/L	2.2	2.8	3.4
x_3	g/100 mL	5	10	15
x_4	day	1	5	9
x_5	rpm	0	50	100

Source: Reprinted from Angellier., H. et al., *Biomacromolecules*, 5, 1545, 2004. With permission.

Note that the coded settings of the three equally-spaced levels of each variable are −1, 0, and 1. A *central composite design* (see Chapter 4 in Khuri and Cornell, 1996) consisting of a one-half fraction of a 2^5 factorial design (runs 1–16), 10 axial points (runs 17–26), and five center-point replications (runs 27–31) was used to measure the response Y at the specified combinations of the levels of the five factors. This particular design is also known as a *face-centered cube* because the axial points are at a distance equal to 1 from the origin. The resulting data set (using the coded settings of the control variables) was given by Angellier et al. (2004) and is reproduced in Table 6.8.

(a) Use the data in Table 6.8 to fit the model,

$$Y = \beta_0 + \sum_{i=1}^{5} \beta_i x_i + \beta_{35} x_3 x_5 + \beta_{44} x_4^2 + \epsilon,$$

where $\epsilon \sim N(0, \sigma^2)$.

TABLE 6.8
Design Settings and Hydrolysis Yield Data

Experimental Run	x_1	x_2	x_3	x_4	x_5	Y (%)
1	−1	−1	−1	−1	1	76.3
2	1	−1	−1	−1	−1	68.1
3	−1	1	−1	−1	−1	47.6
4	1	1	−1	−1	1	26.3
5	−1	−1	1	−1	−1	70.7
6	1	−1	1	−1	1	54.8
7	−1	1	1	−1	1	57.8
8	1	1	1	−1	−1	35.9
9	−1	−1	−1	1	−1	43.9
10	1	−1	−1	1	1	20.3
11	−1	1	−1	1	1	5.4
12	1	1	−1	1	−1	2.8
13	−1	−1	1	1	1	44.8
14	1	−1	1	1	−1	29.3
15	−1	1	1	1	−1	16.7
16	1	1	1	1	1	2.1
17	−1	0	0	0	0	42.3
18	1	0	0	0	0	26.5
19	0	−1	0	0	0	30.4
20	0	1	0	0	0	21.3
21	0	0	−1	0	0	24.1
22	0	0	1	0	0	34.9
23	0	0	0	−1	0	56.4
24	0	0	0	1	0	20.6
25	0	0	0	0	−1	36.0
26	0	0	0	0	1	20.3
27	0	0	0	0	0	37.6
28	0	0	0	0	0	31.8
29	0	0	0	0	0	28.3
30	0	0	0	0	0	29.7
31	0	0	0	0	0	28.9

Source: Reprinted from Angellier, H. et al., *Biomacromolecules*, 5, 1545, 2004. With permission.

(b) Give values of the least-squares estimates of the model's parameters and their standard errors.

(c) Find the values of SS_{Reg} and SS_E, then determine whether or not the model provides a good fit to the data by examining the value of $R^2 = \frac{SS_{Reg}}{SS_{Reg}+SS_E}$, the so-called *coefficient of determination*.

(d) Obtain individual confidence intervals on the model's parameters using a 95% confidence coefficient for each interval.

(e) Obtain Scheffé's simultaneous 95% confidence intervals on the model's parameters.

(f) Test the hypothesis

$$H_0 : \beta_1 + \beta_2 + \beta_4 = 0$$
$$\beta_2 + 3\,\beta_{44} = 2$$

against the alternative hypothesis, H_a:H_0 is not true, and state your conclusion at the $\alpha = 0.05$ level.

(g) For the hypothesis in part (f), compute the power of the test under the alternative hypothesis,

$$H_a : \beta_1 + \beta_2 + \beta_4 = 1,$$
$$\beta_2 + 3\,\beta_{44} = 4$$

given that $\sigma^2 = 1$.

6.20 A study was conducted to predict optimum conditions for microwave-assisted extraction of saponin components from ginseng roots. A central composite design consisting of a 2^2 factorial design, four axial points, and two center-point replications was used to monitor the effects of x_1 = ethanol concentration and x_2 = extraction time on Y = total extract yield. The original and coded settings of x_1 and x_2 are given in the following table:

		Coded Values of Concentration and Time				
Variable	**Unit**	$x_i = -2$	$x_i = -1$	$x_i = 0$	$x_i = 1$	$x_i = 2$
Concentration	%	30	45	60	75	90
Time	S	30	90	150	210	270

The data set was given by Kwon et al. (2003) using the coded settings of concentration and time and is reproduced in Table 6.9.

(a) Use the data to fit the model

$$Y = \beta_0 + \beta_1 x_1 + \beta_2 x_2 + \beta_{11} x_1^2 + \beta_{22} x_2^2 + \beta_{12} x_1 x_2 + \epsilon,$$

where $\epsilon \sim N(0, \sigma^2)$.

TABLE 6.9
Design Settings and Extract Yield Data

Experimental Run	x_1	x_2	Y (%)
1	1	1	21.8
2	1	−1	20.8
3	−1	1	26.3
4	−1	−1	25.2
5	2	0	14.6
6	−2	0	22.2
7	0	2	27.1
8	0	−2	23.6
9	0	0	26.8
10	0	0	26.8

Source: Reprinted from Kwon, J.H. et al., *J. Agric. Food Chem.*, 51, 1807, 2003. With permission.

(b) Find the uniformly minimum variance unbiased estimates of the model's parameters.

(c) Find the uniformly minimum variance unbiased estimate of $\frac{\beta_1+\beta_2+\beta_{12}}{\sigma^2}$.

(d) Obtain a 95% confidence interval on the mean response at (1, 1.5).

(e) Obtain a 95% confidence region on $\boldsymbol{\beta} = (\beta_0, \beta_1, \beta_2, \beta_{11}, \beta_{22}, \beta_{12})'$.

(f) Find the principal axes of the ellipsoid representing the confidence region in part (e).

6.21 Enamines are useful intermediates in organic synthesis. They are usually prepared by condensing the corresponding carbonyl compound (aldehyde or ketone) with a secondary amine under elimination of water. The effects of x_1 = amount of *TiCL4/ketone* (mol/mol) and x_2 = amount of *morpholine/ketone* (mol/mol) on Y, the yield of *enamine* was investigated. A central composite design consisting of a 2^2 factorial design, four axial points, and five center-point replications was used. The original and coded settings of x_1 and x_2 are given in the following table:

	Coded Settings				
Variable	$x_i = -1.414$	$x_i = -1$	$x_i = 0$	$x_i = 1$	$x_i = 1.414$
TiCL4/ketone	0.50	0.57	0.75	0.93	1.00
Morpholine/ketone	3.00	3.70	5.50	7.30	8.00

The corresponding data set was given by Carlson and Carlson (2005) using the coded settings of x_1 and x_2 and is reproduced in Table 6.10.

TABLE 6.10
Design Settings and the Yields of Enamine

Experimental Run	x_1	x_2	Yield (%)
1	−1	−1	73.4
2	1	−1	69.7
3	−1	1	88.7
4	1	1	98.7
5	−1.414	0	76.8
6	1.414	0	84.9
7	0	−1.414	56.6
8	0	1.414	81.3
9	0	0	96.8
10	0	0	96.4
11	0	0	87.5
12	0	0	96.1
13	0	0	90.5

Source: Reprinted from Carlson, R. and Carlson, J.E., *Organ. Process Res. Dev.*, 9, 321, 2005. With permission.

(a) Fit the model

$$Y = \beta_0 + \beta_1 x_1 + \beta_2 x_2 + \beta_{11} x_1^2 + \beta_{22} x_2^2 + \beta_{12} x_1 x_2 + \epsilon,$$

where $\epsilon \sim N(0, \sigma^2)$.

(b) Find the prediction variance $\text{Var}[\hat{Y}(\boldsymbol{x})]$, where $\hat{Y}(\boldsymbol{x})$ is the predicted response at a point $\boldsymbol{x} = (x_1,\ x_2)'$ in the experimental region.

(c) Show that $\text{Var}[\hat{Y}(\boldsymbol{x})]$ is constant at all points that are equidistant from the design center. Thus contours of constant prediction variance are concentric circles centered at the origin within the experimental region.

(d) Find the value of the maximum yield within the experimental region.

(e) Obtain a 95% confidence region on the location of the true maximum yield in part (d).

[Note: A central composite design having the property described in part (c) is said to be *rotatable*. See Khuri and Cornell (1996, Chapter 4).]

7

Less-Than-Full-Rank Linear Models

In this chapter, the matrix $\boldsymbol{X}$ in the model,

$$\boldsymbol{Y} = \boldsymbol{X}\boldsymbol{\beta} + \boldsymbol{\epsilon}, \tag{7.1}$$

does not have a full column rank, as was the case in Chapter 6. Accordingly, (7.1) is labeled as a *less-than-full-rank model*. The analysis of this model in terms of parameter estimation and hypothesis testing is now revisited under the present label.

As in Chapter 6, $\boldsymbol{X}$ is a known matrix of order $n \times p$. Its rank, however, is r, where $r < p$. Consequently, the matrix $\boldsymbol{X}'\boldsymbol{X}$ is no longer nonsingular and formula (6.9) is therefore invalid here. Hence, it cannot be used to estimate $\boldsymbol{\beta}$. This result should not be surprising since the number of linearly independent equations in (6.8) is only r whereas the number of unknown parameters is p, which exceeds r. It is therefore not possible to uniquely estimate all of the elements of $\boldsymbol{\beta}$.

Typical examples of models that are not of full rank include ANOVA (analysis of variance) models, such as crossed (or nested) classification models. For example, the one-way model,

$$Y_{ij} = \mu + \alpha_i + \epsilon_{ij}, \quad i = 1, 2, \ldots, k; \quad j = 1, 2, \ldots, n_i,$$

is not of full rank since, in this case, the matrix $\boldsymbol{X}$ is $\left[\boldsymbol{1}_{n_.} : \oplus_{i=1}^{k}\boldsymbol{1}_{n_i}\right]$, which is of order $n_. \times (k+1)$ and rank k ($n_. = \sum_{i=1}^{k} n_i$), and $\boldsymbol{\beta} = (\mu, \alpha_1, \alpha_2, \ldots, \alpha_k)'$.

7.1 Parameter Estimation

Consider model (7.1), where the error term $\boldsymbol{\epsilon}$ is assumed to have a zero mean vector and a variance–covariance matrix $\sigma^2 \boldsymbol{I}_n$. Using the method of least squares, as was seen earlier in Section 6.1, we obtain the equation

$$\boldsymbol{X}'\boldsymbol{X}\hat{\boldsymbol{\beta}} = \boldsymbol{X}'\boldsymbol{Y}. \tag{7.2}$$

Since $\boldsymbol{X}'\boldsymbol{X}$ is a singular matrix, this equation does not have a unique solution for $\hat{\boldsymbol{\beta}}$. In this case, $\hat{\boldsymbol{\beta}}$ is considered to be just a solution to (7.2), but not an estimator of β because of its nonuniqueness. A solution to (7.2) can be expressed as

$$\hat{\beta} = (X'X)^{-}X'Y, \tag{7.3}$$

where $(X'X)^{-}$ is a generalized inverse of $X'X$. Hence, for a given Y, (7.3) gives infinitely many solutions for (7.2) depending on the choice of $(X'X)^{-}$. For a particular $(X'X)^{-}$, the mean vector and variance–covariance matrix of $\hat{\beta}$ are given by

$$E(\hat{\beta}) = (X'X)^{-}X'X\beta, \tag{7.4}$$

$$\text{Var}(\hat{\beta}) = \sigma^2(X'X)^{-}X'X(X'X)^{-}. \tag{7.5}$$

It should be noted that even though $(X'X)^{-}$ is not unique, the least-squares estimate of the mean response vector, namely, $E(Y) = X\beta$, which is $X\hat{\beta} = X(X'X)^{-}X'Y$, does not depend on the choice of the generalized inverse of $X'X$. The same is true with regard to the error (residual) sum of squares,

$$SS_E = Y'[I_n - X(X'X)^{-}X']Y, \tag{7.6}$$

and the regression sum of squares,

$$SS_{Reg} = Y'X(X'X)^{-}X'Y. \tag{7.7}$$

This is true because the matrix $X(X'X)^{-}X'$, which is idempotent of rank r, is invariant to the choice of $(X'X)^{-}$ (see property (b) in Section 3.7.1).

7.2 Some Distributional Properties

The following properties concerning the distributions of SS_E and SS_{Reg} are similar to those in Section 6.2.1:

(a) If ϵ in model (7.1) has a zero mean vector and a variance–covariance matrix $\sigma^2 I_n$, then $E(MS_E) = \sigma^2$, where $MS_E = \frac{SS_E}{n-r}$ is the error mean square, and r is the rank of X. This follows from the fact that

$$E(SS_E) = \beta'X'[I_n - X(X'X)^{-}X']X\beta + tr\left\{[I_n - X(X'X)^{-}X']\left(\sigma^2 I_n\right)\right\} = (n-r)\sigma^2.$$

(b) If $\epsilon \sim N(\mathbf{0}, \sigma^2 I_n)$, then

(i) $\frac{1}{\sigma^2}SS_E \sim \chi^2_{n-r}$.

(ii) $\frac{1}{\sigma^2}SS_{Reg} \sim \chi^2_r(\theta)$, where $\theta = \frac{1}{\sigma^2}\beta' X' X \beta$.
(iii) SS_E and SS_{Reg} are independent.

Properties (i) and (ii) follow from applying Theorem 5.4 and noting that the matrices $I_n - X(X'X)^- X'$ and $X(X'X)^- X'$ are idempotent of ranks $n - r$ and r, respectively. Furthermore, the noncentrality parameter for $\frac{1}{\sigma^2}SS_E$ is zero, whereas for $\frac{1}{\sigma^2}SS_{Reg}$ it is equal to

$$\begin{aligned}\theta &= \frac{1}{\sigma^2}\beta' X'[X(X'X)^- X']X\beta \\ &= \frac{1}{\sigma^2}\beta' X' X \beta.\end{aligned}$$

Property (iii) is true on the basis of Theorem 5.5 and the fact that

$$[X(X'X)^- X'](\sigma^2 I_n)[I_n - X(X'X)^- X'] = \mathbf{0}.$$

7.3 Reparameterized Model

The model in (7.1) can be replaced by another equivalent model that has a full rank. This can be shown as follows:

Using the Spectral Decomposition Theorem (Theorem 3.4), the matrix $X'X$ can be decomposed as

$$X'X = P\,\mathrm{diag}(\Lambda, \mathbf{0})\,P', \tag{7.8}$$

where

Λ is a diagonal matrix of order $r \times r$ whose diagonal elements are the nonzero eigenvalues of $X'X$
$\mathbf{0}$ is a zero matrix of order $(p - r) \times (p - r)$
P is an orthogonal matrix of orthonormal eigenvectors of $X'X$

Let P be partitioned as $P = [P_1 : P_2]$, where P_1 is of order $p \times r$ whose columns are orthonormal eigenvectors of $X'X$ corresponding to the diagonal elements of Λ, and P_2 is of order $p \times (p - r)$ whose columns are orthonormal eigenvectors of $X'X$ corresponding to the zero eigenvalue, which is of multiplicity $p - r$. From (7.8) we get

$$\begin{bmatrix} P_1' \\ P_2' \end{bmatrix} X'X[P_1 : P_2] = \mathrm{diag}(\Lambda, \mathbf{0}).$$

Hence,

$$P_1' X' X P_1 = \Lambda, \tag{7.9}$$

$$P_2' X' X P_2 = \mathbf{0}. \tag{7.10}$$

From (7.9) and (7.10) we conclude that XP_1, which is of order $n \times r$, is of rank r $\left[rank(XP_1) = rank\left(P_1'X'XP_1\right) = r\right]$, and that $XP_2 = 0$. Now, model (7.1) can be written as

$$\begin{aligned} Y &= XPP'\beta + \epsilon \\ &= [XP_1 : XP_2]\begin{bmatrix} P_1' \\ P_2' \end{bmatrix}\beta + \epsilon \\ &= [XP_1 : 0]\begin{bmatrix} P_1'\beta \\ P_2'\beta \end{bmatrix} + \epsilon \\ &= XP_1P_1'\beta + \epsilon. \end{aligned}$$

Let $\tilde{X} = XP_1$ and $\tilde{\beta} = P_1'\beta$. We then have

$$Y = \tilde{X}\tilde{\beta} + \epsilon. \tag{7.11}$$

We note that (7.11) is a full-rank model since $\tilde{X}$ is of full column rank. In addition, we have the following results which are given by the next three theorems.

Theorem 7.1 The column spaces of X and $\tilde{X}$ are the same.

Proof. From $\tilde{X} = XP_1$, every column of $\tilde{X}$ is a linear combination of the columns of X. Vice versa, $X = \tilde{X}P_1'$ (why?). Hence, every column of X is a linear combination of the columns of $\tilde{X}$. □

Since the column spaces of X and $\tilde{X}$ are identical, models (7.1) and (7.11) are said to be *equivalent*.

Theorem 7.2 The row space of X is identical to the row space of P_1'.

Proof. Formula (7.8) can be written as

$$X'X = P_1\Lambda P_1'. \tag{7.12}$$

Multiplying (7.12) on the left by P_1' and noting that $P_1'P_1 = I_r$, we get

$$P_1'X'X = \Lambda P_1',$$

which implies that $P_1' = \Lambda^{-1}P_1'X'X$. Hence, every row of P_1' is a linear combination of the rows of X. Vice versa, since

$$\begin{aligned} X &= X(X'X)^{-}X'X \\ &= X(X'X)^{-}P_1\Lambda_1P_1', \end{aligned}$$

we conclude that every row of X is a linear combination of the rows of P_1'. Hence, the row spaces of X and P_1' are identical. □

As a consequence of the equivalence of models (7.1) and (7.11), we have the following results given by the following theorem.

Theorem 7.3

(a) The regression and residual sums of squares for models (7.1) and (7.11) are identical.

(b) The least-squares estimates of the mean response vector $E(\boldsymbol{Y})$ from models (7.1) and (7.11) are identical.

Proof.

(a) For model (7.1), $SS_{Reg} = \boldsymbol{Y}'\boldsymbol{X}(\boldsymbol{X}'\boldsymbol{X})^{-}\boldsymbol{X}'\boldsymbol{Y}$, and for model (7.11), the regression sum of squares is

$$\widetilde{SS}_{Reg} = \boldsymbol{Y}'\tilde{\boldsymbol{X}}(\tilde{\boldsymbol{X}}'\tilde{\boldsymbol{X}})^{-1}\tilde{\boldsymbol{X}}'\boldsymbol{Y}. \tag{7.13}$$

Since $\tilde{\boldsymbol{X}} = \boldsymbol{X}\boldsymbol{P}_1$, then by using (7.9),

$$\begin{aligned}\tilde{\boldsymbol{X}}(\tilde{\boldsymbol{X}}'\tilde{\boldsymbol{X}})^{-1}\tilde{\boldsymbol{X}}' &= \boldsymbol{X}\boldsymbol{P}_1\left(\boldsymbol{P}_1'\boldsymbol{X}'\boldsymbol{X}\boldsymbol{P}_1\right)^{-1}\boldsymbol{P}_1'\boldsymbol{X}' \\ &= \boldsymbol{X}\boldsymbol{P}_1\boldsymbol{\Lambda}^{-1}\boldsymbol{P}_1'\boldsymbol{X}'.\end{aligned}$$

But, $\boldsymbol{P}_1\boldsymbol{\Lambda}^{-1}\boldsymbol{P}_1'$ is a generalized inverse of $\boldsymbol{X}'\boldsymbol{X}$ because by (7.12) and the fact that $\boldsymbol{P}_1'\boldsymbol{P}_1 = \boldsymbol{I}_r$,

$$\begin{aligned}\boldsymbol{X}'\boldsymbol{X}\left(\boldsymbol{P}_1\boldsymbol{\Lambda}^{-1}\boldsymbol{P}_1'\right)\boldsymbol{X}'\boldsymbol{X} &= \boldsymbol{P}_1\boldsymbol{\Lambda}\boldsymbol{P}_1'\left(\boldsymbol{P}_1\boldsymbol{\Lambda}^{-1}\boldsymbol{P}_1'\right)\boldsymbol{P}_1\boldsymbol{\Lambda}\boldsymbol{P}_1' \\ &= \boldsymbol{P}_1\boldsymbol{\Lambda}\boldsymbol{P}_1' \\ &= \boldsymbol{X}'\boldsymbol{X}.\end{aligned}$$

Hence,

$$\tilde{\boldsymbol{X}}(\tilde{\boldsymbol{X}}'\tilde{\boldsymbol{X}})^{-1}\tilde{\boldsymbol{X}}' = \boldsymbol{X}(\boldsymbol{X}'\boldsymbol{X})^{-}\boldsymbol{X}'. \tag{7.14}$$

Therefore, $SS_{Reg} = \widetilde{SS}_{Reg}$. We can also conclude that the residual sums of squares for models (7.1) and (7.11) are equal since $SS_E = \boldsymbol{Y}'\boldsymbol{Y} - SS_{Reg}$.

(b) Let $\boldsymbol{\mu} = E(\boldsymbol{Y})$. Then, from (7.1) and (7.11), $\boldsymbol{\mu} = \boldsymbol{X}\boldsymbol{\beta} = \tilde{\boldsymbol{X}}\tilde{\boldsymbol{\beta}}$. The corresponding least-squares estimators are

$$\begin{aligned}\boldsymbol{X}\hat{\boldsymbol{\beta}} &= \boldsymbol{X}(\boldsymbol{X}'\boldsymbol{X})^{-}\boldsymbol{X}'\boldsymbol{Y}, \\ \tilde{\boldsymbol{X}}\hat{\tilde{\boldsymbol{\beta}}} &= \tilde{\boldsymbol{X}}(\tilde{\boldsymbol{X}}'\tilde{\boldsymbol{X}})^{-1}\tilde{\boldsymbol{X}}'\boldsymbol{Y},\end{aligned}$$

respectively. These estimators are equal by (7.14). Note that $\hat{\tilde{\boldsymbol{\beta}}}$ can be expressed as

$$\begin{aligned}\hat{\tilde{\boldsymbol{\beta}}} &= (\tilde{\boldsymbol{X}}'\tilde{\boldsymbol{X}})^{-1}\tilde{\boldsymbol{X}}'\boldsymbol{Y} \\ &= \boldsymbol{\Lambda}^{-1}\tilde{\boldsymbol{X}}'\boldsymbol{Y}, \end{aligned} \tag{7.15}$$

since by (7.9),

$$\tilde{X}'\tilde{X} = P_1'X'XP_1$$
$$= \Lambda. \tag{7.16}$$

□

7.4 Estimable Linear Functions

It was stated earlier in this chapter that the parameter vector β in model (7.1), which has p elements, cannot be estimated in its entirety. Hence, not every linear function of β of the form $a'\beta$ can be estimated. There are, however, certain conditions on a under which $a'\beta$ can be estimated.

Definition 7.1 The linear function, $a'\beta$, where a is a constant vector, is said to be *estimable* if there exists a linear function of Y, the vector of observations in (7.1), of the form $b'Y$ such that $E(b'Y) = a'\beta$.

The next theorem gives a necessary and sufficient condition for the estimability of $a'\beta$.

Theorem 7.4 The linear function $a'\beta$ is estimable if and only if a' belongs to the row space of X in (7.1), that is, $a' = b'X$ for some vector b.

Proof. If $a'\beta$ is estimable, then by Definition 7.1 there exists a vector b such that $E(b'Y) = a'\beta$. Consequently,

$$b'X\beta = a'\beta, \tag{7.17}$$

which must be true for all $\beta \in R^p$. It follows that

$$a' = b'X. \tag{7.18}$$

Hence, a' is a linear combination of the rows of X and therefore belongs to the row space of X. Vice versa, if $a' = b'X$, for some b, then $E(b'Y) = b'X\beta = a'\beta$, which makes $a'\beta$ estimable. □

The following corollary provides a practical method for checking estimability.

Corollary 7.1 The linear function $a'\beta$ is estimable if and only if the matrix

$$X_a = \begin{bmatrix} X \\ a' \end{bmatrix}$$

has the same rank as that of X, or equivalently, if and only if the numbers of nonzero eigenvalues of $X'X$ and $X_a'X_a$ are the same.

Since X is of rank r, then from Theorem 7.4 we can conclude that the number of linearly independent estimable functions of β is equal to r. Furthermore, since the row spaces of X and P_1' are the same, as was seen in Theorem 7.2, $a'\beta$ is estimable if and only if a' belongs to the row space of P_1', that is,

$$a' = \zeta' P_1', \tag{7.19}$$

for some vector ζ. Recall that P_1 is the matrix used in (7.9) whose columns are orthonormal eigenvectors of $X'X$ corresponding to the nonzero eigenvalues of $X'X$. We conclude that the elements of the vector $P_1'\beta$ form a basis for the vector space of all estimable linear functions of β.

7.4.1 Properties of Estimable Functions

Estimable linear functions have several interesting features in the sense that their estimation and tests of significance are carried out in much the same way as in the case of full-rank models in Chapter 6. Their properties are given by the following theorems.

Theorem 7.5 If $a'\beta$ is estimable, then $a'\hat{\beta}$, where $\hat{\beta} = (X'X)^-X'Y$, is invariant to the choice of $(X'X)^-$.

Proof. If $a'\beta$ is estimable, then by Theorem 7.4, $a' = b'X$ for some vector b. Hence,

$$\begin{aligned} a'\hat{\beta} &= a'(X'X)^-X'Y \\ &= b'X(X'X)^-X'Y. \end{aligned}$$

Invariance of $a'\hat{\beta}$ follows from the fact that $X(X'X)^-X$ is invariant to the choice of the generalized inverse of $X'X$ (see property (b) in Section 3.7.1). □

The uniqueness of $a'\hat{\beta}$ for a given Y, when $a'\beta$ is estimable, makes $a'\hat{\beta}$ a full-fledged estimator of $a'\beta$, which was not the case with $\hat{\beta}$. Furthermore, $a'\beta$ and $a'\hat{\beta}$ can be expressed as

$$a'\beta = \zeta'\tilde{\beta}, \tag{7.20}$$

$$a'\hat{\beta} = \zeta'\hat{\tilde{\beta}}, \tag{7.21}$$

where

$\tilde{\beta}$ is the parameter vector in (7.11)
$\hat{\tilde{\beta}}$ is the least-squares estimator of $\tilde{\beta}$ given by (7.15)
ζ' is the vector used in (7.19)

Formula (7.20) is true because from (7.19), $a'\beta = \zeta' P_1' \beta = \zeta' \tilde{\beta}$. As for (7.21), we have

$$\begin{aligned} a'\hat{\beta} &= \zeta' P_1' (X'X)^- X'Y \\ &= \zeta' \Lambda^{-1} P_1' X'X(X'X)^- X'Y, \end{aligned}$$

since from the proof of Theorem 7.2, $P_1' = \Lambda^{-1} P_1' X'X$. Hence,

$$\begin{aligned} a'\hat{\beta} &= \zeta' \Lambda^{-1} P_1' X'Y \\ &= \zeta' \Lambda^{-1} \tilde{X}' Y, \text{ since } \tilde{X} = XP_1, \\ &= \zeta' \hat{\tilde{\beta}}, \text{ by using (7.15).} \end{aligned}$$

Theorem 7.6 (The Gauss–Markov Theorem) Suppose that ϵ in model (7.1) has a zero mean vector and a variance–covariance matrix given by $\sigma^2 I_n$. If $a'\beta$ is estimable, then $a'\hat{\beta} = a'(X'X)^- X'Y$ is the best linear unbiased estimator (BLUE) of $a'\beta$.

Proof. This follows directly from using (7.20) and (7.21) and by applying the Gauss–Markov Theorem (Theorem 6.1) to model (7.11). More specifically, we have that $a'\hat{\beta} = \zeta'\hat{\tilde{\beta}}$ and $\zeta'\hat{\tilde{\beta}}$ is the BLUE of $\zeta'\tilde{\beta}$, which is equal to $a'\beta$ by (7.20). □

Theorem 7.7 Suppose that ϵ in model (7.1) is distributed as $N(\mathbf{0}, \sigma^2 I_n)$, and that $a'\beta$ is estimable. Then,

(a) $a'\hat{\beta}$ is normally distributed with mean $a'\beta$ and variance $a'(X'X)^- a\,\sigma^2$, where $\hat{\beta}$ is given by (7.3).

(b) $a'\hat{\beta}$ and MS_E are independently distributed, where MS_E is the error mean square,

$$MS_E = \frac{1}{n-r} Y'[I_n - X(X'X)^- X']Y \tag{7.22}$$

(c) $a'\hat{\beta}$ and MS_E are uniformly minimum variance unbiased estimators (UMVUE) of $a'\beta$ and σ^2.

Proof.

(a) $a'\hat{\beta} = a'(X'X)^- X'Y = b'X(X'X)^- X'Y$ for some vector b. Since Y is normally distributed, then so is $a'\hat{\beta}$. Its mean is

$$\begin{aligned} E(a'\hat{\beta}) &= b'X(X'X)^- X'X\beta \\ &= b'X\beta \\ &= a'\beta, \end{aligned}$$

and its variance is

$$\begin{aligned}\text{Var}(\boldsymbol{a}'\hat{\boldsymbol{\beta}}) &= \boldsymbol{b}'\boldsymbol{X}(\boldsymbol{X}'\boldsymbol{X})^{-}\boldsymbol{X}'(\sigma^2\boldsymbol{I}_n)\boldsymbol{X}(\boldsymbol{X}'\boldsymbol{X})^{-}\boldsymbol{X}'\boldsymbol{b}\\ &= \boldsymbol{b}'\boldsymbol{X}(\boldsymbol{X}'\boldsymbol{X})^{-}\boldsymbol{X}'\boldsymbol{b}\,\sigma^2\\ &= \boldsymbol{a}'(\boldsymbol{X}'\boldsymbol{X})^{-}\boldsymbol{a}\,\sigma^2.\end{aligned}$$

(b) $\boldsymbol{a}'\hat{\boldsymbol{\beta}}$ and MS_E are independent by applying Theorem 5.6 and the fact that

$$\boldsymbol{a}'(\boldsymbol{X}'\boldsymbol{X})^{-}\boldsymbol{X}'(\sigma^2\boldsymbol{I}_n)[\boldsymbol{I}_n - \boldsymbol{X}(\boldsymbol{X}'\boldsymbol{X})^{-}\boldsymbol{X}'] = \boldsymbol{0}'.$$

(c) By applying Corollary 6.2 to model (7.11), we can assert that $\hat{\tilde{\boldsymbol{\beta}}}$ in (7.15) and

$$\widetilde{MS}_E = \frac{1}{n-r}\boldsymbol{Y}'\left[\boldsymbol{I}_n - \tilde{\boldsymbol{X}}(\tilde{\boldsymbol{X}}'\tilde{\boldsymbol{X}})^{-1}\tilde{\boldsymbol{X}}'\right]\boldsymbol{Y}$$

are complete and sufficient statistics for $\tilde{\boldsymbol{\beta}}$ and σ^2. But, by (7.21), $\boldsymbol{a}'\hat{\boldsymbol{\beta}} = \boldsymbol{\zeta}'\hat{\tilde{\boldsymbol{\beta}}}$, and by (7.14), $MS_E = \widetilde{MS}_E$. Furthermore, $\boldsymbol{\zeta}'\hat{\tilde{\boldsymbol{\beta}}}$ is unbiased for $\boldsymbol{\zeta}'\tilde{\boldsymbol{\beta}}$, and hence for $\boldsymbol{a}'\boldsymbol{\beta}$ by (7.20), and $\widetilde{MS}_E$ is unbiased for σ^2. By the Lehmann–Scheffé Theorem (see Casella and Berger, 2002, p. 369), it follows that $\boldsymbol{a}'\hat{\boldsymbol{\beta}}$ and MS_E are UMVUE of $\boldsymbol{a}'\boldsymbol{\beta}$ and σ^2. □

7.4.2 Testable Hypotheses

The properties described in Section 7.4.1 are now applied to derive tests and confidence intervals concerning estimable linear functions.

Definition 7.2 The hypothesis $H_0 : \boldsymbol{A\beta} = \boldsymbol{m}$ is said to be *testable* if the elements of $\boldsymbol{A\beta}$ are estimable, where $\boldsymbol{A}$ is a matrix of order $s \times p$ and rank s ($\leq r$, the rank of $\boldsymbol{X}$ in model (7.1)), and $\boldsymbol{m}$ is a constant vector.

Lemma 7.1 The hypothesis $H_0 : \boldsymbol{A\beta} = \boldsymbol{m}$ is testable if and only if there exists a matrix $\boldsymbol{S}$ of order $s \times p$ such that $\boldsymbol{A} = \boldsymbol{SX}'\boldsymbol{X}$.

Proof. If $\boldsymbol{A\beta}$ is testable, then the rows of $\boldsymbol{A}$ must belong to the row space of $\boldsymbol{X}$. Thus, there exists a matrix $\boldsymbol{T}$ such that $\boldsymbol{A} = \boldsymbol{TX}$. Hence, $\boldsymbol{A} = \boldsymbol{TX}(\boldsymbol{X}'\boldsymbol{X})^{-}\boldsymbol{X}'\boldsymbol{X} = \boldsymbol{SX}'\boldsymbol{X}$, where $\boldsymbol{S} = \boldsymbol{TX}(\boldsymbol{X}'\boldsymbol{X})^{-}$. Vice versa, if $\boldsymbol{A} = \boldsymbol{SX}'\boldsymbol{X}$ for some matrix $\boldsymbol{S}$, then any row of $\boldsymbol{A}$ is a linear combination of the rows of $\boldsymbol{X}$ implying estimability of $\boldsymbol{A\beta}$ and hence testability of H_0. □

Let us now suppose that $\boldsymbol{A\beta}$ is estimable, where $\boldsymbol{A}$ is $s \times p$ of rank s ($\leq r$). The best linear unbiased estimator of $\boldsymbol{A\beta}$ is $\boldsymbol{A}\hat{\boldsymbol{\beta}} = \boldsymbol{A}(\boldsymbol{X}'\boldsymbol{X})^{-}\boldsymbol{X}'\boldsymbol{Y}$, assuming that $\boldsymbol{\epsilon}$ in (7.1) has a zero mean and a variance–covariance matrix $\sigma^2\boldsymbol{I}_n$. The variance–covariance matrix of $\boldsymbol{A}\hat{\boldsymbol{\beta}}$ is

$$\begin{aligned}\text{Var}(\boldsymbol{A}\hat{\boldsymbol{\beta}}) &= \boldsymbol{A}(\boldsymbol{X}'\boldsymbol{X})^{-}\boldsymbol{X}'\boldsymbol{X}(\boldsymbol{X}'\boldsymbol{X})^{-}\boldsymbol{A}'\,\sigma^2\\ &= \boldsymbol{SX}'\boldsymbol{X}(\boldsymbol{X}'\boldsymbol{X})^{-}\boldsymbol{X}'\boldsymbol{X}(\boldsymbol{X}'\boldsymbol{X})^{-}\boldsymbol{X}'\boldsymbol{XS}'\,\sigma^2\\ &= \boldsymbol{SX}'\boldsymbol{X}(\boldsymbol{X}'\boldsymbol{X})^{-}\boldsymbol{X}'\boldsymbol{XS}'\,\sigma^2\\ &= \boldsymbol{SX}'\boldsymbol{XS}'\,\sigma^2, \qquad (7.23)\end{aligned}$$

where S is the matrix described in Lemma 7.1. Note that $\text{Var}(A\hat{\beta})$ can also be expressed as

$$\begin{aligned}\text{Var}(A\hat{\beta}) &= TX(X'X)^{-}X'X(X'X)^{-}X'T'\,\sigma^2\\ &= TX(X'X)^{-}X'T'\,\sigma^2\\ &= A(X'X)^{-}A'\,\sigma^2,\end{aligned} \tag{7.24}$$

where T is a matrix such that $A = TX$. It is easy to see that $\text{Var}(A\hat{\beta})$ is invariant to the choice of $(X'X)^{-}$ and is a nonsingular matrix. The latter assertion is true because A is of full row rank s and

$$s = rank(A) \leq rank(SX') \leq rank(S) \leq s,$$

since S has s rows. It follows that $rank(SX') = s$. Hence, the matrix $SX'XS'$ is of full rank, which implies that $\text{Var}(A\hat{\beta})$ is nonsingular by (7.23).

A test statistic concerning $H_0 : A\beta = m$ versus $H_a : A\beta \neq m$ can now be obtained, assuming that $A\beta$ is estimable and ϵ in model (7.1) is distributed as $N(\mathbf{0}, \sigma^2 I_n)$. We have that

$$A\hat{\beta} \sim N[A\beta, A(X'X)^{-}A'\,\sigma^2]$$

Hence, under H_0,

$$F = \frac{(A\hat{\beta} - m)'[A(X'X)^{-}A']^{-1}(A\hat{\beta} - m)}{s\,MS_E} \tag{7.25}$$

has the F-distribution with s and $n - r$ degrees of freedom, where MS_E is the error mean square in (7.22). This follows from the fact that $A\hat{\beta}$ is independent of MS_E by Theorem 7.7 and $\frac{1}{\sigma^2}SS_E \sim \chi^2_{n-r}$. The null hypothesis can be rejected at the α-level if $F \geq F_{\alpha,s,n-r}$. The power of this test under the alternative hypothesis $H_a : A\beta = m_a$, where m_a is a given constant vector different from m, is

$$\text{Power} = P[F \geq F_{\alpha,s,n-r} | H_a : A\beta = m_a].$$

Under H_a, F has the noncentral F-distribution $F_{s,n-r}(\theta)$ with the noncentrality parameter

$$\theta = \frac{1}{\sigma^2}(m_a - m)'[A(X'X)^{-}A']^{-1}(m_a - m). \tag{7.26}$$

Hence, the power is given by

$$\text{Power} = P[F_{s,n-r}(\theta) \geq F_{\alpha,s,n-r}].$$

A confidence region on $A\beta$ can also be obtained on the basis of the aforementioned F-distribution. In this case,

$$\frac{(A\hat{\beta} - A\beta)'[A(X'X)^{-}A']^{-1}(A\hat{\beta} - A\beta)}{s\,MS_E}$$

has the F-distribution with s and $n - r$ degrees of freedom. Hence, the $(1 - \alpha)100\%$ confidence region on $A\beta$ is given by

$$\frac{(A\hat{\beta} - A\beta)'[A(X'X)^{-}A']^{-1}(A\hat{\beta} - A\beta)}{s\,MS_E} \leq F_{\alpha,s,n-r}. \tag{7.27}$$

In the special case when $A = a'$, a test statistic concerning $H_0 : a'\beta = m$ versus $H_a : a'\beta \neq m$ is

$$t = \frac{a'\hat{\beta} - m}{[a'(X'X)^{-}a\,MS_E]^{1/2}}, \tag{7.28}$$

which, under H_0, has the t-distribution with $n-r$ degrees of freedom. Accordingly, the $(1 - \alpha)100\%$ confidence interval on $a'\beta$ is given by

$$a'\hat{\beta} \pm [a'(X'X)^{-}a\,MS_E]^{1/2}\, t_{\frac{\alpha}{2},n-r}. \tag{7.29}$$

Example 7.1 Consider the one-way model,

$$Y_{ij} = \mu + \alpha_i + \epsilon_{ij}, \quad i = 1, 2, \ldots, k; \quad j = 1, 2, \ldots, n_i,$$

where
 α_i is a fixed unknown parameter
 the ϵ_{ij}'s are independently distributed as $N(0, \sigma^2)$

This model can be written as in (7.1) with $\beta = (\mu, \alpha_1, \alpha_2, \ldots, \alpha_k)'$ and

$$X = \left[\mathbf{1}_{n.} : \oplus_{i=1}^{k} \mathbf{1}_{n_i}\right],$$

where $n_. = \sum_{i=1}^{k} n_i$ and $\mathbf{1}_{n_i}$ is a vector of ones of order $n_i \times 1$ $(i = 1, 2, \ldots, k)$. The matrix X is of rank k and its row space is spanned by the k vectors,

$$(1, 1, 0, \ldots, 0),\ (1, 0, 1, \ldots, 0), \ldots, (1, 0, 0, \ldots, 0, 1),$$

which are linearly independent. On this basis we have the following results:

Result 1. $\mu + \alpha_i$ $(i = 1, 2, \ldots, k)$ form a basis for all estimable linear functions of β. Hence, $\alpha_{i_1} - \alpha_{i_2}$ is estimable for $i_1 \neq i_2$.

Result 2. μ is nonestimable.

To show this result, let us write μ as $a'\beta$, where $a' = (1, 0, 0, \ldots, 0)$. If μ is estimable, then a' must belong to the row space of X, that is, $a' = b'X$, for some vector b. In this case, we have

$$b'\mathbf{1}_{n.} = 1, \tag{7.30}$$

$$b' \oplus_{i=1}^{k} \mathbf{1}_{n_i} = \mathbf{0}'. \tag{7.31}$$

Equality (7.31) indicates that $\boldsymbol{b}$ is orthogonal to the columns of $\oplus_{i=1}^{k}\mathbf{1}_{n_i}$. Hence, it must be orthogonal to $\mathbf{1}_{n_.}$, which is the sum of the columns of $\oplus_{i=1}^{k}\mathbf{1}_{n_i}$. This contradicts equality (7.30). We therefore conclude that μ is nonestimable.

Result 3. The best linear unbiased estimator (BLUE) of $\alpha_{i_1} - \alpha_{i_2}$, $i_1 \neq i_2$, is $\bar{Y}_{i_1.} - \bar{Y}_{i_2.}$, where $\bar{Y}_{i.} = \frac{1}{n_i}\sum_{j=1}^{n_i} Y_{ij}$, $i = 1, 2, \ldots, k$.

To show this, let us first write $\mu + \alpha_i$ as $\boldsymbol{a}_i'\boldsymbol{\beta}$, where $\boldsymbol{a}_i'$ is a vector with $k+1$ elements; its first element is 1 and the element corresponding to α_i is also 1. Since $\mu + \alpha_i$ is estimable, its BLUE is $\boldsymbol{a}_i'\hat{\boldsymbol{\beta}}$, where

$$\hat{\boldsymbol{\beta}} = (\boldsymbol{X}'\boldsymbol{X})^{-}\boldsymbol{X}'\boldsymbol{Y}$$

$$= \begin{bmatrix} 0 & \mathbf{0}' \\ \mathbf{0} & \boldsymbol{D} \end{bmatrix} \begin{bmatrix} Y_{..} \\ Y_{1.} \\ \cdot \\ \cdot \\ \cdot \\ Y_{k.} \end{bmatrix}, \tag{7.32}$$

where $Y_{..} = \sum_{i=1}^{k}\sum_{j=1}^{n_i} Y_{ij}$, $Y_{i.} = n_i\bar{Y}_{i.}$, and $\boldsymbol{D} = \mathrm{diag}(n_1^{-1}, n_2^{-1}, \ldots, n_k^{-1})$. From (7.32) it follows that $\hat{\mu} = 0$, $\hat{\alpha}_i = \bar{Y}_{i.}$, $i = 1, 2, \ldots, k$. Hence, $\boldsymbol{a}_i'\hat{\boldsymbol{\beta}} = \bar{Y}_{i.}$. Consequently, the BLUE of $\alpha_{i_1} - \alpha_{i_2}$ is $\bar{Y}_{i_1.} - \bar{Y}_{i_2.}$, $i_1 \neq i_2$. Its variance is

$$\mathrm{Var}(\bar{Y}_{i_1.} - \bar{Y}_{i_2.}) = (\frac{1}{n_{i_1}} + \frac{1}{n_{i_2}})\,\sigma^2.$$

The $(1-\alpha)100\%$ confidence interval on $\alpha_{i_1} - \alpha_{i_2}$ is then given by

$$\bar{Y}_{i_1.} - \bar{Y}_{i_2.} \pm \left[\left(\frac{1}{n_{i_1}} + \frac{1}{n_{i_2}}\right) MS_E\right]^{1/2} t_{\alpha/2, n_.-k}$$

Example 7.2 Consider the following two-way crossed classification without interaction model

$$Y_{ijk} = \mu + \alpha_i + \beta_j + \epsilon_{ijk},$$

where α_i and β_j are fixed unknown parameters. The response values are given in the following table:

		B		
		1	**2**	**3**
A	**1**	17 20	15	20
	2	12	—	11 14
	3	6	—	17
	4	9	4 6	19

Writing this model as in (7.1), we have $\boldsymbol{\beta} = (\mu, \alpha_1, \alpha_2, \alpha_3, \alpha_4, \beta_1, \beta_2, \beta_3)'$, and the $\boldsymbol{X}$ matrix is of order 13×8 and rank $r = 6$. The error mean square is $MS_E = 10.0736$ with 7 degrees of freedom.

Suppose that it is desired to test the hypothesis

$$H_0 : \beta_1 = \beta_2 = \beta_3,$$

which is testable since $\beta_1 - \beta_2$ and $\beta_1 - \beta_3$ are estimable. This can be shown in two ways:

(1) The first row of the data contains no missing cells. Hence,

$$E(Y_{ijk}) = \mu + \alpha_i + \beta_j$$

is estimable for $i = 1, j = 1, 2, 3$. Hence, $\beta_1 - \beta_2$ and $\beta_1 - \beta_3$ are estimable.

(2) The hypothesis H_0 can be written as $H_0 : \boldsymbol{A\beta} = \boldsymbol{0}$, where

$$\boldsymbol{A} = \begin{bmatrix} 0 & 0 & 0 & 0 & 0 & 1 & -1 & 0 \\ 0 & 0 & 0 & 0 & 0 & 1 & 0 & -1 \end{bmatrix}. \tag{7.33}$$

Let $\boldsymbol{X}_A$ be the matrix $\boldsymbol{X}$ augmented vertically with $\boldsymbol{A}$, that is,

$$\boldsymbol{X}_A = \begin{bmatrix} \boldsymbol{X} \\ \boldsymbol{A} \end{bmatrix}.$$

It can be verified that the nonzero eigenvalues of $\boldsymbol{X}_A'\boldsymbol{X}_A$ are 1.945, 2.425, 3.742, 5.885, 7.941, 21.063. Hence, the rank of $\boldsymbol{X}_A'\boldsymbol{X}_A$, and therefore the rank of $\boldsymbol{X}_A$, is 6, which is the same as the rank of $\boldsymbol{X}$. By Corollary 7.1, the elements of $\boldsymbol{A\beta}$ are estimable. The BLUE of $\boldsymbol{A\beta}$ is then given by $\boldsymbol{A}\hat{\boldsymbol{\beta}} = (4.6294, -5.1980)'$. Using formula (7.25) with $\boldsymbol{m} = (0, 0)'$ and $s = 2$, we find that the value of the corresponding test statistic is $F = 7.229$ with 2 and 7 degrees of freedom. The corresponding p-value is 0.0198. The power of the test for the alternative hypothesis,

$$H_a : \boldsymbol{A\beta} = (0, 3)',$$

given that $\sigma^2 = 1$ and the level of significance is $\alpha = 0.05$, is given by

$$\begin{aligned} \text{Power} &= P[F_{2,7}(\theta) \geq F_{0.05,2,7}] \\ &= P[F_{2,7}(\theta) \geq 4.74], \end{aligned}$$

where from (7.26) with $\sigma^2 = 1$ and $\boldsymbol{m}_a = (0, 3)'$, $\theta = 24$. Hence, the power value is 0.9422. Note that this value can be easily obtained using the PROBF function in SAS's (2000) PROC IML.

Using now (7.29), the 95% confidence intervals on $\beta_1 - \beta_2$ and $\beta_1 - \beta_3$ are, respectively,

$$\boldsymbol{a}_1'\hat{\boldsymbol{\beta}} \pm [\boldsymbol{a}_1'(\boldsymbol{X}'\boldsymbol{X})^-\boldsymbol{a}_1 MS_E]^{1/2} t_{0.025,7}, \tag{7.34}$$

$$\boldsymbol{a}_2'\hat{\boldsymbol{\beta}} \pm [\boldsymbol{a}_2'(\boldsymbol{X}'\boldsymbol{X})^-\boldsymbol{a}_2 MS_E]^{1/2} t_{0.025,7}, \tag{7.35}$$

where a_i' is the ith row of the matrix A in (7.33), $i = 1, 2$. Making the proper substitutions in (7.34) and (7.35), we get $4.6294 \pm 2.5584(2.365)$ or $(-1.4212, 10.680)$; $-5.1980 \pm 2.0725(2.365)$ or $(-10.0995, -0.2965)$.

7.5 Simultaneous Confidence Intervals on Estimable Linear Functions

Simultaneous confidence intervals on estimable linear functions of β can be derived as follows:

Let $A\beta$ be an estimable linear function of β, where A is $s \times p$ of rank s ($\leq r$, r is the rank of X). Thus, the rows of A belong to the row space of X. Furthermore,

$$P\{[A(\hat{\beta} - \beta)]'[A(X'X)^{-}A']^{-1}[A(\hat{\beta} - \beta)] \leq s\,MS_E F_{\alpha,s,n-r}\} = 1 - \alpha. \quad (7.36)$$

Let $c^2 = s\,MS_E F_{\alpha,s,n-r}$ and $x = [A(X'X)^{-}A']^{-1/2}[A(\hat{\beta} - \beta)]$. Formula (7.36) can be written as

$$P[x'x \leq c^2] = 1 - \alpha. \quad (7.37)$$

Using Lemma 6.1, (7.37) is equivalent to

$$P[\,|\,v'x\,| \leq c\,(v'v)^{1/2},\ \forall\, v \in R^s] = 1 - \alpha,$$

which can be expressed as

$$P[\,|\,d'[A(\hat{\beta} - \beta)]\,| \leq c\,\{d'[A(X'X)^{-}A']d\}^{1/2},\quad \forall\, d \in R^s] = 1 - \alpha, \quad (7.38)$$

where $v'[A(X'X)^{-}A']^{-1/2} = d'$. Letting $\ell' = d'A$ in (7.38), we get

$$P[\,|\,\ell'(\hat{\beta} - \beta)\,| \leq c\,[\ell'(X'X)^{-}\ell]^{1/2},\quad \forall\, \ell' \in \text{row space of } A\,] = 1 - \alpha. \quad (7.39)$$

Formula (7.39) defines simultaneous $(1 - \alpha)$ 100% confidence intervals on all estimable linear functions $\ell'\beta$, where ℓ' belongs to the row space of A, which are given by

$$\ell'\hat{\beta} \pm [s\,MS_E F_{\alpha,s,n-r}\,\ell'(X'X)^{-}\ell]^{1/2}. \quad (7.40)$$

These intervals are known as *Scheffé's simultaneous confidence intervals*. In particular, if $s = r$, then the elements of $A\beta$ form a basis for all estimable linear functions of β, and (7.40) becomes

$$\ell'\hat{\beta} \pm [r\,MS_E F_{\alpha,r,n-r}\,\ell'(X'X)^{-}\ell]^{1/2}, \quad (7.41)$$

where ℓ' belongs to the row space of X. In this case, the row space of A is identical to the row space of X, and (7.41) provides simultaneous $(1 - \alpha)$ 100% confidence intervals on all estimable linear functions of β.

Example 7.3 Consider the one-way model,

$$Y_{ij} = \mu + \alpha_i + \epsilon_{ij}, \quad i = 1, 2, \ldots, k; \quad j = 1, 2, \ldots, n_i,$$

where
α_i is a fixed unknown parameter
the ϵ_{ij}'s are independently distributed as $N(0, \sigma^2)$

As before in Example 7.1, the associated $\boldsymbol{X}$ matrix is

$$\boldsymbol{X} = \left[\mathbf{1}_{n_.} : \oplus_{i=1}^{k} \mathbf{1}_{n_i}\right],$$

where
$n_. = \sum_{i=1}^{k} n_i$
$\boldsymbol{\beta} = (\mu, \alpha_1, \alpha_2, \ldots, \alpha_k)'$

Let $\boldsymbol{A}$ be a $(k-1) \times (k+1)$ matrix of rank $k-1$ of the form $\boldsymbol{A} = [\mathbf{0} : \mathbf{1}_{k-1} : -\boldsymbol{I}_{k-1}]$, where $\boldsymbol{I}_{k-1}$ is the identity matrix of order $(k-1) \times (k-1)$. If $\boldsymbol{\beta} = (\mu, \alpha_1, \alpha_2, \ldots, \alpha_k)'$, then $\boldsymbol{A\beta}$ is given by

$$\boldsymbol{A\beta} = (\alpha_1 - \alpha_2, \alpha_1 - \alpha_3, \ldots, \alpha_1 - \alpha_k)'. \tag{7.42}$$

The vector $\boldsymbol{A\beta}$ is estimable, as was seen in Example 7.1. Hence, the rows of $\boldsymbol{A}$ must belong to the row space of $\boldsymbol{X}$. If $\boldsymbol{\ell}'$ represents any vector in the row space of $\boldsymbol{A}$, then by (7.40), simultaneous $(1-\alpha)100\%$ confidence intervals on all estimable linear functions of the form $\boldsymbol{\ell}'\boldsymbol{\beta}$ are given by

$$\boldsymbol{\ell}'\hat{\boldsymbol{\beta}} \pm [(k-1)MS_E F_{\alpha,k-1,n_.-k}\, \boldsymbol{\ell}'(\boldsymbol{X}'\boldsymbol{X})^{-}\boldsymbol{\ell}]^{1/2}. \tag{7.43}$$

Writing $\boldsymbol{\ell}'$ as $(\ell_0, \ell_1, \ldots, \ell_k)$, we get $\boldsymbol{\ell}'\hat{\boldsymbol{\beta}} = \sum_{i=1}^{k} \ell_i \bar{Y}_{i.}$, and it is easy to show that

$$\boldsymbol{\ell}'(\boldsymbol{X}'\boldsymbol{X})^{-}\boldsymbol{\ell} = \sum_{i=1}^{k} \frac{\ell_i^2}{n_i}.$$

Substituting in (7.43), we get

$$\sum_{i=1}^{k} \ell_i \bar{Y}_{i.} \pm \left[(k-1)MS_E F_{\alpha,k-1,n_.-k} \sum_{i=1}^{k} \frac{\ell_i^2}{n_i}\right]^{1/2}. \tag{7.44}$$

Note that the elements of $\boldsymbol{A\beta}$ in (7.42) are differences between two α_i's, and hence between two treatment means. We now show that the row space of $\boldsymbol{A}$ generates all contrasts among the α_i's (or the k treatment means). By definition, a *contrast* among $\alpha_1, \alpha_2, \ldots, \alpha_k$ is a linear combination of the form $\sum_{i=1}^{k} \lambda_i \alpha_i$ such that $\sum_{i=1}^{k} \lambda_i = 0$. This can also be written as a contrast among the means of the k treatments, namely, $\mu_i = \mu + \alpha_i$ $(i = 1, 2, \ldots, k)$ since $\sum_{i=1}^{k} \lambda_i \alpha_i = \sum_{i=1}^{k} \lambda_i \mu_i$.

Lemma 7.2 The linear combination $\sum_{i=1}^{k} \lambda_i \alpha_i$ is a contrast among $\alpha_1, \alpha_2, \ldots, \alpha_k$ if and only if $\boldsymbol{\ell}' = (0, \lambda_1, \lambda_2, \ldots, \lambda_k)$ belongs to the row space of $\boldsymbol{A}$ in (7.42).

Proof. If $\boldsymbol{\ell}'$ belongs to the row space of $\boldsymbol{A}$, then $\boldsymbol{\ell}' = \boldsymbol{u}'\boldsymbol{A}$ for some vector $\boldsymbol{u} = (u_1, u_2, \ldots, u_{k-1})'$. Then,

$$\begin{aligned} \boldsymbol{\ell}'\,\mathbf{1}_{k+1} &= \boldsymbol{u}'\boldsymbol{A}\mathbf{1}_{k+1} \\ &= 0. \end{aligned} \tag{7.45}$$

From (7.45) it follows that $\sum_{i=1}^{k} \lambda_i = 0$.

Vice versa, if $\sum_{i=1}^{k} \lambda_i \alpha_i$ is a contrast among the α_i's, then

$$(\lambda_1, \lambda_2, \ldots, \lambda_k)\mathbf{1}_k = 0. \tag{7.46}$$

Let $\boldsymbol{A}_1$ be a matrix of order $(k-1) \times k$ obtained by removing the first column of $\boldsymbol{A}$. Then, $\boldsymbol{A}_1\mathbf{1}_k = \mathbf{0}$. Since the rows of $\boldsymbol{A}_1$ are linearly independent, they must form a basis for the orthogonal complement of $\mathbf{1}_k$ in a k-dimensional Euclidean space. From (7.46) we conclude that $(\lambda_1, \lambda_2, \ldots, \lambda_k)$ must be in such an orthogonal complement, that is,

$$(\lambda_1, \lambda_2, \ldots, \lambda_k) = \boldsymbol{h}'\boldsymbol{A}_1,$$

for some vector $\boldsymbol{h}$. Hence,

$$\begin{aligned} \boldsymbol{\ell}' &= (0, \lambda_1, \lambda_2, \ldots, \lambda_k) \\ &= \boldsymbol{h}'\boldsymbol{A}, \end{aligned}$$

which indicates that $\boldsymbol{\ell}'$ belongs to the row space of $\boldsymbol{A}$. □

Corollary 7.2 Any contrast, $\sum_{i=1}^{k} \lambda_i \alpha_i$, among the α_i's is estimable.

Proof. $\sum_{i=1}^{k} \lambda_i \alpha_i = \boldsymbol{\ell}'\boldsymbol{\beta}$, where $\boldsymbol{\ell}' = (0, \lambda_1, \lambda_2, \ldots, \lambda_k)$. By Lemma 7.2, $\boldsymbol{\ell}'$ belongs to the row space of $\boldsymbol{A}$, and hence to the row space of $\boldsymbol{X}$. Thus, $\boldsymbol{\ell}'\boldsymbol{\beta}$ is estimable. □

Using (7.44), Scheffé's simultaneous $(1-\alpha)100\%$ confidence intervals on all contrasts among the α_i's are then given by

$$\sum_{i=1}^{k} \lambda_i \bar{Y}_{i.} \pm \left[(k-1) MS_E F_{\alpha, k-1, n_.-k} \sum_{i=1}^{k} \frac{\lambda_i^2}{n_i} \right]^{1/2}. \tag{7.47}$$

7.5.1 The Relationship between Scheffé's Simultaneous Confidence Intervals and the F-Test Concerning $H_0 : \boldsymbol{A\beta} = \mathbf{0}$

There is an interesting relationship between Scheffé's simultaneous confidence intervals in (7.40) and the F-test concerning the null hypothesis

$$H_0 : \boldsymbol{A\beta} = \mathbf{0}, \tag{7.48}$$

where A is $s \times p$ of rank s ($\leq r$, the rank of X in model (7.1)) whose rows belong to the row space of X. This relationship is given by the following lemma.

Lemma 7.3 The test statistic,

$$F = \frac{(A\hat{\beta})'[A(X'X)^{-}A']^{-1}A\hat{\beta}}{s\,MS_E},$$

concerning the hypothesis $H_0 : A\beta = 0$, is significant at the α-level if and only if there exists an estimable linear function $\ell_0'\beta$, where ℓ_0' belongs to the row space of A, for which the confidence interval in (7.40) does not cover the value zero, that is,

$$|\ \ell_0'\hat{\beta}\ | > [s\,MS_E F_{\alpha,s,n-r}\ell_0'(X'X)^{-}\ell_0]^{1/2}. \tag{7.49}$$

Proof. This follows directly from the fact that formula (7.36) under H_0 is equivalent to

$$P[\,|\ \ell'\hat{\beta}\ | \leq [s\,MS_E F_{\alpha,s,n-r}\ell'(X'X)^{-}\ell]^{1/2}, \quad \forall\, \ell' \in \text{row space of } A\,] = 1 - \alpha, \tag{7.50}$$

which results from using (7.39) with $A\beta$ replaced by the zero vector. In other words,

$$F = \frac{(A\hat{\beta})'[A(X'X)^{-}A']^{-1}A\hat{\beta}}{s\,MS_E} \leq F_{\alpha,s,n-r}, \tag{7.51}$$

if and only if

$$|\ \ell'\hat{\beta}\ | \leq [s\,MS_E F_{\alpha,s,n-r}\ell'(X'X)^{-}\ell]^{1/2}, \tag{7.52}$$

for all ℓ' in the row space of A. Equivalently, the F-test statistic in (7.51) is significant at the α-level, that is,

$$\frac{(A\hat{\beta})'[A(X'X)^{-}A']^{-1}A\hat{\beta}}{s\,MS_E} > F_{\alpha,s,n-r}, \tag{7.53}$$

if and only if

$$|\ \ell_0'\hat{\beta}\ | > [s\,MS_E F_{\alpha,s,n-r}\ell_0'(X'X)^{-}\ell_0]^{1/2} \tag{7.54}$$

for some ℓ_0' in the row space of A. In this case, $\ell_0'\hat{\beta}$ is said to be significantly different from zero at the α-level. □

Corollary 7.3 Consider the one-way model in Example 7.3. The F-test concerning the hypothesis, $H_0 : \alpha_1 = \alpha_2 = \cdots = \alpha_k$, is significant at the α-level if and only if there exists at least one contrast among the α_i's for which the confidence interval in (7.47) does not contain zero.

Proof. Using (7.42), the null hypothesis is equivalent to $H_0 : \boldsymbol{A\beta} = \boldsymbol{0}$, where $\boldsymbol{A}$ is the same matrix used in Example 7.3. By Lemma 7.2, for any $\boldsymbol{\ell}'$ in the row space of $\boldsymbol{A}$, $\boldsymbol{\ell}'\boldsymbol{\beta}$ represents a contrast among the α_i's. Hence, by Lemma 7.3, the F-test concerning H_0 is significant at the α-level if and only if there exists a contrast, $\sum_{i=1}^{k} \lambda_i^0 \alpha_i$, for which the confidence interval in (7.47) does not contain zero, that is,

$$\left| \sum_{i=1}^{k} \lambda_i^0 \bar{Y}_{i.} \right| > \left[(k-1) MS_E F_{\alpha,k-1,n_.-k} \sum_{i=1}^{k} \frac{\lambda_i^{0^2}}{n_i} \right]^{1/2}. \tag{7.55}$$

In this case, $\sum_{i=1}^{k} \lambda_i^0 \bar{Y}_{i.}$ is said to be significantly different from zero. □

7.5.2 Determination of an Influential Set of Estimable Linear Functions

The existence of a significant estimable linear function, whenever the F-test concerning the hypothesis in (7.48) is significant, was established in Section 7.5.1. In the present section, we show how such a linear function can be found. This was initially demonstrated by Khuri (1993).

Consider again the inequality in (7.49). Since $\boldsymbol{\ell}_0'$ belongs to the row space of $\boldsymbol{A}$, $\boldsymbol{\ell}_0'\boldsymbol{\beta}$ can be written as $\boldsymbol{t}_0'\boldsymbol{A\beta}$, where $\boldsymbol{t}_0$ is some vector in R^s, the s-dimensional Euclidean space. It is easy to see that inequality (7.49) is equivalent to

$$\sup_{\boldsymbol{t}\in R^s,\, \boldsymbol{t}\neq \boldsymbol{0}} \left\{ \frac{|\, \boldsymbol{t}'\boldsymbol{A}\hat{\boldsymbol{\beta}}\,|}{[\boldsymbol{t}'\boldsymbol{A}(\boldsymbol{X}'\boldsymbol{X})^-\boldsymbol{A}'\boldsymbol{t}]^{1/2}} \right\} > (s\, MS_E F_{\alpha,s,n-r})^{1/2}. \tag{7.56}$$

Note that $|\, \boldsymbol{t}'\boldsymbol{A}\hat{\boldsymbol{\beta}}\, | = (\boldsymbol{t}'\boldsymbol{A}\hat{\boldsymbol{\beta}}\hat{\boldsymbol{\beta}}'\boldsymbol{A}'\boldsymbol{t})^{1/2}$. Hence, (7.56) can be expressed as

$$\sup_{\boldsymbol{t}\in R^s,\, \boldsymbol{t}\neq \boldsymbol{0}} \left\{ \frac{\boldsymbol{t}'\boldsymbol{G}_1\boldsymbol{t}}{\boldsymbol{t}'\boldsymbol{G}_2\boldsymbol{t}} \right\} > s\, MS_E F_{\alpha,s,n-r}, \tag{7.57}$$

where

$$\boldsymbol{G}_1 = \boldsymbol{A}\hat{\boldsymbol{\beta}}\hat{\boldsymbol{\beta}}'\boldsymbol{A}'$$
$$\boldsymbol{G}_2 = \boldsymbol{A}(\boldsymbol{X}'\boldsymbol{X})^-\boldsymbol{A}'$$

Since $\boldsymbol{G}_2$ is positive definite, then by Theorem 3.11,

$$\sup_{\boldsymbol{t}\in R^s,\, \boldsymbol{t}\neq \boldsymbol{0}} \left\{ \frac{\boldsymbol{t}'\boldsymbol{G}_1\boldsymbol{t}}{\boldsymbol{t}'\boldsymbol{G}_2\boldsymbol{t}} \right\} = e_{\max}\left(\boldsymbol{G}_2^{-1}\boldsymbol{G}_1\right), \tag{7.58}$$

where $e_{\max}\left(\boldsymbol{G}_2^{-1}\boldsymbol{G}_1\right)$ is the largest eigenvalue of $\boldsymbol{G}_2^{-1}\boldsymbol{G}_1$. By a cyclic permutation of the matrices in $\boldsymbol{G}_2^{-1}\boldsymbol{G}_1$, this eigenvalue is equal to $\hat{\boldsymbol{\beta}}'\boldsymbol{A}'[\boldsymbol{A}(\boldsymbol{X}'\boldsymbol{X})^-\boldsymbol{A}']^{-1}\boldsymbol{A}\hat{\boldsymbol{\beta}}$ (see property (f) in Section 3.8), which is the numerator sum of squares of the F-test statistic in (7.53) for testing $H_0 : \boldsymbol{A\beta} = \boldsymbol{0}$.

Now, let $\boldsymbol{t}^*$ be an eigenvector of $\boldsymbol{G}_2^{-1}\boldsymbol{G}_1$ corresponding to $e_{\max}\left(\boldsymbol{G}_2^{-1}\boldsymbol{G}_1\right)$. Then,

$$\frac{\boldsymbol{t}^{*'}\boldsymbol{G}_1\boldsymbol{t}^*}{\boldsymbol{t}^{*'}\boldsymbol{G}_2\boldsymbol{t}^*} = e_{\max}\left(\boldsymbol{G}_2^{-1}\boldsymbol{G}_1\right), \tag{7.59}$$

which follows from the fact that $\boldsymbol{t}^*$ satisfies the equation

$$\left[\boldsymbol{G}_1 - e_{\max}\left(\boldsymbol{G}_2^{-1}\boldsymbol{G}_1\right)\boldsymbol{G}_2\right]\boldsymbol{t}^* = \boldsymbol{0}.$$

From (7.58) and (7.59) we conclude that $\frac{\boldsymbol{t}'\boldsymbol{G}_1\boldsymbol{t}}{\boldsymbol{t}'\boldsymbol{G}_2\boldsymbol{t}}$ attains its supremum when $\boldsymbol{t} = \boldsymbol{t}^*$. The vector $\boldsymbol{t}^*$ can be chosen equal to $\boldsymbol{G}_2^{-1}\boldsymbol{A}\hat{\boldsymbol{\beta}}$ since

$$\begin{aligned}
\boldsymbol{G}_2^{-1}\boldsymbol{G}_1\left(\boldsymbol{G}_2^{-1}\boldsymbol{A}\hat{\boldsymbol{\beta}}\right) &= \boldsymbol{G}_2^{-1}(\boldsymbol{A}\hat{\boldsymbol{\beta}}\hat{\boldsymbol{\beta}}'\boldsymbol{A}')\left(\boldsymbol{G}_2^{-1}\boldsymbol{A}\hat{\boldsymbol{\beta}}\right) \\
&= \left(\hat{\boldsymbol{\beta}}'\boldsymbol{A}'\boldsymbol{G}_2^{-1}\boldsymbol{A}\hat{\boldsymbol{\beta}}\right)\boldsymbol{G}_2^{-1}\boldsymbol{A}\hat{\boldsymbol{\beta}} \\
&= e_{\max}\left(\boldsymbol{G}_2^{-1}\boldsymbol{G}_1\right)\boldsymbol{G}_2^{-1}\boldsymbol{A}\hat{\boldsymbol{\beta}},
\end{aligned}$$

which indicates that $\boldsymbol{G}_2^{-1}\boldsymbol{A}\hat{\boldsymbol{\beta}}$ is an eigenvector of $\boldsymbol{G}_2^{-1}\boldsymbol{G}_1$ for the eigenvalue $e_{\max}\left(\boldsymbol{G}_2^{-1}\boldsymbol{G}_1\right)$.

From (7.56) it can be concluded that if the F-test is significant at the α-level, then

$$\mid \boldsymbol{t}^{*'}\boldsymbol{A}\hat{\boldsymbol{\beta}} \mid > (s\,MS_E F_{\alpha,s,n-r})^{1/2}[\boldsymbol{t}^{*'}\boldsymbol{A}(\boldsymbol{X}'\boldsymbol{X})^{-}\boldsymbol{A}'\boldsymbol{t}^*]^{1/2}.$$

This shows that $\boldsymbol{\ell}^{*'}\hat{\boldsymbol{\beta}} = \boldsymbol{t}^{*'}\boldsymbol{A}\hat{\boldsymbol{\beta}}$ is significantly different from zero. We have therefore identified an estimable linear function, $\boldsymbol{\ell}^{*'}\boldsymbol{\beta}$, that satisfies inequality (7.49).

Let us now determine the elements of $\boldsymbol{A}\hat{\boldsymbol{\beta}}$ which contribute sizably to the significance of the F-test. For this purpose, let us express $\boldsymbol{t}^{*'}\boldsymbol{A}\hat{\boldsymbol{\beta}}$ as

$$\boldsymbol{t}^{*'}\boldsymbol{A}\hat{\boldsymbol{\beta}} = \sum_{i=1}^{s} t_i^*\hat{\gamma}_i, \tag{7.60}$$

where t_i^* and $\hat{\gamma}_i$ are the ith elements of $\boldsymbol{t}^*$ and $\hat{\boldsymbol{\gamma}} = \boldsymbol{A}\hat{\boldsymbol{\beta}}$, respectively ($i = 1, 2, \ldots, s$). Dividing $\hat{\gamma}_i$ by its estimated standard error $\hat{\kappa}_i$, which is equal to the square root of the ith diagonal element of $\boldsymbol{A}(\boldsymbol{X}'\boldsymbol{X})^{-}\boldsymbol{A}'MS_E$, formula (7.60) can be written as

$$\boldsymbol{t}^{*'}\boldsymbol{A}\hat{\boldsymbol{\beta}} = \sum_{i=1}^{s} \omega_i\hat{\tau}_i, \tag{7.61}$$

where

$\hat{\tau}_i = \hat{\gamma}_i/\hat{\kappa}_i$

$\omega_i = t_i^* \hat{\kappa}_i, i = 1, 2, \ldots, s$

Large values of $|\ \omega_i\ |$ identify those elements of $A\hat{\beta}$ that are influential contributors to the significance of the F-test.

Example 7.4 Consider once more the one-way model of Example 7.3. By Corollary 7.3, if the F-test concerning the hypothesis $A\beta = 0$, where $A\beta$ is given in (7.42), is significant at the α-level, then there exists a contrast among $\bar{Y}_{1.}, \bar{Y}_{2.}, \ldots, \bar{Y}_{k.}$ that is significantly different from zero [see inequality (7.55)]. Using the procedure described earlier in Section 7.5.2, it is now possible to find such a contrast. It can be verified that in this example,

$$\begin{aligned} G_2 &= A(X'X)^- A' \\ &= [0 : 1_{k-1} : -I_{k-1}](X'X)^- [0 : 1_{k-1} : -I_{k-1}]' \\ &= \frac{1}{n_1} J_{k-1} + \text{diag}\left(\frac{1}{n_2}, \frac{1}{n_3}, \ldots, \frac{1}{n_k}\right). \end{aligned}$$

Hence,

$$G_2^{-1} = \text{diag}(n_2, n_3, \ldots, n_k) - \frac{1}{n_.}(n_2, n_3, \ldots, n_k)'(n_2, n_3, \ldots, n_k),$$

where $n_. = \sum_{i=1}^{k} n_i$. In addition, we have that $\hat{\gamma} = A\hat{\beta} = (\bar{Y}_{1.} - \bar{Y}_{2.}, \bar{Y}_{1.} - \bar{Y}_{3.}, \ldots, \bar{Y}_{1.} - \bar{Y}_{k.})'$. Hence, the ith element, t_i^*, of $t^* = G_2^{-1} A\hat{\beta}$ is given by

$$t_i^* = n_{i+1}(\bar{Y}_{1.} - \bar{Y}_{i+1.}) - \frac{n_{i+1}}{n_.} \sum_{j=2}^{k} n_j(\bar{Y}_{1.} - \bar{Y}_{j.}), \quad i = 1, 2, \ldots, k-1.$$

The variance–covariance matrix of $\hat{\gamma}$ is estimated by

$$\begin{aligned} \widehat{\text{Var}}(\hat{\gamma}) &= A(X'X)^- A'\, MS_E \\ &= G_2\, MS_E \\ &= \left[\frac{1}{n_1} J_{k-1} + \text{diag}\left(\frac{1}{n_2}, \frac{1}{n_3}, \ldots, \frac{1}{n_k}\right)\right] MS_E. \end{aligned}$$

Hence, the estimated standard error, $\hat{\kappa}_i$, of the ith element of $\hat{\gamma}$ is of the form

$$\hat{\kappa}_i = \left[\left(\frac{1}{n_1} + \frac{1}{n_{i+1}}\right) MS_E\right]^{1/2}, \quad i = 1, 2, \ldots, k-1.$$

Thus, by (7.61), large values of $|\ \omega_i\ | = |\ t_i^*\ |\ \hat{\kappa}_i$ $(i = 1, 2, \ldots, k-1)$ identify those elements of $\hat{\gamma} = (\bar{Y}_{1.} - \bar{Y}_{2.}, \bar{Y}_{1.} - \bar{Y}_{3.}, \ldots, \bar{Y}_{1.} - \bar{Y}_{k.})'$ that contribute sizably to the rejection of $H_0 : \alpha_1 = \alpha_2 = \cdots = \alpha_k$ when the F-test is significant. In particular, if the data set is balanced, that is, $n_i = m$ for $i = 1, 2, \ldots, k$, then it can be shown that

$$| \omega_i | = (2\, m MS_E)^{1/2} \mid \bar{Y}_{i+1.} - \bar{Y}_{..} \mid, \quad i = 1, 2, \ldots, k-1,$$

where $\bar{Y}_{..} = \frac{1}{k} \sum_{i=1}^{k} \bar{Y}_{i.}$.

Note that the elements of $\hat{\gamma} = A\hat{\beta}$ in this example are pairwise differences among the $\bar{Y}_{i.}$'s. None of these differences may be significantly different from zero, even if the F-test is significant. However, a combination of such differences, particularly those that correspond to large values of $| \omega_i |$, will have a significant effect, if the F-test results in the rejection of $H_0 : \alpha_1 = \alpha_2 = \cdots = \alpha_k$.

7.5.3 Bonferroni's Intervals

So far, emphasis has been placed on getting simultaneous confidence intervals on all contrasts (or linear functions) involving the treatment means in the case of the one-way model. In some situations, however, simultaneous confidence intervals on only a fixed number, ν, of contrasts may be of interest. Let $\phi_j = \sum_{i=1}^{k} \lambda_{ij} \mu_i$ $(j = 1, 2, \ldots, \nu)$ be such contrasts. An unbiased estimator of ϕ_j is $\hat{\phi}_j = \sum_{i=1}^{k} \lambda_{ij} \bar{Y}_{i.}$ whose variance is

$$\text{Var}(\hat{\phi}_j) = \sigma^2 \sum_{i=1}^{k} \frac{\lambda_{ij}^2}{n_i}, \quad j = 1, 2, \ldots, \nu.$$

The $(1 - \alpha)$ 100% confidence interval on ϕ_j is then given by

$$\hat{\phi}_j \pm \left(MS_E \sum_{i=1}^{k} \frac{\lambda_{ij}^2}{n_i} \right)^{1/2} t_{\frac{\alpha}{2}, n_. - k}, \quad j = 1, 2, \ldots, \nu.$$

Let A_j denote the event that occurs when this interval contains ϕ_j $(j = 1, 2, \ldots, \nu)$. The so-called *Bonferroni inequality* can be used to obtain a lower bound on the probability $P\left(\bigcap_{j=1}^{\nu} A_j\right)$, namely,

$$P\left(\bigcap_{j=1}^{\nu} A_j\right) \geq 1 - \nu\, \alpha.$$

This inequality results from noting that

$$P\left(\bigcap_{j=1}^{\nu} A_j\right) = 1 - P\left(\bigcup_{j=1}^{\nu} A_j^c\right)$$
$$\geq 1 - \sum_{j=1}^{\nu} P\left(A_j^c\right)$$
$$= 1 - \nu\,\alpha.$$

Thus, the aforementioned intervals provide simultaneous coverage of $\phi_1, \phi_2, \ldots, \phi_\nu$ with a joint coverage probability greater than or equal to $1 - \tilde{\alpha}$, where $\tilde{\alpha} = \nu\,\alpha$. Such intervals are called *Bonferroni's intervals*. It should be noted that the consideration of too many contrasts causes the corresponding intervals to be long and therefore not very desirable.

7.5.4 *Šidák*'s Intervals

An alternative set of intervals for a fixed number, ν, of contrasts for the one-way model case can be obtained on the basis of the following result by *Šidák* (1967, Corollary 2):

Theorem 7.8 Let $\mathbf{Z} = (Z_1, Z_2, \ldots, Z_\kappa)'$ have a multivariate normal distribution with a zero mean vector and a variance–covariance matrix, $\mathbf{\Sigma}$. Let ψ be a positive random variable independent of $\mathbf{Z}$. Then, for any positive constants, $\delta_1, \delta_2, \ldots, \delta_\kappa$,

$$P\left[\frac{|\,Z_1\,|}{\psi} \leq \delta_1, \frac{|\,Z_2\,|}{\psi} \leq \delta_2, \ldots, \frac{|\,Z_\kappa\,|}{\psi} \leq \delta_\kappa\right] \geq \prod_{j=1}^{\kappa} P\left[\frac{|\,Z_j\,|}{\psi} \leq \delta_j\right].$$

Consider now the contrasts, $\phi_j = \sum_{i=1}^{k} \lambda_{ij}\mu_i$, $j = 1, 2, \ldots, \nu$. We assume that these contrasts are linearly independent in the sense that the matrix,

$$\mathbf{\Lambda}_\phi = [\boldsymbol{\lambda}_1 : \boldsymbol{\lambda}_2 : \ldots : \boldsymbol{\lambda}_\nu]'$$

is of full row rank, where $\boldsymbol{\lambda}_j = (\lambda_{1j}, \lambda_{2j}, \ldots, \lambda_{kj})'$, $j = 1, 2, \ldots, \nu$. Then, the random vector $\mathbf{\Lambda}_\phi \bar{\mathbf{Y}}$, where $\bar{\mathbf{Y}} = (\bar{Y}_{1.}, \bar{Y}_{2.}, \ldots, \bar{Y}_{k.})'$ is normally distributed with a variance–covariance matrix, $\sigma^2\,\mathbf{\Lambda}_\phi\,\text{diag}(\frac{1}{n_1}, \frac{1}{n_2}, \ldots, \frac{1}{n_k})\mathbf{\Lambda}'_\phi$. Let us now proceed to apply Theorem 7.8 to

$$\mathbf{Z} = \mathbf{\Delta}^{-1/2}\mathbf{\Lambda}_\phi(\bar{\mathbf{Y}} - \boldsymbol{\mu}),$$

where

$\boldsymbol{\mu} = (\mu_1, \mu_2, \ldots, \mu_k)'$

$\mathbf{\Delta} = \text{diag}(\Delta_{11}, \Delta_{22}, \ldots, \Delta_{\nu\nu})$

Δ_{jj} is the jth diagonal element of $\mathbf{\Lambda}_\phi\ \text{diag}\left(\frac{1}{n_1}, \frac{1}{n_2}, \ldots, \frac{1}{n_k}\right)\mathbf{\Lambda}'_\phi$, $j = 1, 2, \ldots, \nu$

Note that $\boldsymbol{Z}$ is normally distributed with a zero mean vector and a variance–covariance matrix,

$$\boldsymbol{\Sigma} = \sigma^2 \boldsymbol{\Delta}^{-1/2} \boldsymbol{\Lambda}_{\phi} \,\text{diag}\left(\frac{1}{n_1}, \frac{1}{n_2}, \ldots, \frac{1}{n_k}\right) \boldsymbol{\Lambda}'_{\phi}\, \boldsymbol{\Delta}^{-1/2}.$$

Choosing $\psi = \sqrt{MS_E}$, where MS_E is the error (residual) mean square with $n_. - k$ degrees of freedom, we find that

$$\frac{Z_j}{\psi} = \frac{\hat{\phi}_j - \phi_j}{[\sum_{i=1}^{k} \lambda_{ij}^2 / n_i\, MS_E]^{1/2}}, \quad j = 1, 2, \ldots, \nu,$$

has the t-distribution with $n_. - k$ degrees of freedom, where $\hat{\phi}_j = \sum_{i=1}^{k} \lambda_{ij} \bar{Y}_{i.}$. Thus, by choosing $\delta_j = t_{\alpha/2, n_. - k}$, we get

$$P\left(\frac{|\, Z_j \,|}{\psi} \leq \delta_j\right) = 1 - \alpha, \quad j = 1, 2, \ldots, \nu.$$

Hence, on the basis of Theorem 7.8, we can write

$$P\left[\frac{|\, Z_1 \,|}{\psi} \leq \delta_1, \frac{|\, Z_2 \,|}{\psi} \leq \delta_2, \ldots, \frac{|\, Z_\nu \,|}{\psi} \leq \delta_\nu\right] \geq (1 - \alpha)^{\nu}.$$

Consequently, the intervals,

$$\hat{\phi}_j \pm \left(\sum_{i=1}^{k} \frac{\lambda_{ij}^2}{n_i} MS_E\right)^{1/2} t_{\alpha/2, n_. - k}, \quad j = 1, 2, \ldots, \nu,$$

provide a joint coverage of $\phi_1, \phi_2, \ldots, \phi_\nu$ with a probability greater than or equal to $1 - \alpha_s$, where α_s is such that $1 - \alpha_s = (1 - \alpha)^{\nu}$. These are called *Šidák's intervals*. We note that these intervals are of the same form as Bonferroni's intervals, except that in the Bonferroni case, the joint probability of coverage is greater than or equal to $1 - \tilde{\alpha}$ ($\tilde{\alpha} = \nu\, \alpha$) instead of $(1 - \alpha)^{\nu}$ in the *Šidák* case. Since when $\tilde{\alpha} = \alpha_s$, and for $0 < \tilde{\alpha} < 1$ and $\nu \geq 1$, we must have

$$\frac{\tilde{\alpha}}{\nu} \leq 1 - (1 - \tilde{\alpha})^{1/\nu},$$

we conclude that

$$t_{[1 - (1 - \tilde{\alpha})^{1/\nu}]/2, n_. - k} \leq t_{\frac{\tilde{\alpha}}{2\nu}, n_. - k}.$$

Thus, when the lower bound on the joint coverage probability using Bonferroni's inequality (that is, $1 - \tilde{\alpha}$) is the same as in the *Šidák* case, *Šidák*'s intervals are shorter than Bonferroni's intervals. A comparison of Scheffé's intervals with Bonferroni's and *Šidák*'s intervals was made by Fuchs and Sampson (1987).

7.6 Simultaneous Confidence Intervals on All Contrasts among the Means with Heterogeneous Group Variances

Consider again the one-way model,

$$Y_{ij} = \mu + \alpha_i + \epsilon_{ij}, \quad i = 1, 2, \ldots, k; \quad j = 1, 2, \ldots, n_i, \tag{7.62}$$

used earlier in Example 7.3. Here, the ϵ_{ij}'s are assumed to be independently distributed as $N(0, \sigma_i^2)$, $i = 1, 2, \ldots, k$. Note that the error variances are not necessarily equal for all the k groups. The purpose of this section is to develop simultaneous confidence intervals on all contrasts of the form $\sum_{i=1}^{k} \lambda_i \mu_i$, where $\sum_{i=1}^{k} \lambda_i = 0$ and $\mu_i = \mu + \alpha_i$, $i = 1, 2, \ldots, k$. Scheffé's intervals given in (7.47) are not applicable here because they were developed under the assumption that the error variances were equal. Several methods are now given for the derivation of simultaneous confidence intervals on all contrasts under the assumption of heterogeneous error variances.

7.6.1 The Brown–Forsythe Intervals

According to Brown and Forsythe (1974a), approximate $(1 - \alpha)100\%$ simultaneous confidence intervals on all contrasts, $\sum_{i=1}^{k} \lambda_i \mu_i$, are given by

$$\sum_{i=1}^{k} \lambda_i \bar{Y}_{i.} \pm \left[(k-1) F_{\alpha, k-1, \eta}\right]^{1/2} \left[\sum_{i=1}^{k} \frac{\lambda_i^2 s_i^2}{n_i}\right]^{1/2}, \tag{7.63}$$

where $\bar{Y}_{i.} = \frac{1}{n_i} \sum_{j=1}^{n_i} Y_{ij}$, s_i^2 is the sample variance for the sample data from the ith group (treatment), $i = 1, 2, \ldots, k$, and

$$\eta = \left(\sum_{i=1}^{k} \frac{\lambda_i^2 s_i^2}{n_i}\right)^2 \left[\sum_{i=1}^{k} \frac{\lambda_i^4 s_i^4}{n_i^2 (n_i - 1)}\right]^{-1}. \tag{7.64}$$

Formula (7.63) was also given in Tamhane (1979, p. 473). Note that the Brown–Forsythe procedure gives the following approximate conservative and simultaneous $(1 - \alpha)100\%$ confidence intervals for all pairwise differences, $\mu_i - \mu_j$ $(i, j = 1, 2, \ldots, k;\ i < j)$:

$$\bar{Y}_{i.} - \bar{Y}_{j.} \pm [(k-1) F_{\alpha, k-1, \eta_{ij}}]^{1/2} \left(\frac{s_i^2}{n_i} + \frac{s_j^2}{n_j}\right)^{1/2},$$

where η_{ij} is given by

$$\eta_{ij} = \frac{(s_i^2/n_i + s_j^2/n_j)^2}{\frac{s_i^4/n_i^2}{n_i-1} + \frac{s_j^4/n_j^2}{n_j-1}}. \tag{7.65}$$

(See Tamhane, 1979, p. 473.)

7.6.2 Spjøtvoll's Intervals

Spjøvoll (1972) proposed simultaneous confidence intervals on all linear functions, $\sum_{i=1}^{k} c_i\mu_i$, of the means. The derivation of these intervals is based on the following lemma.

Lemma 7.4 Let A be a positive constant. Then,

$$\sum_{i=1}^{k} \frac{(\bar{Y}_{i.} - \mu_i)^2 n_i}{s_i^2} \leq A^2 \tag{7.66}$$

if and only if

$$\left| \sum_{i=1}^{k} \frac{v_i(\bar{Y}_{i.} - \mu_i)n_i^{1/2}}{s_i} \right| \leq A\,(v'v)^{1/2}, \quad \forall\, v \in R^k, \tag{7.67}$$

where $v = (v_1, v_2, \ldots, v_k)'$.

Proof. This can be easily proved by letting $x = (x_1, x_2, \ldots, x_k)'$, where $x_i = \frac{(\bar{Y}_{i.}-\mu_i)n_i^{1/2}}{s_i}$, $i = 1, 2, \ldots, k$, and then using Lemma 6.1 with c^2 replaced by A^2. □

Letting $\frac{v_i n_i^{1/2}}{s_i} = c_i$, $i = 1, 2, \ldots, k$, inequality (7.67) can be written as

$$\left| \sum_{i=1}^{k} c_i(\bar{Y}_{i.} - \mu_i) \right| < A \left(\sum_{i=1}^{k} \frac{c_i^2 s_i^2}{n_i} \right)^{1/2}, \quad \forall\, c \in R^k, \tag{7.68}$$

where $c = (c_1, c_2, \ldots, c_k)'$. Note that $\sum_{i=1}^{k} c_i^2 s_i^2/n_i$ is an unbiased estimate of $\sum_{i=1}^{k} c_i^2 \sigma_i^2/n_i$, which gives the variance of $\sum_{i=1}^{k} c_i \bar{Y}_{i.}$, the unbiased estimator of $\sum_{i=1}^{k} c_i\mu_i$. It can also be noted that since $\frac{(\bar{Y}_{i.}-\mu_i)n_i^{1/2}}{s_i}$ has the t-distribution with $n_i - 1$ degrees of freedom, $F_i = \frac{(\bar{Y}_{i.}-\mu_i)^2 n_i}{s_i^2}$ has the F-distribution with 1 and $n_i - 1$ degrees of freedom, $i = 1, 2, \ldots, k$. Furthermore, these F variates are independently distributed. Hence, the left-hand side of (7.66) is the sum of k mutually independent F-distributed random variables, the ith of which has one degree of freedom for the numerator and $n_i - 1$ degrees of freedom for the denominator ($i = 1, 2, \ldots, k$). If A^2 is chosen as the upper α-quantile of the distribution of $\sum_{i=1}^{k} F_i$, then by Lemma 7.4,

$$P\left(\sum_{i=1}^{k} F_i \leq A^2\right) = 1 - \alpha, \tag{7.69}$$

if and only if

$$P\left(\left|\sum_{i=1}^{k} c_i(\bar{Y}_{i.} - \mu_i)\right| \leq A\left(\sum_{i=1}^{k} \frac{c_i^2 s_i^2}{n_i}\right)^{1/2}, \ \forall \mathbf{c} \in R^k\right) = 1 - \alpha.$$

It follows that simultaneous $(1 - \alpha)100\%$ confidence intervals on all linear functions, $\sum_{i=1}^{k} c_i\mu_i$, of $\mu_1, \mu_2, \ldots, \mu_k$ are given by

$$\sum_{i=1}^{k} c_i\bar{Y}_{i.} \pm A\left(\sum_{i=1}^{k} \frac{c_i^2 s_i^2}{n_i}\right)^{1/2}. \tag{7.70}$$

The constant A depends on α and the degrees of freedom, $n_i - 1$ ($i = 1, 2, \ldots, k$), but not on any unknown parameters. Spjøtvoll approximated the distribution of $\sum_{i=1}^{k} F_i$ by that of a scaled F-variate of the form $\nu_1 F_{k,\nu_2}$, where ν_1 and ν_2 are determined by equating the mean and variance of $\sum_{i=1}^{k} F_i$ to those of $\nu_1 F_{k,\nu_2}$. Since the F_i's are independent, the mean and variance of $\sum_{i=1}^{k} F_i$ are

$$E\left(\sum_{i=1}^{k} F_i\right) = \sum_{i=1}^{k} E(F_i)$$

$$= \sum_{i=1}^{k} \frac{n_i - 1}{n_i - 3},$$

$$\text{Var}\left(\sum_{i=1}^{k} F_i\right) = \sum_{i=1}^{k} \text{Var}(F_i)$$

$$= \sum_{i=1}^{k} \frac{2\,(n_i - 1)^2(n_i - 2)}{(n_i - 3)^2(n_i - 5)}.$$

These formulas are valid provided that $n_i > 5$ for $i = 1, 2, \ldots, k$. Also, the mean and variance of $\nu_1 F_{k,\nu_2}$ are

$$E(\nu_1 F_{k,\nu_2}) = \frac{\nu_1\nu_2}{\nu_2 - 2},$$

$$\text{Var}(\nu_1 F_{k,\nu_2}) = \frac{2\,\nu_1^2(k + \nu_2 - 2)\nu_2^2}{k(\nu_2 - 2)^2(\nu_2 - 4)}.$$

By equating the corresponding means and variances and solving the resulting equations, we get

$$\nu_2 = \frac{(k-2)\left(\sum_{i=1}^{k} \frac{n_i-1}{n_i-3}\right)^2 + 4k\sum_{i=1}^{k} \frac{(n_i-1)^2(n_i-2)}{(n_i-3)^2(n_i-5)}}{k\sum_{i=1}^{k} \frac{(n_i-1)^2(n_i-2)}{(n_i-3)^2(n_i-5)} - \left(\sum_{i=1}^{k} \frac{n_i-1}{n_i-3}\right)^2}, \tag{7.71}$$

$$\nu_1 = \left(1 - \frac{2}{\nu_2}\right)\sum_{i=1}^{k} \frac{n_i - 1}{n_i - 3}. \tag{7.72}$$

Hence, an approximate value of A is given by

$$A \approx [\nu_1 F_{\alpha,k,\nu_2}]^{1/2}. \tag{7.73}$$

The performance of the Brown–Forsythe and Spjøtvoll intervals was evaluated by Tamhane (1979) and Kaiser and Bowden (1983) using computer simulation. It was found that Spjøvoll's intervals are conservative. The Brown–Forsythe intervals, on the other hand, are liberal, that is, the coverage probability of the intervals is less than the nominal value.

7.6.2.1 The Special Case of Contrasts

Suppose that $\sum_{i=1}^{k} c_i\mu_i$ is a contrast among treatment means, that is, $\sum_{i=1}^{k} c_i = 0$. In this case, the vector $\boldsymbol{c} = (c_1, c_2, \ldots, c_k)'$ must belong to a subset, S_c, of R^k, namely, the orthogonal complement of $\mathbf{1}_k$ in R^k, and is therefore of dimension $k-1$. It is easy to see that

$$P\left[\left|\sum_{i=1}^{k} c_i(\bar{Y}_{i.} - \mu_i)\right| \le A\left(\sum_{i=1}^{k} \frac{c_i^2 s_i^2}{n_i}\right)^{1/2}, \forall\, \boldsymbol{c} \in S_c\right] \ge$$

$$P\left[\left|\sum_{i=1}^{k} c_i(\bar{Y}_{i.} - \mu_i)\right| \le A\left(\sum_{i=1}^{k} \frac{c_i^2 s_i^2}{n_i}\right)^{1/2}, \forall\, \boldsymbol{c} \in R^k\right] = P\left[\sum_{i=1}^{k} \frac{(\bar{Y}_{i.} - \mu_i)^2 n_i}{s_i^2} \le A^2\right]$$

$$= 1 - \alpha.$$

Hence, in this special case, Spjøtvoll's intervals in (7.70) become more conservative with a joint coverage probability greater than or equal to $1 - \alpha$. This was noted in the simulation study by Kaiser and Bowden (1983, p. 81).

7.6.3 Exact Conservative Intervals

Let $\sum_{i=1}^{k} c_i\mu_i$ be any linear function of the means. An exact $(1-\alpha)100\%$ confidence interval on μ_i is (U_{1i}, U_{2i}), where

$$U_{1i} = \bar{Y}_{i.} - \left(\frac{s_i^2}{n_i}\right)^{1/2} t_{\alpha/2,\, n_i-1}, \quad i = 1, 2, \ldots, k, \tag{7.74}$$

$$U_{2i} = \bar{Y}_{i.} + \left(\frac{s_i^2}{n_i}\right)^{1/2} t_{\alpha/2,\, n_i-1}, \quad i = 1, 2, \ldots, k, \tag{7.75}$$

where s_i^2 is the sample variance for the ith group ($i = 1, 2, \ldots, k$). These intervals are independent. Hence, their Cartesian product, namely,

$$C = \otimes_{i=1}^{k}[U_{1i}, U_{2i}], \tag{7.76}$$

gives an exact rectangular confidence region on $\mu = (\mu_1, \mu_2, \ldots, \mu_k)'$ with a confidence coefficient = $(1-\alpha)^k = 1-\alpha^*$, where $\alpha^* = 1-(1-\alpha)^k$.

Let $f(.)$ be any continuous function defined on R^k. If $\mu \in C$, then $f(\mu) \in f(C)$ and hence

$$\min_{x \in C} f(x) \leq f(\mu) \leq \max_{x \in C} f(x). \tag{7.77}$$

It follows that

$$P\left(\min_{x \in C} f(x) \leq f(\mu) \leq \max_{x \in C} f(x),\ \forall \text{ continuous} f\right) \geq P(\mu \in C)$$
$$= 1 - \alpha^*.$$

Thus, the double inequality in (7.77) provides simultaneous confidence intervals on the values of $f(\mu)$ for all continuous functions on R^k with a joint confidence coefficient greater than or equal to $1-\alpha^*$. Recall that a similar argument was used in Section 6.7.2 (see the double inequality in (6.92)). Note that since $f(.)$ is continuous, it must attain its maximum and minimum values at points in C. In particular, if $f(\mu) = \sum_{i=1}^{k} c_i\mu_i$ is a linear function of $\mu_1, \mu_2, \ldots, \mu_k$, then simultaneous confidence intervals on all such linear functions are given by

$$\min_{x \in C} \sum_{i=1}^{k} c_i x_i \leq \sum_{i=1}^{k} c_i\mu_i \leq \max_{x \in C} \sum_{i=1}^{k} c_i x_i, \tag{7.78}$$

where $x = (x_1, x_2, \ldots, x_k)'$. The joint coverage probability is therefore greater than or equal to $1-\alpha^*$. Since $f(x) = \sum_{i=1}^{k} c_i x_i$ is a linear function and the region C is bounded by a finite number of hyperplanes (convex polyhedron), the optimization of $f(x)$ over C can be easily carried out by using the following result from *linear programming*: There exists at least one vertex of C at which $f(\boldsymbol{x})$ attains an absolute maximum, and at least one vertex at which $f(\boldsymbol{x})$ attains an absolute minimum (see Theorem 1.4 in Simonnard, 1966, p. 19). Furthermore, since C is a rectangular region, the absolute minimum and maximum of $f(\boldsymbol{x})$ can actually be obtained as follows: Let T_k be a subset of $\{1, 2, \ldots, k\}$ such that $c_i \neq 0$ for $i \in T_k$, and let T_{k_1} and T_{k_2} be two disjoint subsets of T_k such that $T_k = T_{k_1} \bigcup T_{k_2}$ with $c_i > 0$ for $i \in T_{k_1}$ and $c_i < 0$ for $i \in T_{k_2}$. Then,

$$\begin{aligned}\sum_{i=1}^{k} c_i x_i &= \sum_{i \in T_k} c_i x_i \\ &= \sum_{i \in T_{k_1}} c_i x_i + \sum_{i \in T_{k_2}} c_i x_i.\end{aligned}$$

Hence,

$$\min_{\boldsymbol{x} \in C} \sum_{i=1}^{k} c_i x_i = \sum_{i \in T_{k_1}} c_i U_{1i} + \sum_{i \in T_{k_2}} c_i U_{2i}, \tag{7.79}$$

$$\max_{\boldsymbol{x} \in C} \sum_{i=1}^{k} c_i x_i = \sum_{i \in T_{k_1}} c_i U_{2i} + \sum_{i \in T_{k_2}} c_i U_{1i}, \tag{7.80}$$

where U_{1i} and U_{i2} are defined in (7.74) and (7.75), respectively. Formulas (7.79) and (7.80) can then be used to obtain simultaneous confidence intervals on all linear functions $\sum_{i=1}^{k} c_i \mu_i$ with a joint probability of coverage greater than or equal to $1 - \alpha^*$. In particular, if these linear functions are contrasts among the means, then formulas (7.79) and (7.80) can still be used to obtain conservative confidence intervals on all contrasts.

Example 7.5 An experiment was conducted to study the effect of temperature on the shear strength of an adhesive. Four different temperature settings were applied using a completely randomized design. The data are shown in Table 7.1. Testing equality of the population variances for the four temperatures by using Levene's (1960) test gives a significant result with a p-value $= 0.0216$. We can therefore conclude that the population variances are not all equal (for a description of Levene's test, which is known to be robust to possible nonnormality, see, for example, Ott and Longnecker, 2004, p. 314). This test can be easily carried out in PROC GLM of SAS (2000) by using the statement, MEANS TEMPERATURE/HOVTEST = LEVENE, after the model statement, MODEL Y = TEMPERATURE.

Since temperature is a quantitative factor with four equally-spaced levels, it would be of interest to consider its linear, quadratic, and cubic effects, which

TABLE 7.1
Shear Strength Values

	Temperature (°F)			
	200	**220**	**240**	**260**
	7.99	11.03	9.18	8.59
	8.35	10.87	9.55	7.97
	8.73	11.63	9.94	8.13
	8.87	10.31	9.54	8.51
	8.55	11.04	9.02	8.03
	9.04	11.13	9.17	7.88
	9.86	11.30	9.11	8.01
	9.63	10.51	9.53	8.23
	9.86	10.72	8.63	8.45
	8.90	10.65	8.99	8.30
$\bar{Y}$	8.978	10.919	9.266	8.210
s^2	0.402	0.153	0.140	0.061

can be represented by the contrasts, ϕ_1, ϕ_2, ϕ_3, respectively, of the form

$$\phi_1 = -3\mu_1 - \mu_2 + \mu_3 + 3\mu_4,$$
$$\phi_2 = \mu_1 - \mu_2 - \mu_3 + \mu_4,$$
$$\phi_3 = -\mu_1 + 3\mu_2 - 3\mu_3 + \mu_4,$$

where μ_i is the shear strength mean for the ith level of temperature. These contrasts are said to be *orthogonal* because, since the data set is balanced and the temperature levels are equally-spaced, the sum of the cross products of the coefficients of the corresponding means in any two contrasts is equal to zero. In this case, ϕ_1, ϕ_2, ϕ_3 provide a partitioning of the temperature effect, which has 3 degrees of freedom. Note that the coefficients of the means in ϕ_1, ϕ_2, ϕ_3 can be obtained from standard experimental design books (see, for example, Montgomery, 2005, Table IX, p. 625). If the temperature levels were not equally-spaced, or if the sample sizes were unequal, orthogonal linear, quadratic, and cubic contrasts still exist, but are more difficult to find.

Using the Brown–Forsythe intervals in (7.63) with $1 - \alpha = 0.95$, we get

$$-6.085 < \phi_1 < -1.829,$$
$$-3.821 < \phi_2 < -2.173,$$
$$2.517 < \phi_3 < 5.865.$$

Applying now Spjøtvoll's intervals in (7.70) with $1 - \alpha = 0.95$, where by (7.73), $A \approx [\nu_1 F_{0.05,k,\nu_2}]^{1/2} = [4.408 F_{0.05,4,14}]^{1/2} = 3.704$ (see formulas (7.71) and (7.72)), we get

$$-6.430 < \phi_1 < -1.484,$$
$$-4.015 < \phi_2 < -1.979,$$
$$2.129 < \phi_3 < 6.253.$$

Finally, applying the methodology in Section 7.6.3, we find that the individual $(1-\alpha)100\%$ confidence intervals on μ_i $(i = 1, 2, 3, 4)$, where $1-\alpha = (0.95)^{1/4} = 0.987259$ are

$$8.357 < \mu_1 < 9.599,$$
$$10.536 < \mu_2 < 11.302,$$
$$8.899 < \mu_3 < 9.633,$$
$$7.969 < \mu_4 < 8.451.$$

Using now formulas (7.79) and (7.80), the simultaneous confidence intervals on ϕ_1, ϕ_2, ϕ_3 are given by

$$-7.295 < \phi_1 < -0.619,$$
$$-4.610 < \phi_2 < -1.384,$$
$$1.078 < \phi_3 < 7.304.$$

The joint coverage probability is greater than or equal to 0.95.

As expected, the Brown–Forsythe intervals, being liberal, are shorter than the confidence intervals for the other two methods. Both the Spjøtvoll intervals and those obtained from formulas (7.79) and (7.80) are conservative. The difference is that the former are approximate while the latter are exact. We note that the intervals from all three methods do not contain zero. This indicates that the linear, quadratic, and cubic effects of temperature are all significant.

7.7 Further Results Concerning Contrasts and Estimable Linear Functions

In this section, we provide a geometrical approach for the representation of contrasts, and the interpretation of their orthogonality in the context of the one-way model. In addition, we consider the problem of finding simultaneous confidence intervals for two estimable linear functions and their ratio.

7.7.1 A Geometrical Representation of Contrasts

Schey (1985) used geometrical arguments to explain certain features of contrasts for the one-way model. Consider, for example, the balanced one-way model

$$Y_{ij} = \mu + \alpha_i + \epsilon_{ij}, \quad i = 1, 2, \ldots, k; \quad j = 1, 2, \ldots, n_0,$$

where
α_i is a fixed unknown parameter
the ϵ_{ij}'s are independently distributed as $N(0, \sigma^2)$

Let $\mu_i = \mu + \alpha_i$ $(i = 1, 2, \ldots, k)$. Consider the matrix

$$\boldsymbol{W} = \boldsymbol{I}_k - \frac{1}{k}\boldsymbol{J}_k,$$

where $\boldsymbol{J}_k$ is the matrix of ones of order $k \times k$. It is easy to see that $\boldsymbol{W}$ is idempotent of rank $k-1$ and has therefore $k-1$ eigenvalues equal to 1 and one eigenvalue equal to 0. Using the Spectral Decomposition Theorem (see Theorem 3.4), we can write

$$\boldsymbol{W} = \sum_{i=1}^{k-1} \boldsymbol{p}_i \boldsymbol{p}_i', \tag{7.81}$$

where $\boldsymbol{p}_1, \boldsymbol{p}_2, \ldots, \boldsymbol{p}_{k-1}$ are orthonormal eigenvectors of $\boldsymbol{W}$ corresponding to the eigenvalue 1. Hence,

$$\boldsymbol{W}\boldsymbol{p}_i = \boldsymbol{p}_i, \quad i = 1, 2, \ldots, k-1. \tag{7.82}$$

Let $\boldsymbol{p}_k = \frac{1}{\sqrt{k}}\boldsymbol{1}_k$. This is an eigenvector of $\boldsymbol{W}$ corresponding to the eigenvalue 0 since $\boldsymbol{W}\boldsymbol{p}_k = \boldsymbol{0}$. Furthermore, $\boldsymbol{p}_k$ is orthogonal to $\boldsymbol{p}_1, \boldsymbol{p}_2, \ldots, \boldsymbol{p}_{k-1}$ since by (7.82), $\boldsymbol{p}_i'\boldsymbol{p}_k = \boldsymbol{p}_i'\boldsymbol{W}\boldsymbol{p}_k = 0$, $i = 1, 2, \ldots, k-1$. Consequently, $\boldsymbol{p}_1, \boldsymbol{p}_2, \ldots, \boldsymbol{p}_{k-1}, \boldsymbol{p}_k$ form an orthonormal basis in a k-dimensional Euclidean space. It follows that if $\bar{\boldsymbol{Y}} = (\bar{Y}_{1.}, \bar{Y}_{2.}, \ldots, \bar{Y}_{k.})'$, where $\bar{Y}_{i.} = \frac{1}{n_0}\sum_{j=1}^{n_0} Y_{ij}$, $i = 1, 2, \ldots, k$, then $\bar{\boldsymbol{Y}}$ can be expressed as a linear combination of $\boldsymbol{p}_1, \boldsymbol{p}_2, \ldots, \boldsymbol{p}_k$ of the form

$$\bar{\boldsymbol{Y}} = \sum_{i=1}^{k} \hat{\ell}_i \boldsymbol{p}_i. \tag{7.83}$$

Hence, by (7.82) and the fact that $\boldsymbol{W}\boldsymbol{p}_k = \boldsymbol{0}$,

$$\boldsymbol{W}\bar{\boldsymbol{Y}} = \sum_{i=1}^{k-1} \hat{\ell}_i \boldsymbol{p}_i. \tag{7.84}$$

Thus, $\boldsymbol{W}\bar{\boldsymbol{Y}}$ belongs to the $(k-1)$-dimensional Euclidean space spanned by $\boldsymbol{p}_1, \boldsymbol{p}_2, \ldots, \boldsymbol{p}_{k-1}$, which is the orthogonal complement of $\boldsymbol{p}_k$. From (7.83) we note that

$$\boldsymbol{p}_j'\bar{\boldsymbol{Y}} = \hat{\ell}_j, \quad j = 1, 2, \ldots, k-1. \tag{7.85}$$

Thus, $\hat{\ell}_j$ is a contrast in $\bar{Y}_{1.}, \bar{Y}_{2.}, \ldots, \bar{Y}_{k.}$ since $\boldsymbol{p}_j'\mathbf{1}_k = \sqrt{k}\boldsymbol{p}_j'\boldsymbol{p}_k = 0$ for $j = 1, 2, \ldots, k-1$. Furthermore, $\hat{\ell}_j$ is an unbiased estimate of $\ell_j = \boldsymbol{p}_j'\boldsymbol{\mu}$, which is a contrast in $\mu_1, \mu_2, \ldots, \mu_k$ $(j = 1, 2, \ldots, k-1)$, where $\boldsymbol{\mu} = (\mu_1, \mu_2, \ldots, \mu_k)'$. It is easy to see that $\ell_1, \ell_2, \ldots, \ell_{k-1}$ form a basis for all contrasts among $\mu_1, \mu_2, \ldots, \mu_k$. This follows from the fact that if $\boldsymbol{c}'\boldsymbol{\mu}$ is such a contrast with $\boldsymbol{c}'\mathbf{1}_k = 0$, then $\boldsymbol{c} = \sum_{i=1}^{k-1} d_i \boldsymbol{p}_i$ for some constants $d_1, d_2, \ldots, d_{k-1}$. Hence, $\boldsymbol{c}'\boldsymbol{\mu} = \sum_{i=1}^{k-1} d_i \boldsymbol{p}_i'\boldsymbol{\mu} = \sum_{i=1}^{k-1} d_i \ell_i$. Also, if $\boldsymbol{c}_1'\boldsymbol{\mu}$ and $\boldsymbol{c}_2'\boldsymbol{\mu}$ are two orthogonal contrasts, that is, $\boldsymbol{c}_1'\boldsymbol{c}_2 = 0$, then we can write

$$\boldsymbol{c}_1'\boldsymbol{\mu} = \left(\sum_{i=1}^{k-1} d_{1i}\boldsymbol{p}_i\right)'\boldsymbol{\mu} = \sum_{i=1}^{k-1} d_{1i}\ell_i$$

$$\boldsymbol{c}_2'\boldsymbol{\mu} = \left(\sum_{i=1}^{k-1} d_{2i}\boldsymbol{p}_i\right)'\boldsymbol{\mu} = \sum_{i=1}^{k-1} d_{2i}\ell_i$$

for some constants d_{1i}, d_{2i} $(i = 1, 2, \ldots, k-1)$. Note that $\sum_{i=1}^{k-1} d_{1i}d_{2i} = 0$ due to the fact that

$$\boldsymbol{c}_1'\boldsymbol{c}_2 = \left(\sum_{i=1}^{k-1} d_{1i}\boldsymbol{p}_i'\right)\left(\sum_{i=1}^{k-1} d_{2i}\boldsymbol{p}_i\right) = \sum_{i=1}^{k-1} d_{1i}d_{2i}.$$

In particular, ℓ_i and ℓ_j are orthogonal contrasts for $i \neq j$ $(i, j = 1, 2, \ldots, k-1)$.

The above arguments show that $\boldsymbol{p}_1, \boldsymbol{p}_2, \ldots, \boldsymbol{p}_{k-1}$ generate all contrasts among $\mu_1, \mu_2, \ldots, \mu_k$ and that $\ell_1 = \boldsymbol{p}_1'\boldsymbol{\mu}, \ell_2 = \boldsymbol{p}_2'\boldsymbol{\mu}, \ldots, \ell_{k-1} = \boldsymbol{p}_{k-1}'\boldsymbol{\mu}$ are orthogonal contrasts. Formula (7.84) provides a representation of $\boldsymbol{W}\bar{\boldsymbol{Y}}$ as a linear combination of $\boldsymbol{p}_1, \boldsymbol{p}_2, \ldots, \boldsymbol{p}_{k-1}$ whose coefficients form orthogonal contrasts. In addition, the sum of squares for the treatment effect in the one-way model, that is,

$$SS_\alpha = n_0 \sum_{i=1}^{k} (\bar{Y}_{i.} - \bar{Y}_{..})^2,$$

where $\bar{Y}_{..} = \frac{1}{k}\sum_{i=1}^{k} \bar{Y}_{i.}$, can be expressed as

$$SS_\alpha = n_0\,\bar{\boldsymbol{Y}}'\boldsymbol{W}\bar{\boldsymbol{Y}}. \tag{7.86}$$

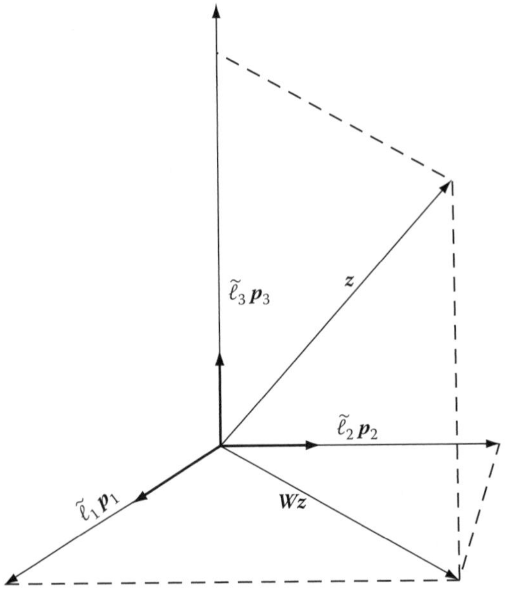

FIGURE 7.1
Orthogonal decomposition of $\boldsymbol{z} = \sqrt{n_0}\bar{\boldsymbol{Y}}$ in the case of three treatment groups. (From Schey, H.M., *Am. Statist.*, 39, 104, 1985. With permission.)

Using formula (7.84) and the fact that $\boldsymbol{W}$ is idempotent, (7.86) can be written as

$$SS_\alpha = n_0 \left(\sum_{i=1}^{k-1} \hat{\ell}_i \boldsymbol{p}_i'\right)\left(\sum_{i=1}^{k-1} \hat{\ell}_i \boldsymbol{p}_i\right)$$
$$= n_0 \sum_{i=1}^{k-1} \hat{\ell}_i^2. \tag{7.87}$$

Note that by (7.85),

$$n_0\, \hat{\ell}_i^2 = (\sqrt{n_0}\, \bar{\boldsymbol{Y}}')\boldsymbol{p}_i \boldsymbol{p}_i'(\sqrt{n_0}\, \bar{\boldsymbol{Y}}).$$

Since $\sqrt{n_0}\, \bar{\boldsymbol{Y}} \sim N(\sqrt{n_0}\, \boldsymbol{\mu}, \sigma^2 \boldsymbol{I}_k)$, $n_0\, \hat{\ell}_i^2$ is distributed as $\sigma^2 \chi_1^2(\theta_i)$, where $\theta_i = n_0\, \boldsymbol{\mu}' \boldsymbol{p}_i \boldsymbol{p}_i' \boldsymbol{\mu}/\sigma^2, i = 1, 2, \ldots, k-1$. Thus, formula (7.87) provides a partitioning of SS_α into $k-1$ independent sums of squares, each having a single degree of freedom, that are associated with the orthogonal contrasts, $\ell_1, \ell_2, \ldots, \ell_{k-1}$. A graphical depiction of these geometric arguments for $k = 3$ is given in Figure 7.1, which shows the orthogonal decomposition of

$$\boldsymbol{z} = \sqrt{n_0}\, \bar{\boldsymbol{Y}}$$
$$= \sum_{i=1}^{3} \tilde{\ell}_i \boldsymbol{p}_i,$$

where $\tilde{\ell}_i = \sqrt{n_0}\,\hat{\ell}_i$, $i = 1, 2, 3$ [see formula (7.83)]. Note that the sum of $\tilde{\ell}_1 \boldsymbol{p}_1$ and $\tilde{\ell}_2 \boldsymbol{p}_2$ is $\boldsymbol{W}\boldsymbol{z}$, and from (7.86), SS_α is the square of the Euclidean norm of $\boldsymbol{W}\boldsymbol{z}$. Furthermore,

$$\begin{aligned}\tilde{\ell}_3^2 &= \boldsymbol{z}'\boldsymbol{z} - \boldsymbol{z}'\boldsymbol{W}\boldsymbol{z} \\ &= \boldsymbol{z}'\left(\frac{1}{3}\boldsymbol{J}_3\right)\boldsymbol{z} \\ &= \frac{n_0}{3}\left(\sum_{i=1}^{3}\bar{Y}_{i.}\right)^2 \\ &= 3\,n_0\bar{Y}_{..}^2.\end{aligned}$$

7.7.2 Simultaneous Confidence Intervals for Two Estimable Linear Functions and Their Ratio

In some experimental situations, one may be interested in estimating two means as well as their ratio. For example, the equivalence of two drugs may be assessed on the basis of the ratio of two treatment means. Also, in quantitative genetics, the *dominance ratio* is to be estimated from *dominance* and *additive* gene effects. These examples, along with others, were reported in Piepho and Emrich (2005) who presented several methods for constructing simultaneous confidence intervals for two means and their ratio. A description of some of these methods is given in Sections 7.7.2.1 through 7.7.2.3. Model (7.1) is considered for this purpose where it is assumed that $\boldsymbol{\epsilon} \sim N(\boldsymbol{0}, \sigma^2\boldsymbol{I}_n)$.

Let $\gamma_1 = \boldsymbol{h}_1'\boldsymbol{\beta}$ and $\gamma_2 = \boldsymbol{h}_2'\boldsymbol{\beta}$ be two estimable linear functions whose ratio, $\frac{\gamma_1}{\gamma_2}$, is denoted by ρ. The objective here is to obtain simultaneous confidence intervals on γ_1, γ_2, and ρ.

7.7.2.1 Simultaneous Confidence Intervals Based on Scheffé's Method

Let $\boldsymbol{\gamma} = (\gamma_1, \gamma_2)'$. Using the methodology outlined in Section 7.4.2, a $(1-\alpha)$ 100% confidence region on $\boldsymbol{\gamma} = \boldsymbol{H}'\boldsymbol{\beta}$, where $\boldsymbol{H} = [\boldsymbol{h}_1 : \boldsymbol{h}_2]$, is given by

$$(\hat{\boldsymbol{\gamma}} - \boldsymbol{\gamma})'[\boldsymbol{H}'(\boldsymbol{X}'\boldsymbol{X})^-\boldsymbol{H}]^{-1}(\hat{\boldsymbol{\gamma}} - \boldsymbol{\gamma}) \le 2\,MS_E\,F_{\alpha,2,n-r}. \qquad (7.88)$$

Then, as in (7.38), Scheffé's simultaneous $(1-\alpha)100\%$ confidence intervals on estimable linear functions of the form $\boldsymbol{k}'\boldsymbol{\gamma}$, for all $\boldsymbol{k}$ in a two-dimensional Euclidean space, are obtained from

$$|\,\boldsymbol{k}'(\hat{\boldsymbol{\gamma}} - \boldsymbol{\gamma})\,| \le [2\,MS_E\,F_{\alpha,2,n-r}\,\boldsymbol{k}'\boldsymbol{H}'(\boldsymbol{X}'\boldsymbol{X})^-\boldsymbol{H}\boldsymbol{k}]^{1/2}, \quad \forall\, \boldsymbol{k} \in R^2. \qquad (7.89)$$

The confidence intervals for $\boldsymbol{\gamma}_1$ and $\boldsymbol{\gamma}_2$ are derived by setting $\boldsymbol{k} = (1, 0)'$ and $(0, 1)'$, respectively. As for the interval for ρ, setting $\boldsymbol{k} = (1, -\rho)'$ in (7.89) and noting that $\boldsymbol{k}'\boldsymbol{\gamma} = \gamma_1 - \rho\,\gamma_2 = 0$, we get

$$|\,\hat{\gamma}_1 - \rho\,\hat{\gamma}_2\,| \le [2\,MS_E\,F_{\alpha,2,n-r}\,(1, -\rho)\boldsymbol{H}'(\boldsymbol{X}'\boldsymbol{X})^-\boldsymbol{H}(1, -\rho)']^{1/2},$$

which is equivalent to

$$(\hat{\gamma}_1 - \rho\,\hat{\gamma}_2)^2 \leq 2\,MS_E\,F_{\alpha,2,n-r}\,(1,-\rho)\boldsymbol{H}'(\boldsymbol{X}'\boldsymbol{X})^{-}\boldsymbol{H}(1,-\rho)'. \tag{7.90}$$

The equality in (7.90) represents a quadratic equation in ρ. Under certain conditions (see Section 6.7.3), this equation yields two real roots for ρ, namely, ρ_1 and ρ_2. In this case, (7.90) provides a confidence interval on ρ given by $[\rho_1, \rho_2]$. This interval, along with the ones for γ_1 and γ_2 will have a joint coverage probability greater than or equal to $1-\alpha$. If, however, the aforementioned conditions are not satisfied, then no confidence interval on ρ can be obtained from (7.90).

7.7.2.2 Simultaneous Confidence Intervals Based on the Bonferroni Inequality

Using the Bonferroni inequality (see Section 7.5.3), simultaneous confidence intervals on γ_1, γ_2, and ρ can be obtained from the inequalities,

$$[(1,0)(\hat{\boldsymbol{\gamma}} - \boldsymbol{\gamma})]^2 \leq \left[MS_E\,F_{\frac{\alpha}{3},1,n-r}\,(1,0)\boldsymbol{H}'(\boldsymbol{X}'\boldsymbol{X})^{-}\boldsymbol{H}(1,0)'\right], \tag{7.91}$$

$$[(0,1)(\hat{\boldsymbol{\gamma}} - \boldsymbol{\gamma})]^2 \leq \left[MS_E\,F_{\frac{\alpha}{3},1,n-r}\,(0,1)\boldsymbol{H}'(\boldsymbol{X}'\boldsymbol{X})^{-}\boldsymbol{H}(0,1)'\right], \tag{7.92}$$

$$[(1,-\rho)(\hat{\boldsymbol{\gamma}} - \boldsymbol{\gamma})]^2 \leq \left[MS_E\,F_{\frac{\alpha}{3},1,n-r}\,(1,-\rho)\boldsymbol{H}'(\boldsymbol{X}'\boldsymbol{X})^{-}\boldsymbol{H}(1,-\rho)'\right], \tag{7.93}$$

respectively. Note that in (7.93), $(1,-\rho)\boldsymbol{\gamma} = 0$. The joint coverage probability for the intervals derived from (7.91) through (7.93) is greater than or equal to $1-\alpha$. Note also that, as in (7.90), the equality in (7.93) can, under certain conditions, yield two real roots for ρ, which provide a confidence interval for ρ.

The Scheffé and Bonferroni intervals differ only in the values of $2\,F_{\alpha,2,n-r}$ and $F_{\frac{\alpha}{3},1,n-r}$, respectively. Piepho and Emrich (2005) made a comparison between such values. They noted that the Bonferroni intervals may be shorter than Scheffé's intervals in some cases, particularly when $n-r$ is large and α is small.

7.7.2.3 Conservative Simultaneous Confidence Intervals

The method described in Section 7.6.3 can be used to obtain simultaneous confidence intervals on γ_1, γ_2, and ρ. This is accomplished by considering the confidence region in (7.88) for $\boldsymbol{\gamma}$, which we denote by C. Then, for any continuous function $f(.)$ defined on C, we have

$$\begin{aligned} P[\min_{x\in C} f(\boldsymbol{x}) \leq f(\boldsymbol{\gamma}) \leq \max_{x\in C} f(\boldsymbol{x}),\ \forall\,\text{continuous}\,f\,] &\geq P(\boldsymbol{\gamma} \in C) \\ &= 1-\alpha. \end{aligned}$$

Choosing $f(\boldsymbol{\gamma}) = \gamma_i$ for $i = 1, 2$ gives confidence intervals for γ_1 and γ_2. These intervals are in fact the projections of C on the x_1 and x_2 axes, which correspond to γ_1 and γ_2, respectively. As for $\rho = \frac{\gamma_1}{\gamma_2}$, choosing $f(\gamma) = \frac{\gamma_1}{\gamma_2}$ gives the interval

$$\min_{x \in \mathcal{C}} \left(\frac{x_1}{x_2} \right) \leq \rho \leq \max_{x \in \mathcal{C}} \left(\frac{x_1}{x_2} \right), \tag{7.94}$$

provided that $\mathcal{C}$ does not include points on the x_1 axis. The joint coverage probability for all three intervals is greater than or equal to $1 - \alpha$. Note that the end points in (7.94), denoted by ρ_ℓ and ρ_u, respectively, are the slopes of the two tangent lines to $\mathcal{C}$ as shown in Figure 7.2.

One numerical example presented by Piepho and Emrich (2005) to illustrate their methods (of deriving the simultaneous confidence intervals) concerned the birth weight data discussed by Nolan, and Speed (2000). The data were obtained by studying the effect of smoking status of mothers on the birth weights of their babies. The weight of the mother was used as a covariate in the following analysis of covariance model,

$$Y_{ij} = \mu_i + \beta(x_{ij} - \bar{x}_{..}) + \epsilon_{ij},$$

where

Y_{ij} denotes the birth weight of the jth baby in the ith group (i takes the value 1 if the mother is a nonsmoker, and the value 2 if she is a smoker)
x_{ij} is the weight of the mother of the jth baby in the ith group
μ_i is the mean of the ith group
β is a slope parameter for the covariate
$\bar{x}_{..}$ is the average of the x_{ij}'s
$\epsilon_{ij} \sim N(0, \sigma^2)$

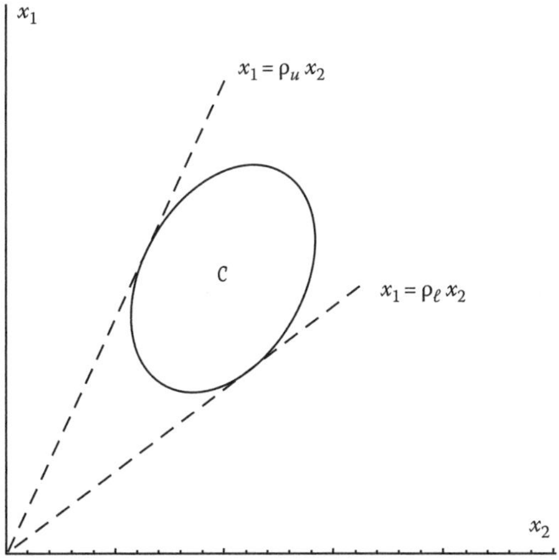

FIGURE 7.2
Lower (ρ_ℓ) and upper (ρ_u) bounds on ρ derived from the confidence region $\mathcal{C}$.

Simultaneous confidence intervals for μ_1, μ_2, and $\rho = \frac{\mu_1}{\mu_2}$ using the aforementioned data were given in Piepho and Emrich (2005, Table 7).

Exercises

7.1 Consider the following data set:

		B			
		1	2	3	4
A	1	16.1	20.5	23.7	—
	2	14.6	—	21.8	—
	3	10.9	12.3	19.7	18.1
	4	—	—	—	6.4

The corresponding model is

$$Y_{ijk} = \mu + \alpha_i + \beta_j + \epsilon_{ijk},$$

where α_i represents the effect of the ith level of factor A, β_j represents the effect of the jth level of factor B, and $\epsilon_{ijk} \sim N(0, \sigma^2)$.

(a) Show that the hypothesis $H_0 : \alpha_1 = \alpha_2 = \alpha_3 = \alpha_4$ is testable. Test H_0 at the 5% level.

(b) Show that the hypothesis $H_0 : \beta_1 = \beta_2 = \beta_3 = \beta_4$ is testable. Test H_0 at the 5% level.

(c) Show that $\mu + \alpha_2 + \frac{1}{4}\sum_{j=1}^{4} \beta_j$ is estimable and find its BLUE.

(d) Show that $\phi = \sum_{i=1}^{4} \lambda_i \alpha_i$ is estimable if $\sum_{i=1}^{4} \lambda_i = 0$. What is its BLUE?

7.2 Find the power of the test in part (a) of Exercise 7.1 assuming that $\sigma^2 = 1$ and that under the alternative hypothesis, $\alpha_1 - \alpha_2 = 2$, $\alpha_1 - \alpha_3 = 0.50$, $\alpha_1 - \alpha_4 = 0$.

7.3 Consider again Exercise 7.1. Write the model in the general form, $Y = X\beta + \epsilon$.

(a) Provide a reparameterization of this model, as was done in Section 7.3, so that it is written as a full-rank model.

(b) Verify that the two models have the same regression and residual sums of squares.

7.4 Consider the model

$$Y_{ijk} = \mu + \alpha_i + \beta_j + \epsilon_{ijk},$$

where $\epsilon_{ijk} \sim N(0, \sigma^2)$. This model is analyzed using the following data set:

		B			
		1	**2**	**3**	**4**
A	**1**	12	—	24	30
	2	—	—	6	11
	3	21	15	16	—

(a) What is the number of basic linearly independent estimable linear functions for this data set? Specify a set of estimable linear functions that form a basis for the space of all estimable linear functions for this model.

(b) Is $\mu_{12} = \mu + \alpha_1 + \beta_2$ estimable? Why or why not?

(c) Show that $\frac{1}{4}\sum_{j=1}^{4}(\mu + \alpha_i + \beta_j)$ is estimable for $i = 1, 2, 3$ and find its BLUE.

(d) Let $\phi_j = \frac{1}{3}\sum_{i=1}^{3}(\mu + \alpha_i + \beta_j)$. Is the hypothesis $H_0 : \phi_1 = \phi_2 = \phi_3 = \phi_4$ testable? Why or why not? If testable, what is the corresponding test statistic?

7.5 Consider the one-way model,

$$Y_{ij} = \mu + \alpha_i + \epsilon_{ij}, \quad i = 1, 2, 3; \quad j = 1, 2, \ldots, n_i,$$

where the ϵ_{ij}'s are independently distributed as $N(0, \sigma^2)$.

(a) Derive a test statistic for testing the hypothesis,

$$H_0 : \frac{1}{5}(\mu + \alpha_1) = \frac{1}{10}(\mu + \alpha_2) = \frac{1}{15}(\mu + \alpha_3)$$

at the $\alpha = 0.05$ level.

(b) Give an expression for the power of the test in part (a) under the alternative hypothesis

$$H_a : \frac{1}{5}(\mu + \alpha_1) - \frac{1}{10}(\mu + \alpha_2) = 1.0,$$
$$\frac{1}{5}(\mu + \alpha_1) - \frac{1}{15}(\mu + \alpha_3) = 2.5,$$

assuming that $\sigma^2 = 1.0$.

7.6 Consider the one-way model,

$$Y_{ij} = \mu + \alpha_i + \epsilon_{ij}, \quad i = 1, 2, \ldots, k; \quad j = 1, 2, \ldots, n_i.$$

Let ϵ be the error vector whose elements are the ϵ_{ij}'s. Suppose that $\epsilon \sim N(\mathbf{0}, \mathbf{\Sigma})$, where

$$\mathbf{\Sigma} = \sigma^2(1-\rho)\mathbf{I}_{n_.} + \sigma^2\rho\,\mathbf{J}_{n_.},$$

where $0 < |\,\rho\,| < 1$, $n_. = \sum_{i=1}^{k} n_i$, $\mathbf{I}_{n_.}$ and $\mathbf{J}_{n_.}$ are, respectively, the identity matrix and matrix of ones of order $n_. \times n_.$.

(a) Show that SS_α and SS_E are independent, where SS_α and SS_E are the usual sums of squares for the α_i's and the residual, respectively, from the ANOVA table for this one-way model.

(b) Show that $\frac{1}{\sigma^2(1-\rho)} SS_\alpha$ has the noncentral chi-squared distribution with $k-1$ degrees of freedom.

(c) Show that $\frac{1}{\sigma^2(1-\rho)} SS_E$ has the central chi-squared distribution with $n_. - k$ degrees of freedom.

(d) Develop a test statistic for the hypothesis $H_0 : \alpha_1 = \alpha_2 = \ldots = \alpha_k$.

7.7 Consider the same model as in Exercise 7.6, except that the ϵ_{ij}'s are now assumed to be independently distributed as normal variables with zero means and variances σ_i^2 $(i = 1, 2, \ldots, k)$. It is assumed that the σ_i^2's are known, but are not equal. Let $\psi = \sum_{i=1}^{k} c_i\mu_i$, where $\mu_i = \mu + \alpha_i$ and $c_1, c_2, \ldots, c_k$ are constants.

(a) Show that the $(1-\alpha)100\%$ simultaneous confidence intervals on all linear functions of the μ_i's of the form ψ are given by

$$\sum_{i=1}^{k} c_i\bar{Y}_{i.} \pm \left[\chi^2_{\alpha,k} \sum_{i=1}^{k} \frac{c_i^2\sigma_i^2}{n_i}\right]^{1/2},$$

where $\bar{Y}_{i.} = \frac{1}{n_i}\sum_{j=1}^{n_i} Y_{ij}$ $(i = 1, 2, \ldots, k)$ and $\chi^2_{\alpha,k}$ is the upper α-quantile of the chi-squared distribution with k degrees of freedom.

(b) Suppose now that the σ_i^2's are unknown. Show how you can obtain an exact $(1-\alpha)100\%$ confidence region on the vector $(\sigma_1^2, \sigma_2^2, \ldots, \sigma_k^2)'$.

7.8 Consider model (7.1) with the following restrictions on the elements of β:

$$\mathbf{T}\mathbf{X}\beta = \mathbf{d},$$

where

T is a known matrix such that TX is of full row rank

d is a known vector in the column space of TX

(a) Find a matrix A and a vector c such that $AY + c$ is an unbiased estimator of $X\beta$.

(b) Does it follow from part (a) that $AX = X$ and $c = 0$? Why or why not?

[Note: Numerical examples concerning parts (a) and (b) can be found in McCulloch and Searle (1995).]

7.9 Twenty four batteries of a particular brand were used in an experiment to determine the effect of temperature on the life span of a battery. Four temperature settings were selected and six batteries were tested at each setting. The measured response was the life of a battery (in hours). The following data were obtained:

Temperature (°F)			
20	**50**	**80**	**110**
130	120	100	90
150	110	95	85
145	105	115	80
120	113	112	92
160	109	92	79
155	114	88	82

(a) Test the polynomial effects (linear, quadratic, and cubic) of temperature at the $\alpha = 0.05$ level.

(b) Let ϕ_1, ϕ_2, ϕ_3 denote the contrasts representing the linear, quadratic, and cubic effects of temperature, respectively. Obtain 95% simultaneous confidence intervals on ϕ_1, ϕ_2, ϕ_3 using Scheffé's procedure. What is the joint coverage probability of these intervals?

7.10 Three repeated measurements were taken from each of n patients undergoing a certain treatment. One measurement was taken after 1 week of administering the treatment, the second measurement was taken after 2 weeks, and the third measurement was taken after 3 weeks. Let Y_i denote the vector of observations obtained from the n patients in week $i\,(= 1, 2, 3)$. Consider the models,

$$Y_i = \beta_{i0}\mathbf{1}_n + X_i\beta_i + \epsilon_i, \quad i = 1, 2, 3,$$

where

X_i is of order $n \times p$ and rank p

ϵ_i is a random error vector

These models can be written as a single multivariate linear model of the form,

$$Y = 1_n \beta_0' + XB + \varepsilon,$$

where

$$\begin{aligned} Y &= [y_1 : y_2 : y_3] \\ \beta_0' &= (\beta_{10}, \beta_{20}, \beta_{30}) \\ X &= [X_1 : X_2 : X_3] \\ B &= \bigoplus_{i=1}^{3} \beta_i \\ \varepsilon &= [\epsilon_1 : \epsilon_2 : \epsilon_3]. \end{aligned}$$

It is assumed that the rows of ε are independently distributed as $N(\mathbf{0}, \boldsymbol{\Sigma})$, where $\boldsymbol{\Sigma}$ is an unknown variance–covariance matrix of order 3×3. The matrix X is of rank ρ, $p \leq \rho \leq 3p$. It is also assumed that $n - \rho \geq 3$.

Consider testing the hypothesis

$$H_0 : \beta_1 = \beta_2 = \beta_3$$

against the alternative hypothesis

$$H_a : \text{at least two } \beta_i's \text{ are not equal.}$$

Show that a test statistic for testing H_0 is given by $e_{\max}\left(S_h S_e^{-1}\right)$, the largest eigenvalue of $S_h S_e^{-1}$, where

$$\begin{aligned} S_h &= C'Y'Z(Z'Z)^- G'[G(Z'Z)^- G']^{-1} G(Z'Z)^- Z'YC, \\ S_e &= C'Y'[I_n - Z(Z'Z)^- Z']YC, \end{aligned}$$

where

$$C = \begin{bmatrix} 1 & 1 \\ -1 & 0 \\ 0 & -1 \end{bmatrix},$$

$$\begin{aligned} Z &= [1_n : X], \\ G &= [\mathbf{0} : W]. \end{aligned}$$

The zero vector in G has p elements and $W = 1_3' \otimes I_p$.

[Hint: Multiply the two sides of the multivariate linear model on the right by $\boldsymbol{C}$ to get $\boldsymbol{YC} = \boldsymbol{1}_n\boldsymbol{\beta}_0'\boldsymbol{C} + \boldsymbol{XBC} + \boldsymbol{\varepsilon C}$. The rows of $\boldsymbol{\varepsilon C}$ are independently distributed as $N(\boldsymbol{0}, \boldsymbol{\Sigma}_c)$, where $\boldsymbol{\Sigma}_c = \boldsymbol{C}'\boldsymbol{\Sigma C}$. The null hypothesis can then be written as $H_0 : \boldsymbol{WBC} = \boldsymbol{0}$. This hypothesis is testable if the rows of $\boldsymbol{G}$ are spanned by the rows of $[\boldsymbol{1}_n : \boldsymbol{X}]$. Multiply now the above model on the right by an arbitrary vector $\boldsymbol{a} = (a_1, a_2)'$ to get the univariate model $\boldsymbol{Y}_a = \boldsymbol{Z}\boldsymbol{\zeta}_a + \boldsymbol{\epsilon}_a$, where $\boldsymbol{Y}_a = \boldsymbol{YCa}$, $\boldsymbol{\zeta}_a = [\boldsymbol{C}'\boldsymbol{\beta}_0 : \boldsymbol{C}'\boldsymbol{B}']'\boldsymbol{a}$, $\boldsymbol{\epsilon}_a = \boldsymbol{\varepsilon Ca}$. Note that $\boldsymbol{\epsilon}_a \sim N(\boldsymbol{0}, \sigma_a^2\boldsymbol{I}_n)$, where $\sigma_a^2 = \boldsymbol{a}'\boldsymbol{C}'\boldsymbol{\Sigma C a}$, and the null hypothesis H_0 is equivalent to $H_0(\boldsymbol{a}) : \boldsymbol{WBCa} = \boldsymbol{0}$, for all $\boldsymbol{a} \neq \boldsymbol{0}$, which can be expressed as $H_0(\boldsymbol{a}) : \boldsymbol{G}\boldsymbol{\zeta}_a = \boldsymbol{0}$, for all $\boldsymbol{a} \neq \boldsymbol{0}$. Note that $\boldsymbol{G}$ is of rank p and is therefore of full row rank. Use now the univariate model to get the following test statistic for testing $H_0(\boldsymbol{a})$:

$$R(\boldsymbol{a}) = \frac{\boldsymbol{Y}_a{}'\boldsymbol{Z}(\boldsymbol{Z}'\boldsymbol{Z})^{-}\boldsymbol{G}'[\boldsymbol{G}(\boldsymbol{Z}'\boldsymbol{Z})^{-}\boldsymbol{G}']^{-1}\boldsymbol{G}(\boldsymbol{Z}'\boldsymbol{Z})^{-}\boldsymbol{Z}'\boldsymbol{Y}_a}{\boldsymbol{Y}_a{}'[\boldsymbol{I}_n - \boldsymbol{Z}(\boldsymbol{Z}'\boldsymbol{Z})^{-}\boldsymbol{Z}']\boldsymbol{Y}_a}$$
$$= \frac{\boldsymbol{a}'\boldsymbol{S}_h\boldsymbol{a}}{\boldsymbol{a}'\boldsymbol{S}_e\boldsymbol{a}},$$

where large values of $R(\boldsymbol{a})$ are significant. Deduce from this information that $e_{\max}\left(\boldsymbol{S}_h\boldsymbol{S}_e^{-1}\right)$ is a test statistic for H_0.]

[Note: More details concerning this exercise and other related issues can be found in Khuri (1986).]

7.11 Consider the one-way model given in (7.62), where $\text{Var}(\epsilon_{ij}) = \sigma^2\lambda_i^2$, $i = 1, 2, \ldots, k$; $j = 1, 2, \ldots, n_i$, where σ^2 is unknown and the λ_i^2's are known.

(a) Show that the ordinary least-squares and the best linear unbiased estimates of $\mu + \alpha_i$ are identical ($i = 1, 2, \ldots, k$).

(b) Find an unbiased estimate of σ^2.

7.12 Consider model (7.1) where $\boldsymbol{\epsilon} \sim N(\boldsymbol{0}, \sigma^2\boldsymbol{I}_n)$. Let $\psi_1, \psi_2, \ldots, \psi_q$ be a set of linearly independent estimable functions of $\boldsymbol{\beta}$. The vector $\hat{\boldsymbol{\psi}}$ denotes the BLUE of $\boldsymbol{\psi} = (\psi_1, \psi_2, \ldots, \psi_q)'$. Its variance–covariance matrix is $\sigma^2\boldsymbol{Q}^{-1}$.

(a) Give an expression for the $(1 - \alpha)100\%$ confidence region on $\boldsymbol{\psi}$.

(b) Suppose that $\boldsymbol{\psi}$ is partitioned as $\boldsymbol{\psi} = [\boldsymbol{\psi}_A' : \boldsymbol{\psi}_B']'$, where $\boldsymbol{\psi}_A$ consists of the first h elements of $\boldsymbol{\psi}$ ($h < q$) and $\boldsymbol{\psi}_B$ consists of the remaining elements. Suppose that $\boldsymbol{Q}$ is partitioned accordingly as

$$\boldsymbol{Q} = \begin{bmatrix} \boldsymbol{Q}_{11} & \boldsymbol{Q}_{12} \\ \boldsymbol{Q}_{12}' & \boldsymbol{Q}_{22} \end{bmatrix}.$$

Show that

$$(\boldsymbol{\psi}_A - \hat{\boldsymbol{\psi}}_A)'\left(\boldsymbol{Q}_{11} - \boldsymbol{Q}_{12}\boldsymbol{Q}_{22}^{-1}\boldsymbol{Q}_{12}'\right)(\boldsymbol{\psi}_A - \hat{\boldsymbol{\psi}}_A) \leq (\boldsymbol{\psi} - \hat{\boldsymbol{\psi}})'\boldsymbol{Q}(\boldsymbol{\psi} - \hat{\boldsymbol{\psi}}),$$

where $\hat{\psi}_A$ is the BLUE of ψ_A.

[Hint: Show that the left-hand side of the above inequality is the minimum value of the right-hand side when the latter is minimized with respect to ψ_B.]

(c) What can be said about the probability that

$$(\psi_A - \hat{\psi}_A)' \left(Q_{11} - Q_{12}Q_{22}^{-1}Q_{12}'\right) (\psi_A - \hat{\psi}_A) \leq q\, MS_E\, F_{\alpha,q,n-r},$$

where MS_E is the error (residual) mean square for model (7.1) and r is the rank of X?

[Hint: Use parts (a) and (b).]

7.13 Let β in model (7.1) be partitioned as $\beta = [\beta_1' : \beta_2']'$, where β_1 is a vector of q elements ($q < p$). By definition, β_1 is testable if there exists a linear function $a'\beta_1$ that is estimable for some vector a. Thus β_1 is testable if and only if $[a' : 0'] = \nu' P_1'$ for some vector ν, where 0 is a zero vector of dimension $p - q$ and P_1 is a matrix of order $p \times r$ whose columns are orthonormal eigenvectors of $X'X$ corresponding to its nonzero eigenvalues, with r being the rank of X [see Section 7.4 and in particular equality (7.19)]. Now, let us partition P_1' as $P_1' = [P_{11}' : P_{12}']$, where P_{11}' and P_{12}' are submatrices of orders $r \times q$ and $r \times (p - q)$, respectively.

(a) Show that β_1 is testable if and only if ν is an eigenvector of $P_{12}'P_{12}$ with a zero eigenvalue such that $P_{11}\nu \neq 0$.

(b) Consider the corresponding reparameterized model in (7.11). Let $\hat{\tilde{\beta}}$ be the BLUE of $\tilde{\beta}$ in this model. Show that $\nu'\hat{\tilde{\beta}}$ is an unbiased estimator of $a'\beta_1$.

7.14 Consider the null hypothesis $H_0 : A\beta = 0$ in Section 7.5.1. Let W denote the set consisting of all vectors ℓ in the row space of A such that $\ell'(X'X)^{-}\ell = 1$. Hence, $\text{Var}(\ell'\hat{\beta}) = \sigma^2$, where σ^2 is the error variance.

(a) Show that W is a nonempty set.

(b) Suppose that $\ell'\hat{\beta}$ is significantly different from zero for some $\ell \in W$. Thus, by inequality (7.54),

$$|\, \ell'\hat{\beta} \,| > (s\, MS_E\, F_{\alpha,s,n-r})^{1/2}.$$

Show that there exists a vector $\ell \in W$ for which $\ell'\hat{\beta}$ is significantly different from zero if and only if

$$\sup_{u \in W} |\, u'\hat{\beta} \,| > (s\, MS_E\, F_{\alpha,s,n-r})^{1/2}.$$

(c) Show that

$$\sup_{u \in W} | u'\hat{\beta} | \leq \{\hat{\beta}' A'[A(X'X)^{-}A']^{-1}A\hat{\beta}\}^{1/2}.$$

7.15 Consider the hypothesis $H_0 : A\beta = m$, where A is $s \times p$ of rank s, but $A\beta$ is nonestimable. Hence, H_0 is not testable.

(a) Obtain the equations that result from minimizing $(Y - X\beta)'(Y - X\beta)$, subject to the restrictions $A\beta = m$, using the method of Lagrange multipliers.

(b) Show that these equations are inconsistent, that is, they produce no solution for $\hat{\beta}$.

(c) Conclude from part (b) that there is no test of the hypothesis H_0.

[Note: This demonstrates that when $A\beta$ is not estimable, no test exists for H_0, which explains the need for estimability in testable hypotheses.]

7.16 Consider the one-way model, $Y_{ij} = \mu + \alpha_i + \epsilon_{ij}$. The hypothesis $H_0 : \mu = 0$ is not testable since μ is nonestimable (see Result 2, Example 7.1). Let SS_E be the residual sum of squares for the model, and let SS_E^0 be the residual sum of squares under H_0. Show that $SS_E^0 = SS_E$.

[Note: This also demonstrates that no test exists for a nontestable hypothesis.]

7.17 Consider model (7.1), where it is assumed that $E(\epsilon) = \mathbf{0}$ and $\text{Var}(\epsilon) = \Sigma = (\sigma_{ij})$. Let

$$MS_E = \frac{1}{n-r} Y'[I_n - X(X'X)^{-}X']Y.$$

(a) Show that

$$E(MS_E) \leq \frac{1}{n-r} \sum_{i=1}^{n} \sigma_{ii}.$$

(b) Show that the upper bound in part (a) cannot be attained unless $X = \mathbf{0}$.

7.18 The yield of a certain grain crop depends on the rate of a particular fertilizer. Four equally-spaced levels of the fertilizer factor were used in a completely randomized design experiment. The yield data (in bushels/acre) are given in the following table:

Level of Fertilizer (lb/plot)			
15	**25**	**35**	**45**
18	39	41	41
12	28	37	42
25	40	30	38
27	33	41	48
26	30	45	45

Let ϕ_1, ϕ_2, ϕ_3 denote the contrasts representing the linear, quadratic, and cubic effects of the fertilizer factor (see Example 7.5). Obtain 95% simultaneous confidence intervals on ϕ_1, ϕ_2, ϕ_3 using

(a) The Brown–Forsythe procedure.

(b) The method by Spjøtvoll.

(c) The exact method described in Section 7.6.3.

8

Balanced Linear Models

The purpose of this chapter is to provide a comprehensive coverage of the properties associated with a balanced linear model. In particular, expressions for the expected mean squares and the distributions of the sums of squares in the corresponding analysis of variance (ANOVA) table are developed using these properties under the usual ANOVA assumptions. Tests of hypothesis and confidence intervals concerning certain unknown parameters of the model can then be derived.

8.1 Notation and Definitions

Suppose that in a given experimental situation, several factors are known to affect a certain response variable denoted by Y. The levels of these factors are indexed by subscripts such as $i, j, k, \ldots$. A typical value of the response can then be identified by attaching these subscripts to Y. For example, in a two-factor experiment involving factors A and B, Y_{ijk} denotes the kth value of the response obtained under level i of A and level j of B.

Definition 8.1 (Balanced data and models). A data set is said to be *balanced* if the range of any one subscript of the response Y does not depend on the values of the other subscripts of Y. A linear model used to analyze a balanced data set is referred to as a *balanced linear model*.

For example, the two-way crossed classification model, which is given by

$$Y_{ijk} = \mu + \alpha_i + \beta_j + (\alpha\beta)_{ij} + \epsilon_{ijk}, \tag{8.1}$$

where $i = 1, 2, \ldots, a$; $j = 1, 2, \ldots, b$; $k = 1, 2, \ldots, n$, is balanced since the range of any of the three subscripts of Y does not depend on the values of the other two subscripts. Accordingly, the corresponding data set consisting of the values of Y_{ijk} is said to be balanced. Note that in model (8.1), if A and B are the factors under consideration, then α_i denotes the effect of level i of factor A, β_j denotes the effect of level j of factor B, $(\alpha\beta)_{ij}$ denotes their interaction effect, and ϵ_{ijk} is a random experimental error.

Definition 8.2 (Crossed and nested factors). Factors A and B are said to be *crossed* if every level of one factor is used in combination with every level of the other factor.

For example, five rates of a potassium fertilizer (factor A) and four rates of a phosphorus fertilizer (factor B) are combined to produce $5\times4 = 20$ treatment combinations. The treatments are allocated at random to 60 identical plots (experimental units) such that each treatment is assigned to each of three plots. In this case, the corresponding model is the one given in (8.1), where Y_{ijk} denotes, for example, the yield of a vegetable crop from the kth plot that receives level i of A and level j of B, $i = 1,2,3,4,5$; $j = 1,2,3,4$; $k = 1,2,3$.

On the other hand, factor B is said to be *nested* within factor A if the levels of B used with a given level of A are different from those used with other levels of A. In this case, B is called the *nested factor* and A is referred to as the *nesting factor*. There is therefore a strict hierarchy in a given nesting relationship. This is not the case when A and B are crossed, and hence no nesting relationship exists between the two factors. A nested classification is sometimes referred to as a *hierarchical classification*.

For example, in an industrial experiment, 12 batches (factor B) of raw material are randomly selected from the warehouse of each of three suppliers (factor A), and three determinations of purity of the raw material are obtained from each batch. In this experiment, B is nested within A, and the corresponding model is written as

$$Y_{ijk} = \mu + \alpha_i + \beta_{ij} + \epsilon_{ijk}, \tag{8.2}$$

where

α_i denotes the effect of the ith supplier ($i = 1,2,3$)

β_{ij} denotes the effect of the jth batch ($j = 1,2,\ldots,12$) obtained from the ith supplier

ϵ_{ijk} is a random experimental error associated with the kth ($k = 1,2,3$) measurement from the jth batch nested within the ith supplier

We note that the identification of the nested effect of B requires two subscripts, i and j, since a particular level of B (for example, the second batch) is only defined after identifying the level of A nesting it. Thus, we can refer, for example, to the second batch obtained from the third supplier. Consequently, if B is nested within A, then whenever j appears as a subscript to an effect in the model, subscript i must also appear.

Let us now consider another example which involves both nested and crossed factors. It concerns a clinical study where three patients are assigned at random to each of three clinics. Three measurements (blood pressure readings) were obtained from each patient on each of four consecutive days. In this case, if A, B, C denote the clinics, patients, and days factors, respectively, then B is obviously nested within A, but A (and hence B) is crossed with C. The corresponding model is therefore of the form

$$Y_{ijkl} = \mu + \alpha_i + \beta_{ij} + \gamma_k + (\alpha\gamma)_{ik} + (\beta\gamma)_{ijk} + \epsilon_{ijkl}, \tag{8.3}$$

where

α_i denotes the effect of the ith clinic ($i = 1, 2, 3$)

β_{ij} denotes the effect of the jth patient ($j = 1, 2, 3$), which is nested within the ith clinic

γ_k denotes the effect of the kth day ($k = 1, 2, 3, 4$)

$(\alpha\gamma)_{ik}$ is the interaction effect between clinics and days

$(\beta\gamma)_{ijk}$ represents the interaction effect between days and patients within clinics

ϵ_{ijkl} is a random experimental error associated with the lth measurement obtained from the jth patient in clinic i on day k ($l = 1, 2, 3$)

Definition 8.3 (Crossed and nested subscripts). Let A and B be two given factors indexed by i and j, respectively. If A and B are crossed, then i and j are said to be *crossed subscripts*. This fact is denoted by writing $(i)(j)$ where i and j are separated by two pairs of parentheses. If, however, B is nested within A, then j is said to be *nested* within i, and this fact is denoted by writing $i : j$ where a colon separates i and j with j appearing to the right of the colon.

Definition 8.4 (Population structure). The *population structure* associated with a given experiment is a complete description of the nesting and nonnesting (crossed) relationships that exist among the factors considered in the experiment.

For example, for model (8.1), the population structure is $[(i)(j)] : k$, where subscripts i and j are crossed and subscript k is nested within both i and j. Note that a pair of square brackets is used to separate k from the combination $(i)(j)$. The population structure for model (8.2) is $(i : j : k)$, which indicates that k is nested within j and the latter is nested within i. As for model (8.3), the corresponding population structure is $[(i : j)(k)] : l$. This clearly shows that j is nested within i and k is crossed with i, and hence with j. Subscript l is nested within i, j, and k.

As will be seen later, the population structure plays an important role in setting up the complete model and the corresponding ANOVA table for the experiment under investigation.

Definition 8.5 (Partial mean). A *partial mean* of a response Y is obtained by averaging Y over the entire range of values of a particular subset of its set of subscripts. A partial mean is denoted by the same symbol as the one used for the response, except that the subscripts that have been averaged out are omitted.

Definition 8.6 (Admissible mean). A partial mean is *admissible* if whenever a nested subscript appears, all the subscripts that nest it appear also.

Definition 8.7 (Rightmost-bracket subscripts). The subscripts of an admissible mean which nest no other subscripts of that mean are said to constitute the set of subscripts belonging to the *rightmost bracket* of the admissible mean.

TABLE 8.1
Population Structures and Admissible Means

Population Structure	Admissible Means
$[(i)(j)]:k$	$Y, Y_{(i)}, Y_{(j)}, Y_{(ij)}, Y_{ij(k)}$
$(i:j:k)$	$Y, Y_{(i)}, Y_{i(j)}, Y_{ij(k)}$
$[(i:j)(k)]:l$	$Y, Y_{(i)}, Y_{(k)}, Y_{i(j)}, Y_{(ik)}, Y_{i(jk)}, Y_{ijk(l)}$

The grouping of these subscripts is indicated by using a pair of parentheses. The remaining subscripts, if any, of the admissible mean are called *nonrightmost-bracket* subscripts and are placed before the rightmost bracket. If the sets of rightmost- and nonrightmost-bracket subscripts are both empty, as is case with the overall mean of the response, then the corresponding admissible mean is denoted by just Y.

For example, the admissible means corresponding to the three population structures, $[(i)(j)]:k$, $(i:j:k)$, and $[(i:j)(k)]:l$ are given in Table 8.1. The identification of the rightmost bracket can also be extended from the admissible means to their corresponding effects in a given model. For example, models (8.1), (8.2), and (8.3) can now be expressed as

$$Y_{ijk} = \mu + \alpha_{(i)} + \beta_{(j)} + (\alpha\beta)_{(ij)} + \epsilon_{ij(k)} \tag{8.4}$$

$$Y_{ijk} = \mu + \alpha_{(i)} + \beta_{i(j)} + \epsilon_{ij(k)} \tag{8.5}$$

$$Y_{ijkl} = \mu + \alpha_{(i)} + \beta_{i(j)} + \gamma_{(k)} + (\alpha\gamma)_{(ik)} + (\beta\gamma)_{i(jk)} + \epsilon_{ijk(l)}. \tag{8.6}$$

As will be seen later, such identification is instrumental in determining the degrees of freedom, sums of squares, and the expected mean squares in the corresponding ANOVA table. We shall therefore use this identification scheme in future balanced ANOVA models.

Definition 8.8 (Component). A *component* associated with an admissible mean is a linear combination of admissible means obtained by selecting all those admissible means that are yielded by the mean under consideration when some, all, or none of its rightmost-bracket subscripts are omitted in all possible ways. Whenever an odd number of subscripts is omitted, the resulting admissible mean is given a negative sign, and whenever an even number of subscripts is omitted, the mean is given a positive sign (the number zero is considered even).

For example, the components corresponding to the admissible means for each of the three population structures in Table 8.1 are displayed in Table 8.2. It is easy to see that if the rightmost bracket of an admissible mean has k subscripts, then the number of admissible means in the corresponding component is equal to 2^k. Furthermore, it can be noted that for any population structure, the sum of all components associated with its admissible means is identical to the response Y. For example, in Table 8.2, the sum of all the

TABLE 8.2
Admissible Means and Corresponding Components

Population Structure	Admissible Mean	Component
$[(i)(j)]:k$	Y	Y
	$Y_{(i)}$	$Y_{(i)} - Y$
	$Y_{(j)}$	$Y_{(j)} - Y$
	$Y_{(ij)}$	$Y_{(ij)} - Y_{(i)} - Y_{(j)} + Y$
	$Y_{ij(k)}$	$Y_{ij(k)} - Y_{(ij)}$
$(i:j:k)$	Y	Y
	$Y_{(i)}$	$Y_{(i)} - Y$
	$Y_{i(j)}$	$Y_{i(j)} - Y_{(i)}$
	$Y_{ij(k)}$	$Y_{ij(k)} - Y_{i(j)}$
$[(i:j)(k)]:l$	Y	Y
	$Y_{(i)}$	$Y_{(i)} - Y$
	$Y_{(k)}$	$Y_{(k)} - Y$
	$Y_{i(j)}$	$Y_{i(j)} - Y_{(i)}$
	$Y_{(ik)}$	$Y_{(ik)} - Y_{(i)} - Y_{(k)} + Y$
	$Y_{i(jk)}$	$Y_{i(jk)} - Y_{i(j)} - Y_{(ik)} + Y_{(i)}$
	$Y_{ijk(l)}$	$Y_{ijk(l)} - Y_{i(jk)}$

components for each of the first two population structures is equal to Y_{ijk}, and for the third population structure, the sum of the components is equal to Y_{ijkl}. Such a relationship leads to a complete linear model for the response under consideration in terms of all the effects which can be derived from the corresponding population structure. For example, for each population structure, the components shown in Table 8.2 are in a one-to-one correspondence with the corresponding effects in models (8.4), (8.5), and (8.6), respectively. Thus, knowledge of the population structure in a given experimental situation is instrumental in identifying the complete model. A general representation of such a model is given in the next section.

8.2 The General Balanced Linear Model

Let $\theta = \{k_1, k_2, \ldots, k_s\}$ be a complete set of subscripts that identify a typical response Y, where $k_j = 1, 2, \ldots, a_j$ $(j = 1, 2, \ldots, s)$. We note that the corresponding data set is balanced since the range of any one subscript, for example k_j, does not depend on the values of the remaining subscripts $(j = 1, 2, \ldots, s)$. The total number of observations in the data set, denoted by N, is $N = \prod_{j=1}^{s} a_j$.

Suppose that for a given population structure, the number of admissible means is $\nu + 2$. The ith admissible mean is denoted by $Y_{\theta_i(\bar{\theta}_i)}$, where $\bar{\theta}_i$ is the set of rightmost-bracket subscripts and θ_i is the set of nonrightmost-bracket subscripts ($i = 0, 1, 2, \ldots, \nu+1$). For $i = 0$, both θ_i and $\bar{\theta}_i$ are empty. For some other admissible means, θ_i may also be empty, as was seen earlier in Table 8.1. The set consisting of all subscripts belonging to both θ_i and $\bar{\theta}_i$ is denoted by ψ_i. The complement of ψ_i with respect to θ is denoted by ψ_i^c ($i = 0, 1, \ldots, \nu+1$). Note that $\psi_i = \theta$ when $i = \nu+1$. The ith component corresponding to the ith admissible mean is denoted by $C_{\theta_i(\bar{\theta}_i)}(Y)$ ($i = 0, 1, \ldots, \nu+1$).

The general form of a balanced linear model can be expressed as

$$Y_\theta = \sum_{i=0}^{\nu+1} g_{\theta_i(\bar{\theta}_i)}, \tag{8.7}$$

where $g_{\theta_i(\bar{\theta}_i)}$ denotes the ith effect in the model. Note that for $i = 0$, $g_{\theta_i(\bar{\theta}_i)}$ is the grand mean μ and for $i = \nu+1$, $g_{\theta_i(\bar{\theta}_i)}$ is the experimental error term. A general expression for the ith component, $C_{\theta_i(\bar{\theta}_i)}$, is given by

$$C_{\theta_i(\bar{\theta}_i)}(Y) = \sum_{j=0}^{\nu+1} \lambda_{ij} Y_{\theta_j(\bar{\theta}_j)}, \quad i = 0, 1, \ldots, \nu+1, \tag{8.8}$$

where $\lambda_{ij} = -1, 0, 1$. The values -1 and 1 are obtained whenever an odd number or an even number of subscripts are omitted from $\bar{\theta}_i$, respectively. The value $\lambda_{ij} = 0$ is used for those admissible means that are not obtained by deleting subscripts from $\bar{\theta}_i$.

Model (8.7) can be written in vector form as

$$\boldsymbol{Y} = \sum_{i=0}^{\nu+1} \boldsymbol{H}_i \boldsymbol{\beta}_i, \tag{8.9}$$

where

$\boldsymbol{Y}$ denotes the vector of N observations

$\boldsymbol{\beta}_i$ is a vector consisting of the elements of $g_{\theta_i(\bar{\theta}_i)}$ ($i = 0, 1, \ldots, \nu+1$)

The matrix $\boldsymbol{H}_i$ is expressible as a direct (Kronecker) product of matrices of the form

$$\boldsymbol{H}_i = \otimes_{j=1}^{s} \boldsymbol{L}_{ij}, \quad i = 0, 1, \ldots, \nu+1, \tag{8.10}$$

where for each i ($= 0, 1, \ldots, \nu+1$), a one-to-one correspondence exists between the matrices $\boldsymbol{L}_{ij}$ ($j = 1, 2, \ldots, s$) and the elements of $\theta = \{k_1, k_2, \ldots, k_s\}$ such that

$$\boldsymbol{L}_{ij} = \begin{cases} \boldsymbol{I}_{a_j}, & k_j \in \psi_i \\ \boldsymbol{1}_{a_j}, & k_j \in \psi_i^c \end{cases}; \ i = 0, 1, \ldots, \nu+1; \ j = 1, 2, \ldots, s, \tag{8.11}$$

where, if we recall, ψ_i is the set of subscripts associated with the ith effect, ψ_i^c is its complement with respect to θ ($i = 0, 1, \ldots, \nu+1$), and $\boldsymbol{I}_{a_j}$ and $\boldsymbol{1}_{a_j}$ are, respectively, the identity matrix of order $a_j \times a_j$ and the $a_j \times 1$ vector of ones $(j = 1, 2, \ldots, s)$. Thus, whenever a subscript of $\theta = \{k_1, k_2, \ldots, k_s\}$ belongs to ψ_i, the corresponding $\boldsymbol{L}_{ij}$ matrix in (8.10) is equal to $\boldsymbol{I}_{a_j}$, otherwise, it is equal to $\mathbf{1}_{a_j}$ $(j = 1, 2, \ldots, s)$. Note that

$$\boldsymbol{H}_i'\boldsymbol{H}_i = b_i\boldsymbol{I}_{c_i}, \quad i = 0, 1, \ldots, \nu + 1, \tag{8.12}$$

where

$$b_i = \begin{cases} \prod_{k_j \in \psi_i^c} a_j, & \text{if } \psi_i^c \neq \emptyset \\ 1, & \text{if } \psi_i^c = \emptyset \end{cases} \; ; \; i = 0, 1, \ldots, \nu + 1$$

$$c_i = \begin{cases} \prod_{k_j \in \psi_i} a_j, & \text{if } \psi_i \neq \emptyset \\ 1, & \text{if } \psi_i = \emptyset \end{cases} \; ; \; i = 0, 1, \ldots, \nu + 1 \tag{8.13}$$

Here
 c_i is the number of columns of $\boldsymbol{H}_i$
 b_i is the number of ones in a typical column of $\boldsymbol{H}_i$ $(i = 0, 1, \ldots, \nu + 1)$
 $\emptyset$ denotes the empty set

Thus, $b_i c_i = N$ for $i = 0, 1, \ldots, \nu + 1$.

Example 8.1 Consider model (8.4), where $i = 1, 2, 3, 4$; $j = 1, 2, 3$; $k = 1, 2$. In vector form, the model is written as

$$\boldsymbol{Y} = \boldsymbol{H}_0\mu + \boldsymbol{H}_1\boldsymbol{\alpha} + \boldsymbol{H}_2\boldsymbol{\beta} + \boldsymbol{H}_3(\boldsymbol{\alpha\beta}) + \boldsymbol{H}_4\boldsymbol{\epsilon},$$

where $\boldsymbol{\alpha}$, $\boldsymbol{\beta}$, $(\boldsymbol{\alpha\beta})$ contain the elements of $\alpha_{(i)}$, $\beta_{(j)}$, $(\alpha\beta)_{(ij)}$, respectively, and $\boldsymbol{H}_0$, $\boldsymbol{H}_1$, $\boldsymbol{H}_2$, $\boldsymbol{H}_3$, $\boldsymbol{H}_4$ are given by

$$\begin{aligned} \boldsymbol{H}_0 &= \mathbf{1}_4 \otimes \mathbf{1}_3 \otimes \mathbf{1}_2 \\ \boldsymbol{H}_1 &= \boldsymbol{I}_4 \otimes \mathbf{1}_3 \otimes \mathbf{1}_2 \\ \boldsymbol{H}_2 &= \mathbf{1}_4 \otimes \boldsymbol{I}_3 \otimes \mathbf{1}_2 \\ \boldsymbol{H}_3 &= \boldsymbol{I}_4 \otimes \boldsymbol{I}_3 \otimes \mathbf{1}_2 \\ \boldsymbol{H}_4 &= \boldsymbol{I}_4 \otimes \boldsymbol{I}_3 \otimes \boldsymbol{I}_2. \end{aligned}$$

Example 8.2 Consider model (8.3), which now can be written in a manner that displays the sets of rightmost-bracket subscripts,

$$Y_{ijkl} = \mu + \alpha_{(i)} + \beta_{i(j)} + \gamma_{(k)} + (\alpha\gamma)_{(ik)} + (\beta\gamma)_{i(jk)} + \epsilon_{ijk(l)},$$

where $i = 1, 2, 3$; $j = 1, 2, 3$; $k = 1, 2, 3, 4$; and $l = 1, 2, 3$. Its vector form is

$$\boldsymbol{Y} = \boldsymbol{H}_0\mu + \boldsymbol{H}_1\boldsymbol{\alpha} + \boldsymbol{H}_2\boldsymbol{\beta} + \boldsymbol{H}_3\boldsymbol{\gamma} + \boldsymbol{H}_4(\boldsymbol{\alpha\gamma}) + \boldsymbol{H}_5(\boldsymbol{\beta\gamma}) + \boldsymbol{H}_6\boldsymbol{\epsilon},$$

where $\boldsymbol{\alpha}$, $\boldsymbol{\beta}$, $\boldsymbol{\gamma}$, $(\boldsymbol{\alpha\gamma})$, and $(\boldsymbol{\beta\gamma})$ contain the elements of $\alpha_{(i)}$, $\beta_{i(j)}$, $\gamma_{(k)}$, $(\alpha\gamma)_{(ik)}$, $(\beta\gamma)_{i(jk)}$, respectively, and

$$\begin{aligned}
H_0 &= \mathbf{1}_3 \otimes \mathbf{1}_3 \otimes \mathbf{1}_4 \otimes \mathbf{1}_3 \\
H_1 &= I_3 \otimes \mathbf{1}_3 \otimes \mathbf{1}_4 \otimes \mathbf{1}_3 \\
H_2 &= I_3 \otimes I_3 \otimes \mathbf{1}_4 \otimes \mathbf{1}_3 \\
H_3 &= \mathbf{1}_3 \otimes \mathbf{1}_3 \otimes I_4 \otimes \mathbf{1}_3 \\
H_4 &= I_3 \otimes \mathbf{1}_3 \otimes I_4 \otimes \mathbf{1}_3 \\
H_5 &= I_3 \otimes I_3 \otimes I_4 \otimes \mathbf{1}_3 \\
H_6 &= I_3 \otimes I_3 \otimes I_4 \otimes I_3.
\end{aligned}$$

8.3 Properties of Balanced Models

Linear models associated with balanced data have very interesting properties. Some of these properties are listed here as lemmas without proofs (see Zyskind, 1962; Khuri, 1982 for more details).

Lemma 8.1 For any component associated with model (8.7), the sum of values of the component is zero when the summation is taken over any subset of subscripts in its rightmost bracket. Thus, if $C_{\theta_i(\bar{\theta}_i)}(Y)$ is the ith component, and if $\bar{\theta}_i \neq \emptyset$, then

$$\sum_{\tau_i \subset \bar{\theta}_i} C_{\theta_i(\bar{\theta}_i)}(Y) = 0,$$

where τ_i denotes a subset of subscripts in $\bar{\theta}_i$.

Definition 8.9 The sum of squares associated with the ith effect in model (8.7) is expressed as a quadratic form $\boldsymbol{Y}'\boldsymbol{P}_i\boldsymbol{Y}(i = 0, 1, \ldots, \nu + 1)$ defined as

$$\boldsymbol{Y}'\boldsymbol{P}_i\boldsymbol{Y} = \sum_{\theta} [C_{\theta_i(\bar{\theta}_i)}(Y)]^2, \quad i = 0, 1, \ldots, \nu + 1. \tag{8.14}$$

For example, for the two-way model in Example 8.1, the sums of squares for the various effects in the model are (see Table 8.2)

$$\begin{aligned}
\boldsymbol{Y}'\boldsymbol{P}_0\boldsymbol{Y} &= \sum_{i=1}^{4}\sum_{j=1}^{3}\sum_{k=1}^{2} Y^2 \\
&= 24\, Y^2
\end{aligned}$$

$$Y'P_1Y = \sum_{i=1}^{4}\sum_{j=1}^{3}\sum_{k=1}^{2}[Y_{(i)} - Y]^2$$

$$= 6\sum_{i=1}^{4}[Y_{(i)} - Y]^2$$

$$Y'P_2Y = \sum_{i=1}^{4}\sum_{j=1}^{3}\sum_{k=1}^{2}[Y_{(j)} - Y]^2$$

$$= 8\sum_{j=1}^{3}[Y_{(j)} - Y]^2$$

$$Y'P_3Y = \sum_{i=1}^{4}\sum_{j=1}^{3}\sum_{k=1}^{2}[Y_{(ij)} - Y_{(i)} - Y_{(j)} + Y]^2$$

$$= 2\sum_{i=1}^{4}\sum_{j=1}^{3}[Y_{(ij)} - Y_{(i)} - Y_{(j)} + Y]^2$$

$$Y'P_4Y = \sum_{i=1}^{4}\sum_{j=1}^{3}\sum_{k=1}^{2}[Y_{ij(k)} - Y_{(ij)}]^2.$$

Lemma 8.2 The sum of squares associated with the ith effect in model (8.7) can be written as

$$Y'P_iY = \sum_{j=0}^{\nu+1}\lambda_{ij}\sum_{\theta}Y^2_{\theta_j(\bar{\theta}_j)}, \quad i = 0, 1, \ldots, \nu+1, \tag{8.15}$$

where λ_{ij} is the coefficient of $Y_{\theta_j(\bar{\theta}_j)}$ in the expression given by (8.8) which defines the ith component ($i, j = 0, 1, \ldots, \nu+1$). Thus, the ith sum of squares is a linear combination of the sums of squares (over the set of all subscripts, θ) of the admissible means that make up the ith component. This linear combination has the same coefficients (that is, the λ_{ij}'s) as those used to define the ith component in terms of its admissible means.

For example, the sums of squares shown earlier for the two-way model can be expressed as

$$Y'P_0Y = 24\,Y^2$$

$$Y'P_1Y = 6\sum_{i=1}^{4}Y^2_{(i)} - 24\,Y^2$$

$$Y'P_2Y = 8\sum_{j=1}^{3}Y^2_{(j)} - 24\,Y^2$$

$$Y'P_3Y = 2\sum_{i=1}^{4}\sum_{j=1}^{3} Y_{(ij)}^2 - 6\sum_{i=1}^{4} Y_{(i)}^2 - 8\sum_{j=1}^{3} Y_{(j)}^2 + 24\,Y^2$$

$$Y'P_4Y = \sum_{i=1}^{4}\sum_{j=1}^{3}\sum_{k=1}^{2} Y_{ij(k)}^2 - 2\sum_{i=1}^{4}\sum_{j=1}^{3} Y_{(ij)}^2.$$

Lemma 8.3 The matrix P_i in formula (8.14) satisfies the following properties:

(a) P_i is idempotent ($i = 0, 1, \ldots, \nu + 1$)

(b) $P_iP_j = 0$ for $i \neq j$ ($i,\ j = 0,\ 1, \ldots,\ \nu + 1$)

(c) $\sum_{i=0}^{\nu+1} P_i = I_N$.

Lemma 8.4 The matrix P_i in formula (8.14) is expressible as

$$P_i = \sum_{j=0}^{\nu+1} \lambda_{ij}\frac{A_j}{b_j}, \quad i = 0, 1, \ldots, \nu + 1, \tag{8.16}$$

where $A_j = H_jH_j'$ ($j = 0, 1, \ldots, \nu + 1$), the λ_{ij}'s are the same coefficients as in (8.8), and b_j is the number defined in (8.13).

Proof. Let Y_i denote the vector consisting of all the values of the ith admissible mean, $Y_{\theta_i(\bar{\theta}_i)}$ ($i = 0, 1, \ldots, \nu + 1$). Then,

$$Y_i = \frac{1}{b_i}H_i'Y, \quad i = 0, 1, \ldots, \nu + 1. \tag{8.17}$$

Using formulas (8.15) and (8.17) we can write

$$\begin{aligned} Y'P_iY &= \sum_{j=0}^{\nu+1} \lambda_{ij} \sum_{\theta} Y_{\theta_j(\bar{\theta}_j)}^2 \\ &= \sum_{j=0}^{\nu+1} \lambda_{ij}\, b_j \sum_{\psi_j} Y_{\theta_j(\bar{\theta}_j)}^2 \\ &= \sum_{j=0}^{\nu+1} \lambda_{ij}\, b_j \left(\frac{Y'A_jY}{b_j^2}\right) \\ &= Y'\left(\sum_{j=0}^{\nu+1} \lambda_{ij}\frac{A_j}{b_j}\right)Y. \end{aligned} \tag{8.18}$$

From (8.18) we conclude (8.16). □

Lemma 8.5 Consider formula (8.16). Then,

$$\frac{\boldsymbol{A}_j}{b_j} = \sum_{\psi_i \subset \psi_j} \boldsymbol{P}_i, \; j = 0, 1, \ldots, \nu + 1, \tag{8.19}$$

where the summation in (8.19) extends over all those i for which $\psi_i \subset \psi_j$ for a given j ($i, j = 0, 1, \ldots, \nu + 1$).

Proof. Formula (8.16) is derived from formula (8.8) by replacing $C_{\theta_i(\bar{\theta}_i)}(Y)$ with $\boldsymbol{P}_i$ and $Y_{\theta_j(\bar{\theta}_j)}$ with $\boldsymbol{A}_j/b_j$. It is easy to see that the sum of all components whose sets of subscripts are contained inside ψ_j is equal to $Y_{\theta_j(\bar{\theta}_j)}$, that is,

$$Y_{\theta_j(\bar{\theta}_j)} = \sum_{\psi_i \subset \psi_j} C_{\theta_i(\bar{\theta}_i)}(Y), \; j = 0, 1, \ldots, \nu + 1.$$

If we replace $Y_{\theta_j(\bar{\theta}_j)}$ by $\boldsymbol{A}_j/b_j$ and $C_{\theta_i(\bar{\theta}_i)}(Y)$ by $\boldsymbol{P}_i$, we get (8.19). □

Lemma 8.6 If $\boldsymbol{A}_j = \boldsymbol{H}_j\boldsymbol{H}_j'$, then

$$\boldsymbol{A}_j\boldsymbol{P}_i = \kappa_{ij}\boldsymbol{P}_i, \; i, j = 0, 1, \ldots, \nu + 1, \tag{8.20}$$

where

$$\kappa_{ij} = \begin{cases} b_j, & \psi_i \subset \psi_j \\ 0, & \psi_i \not\subset \psi_j \end{cases} ; \; i, j = 0, 1, \ldots, \nu + 1. \tag{8.21}$$

Proof. Multiplying the two sides of (8.19) on the right by $\boldsymbol{P}_i$, we get

$$\boldsymbol{A}_j\boldsymbol{P}_i = b_j \left(\sum_{\psi_\ell \subset \psi_j} \boldsymbol{P}_\ell \right) \boldsymbol{P}_i, \; i, j = 0, 1, \ldots, \nu + 1.$$

If $\psi_i \subset \psi_j$, then

$$\left(\sum_{\psi_\ell \subset \psi_j} \boldsymbol{P}_\ell \right) \boldsymbol{P}_i = \boldsymbol{P}_i,$$

since $\boldsymbol{P}_i$ is idempotent and $\boldsymbol{P}_i\boldsymbol{P}_\ell = \boldsymbol{0}$ if $i \neq \ell$. If, however, $\psi_i \not\subset \psi_j$, then

$$\left(\sum_{\psi_\ell \subset \psi_j} \boldsymbol{P}_\ell \right) \boldsymbol{P}_i = \boldsymbol{0},$$

by the orthogonality of the P_i's. We conclude that $A_j P_i = \kappa_{ij} P_i$, where κ_{ij} is defined by (8.21). □

Lemma 8.7 Let m_i be the rank of P_i in (8.14). Then, m_i is the same as the number of degrees of freedom for the ith effect in model (8.7), and is equal to

$$m_i = \left[\prod_{k_j \in \theta_i} a_j\right]\left[\prod_{k_j \in \bar{\theta}_i} (a_j - 1)\right], \quad i = 0, 1, \ldots, \nu + 1, \tag{8.22}$$

where $\bar{\theta}_i$ and θ_i are, respectively, the rightmost- and nonrightmost-bracket subscripts for the ith effect ($i = 0, 1, \ldots, \nu + 1$).

Example 8.3 Consider again the model used in Example 8.2. If P_i is the matrix associated with the ith sum of squares for this model ($i = 0, 1, \ldots, 6$), then using formula (8.16) and the fact that $a_1 = 3$, $a_2 = 3$, $a_3 = 4$, $a_4 = 3$, we have (see the corresponding components in Table 8.2)

$$\begin{aligned}
P_0 &= \frac{1}{108} A_0 \\
P_1 &= \frac{1}{36} A_1 - \frac{1}{108} A_0 \\
P_2 &= \frac{1}{12} A_2 - \frac{1}{36} A_1 \\
P_3 &= \frac{1}{27} A_3 - \frac{1}{108} A_0 \\
P_4 &= \frac{1}{9} A_4 - \frac{1}{36} A_1 - \frac{1}{27} A_3 + \frac{1}{108} A_0 \\
P_5 &= \frac{1}{3} A_5 - \frac{1}{12} A_2 - \frac{1}{9} A_4 + \frac{1}{36} A_1 \\
P_6 &= A_6 - \frac{1}{3} A_5,
\end{aligned}$$

where

$$\begin{aligned}
A_0 &= H_0 H_0' = J_3 \otimes J_3 \otimes J_4 \otimes J_3 \\
A_1 &= H_1 H_1' = I_3 \otimes J_3 \otimes J_4 \otimes J_3 \\
A_2 &= H_2 H_2' = I_3 \otimes I_3 \otimes J_4 \otimes J_3 \\
A_3 &= H_3 H_3' = J_3 \otimes J_3 \otimes I_4 \otimes J_3 \\
A_4 &= H_4 H_4' = I_3 \otimes J_3 \otimes I_4 \otimes J_3 \\
A_5 &= H_5 H_5' = I_3 \otimes I_3 \otimes I_4 \otimes J_3 \\
A_6 &= H_6 H_6' = I_3 \otimes I_3 \otimes I_4 \otimes I_3.
\end{aligned}$$

Solving for $A_0, A_1, \ldots, A_6$ in terms of $\boldsymbol{P}_0, \boldsymbol{P}_1, \ldots, \boldsymbol{P}_6$, we get

$$A_0 = 108\,\boldsymbol{P}_0$$
$$A_1 = 36(\boldsymbol{P}_0 + \boldsymbol{P}_1)$$
$$A_2 = 12(\boldsymbol{P}_0 + \boldsymbol{P}_1 + \boldsymbol{P}_2)$$
$$A_3 = 27(\boldsymbol{P}_0 + \boldsymbol{P}_3)$$
$$A_4 = 9(\boldsymbol{P}_0 + \boldsymbol{P}_1 + \boldsymbol{P}_3 + \boldsymbol{P}_4)$$
$$A_5 = 3(\boldsymbol{P}_0 + \boldsymbol{P}_1 + \boldsymbol{P}_2 + \boldsymbol{P}_3 + \boldsymbol{P}_4 + \boldsymbol{P}_5)$$
$$A_6 = \sum_{i=0}^{6} \boldsymbol{P}_i.$$

These equalities provide a verification of formula (8.19). Using formula (8.20), it can be verified that, for example,

$$A_1\boldsymbol{P}_1 = 36\,\boldsymbol{P}_1, \quad A_1\boldsymbol{P}_2 = \boldsymbol{0}, \quad A_2\boldsymbol{P}_1 = 12\,\boldsymbol{P}_1$$
$$A_2\boldsymbol{P}_2 = 12\,\boldsymbol{P}_2, \quad A_3\boldsymbol{P}_3 = 27\,\boldsymbol{P}_3, \quad A_3\boldsymbol{P}_4 = \boldsymbol{0}$$
$$A_4\boldsymbol{P}_5 = \boldsymbol{0}, \quad A_4\boldsymbol{P}_1 = 9\,\boldsymbol{P}_1, \quad A_4\boldsymbol{P}_2 = \boldsymbol{0}$$
$$A_4\boldsymbol{P}_3 = 9\,\boldsymbol{P}_3, \quad A_5\boldsymbol{P}_1 = 3\,\boldsymbol{P}_1, \quad A_5\boldsymbol{P}_2 = 3\,\boldsymbol{P}_2$$
$$A_5\boldsymbol{P}_6 = \boldsymbol{0}, \quad A_6\boldsymbol{P}_5 = \boldsymbol{P}_5.$$

8.4 Balanced Mixed Models

Model (8.7) may contain some random effects besides the experimental error. If all the effects in the model are randomly distributed, except for the term corresponding to $i = 0$, then the model is called a *random-effects model* (or just a *random model*). If all the effects in the model are represented by fixed unknown parameters, except for the experimental error term corresponding to $i = \nu + 1$, then the model is called a *fixed-effects model* (or just a *fixed model*). If, however, the model has fixed effects (besides the term corresponding to $i = 0$) and at least one random effect (besides the experimental error), then it is called a *mixed-effects model* (or just a *mixed model*).

The purpose of this section is to discuss the distribution of the sums of squares in the ANOVA table corresponding to a general balanced model, as in (8.7), when the model can be fixed, random, or mixed. Without loss of generality, we consider that the effects associated with $i = 0, 1, \ldots, \nu - p$ are fixed and those corresponding to $i = \nu - p + 1, \nu - p + 2, \ldots, \nu + 1$ are random, where p is a nonnegative integer not exceeding ν. The model is fixed if $p = 0$; random if $p = \nu$, otherwise, it is mixed if $0 < p < \nu$. Model (8.9) can then be written as

$$\boldsymbol{Y} = \boldsymbol{X}\boldsymbol{g} + \boldsymbol{Z}\boldsymbol{h}, \tag{8.23}$$

where
$Xg = \sum_{i=0}^{\nu-p} H_i\beta_i$ is the fixed portion of the model
$Zh = \sum_{i=\nu-p+1}^{\nu+1} H_i\beta_i$ is its random portion

Thus, for $i = 0, 1, \ldots, \nu - p$, β_i is a vector of fixed unknown parameters, and for $i = \nu - p + 1, \nu - p + 2, \ldots, \nu + 1$, β_i is a random vector. We assume that $\beta_{\nu-p+1}, \beta_{\nu-p+2}, \ldots, \beta_{\nu+1}$ are mutually independent and normally distributed such that

$$\beta_i \sim N(\mathbf{0}, \sigma_i^2 I_{c_i}),\ i = \nu - p + 1, \nu - p + 2, \ldots, \nu + 1, \tag{8.24}$$

where c_i is given in (8.13) and represents the number of columns of H_i. Under these assumptions, $E(Y) = Xg$, and the variance–covariance matrix, Σ, of Y is of the form

$$\Sigma = \sum_{i=\nu-p+1}^{\nu+1} \sigma_i^2 A_i, \tag{8.25}$$

where $A_i = H_i H_i' \ (i = \nu - p + 1, \nu - p + 2, \ldots, \nu + 1)$.

8.4.1 Distribution of Sums of Squares

The distribution of the sums of squares associated with the various effects in the balanced mixed model defined earlier can be easily derived using the properties outlined in Section 8.3. This is based on the following theorem.

Theorem 8.1 Let $Y'P_iY$ be the sum of squares associated with the ith effect for model (8.23), $i = 0, 1, \ldots, \nu + 1$. Then, under the assumptions mentioned earlier concerning the model's random effects,

(a) $Y'P_0Y, Y'P_1Y, \ldots, Y'P_{\nu+1}Y$ are mutually independent.

(b)
$$\frac{Y'P_iY}{\delta_i} \sim \chi^2_{m_i}(\lambda_i),\ i = 0, 1, \ldots, \nu + 1, \tag{8.26}$$

where m_i is the number of degrees of freedom for the ith effect (which is the same as the rank of P_i), λ_i is the noncentrality parameter given by

$$\lambda_i = \frac{g'X'P_iXg}{\delta_i},\ i = 0, 1, \ldots, \nu + 1, \tag{8.27}$$

and

$$\delta_i = \sum_{j \in W_i} b_j\sigma_j^2,\ i = 0, 1, \ldots, \nu + 1, \tag{8.28}$$

where b_j is defined in formula (8.13) and W_i is the set

$$W_i = \{j : \nu - p + 1 \le j \le \nu + 1 \mid \psi_i \subset \psi_j\},\ i = 0, 1, \ldots, \nu + 1, \tag{8.29}$$

where, if we recall, ψ_i is the set of subscripts that identifies the ith effect (see Section 8.2).

(c) $$E(\boldsymbol{Y}'\boldsymbol{P}_i\boldsymbol{Y}) = \boldsymbol{g}'\boldsymbol{X}'\boldsymbol{P}_i\boldsymbol{X}\boldsymbol{g} + m_i\,\delta_i, \; i = 0, 1, \ldots, \nu + 1. \tag{8.30}$$

(d) The noncentrality parameter λ_i in (b) is equal to zero if the ith effect is random, that is, for $i = \nu - p + 1, \nu - p + 2, \ldots, \nu + 1$. Thus, for such an effect, $\frac{1}{\delta_i}\boldsymbol{Y}'\boldsymbol{P}_i\boldsymbol{Y}$ has the central chi-squared distribution with m_i degrees of freedom and $E(\boldsymbol{Y}'\boldsymbol{P}_i\boldsymbol{Y}) = m_i\delta_i$.

Proof.

(a) Using formulas (8.20) and (8.25), we can write

$$\begin{aligned}\boldsymbol{\Sigma}\boldsymbol{P}_i &= \left(\sum_{j=\nu-p+1}^{\nu+1} \sigma_j^2 \boldsymbol{A}_j\right)\boldsymbol{P}_i \\ &= \left(\sum_{j=\nu-p+1}^{\nu+1} \kappa_{ij}\,\sigma_j^2\right)\boldsymbol{P}_i, \; i = 0, 1, \ldots, \nu + 1.\end{aligned}$$

But, by (8.21),

$$\sum_{j=\nu-p+1}^{\nu+1} \kappa_{ij}\,\sigma_j^2 = \sum_{j \in W_i} b_j \sigma_j^2,$$

where W_i is the set defined in (8.29). Hence,

$$\boldsymbol{\Sigma}\boldsymbol{P}_i = \delta_i\boldsymbol{P}_i, \; i = 0, 1, \ldots, \nu + 1. \tag{8.31}$$

From (8.31) and Lemma 8.3 (b), it follows that

$$\begin{aligned}\boldsymbol{P}_{i_1}\boldsymbol{\Sigma}\boldsymbol{P}_{i_2} &= \boldsymbol{P}_{i_1}(\delta_{i_2}\boldsymbol{P}_{i_2}) \\ &= \boldsymbol{0}, \; i_1 \neq i_2;\, i_1, i_2 = 0, 1, \ldots, \nu + 1.\end{aligned} \tag{8.32}$$

Now, since $\boldsymbol{P}_i$ is symmetric of rank m_i, there exists a matrix $\boldsymbol{L}_i$ of order $N \times m_i$ and rank m_i such that $\boldsymbol{P}_i = \boldsymbol{L}_i\boldsymbol{L}_i'$ $(i = 0, 1, \ldots, \nu + 1)$ (see Corollary 3.1). Hence, from (8.32) we obtain

$$\boldsymbol{L}_{i_1}'\boldsymbol{\Sigma}\boldsymbol{L}_{i_2} = \boldsymbol{0}, \quad i_1 \neq i_2;\, i_1, i_2 = 0, 1, \ldots, \nu + 1, \tag{8.33}$$

since $\boldsymbol{L}_i$ is of full column rank $(i = 0, 1, \ldots, \nu + 1)$. Let $\boldsymbol{L} = [\boldsymbol{L}_0 : \boldsymbol{L}_1 : \ldots : \boldsymbol{L}_{\nu+1}]'$. This matrix is of full row rank because each $\boldsymbol{L}_i'$ $(i = 0, 1, \ldots, \nu+1)$ is of full row rank and the rows of $\boldsymbol{L}_{i_1}'$ are linearly independent of the rows of $\boldsymbol{L}_{i_2}'$ $(i_1 \neq i_2;\, i_1, i_2 = 0, 1, \ldots, \nu + 1)$. The latter assertion follows from the fact that if the rows of $\boldsymbol{L}_{i_1}'$ are not linearly independent of the rows of $\boldsymbol{L}_{i_2}'$, then there exists a matrix $\boldsymbol{M}$ such that $\boldsymbol{L}_{i_1}' = \boldsymbol{M}\boldsymbol{L}_{i_2}'$, which, by (8.33), implies that $\boldsymbol{M} = \boldsymbol{0}$ since $\boldsymbol{L}_{i_2}'\boldsymbol{\Sigma}\boldsymbol{L}_{i_2}$ is nonsingular. This results in

$L'_{i_1} = \mathbf{0}$, which is a contradiction. We conclude that $L'_0 Y, L'_1 Y, \ldots, L'_{\nu+1} Y$ form a partitioning of the multivariate normal vector LY and are therefore normally distributed. Furthermore, they are uncorrelated because of (8.33). Hence, by Corollary 4.2, they are also mutually independent. We conclude that $Y'P_0Y, Y'P_1Y, \ldots, Y'P_{\nu+1}Y$ are mutually independent since $P_i = L_i L'_i$ $(i = 0, 1, \ldots, \nu + 1)$.

(b) From (8.31) we have that

$$\frac{P_i}{\delta_i}\Sigma = P_i, \ i = 0, 1, \ldots, \nu + 1.$$

Since P_i is idempotent [see Lemma 8.3 (a)] of rank m_i, $\frac{1}{\delta_i} Y'P_iY \sim \chi^2_{m_i}(\lambda_i)$ by Theorem 5.4. The noncentrality parameter λ_i is given by (8.27) since $E(Y) = Xg$.

(c) Applying Theorem 5.2 to $Y'P_iY$, we obtain

$$\begin{aligned} E(Y'P_iY) &= g'X'P_iXg + tr(P_i\Sigma) \\ &= g'X'P_iXg + tr(\delta_i P_i), \quad \text{by (8.31)}, \\ &= g'X'P_iXg + m_i\delta_i. \end{aligned}$$

(d) From model (8.23) we have

$$P_iX = P_i[H_0 : H_1 : \ldots : H_{\nu-p}], \ i = 0, 1, \ldots, \nu + 1. \tag{8.34}$$

If the ith effect is random (i.e., for $i = \nu - p + 1, \nu - p + 2, \ldots, \nu + 1$), then $P_iA_j = \mathbf{0}$ for $j = 0, 1, \ldots, \nu - p$. This follows from (8.20) and (8.21) since ψ_i is not a subset of ψ_j (a set of subscripts, ψ_i, associated with a random effect cannot be a subset of ψ_j for a fixed effect). Hence, $P_iH_j = \mathbf{0}$ by the fact that $P_iH_jH'_j = \mathbf{0}$ and $H'_jH_j = b_j I_{c_j}$ [see formula (8.12)], $j = 0, 1, \ldots, \nu - p$. From (8.34) we conclude that $P_iX = \mathbf{0}$. Thus, the noncentrality parameter in (8.27) must be equal to zero if the ith effect is random $(i = \nu - p + 1, \nu - p + 2, \ldots, \nu + 1)$. □

8.4.2 Estimation of Fixed Effects

In this section, we show how to derive estimates of estimable linear functions of g, the vector of fixed effects in model (8.23). The following theorem identifies a basis for all such functions.

Theorem 8.2 Let P_i be the same matrix as in Theorem 8.1 $(i = 0, 1, \ldots, \nu + 1)$. Then,

(a) For the ith fixed effect in model (8.23), $rank(P_iX) = m_i$, where $m_i = rank(P_i)$, $i = 0, 1, \ldots, \nu - p$.

(b) $rank(X) = \sum_{i=0}^{\nu-p} m_i$.

(c) $P_0Xg, P_1Xg, \ldots, P_{\nu-p}Xg$ are linearly independent.

(d) Any estimable linear function of g can be written as the sum of linear functions of $P_0Xg, P_1Xg, \ldots, P_{\nu-p}Xg$.

Proof.

(a) Recall from the proof of part (d) of Theorem 8.1 that if the ith effect is random, that is, for $i = \nu - p + 1, \nu - p + 2, \ldots, \nu + 1$, then $P_iX = 0$. Let us now consider P_iX for the ith fixed effect ($i = 0, 1, \ldots, \nu - p$). We have that

$$
\begin{aligned}
rank(P_iX) &= rank\{P_i[H_0 : H_1 : \ldots : H_{\nu-p}]\} \\
&= rank\{P_i[H_0 : H_1 : \ldots : H_{\nu-p}][H_0 : H_1 : \ldots : H_{\nu-p}]'P_i\} \\
&= rank\left[P_i\left(\sum_{j=0}^{\nu-p} A_j\right)P_i\right] \\
&= rank\left[P_i\sum_{j=0}^{\nu-p} \kappa_{ij}P_i\right], \quad \text{using formula (8.20)} \\
&= rank\left[\left(\sum_{j=0}^{\nu-p} \kappa_{ij}\right)P_i\right], \quad \text{since } P_i \text{ is idempotent,} \\
&= rank(P_i) = m_i, \; i = 0, 1, \ldots, \nu - p.
\end{aligned}
$$

Note that $\sum_{j=0}^{\nu-p} \kappa_{ij} > 0$ since $\sum_{j=0}^{\nu-p} \kappa_{ij} \geq \kappa_{ii}$ for $i = 0, 1, \ldots, \nu - p$, and $\kappa_{ii} = b_i > 0$ by formula (8.21).

(b) Recall from Lemma 8.3 (c) that $\sum_{i=0}^{\nu+1} P_i = I_N$. Hence,

$$
\begin{aligned}
X &= \left(\sum_{i=0}^{\nu+1} P_i\right)X \\
&= \sum_{i=0}^{\nu-p} P_iX, \text{ since } P_iX = 0, \text{ for } i = \nu - p + 1, \nu - p + 2, \ldots, \nu + 1.
\end{aligned}
$$

Therefore,

$$
\begin{aligned}
rank(X) &= rank\left(\sum_{i=0}^{\nu-p} P_iX\right) \\
&= rank\left(\sum_{i=0}^{\nu-p} P_iXX'\sum_{i=0}^{\nu-p} P_i\right)
\end{aligned}
$$

$$= rank\left[\left(\sum_{i=0}^{\nu-p} \boldsymbol{P}_i\right)\left(\sum_{j=0}^{\nu-p} \boldsymbol{A}_j\right)\left(\sum_{k=0}^{\nu-p} \boldsymbol{P}_k\right)\right]$$

$$= rank\left[\left(\sum_{i=0}^{\nu-p} \boldsymbol{P}_i\right)\sum_{j=0}^{\nu-p}\sum_{k=0}^{\nu-p} \boldsymbol{A}_j\boldsymbol{P}_k\right]$$

$$= rank\left[\left(\sum_{i=0}^{\nu-p} \boldsymbol{P}_i\right)\sum_{j=0}^{\nu-p}\sum_{k=0}^{\nu-p} \kappa_{kj}\boldsymbol{P}_k\right], \text{ by (8.20)}$$

$$= rank\left(\sum_{i=0}^{\nu-p}\sum_{j=0}^{\nu-p}\sum_{k=0}^{\nu-p} \kappa_{kj}\boldsymbol{P}_i\boldsymbol{P}_k\right)$$

$$= rank\left(\sum_{i=0}^{\nu-p}\sum_{j=0}^{\nu-p} \kappa_{ij}\boldsymbol{P}_i\right)$$

$$= rank\left[\sum_{i=0}^{\nu-p}\left(\sum_{j=0}^{\nu-p} \kappa_{ij}\right)\boldsymbol{P}_i\right].$$

Let $\kappa_{i.} = \sum_{j=0}^{\nu-p} \kappa_{ij}$. Then, $\kappa_{i.} > 0$ for $i = 0, 1, \ldots, \nu - p$, as was seen earlier. We therefore have

$$rank(\boldsymbol{X}) = rank\left(\sum_{i=0}^{\nu-p} \kappa_{i.}\boldsymbol{P}_i\right)$$

$$= rank\left(\kappa_{0.}^{1/2}\boldsymbol{P}_0 : \kappa_{1.}^{1/2}\boldsymbol{P}_1 : \ldots : \kappa_{\nu-p.}^{1/2}\boldsymbol{P}_{\nu-p}\right)$$

$$= \sum_{i=0}^{\nu-p} m_i,$$

since $rank(\boldsymbol{P}_i) = m_i$ $(i = 0, 1, \ldots, \nu - p)$ and the columns of $\boldsymbol{P}_i$ are linearly independent of the columns of $\boldsymbol{P}_{i'}$ $(i \neq i')$. The latter assertion follows from Lemma 8.3.

(c) Suppose that $\boldsymbol{P}_0\boldsymbol{X}\boldsymbol{g}, \boldsymbol{P}_1\boldsymbol{X}\boldsymbol{g}, \ldots, \boldsymbol{P}_{\nu-p}\boldsymbol{X}\boldsymbol{g}$ are linearly dependent. Then, there exist constants $\zeta_0, \zeta_1, \ldots, \zeta_{\nu-p}$, not all equal to zero, such that

$$\sum_{i=0}^{\nu-p} \zeta_i\boldsymbol{P}_i\boldsymbol{X}\boldsymbol{g} = \boldsymbol{0}, \ \forall \boldsymbol{g}.$$

Hence,

$$\sum_{i=0}^{\nu-p} \zeta_i\boldsymbol{P}_i\boldsymbol{X} = \boldsymbol{0}.$$

Multiplying both sides on the right by X', we get

$$\sum_{i=0}^{\nu-p} \zeta_i P_i X X' = \mathbf{0}.$$

This implies that

$$\sum_{i=0}^{\nu-p} \zeta_i P_i \sum_{j=0}^{\nu-p} A_j = \mathbf{0},$$

that is,

$$\sum_{i=0}^{\nu-p} \sum_{j=0}^{\nu-p} \zeta_i P_i A_j = \mathbf{0},$$

or equivalently,

$$\sum_{i=0}^{\nu-p} \sum_{j=0}^{\nu-p} \zeta_i \, \kappa_{ij} \, P_i = \mathbf{0}, \text{ by (8.20).}$$

Thus,

$$\sum_{i=0}^{\nu-p} \zeta_i \, \kappa_{i.} \, P_i = \mathbf{0}.$$

Multiplying now both sides of this equation on the right by P_j ($j = 0, 1, \ldots, \nu - p$), we get (using Lemma 8.3)

$$\zeta_j \, \kappa_{j.} \, P_j = \mathbf{0}, \; j = 0, 1, \ldots, \nu - p.$$

We conclude that $\zeta_j = 0$ for $j = 0, 1, \ldots, \nu - p$, a contradiction. Therefore, $P_0 X g, P_1 X g, \ldots, P_{\nu-p} X g$ must be linearly independent.

(d) We recall from the proof of part (b) that $X = \sum_{i=0}^{\nu-p} P_i X$. Thus,

$$X g = \sum_{i=0}^{\nu-p} P_i X g.$$

Hence, $E(Y) = Xg$ is the sum of $P_0 X g, P_1 X g, \ldots, P_{\nu-p} X g$. If $Q' g$ is a vector of estimable linear functions of g, then there exists a matrix T' such that $Q' = T' X$ (see Section 7.4). Consequently,

$$\begin{aligned} Q' g &= T' X g \\ &= \sum_{i=0}^{\nu-p} T' P_i X g. \end{aligned}$$

Thus, $Q'g$ is the sum of linear functions of $P_0Xg, P_1Xg, \ldots, P_{\nu-p}Xg$. Note that an unbiased estimator of $Q'g$ is given by

$$\widehat{Q'g} = \sum_{i=0}^{\nu-p} T'P_iY. \qquad \square$$

Theorem 8.3 Let $Q'g$ be a vector of estimable linear functions of g. Then, the best linear unbiased estimator of $Q'g$ is given by $Q'(X'X)^{-}X'Y$.

The proof of this theorem requires the following lemmas:

Lemma 8.8 $P_i\Sigma = \Sigma P_i$ for $i = 0, 1, \ldots, \nu + 1$, where Σ is the variance-covariance matrix in (8.25).

Proof. This follows directly from formula (8.31), where $\Sigma P_i = \delta_i P_i$ ($i = 0, 1, \ldots, \nu + 1$). Hence, $\Sigma P_i = P_i\Sigma$. $\square$

Lemma 8.9 $A_i\Sigma = \Sigma A_i$ for $i = 0, 1, \ldots, \nu + 1$, where $A_i = H_iH_i'$ and Σ is the same as in Lemma 8.8.

Proof. Using Lemma 8.5, we can write

$$A_i = b_i \sum_{\psi_j \subset \psi_i} P_j, \; i = 0, 1, \ldots, \nu + 1.$$

Hence,

$$\begin{aligned} A_i\Sigma &= b_i \left(\sum_{\psi_j \subset \psi_i} P_j \right) \Sigma \\ &= b_i \sum_{\psi_j \subset \psi_i} (\Sigma P_j), \text{ by Lemma 8.8}, \\ &= \Sigma \left(b_i \sum_{\psi_j \subset \psi_i} P_j \right) \\ &= \Sigma A_i. \qquad \square \end{aligned}$$

Lemma 8.10 There exists a nonsingular matrix F such that $\Sigma X = XF$.

Proof. $\Sigma X = \Sigma[H_0 : H_1 : \ldots : H_{\nu-p}]$. From Lemma 8.9, $A_i\Sigma = \Sigma A_i$, $i = 0, 1, \ldots, \nu + 1$. Then,

$$H_iH_i'\Sigma = \Sigma H_iH_i', \; i = 0, 1, \ldots, \nu + 1.$$

Multiplying both sides on the right by H_i and recalling from (8.12) that $H_i'H_i = b_iI_{c_i}$ ($i = 0, 1, \ldots, \nu + 1$), we get

$$H_iH_i'\Sigma H_i = b_i\Sigma H_i, \; i = 0, 1, \ldots, \nu + 1.$$

Hence,

$$\begin{aligned}\boldsymbol{\Sigma H}_i &= \frac{1}{b_i}\boldsymbol{H}_i\boldsymbol{H}_i'\boldsymbol{\Sigma H}_i \\ &= \boldsymbol{H}_i\boldsymbol{G}_i, \ i = 0, 1, \ldots, \nu + 1,\end{aligned}$$

where $\boldsymbol{G}_i = \frac{1}{b_i}\boldsymbol{H}_i'\boldsymbol{\Sigma H}_i$. Note that $\boldsymbol{G}_i$ is a $c_i \times c_i$ matrix of rank c_i and is therefore nonsingular. It follows that

$$\begin{aligned}\boldsymbol{\Sigma X} &= [\boldsymbol{\Sigma H}_0 : \boldsymbol{\Sigma H}_1 : \ldots : \boldsymbol{\Sigma H}_{\nu-p}] \\ &= [\boldsymbol{H}_0\boldsymbol{G}_0 : \boldsymbol{H}_1\boldsymbol{G}_1 : \ldots : \boldsymbol{H}_{\nu-p}\boldsymbol{G}_{\nu-p}] \\ &= [\boldsymbol{H}_0 : \boldsymbol{H}_1 : \ldots : \boldsymbol{H}_{\nu-p}]\bigoplus_{i=0}^{\nu-p}\boldsymbol{G}_i \\ &= \boldsymbol{XF},\end{aligned}$$

where $\boldsymbol{F} = \bigoplus_{i=0}^{\nu-p}\boldsymbol{G}_i$, which is nonsingular. □

Proof of Theorem 8.3 The best linear unbiased estimator (BLUE) of $\boldsymbol{Q}'\boldsymbol{g}$ is $\boldsymbol{Q}'(\boldsymbol{X}'\boldsymbol{\Sigma}^{-1}\boldsymbol{X})^-\boldsymbol{X}'\boldsymbol{\Sigma}^{-1}\boldsymbol{Y}$. By Lemma 8.10, $\boldsymbol{\Sigma X} = \boldsymbol{XF}$ for some nonsingular matrix $\boldsymbol{F}$. Hence,

$$\begin{aligned}\boldsymbol{X} &= \boldsymbol{\Sigma X F}^{-1} \\ \boldsymbol{X}'\boldsymbol{\Sigma}^{-1} &= (\boldsymbol{F}^{-1})'\boldsymbol{X}'.\end{aligned}$$

Consequently,

$$(\boldsymbol{X}'\boldsymbol{\Sigma}^{-1}\boldsymbol{X})^-\boldsymbol{X}'\boldsymbol{\Sigma}^{-1}\boldsymbol{Y} = [(\boldsymbol{F}^{-1})'\boldsymbol{X}'\boldsymbol{X}]^-(\boldsymbol{F}^{-1})'\boldsymbol{X}'\boldsymbol{Y}. \tag{8.35}$$

Note that $[(\boldsymbol{F}^{-1})'\boldsymbol{X}'\boldsymbol{X}]^-$ can be chosen equal to $(\boldsymbol{X}'\boldsymbol{X})^-\boldsymbol{F}'$ since

$$(\boldsymbol{F}^{-1})'\boldsymbol{X}'\boldsymbol{X}(\boldsymbol{X}'\boldsymbol{X})^-\boldsymbol{F}'(\boldsymbol{F}^{-1})'\boldsymbol{X}'\boldsymbol{X} = (\boldsymbol{F}^{-1})'\boldsymbol{X}'\boldsymbol{X}.$$

From (8.35) it follows that

$$\begin{aligned}(\boldsymbol{X}'\boldsymbol{\Sigma}^{-1}\boldsymbol{X})^-\boldsymbol{X}'\boldsymbol{\Sigma}^{-1}\boldsymbol{Y} &= (\boldsymbol{X}'\boldsymbol{X})^-\boldsymbol{F}'(\boldsymbol{F}^{-1})'\boldsymbol{X}'\boldsymbol{Y} \\ &= (\boldsymbol{X}'\boldsymbol{X})^-\boldsymbol{X}'\boldsymbol{Y}.\end{aligned}$$

This implies that the BLUE of $\boldsymbol{Q}'\boldsymbol{g}$ is $\boldsymbol{Q}'(\boldsymbol{X}'\boldsymbol{X})^-\boldsymbol{X}'\boldsymbol{Y}$. □

Theorem 8.3 shows that the BLUE of $\boldsymbol{Q}'\boldsymbol{g}$ can be obtained without the need to know $\boldsymbol{\Sigma}$, which is usually unknown. Furthermore, this estimator is the same as the ordinary least-squares estimator of $\boldsymbol{Q}'\boldsymbol{g}$.

Theorem 8.4 The variance–covariance matrix of the best linear unbiased estimator of $\boldsymbol{Q}'\boldsymbol{g}$ in Theorem 8.3 is given by

$$\text{Var}[\boldsymbol{Q}'(\boldsymbol{X}'\boldsymbol{X})^-\boldsymbol{X}'\boldsymbol{Y}] = \boldsymbol{T}'\left(\sum_{i=0}^{\nu-p}\delta_i\boldsymbol{P}_i\right)\boldsymbol{T},$$

where $\boldsymbol{T}$ is such that $\boldsymbol{Q}' = \boldsymbol{T}'\boldsymbol{X}$, and δ_i is given in (8.28) $(i = 0, 1, \ldots, \nu - p)$.

The proof of this theorem depends on the following lemma.

Lemma 8.11 $X(X'X)^{-}X' = \sum_{i=0}^{\nu-p} P_i$.

Proof. From the proof of part (b) of Theorem 8.2, we have that $X = \left(\sum_{i=0}^{\nu-p} P_i\right) X$.

Furthermore, $X(X'X)^{-}X'X = X$. Hence,

$$\left[X(X'X)^{-}X' - \sum_{i=0}^{\nu-p} P_i\right] X = 0. \tag{8.36}$$

We may also recall that $P_i X = 0$ for $i = \nu-p+1, \nu-p+2, \ldots, \nu+1$. On this basis and using the orthogonality of the P_i's, we can write

$$\left[X(X'X)^{-}X' - \sum_{i=0}^{\nu-p} P_i\right] \sum_{j=\nu-p+1}^{\nu+1} P_j = 0. \tag{8.37}$$

Note that the column space of $\sum_{j=\nu-p+1}^{\nu+1} P_j$, which is of dimension $= \sum_{j=\nu-p+1}^{\nu+1} m_j$, is the orthogonal complement in R^N of the column space of X, which is of dimension $= rank(X) = \sum_{i=0}^{\nu-p} m_i$. From (8.36) and (8.37) we therefore conclude that

$$X(X'X)^{-}X' - \sum_{i=0}^{\nu-p} P_i = 0. \qquad \square$$

Proof of Theorem 8.4 We have that $Q' = T'X$. Hence,

$$\begin{aligned}
\text{Var}[Q'(X'X)^{-}X'Y] &= \text{Var}[T'X(X'X)^{-}X'Y] \\
&= T'X(X'X)^{-}X'\left(\sum_{j=\nu-p+1}^{\nu+1} \sigma_j^2 A_j\right) X(X'X)^{-}X'T, \text{ using (8.25)}, \\
&= T'\left(\sum_{i=0}^{\nu-p} P_i\right)\left(\sum_{j=\nu-p+1}^{\nu+1} \sigma_j^2 A_j\right)\left(\sum_{i=0}^{\nu-p} P_i\right) T, \text{ by Lemma 8.11}, \\
&= T'\left(\sum_{i=0}^{\nu-p}\sum_{j=\nu-p+1}^{\nu+1} \sigma_j^2 P_i A_j\right)\left(\sum_{i=0}^{\nu-p} P_i\right) T \\
&= T'\left(\sum_{i=0}^{\nu-p}\sum_{j=\nu-p+1}^{\nu+1} \kappa_{ij}\, \sigma_j^2 P_i\right)\left(\sum_{i=0}^{\nu-p} P_i\right) T, \text{ by Lemma 8.6}, \\
&= T'\left(\sum_{i=0}^{\nu-p} \delta_i P_i\right)\left(\sum_{i=0}^{\nu-p} P_i\right) T, \text{ by (8.28)}.
\end{aligned} \tag{8.38}$$

Note that (8.38) follows from the fact that

$$\sum_{j=\nu-p+1}^{\nu+1} \kappa_{ij}\, \sigma_j^2 = \sum_{j\in W_i} b_j\, \sigma_j^2$$
$$= \delta_i .$$

Using (8.38) and Lemma 8.3, we get

$$\text{Var}[\boldsymbol{Q}'(\boldsymbol{X}'\boldsymbol{X})^{-}\boldsymbol{X}'\boldsymbol{Y}] = \boldsymbol{T}'\left(\sum_{i=0}^{\nu-p} \delta_i\, \boldsymbol{P}_i\right)\boldsymbol{T}. \tag{8.39}$$

□

Example 8.4 Consider the balanced mixed two-way crossed classification model

$$Y_{ijk} = \mu + \alpha_{(i)} + \beta_{(j)} + (\alpha\beta)_{(ij)} + \epsilon_{ij(k)}, \tag{8.40}$$

where $\alpha_{(i)}$ is a fixed unknown parameter and $\beta_{(j)}$, $(\alpha\beta)_{(ij)}$, and $\epsilon_{ij(k)}$ are randomly distributed as $N(0, \sigma^2_{\beta})$, $N(0, \sigma^2_{\alpha\beta})$, and $N(0, \sigma^2_{\epsilon})$, respectively. All random effects are independent. The data set used with model (8.40) is given in Table 8.3. The corresponding ANOVA table is shown in Table 8.4. Note that the expected mean squares were obtained by applying formula (8.30). More specifically, if we number the effects in model (8.40) as 0, 1, 2, 3, and 4, respectively, then

$$E(MS_A) = \boldsymbol{g}'\boldsymbol{X}'\boldsymbol{P}_1\boldsymbol{X}\boldsymbol{g} + \delta_1$$
$$E(MS_B) = \delta_2$$
$$E(MS_{AB}) = \delta_3$$
$$E(MS_E) = \delta_4,$$

where $\delta_1 = 3\,\sigma^2_{\alpha\beta} + \sigma^2_{\epsilon}$, $\delta_2 = 6\,\sigma^2_{\beta} + 3\,\sigma^2_{\alpha\beta} + \sigma^2_{\epsilon}$, $\delta_3 = 3\,\sigma^2_{\alpha\beta} + \sigma^2_{\epsilon}$, and $\delta_4 = \sigma^2_{\epsilon}$, as can be seen from applying formula (8.28). Furthermore, $\boldsymbol{X} = [\boldsymbol{H}_0 : \boldsymbol{H}_1]$,

TABLE 8.3
Data Set for Model (8.40)

	B		
A	**1**	**2**	**3**
1	74	71	99
	64	68	104
	60	75	93
2	99	108	114
	98	110	111
	107	99	108

TABLE 8.4
ANOVA Table for Model (8.40)

Source	DF	SS	MS	E(MS)	F	p-Value
A	1	3362.00	3362.00	$9\sum_{i=1}^{2}\alpha_{(i)}^2 + 3\sigma_{\alpha\beta}^2 + \sigma_\epsilon^2$	13.26	0.0678
B	2	1476.33	738.17	$6\sigma_\beta^2 + 3\sigma_{\alpha\beta}^2 + \sigma_\epsilon^2$	2.91	0.2556
$A * B$	2	507.00	253.50	$3\sigma_{\alpha\beta}^2 + \sigma_\epsilon^2$	9.37	0.0035
Error	12	324.67	27.06	σ_ϵ^2		

where $\boldsymbol{H}_0 = \boldsymbol{1}_2 \otimes \boldsymbol{1}_3 \otimes \boldsymbol{1}_3 = \boldsymbol{1}_{18}$, $\boldsymbol{H}_1 = \boldsymbol{I}_2 \otimes \boldsymbol{1}_3 \otimes \boldsymbol{1}_3$, $\boldsymbol{g} = [\mu, \alpha_{(1)}, \alpha_{(2)}]'$, and from (8.16), $\boldsymbol{P}_1$ is given by

$$\boldsymbol{P}_1 = \frac{1}{9}\boldsymbol{I}_2 \otimes \boldsymbol{J}_3 \otimes \boldsymbol{J}_3 - \frac{1}{18}\boldsymbol{J}_2 \otimes \boldsymbol{J}_3 \otimes \boldsymbol{J}_3. \tag{8.41}$$

A more direct way to compute $\boldsymbol{g}'\boldsymbol{X}'\boldsymbol{P}_1\boldsymbol{X}\boldsymbol{g}$ is described as follows:

We note that $\boldsymbol{g}'\boldsymbol{X}'\boldsymbol{P}_1\boldsymbol{X}\boldsymbol{g}$ is the same as $SS_A = \boldsymbol{Y}'\boldsymbol{P}_1\boldsymbol{Y}$, except that $\boldsymbol{Y}$ is replaced by $\boldsymbol{X}\boldsymbol{g}$, which is the mean of $\boldsymbol{Y}$. Therefore, to compute $\boldsymbol{g}'\boldsymbol{X}'\boldsymbol{P}_1\boldsymbol{X}\boldsymbol{g}$, it is sufficient to use the formula for SS_A, namely,

$$SS_A = 9\sum_{i=1}^{2}[Y_{(i)} - Y]^2, \tag{8.42}$$

and then replace $Y_{(i)}$ and Y with their expected values, respectively. Using model (8.40), these expected values are

$$E[Y_{(i)}] = \mu + \alpha_{(i)}$$
$$E(Y) = \mu,$$

since $\sum_{i=1}^{2}\alpha_{(i)} = 0$. Hence,

$$\boldsymbol{g}'\boldsymbol{X}'\boldsymbol{P}_1\boldsymbol{X}\boldsymbol{g} = 9\sum_{i=1}^{2}\alpha_{(i)}^2.$$

This short-cut to computing $\boldsymbol{g}'\boldsymbol{X}'\boldsymbol{P}_1\boldsymbol{X}\boldsymbol{g}$ can be applied in general to any fixed effect in a balanced mixed linear model: If the ith effect is fixed, then

$$\boldsymbol{g}'\boldsymbol{X}'\boldsymbol{P}_i\boldsymbol{X}\boldsymbol{g} = \sum_{\theta} g_{\theta_i(\bar{\theta}_i)}^2, \tag{8.43}$$

where, if we recall from Section 8.2, θ is the complete set of subscripts that identify the response Y in the model under consideration, and $g_{\theta_i(\bar{\theta}_i)}$ denotes the ith effect in the model [see model (8.7)].

Now, suppose that it is of interest to estimate $\alpha_{(1)} - \alpha_{(2)}$, which is estimable since $\alpha_{(1)} - \alpha_{(2)} = (\mu + \alpha_{(1)}) - (\mu + \alpha_{(2)})$ and both $\mu + \alpha_{(1)}$ and $\mu + \alpha_{(2)}$ are estimable. Using Theorem 8.3, the BLUE of $\alpha_{(1)} - \alpha_{(2)}$ is given by $\hat{\alpha}_{(1)} - \hat{\alpha}_{(2)}$, where $\hat{\alpha}_{(1)}$ and $\hat{\alpha}_{(2)}$ are obtained from

$$\begin{aligned}\hat{g} &= [\hat{\mu}, \hat{\alpha}_{(1)}, \hat{\alpha}_{(2)}]' \\ &= (X'X)^{-}X'Y \\ &= [0, Y_{(1)}, Y_{(2)}]' \\ &= (0, 78.667, 106.00)'.\end{aligned}$$

Thus, $\hat{\alpha}_{(1)} - \hat{\alpha}_{(2)} = Y_{(1)} - Y_{(2)} = -27.333$. Furthermore, since

$$Y_{(i)} = \mu + \alpha_{(i)} + \frac{1}{3}\sum_{j=1}^{3}\beta_{(j)} + \frac{1}{3}\sum_{j=1}^{3}(\alpha\beta)_{(ij)} + \frac{1}{9}\sum_{k=1}^{3}\sum_{j=1}^{3}\epsilon_{ij(k)},$$

then

$$\mathrm{Var}[Y_{(1)} - Y_{(2)}] = \frac{2}{9}(3\,\sigma^2_{\alpha\beta} + \sigma^2_{\epsilon}). \tag{8.44}$$

Hence, the 95% confidence interval on $\alpha_{(1)} - \alpha_{(2)}$ is given by

$$\begin{aligned}Y_{(1)} - Y_{(2)} \pm \left(\frac{2}{9}MS_{AB}\right)^{1/2} t_{0.025,2} &= -27.33 \pm \left[\frac{2}{9}(253.50)\right]^{1/2} (4.303) \\ &= -27.33 \pm 32.296,\end{aligned}$$

which yields the interval $(-59.626, 4.966)$. This indicates that, at the 5% level, no significant difference can be detected between the two means of factor A. This coincides with the outcome of the F-test for A from Table 8.4 (p-value = 0.0678).

Note that (8.44) could have been derived using formula (8.39) as follows: Let us write $\alpha_{(1)} - \alpha_{(2)}$ as $q'g$, where $q' = (0, 1, -1)$, $q' = t'X$, and the elements of t are all zero, except for the first and the 10th elements, which are equal to 1 and -1, respectively. Applying formula (8.39), we get

$$\mathrm{Var}[Y_{(1)} - Y_{(2)}] = t'(\delta_0 P_0 + \delta_1 P_1)t,$$

where $\delta_0 = 6\,\sigma^2_{\beta} + 3\,\sigma^2_{\alpha\beta} + \sigma^2_{\epsilon}$, $\delta_1 = 3\,\sigma^2_{\alpha\beta} + \sigma^2_{\epsilon}$, $P_0 = \frac{1}{18}J_2 \otimes J_3 \otimes J_3$, and P_1 is given in (8.41). It follows that $t'P_0 = 0'$ and $t'P_1 t = \frac{2}{9}$. Hence,

$$\mathrm{Var}[Y_{(1)} - Y_{(2)}] = \frac{2}{9}(3\,\sigma^2_{\alpha\beta} + \sigma^2_{\epsilon}).$$

This gives the same result as in (8.44).

8.5 Complete and Sufficient Statistics

Consider model (8.23) under the same assumptions made earlier in Section 8.4 concerning the random effects in the model. The density function of Y is given by

$$f(y, \Delta) = \frac{1}{(2\pi)^{n/2}[\det(\Sigma)]^{1/2}} \exp\left[-\frac{1}{2}(y - Xg)'\Sigma^{-1}(y - Xg)\right], \qquad (8.45)$$

where Σ is the variance–covariance matrix in (8.25), and Δ denotes the vector of all model's unknown parameters, including the elements of Σ. The purpose of this section is to find complete and sufficient statistics for model (8.23).

Lemma 8.12 Let P_i be the matrix associated with the ith sum of squares for model (8.23) ($i = 0, 1, \ldots, \nu + 1$). Then, P_i can be written as $P_i = Q_i Q_i'$, where Q_i is an $N \times m_i$ matrix of rank m_i such that $Q_i' Q_i = I_{m_i}$. Furthermore, if

$$Q^* = [Q_0 : Q_1 : \ldots : Q_{\nu+1}], \qquad (8.46)$$

then, Q^* is an orthogonal matrix that diagonalizes Σ, that is,

$$Q^{*'} \Sigma Q^* = \Lambda, \qquad (8.47)$$

where Λ is a diagonal matrix of the form

$$\Lambda = \bigoplus_{i=0}^{\nu+1} (\delta_i I_{m_i}),$$

and δ_i is given in (8.28).

Proof. The matrix P_i is idempotent of order $N \times N$ and rank m_i ($i = 0, 1, \ldots, \nu + 1$). Hence, by applying the Spectral Decomposition Theorem (Theorem 3.4), P_i can be expressed as

$$P_i = Q_i Q_i', \; i = 0, 1, \ldots, \nu + 1,$$

where Q_i is of order $N \times m_i$ whose columns are orthonormal eigenvectors of P_i corresponding to the eigenvalue 1. Thus, $Q_i' Q_i = I_{m_i}$. Define the matrix Q^* as in (8.46). Then, it is easy to verify that $Q^* Q^{*'} = \sum_{i=0}^{\nu+1} P_i = I_N$, and $Q^{*'} Q^* = \bigoplus_{i=0}^{\nu+1} I_{m_i} = I_N$. Hence, Q^* is orthogonal. Furthermore, $Q^{*'} \Sigma Q^*$ can be partitioned as

$$Q^{*'} \Sigma Q^* = (\Lambda_{ij}),$$

where $\Lambda_{ij} = Q_i' \Sigma Q_j$, $i, j = 0, 1, \ldots, \nu + 1$. We now show that $\Lambda_{ij} = \mathbf{0}$ for $i \neq j$ and $\Lambda_{ii} = \delta_i I_{m_i}$, $i = 0, 1, \ldots, \nu + 1$.

We have that for $i \neq j$,

$$\begin{aligned} P_i \Sigma P_j &= \Sigma P_i P_j, \text{ by Lemma 8.8} \\ &= \mathbf{0}, \text{ by Lemma 8.3(b).} \end{aligned}$$

Hence,

$$Q_i Q_i' \Sigma Q_j Q_j' = \mathbf{0}, \; i \neq j. \qquad (8.48)$$

Upon multiplying the two sides of (8.48) by Q_i' on the left and by Q_j on the right, we get

$$Q_i'\Sigma Q_j = \mathbf{0},\ i \neq j. \tag{8.49}$$

Also, from (8.31), we have

$$\begin{aligned} P_i \Sigma P_i &= P_i\,(\delta_i\, P_i) \\ &= \delta_i\, P_i,\ i = 0, 1, \ldots, \nu + 1. \end{aligned}$$

Hence,

$$Q_i Q_i'\, \Sigma\, Q_i Q_i' = \delta_i\, Q_i Q_i',$$

which implies that

$$Q_i'\, \Sigma\, Q_i = \delta_i\, I_{m_i},\ i = 0, 1, \ldots, \nu + 1. \tag{8.50}$$

From (8.49) and (8.50) we conclude that

$$\begin{aligned} \Lambda_{ij} &= \mathbf{0},\ i \neq j \\ \Lambda_{ii} &= \delta_i\, I_{m_i},\ i = 0, 1, \ldots, \nu + 1. \end{aligned}$$

Thus,

$$\begin{aligned} Q^{*'}\, \Sigma\, Q^* &= (\Lambda_{ij}) \\ &= \bigoplus_{i=0}^{\nu+1} \Lambda_{ii} \\ &= \Lambda, \end{aligned}$$

where

$$\Lambda = \bigoplus_{i=0}^{\nu+1} (\delta_i\, I_{m_i}). \tag{8.51}$$

□

Lemma 8.13 Consider the density function in (8.45). Then,

$$\begin{aligned} (y - Xg)'\Sigma^{-1}(y - Xg) = \sum_{i=0}^{\nu-p} \frac{1}{\delta_i}\, y'P_i y\, - 2 \sum_{i=0}^{\nu-p} \frac{1}{\delta_i}\, y'P_i X g \\ + g'X' \left(\sum_{i=0}^{\nu-p} \frac{1}{\delta_i}\, P_i \right) Xg + \sum_{i=\nu-p+1}^{\nu+1} \frac{1}{\delta_i}\, y'P_i y. \end{aligned}$$

Proof. Using formulas (8.47) and (8.51), Σ^{-1} can be expressed as

$$\Sigma^{-1} = Q^* \left[\bigoplus_{i=0}^{\nu+1} \left(\frac{1}{\delta_i} I_{m_i} \right) \right] Q^{*'}.$$

Hence,

$$
\begin{aligned}
(y - Xg)'\Sigma^{-1}(y - Xg) &= (y - Xg)'Q^* \left[\bigoplus_{i=0}^{\nu+1}\left(\frac{1}{\delta_i} I_{m_i}\right)\right] Q^{*'}(y - Xg) \\
&= (y - Xg)'\left[\sum_{i=0}^{\nu+1}\left(\frac{1}{\delta_i} Q_i Q_i^{'}\right)\right](y - Xg) \\
&= (y - Xg)'\left(\sum_{i=0}^{\nu+1}\frac{1}{\delta_i} P_i\right)(y - Xg) \\
&= (y - Xg)'\left(\sum_{i=0}^{\nu-p}\frac{1}{\delta_i} P_i\right)(y - Xg) \\
&\quad + (y - Xg)'\left(\sum_{i=\nu-p+1}^{\nu+1}\frac{1}{\delta_i} P_i\right)(y - Xg).
\end{aligned}
$$

Since $P_i X = 0$ for $i = \nu - p + 1, \ldots, \nu + 1$, then

$$
\begin{aligned}
(y - Xg)'\Sigma^{-1}(y - Xg) &= \sum_{i=0}^{\nu-p}\frac{1}{\delta_i} y'P_i y - 2\sum_{i=0}^{\nu-p}\frac{1}{\delta_i} y'P_i Xg + g'X'\left(\sum_{i=0}^{\nu-p}\frac{1}{\delta_i} P_i\right)Xg \\
&\quad + \sum_{i=\nu-p+1}^{\nu+1}\frac{1}{\delta_i} y'P_i y. \qquad (8.52)
\end{aligned}
$$

□

Theorem 8.5 Consider model (8.23) with the associated density function in (8.45). Then, $Q_0'Y, Q_1'Y, \ldots, Q_{\nu-p}'Y$; $Y'P_{\nu-p+1}Y, Y'P_{\nu-p+2}Y, \ldots, Y'P_{\nu+1}Y$ are complete and sufficient statistics for model (8.23), where Q_i and P_j are the same matrices as in Lemma 8.12 ($i = 0, 1, \ldots, \nu - p$; $j = \nu - p + 1, \nu - p + 2, \ldots, \nu + 1$).

Proof. Using formulas (8.47) and (8.52), the density function in (8.45) can be expressed as

$$
\begin{aligned}
f(y, \Delta) &= \frac{1}{(2\pi)^{n/2}\prod_{i=0}^{\nu+1}\delta_i^{m_i/2}}\exp\left\{-\frac{1}{2}\sum_{i=0}^{\nu-p}\frac{1}{\delta_i}[y'P_i y - 2y'P_i Xg\right. \\
&\quad \left. + g'X'P_i Xg] - \frac{1}{2}\sum_{i=\nu-p+1}^{\nu+1}\frac{1}{\delta_i} y'P_i y\right\}. \qquad (8.53)
\end{aligned}
$$

Since $P_i = Q_i Q_i', i = 0, 1, \ldots, \nu + 1$, then by the Factorization Theorem (Theorem 6.2), $Q_i'Y$ ($i = 0, 1, \ldots, \nu - p$) and $Y'P_iY$($i = \nu - p + 1, \nu - p +$

$2, \ldots, \nu+1$) form a set of sufficient statistics. These statistics are also complete. The last assertion is true by Theorem 15.16 in Arnold (1981, p. 271), which states that if

(a) $\tilde{Q}'Y$, where $\tilde{Q} = [Q_0 : Q_1 : \ldots : Q_{\nu-p}]$, and $Y'P_iY (i = \nu - p + 1, \nu - p + 2, \ldots, \nu + 1)$ are mutually independent (which they are here since $P_i\Sigma P_j = P_iP_j\Sigma = \mathbf{0}$ for $i \neq j$, hence, $Q_i'\Sigma P_j = \mathbf{0}$ for $i = 0, 1, \ldots, \nu - p$; $j = \nu - p + 1, \nu - p + 2, \ldots, \nu + 1$),

(b) $\tilde{Q}'Y \sim N(\tilde{Q}'Xg, \tilde{Q}'\Sigma\tilde{Q})$, and $\frac{1}{\delta_i} Y'P_iY \sim \chi^2_{m_i}$, $i = \nu - p + 1, \nu - p + 2, \ldots, \nu + 1$, which is true by Theorem 8.1, where $\tilde{Q}'\Sigma\tilde{Q}$ is a function of $\delta_{\nu-p+1}, \delta_{\nu-p+2}, \ldots, \delta_{\nu+1}$, and

(c) the elements of $\tilde{Q}'Xg$ and $\delta_{\nu-p+1}, \delta_{\nu-p+2}, \ldots, \delta_{\nu+1}$ are unrelated and their ranges contain open rectangles,

then $\tilde{Q}'Y$ and $Y'P_iY$ $(i = \nu-p+1, \nu-p+2, \ldots, \nu+1)$ are complete statistics. □

Example 8.5 Consider a general mixed two-way model as the one used in Example 8.4, where $\alpha_{(i)}$ is fixed and $\beta_{(j)}$ is random ($i = 1, 2, \ldots, a$; $j = 1, 2, \ldots, b$; $k = 1, 2, \ldots, n$). In this case, $Q_0'Y$, $Q_1'Y$, $Y'P_2Y$, $Y'P_3Y$, $Y'P_4Y$ are complete and sufficient statistics, where Q_0 and Q_1 are such that $P_0 = Q_0Q_0'$, $P_1 = Q_1Q_1'$. Here, $P_0 = \frac{1}{abn}H_0H_0'$, $P_1 = \frac{1}{bn}H_1H_1' - \frac{1}{abn}H_0H_0'$, where $H_0 = \mathbf{1}_a \otimes \mathbf{1}_b \otimes \mathbf{1}_n$, $H_1 = I_a \otimes \mathbf{1}_b \otimes \mathbf{1}_n$, and P_1, P_2, P_3, P_4 are the matrices giving the sums of squares for factors A, B, the interaction $A * B$, and the error term, respectively, in the model. Consider the following lemma.

Lemma 8.14 The row space of $\tilde{Q}' = [Q_0 : Q_1]'$ is the same as the row space of H_1'.

Proof. We have that

$$\begin{aligned} Q_0' &= Q_0'P_0, \text{ since } Q_0'Q_0 = 1, \\ &= \frac{1}{abn} Q_0'H_0H_0' \\ &= \frac{1}{abn} Q_0'H_0\mathbf{1}_a'H_1', \text{ since } H_0' = \mathbf{1}_a'H_1', \end{aligned} \tag{8.54}$$

and

$$\begin{aligned} Q_1' &= Q_1'P_1, \text{ since } Q_1'Q_1 = I_{a-1}, \\ &= Q_1'\left[\frac{1}{bn}H_1H_1' - \frac{1}{abn}H_0H_0'\right] \\ &= Q_1'\left[\frac{1}{bn}H_1 - \frac{1}{abn}H_0\mathbf{1}_a'\right]H_1'. \end{aligned} \tag{8.55}$$

Formulas (8.54) and (8.55) indicate that the row space of $\tilde{Q}'$ is contained within the row space of H_1'.

Vice versa, we have that

$$Q_1Q_1' + Q_0Q_0' = \frac{1}{bn}H_1H_1'.$$

Hence, since $H_1'H_1 = bnI_a$,

$$H_1' = H_1'[Q_1Q_1' + Q_0Q_0'] = H_1'\tilde{Q}\tilde{Q}',$$

which indicates that the row space of H_1' is contained within the row space of $\tilde{Q}'$. We can therefore conclude that the row spaces of $\tilde{Q}'$ and H_1' are identical. □

On the basis of Lemma 8.14, there exists a nonsingular matrix U such that $H_1' = U\tilde{Q}'$. Since the elements of $\tilde{Q}'Y$ together with $Y'P_2Y, Y'P_3Y$, and $Y'P_4Y$ are complete and sufficient statistics for the model by Theorem 8.5, then so are the elements of $H_1'Y$ and $Y'P_2Y, Y'P_3Y, Y'P_4Y$. But, $H_1'Y = bn[\bar{Y}_{(1)}, \bar{Y}_{(2)}, \ldots, \bar{Y}_{(a)}]$, where $\bar{Y}_{(i)}$ is the sample mean of level i of factor A $(i = 1, 2, \ldots, a)$. We finally conclude that $\bar{Y}_{(1)}, \bar{Y}_{(2)}, \ldots, \bar{Y}_{(a)}$; $Y'P_2Y, Y'P_3Y$, and $Y'P_4Y$ are complete and sufficient statistics for the mixed two-way model.

8.6 ANOVA Estimation of Variance Components

Analysis of variance (ANOVA) estimates of the variance components for model (8.23) are obtained by equating the mean squares of the random effects to their corresponding expected values, then solving the resulting equations. Such estimates are obviously unbiased. Furthermore, under the same assumptions made earlier in Section 8.4 concerning the distribution of the random effects, these unbiased estimates are functions of the mean squares of the random effects, which along with $Q_0'Y, Q_1'Y, \ldots, Q_{\nu-p}'Y$ are complete and sufficient statistics for the model by Theorem 8.5. Consequently, and on the basis of the Lehmann–Scheffé Theorem (see, for example, Casella and Berger, 2002, p. 369; Graybill, 1976, Theorem 2.7.7, p. 78), we conclude that the ANOVA estimators are *uniformly minimum variance unbiased estimators* (UMVUE) of the variance components.

8.6.1 The Probability of a Negative ANOVA Estimator

One of the undesirable features of the ANOVA estimators of variance components is the possibility that they can be negative. We now show how to compute the probability that such estimators can be negative.

Let $\hat{\sigma}^2_{\nu-p+1}, \hat{\sigma}^2_{\nu-p+2}, \ldots, \hat{\sigma}^2_{\nu+1}$ denote the ANOVA estimators of $\sigma^2_{\nu-p+1}, \sigma^2_{\nu-p+2}, \ldots, \sigma^2_{\nu+1}$. Then, $\hat{\sigma}^2_i$ can be written as

$$\hat{\sigma}^2_i = \sum_{j=\nu-p+1}^{\nu+1} \eta_{ij} MS_j, \; i = \nu - p + 1, \nu - p + 2, \ldots, \nu + 1, \quad (8.56)$$

where $MS_j = \frac{1}{m_j} \boldsymbol{Y}' \boldsymbol{P}_j \boldsymbol{Y}$ and the η_{ij}'s are known constants. Under the assumptions made earlier in Section 8.4 concerning the distribution of the random effects, $\frac{1}{\delta_j} \boldsymbol{Y}' \boldsymbol{P}_j \boldsymbol{Y} \sim \chi^2_{m_j}$ $(j = \nu - p + 1, \nu - p + 2, \ldots, \nu + 1)$. Hence, $\hat{\sigma}^2_i$ can be written as

$$\hat{\sigma}^2_i = \sum_{j=\nu-p+1}^{\nu+1} \eta_{ij} \frac{\delta_j}{m_j} \chi^2_{m_j}, \; i = \nu - p + 1, \nu - p + 2, \ldots, \nu + 1. \quad (8.57)$$

Some of the η_{ij} coefficients may be equal to zero. Let κ_i be the number of nonzero η_{ij} in (8.57), and let the nonzero values of $\eta_{ij} \frac{\delta_j}{m_j}$ be denoted by η^*_{ij} $(i = \nu - p + 1, \nu - p + 2, \ldots, \nu + 1; j = 1, 2, \ldots, \kappa_i)$. If the corresponding values of m_j in (8.57) are denoted by $m^*_{i1}, m^*_{i2}, \ldots, m^*_{i\kappa_i}$, then (8.57) can be expressed as

$$\hat{\sigma}^2_i = \sum_{j=1}^{\kappa_i} \eta^*_{ij} \chi^2_{m^*_{ij}}, \; i = \nu - p + 1, \nu - p + 2, \ldots, \nu + 1. \quad (8.58)$$

The exact probability of a negative $\hat{\sigma}^2_i$ can then be obtained by applying a formula given by Imhof [1961, formula (3.2)], namely,

$$P(\hat{\sigma}^2_i < 0) = \frac{1}{2} - \frac{1}{\pi} \int_0^{\infty} \frac{\sin[\tau_i(u)]}{u \, \rho_i(u)} \, du,$$

where

$$\tau_i(u) = \frac{1}{2} \sum_{j=1}^{\kappa_i} m^*_{ij} \arctan(\eta^*_{ij} u)$$

$$\rho_i(u) = \prod_{j=1}^{\kappa_i} (1 + \eta^{*2}_{ij} u^2)^{m^*_{ij}/4}.$$

(see also Khuri, 1994, p. 903). A computer program written in FORTRAN was given by Koerts and Abrahamse (1969) to implement Imhof's (1961) procedure.

Alternatively, by using the representation of $\hat{\sigma}^2_i$ in (8.57) as a linear combination of mutually independent chi-squared variates, the exact value of $P(\hat{\sigma}^2_i < 0)$ can be more easily obtained by utilizing Davies' (1980) algorithm (see Section 5.6).

Example 8.6 Consider the random model

$$Y_{ijkl} = \mu + \alpha_{(i)} + \beta_{(j)} + (\alpha\beta)_{(ij)} + \gamma_{j(k)} + (\alpha\gamma)_{j(ik)} + \epsilon_{ijk(l)},$$

$i = 1,2,3;\ j = 1,2,\ldots,5;\ k = 1,2,3,4;$ and $l = 1,2$, where $\alpha_{(i)} \sim N(0,\sigma^2_\alpha)$, $\beta_{(j)} \sim N(0,\sigma^2_\beta)$, $(\alpha\beta)_{(ij)} \sim N(0,\sigma^2_{\alpha\beta})$, $\gamma_{j(k)} \sim N(0,\sigma^2_{\gamma(\beta)})$, $(\alpha\gamma)_{j(ik)} \sim N(0,\sigma^2_{\alpha\gamma(\beta)})$, $\epsilon_{ijk(l)} \sim N(0,\sigma^2_\epsilon)$. In this case,

$$\begin{aligned}
\delta_1 &= E(MS_1) = 40\,\sigma^2_\alpha + 8\,\sigma^2_{\alpha\beta} + 2\,\sigma^2_{\alpha\gamma(\beta)} + \sigma^2_\epsilon\\
\delta_2 &= E(MS_2) = 24\,\sigma^2_\beta + 8\,\sigma^2_{\alpha\beta} + 6\,\sigma^2_{\gamma(\beta)} + 2\,\sigma^2_{\alpha\gamma(\beta)} + \sigma^2_\epsilon\\
\delta_3 &= E(MS_3) = 8\,\sigma^2_{\alpha\beta} + 2\,\sigma^2_{\alpha\gamma(\beta)} + \sigma^2_\epsilon\\
\delta_4 &= E(MS_4) = 6\,\sigma^2_{\gamma(\beta)} + 2\,\sigma^2_{\alpha\gamma(\beta)} + \sigma^2_\epsilon\\
\delta_5 &= E(MS_5) = 2\,\sigma^2_{\alpha\gamma(\beta)} + \sigma^2_\epsilon\\
\delta_6 &= E(MS_6) = \sigma^2_\epsilon,
\end{aligned}$$

where $MS_1, MS_2, \ldots, MS_6$ are the mean squares associated with $\alpha_{(i)}$, $\beta_{(j)}$, $(\alpha\beta)_{(ij)}$, $\gamma_{j(k)}$, $(\alpha\gamma)_{j(ik)}$, and $\epsilon_{ijk(l)}$, respectively, with $m_1 = 2$, $m_2 = 4$, $m_3 = 8$, $m_4 = 15$, $m_5 = 30$, and $m_6 = 60$ degrees of freedom.

Consider now the ANOVA estimator of σ^2_β, which is given by

$$\hat{\sigma}^2_\beta = \frac{1}{24}\,(MS_2 + MS_5 - MS_3 - MS_4).$$

In this case, $\beta_{(j)}$ is the second random effect in the model, and the values of η^*_{2j} and m^*_{2j} in formula (8.58) are given by

$$\begin{aligned}
\eta^*_{21} &= \frac{1}{24}\,\frac{\delta_2}{4}, \quad m^*_{21} = 4\\
\eta^*_{22} &= \frac{1}{24}\,\frac{\delta_5}{30}, \quad m^*_{22} = 30\\
\eta^*_{23} &= -\frac{1}{24}\,\frac{\delta_3}{8}, \quad m^*_{23} = 8\\
\eta^*_{24} &= -\frac{1}{24}\,\frac{\delta_4}{15}, \quad m^*_{24} = 15.
\end{aligned}$$

Note that the values of $P(\hat{\sigma}^2_\beta < 0)$ depend on the following ratios of variance components: $\frac{\sigma^2_\beta}{\sigma^2_\epsilon}$, $\frac{\sigma^2_{\gamma(\beta)}}{\sigma^2_\epsilon}$, $\frac{\sigma^2_{\alpha\beta}}{\sigma^2_\epsilon}$, and $\frac{\sigma^2_{\alpha\gamma(\beta)}}{\sigma^2_\epsilon}$. By assigning specific values to these ratios, we can compute the probability that $\hat{\sigma}^2_\beta < 0$ with the help of Davies' (1980) algorithm. Table 8.5 gives some of such values.

We note that large values of $P(\hat{\sigma}^2_\beta < 0)$ are associated with small values of $\sigma^2_\beta/\sigma^2_\epsilon$, which is an expected result.

TABLE 8.5
Values of $P(\hat{\sigma}^2_\beta < 0)$

$\sigma^2_\beta/\sigma^2_\epsilon$	$\sigma^2_{\gamma(\beta)}/\sigma^2_\epsilon$	$\sigma^2_{\alpha\beta}/\sigma^2_\epsilon$	$\sigma^2_{\alpha\gamma(\beta)}/\sigma^2_\epsilon$	$P(\hat{\sigma}^2_\beta < 0)$
0.1	0.5	0.5	0.5	0.447830
0.1	0.5	0.5	2.0	0.466313
0.1	0.5	2.5	2.0	0.509234
0.1	3.0	2.5	2.0	0.537932
0.5	0.5	2.5	0.5	0.375784
0.5	3.0	0.5	0.5	0.376904
0.5	3.0	2.5	0.5	0.441052
2.0	0.5	0.5	0.5	0.044171
2.0	3.0	0.5	0.5	0.148058
5.0	0.5	0.5	2.0	0.016957
5.0	0.5	2.5	0.5	0.052385

8.7 Confidence Intervals on Continuous Functions of the Variance Components

In this section, we discuss a procedure for constructing exact, but conservative, simultaneous confidence intervals on all continuous functions of the variance components in a balanced mixed model situation.

Consider model (8.23) under the same assumptions regarding the distribution of the random effects as was described in Section 8.4. Let $\varphi(.)$ be a continuous function of the expected mean squares of the random effects, namely, $\delta_{\nu-p+1}, \delta_{\nu-p+2}, \ldots, \delta_{\nu+1}$, and hence of the variance components, $\sigma^2_{\nu-p+1}, \sigma^2_{\nu-p+2}, \ldots, \sigma^2_{\nu+1}$. Recall that for such mean squares, $m_i\, MS_i/\delta_i \sim \chi^2_{m_i}$. Hence, a $(1-\alpha)$ 100% confidence interval on δ_i is given by

$$B_i = \left[\frac{m_i\, MS_i}{\chi^2_{\alpha/2,\, m_i}},\ \frac{m_i\, MS_i}{\chi^2_{1-\alpha/2,\, m_i}}\right],\ i = \nu - p + 1, \nu - p + 2, \ldots, \nu + 1. \tag{8.59}$$

Since the MS_i's are mutually independent, a rectangular confidence region on the values of δ_i $(i = \nu - p + 1, \nu - p + 2, \ldots, \nu + 1)$ with a confidence coefficient $1 - \alpha^*$ can be written as

$$B_* = \times_{i=\nu-p+1}^{\nu+1} B_i, \tag{8.60}$$

where B_* is the Cartesian product of the B_i's and $1 - \alpha^* = (1-\alpha)^{p+1}$.

Note that some of the δ_i's may be subject to certain linear inequality constraints. For example, in a balanced nested random model, we have that $\delta_1 \geq \delta_2 \geq \ldots \geq \delta_{\nu+1}$. Let $\mathcal{R}$ be a subset of the $(\nu+1)$-dimensional Euclidean space which consists of all points whose coordinates are subject to the same linear inequality constraints as those among the corresponding δ_i's. This subset is called the *region of definition* for functions of the δ_i's. Let $\boldsymbol{\delta}$ be a vector whose elements consist of the δ_i's. Since $\boldsymbol{\delta} \in \mathcal{R}$, the confidence region B_* in (8.60) is considered meaningful only if it intersects with $\mathcal{R}$, that is, if $\mathcal{R} \cap B_* \neq \emptyset$, the empty set. Let H_* be the family of $1 - \alpha^*$ confidence regions as defined by (8.60), and let $\tilde{H}_* \subset H_*$ be defined as

$$\tilde{H}_* = \{B_* \in H_* \,|\, \mathcal{R} \cap B_* \neq \emptyset\}.$$

Then, the probability of coverage for $\tilde{H}_*$ is greater than or equal to $1 - \alpha^*$. This is true because

$$\begin{aligned} P[\,\boldsymbol{\delta} \in B_* \,|\, B_* \in \tilde{H}_*\,] &= \frac{P[\,\boldsymbol{\delta} \in B_*\,]}{P[\,B_* \in \tilde{H}_*\,]} \\ &\geq P[\,\boldsymbol{\delta} \in B_*\,] \\ &= 1 - \alpha^*. \end{aligned}$$

Furthermore, $\boldsymbol{\delta} \in B_*$ if and only if $\boldsymbol{\delta} \in \tilde{B}_*$, where $\tilde{B}_* = \mathcal{R} \cap B_*$ since $\boldsymbol{\delta} \in \mathcal{R}$. It follows that

$$P[\,\boldsymbol{\delta} \in \tilde{B}_* \,|\, B_* \in \tilde{H}_*\,] \geq 1 - \alpha^*.$$

Thus, by truncating each $B_* \in \tilde{H}_*$ to $\tilde{B}_*$ we obtain a family of confidence regions for $\boldsymbol{\delta}$, each member of which is contained inside $\mathcal{R}$. This is called the *truncated* $\tilde{H}_*$ *family*. The confidence region $\tilde{B}_*$ will be used to obtain simultaneous confidence intervals on all continuous functions of the variance components. This is shown in the next lemma.

Lemma 8.15 Let $\mathcal{F}$ be a family of continuous functions of $\delta_{\nu-p+1}, \delta_{\nu-p+2}, \ldots, \delta_{\nu+1}$, and hence of the variance components. Then, the probability, $P[\varphi(\boldsymbol{\delta}) \in \varphi(\tilde{B}_*), \forall\, \varphi \in \mathcal{F}]$, is greater than or equal to $1 - \alpha^*$

Proof. Let φ be any continuous function in $\mathcal{F}$. If $\boldsymbol{\delta} \in \tilde{B}_*$, where $B_* \in \tilde{H}_*$, then $\varphi(\boldsymbol{\delta}) \in \varphi(\tilde{B}_*)$. Hence,

$$\begin{aligned} P[\,\varphi(\boldsymbol{\delta}) \in \varphi(\tilde{B}_*)\,] &\geq P[\,\boldsymbol{\delta} \in \tilde{B}_*\,] \\ &\geq 1 - \alpha^*. \end{aligned} \quad (8.61)$$

Since inequality (8.61) is true for all $\varphi \in \mathcal{F}$, we conclude that

$$P[\,\varphi(\boldsymbol{\delta}) \in \varphi(\tilde{B}_*),\ \forall\, \varphi \in \mathcal{F}\,] \quad (8.62)$$

is greater than or equal to $1 - \alpha^*$. □

From Lemma 8.15 it follows that for any $\varphi \in \mathcal{F}$, a confidence interval for $\varphi(\boldsymbol{\delta})$ with a confidence coefficient of at least $1 - \alpha^*$ is given by the range of values of the function φ over $\tilde{B}_*$, that is,

$$[\,\min_{x \in \tilde{B}_*} \varphi(x),\ \max_{x \in \tilde{B}_*} \varphi(x)\,]. \quad (8.63)$$

Note that since φ is continuous and $\tilde{B}_*$ is a closed and bounded set in the $(p+1)$-dimensional Euclidean space, φ must attain its maximum and minimum values in $\tilde{B}_*$. Thus, the family of intervals in (8.63) provides simultaneous confidence intervals on all continuous functions of the δ_i's and hence on the variance components. These intervals are conservative since according to (8.62), the actual coverage probability is greater than or equal to $1-\alpha^*$.

8.7.1 Confidence Intervals on Linear Functions of the Variance Components

Let Φ be a linear function of the variance components, which can be expressed in terms of the δ_i's as

$$\Phi = \sum_{i=\nu-p+1}^{\nu+1} h_i \delta_i, \tag{8.64}$$

where the h_i's are known constants. Such a function is obviously continuous and can be regarded as a special case of the functions considered earlier. The interval in (8.63) can then be applied to obtain simultaneous confidence intervals on all linear functions of the form (8.64). In this case, the function $\varphi(\boldsymbol{x})$ in (8.63) is of the form $\varphi(\boldsymbol{x}) = \sum_{i=\nu-p+1}^{\nu+1} h_i x_i$, where $\boldsymbol{x} = (x_{\nu-p+1}, x_{\nu-p+2}, \ldots, x_{\nu+1})'$. Thus, the intervals,

$$\left[\min_{\boldsymbol{x}\in\tilde{B}_*} \sum_{i=\nu-p+1}^{\nu+1} h_i x_i,\ \max_{\boldsymbol{x}\in\tilde{B}_*} \sum_{i=\nu-p+1}^{\nu+1} h_i x_i\right], \tag{8.65}$$

provide simultaneous confidence intervals on all linear functions of the form shown in (8.64) with a joint confidence coefficient greater than or equal to $1-\alpha^* = (1-\alpha)^{p+1}$. Such intervals can be very conservative, especially for large values of p.

In some cases, we may be interested in a particular linear function Φ of the form in (8.64) where some of the h_i coefficients may be equal to zero. Let S be a subset of $\{\nu-p+1, \nu-p+2, \ldots, \nu+1\}$ consisting of $n(S)$ elements such that $h_i \neq 0$ for $i \in S$. Thus, $\sum_{i=\nu-p+1}^{\nu+1} h_i\delta_i = \sum_{i\in S} h_i\delta_i$. Let $\mathcal{R}_s$ and B_{*s} be the counterparts of $\mathcal{R}$ and B_*, respectively, based only on the elements of S. Then, the interval

$$\left[\min_{\boldsymbol{x}\in\tilde{B}_{*s}} \sum_{i\in S} h_i x_i,\ \max_{\boldsymbol{x}\in\tilde{B}_{*s}} \sum_{i\in S} h_i x_i\right] \tag{8.66}$$

contains Φ with a coverage probability greater than or equal to $1-\alpha_s^*$, where $\boldsymbol{x} = \{x_i;\ i\in S\}$, $\tilde{B}_{*s} = \mathcal{R}_s \cap B_{*s}$, and $\alpha_s^* = 1-(1-\alpha)^{n(S)}$.

Since $\sum_{i\in S} h_i x_i$ is a linear function and the region $\tilde{B}_{*s}$ is bounded by a finite number of hyperplanes (convex polyhedron) in the $n(S)$-dimensional

Euclidean space, the optimization of $\sum_{i\in S} h_i x_i$ over $\tilde{B}_{*s}$ can be conducted by using the *simplex method* of linear programming (see, for example, Simonnard, 1966). In particular, we have the following result from Simonnard (1966, p. 19):

Lemma 8.16 Let $\tilde{B}_{*s}$ be a convex polyhedron on which the linear function $\varphi(x) = \sum_{i\in S} h_i x_i$ is defined. Then, there exists at least one vertex of $\tilde{B}_{*s}$ at which φ attains an absolute maximum, and at least one vertex at which φ attains an absolute minimum.

Thus, by evaluating $\varphi(x) = \sum_{i\in S} h_i x_i$ at the vertices of $\tilde{B}_{*s}$ we can arrive at the values of the optima of φ needed to obtain the interval in (8.66). In particular, if $B_{*s} \subset \mathcal{R}_s$, that is, $\tilde{B}_{*s} = B_{*s}$, then such optima are given by

$$\min_{x\in B_{*s}} \sum_{i\in S} h_i x_i = \sum_{i\in S_1} h_i u_i + \sum_{i\in S_2} h_i v_i \tag{8.67}$$

$$\max_{x\in B_{*s}} \sum_{i\in S} h_i x_i = \sum_{i\in S_1} h_i v_i + \sum_{i\in S_2} h_i u_i, \tag{8.68}$$

where, from (8.59),

$$u_i = \frac{m_i MS_i}{\chi^2_{\alpha/2,\, m_i}}, \quad i \in S \tag{8.69}$$

$$v_i = \frac{m_i MS_i}{\chi^2_{1-\alpha/2,\, m_i}}, \quad i \in S, \tag{8.70}$$

S_1 and S_2 are two disjoint subsets of S such that $S = S_1 \cup S_2$ with $h_i > 0$ for $i \in S_1$ and $h_i < 0$ for $i \in S_2$, and $\alpha = 1 - (1 - \alpha_s^*)^{1/n(S)}$. Thus, in this case, $\sum_{i\in S} h_i x_i$ attains its maximum value at the vertex of B_{*s} whose coordinates are determined by v_i for $i \in S_1$ and u_i for $i \in S_2$; and its minimum value is attained at the vertex whose coordinates are determined by u_i for $i \in S_1$ and v_i for $i \in S_2$.

Further details concerning simultaneous confidence intervals on functions of variance components can be found in Khuri (1981).

Example 8.7 Anderson and Bancroft (1952, p. 323) reported an example given by Crump (1946) of a series of genetic experiments on the number of eggs laid by each of 12 females from 25 races (factor B) of the common fruitfly (*Drosophila melanogaster*) on the fourth day of laying. The whole exsperiment was carried out four times (factor A). The corresponding ANOVA table for a random two-way model with interaction is given in Table 8.6 (see also Khuri, 1981, Section 2.3).

Let us consider setting up a 90% confidence interval on σ^2_α. From the ANOVA table we have $\sigma^2_\alpha = \frac{1}{300}(\delta_1 - \delta_3)$, where $\delta_1 = E(MS_A)$, $\delta_3 = E(MS_{AB})$. Thus,

$$\Phi = \frac{1}{300}(\delta_1 - \delta_3). \tag{8.71}$$

TABLE 8.6
ANOVA Table for Example 8.7

Source	DF	MS	E(MS)
A (experiments)	3	46,659	$\delta_1 = 300\,\sigma^2_{\alpha} + 12\,\sigma^2_{\alpha\beta} + \sigma^2_{\epsilon}$
B (races)	24	3243	$\delta_2 = 48\,\sigma^2_{\beta} + 12\,\sigma^2_{\alpha\beta} + \sigma^2_{\epsilon}$
A^*B	72	459	$\delta_3 = 12\sigma^2_{\alpha\beta} + \sigma^2_{\epsilon}$
Error	1100	231	$\delta_4 = \sigma^2_{\epsilon}$

In this case, we need to construct a 90% confidence region on $(\delta_1,\ \delta_3)'$, namely, the Cartesian product, $B_{*s} = B_1 \times B_3$, where from (8.59) and Table 8.6,

$$B_1 = \left[\frac{3\,(46,659)}{\chi^2_{\alpha/2,3}},\ \frac{3\,(46,659)}{\chi^2_{1-\alpha/2,3}}\right] \tag{8.72}$$

$$B_3 = \left[\frac{72\,(459)}{\chi^2_{\alpha/2,72}},\ \frac{72\,(459)}{\chi^2_{1-\alpha/2,72}}\right]. \tag{8.73}$$

Here, $(1-\alpha)^2 = 0.90$, that is, $\alpha = 0.0513$. Substituting in (8.72) and (8.73), we get

$$B_1 = [15053.0945,\ 637048.4606]$$
$$B_3 = [340.0268,\ 654.0737].$$

In this example, the region, $\mathcal{R}_s$, of definition for functions of only δ_1 and δ_3 is

$$\mathcal{R}_s = \{(x_1, x_3) | x_1 \geq x_3 \geq 0\}.$$

We note that $B_{*s} = B_1 \times B_3$ is contained inside $\mathcal{R}_s$. Thus, $\tilde{B}_{*s} = B_{*s}$. Consequently, the confidence interval on σ^2_{α} is obtained by optimizing the function $\varphi(\boldsymbol{x}) = \frac{1}{300}(x_1 - x_3)$ over B_{*s}, where $\boldsymbol{x} = (x_1,\ x_3)'$. By applying formulas (8.67) and (8.68), this interval is given by

$$[\min_{\boldsymbol{x}\in B_{*s}} \varphi(\boldsymbol{x}),\ \max_{\boldsymbol{x}\in B_{*s}} \varphi(\boldsymbol{x})] = [(u_1 - v_3)/300,\ (v_1 - u_3)/300], \tag{8.74}$$

where from (8.69) and (8.70),

$$u_1 = \frac{3\,MS_A}{\chi^2_{\alpha/2,3}}$$
$$v_1 = \frac{3\,MS_A}{\chi^2_{1-\alpha/2,3}}$$
$$u_3 = \frac{72\,MS_{AB}}{\chi^2_{\alpha/2,72}}$$
$$v_3 = \frac{72\,MS_{AB}}{\chi^2_{1-\alpha/2,72}}.$$

The confidence coefficient for this interval is therefore greater than or equal to 0.90, that is,

$$P\left[\frac{u_1 - v_3}{300} < \sigma_\alpha^2 < \frac{v_1 - u_3}{300}\right] \geq 0.90. \tag{8.75}$$

Using the corresponding entries in Table 8.6, the interval in (8.74) is equal to [47.9967, 2122.3614].

Since $\frac{3MS_A}{\delta_1} \sim \chi_3^2$ independently of $\frac{72MS_{AB}}{\delta_3} \sim \chi_{72}^2$, the exact coverage probability for the interval in (8.74) can be written as (see formula (8.75))

$$\begin{aligned} P[u_1 - v_3 &< \delta_1 - \delta_3 < v_1 - u_3] \\ &= P\left[\frac{\delta_1 F_1}{\chi^2_{\alpha/2,3}} - \frac{\delta_3 F_3}{\chi^2_{1-\alpha/2,72}} < \delta_1 - \delta_3 < \frac{\delta_1 F_1}{\chi^2_{1-\alpha/2,3}} - \frac{\delta_3 F_3}{\chi^2_{\alpha/2,72}}\right] \\ &= P\left[\frac{F_1}{\chi^2_{\alpha/2,3}} - \frac{\delta_3}{\delta_1}\frac{F_3}{\chi^2_{1-\alpha/2,72}} < 1 - \frac{\delta_3}{\delta_1} < \frac{F_1}{\chi^2_{1-\alpha/2,3}} - \frac{\delta_3}{\delta_1}\frac{F_3}{\chi^2_{\alpha/2,72}}\right], \end{aligned} \tag{8.76}$$

where $F_1 = 3MS_A/\delta_1$, $F_3 = 72MS_{AB}/\delta_3$ are independently distributed such that $F_1 \sim \chi_3^2$ and $F_3 \sim \chi_{72}^2$. Hence, for a given value of δ_3/δ_1, the exact probability in (8.76) can be evaluated by using a double integration computer program, or by applying Davies' (1980) algorithm mentioned in Section 5.6. Table 8.7 gives values of this probability for several values of δ_3/δ_1. We note that for large values of δ_3/δ_1 (≤ 1), the exact probability is sizably larger than 0.90. To remedy this situation, we can consider reducing the 90% confidence coefficient.

TABLE 8.7
Exact Coverage Probability in (8.76) with a Minimum 90% Coverage

δ_3/δ_1	Exact Coverage Probability
0.05	0.9537
0.10	0.9561
0.20	0.9603
0.30	0.9639
0.40	0.9670
0.50	0.9696
0.60	0.9719
0.70	0.9740
0.80	0.9758
0.90	0.9774
1.00	0.9789

8.8 Confidence Intervals on Ratios of Variance Components

In some experimental situations, certain functions of the variance components, particularly ratios thereof, may be of greater interest than the components themselves. For example, in animal breeding experiments, estimation of functions, such as *heritability*, which is a ratio involving several variance components, is of paramount interest to breeders in order to assess the potential for genetic improvement of a certain breed of animals. Confidence intervals on such functions can therefore be very useful in making statistical inferences about these functions.

Harville and Fenech (1985) obtained an exact confidence interval on a heritability parameter, which is used in animal and plant breeding problems. Burdick and Graybill (1992, Section 3.4) provided a review of methods for constructing confidence intervals on ratios of variance components. Broemeling (1969) proposed simultaneous confidence intervals on particular forms of such ratios. The latter intervals are helpful in assessing the relative measures of variability of various effects with respect to the experimental error variance in the model. In this section, we describe the development of Broemeling's (1969) confidence intervals.

Consider model (8.23) under the same assumptions made earlier in Section 8.4 with regard to the the distribution of the random effects. Of interest here is the derivation of simultaneous confidence intervals on the ratios $\sigma_i^2/\sigma_\epsilon^2$ $(i = \nu - p + 1, \nu - p + 2, \ldots, \nu)$, where $\sigma_\epsilon^2 = \sigma_{\nu+1}^2$ is designated as the experimental error variance component.

Let MS_i be the mean square associated with the ith random effect $(i = \nu - p + 1, \ldots, \nu)$ and let MS_E be the error mean square. On the basis of Theorem 8.1, we have

$$P\left[\frac{\delta_i}{\sigma_\epsilon^2}\frac{MS_E}{MS_i} \le F_{\alpha_i, m_e, m_i}\right] = 1 - \alpha_i, \; i = \nu - p + 1, \nu - p + 2, \ldots, \nu, \quad (8.77)$$

where, if we recall, $\delta_i = E(MS_i)$, m_i and m_e are the degrees of freedom for MS_i and MS_E, respectively. The following lemma, which is due to Kimball (1951), is now needed.

Lemma 8.17 Let $X_1, X_2, \ldots, X_r, X_{r+1}$ be random variables, mutually independent and distributed as chi-squared variates with $n_1, n_2, \ldots, n_r, n_{r+1}$ degrees of freedom, respectively, then

$$P\left[\frac{X_{r+1}/n_{r+1}}{X_i/n_i} \le F_{\alpha_i, n_{r+1}, n_i};\, i = 1, 2, \ldots, r\right]$$
$$\ge \prod_{i=1}^{r} P\left[\frac{X_{r+1}/n_{r+1}}{X_i/n_i} \le F_{\alpha_i, n_{r+1}, n_i}\right] = \prod_{i=1}^{r}(1 - \alpha_i). \quad (8.78)$$

Applying Lemma 8.17 to (8.77) and recalling $m_i MS_i/\delta_i \sim \chi^2_{m_i}$ independently from $m_e MS_E/\sigma^2_\epsilon \sim \chi^2_{m_e}$ $(i = \nu - p + 1,\ \nu - p + 2, \ldots,\ \nu)$, we obtain

$$P\left[\frac{\delta_i}{\sigma^2_\epsilon}\frac{MS_E}{MS_i} \leq F_{\alpha_i, m_e, m_i};\ i = \nu - p + 1, \nu - p + 2, \ldots, \nu\right] \geq 1 - \alpha, \qquad (8.79)$$

where $1 - \alpha = \prod_{i=\nu-p+1}^{\nu}(1 - \alpha_i)$. This inequality determines a conservative confidence region with a confidence coefficient greater than or equal to $1 - \alpha$ for the ratios $\delta_i/\sigma^2_\epsilon$, $i = \nu - p + 1,\ \nu - p + 2, \ldots,\ \nu$. Recall from Theorem 8.1 that

$$\begin{aligned}\delta_i &= \sum_{j=\nu-p+1}^{\nu+1} \kappa_{ij}\, \sigma^2_j \\ &= \sigma^2_\epsilon + \sum_{j=\nu-p+1}^{\nu} \kappa_{ij}\, \sigma^2_j, \quad i = \nu - p + 1, \nu - p + 2, \ldots, \nu.\end{aligned}$$

Hence, the inequality in (8.79) defines a conservative confidence region on the variance ratios, $\gamma_{\nu-p+1}, \gamma_{\nu-p+2}, \ldots, \gamma_\nu$, where $\gamma_i = \sigma^2_i/\sigma^2_\epsilon$, $i = \nu - p + 1$, $\nu - p + 2, \ldots,\ \nu$, of the form

$$\mathcal{K} = \left\{\gamma : 1 + \sum_{j=\nu-p+1}^{\nu} \kappa_{ij}\,\gamma_j \leq \frac{MS_i}{MS_E} F_{\alpha_i, m_e, m_i};\ i = \nu - p + 1, \nu - p + 2, \ldots, \nu\right\}, \qquad (8.80)$$

where $\gamma = (\gamma_{\nu-p+1}, \gamma_{\nu-p+2}, \ldots, \gamma_\nu)'$. The confidence coefficient for the region $\mathcal{K}$ is greater than or equal to $1 - \alpha$. Note that $\mathcal{K}$ is a region bounded by the hyperplanes,

$$1 + \sum_{j=\nu-p+1}^{\nu} \kappa_{ij}\,\gamma_j = \frac{MS_i}{MS_E} F_{\alpha_i, m_e, m_i}, \quad i = \nu - p + 1, \nu - p + 2, \ldots, \nu.$$

Broemeling's (1969) simultaneous confidence intervals on the γ_i's are obtained by taking the orthogonal projections of $\mathcal{K}$ on the coordinate axes corresponding to $\gamma_{\nu-p+1}, \gamma_{\nu-p+2}, \ldots, \gamma_\nu$. This follows from the fact that

$$P[\gamma_i \in Pr_i(\mathcal{K});\ i = \nu - p + 1, \nu - p + 2, \ldots, \nu] \geq P[\gamma \in \mathcal{K}] \geq 1 - \alpha,$$

where $Pr_i(\mathcal{K})$ denotes the orthogonal projection of $\mathcal{K}$ on the γ_i-axis. Hence, the joint coverage probability of the intervals $Pr_i(\mathcal{K})$, $i = \nu - p + 1,\ \nu - p + 2, \ldots,\ \nu$, is greater than or equal to $1 - \alpha$. Broemeling's confidence intervals on the γ_i's are therefore conservative. The exact confidence coefficient associated with the confidence region $\mathcal{K}$ in (8.80) was derived by Sahai and Anderson (1973) in terms of a fairly complicated multidimensional integral.

TABLE 8.8
ANOVA Table for Example 8.8

Source	DF	SS	MS	E(MS)
A	$a-1$	SS_A	MS_A	$\delta_1 = bn\,\sigma^2_\alpha + n\,\sigma^2_{\beta(\alpha)} + \sigma^2_\epsilon$
$B(A)$	$a(b-1)$	$SS_{B(A)}$	$MS_{B(A)}$	$\delta_2 = n\,\sigma^2_{\beta(\alpha)} + \sigma^2_\epsilon$
Error	$ab(n-1)$	SS_E	MS_E	$\delta_3 = \sigma^2_\epsilon$

Note that since the coefficients of γ_j in (8.80) are nonnegative, the orthogonal projections of $\mathcal{K}$ on the γ_i axes are bounded from below by zero. Hence, Broemeling's intervals are one sided.

Example 8.8 Consider the two-fold nested random model

$$Y_{ijk} = \mu + \alpha_{(i)} + \beta_{i(j)} + \epsilon_{ij(k)},\ i = 1, 2, \ldots, a;\ j = 1, 2, \ldots, b;\ k = 1, 2, \ldots, n,$$

where $\alpha_{(i)} \sim N(0, \sigma^2_\alpha)$, $\beta_{i(j)} \sim N(0, \sigma^2_{\beta(\alpha)})$, $\epsilon_{ij(k)} \sim N(0, \sigma^2_\epsilon)$. All random effects are independent. The corresponding ANOVA table is Table 8.8.

In this case, the confidence region $\mathcal{K}$ in (8.80) is of the form

$$\mathcal{K} = \left\{\gamma : 1 + bn\gamma_1 + n\gamma_2 \leq \frac{MS_A}{MS_E} F_{\alpha_1, ab(n-1), a-1},\ 1 + n\gamma_2 \right.$$
$$\left. \leq \frac{MS_{B(A)}}{MS_E} F_{\alpha_2, ab(n-1), a(b-1)} \right\}, \qquad (8.81)$$

where $\gamma_1 = \sigma^2_\alpha / \sigma^2_\epsilon$, $\gamma_2 = \sigma^2_{\beta(\alpha)}/\sigma^2_\epsilon$, and $1 - \alpha = (1 - \alpha_1)(1 - \alpha_2)$. This region is shown in Figure 8.1.

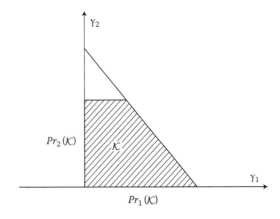

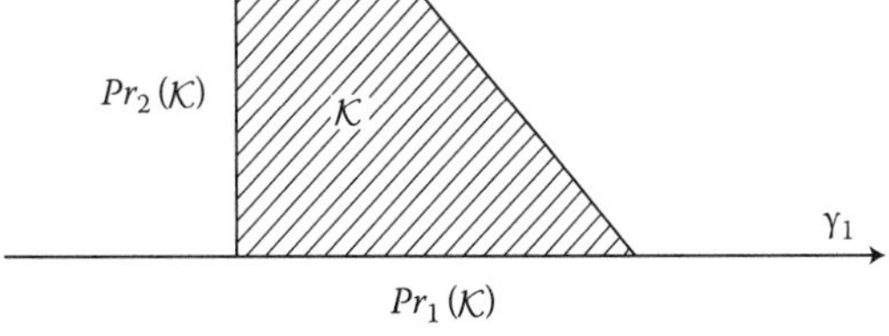

FIGURE 8.1
The confidence region, $\mathcal{K}$, in (8.81).

The projections of $\mathcal{K}$ on the γ_1 and γ_2 axes are given by $[0,\ Pr_1(\mathcal{K})]$, $[0,\ Pr_2(\mathcal{K})]$, respectively, where

$$Pr_1(\mathcal{K}) = \frac{1}{bn}\left[\frac{MS_A}{MS_E}F_{\alpha_1,ab(n-1),a-1} - 1\right]$$

$$Pr_2(\mathcal{K}) = \min\left\{\frac{1}{n}\left[\frac{MS_A}{MS_E}F_{\alpha_1,ab(n-1),a-1} - 1\right],\right.$$
$$\left.\frac{1}{n}\left[\frac{MS_{B(A)}}{MS_E}F_{\alpha_2,ab(n-1),a(b-1)} - 1\right]\right\}.$$

Hence, simultaneous confidence intervals on $\sigma^2_\alpha/\sigma^2_\epsilon$, $\sigma^2_{\beta(\alpha)}/\sigma^2_\epsilon$ with a confidence coefficient greater than or equal to $1 - \alpha$ are given by

$$0 < \frac{\sigma^2_\alpha}{\sigma^2_\epsilon} < Pr_1(\mathcal{K})$$

$$0 < \frac{\sigma^2_{\beta(\alpha)}}{\sigma^2_\epsilon} < Pr_2(\mathcal{K}).$$

Note that the procedure outlined in Section 8.7 can also be applied to obtain simultaneous confidence intervals on ratios of variance components (see Khuri, 1981, pp. 880–882). In particular, it can be used to derive confidence intervals on $\sigma^2_\alpha/\sigma^2_\epsilon$, $\sigma^2_{\beta(\alpha)}/\sigma^2_\epsilon$ in Example 8.8.

Exercises

8.1 Consider the following model

$$Y_{ijkl} = \mu + \alpha_{(i)} + \beta_{i(j)} + \gamma_{ij(k)} + \epsilon_{ijk(l)},$$

$i = 1,2,\ldots,a;\ j = 1,2,\ldots,b;\ k = 1,2,\ldots,c;\ l = 1,2,\ldots,n,$ where $\alpha_{(i)}$ is fixed and $\beta_{i(j)}$ and $\gamma_{ij(k)}$ are normally distributed as $N(0,\sigma^2_{\beta(\alpha)})$, $N(0,\sigma^2_{\gamma(\alpha\beta)})$, respectively. All random effects are mutually independent and independent of $\epsilon_{ijk(l)} \sim N(0,\sigma^2_\epsilon)$.

(a) Write down the corresponding population structure.

(b) Give the expected mean squares for all the effects in the corresponding ANOVA table.

(c) Let SS_A be the sum of squares associated with the fixed effect. What distribution does SS_A have?

(d) Let $\hat{\sigma}^2_{\beta(\alpha)}$ denote the ANOVA estimator of $\sigma^2_{\beta(\alpha)}$. Give an expression for computing the probability that $\hat{\sigma}^2_{\beta(\alpha)} < 0$.

(e) Set up a $(1 - \alpha)100\%$ confidence interval on $\mu_1 - \mu_2$, where μ_1 and μ_2 are the means of levels 1 and 2 of the fixed-effects factor (assume that $a \geq 2$).

8.2 Consider the population structure $\{[(i)(j)] : k\} : l$, where $i = 1, 2, 3$, $j = 1, 2, 3, 4$; $k = 1, 2, 3$; $l = 1, 2$. Let α, β, and γ denote the effects associated with subscripts i, j, and k, respectively. It is assumed that $\alpha_{(i)}$ is fixed, but all the remaining effects in the model are random. The usual assumptions made earlier in Section 8.4 regarding the random effects can be considered to be valid here.

(a) Obtain the complete ANOVA table, including the expected mean squares.

(b) Give test statistics for testing the significance of all the effects in the ANOVA table.

(c) Give an expression for the power of the test concerning the α-effect corresponding to a 5% level of significance (assume a particular alternative hypothesis).

8.3 Two drugs were compared in a multi-center study at 53 research centers. At each center, the two drugs were assigned to 14 subjects. One of the objectives of this study was to assess the side effects of the drugs. At tri-weekly intervals following treatment, subjects returned to the clinics for measurement of several critical signs, including sitting heart rate.

(a) Determine the appropriate population structure for this experiment.

(b) Give the complete model and set up the ANOVA table.

8.4 A manufacturer wants to investigate the variation of the quality of a product with regard to *type A* pre-production processes and *type B* pre-production processes. Factor A has 4 levels and each level has 5 sublevels. Factor B has 4 levels and each level has 6 sublevels. Each sublevel of each level of the A-factor is combined with each sublevel of each level of the B-factor. The same number of replications (3) is available for each sublevel combination.

(a) Write the complete model for this experiment.

(b) Give expressions for the expected mean squares assuming that the effects of A and B are fixed, while the remaining effects are random (the sublevels are chosen at random).

8.5 In an investigation of the can-making properties of tin plate, two methods of annealing were studied. Three coils were selected at random out of a supposedly very large population of coils made by each of these two methods. From each coil, samples were taken from two particular locations, namely, the head and tail of each coil. From each sample, two sets of cans were made up independently and from each set an estimate of the can life was obtained. The data are given in the following table.

		Annealing Method (i)					
		1			**2**		
		Coil (j)			**Coil (j)**		
Location (k)	**Replication (l)**	**1**	**2**	**3**	**4**	**5**	**6**
1	1	288	355	329	310	303	299
	2	295	369	343	282	321	328
2	1	278	336	320	288	302	289
	2	272	342	315	287	297	284

(a) Write down the complete model.

(b) Obtain the expected mean square values and the degrees of freedom for all the effects in the model.

(c) Compute all sums of squares in the ANOVA table and provide appropriate test statistics (assuming that the assumptions made with regard to the random effects, as outlined in Section 8.4, are valid here).

8.6 Consider again Exercise 8.5.

(a) Obtain a 95% confidence interval on $\mu_1 - \mu_2$, where μ_1 and μ_2 are the means of annealing methods 1 and 2, respectively.

(b) Let $\sigma^2_{\beta(\alpha)}$, $\sigma^2_{\delta\beta(\alpha)}$, σ^2_ϵ denote the variance components for coil(method), location*coil(method), and the error term, respectively. Obtain a 95% confidence interval on $\sigma^2_{\beta(\alpha)}/(2\,\sigma^2_{\delta\beta(\alpha)} + \sigma^2_\epsilon)$.

(c) If the hypothesis concerning the location effect is tested at the $\alpha = 0.05$ level, find the power of the F-test given that

$$\frac{\sum_{k=1}^{2} \delta^2_{(k)}}{2\,\sigma^2_{\delta\beta(\alpha)} + \sigma^2_\epsilon} = 0.10,$$

where $\delta_{(k)}$ denotes the effect of location k ($= 1, 2$).

(d) Let $\hat{\sigma}^2_{\beta(\alpha)}$ denote the ANOVA estimator of $\sigma^2_{\beta(\alpha)}$, that is,

$$\hat{\sigma}^2_{\beta(\alpha)} = \frac{1}{4}\,[MS_{coil(method)} - MS_{location*coil(method)}].$$

Find the probability that $\hat{\sigma}^2_{\beta(\alpha)} < 0$.

8.7 Consider the balanced one-way model,

$$Y_{ij} = \mu + \alpha_{(i)} + \epsilon_{i(j)}, \; i = 1, 2, \ldots, k; \; j = 1, 2, \ldots, n,$$

where $\alpha_{(i)}$ is a fixed unknown parameter and the $\epsilon_{i(j)}$'s are independently distributed as $N(0, \sigma^2_\epsilon)$. Let $SS_A = n \sum_{i=1}^{k}(Y_{(i)} - Y)^2$ be the sum of squares associated with $\alpha_{(i)}$ (for the treatment effect).

(a) Express SS_A as a quadratic form in $\bar{Y} = (Y_{(1)}, Y_{(2)}, \ldots, Y_{(k)})'$.

(b) Partition SS_A into $k - 1$ mutually independent sums of squares with one degree of freedom each.

(c) Deduce that the one-degree-of-freedom sums of squares in part (b) represent sums of squares of orthogonal contrasts among the true means of the treatment effect.

8.8 An experiment was conducted to monitor the chemical content of a large tank. Eight liquid samples (factor *A*) were randomly selected from the tank over time. A random sample of four operators (factor *B*) was used to measure the chemical content of each sample. Each operator measured the acid concentration of each sample $n = 2$ times [A description of this experiment was given in Burdick and Larsen (1997)]. The corresponding ANOVA table is shown below (the usual assumptions concerning the random effects are the same as in Section 8.4)

Source	**DF**	*SS*	*MS*
A	7	356.769	50.967
B	3	38.157	12.719
$A * B$	21	17.052	0.812
Error	32	13.280	0.415

(a) Test the significance of *A*, *B*, and their interaction $A * B$ at the $\alpha = 0.05$ level.

(b) Obtain a 95% confidence interval on $(\sigma^2_\epsilon + \sigma^2_{\alpha\beta})/\sigma^2_\epsilon$, where σ^2_ϵ and $\sigma^2_{\alpha\beta}$ are the variance components for the experimental error term and the interaction $A * B$, respectively.

(c) Obtain simultaneous confidence intervals on $\sigma^2_\alpha/\sigma^2_\epsilon$, $\sigma^2_\beta/\sigma^2_\epsilon$, and $\sigma^2_{\alpha\beta}/\sigma^2_\epsilon$, where σ^2_α and σ^2_β are the variance components for *A* and *B*, respectively, with a joint coverage probability greater than or equal to 0.95.

8.9 Consider the same one-way model as in Exercise 8.7, except that $\alpha_{(i)}$ is now considered to be randomly distributed as $N(0, \sigma^2_\alpha)$ independently of $\epsilon_{i(j)}$, which has the normal distribution $N(0, \sigma^2_\epsilon)$.

(a) Obtain individual confidence intervals on $\theta_1 = n\,\sigma^2_\alpha + \sigma^2_\epsilon$ and $\theta_2 = \sigma^2_\alpha/\sigma^2_\epsilon$.

(b) Use Bonferroni's inequality and the intervals in part (a) to obtain simultaneous confidence intervals on θ_1 and θ_2 with a joint coverage probability greater than or equal to 0.90.

(c) Use part (b) to obtain a confidence region on $(\sigma^2_\epsilon, \sigma^2_\alpha)$ with a confidence coefficient greater than or equal to 0.90.

(d) Use part (c) to obtain simultaneous confidence intervals on σ^2_ϵ, σ^2_α with a joint coverage probability greater than or equal to 0.90.

8.10 Consider the following ANOVA table for a balanced random two-fold nested model (the assumptions concerning the random effects are the same as in Section 8.4).

Source	DF	*MS*	*E(MS)*
A	11	3.5629	$15\,\sigma^2_\alpha + 3\,\sigma^2_{\beta(\alpha)} + \sigma^2_\epsilon$
$B(A)$	48	1.2055	$3\,\sigma^2_{\beta(\alpha)} + \sigma^2_\epsilon$
Error	120	0.6113	σ^2_ϵ

(a) Use the methodology described in Section 8.7.1 to obtain an exact confidence interval on σ^2_α with a confidence coefficient greater than or equal to 0.90.

(b) Use the methodology described in Section 8.8 to obtain simultaneous confidence intervals on $\sigma^2_\alpha/\sigma^2_\epsilon$, $\sigma^2_{\beta(\alpha)}/\sigma^2_\epsilon$ with a confidence coefficient greater than or equal to 0.90.

9

The Adequacy of Satterthwaite's Approximation

In the analysis of a balanced mixed model, it may not be possible to obtain an exact F-test concerning a certain hypothesis from the corresponding ANOVA table. This occurs when no single mean square exists that can serve as an "error term" in the denominator of the test's F-ratio. A common approach to this problem is to concoct a "synthetic error term" which consists of a linear combination of mean squares of random effects. It is also possible to construct an alternative test statistic by synthesizing both the numerator and the denominator of the F-ratio, that is, by creating two linear combinations of mean squares, one for the numerator and the other for the denominator. The choice of these linear combinations is based on requiring the numerator and denominator to have the same expected value under the null hypothesis to be tested. Under the alternative hypothesis, the expected value of the numerator exceeds that of the denominator by a positive constant. Each linear combination of mean squares of random effects is usually approximately represented as a scalar multiple of a chi-squared random variable whose number of degrees of freedom is estimated using the so-called *Satterthwaite's formula*. This yields an F- ratio which has an approximate F-distribution. The whole process leading up to this approximation is referred to as *Satterthwaite's approximation*.

In this chapter, we investigate the adequacy of the approximation of such linear combinations of mean squares with the chi-squared distribution. In addition, a measure will be developed to quantify the closeness of this approximation.

9.1 Satterthwaite's Approximation

Consider the balanced mixed model given in Section 8.4 [model (8.23)] under the same assumptions made earlier concerning the model's random effects. Let $MS_1, MS_2, \ldots, MS_k$ denote the mean squares for a set of random effects. For the sake of simplicity, the random effects have been renumbered so that $MS_1, MS_2, \ldots, MS_k$ form a subset of the entire set of random effects in the model. We recall that the MS_i's are mutually independent and that $m_i MS_i/\delta_i \sim \chi^2_{m_i}$, where m_i is the number of degrees of freedom for MS_i with $\delta_i = E(MS_i)$, $i = 1, 2, \ldots, k$ (see Theorem 8.1).

Let MS^* be defined as the linear combination,

$$MS^* = \sum_{i=1}^{k} a_i\, MS_i, \tag{9.1}$$

where $a_1, a_2, \ldots, a_k$ are known nonzero constants. Satterthwaite (1941,1946) suggested that MS^* can be distributed approximately as $(\delta^*/\nu)\,\chi^2_\nu$, where $\delta^* = E(MS^*)$ and ν is given by the formula

$$\nu = \frac{\left(\sum_{i=1}^{k} a_i\,\delta_i\right)^2}{\sum_{i=1}^{k}(a_i\,\delta_i)^2/m_i}. \tag{9.2}$$

This formula is derived in the following theorem.

Theorem 9.1 Let $MS_1, MS_2, \ldots, MS_k$ be mutually independent mean squares such that $m_i MS_i/\delta_i \sim \chi^2_{m_i}$, where $\delta_i = E(MS_i)$, $i = 1, 2, \ldots, k$. Let MS^* be defined as in (9.1). Then, $\nu\, MS^*/\delta^*$ has the approximate chi-squared distribution with ν degrees of freedom, where $\delta^* = E(MS^*)$ and ν is given by formula (9.2).

Proof. Let us represent MS^* approximately as a scalar multiple of a chi-squared random variable of the form

$$MS^* \approx a\,\chi^2_\nu, \tag{9.3}$$

where a and ν are constants to be determined so that the two sides in (9.3) have identical means and variances. Of course, this is feasible only if MS^* is positive. Note that

$$\delta^* = E(MS^*) = \sum_{i=1}^{k} a_i\,\delta_i, \tag{9.4}$$

and

$$\begin{aligned}
\text{Var}\left(\sum_{i=1}^{k} a_i\,MS_i\right) &= \text{Var}\left(\sum_{i=1}^{k} a_i\,\frac{\delta_i}{m_i}\,\chi^2_{m_i}\right)\\
&= \sum_{i=1}^{k}\left(\frac{a_i\,\delta_i}{m_i}\right)^2 \text{Var}\left(\chi^2_{m_i}\right)\\
&= \sum_{i=1}^{k}\left(\frac{a_i\,\delta_i}{m_i}\right)^2 (2\,m_i)\\
&= 2\sum_{i=1}^{k}\frac{a_i^2\,\delta_i^2}{m_i}.
\end{aligned} \tag{9.5}$$

Equating the means and variances of the two sides of formula (9.3), we obtain

$$\sum_{i=1}^{k} a_i\,\delta_i = a\,\nu, \tag{9.6}$$

$$2\sum_{i=1}^{k} \frac{a_i^2\,\delta_i^2}{m_i} = 2\,a^2\,\nu. \tag{9.7}$$

Solving (9.6) and (9.7) for a and ν, we get

$$a = \frac{\sum_{i=1}^{k} a_i^2\,\delta_i^2/m_i}{\sum_{i=1}^{k} a_i\,\delta_i}, \tag{9.8}$$

$$\nu = \frac{\left(\sum_{i=1}^{k} a_i\,\delta_i\right)^2}{\sum_{i=1}^{k} a_i^2\,\delta_i^2/m_i}. \tag{9.9}$$

From (9.3) and (9.6) we then have

$$MS^* \approx \frac{1}{\nu}\left(\sum_{i=1}^{k} a_i\,\delta_i\right)\chi^2_{\nu}, \tag{9.10}$$

that is,

$$\frac{\nu\,MS^*}{\sum_{i=1}^{k} a_i\,\delta_i} \;\; \underset{\sim}{approx.} \;\; \chi^2_{\nu}.$$

□

In practice, $\delta_1, \delta_2, \ldots, \delta_k$ are unknown and are usually estimated by their unbiased estimates, namely, $MS_1, MS_2, \ldots, MS_k$, respectively. Substituting these estimates in formula (9.9), we get

$$\hat{\nu} = \frac{\left(\sum_{i=1}^{k} a_i\,MS_i\right)^2}{\sum_{i=1}^{k} a_i^2\,MS_i^2/m_i}, \tag{9.11}$$

which serves as an estimate of ν. Formula (9.11) is known as *Satterthwaite's formula*. We then have

$$\frac{\hat{\nu}\,MS^*}{\delta^*} \;\; \underset{\sim}{approx.} \;\; \chi^2_{\hat{\nu}}. \tag{9.12}$$

Satterthwaite (1946) cautioned about using his formula when one or more of the a_i's in (9.1) are negative. In this case, MS^* can be negative with a possibly large probability, and the approximate distribution in (9.12) will therefore become rather poor.

Satterthwaite's approximation will be used to obtain an approximate F-statistic for testing a hypothesis concerning a fixed or random effect in a given balanced mixed model when no single mean square exists in the corresponding ANOVA table that can be used as an "error term." This approach was adopted by Cochran (1951).

9.1.1 A Special Case: The Behrens–Fisher Problem

The *Behrens–Fisher problem* concerns the comparison of two means from normal populations with unknown variances, which are assumed to be unequal. If $\bar{Y}_1$ and $\bar{Y}_2$ are two independent sample means based on n_1 and n_2 observations from normal populations, $N(\mu_1, \sigma_1^2)$, $N(\mu_2, \sigma_2^2)$, respectively, $\sigma_1^2 \neq \sigma_2^2$, then an approximate t-test for the equality of the population means, μ_1 and μ_2, is given by

$$t = \frac{\bar{Y}_1 - \bar{Y}_2}{\left(\frac{s_1^2}{n_1} + \frac{s_2^2}{n_2}\right)^{1/2}}, \tag{9.13}$$

where s_1^2 and s_2^2 are the corresponding sample variances. Note that the denominator in (9.13) is an estimate of the standard deviation of $\bar{Y}_1 - \bar{Y}_2$, namely, $\left(\frac{\sigma_1^2}{n_1} + \frac{\sigma_2^2}{n_2}\right)^{1/2}$ since $E(s_i^2) = \sigma_i^2$ $(i = 1, 2)$. Using Satterthwaite's formula (9.11), the number of degrees of freedom associated with $\frac{s_1^2}{n_1} + \frac{s_2^2}{n_2}$ is approximately given by

$$\hat{\nu} = \frac{\left(\frac{s_1^2}{n_1} + \frac{s_2^2}{n_2}\right)^2}{\frac{\left(\frac{s_1^2}{n_1}\right)^2}{n_1 - 1} + \frac{\left(\frac{s_2^2}{n_2}\right)^2}{n_2 - 1}}. \tag{9.14}$$

It can be shown (see Gaylor and Hopper, 1969, p. 693) that

$$\min(n_1 - 1, n_2 - 1) \leq \hat{\nu} \leq n_1 + n_2 - 2.$$

On the basis of formula (9.12) we can then write

$$\frac{\hat{\nu}}{\frac{\sigma_1^2}{n_1} + \frac{\sigma_2^2}{n_2}} \left(\frac{s_1^2}{n_1} + \frac{s_2^2}{n_2}\right) \underset{approx.}{\sim} \chi^2_{\hat{\nu}}. \tag{9.15}$$

From (9.13) and (9.15) we conclude that the test statistic t in (9.13) has, under the null hypothesis $H_0 : \mu_1 = \mu_2$, the approximate t-distribution with $\hat{\nu}$ degrees of freedom. Using this fact, an approximate $(1 - \alpha)$ 100% confidence interval on $\mu_1 - \mu_2$ is given by

$$\bar{Y}_1 - \bar{Y}_2 \pm \left(\frac{s_1^2}{n_1} + \frac{s_2^2}{n_2}\right)^{1/2} t_{\frac{\alpha}{2}, \hat{\nu}}.$$

Example 9.1 A consumer group was interested in examining consistency of prices of a variety of food items sold in large supermarkets. The study was conducted in a random sample of four standard metropolitan areas (factor A). Three supermarkets (factor B) were randomly selected in each of the four areas. Four food items (factor C) were randomly chosen for the study. The prices of these items (in dollars) were recorded for a random sample of 3 months. One record was obtained per month from each supermarket for each food item. The resulting data set is shown in Table 9.1, which is a modification of the original unbalanced data given in Khuri, Mathew, and Sinha (1998, p. 139). The population structure is $[(i:j)(k)]:l$, where i, j, k, l are subscripts associated with factors A, B, C, and the replications, respectively. The corresponding complete model is therefore of the form

$$Y_{ijkl} = \mu + \alpha_{(i)} + \beta_{i(j)} + \gamma_{(k)} + (\alpha\gamma)_{(ik)} + (\beta\gamma)_{i(jk)} + \epsilon_{ijk(l)}, \tag{9.16}$$

where

$\alpha_{(i)}$ is the effect of the ith metropolitan area ($i = 1, 2, 3, 4$)

$\beta_{i(j)}$ is the effect of the jth supermarket within the ith metropolitan area ($j = 1, 2, 3$)

$\gamma_{(k)}$ is the effect of the kth food item ($k = 1, 2, 3, 4$)

All the effects in model (9.16) are random and are assumed to be mutually independent and normally distributed with zero means and variances σ^2_α, $\sigma^2_{\beta(\alpha)}$, σ^2_γ, $\sigma^2_{\alpha\gamma}$, $\sigma^2_{\beta(\alpha)\gamma}$, σ^2_ϵ, respectively. The corresponding ANOVA table is given in Table 9.2. We note that the F-test statistics for testing $H_0: \sigma^2_{\beta(\alpha)} = 0$, $H_0: \sigma^2_\gamma = 0$, $H_0: \sigma^2_{\alpha\gamma} = 0$, and $H_0: \sigma^2_{\beta(\alpha)\gamma} = 0$ are $F = \frac{MS_{B(A)}}{MS_{C*B(A)}}$, $F = \frac{MS_C}{MS_{A*C}}$, $F = \frac{MS_{A*C}}{MS_{C*B(A)}}$, and $F = \frac{MS_{C*B(A)}}{MS_E}$, respectively. However, for testing the hypothesis, $H_0: \sigma^2_\alpha = 0$, no single mean square exists in the ANOVA table that can be used as an "error term." This follows from the fact that under H_0,

$$E(MS_A) = 12\,\sigma^2_{\beta(\alpha)} + 9\,\sigma^2_{\alpha\gamma} + 3\,\sigma^2_{\beta(\alpha)\gamma} + \sigma^2_\epsilon, \tag{9.17}$$

and no mean square (for any of the remaining random effects) has such an expected value. One approach to this problem is to consider the test statistic,

$$F_1 = \frac{MS_A}{MS_{B(A)} + MS_{A*C} - MS_{C*B(A)}}, \tag{9.18}$$

since both numerator and denominator have the same expected value under $H_0: \sigma^2_\alpha = 0$ and differ only by $36\,\sigma^2_\alpha$ when $\sigma^2_\alpha \neq 0$. The denominator in (9.18), being a linear combination of several mean squares, is referred to as a *synthetic* (artificial) "error term" for testing H_0. Its number of degrees of freedom can be approximately obtained by applying Satterthwaite's formula

TABLE 9.1
Prices of Food Items in Large Supermarkets

A	*B*	Food Item (*C*)			
Area	**Supermarket**	**1**	**2**	**3**	**4**
1	1	3.15	5.70	1.30	6.12
		3.15	5.68	1.29	6.14
		3.18	5.70	1.29	6.16
	2	3.28	5.75	1.27	6.18
		3.24	5.72	1.25	6.16
		3.26	5.71	1.26	6.15
	3	3.19	5.65	1.21	6.10
		3.18	5.61	1.21	6.11
		3.16	5.59	1.20	6.12
2	1	3.30	5.80	1.51	6.20
		3.28	5.82	1.51	6.20
		3.27	5.80	1.52	6.21
	2	3.25	5.82	1.49	6.24
		3.23	5.79	1.47	6.22
		3.21	5.78	1.45	6.20
	3	3.32	5.72	1.46	6.26
		3.30	5.74	1.45	6.23
		3.30	5.71	1.43	6.20
3	1	3.29	5.79	1.57	6.30
		3.28	5.79	1.56	6.28
		3.31	5.78	1.58	6.31
	2	3.35	5.81	1.50	6.29
		3.32	5.80	1.49	6.28
		3.31	5.80	1.49	6.27
	3	3.24	5.72	1.58	6.32
		3.26	5.69	1.55	6.32
		3.23	5.70	1.54	6.30
4	1	3.14	5.50	1.20	6.08
		3.14	5.49	1.22	6.08
		3.12	5.48	1.22	6.07
	2	3.18	5.55	1.18	6.06
		3.18	5.55	1.18	6.04
		3.17	5.53	1.17	6.02
	3	3.20	5.59	1.21	6.12
		3.18	5.56	1.22	6.11
		3.16	5.53	1.21	6.11

TABLE 9.2
ANOVA Table for Model (9.16)

Source	DF	*MS*	*E(MS)*	*F*	*p*-Value
A	3	0.4038	$36\,\sigma^2_{\alpha}+12\,\sigma^2_{\beta(\alpha)}+9\,\sigma^2_{\alpha\gamma}+3\,\sigma^2_{\beta(\alpha)\gamma}+\sigma^2_{\epsilon}$	13.15	0.00023
$B(A)$	8	0.0069	$12\,\sigma^2_{\beta(\alpha)}+3\,\sigma^2_{\beta(\alpha)\gamma}+\sigma^2_{\epsilon}$	1.88	0.1111
C	3	180.9015	$36\,\sigma^2_{\gamma}+9\,\sigma^2_{\alpha\gamma}+3\,\sigma^2_{\beta(\alpha)\gamma}+\sigma^2_{\epsilon}$	7505.04	<0.0001
$A*C$	9	0.0241	$9\,\sigma^2_{\alpha\gamma}+3\,\sigma^2_{\beta(\alpha)\gamma}+\sigma^2_{\epsilon}$	6.53	0.0001
$C*B(A)$	24	0.0037	$3\,\sigma^2_{\beta(\alpha)\gamma}+\sigma^2_{\epsilon}$	15.41	<0.0001
Error	96	0.0002	σ^2_{ϵ}		

(9.11), which, on the basis of Table 9.2, gives the value

$$\hat{\nu} = \frac{[MS_{B(A)} + MS_{A*C} - MS_{C*B(A)}]^2}{\frac{[MS_{B(A)}]^2}{8} + \frac{[MS_{A*C}]^2}{9} + \frac{[-MS_{C*B(A)}]^2}{24}}$$
$$= 10.513 \approx 11. \quad (9.19)$$

The test statistic F_1 in (9.18) has the value $F_1 = 14.77$. The corresponding p-value (based on 3 and $\hat{\nu}$ degrees of freedom) is 0.0004.

One disadvantage of using F_1 is that the linear combination of mean squares in the denominator contains a negative coefficient. It can therefore have a positive probability of being negative, which, of course, is undesirable (see Section 8.6.1). It may be recalled that Satterthwaite (1946) cautioned about the use of his formula under these circumstances.

An alternative approach to testing $H_0 : \sigma^2_{\alpha} = 0$ is to choose another F-ratio whose numerator and denominator can both be synthesized (that is, each consists of a positive linear combination of mean squares) and have equal expected values under H_0. For example, we can consider the test statistic,

$$F_2 = \frac{MS_A + MS_{C*B(A)}}{MS_{B(A)} + MS_{A*C}}. \quad (9.20)$$

Under H_0, the expected values of the numerator and denominator are equal to $12\,\sigma^2_{\beta(\alpha)} + 9\,\sigma^2_{\alpha\gamma} + 6\,\sigma^2_{\beta(\alpha)\gamma} + 2\,\sigma^2_{\epsilon}$, but, under $H_a : \sigma^2_{\alpha} \neq 0$, the expected value of the numerator exceeds that of the denominator by $36\,\sigma^2_{\alpha}$. In addition, F_2 avoids the problem of having negative coefficients in the linear combinations of mean squares. Note that the numerator and denominator of F_2 are independent, and on the basis of Theorem 9.1, F_2 is distributed approximately as $F_{\hat{\nu}_1,\hat{\nu}_2}$, where by formula (9.11),

$$\hat{\nu}_1 = \frac{[MS_A + MS_{C*B(A)}]^2}{\frac{[MS_A]^2}{3} + \frac{[MS_{C*B(A)}]^2}{24}}$$
$$= 3.055 \approx 3,$$
$$\hat{\nu}_2 = \frac{[MS_{B(A)} + MS_{A*C}]^2}{\frac{[MS_{B(A)}]^2}{8} + \frac{[MS_{A*C}]^2}{9}}$$
$$= 13.63 \approx 14.$$

The value of F_2 in (9.20) is $F_2 = 13.15$ and the corresponding p-value (with 3 and 14 degrees of freedom) is approximately equal to 0.00023. This value along with $F_2 = 13.15$ are the entries corresponding to A in Table 9.2.

Statistics similar to F_1 and F_2 were compared, using computer simulation and on the basis of the probability of Type I error and power values, by Hudson and Krutchkoff (1968) and Davenport and Webster (1973). They noted that the statistics do about equally well for testing $H_0 : \sigma^2_\alpha = 0$ if the degrees of freedom of the mean squares in F_1 and F_2 are not too small and the nuisance parameters (the variance components for the other random effects) are not all negligible. In the remaining cases neither statistic does well, but F_2 is better than F_1 in its approximation to the nominal level of significance and in terms of power.

Estimators other than the one in (9.11) for the degrees of freedom ν were suggested by Ames and Webster (1991). Myers and Howe (1971) adopted a different approach to the standard practice of assigning approximate degrees of freedom (based on Satterthwaite's formula) to the ratio of synthetic mean squares, where the numerator and denominator are treated separately as approximate chi-squared statistics (as in the case of F_2). Their approach was to approximate the distribution of the ratio directly as an F-statistic. Davenport (1975) compared the two approaches. His empirical studies indicated that the probability of Type I error using the the Myers–Howe procedure was greater than that using the standard approach based on Satterthwaite's approximation. Hence, the Myers–Howe procedure was not believed to be an overall improvement over the Satterthwaite procedure.

9.2 Adequacy of Satterthwaite's Approximation

In this section, we examine Satterthwaite's approximation of the distribution of a nonnegative linear combination of independent mean squares. A necessary and sufficient condition for this approximation to be exact will be presented for the case of a general balanced mixed model. The initial work leading up to the development of the methodology in this section was established in Khuri (1995a).

Consider the balanced mixed model (8.23) under the same assumptions concerning the distributions of the random effects as in Section 8.4. Let MS^* be a linear combination of mean squares as in (9.1), and it is here assumed that $a_i > 0$ for $i = 1, 2, \ldots, k$. Recall that

$$MS_i = \frac{1}{m_i} \boldsymbol{Y}'\boldsymbol{P}_i\boldsymbol{Y}, \tag{9.21}$$

where

$\boldsymbol{P}_i$ is the idempotent matrix described in Lemma 8.4
m_i is the corresponding number of degrees of freedom ($i = 1, 2, \ldots, k$)

Formula (9.1) can then be written as

$$MS^* = \boldsymbol{Y}'\boldsymbol{B}\boldsymbol{Y}, \tag{9.22}$$

where

$$\boldsymbol{B} = \sum_{i=1}^{k} \frac{a_i}{m_i} \boldsymbol{P}_i. \tag{9.23}$$

We note that $\boldsymbol{B}$ is positive semidefinite since $a_i > 0$ for $i = 1, 2, \ldots, k$. Thus, according to Theorem 9.1,

$$\frac{\nu MS^*}{\delta^*} \underset{approx.}{\sim} \chi^2_{\nu}, \tag{9.24}$$

where δ^* and ν are given by formulas (9.4) and (9.9), respectively. Using (9.22) in (9.24), we obtain

$$\frac{\nu MS^*}{\delta^*} = \boldsymbol{Y}'\boldsymbol{C}\boldsymbol{Y}, \tag{9.25}$$

where

$$\boldsymbol{C} = \frac{\nu}{\delta^*}\boldsymbol{B}.$$

Consider now the following theorem.

Theorem 9.2 A necessary and sufficient condition for $\frac{\nu MS^*}{\delta^*}$ in (9.25) to have a central chi-squared distribution is

$$\frac{a_i \delta_i}{m_i} = \frac{\delta^*}{\nu}, \quad i = 1, 2, \ldots, k, \tag{9.26}$$

where $\delta_i = E(MS_i)$, $i = 1, 2, \ldots, k$.

Proof. Following Theorem 5.4,

$$\frac{\nu MS^*}{\delta^*} = \boldsymbol{Y}'\boldsymbol{C}\boldsymbol{Y},$$

has the chi-squared distribution if and only if

$$\boldsymbol{C}\boldsymbol{\Sigma}\boldsymbol{C}\boldsymbol{\Sigma} = \boldsymbol{C}\boldsymbol{\Sigma}, \tag{9.27}$$

where $\boldsymbol{\Sigma}$ is the variance–covariance matrix of $\boldsymbol{Y}$ given by

$$\boldsymbol{\Sigma} = \sum_{i=\nu-p+1}^{\nu+1} \sigma_i^2 \boldsymbol{A}_i, \tag{9.28}$$

and $\boldsymbol{A}_i = \boldsymbol{H}_i\boldsymbol{H}_i'$ [see formula (8.25)]. Equality (9.27) is identical to

$$\boldsymbol{C}\boldsymbol{\Sigma}\boldsymbol{C} = \boldsymbol{C} \tag{9.29}$$

since $\boldsymbol{\Sigma}$ is nonsingular. Recall from Section 8.4 that

$$E(\boldsymbol{Y}) = \boldsymbol{X}\boldsymbol{g} = \sum_{i=0}^{\nu-p} \boldsymbol{H}_i\boldsymbol{\beta}_i, \tag{9.30}$$

and from Theorem 8.1(d), the noncentrality parameter for a random effect is equal to zero. Thus,

$$\boldsymbol{g}'\boldsymbol{X}'\boldsymbol{C}\boldsymbol{X}\boldsymbol{g} = 0, \tag{9.31}$$

which indicates that the noncentrality parameter for $\boldsymbol{Y}'\boldsymbol{C}\boldsymbol{Y}$ is also equal to zero. Condition (9.29) is then necessary and sufficient for $\boldsymbol{Y}'\boldsymbol{P}\boldsymbol{Y}$ to have the central chi-squared distribution. Furthermore, $\boldsymbol{C}\boldsymbol{\Sigma}\boldsymbol{C}$ in (9.29) can be expressed as

$$\begin{aligned}\boldsymbol{C}\boldsymbol{\Sigma}\boldsymbol{C} &= \left(\frac{\nu}{\delta^*}\sum_{i=1}^{k}\frac{a_i}{m_i}\boldsymbol{P}_i\right)\boldsymbol{\Sigma}\left(\frac{\nu}{\delta^*}\sum_{j=1}^{k}\frac{a_j}{m_j}\boldsymbol{P}_j\right)\\ &= \frac{\nu^2}{\delta^{*2}}\sum_{i=1}^{k}\frac{a_i^2}{m_i^2}\,\delta_i\,\boldsymbol{P}_i.\end{aligned} \tag{9.32}$$

This is true because

$$\begin{aligned}\boldsymbol{P}_i\boldsymbol{\Sigma}\boldsymbol{P}_j &= \boldsymbol{P}_i\left(\sum_{l=\nu-p+1}^{\nu+1}\sigma_l^2\boldsymbol{A}_l\right)\boldsymbol{P}_j\\ &= \left(\sum_{l=\nu-p+1}^{\nu+1}\kappa_{il}\,\sigma_l^2\,\boldsymbol{P}_i\right)\boldsymbol{P}_j, \quad \text{[see formula (8.20)]}\\ &= (\delta_i\,\boldsymbol{P}_i)\boldsymbol{P}_j, \quad \text{by formulas (8.21) and (8.28).}\end{aligned}$$

Hence,

$$\boldsymbol{P}_i \boldsymbol{\Sigma} \boldsymbol{P}_j = \begin{cases} \delta_i \boldsymbol{P}_i, & i = j \\ \boldsymbol{0}, & i \neq j \end{cases} \quad i, j = 1, 2, \ldots, k. \tag{9.33}$$

Now, if condition (9.26) is true, then

$$\begin{aligned} \boldsymbol{C}\boldsymbol{\Sigma}\boldsymbol{C} &= \frac{\nu}{\delta^*} \sum_{i=1}^{k} \frac{a_i}{m_i} \boldsymbol{P}_i \\ &= \boldsymbol{C}. \end{aligned}$$

Vice versa, if condition (9.29) is valid, then from (9.32),

$$\frac{\nu^2}{\delta^{*2}} \sum_{i=1}^{k} \frac{a_i^2}{m_i^2} \delta_i \boldsymbol{P}_i = \frac{\nu}{\delta^*} \sum_{i=1}^{k} \frac{a_i}{m_i} \boldsymbol{P}_i,$$

which implies that

$$\frac{\nu^2}{\delta^{*2}} \frac{a_i^2}{m_i^2} \delta_i = \frac{\nu}{\delta^*} \frac{a_i}{m_i}, \quad i = 1, 2, \ldots, k, \tag{9.34}$$

by the linear independence of $\boldsymbol{P}_1, \boldsymbol{P}_2, \ldots, \boldsymbol{P}_k$. Equality (9.34) gives rise to condition (9.26). □

Corollary 9.1 A necessary and sufficient condition for $\frac{\nu MS^*}{\delta^*}$ in (9.25) to have a central chi-squared distribution is

$$\frac{a_1 \delta_1}{m_1} = \frac{a_2 \delta_2}{m_2} = \ldots = \frac{a_k \delta_k}{m_k} \tag{9.35}$$

Proof. Obviously, (9.35) follows from (9.26). Let us now suppose that (9.35) is true. Let ρ denote the common value of $\frac{a_i \delta_i}{m_i}$ $(i = 1, 2, \ldots, k)$. Then,

$$\frac{a_i^2 \delta_i^2}{m_i^2} = \rho^2, \quad i = 1, 2, \ldots, k.$$

This can be expressed as

$$\begin{aligned} \frac{a_i^2 \delta_i^2}{m_i} &= \rho^2 m_i, \quad i = 1, 2, \ldots, k \\ &= \rho\, (a_i \delta_i), \quad i = 1, 2, \ldots, k. \end{aligned}$$

Thus,

$$\sum_{i=1}^{k} \frac{a_i^2 \delta_i^2}{m_i} = \rho \sum_{i=1}^{k} a_i \delta_i.$$

We conclude that

$$\rho = \frac{\sum_{i=1}^{k} \frac{a_i^2\,\delta_i^2}{m_i}}{\sum_{i=1}^{k} a_i\,\delta_i}$$
$$= \frac{\delta^*}{\nu}, \text{ by applying (9.4) and (9.9).}$$

Hence,

$$\frac{a_i\,\delta_i}{m_i} = \frac{\delta^*}{\nu}, \quad i = 1, 2, \ldots, k.$$

which establishes the validity of (9.26). □

9.2.1 Testing Departure from Condition (9.35)

Since $\delta_1, \delta_2, \ldots, \delta_k$ are unknown, it is not possible to determine if the necessary and sufficient condition in (9.35) is true or not. We can, however, use the data vector $\boldsymbol{Y}$ to determine if there is a significant departure from this condition. This amounts to finding a statistic for testing the hypothesis

$$H_0 : \frac{a_1\,\delta_1}{m_1} = \frac{a_2\,\delta_2}{m_2} = \cdots = \frac{a_k\,\delta_k}{m_k} \tag{9.36}$$

against the alternative hypothesis that at least one of the above equalities is not valid. For this purpose, we define z_i as

$$z_i = \frac{1}{\sqrt{m_i}}\boldsymbol{P}_i\boldsymbol{Y}, \quad i = 1, 2, \ldots, k.$$

It is easy to see that z_i has the singular normal distribution with a zero mean vector and a variance–covariance matrix given by

$$\mathrm{Var}(z_i) = \frac{\delta_i}{m_i}\boldsymbol{P}_i, \quad i = 1, 2, \ldots, k. \tag{9.37}$$

This is true because

$$E(z_i) = \frac{1}{\sqrt{m_i}}\boldsymbol{P}_i\boldsymbol{X}\boldsymbol{g}, \quad i = 1, 2, \ldots, k$$
$$= \frac{1}{\sqrt{m_i}}\boldsymbol{P}_i\sum_{j=0}^{\nu-p}\boldsymbol{H}_j\boldsymbol{\beta}_j$$
$$= \boldsymbol{0}, \quad i = 1, 2, \ldots, k,$$

since $\boldsymbol{P}_i\boldsymbol{H}_j = \boldsymbol{0}$ ($i = 1, 2, \ldots, k; j = 0, 1, \ldots, \nu - p$), as was seen earlier [see the proof of Theorem 8.1(d)]. Furthermore, using (9.33), we have

$$\mathrm{Var}(z_i) = \frac{1}{m_i}\boldsymbol{P}_i\boldsymbol{\Sigma}\boldsymbol{P}_i$$
$$= \frac{\delta_i}{m_i}\boldsymbol{P}_i, \quad i = 1, 2, \ldots, k.$$

Now, let ζ_i be defined as $\zeta_i = \boldsymbol{G}\boldsymbol{z}_i$ $(i = 1, 2, \ldots, k)$, where $\boldsymbol{G}$ is an orthogonal matrix that simultaneously diagonalizes $\boldsymbol{P}_0, \boldsymbol{P}_1, \ldots, \boldsymbol{P}_{\nu+1}$. Such a matrix exists by the fact that $\boldsymbol{P}_i\boldsymbol{P}_j = \boldsymbol{0}$, $i \neq j$, thus $\boldsymbol{P}_i\boldsymbol{P}_j = \boldsymbol{P}_j\boldsymbol{P}_i$ for $i, j = 1, 2, \ldots, k$ (see Theorem 3.9). The actual determination of the matrix $\boldsymbol{G}$ is shown in Appendix 9.A. Note that ζ_i has the singular normal distribution with a zero mean vector and a variance–covariance matrix of the form

$$\begin{aligned}\mathrm{Var}(\zeta_i) &= \boldsymbol{G}\,\frac{\delta_i}{m_i}\,\boldsymbol{P}_i\boldsymbol{G}' \\ &= \frac{\delta_i}{m_i}\,\boldsymbol{\Lambda}_i, \quad i = 1, 2, \ldots, k,\end{aligned} \tag{9.38}$$

where $\boldsymbol{\Lambda}_i$ is a diagonal matrix whose diagonal elements are the eigenvalues of $\boldsymbol{P}_i$, which is idempotent of *rank* m_i. Hence, $\boldsymbol{\Lambda}_i$ has m_i diagonal elements equal to 1 and the remaining elements are equal to 0. Furthermore, the ζ_i's are uncorrelated because for $i \neq j$,

$$\begin{aligned}\mathrm{Cov}(\zeta_i, \zeta_j) &= \boldsymbol{G}\,\mathrm{Cov}(\boldsymbol{z}_i, \boldsymbol{z}_j)\,\boldsymbol{G}' \\ &= \frac{1}{\sqrt{m_i m_j}}\,\boldsymbol{G}\boldsymbol{P}_i\boldsymbol{\Sigma}\boldsymbol{P}_j\boldsymbol{G}' \\ &= \boldsymbol{0}, \quad \text{by (9.33).}\end{aligned}$$

Let us now define ζ_i^* to be the vector consisting of the m_i elements of ζ_i that have variance $\frac{\delta_i}{m_i}$, $i = 1, 2, \ldots, k$ [see formula (9.38)]. Then, the vectors $\zeta_1^*, \zeta_2^*, \ldots, \zeta_k^*$ are normally distributed as $N(\boldsymbol{0}, \frac{\delta_i}{m_i}\boldsymbol{I}_{m_i})$, $i = 1, 2, \ldots, k$, and are also uncorrelated. Since these vectors form a partitioning of a linear transformation of $\boldsymbol{Y}$ that has the multivariate normal distribution, they must be mutually independent by Corollary 4.2. Let $\boldsymbol{v}_i = \sqrt{a_i}\,\zeta_i^*$ $(i = 1, 2, \ldots, k)$. Then, $\boldsymbol{v}_i$ has the normal distribution with a zero mean vector and a variance–covariance matrix $\frac{a_i\delta_i}{m_i}\boldsymbol{I}_{m_i}$, $i = 1, 2, \ldots, k$. It follows that the elements of $\boldsymbol{v}_1, \boldsymbol{v}_2, \ldots, \boldsymbol{v}_k$ form independent random samples of sizes $m_1, m_2, \ldots, m_k$, respectively, from k normally distributed populations with variances given by the values of $\frac{a_i\delta_i}{m_i}$ $(i = 1, 2, \ldots, k)$. The hypothesis H_0 in (9.36) is therefore equivalent to a hypothesis concerning homogeneity of variances of k normally distributed populations.

There are several procedures for testing homogeneity of population variances. Two such procedures are mentioned here

(a) Bartlett's test (Bartlett, 1937)

Bartlett's test statistic is given by (see, for example, Brownlee, 1965, Section 9.5)

$$T_1 = \frac{1}{d}\left[(m_. - k)\log\left(s^2\right) - \sum_{i=1}^{k}(m_i - 1)\log\left(s_i^2\right)\right], \tag{9.39}$$

where $m_. = \sum_{i=1}^{k} m_i$, s_i^2 is the sample variance for the elements in v_i $(i = 1, 2, \ldots, k)$, s^2 is the pooled sample variance,

$$s^2 = \frac{1}{m_. - k} \sum_{i=1}^{k} (m_i - 1)s_i^2,$$

and d is given by

$$d = 1 + \frac{1}{3(k-1)} \left[\left(\sum_{i=1}^{k} \frac{1}{m_i - 1} \right) - \frac{1}{m_. - k} \right].$$

Under H_0, T_1 has approximately the chi-squared distribution with $k - 1$ degrees of freedom. This hypothesis is rejected at the approximate α-level if $T_1 \geq \chi^2_{\alpha,k-1}$.

Note that in our particular application, the means of $v_1, v_2, \ldots, v_k$ are known since they are equal to zero. In this case, it would be more appropriate to test H_0 using the following statistic:

$$T_2 = -\sum_{i=1}^{k} m_i \log \left(\frac{\hat{\theta}_i^2}{\hat{\theta}^2} \right),$$

where $\hat{\theta}_i^2 = \frac{1}{m_i} v_i' v_i$, $i = 1, 2, \ldots, k$, $\hat{\theta}^2 = \frac{1}{m_.} \sum_{i=1}^{k} m_i \hat{\theta}_i^2$. This statistic is derived by applying the likelihood ratio procedure on which Bartlett's test is based. Under H_0, T_2 has approximately the chi-squared distribution with $k-1$ degrees of freedom. The exact null distribution of T_2 was obtained by Nagarsenker (1984).

Bartlett's test is adequate when the underlying distribution of the data is normal. However, this test can be quite sensitive to nonnormality of the data. In this case, Levene's (1960) test is preferred.

(b) Levene's (1960) test

This test is a widely used homogeneity of variance test because it is much more robust to nonnormality. Levene (1960) suggested using the one-way analysis of variance on a transformation on the elements of $v_1, v_2, \ldots, v_k$. Several transformations were proposed by him, including $\mid v_{ij} - \bar{v}_{i.} \mid$, $(v_{ij} - \bar{v}_{i.})^2$, $\log \mid v_{ij} - \bar{v}_{i.} \mid$, where v_{ij} is the jth element of v_i and $\bar{v}_{i.} = \frac{1}{m_i} \sum_{j=1}^{m_i} v_{ij}$ $(i = 1, 2, \ldots, k;\ j = 1, 2, \ldots, m_i)$. Another variation suggested by Brown and Forsythe (1974b) is to replace $\bar{v}_{i.}$ in $\mid v_{ij} - \bar{v}_{i.} \mid$ with the median of the ith group. (For more details, see Conover, Johnson and Johnson, 1981, p. 355.) For example, using the square transformation, we get the test statistic,

$$L = \frac{\sum_{i=1}^{k} m_i (\bar{u}_{i.} - \bar{u}_{..})^2/(k-1)}{\sum_{i=1}^{k} \sum_{j=1}^{m_i} (u_{ij} - \bar{u}_{i.})^2/(m_. - k)},$$

where $u_{ij} = (v_{ij} - \bar{v}_{i.})^2$, $\bar{u}_{i.} = \frac{1}{m_i}\sum_{j=1}^{m_i} u_{ij}$, $\bar{u}_{..} = \frac{1}{m_.}\sum_{i=1}^{k} m_i\,\bar{u}_{i.}$. Under H_0, L has approximately the F-distribution with $k-1$ and $m_. - k$ degrees of freedom. The test is significant at the approximate α-level if $L \geq F_{\alpha,k-1,m_.-k}$.

Levene's test can be easily implemented in SAS (2000) by using the statement "MEANS GROUP/HOVTEST = LEVENE" in PROC GLM, where "GROUP" represents a k-level factor in a one-way model. The associated design is completely randomized consisting of k samples where the elements in the ith sample make up the vector v_i $(i = 1, 2, \ldots, k)$. It should be noted that in SAS, groups with fewer than three observations are dropped from Levene's test. This occurs whenever $m_i = 2$ for some i.

The rejection of the null hypothesis H_0 at a small level of significance gives an indication of a possibly inadequate Satterthwaite's approximation.

Example 9.2 Consider the balanced random two-way model,

$$Y_{ijk} = \mu + \alpha_{(i)} + \beta_{(j)} + (\alpha\beta)_{(ij)} + \epsilon_{ij(k)}, \quad i = 1,2,3; \quad j = 1,2,3,4; \quad k = 1,2,3, \tag{9.40}$$

where $\alpha_{(i)} \sim N(0, \sigma^2_\alpha)$, $\beta_{(j)} \sim N(0, \sigma^2_\beta)$, $(\alpha\beta)_{(ij)} \sim N(0, \sigma^2_{\alpha\beta})$, and $\epsilon_{ij(k)} \sim N(0, \sigma^2_\epsilon)$; all random effects are independent. The corresponding ANOVA table is shown in Table 9.3.

Let ϕ denote the total variation, that is, $\phi = \sigma^2_\alpha + \sigma^2_\beta + \sigma^2_{\alpha\beta} + \sigma^2_\epsilon$, which can be expressed in terms of the δ_i's $(i = 1, 2, 3, 4)$ as

$$\phi = \frac{1}{12}\,\delta_1 + \frac{1}{9}\,\delta_2 + \frac{5}{36}\,\delta_3 + \frac{2}{3}\,\delta_4.$$

An unbiased estimate of ϕ is given by

$$\hat{\phi} = \frac{1}{12}\,MS_A + \frac{1}{9}\,MS_B + \frac{5}{36}\,MS_{AB} + \frac{2}{3}\,MS_E. \tag{9.41}$$

The null hypothesis H_0 in (9.36) has the form,

$$H_0 : \frac{\delta_1}{24} = \frac{\delta_2}{27} = \frac{5\,\delta_3}{216} = \frac{\delta_4}{36}. \tag{9.42}$$

TABLE 9.3
ANOVA Table for Model (9.40)

Source	DF	SS	MS	E(MS)
A	$m_1 = 2$	SS_A	MS_A	$\delta_1 = 12\,\sigma^2_\alpha + 3\,\sigma^2_{\alpha\beta} + \sigma^2_\epsilon$
B	$m_2 = 3$	SS_B	MS_B	$\delta_2 = 9\,\sigma^2_\beta + 3\,\sigma^2_{\alpha\beta} + \sigma^2_\epsilon$
$A * B$	$m_3 = 6$	SS_{AB}	MS_{A*B}	$\delta_3 = 3\,\sigma^2_{\alpha\beta} + \sigma^2_\epsilon$
Error	$m_4 = 24$	SS_E	MS_E	$\delta_4 = \sigma^2_\epsilon$

Since $\frac{SS_A}{\delta_1} \sim \chi^2_2$, $\frac{SS_B}{\delta_2} \sim \chi^2_3$, $\frac{SS_{AB}}{\delta_3} \sim \chi^2_6$, $\frac{SS_E}{\delta_4} \sim \chi^2_{24}$, $\hat{\phi}$ in (9.41) can be expressed as a linear combination of mutually independent central chi-squared variates of the form

$$\hat{\phi} = \frac{\delta_1}{24}\chi^2_2 + \frac{\delta_2}{27}\chi^2_3 + \frac{5\,\delta_3}{216}\chi^2_6 + \frac{\delta_4}{36}\chi^2_{24}. \tag{9.43}$$

By Theorem 9.1, the approximate distribution of $\frac{\nu\hat{\phi}}{\phi}$ is χ^2_ν, that is, $\hat{\phi} \underset{approx.}{\sim} \frac{\phi}{\nu}\chi^2_\nu$, where ν is given by [see formula (9.2)]

$$\nu = \frac{\left(\frac{1}{12}\delta_1 + \frac{1}{9}\delta_2 + \frac{5}{36}\delta_3 + \frac{2}{3}\delta_4\right)^2}{\frac{\left(\frac{1}{12}\delta_1\right)^2}{2} + \frac{\left(\frac{1}{9}\delta_2\right)^2}{3} + \frac{\left(\frac{5}{36}\delta_3\right)^2}{6} + \frac{\left(\frac{2}{3}\delta_4\right)^2}{24}}. \tag{9.44}$$

It would be of interest here to contrast the result of the test concerning H_0 in (9.42) with deviations of the quantiles of the exact distribution of $\hat{\phi}$ in (9.43) from those of $\frac{\phi}{\nu}\chi^2_\nu$. For this purpose, certain values are selected for the variance components and μ in model (9.40). For example, the following values are selected: $\sigma^2_\alpha = 30$, $\sigma^2_\beta = 20$, $\sigma^2_{\alpha\beta} = 10$, $\sigma^2_\epsilon = 1$, and $\mu = 20.5$. A random vector $\boldsymbol{Y}$ of order 36×1 can then be generated from $N(\mu\mathbf{1}_{36}, \boldsymbol{\Sigma})$, where

$$\boldsymbol{\Sigma} = \sigma^2_\alpha(\boldsymbol{I}_3 \otimes \boldsymbol{J}_4 \otimes \boldsymbol{J}_3) + \sigma^2_\beta(\boldsymbol{J}_3 \otimes \boldsymbol{I}_4 \otimes \boldsymbol{J}_3) + \sigma^2_{\alpha\beta}(\boldsymbol{I}_3 \otimes \boldsymbol{I}_4 \otimes \boldsymbol{J}_3) + \sigma^2_\epsilon(\boldsymbol{I}_3 \otimes \boldsymbol{I}_4 \otimes \boldsymbol{I}_3).$$

The elements of the generated response vector, $\boldsymbol{Y}$, are given in Table 9.4. Using the data in Table 9.4, the corresponding value of Bartlett's test statistic is $T_1 = 43.952$ with 3 degrees of freedom and an approximate p-value less than 0.0001 (the value of the test statistic T_2 is 91.158). In addition, the value of Levene's test statistic is $L = 14.98$ with an approximate p-value less than 0.0001. Thus all tests concerning H_0 in (9.42) are highly significant. This gives a

TABLE 9.4
Generated Data for Model (9.40)

	B			
A	1	2	3	4
1	11.729	19.369	6.107	8.190
	11.910	19.356	4.086	9.835
	14.460	20.772	5.451	9.547
2	30.374	24.983	18.423	19.366
	28.007	27.573	19.196	18.148
	26.612	27.205	19.236	17.840
3	14.874	7.360	4.560	15.947
	13.964	9.163	2.610	15.829
	17.203	6.879	3.827	16.222

TABLE 9.5
Exact and Approximate Quantiles of $\hat{\phi}$

p	Exact pth Quantile	Approximate pth Quantile
0.0040	8.0	4.892
0.0093	10.0	6.968
0.0356	15.0	12.436
0.0791	20.0	18.011
0.2000	30.0	29.100
0.3371	40.0	39.996
0.4046	45.0	45.366
0.5290	55.0	55.951
0.6800	70.0	71.464
0.7566	80.0	81.579
0.8411	95.0	96.436
0.8975	110.0	110.951
0.9344	125.0	125.158
0.9692	150.0	148.247
0.9805	165.0	161.793
0.9876	180.0	175.142
0.9933	200.0	192.674

strong indication of an inadequate Satterthwaite's approximation concerning the distribution of $\hat{\phi}$ in (9.41).

Let us now compare some quantiles of the exact distribution of $\hat{\phi}$ in (9.43) with those of the approximate $\frac{\phi}{\nu}\chi^2_\nu$ distribution, where ν is given by (9.44). Using the values of the variance components mentioned earlier, we get $\nu = 5.18853$ and $\phi = 61$. Note that the quantiles of the exact distribution of $\hat{\phi}$ can be obtained by using the representation in (9.43) and then applying Davies' (1980) algorithm, which was mentioned in Section 5.6. Table 9.5 gives some exact and approximate quantiles of $\hat{\phi}$ corresponding to several probability values.

We note that there is a pronounced difference between the exact and approximate quantiles for small values of p (< 0.04) as well as large values (> 0.98), that is, in the the lower and upper tail areas of the distribution of $\hat{\phi}$ [$p = P(\hat{\phi} < q_p)$, where q_p is the pth quantile of $\hat{\phi}$].

9.3 Measuring the Closeness of Satterthwaite's Approximation

In the previous section, the adequacy of Satterthwaite's approximation was formulated as a test of hypothesis using the data vector $\boldsymbol{Y}$. In the present section, a measure is provided to quantify the closeness of this approximation. The measure was initially developed by Khuri (1995b). As before, Satterthwaite's approximation is considered in conjunction with the distribution of MS^*, the linear combination of mean squares in (9.1), where $a_i > 0$ for $i = 1, 2, \ldots, k$. Furthermore, the mean squares are associated with the balanced mixed model (8.23) under the same distributional assumptions concerning the random effects as was described earlier in Section 8.4.

As will be seen later in this section, the purpose of the aforementioned measure is to provide an index that can be effectively used in theoretical investigations to determine the experimental conditions under which Satterthwaite's approximation will be inadequate. This leads to a better understanding of the causes that contribute to such inadequacy. Thus, through this measure, it will be possible to select a design that can enhance the adequacy of the approximation before collecting any data on the response, Y.

The following theorem is instrumental in the development of this measure.

Theorem 9.3 Let MS^* be expressed as in (9.22), that is, $MS^* = \boldsymbol{Y}'\boldsymbol{B}\boldsymbol{Y}$, where $\boldsymbol{Y} \sim N(\boldsymbol{X}\boldsymbol{g}, \boldsymbol{\Sigma})$; $\boldsymbol{X}\boldsymbol{g}$ is the mean of $\boldsymbol{Y}$ according to model (8.23) and $\boldsymbol{\Sigma}$ is the variance–covariance matrix of $\boldsymbol{Y}$ given by formula (9.28). If r is the rank of $\boldsymbol{B}$, then

(a)
$$\frac{r\sum_{i=1}^{r}\tau_i^2}{\left(\sum_{i=1}^{r}\tau_i\right)^2} \geq 1,$$
where $\tau_1, \tau_2, \ldots, \tau_r$ are the nonzero eigenvalues of $\boldsymbol{B}\boldsymbol{\Sigma}$.

(b) $\boldsymbol{Y}'\boldsymbol{B}\boldsymbol{Y}$ is distributed as a scaled chi-squared variate if and only if
$$\frac{r\sum_{i=1}^{r}\tau_i^2}{\left(\sum_{i=1}^{r}\tau_i\right)^2} = 1. \tag{9.45}$$

Proof.

(a) This is obvious by the *Cauchy–Schwarz inequality*,
$$\left(\sum_{i=1}^{r}\tau_i\right)^2 \leq r\sum_{i=1}^{r}\tau_i^2. \tag{9.46}$$

(b) If $\boldsymbol{Y}'\boldsymbol{B}\boldsymbol{Y}$ is distributed as a scaled chi-squared variate, then $\boldsymbol{B}\boldsymbol{\Sigma}$ must be a scalar multiple of an idempotent matrix by Theorem 5.4. Hence, $\tau_1, \tau_2, \ldots, \tau_r$ must be equal which implies (9.45). Vice versa, suppose now that condition (9.45) is valid. In general, it is known that equality in the Cauchy–Schwarz inequality, $\left(\sum_{i=1}^{n} a_i b_i\right)^2 \leq \sum_{i=1}^{n} a_i^2 \sum_{i=1}^{n} b_i^2$, holds if and only if $a_i = k\, b_i$, $i = 1, 2, \ldots, n$, where k is a constant. Applying this fact to the inequality in (9.46), we conclude that equality holds if and only if $\tau_1 = \tau_2 = \ldots = \tau_r$ (in this case, $a_i = 1, b_i = \tau_i, i = 1, 2, \ldots, r$). Thus, if condition (9.45) is true, then $\boldsymbol{B}\boldsymbol{\Sigma}$ must be a scalar multiple of an idempotent matrix, which indicates that $\boldsymbol{Y}'\boldsymbol{B}\boldsymbol{Y}$ is distributed as a scalar multiple of a chi-squared variate. □

From Theorem 9.3 we conclude that Satterthwaite's approximation is exact if and only if condition (9.45) is true. Thus, the size of the ratio, $r\left(\sum_{i=1}^{r}\tau_i^2\right)/\left(\sum_{i=1}^{r}\tau_i\right)^2$, as compared to 1, is the determining factor in evaluating the closeness of Satterthwaite's approximation. Let us therefore define the function $\lambda(\boldsymbol{\tau})$ as

$$\lambda(\boldsymbol{\tau}) = \frac{r\sum_{i=1}^{r}\tau_i^2}{\left(\sum_{i=1}^{r}\tau_i\right)^2}, \tag{9.47}$$

where $\boldsymbol{\tau} = (\tau_1, \tau_2, \ldots, \tau_r)'$. Then, by Theorem 9.3, $\lambda(\boldsymbol{\tau}) \geq 1$ and equality is achieved if and only if the τ_i's are equal, that is, if and only if Satterthwaite's approximation is exact. Note that $\lambda(\boldsymbol{\tau}) \leq r$. This is true by the fact that the τ_i's are positive since $\boldsymbol{B}$ is positive semidefinite and hence the nonzero eigenvalues of $\boldsymbol{B\Sigma}$ must be positive (they are the same as the nonzero eigenvalues of $\boldsymbol{\Sigma}^{1/2}\boldsymbol{B}\boldsymbol{\Sigma}^{1/2}$, which is positive semidefinite). The function $\lambda(\boldsymbol{\tau})$ will therefore be utilized to develop a measure for the adequacy of this approximation.

Without any loss of generality, we consider that τ_1 and τ_r are the largest and the smallest of the τ_i's, respectively. The function $\lambda(\boldsymbol{\tau})$ in (9.47) can then be expressed as

$$\lambda(\boldsymbol{\tau}) = \frac{r\left(1 + \kappa^2 + \sum_{i=1}^{r-2}\kappa_i^2\right)}{\left(1 + \kappa + \sum_{i=1}^{r-2}\kappa_i\right)^2}, \tag{9.48}$$

where $\kappa = \frac{\tau_r}{\tau_1}$ and $\kappa_i = \frac{\tau_{i+1}}{\tau_1}$, $i = 1, 2, \ldots, r-2$. Note that $0 < \kappa \leq \kappa_i \leq 1$, $i = 1, 2, \ldots, r-2$. For a fixed κ, $\lambda(\boldsymbol{\tau})$ is defined over the rectangular region,

$$\mathcal{R}_\kappa = \{(\kappa_1, \kappa_2, \ldots, \kappa_{r-2}) \,|\, 0 < \kappa \leq \kappa_i \leq 1,\ i = 1, 2, \ldots, r-2\}, \tag{9.49}$$

which is a closed and bounded subset of the $(r-2)$-dimensional Euclidean space. Since $\lambda(\boldsymbol{\tau})$ is a continuous function of $\kappa_1, \kappa_2, \ldots, \kappa_{r-2}$ over $\mathcal{R}_\kappa$, it must attain its minimum and maximum values over $\mathcal{R}_\kappa$ at some points in $\mathcal{R}_\kappa$. Let us denote the maximum of $\lambda(\boldsymbol{\tau})$ by $\lambda_{\max}(\kappa)$. Thus, $1 \leq \lambda_{\max}(\kappa) \leq r$. It is easy to see that $\lambda_{\max}(\kappa)$ is a monotone decreasing function of κ over the interval $0 < \kappa \leq 1$. Consequently, the supremum of $\lambda_{\max}(\kappa)$ over $0 < \kappa \leq 1$, denoted by λ_{sup}, is the limit of $\lambda_{\max}(\kappa)$ as $\kappa \to 0$, that is,

$$\lambda_{sup} = \lim_{\kappa \to 0}\lambda_{\max}(\kappa). \tag{9.50}$$

Note that $1 \leq \lambda_{sup} \leq r$.

Lemma 9.1 $\lambda_{sup} = 1$ if and only if Satterthwaite's approximation is exact.

Proof. If Satterthwaite's approximation is exact, then $\lambda(\boldsymbol{\tau}) = 1$, as was seen earlier. Hence, $\lambda_{sup} = 1$. Vice versa, if $\lambda_{sup} = 1$, then by the fact that $1 \leq \lambda(\boldsymbol{\tau})$, we have $1 \leq \lambda(\boldsymbol{\tau}) \leq \lambda_{\max}(\kappa) \leq \lambda_{sup} = 1$. Thus, $\lambda(\boldsymbol{\tau}) = 1$, which implies that Satterthwaite's approximation is exact. □

On the basis of the above arguments, the value of λ_{sup} can be used as a measure of closeness of Satterthwaite's approximation. Since $1 \le \lambda_{sup} \le r$, a value of λ_{sup} close to its lower bound, 1, indicates a close approximation.

9.3.1 Determination of λ_{sup}

Let us first obtain an expression for $\lambda_{\max}(\kappa)$, the maximum of $\lambda(\boldsymbol{\tau})$ in (9.48) over the region $\mathcal{R}_\kappa$ in (9.49) for a fixed κ. The development of this expression is based on the following two theorems established by Thibaudeau and Styan (1985, Theorems 2.2 and 2.3) (see also Khuri, 1995b, Theorems 3.1 and 3.2).

Theorem 9.4 For a fixed κ, the maximum of the function $\lambda(\boldsymbol{\tau})$ in (9.48) over the region $\mathcal{R}_\kappa$ in (9.49) always occurs at one or more of the extreme points of $\mathcal{R}_\kappa$.

Theorem 9.5 For a fixed κ, the function $\lambda(\boldsymbol{\tau})$, restricted to the region $\mathcal{R}_\kappa$, attains its maximum if and only if $u-1$ of the κ_i's in (9.48), $i = 1, 2, \ldots, r-2$, are equal to κ and the remaining $r-u-1$ of the κ_i's are equal to 1, where u denotes the value of $t = 1, 2, \ldots, r-1$ that minimizes the function $f(t)$ defined by

$$f(t) = \frac{r - t + \kappa t}{(r - t + \kappa^2 t)^{1/2}}, \quad t = 1, 2, \ldots, r-1. \tag{9.51}$$

In this case, the maximum of $\lambda(\boldsymbol{\tau})$ over the region $\mathcal{R}_\kappa$ is given by

$$\lambda_{\max}(\kappa) = \frac{r\,(r - u + \kappa^2 u)}{(r - u + \kappa u)^2}. \tag{9.52}$$

The value of u depends on the values of κ and r as shown in Table 9.6. Note that in this table, a_h is defined as

$$a_h = \frac{h(1+h)}{(r-h)(r-1-h)}, \quad h = 1, 2, \ldots, \left[\frac{r-1}{2}\right], \tag{9.53}$$

where $[\frac{r-1}{2}]$ denotes the greatest integer less than or equal to $\frac{r-1}{2}$. Note also that $0 < a_{h-1} < a_h \le 1; h = 2, 3, \ldots, [\frac{r-1}{2}]$, and $a_{[\frac{r-1}{2}]} = 1$ if and only if r is odd.

Having established $\lambda_{\max}(\kappa)$ as in (9.52), λ_{sup} is obtained by taking the limit of $\lambda_{\max}(\kappa)$ as κ tends to zero.

9.4 Examples

Three examples are presented in this section to demonstrate the utility of the measure described in the previous section.

TABLE 9.6
Values of u That Minimize $f(t)$ in (9.51)

K	u
$\left(0, a_1^{1/2}\right]$	$r-1$
$\left(a_1^{1/2}, a_2^{1/2}\right]$	$r-2$
$\left(a_2^{1/2}, a_3^{1/2}\right]$	$r-3$
$\vdots$	$\vdots$
$\left(a^{1/2}_{\left[\frac{1}{2}(r-1)-1\right]}, a^{1/2}_{\left[\frac{1}{2}(r-1)\right]}\right]$	$\left[\frac{r}{2}\right]+1$
$\left(a^{1/2}_{\left[\frac{1}{2}(r-1)\right]}, 1\right]$	$\left[\frac{r}{2}\right]=\frac{r}{2}$

$a_h = [h(1+h)]/[(r-h)(r-1-h)], h = 1, 2, \ldots, \left[\frac{r-1}{2}\right]$.
The last line holds only when r is even; when r is odd then $a_{\left[\frac{1}{2}(r-1)\right]} = 1$. This is an adaptation of Thibaudeau and Styan (1985, Table 1).

9.4.1 The Behrens–Fisher Problem

This problem was mentioned in Section 9.1.1. It concerns testing the null hypothesis $H_0 : \mu_1 = \mu_2$ against the alternative hypothesis $H_a : \mu_1 \neq \mu_2$, where, if we recall, μ_1 and μ_2 are the means of two normal populations whose variances, σ_1^2 and σ_2^2, are unknown, but are assumed to be unequal. The corresponding test statistic is given in (9.13), which, under H_0, has the approximate t-distribution with ν degrees of freedom, where ν is given by

$$\nu = \frac{\left(\frac{\sigma_1^2}{n_1} + \frac{\sigma_2^2}{n_2}\right)^2}{\frac{\left(\frac{\sigma_1^2}{n_1}\right)^2}{n_1-1} + \frac{\left(\frac{\sigma_2^2}{n_2}\right)^2}{n_2-1}},$$

which is a special case of the general formula given in (9.9). An estimate of ν is $\hat{\nu}$ which is given in (9.14).

In this example, $MS^* = \frac{s_1^2}{n_1} + \frac{s_2^2}{n_2}$. Hence, the vector $\boldsymbol{Y}$ and the matrix $\boldsymbol{B}$ in formula (9.22) are $\boldsymbol{Y} = (\boldsymbol{Y}_1' : \boldsymbol{Y}_2')'$ and $\boldsymbol{B} = \text{diag}\left(\frac{1}{n_1}\boldsymbol{B}_1, \frac{1}{n_2}\boldsymbol{B}_2\right)$, where $\boldsymbol{Y}_i$ is the vector of n_i observations in the ith sample and $\boldsymbol{B}_i = \left(\boldsymbol{I}_{n_i} - \frac{1}{n_i}\boldsymbol{J}_{n_i}\right)/(n_i - 1)$, $i = 1, 2$. Thus, the variance–covariance matrix of $\boldsymbol{Y}$ is given by

$$\boldsymbol{\Sigma} = \text{diag}\left(\sigma_1^2\boldsymbol{I}_{n_1}, \sigma_2^2\boldsymbol{I}_{n_2}\right). \tag{9.54}$$

Hence, the matrix $\boldsymbol{B\Sigma}$ is of the form

$$\boldsymbol{B\Sigma} = \text{diag}\left(\frac{1}{n_1}\,\sigma_1^2\,\boldsymbol{B}_1,\ \frac{1}{n_2}\,\sigma_2^2\,\boldsymbol{B}_2\right). \tag{9.55}$$

Note that the rank of $\boldsymbol{B\Sigma}$, which is the same as the rank of $\boldsymbol{B}$, is equal to $r = n_1+n_2-2$. The nonzero eigenvalues of $\boldsymbol{B\Sigma}$ are $\sigma_i^2/[n_i(n_i-1)]$ of multiplicity $n_i - 1$ since $(n_i - 1)\boldsymbol{B}_i$ is idempotent of *rank* $n_i - 1$ $(i = 1, 2)$. The largest and the smallest of these eigenvalues are

$$\tau_1 = \max\left[\frac{\sigma_1^2}{n_1(n_1-1)},\ \frac{\sigma_2^2}{n_2(n_2-1)}\right]$$

$$\tau_r = \min\left[\frac{\sigma_1^2}{n_1(n_1-1)},\ \frac{\sigma_2^2}{n_2(n_2-1)}\right].$$

Hence, $\kappa = \frac{\tau_r}{\tau_1}$ can be written as

$$\kappa = \begin{cases} \dfrac{n_2(n_2-1)}{n_1(n_1-1)}\,\dfrac{\sigma_1^2}{\sigma_2^2}, & \text{if } \dfrac{\sigma_1^2}{n_1(n_1-1)} < \dfrac{\sigma_2^2}{n_2(n_2-1)} \\[2ex] \dfrac{n_1(n_1-1)}{n_2(n_2-1)}\,\dfrac{\sigma_2^2}{\sigma_1^2}, & \text{if } \dfrac{\sigma_2^2}{n_2(n_2-1)} < \dfrac{\sigma_1^2}{n_1(n_1-1)}. \end{cases}$$

It is easy to show that $\lambda(\boldsymbol{\tau})$ in (9.47) can be expressed as

$$\lambda(\boldsymbol{\tau}) = \begin{cases} \dfrac{(n_1+n_2-2)\,[n_2-1\,+\,(n_1-1)\,\kappa^2]}{[n_2-1\,+\,(n_1-1)\,\kappa]^2}, & \text{if } \dfrac{\sigma_1^2}{n_1(n_1-1)} < \dfrac{\sigma_2^2}{n_2(n_2-1)} \\[2ex] \dfrac{(n_1+n_2-2)\,[n_1-1\,+\,(n_2-1)\,\kappa^2]}{[n_1-1\,+\,(n_2-1)\,\kappa]^2}, & \text{if } \dfrac{\sigma_2^2}{n_2(n_2-1)} < \dfrac{\sigma_1^2}{n_1(n_1-1)}. \end{cases}$$

In this case, $\lambda_{\max}(\kappa) = \lambda(\boldsymbol{\tau})$. We note that $\lambda(\boldsymbol{\tau})$ is a monotone decreasing function of κ over the interval $0 < \kappa \leq 1$. Since small values of $\lambda_{\max}(\kappa)$ are needed for an adequate approximation, a large value of κ (close to 1) will therefore be desirable. Thus, the adequacy of the approximation depends on how close $\sigma_1^2/[n_1(n_1-1)]$ is to $\sigma_2^2/[n_2(n_2-1)]$. Furthermore, the supremum of $\lambda(\boldsymbol{\tau})$ over the interval $0 < \kappa \leq 1$ is equal to its limit as $\kappa \to 0$, that is,

$$\begin{aligned} \lambda_{sup} &= \max\left[\frac{n_1+n_2-2}{n_2-1},\ \frac{n_1+n_2-2}{n_1-1}\right] \\ &= 1 + \max\left[\frac{n_1-1}{n_2-1},\ \frac{n_2-1}{n_1-1}\right]. \end{aligned} \tag{9.56}$$

Formula (9.56) indicates that λ_{sup} increases, resulting in a worsening of Satterthwaite's approximation, as the discrepancy in the degrees of freedom for the two samples increases. This conclusion is consistent with the simulation results given by Burdick and Graybill (1984, p. 133), Davenport and Webster (1972, p. 556), and Hudson and Krutchkoff (1968, p. 433). It can also be noted from (9.56) that $\lambda_{sup} \geq 2$, and equality is attained if and only if $n_1 = n_2$. This implies that Satterthwaite's approximation cannot be exact [that is, the statistic in (9.13) cannot have the exact t-distribution], even if $n_1 = n_2$.

9.4.2 A Confidence Interval on the Total Variation

Consider the balanced random two-way without interaction model,

$$Y_{ij} = \mu + \alpha_{(i)} + \beta_{(j)} + \epsilon_{(ij)}, \quad i = 1, 2, 3, 4; \quad j = 1, 2, 3, 4,$$

where $\alpha_{(i)} \sim N(0, \sigma^2_\alpha)$ is independent of $\beta_{(j)} \sim N(0, \sigma^2_\beta)$ and both are independent of $\epsilon_{(ij)} \sim N(0, \sigma^2_\epsilon)$. The corresponding ANOVA table is

Source	DF	MS	$E(MS)$
A	3	MS_A	$\delta_1 = 4\,\sigma^2_\alpha + \sigma^2_\epsilon$
B	3	MS_B	$\delta_2 = 4\,\sigma^2_\beta + \sigma^2_\epsilon$
Error	9	MS_E	$\delta_3 = \sigma^2_\epsilon$

The total variation is

$$\phi = \sigma^2_\alpha + \sigma^2_\beta + \sigma^2_\epsilon,$$

which can be expressed as

$$\phi = \frac{1}{4}(\delta_1 + \delta_2 + 2\,\delta_3).$$

Using Satterthwaite's approximation, an approximate $(1 - \alpha)$ 100% confidence interval on ϕ is given by

$$\left[\frac{\hat{\nu}\,\hat{\phi}}{\chi^2_{\alpha/2,\,\hat{\nu}}}, \frac{\hat{\nu}\,\hat{\phi}}{\chi^2_{1-\alpha/2,\,\hat{\nu}}}\right],$$

where

$$\hat{\phi} = \frac{1}{4}(MS_A + MS_B + 2\,MS_E), \tag{9.57}$$

and

$$\hat{\nu} = \frac{(MS_A + MS_B + 2\,MS_E)^2}{\frac{(MS_A)^2}{3} + \frac{(MS_B)^2}{3} + \frac{4\,(MS_E)^2}{9}}.$$

To determine the closeness of Satterthwaite's approximation in this example, let us express formula (9.57) as

$$\hat{\phi} = \boldsymbol{Y}'\boldsymbol{B}\boldsymbol{Y},$$

where $\boldsymbol{Y}$ is the data vector (consisting of 16 observations), and $\boldsymbol{B}$ is given by

$$\boldsymbol{B} = \frac{1}{4}\left[\frac{1}{3}\boldsymbol{P}_1 + \frac{1}{3}\boldsymbol{P}_2 + \frac{2}{9}\boldsymbol{P}_3\right].$$

Here, $\boldsymbol{P}_1$, $\boldsymbol{P}_2$, and $\boldsymbol{P}_3$ are the idempotent matrices associated with the sums of squares corresponding to $\alpha_{(i)}$, $\beta_{(j)}$, and the error term, respectively [see formula (9.21)]. Using (9.28), the variance–covariance matrix of $\boldsymbol{Y}$ is

$$\boldsymbol{\Sigma} = \sigma^2_{\alpha}\boldsymbol{A}_1 + \sigma^2_{\beta}\boldsymbol{A}_2 + \sigma^2_{\epsilon}\boldsymbol{A}_3,$$

where $\boldsymbol{A}_1 = \boldsymbol{I}_4 \otimes \boldsymbol{J}_4$, $\boldsymbol{A}_2 = \boldsymbol{J}_4 \otimes \boldsymbol{I}_4$, $\boldsymbol{A}_3 = \boldsymbol{I}_4 \otimes \boldsymbol{I}_4$. Thus,

$$\begin{aligned}\boldsymbol{B\Sigma} &= \frac{4\,\sigma^2_{\alpha}}{12}\boldsymbol{P}_1 + \frac{\sigma^2_{\epsilon}}{12}\boldsymbol{P}_1 + \frac{4\,\sigma^2_{\beta}}{12}\boldsymbol{P}_2 + \frac{\sigma^2_{\epsilon}}{12}\boldsymbol{P}_2 + \frac{\sigma^2_{\epsilon}}{18}\boldsymbol{P}_3\\ &= \frac{\delta_1}{12}\boldsymbol{P}_1 + \frac{\delta_2}{12}\boldsymbol{P}_2 + \frac{\delta_3}{18}\boldsymbol{P}_3.\end{aligned}$$

This follows from an application of formula (8.20). Since $\boldsymbol{P}_1, \boldsymbol{P}_2, \boldsymbol{P}_3$ are simultaneously diagonalizable and are idempotent of ranks 3, 3, 9, respectively, we conclude that the nonzero eigenvalues of $\boldsymbol{B\Sigma}$ are $\frac{\delta_1}{12}$ of multiplicity 3, $\frac{\delta_2}{12}$ of multiplicity 3, and $\frac{\delta_3}{18}$ of multiplicity 9. Thus, the rank of $\boldsymbol{B\Sigma}$ is $r = 3 + 3 + 9 = 15$. The largest and the smallest of these eigenvalues are

$$\begin{aligned}\tau_1 &= \max\left[\frac{\delta_1}{12}, \frac{\delta_2}{12}, \frac{\delta_3}{18}\right]\\ &= \frac{1}{12}\max(\delta_1, \delta_2), \qquad (9.58)\\ \tau_{15} &= \min\left[\frac{\delta_1}{12}, \frac{\delta_2}{12}, \frac{\delta_3}{18}\right]\\ &= \frac{\delta_3}{18}. \qquad (9.59)\end{aligned}$$

The value of $\lambda_{max}(\kappa)$ is determined by applying formula (9.52) with $r = 15$, $\kappa = \tau_{15}/\tau_1$, and u is obtained from Table 9.6. Note that in this table, $a_1^{1/2} = 0.105$, $a_2^{1/2} = 0.196$, $a_3^{1/2} = 0.302$, $a_4^{1/2} = 0.426$, $a_5^{1/2} = 0.577$, $a_6^{1/2} = 0.764$, $a_7^{1/2} = 1$, as can be seen from applying formula (9.53). Consequently, using formula (9.52) and Table 9.6, we get the following expression for $\lambda_{max}(\kappa)$:

$$\begin{aligned}\lambda_{max}(\kappa) &= \frac{15(1+14\,\kappa^2)}{(1+14\,\kappa)^2}, \quad 0 < \kappa \leq 0.105\\ &= \frac{15(2+13\,\kappa^2)}{(2+13\,\kappa)^2}, \quad 0.105 < \kappa \leq 0.196\\ &= \frac{15(3+12\,\kappa^2)}{(3+12\,\kappa)^2}, \quad 0.196 < \kappa \leq 0.302\\ &= \frac{15(4+11\kappa^2)}{(4+11\,\kappa)^2}, \quad 0.302 < \kappa \leq 0.426\end{aligned}$$

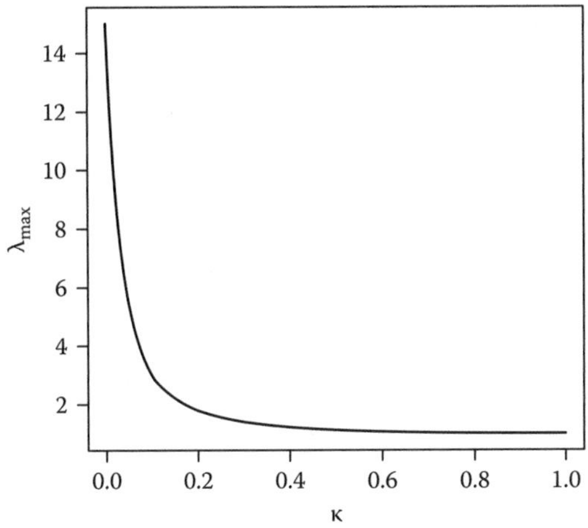

FIGURE 9.1
The graph of $\lambda_{max}(\kappa)$.

$$= \frac{15(5+10\,\kappa^2)}{(5+10\,\kappa)^2}, \quad 0.426 < \kappa \leq 0.577$$

$$= \frac{15(6+9\,\kappa^2)}{(6+9\,\kappa)^2}, \quad 0.577 < \kappa \leq 0.764$$

$$= \frac{15(7+8\,\kappa^2)}{(7+8\,\kappa)^2}, \quad 0.764 < \kappa \leq 1.$$

The graph of $\lambda_{max}(\kappa)$ is given in Figure 9.1. The value of λ_{sup} is the limit of $\lambda_{max}(\kappa)$ as $\kappa \to 0$, that is, $\lambda_{sup} = 15$. Note that since $\kappa = \tau_{15}/\tau_1$, then from (9.58) and (9.59),

$$\begin{aligned}\kappa &= \frac{2}{3}\left[\max\left(\frac{\delta_1}{\delta_3}, \frac{\delta_2}{\delta_3}\right)\right]^{-1} \\ &= \frac{2}{3}\left[1 + 4\max\left(\frac{\sigma^2_\alpha}{\sigma^2_\epsilon}, \frac{\sigma^2_\beta}{\sigma^2_\epsilon}\right)\right]^{-1} \\ &< \frac{2}{3}.\end{aligned} \quad (9.60)$$

Since $\lambda_{max}(\kappa)$ is a monotone decreasing function of κ, and small values of $\lambda_{max}(\kappa)$ are desirable for an adequate approximation, we conclude from (9.60) that large values of σ^2_α or σ^2_β (as compared to σ^2_ϵ) can worsen Satterthwaite's approximation. This agrees with the simulation results reported by Burrdick and Graybill (1984, p. 134).

9.4.3 A Linear Combination of Mean Squares

This section gives an extension of the results in Section 9.4.2. Let us again consider the linear combination, MS^*, of mean squares in formula (9.1), in conjunction with a balanced random model. All the random effects satisfy the same assumptions regarding normality, independence, and equality of variances, as was described in Section 8.4. Let us also express MS^* as a quadratic form, $\boldsymbol{Y}'\boldsymbol{B}\boldsymbol{Y}$, where $\boldsymbol{B}$ is given by formula (9.23). The variance–covariance matrix of $\boldsymbol{Y}$ is

$$\boldsymbol{\Sigma} = \sum_{j=1}^{\nu+1} \sigma_j^2 \boldsymbol{A}_j. \tag{9.61}$$

Then, by an application of formula (8.20), it is easy to show that

$$\begin{aligned}\boldsymbol{B\Sigma} &= \left(\sum_{i=1}^{k} \frac{a_i}{m_i} \boldsymbol{P}_i\right)\left(\sum_{j=1}^{\nu+1} \sigma_j^2 \boldsymbol{A}_j\right)\\ &= \sum_{i=1}^{k} \frac{a_i}{m_i}\left(\sum_{j=1}^{\nu+1} \kappa_{ij}\sigma_j^2\right)\boldsymbol{P}_i\\ &= \sum_{i=1}^{k} \frac{a_i}{m_i}\left(\sum_{j\in W_i} b_j\,\sigma_j^2\right)\boldsymbol{P}_i\\ &= \sum_{i=1}^{k} \frac{a_i\,\delta_i}{m_i}\boldsymbol{P}_i. \quad \text{[see (8.28)]}\end{aligned}$$

Since $\boldsymbol{P}_i$ is idempotent of *rank* m_i, and the $\boldsymbol{P}_i$'s are simultaneously diagonalizable, the nonzero eigenvalues of $\boldsymbol{B\Sigma}$ are given by $\frac{a_i\,\delta_i}{m_i}$ with multiplicity m_i $(i = 1, 2, \ldots, k)$. Hence,

$$\kappa = \frac{\min_i\left\{\frac{a_i\,\delta_i}{m_i}\right\}}{\max_i\left\{\frac{a_i\,\delta_i}{m_i}\right\}}.$$

If the values of $\frac{a_i\,\delta_i}{m_i}$ are highly variable, then κ will be small, which results in a large value of $\lambda_{\max}(\kappa)$ leading to unsatisfactory Satterthwaite's approximation. This conclusion agrees with the results given in Section 9.2, where the equality of $\frac{a_i\delta_i}{m_i}$, for $i = 1, 2, \ldots, k$, was a necessary and sufficient condition for Satterthwaite's approximation to be exact, that is, for MS^* to have a scaled chi-squared distribution (see Corollary 9.1).

Appendix 9.A: Determination of the Matrix G in Section 9.2.1

From Lemma 8.12, we have the orthogonal matrix

$$Q^* = [Q_0 : Q_1 : \ldots : Q_{\nu+1}],$$

which is defined in formula (8.46). The m_i columns of Q_i are orthonormal eigenvectors of P_i corresponding to the eigenvalue 1 such that $P_i = Q_i Q_i'$ and $Q_i' Q_i = I_{m_i}$ $(i = 0, 1, \ldots, \nu + 1)$. Let $G = Q^{*'}$. It is easy to see that

$$\begin{aligned} G P_i G' &= Q^{*'} Q_i Q_i' Q^* \\ &= \Lambda_i, \quad i = 0, 1, \ldots, \nu + 1, \end{aligned}$$

where Λ_i is a block-diagonal matrix whose diagonal blocks are all zero, except for the ith block, which is equal to I_{m_i} $(i = 0, 1, \ldots, \nu + 1)$. It follows that G diagonalizes $P_0, P_1, \ldots, P_{\nu+1}$ simultaneously. □

Exercises

9.1 Davenport and Webster (1973) reported an experiment where three uncalibrated thermometers (factor A) were randomly drawn from a large stock and three analysts (factor B) were randomly chosen from a large number of analysts. Each analyst used each thermometer to determine the melting point of a homogeneous sample of hydroquinone following a specified procedure. This was repeated on three separate weeks (factor C). The results are given in the following table. The same data set was initially given in Johnson and Leone (1964, pp. 239–240).

		Thermometer (A)		
Analyst (B)	**Week (C)**	**I**	**II**	**III**
1	1	174.0	173.0	171.5
	2	173.5	173.5	172.5
	3	174.5	173.0	173.0
2	1	173.0	172.0	171.0
	2	173.0	173.0	172.0
	3	173.5	173.5	171.5
3	1	173.5	173.0	173.0
	2	173.0	173.5	173.0
	3	173.0	172.5	172.5

Source: Davenport, J.M. and Webster, J.T., *Technometrics*, 15, 779, 1973. With permission.

The model considered for this experiment is

$$Y_{ijk} = \mu + \alpha_{(i)} + \beta_{(j)} + \gamma_{(k)} + (\alpha\beta)_{(ij)} + (\beta\gamma)_{(jk)} + \epsilon_{(ijk)},$$

where $\alpha_{(i)}$, $\beta_{(j)}$, and $\gamma_{(k)}$ denote the effects associated with factors A, B, and C, respectively. All effects are considered random under the usual assumptions of normality, independence, and equal variances. Note that the interaction $(\alpha\gamma)_{(ik)}$ was not considered here because the characteristic of a thermometer was not expected to vary from one week to another.

(a) Test the significance of $\sigma^2_{\alpha\beta}$ and $\sigma^2_{\beta\gamma}$, the two interaction variance components. Let $\alpha = 0.05$.

(b) Apply Satterthwaite's procedure to test the significance of σ^2_{β}, the variance component associated with factor B. Let $\alpha = 0.05$.

(c) Obtain an approximate 95% confidence interval on σ^2_{β} using Satterthwaite's approximation.

(d) Compute the actual level of significance of the approximate F-test concerning the hypothesis $H_0 : \sigma^2_{\beta} = 0$ versus $H_a : \sigma^2_{\beta} > 0$ at the nominal 5% level, given that $\sigma^2_{\alpha\beta} = 0.05$, $\sigma^2_{\beta\gamma} = 0.10$, and $\sigma^2_{\epsilon} = 0.01$.

9.2 Consider the ANOVA table for Exercise 8.10.

(a) Use the methodology described in Section 8.7.1 to obtain an exact, but conservative, confidence interval on $\sigma^2_{\beta(\alpha)}$ with a coverage probability greater than or equal to 0.90.

(b) Obtain an approximate 90% confidence interval on $\sigma^2_{\beta(\alpha)}$ using Satterthwaite's approximation.

(c) Compare the actual coverage probabilities for the intervals in parts (a) and (b) given that $\sigma^2_{\beta(\alpha)} = 0.35$ and $\sigma^2_{\epsilon} = 0.25$.

9.3 Consider the expression for $\hat{\nu}$ in formula (9.14).

(a) Show that $\min(n_1 - 1, n_2 - 1) \leq \hat{\nu} \leq n_1 + n_2 - 2$.

(b) Under what condition would the upper bound in the double inequality in part (a) be attained?
[Hint: See Gaylor and Hopper (1969, p. 693).]

9.4 Consider the ANOVA table for Example 9.1 (Table 9.2) and the corresponding test statistics, F_1 and F_2, given by formulas (9.18) and (9.20), respectively, for testing $H_0 : \sigma^2_{\alpha} = 0$ versus $H_a : \sigma^2_{\alpha} > 0$. Let $\alpha = 0.05$ be the nominal level of significance for both tests.

Compare the actual Type I error rates for both tests given that $\sigma^2_{\beta(\alpha)} = 0.10$, $\sigma^2_{\alpha\gamma} = 1.5$, $\sigma^2_{\beta(\alpha)\gamma} = 1.7$, and $\sigma^2_{\epsilon} = 1$.

9.5 Consider again the data set for Example 9.1 and the test statistic F_2 in (9.20).

Use the methodology described in Section 9.2.1 to determine the adequacy of Satterthwaite's approximation of the distributions of both numerator and denominator of F_2 (give the values of Bartlett's and Levene's test statistics in each case).

9.6 Let $X_1, X_2, \ldots, X_n$ be mutually independent random variables such that $X_i \sim \chi^2_1$ $(i = 1, 2, \ldots, n)$. Let $a_1, a_2, \ldots, a_n$ be positive constants. Show that

$$P\left(\sum_{i=1}^{n} a_i X_i < c\right) \le P(a\,X < c),$$

where c is any positive constant, $a = (\prod_{i=1}^{n} a_i)^{1/n}$, and $X \sim \chi^2_n$.

9.7 Let $X_1, X_2, \ldots, X_k$ be mutually independent random variables such that $X_i \sim \chi^2_{n_i}$ $(i = 1, 2, \ldots, k)$. Let $a_1, a_2, \ldots, a_k$ be positive constants. Show that

$$P\left(\sum_{i=1}^{k} a_i X_i < c\right) \le P(aX < c),$$

where c is any positive constant, $a = \left(\prod_{i=1}^{k} a_i^{n_i}\right)^{1/n}$, $n = \sum_{i=1}^{k} n_i$, and $X \sim \chi^2_n$.

[Hint: This is an extension of Exercise 9.6. It can be reduced to the previous case by decomposing X_i $(i = 1, 2, \ldots, k)$ into n_i mutually independent χ^2 variates with one degree of freedom each.]

9.8 Consider a special case of Exercise 9.7 where $k = 3$ and $X_1 \sim \chi^2_3$, $X_2 \sim \chi^2_2$, $X_3 \sim \chi^2_5$.

(a) Approximate the distribution of $X_1 + 5\,X_2 + 3\,X_3$ using the method of Satterthwaite.

(b) Use part (a) to find an approximate value of $P\,(X_1 + 5\,X_2 + 3\,X_3 < 46.8)$.

(c) Compare the result from part (b) with the upper bound, $P\,(a\,X < c)$, in Exercise 9.7, where $c = 46.8$.

9.9 Consider the two-way model and corresponding ANOVA table (Table 9.3) for Example 9.2. Let $\hat{\phi}$ be the linear combination of mean squares given in (9.41), that is,

$$\hat{\phi} = \frac{1}{12} MS_A + \frac{1}{9} MS_B + \frac{5}{36} MS_{AB} + \frac{2}{3} MS_E.$$

(a) Use the methodology described in Section 9.3 to develop a measure to assess the closeness of Satterthwaite's approximation of the distribution of $\hat{\phi}$.

(b) Provide a plot of $\lambda_{max}(\kappa)$, $0 < \kappa \leq 1$.

[Hint: Use formula (9.52) and Table 9.6.]

10

Unbalanced Fixed-Effects Models

We recall from Chapter 8 that a data set consisting of the values of some response, Y, is said to be *balanced* if the range of any one subscript of Y does not depend on the values of the other subscripts. If this condition is not satisfied by the data set, then it is said to be *unbalanced*. Thus, we may have unequal numbers of observations in the subclasses of the data, with possibly some subclasses containing no observations at all. In the latter case, we say that we have *empty subclasses* or *empty cells*. For example, in Table 9.1 which gives the prices of a variety of food items sold in large supermarkets (see Example 9.1), we have a total of 48 cells each containing three observations (prices recorded for a total of three months). If, for example, for supermarket 2 in area 1, the price of food item 2 was recorded only twice, then we end up with a data set that violates the condition stated earlier. It is also possible that no information at all was recorded regarding the price of such an item in that supermarket. In this case, the cell (1, 2, 2) corresponding to area 1, supermarket 2, and food item 2 will be empty.

In contrast to balanced data, the analysis of unbalanced data is much more involved. The main difficulty stems from the fact that in the case of unbalanced data, the partitioning of the total sum of squares can be made in a variety of ways; hence there is no unique way to write the ANOVA table as is the case with balanced data. Furthermore, the nice properties we saw earlier for balanced data in Chapter 8 are no longer applicable to unbalanced data. This makes it very difficult to develop a unified approach for the treatment of unbalanced data. It is therefore not surprising that such data are sometimes labeled as *messy*. Linear models representing unbalanced data are referred to as *unbalanced models*.

In this chapter, we consider the analysis of some unbalanced fixed-effects models, that is, models having only fixed effects except for the error term. The methodology described here depends on a particular notation called the *R-Notation*, which will be defined in the next section.

10.1 The *R*-Notation

Consider the model

$$Y = X\beta + \epsilon, \tag{10.1}$$

where

X is a matrix of order $n \times p$

β is a vector of fixed unknown parameters

ϵ is an experimental error vector assumed to have a zero mean and a variance–covariance matrix $\sigma^2 I_n$

Suppose that X is partitioned as $X = [X_1 : X_2]$, where X_1 and X_2 are of orders $n \times p_1$ and $n \times p_2$, respectively. The vector β is partitioned accordingly as $\beta = (\beta_1' : \beta_2')'$. Model (10.1) can then be written as

$$Y = X_1\beta_1 + X_2\beta_2 + \epsilon. \tag{10.2}$$

The regression sum of squares for the full model, that is, model (10.1) is $SS_{Reg} = Y'X(X'X)^{-}X'Y$, as was seen in formula (7.7). Let this sum of squares be denoted by $R(\beta)$, or, equivalently, $R(\beta_1, \beta_2)$. We thus have

$$R(\beta_1, \beta_2) = Y'X(X'X)^{-}X'Y. \tag{10.3}$$

Let us now consider the reduced model,

$$Y = X_1\beta_1 + \epsilon. \tag{10.4}$$

Its regression sum of squares is likewise denoted by $R(\beta_1)$ and is given by

$$R(\beta_1) = Y'X_1(X_1'X_1)^{-}X_1'Y. \tag{10.5}$$

The difference between the regression sums of squares in (10.3) and (10.5) is denoted by $R(\beta_2 \mid \beta_1)$. We thus have

$$\begin{aligned} R(\beta_2 \mid \beta_1) &= R(\beta_1, \beta_2) - R(\beta_1) \\ &= Y'[X(X'X)^{-}X' - X_1(X_1'X_1)^{-}X_1']Y. \end{aligned} \tag{10.6}$$

We note that $R(\beta_2 \mid \beta_1)$ represents the increase in the regression sum of squares which results from adding β_2 to a model that contains only β_1. In this case, β_2 is said to be adjusted for β_1, or that β_2 is added after β_1. Thus, the equality,

$$R(\beta_1, \beta_2) = R(\beta_1) + R(\beta_2 \mid \beta_1), \tag{10.7}$$

provides a partitioning of the regression sum of squares for the full model into $R(\beta_1)$ and $R(\beta_2 \mid \beta_1)$.

Formula (10.7) can be easily generalized whenever β is partitioned into k subvectors $\beta_1, \beta_2, \ldots, \beta_k$ so that

$$R(\beta) = R(\beta_1) + R(\beta_2 \mid \beta_1) + \cdots + R(\beta_k \mid \beta_1, \beta_2, \ldots, \beta_{k-1}),$$

where for $i = 3, 4, \ldots, k$, the ith R-expression on the right-hand side represents the increase in the regression sum of squares which results from adding β_i to a model that includes $\beta_1, \ldots, \beta_{i-1}$.

For example, for the model

$$Y_{ij} = \mu + \alpha_{(i)} + \beta_{(j)} + \epsilon_{(ij)},$$

the regression sum of squares for the full model, denoted by $R(\mu, \alpha, \beta)$, can be partitioned as

$$R(\mu, \alpha, \beta) = R(\mu) + R(\alpha \mid \mu) + R(\beta \mid \mu, \alpha),$$

where $R(\mu)$ is the regression sum of squares for a model that contains only μ, $R(\alpha \mid \mu) = R(\mu, \alpha) - R(\mu)$, with $R(\mu, \alpha)$ being the regression sum of squares for a model that contains only μ and α, and $R(\beta \mid \mu, \alpha) = R(\mu, \alpha, \beta) - R(\mu, \alpha)$.

The following theorem can be easily proved on the basis of the general principles outlined in Chapter 5 (see Exercise 10.1).

Theorem 10.1 Consider model (10.2) where it is assumed that $\boldsymbol{\epsilon} \sim N(\mathbf{0}, \sigma^2 \boldsymbol{I}_n)$. Let $r = rank(\boldsymbol{X})$ and $r_1 = rank(\boldsymbol{X}_1)$. Then,

(a) $\frac{1}{\sigma^2} R(\boldsymbol{\beta}_2 \mid \boldsymbol{\beta}_1)$ has the noncentral chi-squared distribution with $r - r_1$ degrees of freedom and a noncentrality parameter given by

$$\lambda = \frac{1}{\sigma^2} \boldsymbol{\beta}_2'[\boldsymbol{X}_2'\boldsymbol{X}_2 - \boldsymbol{X}_2'\boldsymbol{X}_1(\boldsymbol{X}_1'\boldsymbol{X}_1)^{-}\boldsymbol{X}_1'\boldsymbol{X}_2]\boldsymbol{\beta}_2.$$

(b) $R(\boldsymbol{\beta}_1)$, $R(\boldsymbol{\beta}_2 \mid \boldsymbol{\beta}_1)$, and SS_E are mutually independent, where SS_E is the error (residual) sum of squares for model (10.2).

(c) $E[R(\boldsymbol{\beta}_2 \mid \boldsymbol{\beta}_1)] = \sigma^2 \lambda + \sigma^2 (r - r_1)$, where λ is the noncentrality parameter in (a).

It can be seen on the basis of Theorem 10.1 that the F-statistic,

$$F = \frac{R(\boldsymbol{\beta}_2 \mid \boldsymbol{\beta}_1)/(r - r_1)}{SS_E/(n - r)}, \tag{10.8}$$

with $r - r_1$ and $n - r$ degrees of freedom, can be used to test the null hypothesis

$$H_0 : [\boldsymbol{I}_n - \boldsymbol{X}_1(\boldsymbol{X}_1'\boldsymbol{X}_1)^{-}\boldsymbol{X}_1']\boldsymbol{X}_2\boldsymbol{\beta}_2 = \mathbf{0}. \tag{10.9}$$

(see Exercise 10.2). In particular, if $\boldsymbol{X}$ is of full column rank, then this null hypothesis is reduced to $H_0 : \boldsymbol{\beta}_2 = \mathbf{0}$.

There are particular R-expressions used in testing certain hypotheses for model (10.1). These expressions represent sums of squares known as Type I, Type II, and Type III sums of squares.

Definition 10.1 A Type I sum of squares (S.S.) for an effect, u, in the model is $R(u \mid v_u)$, where v_u represents all the effects preceding u in the model. Type I sums of squares are obtained by adding one effect at a time to the model until all effects have been added. For this reason, such sums of squares are described as being *sequential*.

For example, Type I sums of squares for the α- and β-effects in the model, $Y_{ij} = \mu + \alpha_{(i)} + \beta_{(j)} + \epsilon_{(ij)}$, $i = 1, 2, \ldots a$; $j = 1, 2, \ldots, b$ are $R(\alpha \mid \mu)$ and $R(\beta \mid \mu, \alpha)$, respectively. If, however, the same model is written as $Y_{ij} = \mu + \beta_{(j)} + \alpha_{(i)} + \epsilon_{(ij)}$, then the corresponding Type I sums of squares for β and α are $R(\beta \mid \mu)$ and $R(\alpha \mid \mu, \beta)$, respectively. Thus, Type I sums of squares depend on the ordering of the effects in the model. Note that the Type I S.S. for μ is $R(\mu) = (\sum_{i=1}^{a} \sum_{j=1}^{b} Y_{ij})^2/(ab)$, which is usually referred to as *the correction term*. We also note that $R(\mu)$, $R(\alpha \mid \mu)$, and $R(\beta \mid \mu, \alpha)$ provide a partitioning of the regression sum of squares, $R(\mu, \alpha, \beta)$, for the model.

Definition 10.2 A Type II sum of squares (S.S.) for an effect, u, in the model is adjusted for all the other effects in the model, except for those that contain u (i.e., not adjusted for interactions involving u, or effects nested within u).

For example, for the model, $Y_{ij} = \mu + \alpha_{(i)} + \beta_{(j)} + \epsilon_{(ij)}$, the Type II sums of squares for the α- and β-effects are $R(\alpha \mid \mu, \beta)$, $R(\beta \mid \mu, \alpha)$, respectively. Also, for the model, $Y_{ijk} = \mu + \alpha_{(i)} + \beta_{(j)} + (\alpha\beta)_{(ij)} + \epsilon_{ij(k)}$ $(i = 1, 2, \ldots, a; j = 1, 2, \ldots, b; k = 1, 2, \ldots, n_{ij})$, the Type II sums of squares for the α- and β-effects, and the $(\alpha\beta)$ interaction are $R(\alpha \mid \mu, \beta)$, $R(\beta \mid \mu, \alpha)$, and $R(\alpha\beta \mid \mu, \alpha, \beta)$, respectively. Note that Type II sums of squares do not necessarily add up to the total regression sum of squares for the model, but are invariant to the ordering of the effects in the model.

Definition 10.3 A Type III sum of squares (S.S.) for an effect, u, is, in principle, obtained by adjusting u for all the other effects in the model.

This definition makes sense when the model in (10.1) is of full rank (as in a regression model). However, for a less-than-full-rank model and in the absence of any constraints on the model parameters, this definition may produce a value equal to zero for some of the effects in the model, as will be seen later in Section 10.3. Type III sums of squares are also called *partial sums of squares*.

10.2 Two-Way Models without Interaction

Consider the two-way without interaction model,

$$Y_{ijk} = \mu + \alpha_{(i)} + \beta_{(j)} + \epsilon_{ij(k)}, \quad i = 1, 2, \ldots, a; \ j = 1, 2, \ldots, b; \ k = 0, 1, \ldots, n_{ij}, \tag{10.10}$$

where μ, $\alpha_{(i)}$, and $\beta_{(j)}$ are unknown parameters with the latter two representing the effects of the levels of two factors, denoted by A and B, respectively, and the elements of $\epsilon_{ij(k)}$ are independently distributed as $N(0, \sigma^2)$. We assume that $n_{..} > a + b - 1$, where $n_{..}$ is the total number of observations. Note that n_{ij} can be zero for some (i, j) indicating the possibility of some empty (or missing) cells. This model can be written in vector form as in (10.1). In this case, the matrix $\boldsymbol{X}$ is of order $n_{..} \times (a + b + 1)$ of the form, $\boldsymbol{X} = [\boldsymbol{1}_{n_{..}} : \boldsymbol{H}_1 : \boldsymbol{H}_2]$, where $\boldsymbol{H}_1 = \oplus_{i=1}^{a} \boldsymbol{1}_{n_{i.}}$, $\boldsymbol{H}_2 = [\oplus_{j=1}^{b} \boldsymbol{1}'_{n_{1j}} : \oplus_{j=1}^{b} \boldsymbol{1}'_{n_{2j}} : \ldots : \oplus_{j=1}^{b} \boldsymbol{1}'_{n_{aj}}]'$, and $n_{i.} = \sum_{j=1}^{b} n_{ij}$. Note that $\boldsymbol{X}$ is of rank $a+b-1$. This results from the fact that the a columns of $\boldsymbol{X}$ corresponding to the $\alpha_{(i)}$'s add up to the first column of $\boldsymbol{X}$, which is the column of ones. The same applies to the b columns of $\boldsymbol{X}$ corresponding to the $\beta_{(j)}$'s. Furthermore, the vector $\boldsymbol{\beta}$ in (10.1) consists of μ, the $\alpha_{(i)}$'s, and the $\beta_{(j)}$'s in model (10.10).

10.2.1 Estimable Linear Functions for Model (10.10)

Let $\mu_{ij} = \mu + \alpha_{(i)} + \beta_{(j)}$ be the mean of the (i, j)th cell. If cell (i, j) is nonempty, then μ_{ij} is estimable. Since the rank of $\boldsymbol{X}$ is $a + b - 1$, the number of linearly independent estimable linear functions of the parameter vector in the model must be equal to $a + b - 1$. Furthermore, if for a given j, $\mu_{ij} - \mu_{i'j}$ is estimable for some $i \neq i'$, then so is $\alpha_{(i)} - \alpha_{(i')}$. Likewise, if for a given i, $\mu_{ij} - \mu_{ij'}$ is estimable for some $j \neq j'$, then $\beta_{(j)} - \beta_{(j')}$ is also estimable.

Lemma 10.1 Suppose that the pattern of the two-way data is such that $\alpha_{(i)} - \alpha_{(i')}$ and $\beta_{(j)} - \beta_{(j')}$ are estimable for all $i \neq i'$ and $j \neq j'$ for model (10.10). If (i_0, j_0) is a nonempty cell, then

(a) All cell means, μ_{ij}, in the model are estimable.

(b) $\mu_{i_0 j_0}$, $\alpha_{(i)} - \alpha_{(i')}$, and $\beta_{(j)} - \beta_{(j')}$, for all $i \neq i'$ and $j \neq j'$, form a basis for the space of all estimable linear functions of the parameters in the model.

Proof.

(a) For any $(i, j) \neq (i_0, j_0)$, μ_{ij} can be written as

$$\begin{aligned}\mu_{ij} &= \mu + \alpha_{(i)} + \beta_{(j)} \\ &= \mu + \alpha_{(i_0)} + \beta_{(j_0)} + \alpha_{(i)} - \alpha_{(i_0)} + \beta_{(j)} - \beta_{(j_0)} \\ &= \mu_{i_0 j_0} + \alpha_{(i)} - \alpha_{(i_0)} + \beta_{(j)} - \beta_{(j_0)}.\end{aligned}$$

The right-hand side is the sum of estimable functions, hence μ_{ij} is estimable for all (i, j).

(b) This is true since all these functions are estimable in addition to being linearly independent, and their number is equal to $a + b - 1$, the rank of $\boldsymbol{X}$. □

In general, if $A\beta$ is an estimable linear function of the parameter vector β, where A is a known matrix of order $s \times p$ and rank s ($s \leq a + b - 1$ and $p = a + b + 1$), then by the Gauss–Markov Theorem (Theorem 7.6), its best linear unbiased estimator (BLUE) is given by $A\hat{\beta}$, where $\hat{\beta} = (X'X)^{-}X'Y$ and Y is the vector of $n_{..}$ observations from model (10.10). It follows that under the estimability condition of Lemma 10.1, the BLUE of μ_{ij} is

$$\hat{\mu}_{ij} = \hat{\mu} + \hat{\alpha}_{(i)} + \hat{\beta}_{(j)}, \quad i = 1, 2, \ldots, a; j = 1, 2, \ldots, b, \tag{10.11}$$

where $\hat{\mu}$, $\hat{\alpha}_{(i)}$ ($i = 1, 2, \ldots, a$) and $\hat{\beta}_{(j)}$ ($j = 1, 2, \ldots, b$) are the $1 + a + b$ elements of $\hat{\beta}$. Furthermore, $\hat{\alpha}_{(i)} - \hat{\alpha}_{(i')}$ and $\hat{\beta}_{(j)} - \hat{\beta}_{(j')}$ are the BLUEs of $\alpha_{(i)} - \alpha_{(i')}$ and $\beta_{(j)} - \beta_{(j')}$, respectively. In particular, the following linear functions are estimable:

$$\frac{1}{b}\sum_{j=1}^{b} \mu_{ij} = \mu + \alpha_{(i)} + \frac{1}{b}\sum_{j=1}^{b} \beta_{(j)} \tag{10.12}$$

$$\frac{1}{a}\sum_{i=1}^{a} \mu_{ij} = \mu + \frac{1}{a}\sum_{i=1}^{a} \alpha_{(i)} + \beta_{(j)}, \tag{10.13}$$

and their BLUEs are $\frac{1}{b}\sum_{j=1}^{b} \hat{\mu}_{ij}$ and $\frac{1}{a}\sum_{i=1}^{a} \hat{\mu}_{ij}$, respectively. By definition, $\frac{1}{b}\sum_{j=1}^{b} \mu_{ij}$ is the *least-squares mean* for row i ($i = 1, 2, \ldots, a$), denoted by $LSM(\alpha_{(i)})$, and $\frac{1}{a}\sum_{i=1}^{a} \mu_{ij}$ is the least-squares mean for column j, denoted by $LSM(\beta_{(j)})$. These means are also called *population marginal means* (see Searle, Speed, and Milliken, 1980).

The least-squares means should not be confused with the *weighted means* of the cell means in row i and column j, namely, $\frac{1}{n_{i.}}\sum_{j=1}^{b} n_{ij}\mu_{ij}$ and $\frac{1}{n_{.j}}\sum_{i=1}^{a} n_{ij}\mu_{ij}$, respectively, where $n_{.j} = \sum_{i=1}^{a} n_{ij}$ ($i = 1, 2, \ldots, a$; $j = 1, 2, \ldots, b$). If the data set is balanced, then the least-squares means are equal to the corresponding weighted means.

10.2.2 Testable Hypotheses for Model (10.10)

In general, if $A\beta$ is an estimable linear function of β, where, as before, A is a known matrix of order $s \times p$ and rank s ($\leq a + b - 1$), then the hypothesis,

$$H_0 : A\beta = m, \tag{10.14}$$

is testable (see Definition 7.2), where m is a known constant vector. Using the methodology in Section 7.4.2 and under the assumption that ϵ in model (10.1) is distributed as $N(0, \sigma^2 I_{n_{..}})$, the test statistic for H_0 is given by the F-ratio

$$F = \frac{(A\hat{\beta} - m)'[A(X'X)^{-}A']^{-1}(A\hat{\beta} - m)}{s\,MS_E}, \tag{10.15}$$

where $MS_E = SS_E/(n_{..} - a - b + 1)$ is the error (residual) mean square. Under H_0, F has the F-distribution with s and $n_{..} - a - b + 1$ degrees of freedom, and H_0 is rejected at the α-level if $F \geq F_{\alpha, s, n_{..}-a-b+1}$.

Let us now consider model (10.10) under the assumption that the $\epsilon_{ij(k)}$'s are independently distributed as $N(0, \sigma^2)$. If the estimability condition stated in Lemma 10.1 is valid for a given data set, then all the cell means, namely μ_{ij}, for $i = 1, 2, \ldots, a$; $j = 1, 2, \ldots, b$, are estimable. There are two types of hypotheses that can be tested concerning the levels of factors A and B. The development of these hypotheses is based on the following lemmas.

Lemma 10.2 $R(\mu)$, $R(\alpha \mid \mu)$, and $R(\beta \mid \mu, \alpha)$ are mutually independent and are independent of $MS_E = SS_E/(n_{..} - a - b + 1)$, the error mean square for model (10.10). Furthermore, $R(\alpha \mid \mu)/\sigma^2$ and $R(\beta \mid \mu, \alpha)/\sigma^2$ are distributed as chi-squared variates with $a - 1$ and $b - 1$ degrees of freedom, respectively, and $SS_E/\sigma^2 \sim \chi^2_{n_{..}-a-b+1}$.

Proof. This follows directly from applying Example 5.6 since

$$R(\mu) + R(\alpha \mid \mu) + R(\beta \mid \mu, \alpha) + SS_E = \boldsymbol{Y}'\boldsymbol{Y}. \qquad \square$$

It can similarly be shown that $R(\beta \mid \mu)$, $R(\alpha \mid \mu, \beta)$, and SS_E are mutually independent, and that $R(\beta \mid \mu)/\sigma^2$ and $R(\alpha \mid \mu, \beta)/\sigma^2$ are distributed as chi-squared variates with $b - 1$ and $a - 1$ degrees of freedom, respectively [Here, $R(\mu)$, $R(\beta \mid \mu)$, $R(\alpha \mid \mu, \beta)$, and SS_E provide a partitioning of $\boldsymbol{Y}'\boldsymbol{Y}$].

Lemma 10.3 The noncentrality parameter of $R(\alpha \mid \mu)/\sigma^2$ is equal to zero if and only if the values of

$$\lambda_i = \alpha_{(i)} + \frac{1}{n_{i.}} \sum_{j=1}^{b} n_{ij}\, \beta_{(j)}, \; i = 1, 2, \ldots, a, \tag{10.16}$$

are equal for all i.

Proof. We have that

$$\begin{aligned} R(\alpha \mid \mu) &= R(\mu, \alpha) - R(\mu) \\ &= \boldsymbol{Y}' \left[\bigoplus_{i=1}^{a} \left(\frac{1}{n_{i.}} \boldsymbol{J}_{n_{i.}} \right) - \frac{1}{n_{..}} \boldsymbol{J}_{n_{..}} \right] \boldsymbol{Y}. \end{aligned}$$

The matrix, $\bigoplus_{i=1}^{a} \left(\frac{1}{n_{i.}} \boldsymbol{J}_{n_{i.}} \right) - \frac{1}{n_{..}} \boldsymbol{J}_{n_{..}}$, is idempotent of rank $a - 1$. The noncentrality parameter, θ, of $R(\alpha \mid \mu)/\sigma^2$ is

$$\theta = \frac{1}{\sigma^2} \boldsymbol{\mu}' \left[\bigoplus_{i=1}^{a} \left(\frac{1}{n_{i.}} \boldsymbol{J}_{n_{i.}} \right) - \frac{1}{n_{..}} \boldsymbol{J}_{n_{..}} \right] \boldsymbol{\mu}, \tag{10.17}$$

where μ is the expected value of $\boldsymbol{Y}$ given by

$$\boldsymbol{\mu} = \mu \mathbf{1}_{n_{..}} + \boldsymbol{H}_1(\alpha_1, \alpha_2, \ldots, \alpha_a)' + \boldsymbol{H}_2(\beta_1, \beta_2, \ldots, \beta_b)',$$

and $\boldsymbol{H}_1$ and $\boldsymbol{H}_2$ are the matrices defined in the beginning of Section 10.2.

Now, $\theta = 0$ if and only if

$$\left[\bigoplus_{i=1}^{a} \left(\frac{1}{n_{i.}} \boldsymbol{J}_{n_{i.}}\right) - \frac{1}{n_{..}} \boldsymbol{J}_{n_{..}}\right] \boldsymbol{\mu} = \mathbf{0}.$$

It is easy to show that

$$\left[\bigoplus_{i=1}^{a} \left(\frac{1}{n_{i.}} \boldsymbol{J}_{n_{i.}}\right) - \frac{1}{n_{..}} \boldsymbol{J}_{n_{..}}\right] \boldsymbol{\mu} = [\boldsymbol{u}_1' : \boldsymbol{u}_2' : \ldots : \boldsymbol{u}_a']', \tag{10.18}$$

where $\boldsymbol{u}_i$ is a column vector of $n_{i.}$ elements ($i = 1, 2, \ldots, a$) of the form

$$\boldsymbol{u}_i = \left[\alpha_{(i)} - \frac{1}{n_{..}} \sum_{j=1}^{a} n_{j.}\, \alpha_{(j)} + \frac{1}{n_{i.}} \sum_{k=1}^{b} n_{ik}\, \beta_{(k)} - \frac{1}{n_{..}} \sum_{l=1}^{b} n_{.l}\, \beta_{(l)}\right] \mathbf{1}_{n_{i.}}, \quad i = 1, 2, \ldots, a.$$

From (10.18) we conclude that $\theta = 0$ if and only if $\boldsymbol{u}_i = \mathbf{0}$ for $i = 1, 2, \ldots, a$, that is, if and only if the values of λ_i in (10.16) are equal for all i. □

It can similarly be shown that the noncentrality parameter of $R(\beta \mid \mu)/\sigma^2$ is equal to zero if and only if the values of

$$\beta_{(j)} + \frac{1}{n_{.j}} \sum_{i=1}^{a} n_{ij}\, \alpha_{(i)}, \quad j = 1, 2, \ldots, b, \tag{10.19}$$

are equal for all j.

Lemma 10.4 The noncentrality parameter of $R(\beta \mid \mu, \alpha)/\sigma^2$ is equal to zero if and only if $\beta_{(1)} = \beta_{(2)} = \ldots = \beta_{(b)}$.

Proof. Let us apply Theorem 10.1 to model (10.10). In this case, $\boldsymbol{X}_1 = [\mathbf{1}_{n_{..}} : \boldsymbol{H}_1]$, $\boldsymbol{X}_2 = \boldsymbol{H}_2$, $\boldsymbol{\beta}_2 = (\beta_1, \beta_2, \ldots, \beta_b)'$, and the noncentrality parameter of $R(\beta \mid \mu, \alpha)/\sigma^2$ is equal to λ, which is given in part (a) of Theorem 10.1. Since the matrix $\boldsymbol{I}_{n_{..}} - \boldsymbol{X}_1(\boldsymbol{X}_1'\boldsymbol{X}_1)^{-}\boldsymbol{X}_1'$ is idempotent, $\lambda = 0$ if and only if

$$[\boldsymbol{I}_{n_{..}} - \boldsymbol{X}_1(\boldsymbol{X}_1'\boldsymbol{X}_1)^{-}\boldsymbol{X}_1']\boldsymbol{X}_2\boldsymbol{\beta}_2 = \mathbf{0}. \tag{10.20}$$

It is easy to show that

$$[\boldsymbol{I}_{n_{..}} - \boldsymbol{X}_1(\boldsymbol{X}_1'\boldsymbol{X}_1)^{-}\boldsymbol{X}_1']\boldsymbol{X}_2\boldsymbol{\beta}_2 = [\boldsymbol{v}_1' : \boldsymbol{v}_2' : \ldots : \boldsymbol{v}_a']', \tag{10.21}$$

where v_i is a column vector of $n_{i.}$ elements $(i = 1, 2, \ldots, a)$ of the form

$$v_i = \begin{bmatrix} \beta_{(1)}\mathbf{1}_{n_{i1}} \\ \beta_{(2)}\mathbf{1}_{n_{i2}} \\ \cdot \\ \cdot \\ \cdot \\ \beta_{(b)}\mathbf{1}_{n_{ib}} \end{bmatrix} - \left(\frac{1}{n_{i.}}\sum_{j=1}^{b} n_{ij}\,\beta_{(j)}\right)\mathbf{1}_{n_{i.}}, \quad i = 1, 2, \ldots, a. \tag{10.22}$$

From formulas (10.20)–(10.22) it follows that $\lambda = 0$ if and only if $\beta_{(1)} = \beta_{(2)} = \cdots = \beta_{(b)}$. □

In a similar manner, it can be shown that the noncentrality parameter of $R(\alpha \mid \mu, \beta)/\sigma^2$ is equal to zero if and only if $\alpha_{(1)} = \alpha_{(2)} = \cdots = \alpha_{(a)}$.

10.2.2.1 Type I Testable Hypotheses

Type I testable hypotheses for model (10.10) are those hypotheses tested by F-ratios that use Type I sums of squares in their numerators. If the model is written so that $\alpha_{(i)}$ appears first followed by $\beta_{(j)}$, then Type I sums of squares for factors A and B are $R(\alpha \mid \mu)$ and $R(\beta \mid \mu, \alpha)$, respectively. The corresponding F-ratios, are

$$F(\alpha \mid \mu) = \frac{R(\alpha \mid \mu)}{(a-1)\,MS_E} \tag{10.23}$$

$$F(\beta \mid \mu, \alpha) = \frac{R(\beta \mid \mu, \alpha)}{(b-1)\,MS_E}. \tag{10.24}$$

If, however, the model is written with $\beta_{(j)}$ appearing first followed by $\alpha_{(i)}$, then Type I sums of squares for factors B and A are $R(\beta \mid \mu)$ and $R(\alpha \mid \mu, \beta)$, respectively, as was seen in Section 10.1, and the corresponding F-ratios are

$$F(\beta \mid \mu) = \frac{R(\beta \mid \mu)}{(b-1)\,MS_E} \tag{10.25}$$

$$F(\alpha \mid \mu, \beta) = \frac{R(\alpha \mid \mu, \beta)}{(a-1)\,MS_E}. \tag{10.26}$$

Now, $F(\alpha \mid \mu)$ and $F(\beta \mid \mu)$ test the hypotheses that their corresponding noncentrality parameters are equal to zero. On the basis of Lemmas 10.2 and 10.3, the hypothesis tested by $F(\alpha \mid \mu)$ is

$$H_0 : \alpha_{(i)} + \frac{1}{n_{i.}}\sum_{j=1}^{b} n_{ij}\,\beta_{(j)} \text{ equal for all } i = 1, 2, \ldots, a. \tag{10.27}$$

Similarly, the hypothesis tested by $F(\beta \mid \mu)$ is

$$H_0 : \beta_{(j)} + \frac{1}{n_{.j}}\sum_{i=1}^{a} n_{ij}\,\alpha_{(i)} \text{ equal for all } j = 1, 2, \ldots, b. \tag{10.28}$$

Using the cell means, μ_{ij}, (10.27) and (10.28) can be written as

$$H_0 : \frac{1}{n_{i.}} \sum_{j=1}^{b} n_{ij}\,\mu_{ij} \text{ equal for all } i = 1, 2, \ldots, a \tag{10.29}$$

$$H_0 : \frac{1}{n_{.j}} \sum_{i=1}^{a} n_{ij}\,\mu_{ij} \text{ equal for all } j = 1, 2, \ldots, b. \tag{10.30}$$

The hypothesis in (10.29) indicates equality of the weighted means of the cell means in rows $1, 2, \ldots, a$, and the one in (10.30) indicates equality of the weighted means of the cell means in columns $1, 2, \ldots, b$. These hypotheses are not desirable for testing the effects of factors A and B since they are data dependent (they depend on the cell frequencies). A hypothesis is supposed to be set up before collecting the data in a given experimental situation. Thus, $F(\alpha \mid \mu)$ and $F(\beta \mid \mu)$ are not recommended F-ratios for testing the effects of A and B. Let us therefore consider the other two F-ratios, namely, $F(\alpha \mid \mu, \beta)$ and $F(\beta \mid \mu, \alpha)$, whose corresponding hypotheses are described in the next section.

10.2.2.2 Type II Testable Hypotheses

Type II testable hypotheses for factors A and B are hypotheses tested by the F-ratios shown in formulas (10.26) and (10.24), respectively. Given that these ratios test that their corresponding noncentrality parameters are equal to zero, we conclude, on the basis of Lemmas 10.2 and 10.4, that these hypotheses are of the form

$$H_0 : \alpha_{(1)} = \alpha_{(2)} = \cdots = \alpha_{(a)} \tag{10.31}$$

$$H_0 : \beta_{(1)} = \beta_{(2)} = \cdots = \beta_{(b)}. \tag{10.32}$$

Using the cell means, μ_{ij}, (10.31) and (10.32) can be written as

$$H_0 : \frac{1}{b} \sum_{j=1}^{b} \mu_{ij} \text{ equal for all } i = 1, 2, \ldots, a \tag{10.33}$$

$$H_0 : \frac{1}{a} \sum_{i=1}^{a} \mu_{ij} \text{ equal for all } j = 1, 2, \ldots, b. \tag{10.34}$$

We recall from Section 10.2.1 that the expressions in (10.33) and (10.34) are the least-squares means for row i and column j, respectively ($i = 1, 2, \ldots, a$; $j = 1, 2, \ldots, b$). Thus, the F-ratios, $F(\alpha \mid \mu, \beta)$ and $F(\beta \mid \mu, \alpha)$, test equality of the $\alpha_{(i)}$'s and of the $\beta_{(j)}$'s, respectively, or equivalently, equality of the least-squares means for the a rows and b columns, respectively. These hypotheses do not depend on the data and, unlike Type I hypotheses, are invariant to the ordering of the effects in model (10.10). Furthermore, such hypotheses are of the same form as the ones tested in a balanced data situation. Consequently, the F-ratios in (10.26) and (10.24) are the recommended test statistics for testing the effects of factors A and B, respectively. The hypothesis in (10.33) is rejected at the α-level if $F(\alpha \mid \mu, \beta) \geq F_{\alpha,a-1,n_{..}-a-b+1}$. Similarly, the hypothesis in (10.34) is rejected at the α-level if $F(\beta \mid \mu, \alpha) \geq F_{\alpha,b-1,n_{..}-a-b+1}$.

Note that the Type I hypothesis for the last effect in the model is identical to its Type II hypothesis, and the corresponding F-ratios are identical. In particular, if the data set is balanced, then Type I and Type II hypotheses and F-ratios are the same.

It should also be noted that, as a follow-up to the Type II tests, if a particular F-test is significant, then any multiple comparisons among the levels of the corresponding factor should be made using the least-squares means of that factor. Thus, multiple comparisons among the weighted means of the cell means (for the a rows and b columns) should not be considered since this amounts to testing using the F-ratios, $F(\alpha \mid \mu)$ and $F(\beta \mid \mu)$, which is undesirable. For example, to compare the least-squares means for rows i and i' $(i \neq i')$, we can consider the null hypothesis, $H_0 : \boldsymbol{a}_i'\boldsymbol{\beta} = \boldsymbol{a}_{i'}'\boldsymbol{\beta}$, where $\boldsymbol{a}_i$ and $\boldsymbol{a}_{i'}$ are known constant vectors so that $\boldsymbol{a}_i'\boldsymbol{\beta} = LSM(\alpha_{(i)})$ and $\boldsymbol{a}_{i'}'\boldsymbol{\beta} = LSM(\alpha_{(i')})$. The corresponding test statistic is

$$t = \frac{(\boldsymbol{a}_i - \boldsymbol{a}_{i'})'\hat{\boldsymbol{\beta}}}{[(\boldsymbol{a}_i - \boldsymbol{a}_{i'})'(\boldsymbol{X}'\boldsymbol{X})^{-}(\boldsymbol{a}_i - \boldsymbol{a}_{i'})\,MS_E]^{1/2}}, \qquad (10.35)$$

where $\hat{\boldsymbol{\beta}} = (\boldsymbol{X}'\boldsymbol{X})^{-}\boldsymbol{X}'\boldsymbol{Y}$. Under H_0, this statistic has the t-distribution with $n_{..} - a - b + 1$ degrees of freedom. The two least-squares means are considered to be significantly different at the α-level if $\mid t \mid \geq t_{\alpha/2,n_{..}-a-b+1}$. A similar t-test can be used to compare the least-squares means for columns j and j' $(j \neq j')$.

Example 10.1 An experiment was conducted to study the effects of three different fats (factor A) and three different additives (factor B) on the specific volume of bread loaves. The resulting data are given in Table 10.1. We note that we have two empty cells, but the estimability condition of Lemma 10.1 is clearly satisfied. Thus all nine cell means are estimable. The error mean squares is $MS_E = 0.7459$ with $n_{..} - a - b + 1 = 17 - 3 - 3 + 1 = 12$ degrees of freedom. Tables 10.2 and 10.3 give the results of the Type I and Type II analyses.

It may be recalled that testing the significance of the effects of factors A and B should be made on the basis of the Type II analysis. Using Table 10.3, we find that the F-ratios for factors A and B are $F(\alpha \mid \mu, \beta) = 2.09$ (p-value $=$ 0.1665) and $F(\beta \mid \mu, \alpha) = 14.73$ (p-value $= 0.0006$), respectively. Thus, the effect of B is significant, but the one for A is not. This means that we have significant differences among the least-squares means for the three additives, but no significant differences can be detected among the least-squares means for the three fats. The values in Tables 10.2 and 10.3 were obtained by using the SAS software (SAS, 2000, PROC GLM) (see Sections 10.4 and 10.5).

TABLE 10.1
Volume of Bread Loaves Data

	B = Additive		
A = Fat	**1**	**2**	**3**
1	6.4 6.2 6.5	7.3 7.1	—
2	5.9 5.1	—	7.1 7.5 6.9
3	4.1 4.2	7.3 6.5 8.1	9.2 8.9

TABLE 10.2
Type I Analysis for Model (10.10)

Source	DF	Type I *SS*	*MS*	*F*	*p*-Value
A	2	0.4706	0.2353	0.32	0.7353
B	2	21.9689	10.9844	14.73	0.0006

TABLE 10.3
Type II Analysis for Model (10.10)

Source	DF	Type II *SS*	*MS*	*F*	*p*-Value
A	2	3.1175	1.5587	2.09	0.1665
B	2	21.9689	10.9844	14.73	0.0006

Note that from Table 10.2, the Type I F-ratios for A and B according to model (10.10) are $F(\alpha \mid \mu) = 0.32$ and $F(\beta \mid \mu, \alpha) = 14.73$, respectively. Table 10.2 does not give the value of $F(\beta \mid \mu)$, which is actually equal to 12.95. To get this value, the SAS model has to be rewritten with the effect of B appearing first followed by the one for A. In any case, $F(\beta \mid \mu)$ and $F(\alpha \mid \mu)$ should not be used to test the effects of B and A, as was mentioned earlier.

The best linear unbiased estimates of the least-squares means for the levels of factors A and B are given by [see also formulas (10.12) and (10.13)]

$$\widehat{LSM}(\alpha_{(i)}) = \hat{\mu} + \hat{\alpha}_{(i)} + \frac{1}{3}\sum_{j=1}^{3}\hat{\beta}_{(j)}, \; i = 1, 2, 3,$$

$$\widehat{LSM}(\beta_{(j)}) = \hat{\mu} + \frac{1}{3}\sum_{i=1}^{3}\hat{\alpha}_{(i)} + \hat{\beta}_{(j)}, \; j = 1, 2, 3,$$

TABLE 10.4
Estimates of the Least-Squares Means

A	$\widehat{LSM}(\alpha_{(i)})$	B	$\widehat{LSM}(\beta_{(j)})$
1	7.5668	1	5.3910
2	6.2587	2	7.0057
3	6.8854	3	8.3142

TABLE 10.5
Pairwise Comparisons among the Least-Squares Means of B

	B		
B	1	2	3
1	•	−3.0347[a] (0.0104)	−5.2178 (0.0002)
2	3.0347 (0.0104)	•	−2.0378 (0.0642)
3	5.2178 (0.0002)	2.0378 (0.0642)	•

[a] t-value for the difference, $\widehat{LSM}(\beta_{(1)}) - \widehat{LSM}(\beta_{(2)})$. The quantity inside parentheses is the p-value.

where $\hat{\mu}$, $\hat{\alpha}_{(i)}$ $(i = 1, 2, 3)$, and $\hat{\beta}_{(j)}$ $(j = 1, 2, 3)$ are the elements of $\hat{\beta} = (X'X)^{-}X'Y$. Using Table 10.1, the actual values of $\widehat{LSM}(\alpha_{(i)})$ and $\widehat{LSM}(\beta_{(j)})$ are shown in Table 10.4. Since the test for factor B is significant, it would be of interest to compare its least-squares means using the t-test described in Section 10.2.2.2. Table 10.5 gives the t-values for the pairwise comparisons among the three levels of B along with their corresponding p-values.

The entries in Tables 10.4 and 10.5 were obtained by using the following statements in SAS's (2000) PROC GLM:

```
PROC GLM;
CLASS A B;
MODEL Y = A B;
LSMEANS A B/TDIFF;
RUN;
```

where "LSMEANS A B" stands for the least-squares means for the levels of factors A and B, and "TDIFF" is an option that requests the t-values for the pairwise comparisons of the least-squares means along with their corresponding p-values.

10.3 Two-Way Models with Interaction

Let us now consider the complete two-way model,

$$Y_{ijk} = \mu + \alpha_{(i)} + \beta_{(j)} + (\alpha\beta)_{(ij)} + \epsilon_{ij(k)}, \tag{10.36}$$

$i = 1, 2, \ldots, a$; $j = 1, 2, \ldots, b$; $k = 1, 2, \ldots, n_{ij}$. Note that $n_{ij} > 0$ for all i, j indicating that the data set contains no empty cells. This is an extension of model (10.10) with the addition of the interaction effect, $(\alpha\beta)_{(ij)}, i = 1, 2, \ldots, a$; $j = 1, 2, \ldots, b$. Thus all the cell means, μ_{ij}, are estimable for all i, j. As before, the $\epsilon_{ij(k)}$'s are assumed to be independently distributed as $N(0, \sigma^2)$.

For this model, the matrix $\boldsymbol{X}$ in (10.1) is of order $n_{..} \times (1 + a + b + ab)$ of the form

$$\boldsymbol{X} = [\boldsymbol{1}_{n_{..}} : \boldsymbol{H}_1 : \boldsymbol{H}_2 : \boldsymbol{H}_3], \tag{10.37}$$

where $\boldsymbol{H}_1$ and $\boldsymbol{H}_2$ are the same as in Section 10.2 and $\boldsymbol{H}_3 = \oplus_{i=1}^{a} \oplus_{j=1}^{b} \boldsymbol{1}_{n_{ij}}$. We note that the rank of $\boldsymbol{X}$ is equal to ab, which is the rank of $\boldsymbol{H}_3$. We assume that $n_{..} > ab$. The vector, $\boldsymbol{\beta}$, of unknown parameters in (10.1) consists in this case of μ, the $\alpha_{(i)}$'s $(i = 1, 2, \ldots, a)$, the $\beta_{(j)}$'s $(j = 1, 2, \ldots, b)$, and the $(\alpha\beta)_{(ij)}$'s.

Since μ_{ij} is estimable for all i and j, all linear functions of μ_{ij} are also estimable. The BLUE of μ_{ij} is $\bar{Y}_{ij.} = \frac{1}{n_{ij}} \sum_{k=1}^{n_{ij}} Y_{ijk}$ $(i = 1, 2, \ldots, a; j = 1, 2, \ldots, b)$. This follows from the fact that $\hat{\boldsymbol{\beta}} = (\boldsymbol{X}'\boldsymbol{X})^{-}\boldsymbol{X}'\boldsymbol{Y}$ gives $\hat{\mu} = 0$, $\hat{\alpha}_{(i)} = 0$ $(i = 1, 2, \ldots, a)$, $\hat{\beta}_{(j)} = 0$ $(j = 1, 2, \ldots, b)$, and $\widehat{(\alpha\beta)}_{(ij)} = \bar{Y}_{ij.}$ $(i = 1, 2, \ldots, a; j = 1, 2, \ldots, b)$. In particular, the following linear functions are estimable:

$$\begin{aligned} LSM(\alpha_{(i)}) &= \frac{1}{b}\sum_{j=1}^{b} \mu_{ij} \\ &= \mu + \alpha_{(i)} + \frac{1}{b}\sum_{j=1}^{b} \beta_{(j)} + \frac{1}{b}\sum_{j=1}^{b} (\alpha\beta)_{(ij)}, \quad i = 1, 2, \ldots, a, \end{aligned} \tag{10.38}$$

$$\begin{aligned} LSM(\beta_{(j)}) &= \frac{1}{a}\sum_{i=1}^{a} \mu_{ij} \\ &= \mu + \frac{1}{a}\sum_{i=1}^{a} \alpha_{(i)} + \beta_{(j)} + \frac{1}{a}\sum_{i=1}^{a} (\alpha\beta)_{(ij)}, \quad j = 1, 2, \ldots, b, \end{aligned} \tag{10.39}$$

$$\begin{aligned} \Theta_{ij,i'j'} &= \mu_{ij} - \mu_{ij'} - \mu_{i'j} + \mu_{i'j'} \\ &= (\alpha\beta)_{(ij)} - (\alpha\beta)_{(ij')} - (\alpha\beta)_{(i'j)} + (\alpha\beta)_{(i'j')}, \\ & i \neq i' = 1, 2, \ldots, a;\ j \neq j' = 1, 2, \ldots, b, \end{aligned} \tag{10.40}$$

$$WM(\alpha_{(i)}) = \frac{1}{n_{i.}}\sum_{j=1}^{b} n_{ij}\mu_{ij},$$

$$= \mu + \alpha_{(i)} + \frac{1}{n_{i.}}\sum_{j=1}^{b} n_{ij}\beta_{(j)} + \frac{1}{n_{i.}}\sum_{j=1}^{b} n_{ij}\,(\alpha\beta)_{(ij)},$$

$$i = 1, 2, \ldots, a, \tag{10.41}$$

$$WM(\beta_{(j)}) = \frac{1}{n_{.j}}\sum_{i=1}^{a} n_{ij}\mu_{ij},$$

$$= \mu + \frac{1}{n_{.j}}\sum_{i=1}^{a} n_{ij}\alpha_{(i)} + \beta_{(j)} + \frac{1}{n_{.j}}\sum_{i=1}^{a} n_{ij}\,(\alpha\beta)_{(ij)},$$

$$j = 1, 2, \ldots, b. \tag{10.42}$$

Here, $LSM(\alpha_{(i)})$ and $LSM(\beta_{(j)})$ are the least-squares means for row i and column j, respectively ($i = 1, 2, \ldots, a$; $j = 1, 2, \ldots, b$), $\Theta_{ij,i'j'}$ is an interaction contrast to be defined later in Section 10.3.1 ($i \neq i'$, $j \neq j'$); $WM(\alpha_{(i)})$ and $WM(\beta_{(j)})$ are the weighted means of the cell means in row i and column j, respectively ($i = 1, 2, \ldots, a$; $j = 1, 2, \ldots, b$).

10.3.1 Tests of Hypotheses

The analysis of model (10.36) with regard to the testing of the main effects of A and B can be carried out using the so-called *method of weighted squares of means* (MWSM), which was introduced by Yates (1934). The following is a description of this method.

Let $X_{ij} = \bar{Y}_{ij.} = \frac{1}{n_{ij}}\sum_{k=1}^{n_{ij}} Y_{ijk}$. The MWSM is based on using the X_{ij}'s in setting up the following sums of squares for A and B:

$$SS_{A\omega} = \sum_{i=1}^{a} \omega_{1i}(\bar{X}_{i.} - \bar{X}_{1\omega})^2 \tag{10.43}$$

$$SS_{B\omega} = \sum_{j=1}^{b} \omega_{2j}(\bar{X}_{.j} - \bar{X}_{2\omega})^2, \tag{10.44}$$

where $\bar{X}_{i.} = \frac{1}{b}\sum_{j=1}^{b} X_{ij}$, $\bar{X}_{.j} = \frac{1}{a}\sum_{i=1}^{a} X_{ij}$, $\omega_{1i} = \left(\frac{1}{b^2}\sum_{j=1}^{b}\frac{1}{n_{ij}}\right)^{-1}$, $\omega_{2j} = \left(\frac{1}{a^2}\sum_{i=1}^{a}\frac{1}{n_{ij}}\right)^{-1}$, $\bar{X}_{1\omega} = \sum_{i=1}^{a}\omega_{1i}\bar{X}_{i.}/\sum_{i=1}^{a}\omega_{1i}$, and $\bar{X}_{2\omega} = \sum_{j=1}^{b}\omega_{2j}\bar{X}_{.j}/\sum_{j=1}^{b}\omega_{2j}$. Note that

$$\text{Var}(\bar{X}_{i.}) = \frac{\sigma^2}{b^2}\sum_{j=1}^{b}\frac{1}{n_{ij}}, \quad i = 1, 2, \ldots, a,$$

$$\text{Var}(\bar{X}_{.j}) = \frac{\sigma^2}{a^2}\sum_{i=1}^{a}\frac{1}{n_{ij}}, \quad j = 1, 2, \ldots, b.$$

Thus, the weights, ω_{1i} and ω_{2j}, used in the sums of squares in (10.43) and (10.44) are equal to $\sigma^2/\text{Var}(\bar{X}_{i.})$ and $\sigma^2/\text{Var}(\bar{X}_{.j})$, respectively. Furthermore, $\bar{X}_{1\omega}$ and $\bar{X}_{2\omega}$ are the weighted averages of the $\bar{X}_{i.}$'s and the $\bar{X}_{.j}$'s, weighted by the ω_{1i}'s and the ω_{2j}'s, respectively.

Theorem 10.2 If the $\epsilon_{ij(k)}$'s in model (10.36) are mutually independent such that $\epsilon_{ij(k)} \sim N(0, \sigma^2)$, then

(a) $SS_{A\omega}/\sigma^2 \sim \chi^2_{a-1}(\lambda_{1\omega})$.

(b) $SS_{B\omega}/\sigma^2 \sim \chi^2_{b-1}(\lambda_{2\omega})$.

(c) The error sum of squares for model (10.36), namely,

$$SS_E = \sum_{i=1}^{a}\sum_{j=1}^{b}\sum_{k=1}^{n_{ij}}(Y_{ijk} - \bar{Y}_{ij.})^2, \tag{10.45}$$

is independent of $SS_{A\omega}$ and $SS_{B\omega}$.

(d) $SS_E/\sigma^2 \sim \chi^2_{n_{..}-ab}$,

where $\lambda_{1\omega}$ and $\lambda_{2\omega}$ are the noncentrality parameters associated with factors A and B, respectively.

Proof.

(a) The $\bar{X}_{i.}$'s are mutually independent and normally distributed such that

$$E(\bar{X}_{i.}) = \frac{1}{b}\sum_{j=1}^{b}\mu_{ij}$$
$$= LSM(\alpha_{(i)}), \quad i = 1, 2, \ldots, a,$$

which is the least-squares mean for level i of A [see (10.38)], and

$$\text{Var}(\bar{X}_{i.}) = \frac{\sigma^2}{\omega_{1i}}, \quad i = 1, 2, \ldots, a.$$

Let $\bar{\boldsymbol{X}} = (\bar{X}_{1.}, \bar{X}_{2.}, \ldots, \bar{X}_{a.})'$. Then, $\bar{\boldsymbol{X}}$ is normally distributed with mean

$$E(\bar{\boldsymbol{X}}) = [LSM(\alpha_{(1)}), LSM(\alpha_{(2)}), \ldots, LSM(\alpha_{(a)})]', \tag{10.46}$$

and a variance–covariance matrix,

$$\text{Var}(\bar{X}) = \sigma^2 \,\text{diag}\left(\omega_{11}^{-1}, \omega_{12}^{-1}, \ldots, \omega_{1a}^{-1}\right). \tag{10.47}$$

Let $\boldsymbol{\omega}_1 = (\omega_{11}, \omega_{12}, \ldots, \omega_{1a})'$. Then, $SS_{A\omega}$ in (10.43) can be written as

$$\begin{aligned} SS_{A\omega} &= \sum_{i=1}^{a} \omega_{1i} \bar{X}_{i.}^2 - \frac{1}{\sum_{i=1}^{a} \omega_{1i}} \left(\sum_{i=1}^{a} \omega_{1i} \bar{X}_{i.} \right)^2 \\ &= \bar{X}' \left[\text{diag}(\omega_{11}, \omega_{12}, \ldots, \omega_{1a}) - \frac{1}{\sum_{i=1}^{a} \omega_{1i}} \boldsymbol{\omega}_1 \boldsymbol{\omega}_1' \right] \bar{X}. \end{aligned} \tag{10.48}$$

Given the fact that $\bar{X}$ is normally distributed with the variance–covariance matrix in (10.47), then by Theorem 5.4, $SS_{A\omega}/\sigma^2 \sim \chi^2_{a-1}(\lambda_{1\omega})$. This is true because the matrix,

$$\frac{1}{\sigma^2} \left[\text{diag}(\omega_{11}, \omega_{12}, \ldots, \omega_{1a}) - \frac{1}{\sum_{i=1}^{a} \omega_{1i}} \boldsymbol{\omega}_1 \boldsymbol{\omega}_1' \right] \times \sigma^2 \text{diag}\left(\omega_{11}^{-1}, \omega_{12}^{-1}, \ldots, \omega_{1a}^{-1}\right) = \boldsymbol{I}_a - \frac{1}{\sum_{i=1}^{a} \omega_{1i}} \boldsymbol{\omega}_1 \boldsymbol{1}_a',$$

is idempotent of rank $a - 1$. The noncentrality parameter $\lambda_{1\omega}$ is

$$\lambda_{1\omega} = \frac{1}{\sigma^2} \left[E(\bar{X}) \right]' \left[\text{diag}(\omega_{11}, \omega_{12}, \ldots, \omega_{1a}) - \frac{1}{\sum_{i=1}^{a} \omega_{1i}} \boldsymbol{\omega}_1 \boldsymbol{\omega}_1' \right] E(\bar{X}). \tag{10.49}$$

(b) This is similar to (a).

(c) This follows from the fact that SS_E is independent of the $\bar{X}_{ij}$'s and is therefore independent of both $SS_{A\omega}$ and $SS_{B\omega}$.

(d) $SS_E = \boldsymbol{Y}' \left[\boldsymbol{I}_{n_{..}} - \oplus_{i=1}^{a} \oplus_{j=1}^{b} \frac{1}{n_{ij}} \boldsymbol{J}_{n_{ij}} \right] \boldsymbol{Y}$, where $\boldsymbol{Y} \sim N(\boldsymbol{X\beta}, \sigma^2 \boldsymbol{I}_{n_{..}})$. The matrix, $\boldsymbol{I}_{n_{..}} - \oplus_{i=1}^{a} \oplus_{j=1}^{b} \frac{1}{n_{ij}} \boldsymbol{J}_{n_{ij}}$, is idempotent of rank $n_{..} - ab$. Hence, SS_E/σ^2 is a chi-squared variate with $n_{..} - ab$ degrees of freedom. Its noncentrality parameter is zero since

$$\left[\boldsymbol{I}_{n_{..}} - \bigoplus_{i=1}^{a} \bigoplus_{j=1}^{b} \frac{1}{n_{ij}} \boldsymbol{J}_{n_{ij}} \right] \boldsymbol{H}_3 = \boldsymbol{0}.$$

Hence, the products of $\boldsymbol{I}_{n_{..}} - \oplus_{i=1}^{a} \oplus_{j=1}^{b} \frac{1}{n_{ij}} \boldsymbol{J}_{n_{ij}}$ with $\boldsymbol{1}_{n_{..}}$, $\boldsymbol{H}_1$, and $\boldsymbol{H}_2$ [see (10.37)] are also zero. Consequently,

$$\boldsymbol{\beta}' \boldsymbol{X}' \left[\boldsymbol{I}_{n_{..}} - \bigoplus_{i=1}^{a} \bigoplus_{j=1}^{b} \frac{1}{n_{ij}} \boldsymbol{J}_{n_{ij}} \right] \boldsymbol{X\beta} = 0. \qquad \square$$

Lemma 10.5

(a) The ratio, $F_A = \frac{SS_{A\omega}}{(a-1)MS_E}$ tests the hypothesis,

$$H_0 : LSM(\alpha_{(i)}) \text{ equal for all } i = 1, 2, \ldots, a. \tag{10.50}$$

Under H_0, F_A has the F-distribution with $a - 1$ and $n_{..} - ab$ degrees of freedom.

(b) The ratio, $F_B = \frac{SS_{B\omega}}{(b-1)MS_E}$ tests the hypothesis,

$$H_0 : LSM(\beta_{(j)}) \text{ equal for all } j = 1, 2, \ldots, b. \tag{10.51}$$

Under H_0, F_B has the F-distribution with $b - 1$ and $n_{..} - ab$ degrees of freedom.

Proof.

(a) The noncentrality parameter of $SS_{A\omega}/\sigma^2$ is $\lambda_{1\omega}$, which is given in (10.49). Note that $\sigma^2\lambda_{1\omega}$ is the same quadratic form as the one in (10.48), except that the mean of $\bar{\mathbf{X}}$ is used instead of $\bar{\mathbf{X}}$. Using (10.43), $\lambda_{1\omega}$ can be written as

$$\lambda_{1\omega} = \frac{1}{\sigma^2}\sum_{i=1}^{a}\omega_{1i}\left[LSM(\alpha_{(i)}) - \frac{1}{\sum_{l=1}^{a}\omega_{1l}}\sum_{l=1}^{a}\omega_{1l}LSM(\alpha_{(l)})\right]^2.$$

Thus, $\lambda_{1\omega} = 0$ if and only if

$$LSM(\alpha_{(1)}) = LSM(\alpha_{(2)}) = \ldots = LSM(\alpha_{(a)}).$$

Under H_0, F_A has the central F-distribution with $a-1$ and $n_{..}-ab$ degrees of freedom by Theorem 10.2 (a, c, d).

(b) This is similar to (a). □

10.3.1.1 Testing the Interaction Effect

Consider the ratio,

$$F(\alpha\beta \mid \mu, \alpha, \beta) = \frac{R(\alpha\beta \mid \mu, \alpha, \beta)}{(a-1)(b-1)MS_E}, \tag{10.52}$$

where

$$R(\alpha\beta \mid \mu, \alpha, \beta) = R(\mu, \alpha, \beta, \alpha\beta) - R(\mu, \alpha, \beta), \tag{10.53}$$

is the increase in the regression sum of squares which results from adding $(\alpha\beta)_{(ij)}$ to model (10.10) giving rise to model (10.36). According to Definition 10.3, $R(\alpha\beta \mid \mu, \alpha, \beta)$ is Type III sum of squares for the interaction effect.

Applying Theorem 10.1 with $X_1 = [\mathbf{1}_{n_{..}} : H_1 : H_2]$, $X_2 = H_3$, $r = ab$, $r_1 = a + b - 1$, we find that

$$\frac{1}{\sigma^2} R(\alpha\beta \mid \mu, \alpha, \beta) \sim \chi^2_{ab-a-b+1}(\lambda_{12}),$$

where

$$\lambda_{12} = \frac{1}{\sigma^2} (\boldsymbol{\alpha\beta})' X_2' [I_{n_{..}} - X_1(X_1'X_1)^- X_1'] X_2 (\boldsymbol{\alpha\beta}), \tag{10.54}$$

and $(\boldsymbol{\alpha\beta}) = [(\alpha\beta)_{(11)}, (\alpha\beta)_{(12)}, \ldots, (\alpha\beta)_{(ab)}]'$. Furthermore, $R(\alpha\beta \mid \mu, \alpha, \beta)$ is independent of SS_E, the error sum of squares in (10.45). Hence, $F(\alpha\beta \mid \mu, \alpha, \beta)$ has the noncentral F-distribution with $(a-1)(b-1)$ and $n_{..} - ab$ degrees of freedom and a noncentrality parameter λ_{12}. By (10.9), $\lambda_{12} = 0$ if and only if

$$[I_{n_{..}} - X_1(X_1'X_1)^- X_1'] X_2 (\boldsymbol{\alpha\beta}) = \mathbf{0}. \tag{10.55}$$

Let us now consider the following lemma:

Lemma 10.6 The noncentrality parameter, λ_{12}, in (10.54) is equal to zero if and only if

$$(\alpha\beta)_{(ij)} = \gamma_{1i} + \gamma_{2j}, \quad i = 1, 2, \ldots, a; \ j = 1, 2, \ldots, b, \tag{10.56}$$

where γ_{1i} and γ_{2j} are constants.

Proof. We have that $X_2 = H_3$. Also, the matrix, $X_1(X_1'X_1)^- X_1'$ can be written as

$$X_1(X_1'X_1)^- X_1' = (H_1 : H_2)[(H_1 : H_2)'(H_1 : H_2)]^- (H_1 : H_2)',$$

since the column vector $\mathbf{1}_{n_{..}}$ in X_1 is the sum of the columns of H_1 (also of H_2) and is therefore linearly dependent on the columns of $(H_1 : H_2)$. Formula (10.55) can then be expressed as

$$H_3 \, (\boldsymbol{\alpha\beta}) = (H_1 : H_2) \, \boldsymbol{\gamma}, \tag{10.57}$$

where $\boldsymbol{\gamma} = [(H_1 : H_2)'(H_1 : H_2)]^- (H_1 : H_2)' H_3 \, (\boldsymbol{\alpha\beta})$. Let $\boldsymbol{\gamma}$ be partitioned as $\boldsymbol{\gamma} = [\boldsymbol{\gamma}_1' : \boldsymbol{\gamma}_2']'$, where $\boldsymbol{\gamma}_1$ and $\boldsymbol{\gamma}_2$ are vectors of a and b elements, respectively. Recall that $H_1 = \oplus_{i=1}^{a} \mathbf{1}_{n_{i.}}$, $H_2 = [\oplus_{j=1}^{b} \mathbf{1}'_{n_{1j}} : \oplus_{j=1}^{b} \mathbf{1}'_{n_{2j}} : \ldots : \oplus_{j=1}^{b} \mathbf{1}'_{n_{aj}}]'$, $H_3 = \oplus_{i=1}^{a} \oplus_{j=1}^{b} \mathbf{1}_{n_{ij}}$. From (10.57) we then have

$$\left[\bigoplus_{i=1}^{a} \bigoplus_{j=1}^{b} \mathbf{1}_{n_{ij}} \right] (\boldsymbol{\alpha\beta}) = \left[\bigoplus_{i=1}^{a} \mathbf{1}_{n_{i.}} \right] \boldsymbol{\gamma}_1 + H_2 \, \boldsymbol{\gamma}_2. \tag{10.58}$$

Each term in (10.58) is a vector of $n_{..}$ elements consisting of a string of a vectors, the ith of which consists of $n_{i.}$ elements ($i = 1, 2, \ldots, a$). Equating the ith of such vectors on the left-hand side to the sum of the corresponding vectors on the right-hand side, we get

$$\left[\bigoplus_{j=1}^{b} \mathbf{1}_{n_{ij}}\right](\boldsymbol{\alpha\beta})_i = \gamma_{1i}\mathbf{1}_{n_{i.}} + \left[\bigoplus_{j=1}^{b} \mathbf{1}_{n_{ij}}\right]\boldsymbol{\gamma}_2, \quad i = 1, 2, \ldots, a, \tag{10.59}$$

where $(\boldsymbol{\alpha\beta})_i$ is the corresponding ith portion of $(\boldsymbol{\alpha\beta})$ such that

$$(\boldsymbol{\alpha\beta})_i = [(\alpha\beta)_{(i1)}, (\alpha\beta)_{(i2)}, \ldots, (\alpha\beta)_{(ib)}]',$$

and γ_{1i} is the ith element of $\boldsymbol{\gamma}_1$ ($i = 1, 2, \ldots, a$). Equating the coefficients of $\mathbf{1}_{n_{ij}}$ on both sides of (10.59), we get

$$(\alpha\beta)_{(ij)} = \gamma_{1i} + \gamma_{2j}, \quad i = 1, 2, \ldots, a; \; j = 1, 2, \ldots, b,$$

where γ_{2j} is the jth element of $\boldsymbol{\gamma}_2$ ($j = 1, 2, \ldots, b$). Formula (10.56) is then proved. □

Lemma 10.6 indicates that $\lambda_{12} = 0$ if and only if model (10.36) is additive, that is, contains no interaction effect. This follows from the fact that (10.56) implies that

$$\begin{aligned}\mu_{ij} &= \mu + \alpha_{(i)} + \beta_{(j)} + (\alpha\beta)_{(ij)} \\ &= \mu + (\alpha_{(i)} + \gamma_{1i}) + (\beta_{(j)} + \gamma_{2j}), \quad i = 1, 2, \ldots, a; \; j = 1, 2, \ldots, b.\end{aligned}$$

Thus, in this case, μ_{ij} is the sum of μ, a constant depending on i only, and another constant depending on j only. Consequently, the F-ratio, $F(\alpha\beta \mid \mu, \alpha, \beta)$, in (10.52) tests the hypothesis that the model is additive. If $\Theta_{ij,i'j'}$ is defined as

$$\Theta_{ij,i'j'} = (\alpha\beta)_{(ij)} - (\alpha\beta)_{(ij')} - (\alpha\beta)_{(i'j)} + (\alpha\beta)_{(i'j')}, \quad \text{for all } i \neq i'; \; j \neq j', \tag{10.60}$$

then, $\Theta_{ij,i'j'} = 0$ if (10.56) is true. Note that $\Theta_{ij,i'j'}$ can also be written as

$$\Theta_{ij,i'j'} = \mu_{ij} - \mu_{ij'} - \mu_{i'j} + \mu_{i'j'}, \quad \text{for all } i \neq i'; \; j \neq j'. \tag{10.61}$$

This is a contrast in the means of the cells (i, j), (i, j'), (i', j), (i', j'), and is referred to as an *interaction contrast*. We can therefore conclude that $F(\alpha\beta \mid \mu, \alpha, \beta)$ tests the hypothesis

$$\begin{aligned}H_0 : \Theta_{ij,i'j'} &= (\alpha\beta)_{(ij)} - (\alpha\beta)_{(ij')} - (\alpha\beta)_{(i'j)} + (\alpha\beta)_{(i'j')} \\ &= \mu_{ij} - \mu_{ij'} - \mu_{i'j} + \mu_{i'j'} \\ &= 0, \quad \text{for all } i \neq i', \; j \neq j'.\end{aligned} \tag{10.62}$$

This is called the *no-interaction hypothesis*. Note that the number of linearly independent contrasts of the form $\Theta_{ij,i'j'}$ is equal to $(a-1)(b-1)$. The test is significant at the α-level if $F(\alpha\beta \mid \mu, \alpha, \beta) \geq F_{\alpha,(a-1)(b-1),n_{..}-ab}$. Since this test statistic utilizes a Type III sum of squares in its numerator, namely, $R(\alpha\beta \mid \mu, \alpha, \beta)$, it is considered a Type III F-ratio for the interaction.

Note that when (10.62) is true,

$$\mu_{ij} - \mu_{i'j} = \mu_{ij'} - \mu_{i'j'}, \quad \text{for all } i \neq i'; j \neq j'.$$

This means that the change in the mean response under one factor (using, for example, levels i and i' of A) is the same for all levels of the other factor (for example, B). In this case, we say that factors A and B do not interact, or that the $A * B$ interaction is zero.

In the event the $A * B$ interaction is significant, testing the main effects of A and B using F_A and F_B, respectively, from Lemma 10.5 may not be meaningful. This is true because comparisons among the least-squares means for A (or B) can be masked (or obscured) by the interaction. More specifically, suppose that we were to compare, for example, $LSM(\alpha_{(i)})$ against $LSM(\alpha_{(i')})$, $i \neq i'$, when $A * B$ is present. Then,

$$LSM(\alpha_{(i)}) - LSM(\alpha_{(i')}) = \frac{1}{b}\sum_{j=1}^{b}(\mu_{ij} - \mu_{i'j}),\ i \neq i'.$$

Since $\mu_{ij} - \mu_{i'j}$ depends on j, some of these differences may be positive for certain values of j and some may be negative for other values of j. As a result, $LSM(\alpha_{(i)}) - LSM(\alpha_{(i')})$ may be small giving the false indication that the two least-squares means are not different, which may not be the case. In this situation, to determine if A has an effect, it would be more meaningful to compare the cell means, $\mu_{1j}, \mu_{2j}, \ldots, \mu_{aj}$, for a fixed j $(=1, 2, \ldots, b)$. Similarly, the cell means of B can be compared for a fixed i $(=1, 2, \ldots, a)$. Such comparisons can be made by using, for example, *Tukey's Studentized range test*, which controls the *experimentwise Type I error rate*, if the factor under consideration is qualitative [Tukey's test requires the means to have equal sample sizes. See, for example, Ott and Longnecker (2004, Section 8.6). For unequal sample sizes, a modified version of Tukey's test, due to Kramer (1956), can be used.]. If, however, the factor is quantitative, then it would be more informative to test the significance of certain orthogonal contrasts among the means of the factor for fixed levels of the other factor. These contrasts represent the so-called *polynomial effects* of the factor (see, for example, Christensen, 1996, Section 7.12).

In case $A * B$ is not significant, it will be meaningful to compare the least-squares means for the levels of A and B, if their effects are significant. If the data set contains no empty cells, as was assumed earlier, then all least-squares means are estimable. Their best linear unbiased estimates (BLUE) are obtained by replacing the parameters in a given least-squares mean by the corresponding elements of $\hat{\boldsymbol{\beta}} = (\boldsymbol{X}'\boldsymbol{X})^{-}\boldsymbol{X}'\boldsymbol{Y}$, where $\boldsymbol{X}$ is the matrix given in (10.37) and $\boldsymbol{\beta}$ is the vector of all model parameters. Hence, the BLUE of $LSM(\alpha_{(i)})$ and $LSM(\beta_{(j)})$ in (10.38) and (10.39), respectively, are

$$\widehat{LSM}(\alpha_{(i)}) = \hat{\mu} + \hat{\alpha}_{(i)} + \frac{1}{b}\sum_{j=1}^{b}[\hat{\beta}_{(j)} + \widehat{\alpha\beta}_{(ij)}], \quad i = 1, 2, \ldots, a,$$

$$\widehat{LSM}(\beta_{(j)}) = \hat{\mu} + \hat{\beta}_{(j)} + \frac{1}{a}\sum_{i=1}^{a}[\hat{\alpha}_{(i)} + \widehat{\alpha\beta}_{(ij)}], \quad j = 1, 2, \ldots, b,$$

where $\hat{\mu}$, the $\hat{\alpha}_{(i)}$'s, $\hat{\beta}_{(j)}$'s, and $\widehat{\alpha\beta}_{(ij)}$'s are the elements of $\hat{\beta} = (X'X)^{-}X'Y$. Since the columns of the matrix $H_3 = \oplus_{i=1}^{a} \oplus_{j=1}^{b} \mathbf{1}_{n_{ij}}$ span the column space of X [see (10.37)], then $\hat{\mu} = 0$, $\hat{\alpha}_{(i)} = 0$ $(i = 1, 2, \ldots, a)$, $\hat{\beta}_{(j)} = 0$ $(j = 1, 2, \ldots, b)$, $\widehat{\alpha\beta}_{(ij)} = \bar{Y}_{ij.}$ $(i = 1, 2, \ldots, a; j = 1, 2, \ldots, b)$, as was seen earlier. Hence,

$$\widehat{LSM}(\alpha_{(i)}) = \frac{1}{b} \sum_{j=1}^{b} \bar{Y}_{ij.}, \quad i = 1, 2, \ldots, a,$$

$$\widehat{LSM}(\beta_{(j)}) = \frac{1}{a} \sum_{i=1}^{a} \bar{Y}_{ij.}, \quad j = 1, 2, \ldots, b.$$

The test statistic for comparing, for example, $LSM(\alpha_{(i)})$ against $LSM(\alpha_{(i')})$, $i \neq i'$, is given by the same t-statistic shown in (10.35), except that, in this case, the error mean square, MS_E, has $n_{..} - ab$ degrees of freedom.

10.3.2 Type III Analysis in SAS

We have seen that the testing of the no-interaction hypothesis in (10.62) was done using the Type III F-ratio, $F(\alpha\beta \mid \mu, \alpha, \beta)$, which can be easily obtained by invoking PROC GLM in SAS. In addition to this ratio, SAS also provides Type III F-ratios for factors A and B. The latter ratios, however, are derived after imposing certain restrictions on the parameters of model (10.36), namely,

$$\begin{aligned} \sum_{i=1}^{a} \alpha_{(i)} &= 0, \\ \sum_{j=1}^{b} \beta_{(j)} &= 0, \\ \sum_{i=1}^{a} (\alpha\beta)_{(ij)} &= 0, \quad \text{for all } j = 1, 2, \ldots, b, \\ \sum_{j=1}^{b} (\alpha\beta)_{(ij)} &= 0, \quad \text{for all } i = 1, 2, \ldots, a. \end{aligned} \tag{10.63}$$

The need for such restrictions stems from the fact that the Type III sums of squares for factors A and B for model (10.36) are actually identically equal to zero. This model is said to be *overparameterized* because the number of its unknown parameters, namely $1 + a + b + ab$, exceeds the rank of X in (10.37), which is equal to ab. For such a model, the Type III sum of squares for A is, by definition,

$$R(\alpha \mid \mu, \beta, \alpha\beta) = R(\mu, \alpha, \beta, \alpha\beta) - R(\mu, \beta, \alpha\beta).$$

But,

$$R(\mu, \alpha, \beta, \alpha\beta) = Y'X(X'X)^{-}X'Y$$
$$= Y'H_3(H_3{}'H_3)^{-1}H_3{}'Y, \quad (10.64)$$

since the column space of X is spanned by the columns of $H_3 = \oplus_{i=1}^{a} \oplus_{j=1}^{b} \mathbf{1}_{n_{ij}}$. Furthermore,

$$R(\mu, \beta, \alpha\beta) = Y'X_\alpha(X_\alpha{}'X_\alpha)^{-}X_\alpha{}'Y, \quad (10.65)$$

where $X_\alpha = [\mathbf{1}_{n_{..}} : H_2 : H_3]$ is obtained from X by removing the a columns corresponding to $\alpha_{(i)}$ in the model. Since the column space of X_α is also spanned by the columns of H_3, then

$$Y'X_\alpha(X_\alpha{}'X_\alpha)^{-}X_\alpha{}'Y = Y'H_3(H_3{}'H_3)^{-1}H_3{}'Y. \quad (10.66)$$

From (10.64)–(10.66) we conclude that

$$R(\alpha \mid \mu, \beta, \alpha\beta) = 0. \quad (10.67)$$

Similarly, it can be shown that

$$R(\beta \mid \mu, \alpha, \alpha\beta) = 0. \quad (10.68)$$

Now, let us reparameterize model (10.36) using the restrictions in (10.63). Since the number of linearly independent equations in (10.63) is equal to $1 + a + b$, the number of linearly independent parameters in model (10.36) under these restrictions is ab, which is equal to the rank of X. Model (10.36) can then be reparameterized and expressed in terms of only ab linearly independent parameters to get the model,

$$Y = X^*\beta^* + \epsilon, \quad (10.69)$$

where the elements of β^* consist of ab linearly independent parameters and X^* is a matrix of order $n_{..} \times ab$ and rank ab. Thus, X^* is of full column rank and (10.69) is therefore a full-rank model. Using this model, the Type III sums of squares for A and B are expressed as $R(\alpha^* \mid \mu^*, \beta^*, \alpha\beta^*)$ and $R(\beta^* \mid \mu^*, \alpha^*, \alpha\beta^*)$, respectively, and are obviously not identically equal to zero. These R-expressions are the Type III sums of squares given by SAS for A and B, respectively.

It can be shown that $R(\alpha^* \mid \mu^*, \beta^*, \alpha\beta^*)$ and $R(\beta^* \mid \mu^*, \alpha^*, \alpha\beta^*)$ are the same as $SS_{A\omega}$ and $SS_{B\omega}$, the sums of squares for A and B in (10.43) and (10.44), respectively, which were derived using the method of weighted squares of means (see, for example, Speed and Hocking, 1976, p. 32; Searle, Speed, and Henderson, 1981, Section 5.2; Searle, 1994, Section 3.1). We can therefore conclude that on the basis of the reparameterized model in (10.69), the Type III F-ratios for A and B, namely,

$$F(\alpha^* \mid \mu^*, \beta^*, \alpha\beta^*) = \frac{R(\alpha^* \mid \mu^*, \beta^*, \alpha\beta^*)}{(a-1)MS_E}$$

$$F(\beta^* \mid \mu^*, \alpha^*, \alpha\beta^*) = \frac{R(\beta^* \mid \mu^*, \alpha^*, \alpha\beta^*)}{(b-1)MS_E},$$

are identical to the F-ratios, F_A and F_B, given in Lemma 10.5. The corresponding hypotheses are the ones described in (10.50) and (10.51) which equate the least-squares means for A and B, respectively.

Thus, in conclusion, the Type III analysis given in SAS can be used to test the significance of factors A, B, and their interaction $A * B$. Note that the Type III F-ratio for $A * B$, which is based on the overparameterized model (10.36), is identical to the one obtained under the reparameterized model (10.69), that is, $F(\alpha\beta \mid \mu, \alpha, \beta) = F(\alpha\beta^* \mid \mu^*, \alpha^*, \beta^*)$. This is true because $R(\alpha\beta \mid \mu, \alpha, \beta) = R(\alpha\beta^* \mid \mu^*, \alpha^*, \beta^*)$.

10.3.3 Other Testable Hypotheses

There are other testable hypotheses concerning the parameters of model (10.36). The values of their test statistics can be obtained through the use of PROC GLM in SAS. These hypotheses, however, are not desirable because they are data dependent. We now give a brief account of these hypotheses and their corresponding test statistics. More details can be found in Searle (1971, Chapter 7) and Speed and Hocking (1976).

(a) Hypotheses tested by $F(\alpha \mid \mu)$ and $F(\beta \mid \mu)$

By definition,

$$F(\alpha \mid \mu) = \frac{R(\alpha \mid \mu)}{(a-1)MS_E} \tag{10.70}$$

$$F(\beta \mid \mu) = \frac{R(\beta \mid \mu)}{(b-1)MS_E}, \tag{10.71}$$

where $R(\alpha \mid \mu) = R(\mu, \alpha) - R(\mu)$ is Type I sum of squares for factor A, if A appears first in the SAS model, $Y = A\ B\ A * B$. Similarly, $R(\beta \mid \mu) = R(\mu, \beta) - R(\mu)$ is Type I sum of squares for factor B, if B appears first in the SAS model. Hence, $F(\alpha \mid \mu)$ and $F(\beta \mid \mu)$ are considered Type I F-ratios.

The statistic, $F(\alpha \mid \mu)$, tests the hypothesis,

$$H_0 : WM(\alpha_{(i)}) \text{ equal for all } i = 1, 2, \ldots, a,$$

where $WM(\alpha_{(i)})$ is the weighted mean of the cell means in row i $(= 1, 2, \ldots, a)$ as shown in formula (10.41). This hypothesis can be written as

$$H_0 : \alpha_{(i)} + \frac{1}{n_{i.}} \sum_{j=1}^{b} n_{ij} \left[\beta_{(j)} + (\alpha\beta)_{(ij)}\right] \text{ equal for all } i = 1, 2, \ldots, a. \tag{10.72}$$

The proof of the testability of this hypothesis by $F(\alpha \mid \mu)$ is similar to the one given in Section 10.2.2.1. Similarly, the ratio, $F(\beta \mid \mu)$, tests the hypothesis,

$$H_0 : WM(\beta_{(j)}) \text{ equal for all } j = 1, 2, \ldots, b,$$

or equivalently, the hypothesis,

$$H_0 : \beta_{(j)} + \frac{1}{n_{.j}} \sum_{i=1}^{a} n_{ij}[\alpha_{(i)} + (\alpha\beta)_{(ij)}] \text{ equal for all } j = 1, 2, \ldots, b, \tag{10.73}$$

where $WM(\beta_{(j)})$ is the weighted mean of the cell means in column j $(= 1, 2, \ldots, b)$ as shown in formula (10.42).

The hypotheses stated in (10.72) and (10.73) concern factors A and B, respectively, but are not desirable because they depend on the cell frequencies and are therefore data dependent.

(b) Hypotheses tested By $F(\alpha \mid \mu, \beta)$ and $F(\beta \mid \mu, \alpha)$

By definition,

$$F(\alpha \mid \mu, \beta) = \frac{R(\alpha \mid \mu, \beta)}{(a-1)MS_E} \tag{10.74}$$

$$F(\beta \mid \mu, \alpha) = \frac{R(\beta \mid \mu, \alpha)}{(b-1)MS_E}, \tag{10.75}$$

where $R(\alpha \mid \mu, \beta) = R(\mu, \alpha, \beta) - R(\mu, \beta)$ and $R(\beta \mid \mu, \alpha) = R(\mu, \alpha, \beta) - R(\mu, \alpha)$ are Type II sums of squares for factors A and B, respectively. Thus, $F(\alpha \mid \mu, \beta)$ and $F(\beta \mid \mu, \alpha)$ are considered Type II F-ratios for A and B, respectively.

The hypotheses tested by the Type II F-ratios are very complicated. According to Searle (1971, pp. 308–310), $F(\alpha \mid \mu, \beta)$ tests the hypothesis,

$$H_0 : \phi_i = 0 \text{ for all } i = 1, 2, \ldots, a, \tag{10.76}$$

where

$$\phi_i = \sum_{j=1}^{b} \left(n_{ij}\, \mu_{ij} - \frac{n_{ij}}{n_{.j}} \sum_{k=1}^{a} n_{kj}\, \mu_{kj} \right), \quad i = 1, 2, \ldots, a,$$

and $F(\beta \mid \mu, \alpha)$ tests the hypothesis,

$$H_0 : \psi_j = 0 \text{ for all } j = 1, 2, \ldots, b, \tag{10.77}$$

where

$$\psi_j = \sum_{i=1}^{a} \left(n_{ij}\, \mu_{ij} - \frac{n_{ij}}{n_{i.}} \sum_{k=1}^{b} n_{ik}\, \mu_{ik} \right), \quad j = 1, 2, \ldots, b.$$

Note that $\sum_{i=1}^{a} \phi_i = 0$ and $\sum_{j=1}^{b} \psi_j = 0$.

The hypotheses in (10.76) and (10.77), which concern factors A and B, respectively, are also undesirable because they are data dependent by being dependent on the cell frequencies. Furthermore, they are not easy to interpret. Perhaps a more palatable formulation of these hypotheses is the following:

$$H_0 : WM(\alpha_{(i)}) = \frac{1}{n_{i.}} \sum_{j=1}^{b} n_{ij}\, WM(\beta_{(j)}), \quad i = 1, 2, \ldots, a, \tag{10.78}$$

$$H_0 : WM(\beta_{(j)}) = \frac{1}{n_{.j}} \sum_{i=1}^{a} n_{ij}\, WM(\alpha_{(i)}), \quad j = 1, 2, \ldots, b, \tag{10.79}$$

where, if we recall, $WM(\alpha_{(i)})$ and $WM(\beta_{(j)})$ are the weighted means of the cell means in row i and column j, respectively [see (10.41) and (10.42)]. The hypothesis in (10.78) states that $WM(\alpha_{(i)})$ is the weighted mean of the $WM(\beta_{(j)})$'s ($i = 1, 2, \ldots, a$), and the one in (10.79) states that $WM(\beta_{(j)})$ is the weighted mean of the $WM(\alpha_{(i)})$'s ($j = 1, 2, \ldots, b$).

In conclusion, Type III analysis concerning model (10.36) is preferred over Types I and II analyses because its hypotheses for A, B, and $A * B$ are not data dependent. Furthermore, they are of the same form as the ones tested when the data set is balanced.

It should be noted that SAS also provides Type IV F-ratios for A, B, and $A * B$. If the data set contains no empty cells, as is the case with model (10.36), then Type III and Type IV F-ratios and hypotheses are identical. The difference between the two can only occur when the data set contains some empty cells. In general, a Type IV hypothesis for an effect, u, is a hypothesis obtained by setting up linearly independent contrasts among the levels of u derived from subsets of nonempty cells. These subsets can be chosen in a variety of ways. Thus, a Type IV hypothesis is not unique when the data set contains some empty cells. By expressing a Type IV hypothesis in the general form given in (10.14), a test statistic can then be obtained by using the F-ratio in (10.15), where, in this case, MS_E is the error mean square for model (10.36) with $n_{..} - q$ degrees of freedom, where q is the number of nonempty cells.

In any case, the analysis of data containing some empty cells is not satisfactory since even the Type III hypotheses become dependent on the pattern of nonempty cells. More specifically, Type III hypotheses for factors A and B, which equate all the least-squares means for A as well as those for B, respectively, cannot be tested since some of the least-squares means will be nonestimable. For example, if row i $(= 1, 2, \ldots, a)$ contains at least one empty cell, then $LSM(\alpha_{(i)})$ is nonsetimable. Similarly, if column j $(= 1, 2, \ldots, b)$ contains at least one empty cell, then $LSM(\beta_{(j)})$ is nonestimable. Consequently, it is not possible to test equality of all the least-squares means for A and B in this case. In addition, the hypothesis tested by $F(\alpha\beta \mid \mu, \alpha, \beta)$ is no longer the no-interaction hypothesis as in (10.62). This is because the interaction contrasts, $\Theta_{ij,i'j'}$, in (10.60) are not all estimable when some cells are empty (see Searle, 1971, p. 311). In fact, $F(\alpha\beta \mid \mu, \alpha, \beta)$ has, in this case, only $q - a - b + 1$ degrees of freedom for the numerator instead of $ab - a - b + 1$. Hence, a complete analysis of the data cannot be carried out in this case. For this reason, we have not considered the analysis of model (10.36) in situations involving missing data. The interested reader is referred to Searle (1971, 1987), Speed, Hocking, and Hackney (1978), and Milliken and Johnson (1984) for additional details concerning this topic.

10.4 Higher-Order Models

The analysis of a fixed-effects model in the general unbalanced case, with no missing data in the subclasses, can be done by first identifying certain hypotheses of interest concerning the model's parameters. Any of these hypotheses is then tested using an appropriate F-ratio as in (10.15) (the usual assumptions of normality, independence, and equality of error variances are considered valid). A more convenient way to accomplish this is to do the analysis by relying on PROC GLM in SAS. The following is a brief overview of what can be gleaned from the SAS output that may be helpful in the analysis of the model.

(a) **The E option**

The E option in the model statement of PROC GLM gives the general form of all estimable linear functions of $\boldsymbol{\beta}$ for a general model as the one shown in (10.1). More specifically, if $\boldsymbol{L}$ denotes any given constant vector, the linear function, $\boldsymbol{L}'(\boldsymbol{X}'\boldsymbol{X})^{-}\boldsymbol{X}'\boldsymbol{X}\boldsymbol{\beta}$, is estimable. In fact, it is easy to show that any linear function of $\boldsymbol{\beta}$ is estimable if and only if it can be expressed as $\boldsymbol{L}'(\boldsymbol{X}'\boldsymbol{X})^{-}\boldsymbol{X}'\boldsymbol{X}\boldsymbol{\beta}$ for some vector $\boldsymbol{L}$. The elements of $\boldsymbol{L}'(\boldsymbol{X}'\boldsymbol{X})^{-}\boldsymbol{X}'\boldsymbol{X}$ are given in the SAS output as a result of invoking the E option. It is interesting to note here that the number of elements of $\boldsymbol{L}$ that appear in $\boldsymbol{L}'(\boldsymbol{X}'\boldsymbol{X})^{-}\boldsymbol{X}'\boldsymbol{X}\boldsymbol{\beta}$ is actually equal to the rank of $\boldsymbol{X}$. Furthermore, the coefficients of the elements of $\boldsymbol{L}$ in this linear combination

TABLE 10.6
General Form of Estimable Functions

Effect	Coefficients
μ	L_1
$\alpha_{(1)}$	L_2
$\alpha_{(2)}$	L_3
$\alpha_{(3)}$	$L_1 - L_2 - L_3$
$\beta_{(1)}$	L_5
$\beta_{(2)}$	L_6
$\beta_{(3)}$	$L_1 - L_5 - L_6$

are all estimable and form a basis for the space of all estimable linear functions of $\boldsymbol{\beta}$.

For example, consider model (10.10) along with the data set from Example 10.1. In this case, $\boldsymbol{\beta} = (\mu, \alpha_{(1)}, \alpha_{(2)}, \alpha_{(3)}, \beta_{(1)}, \beta_{(2)}, \beta_{(3)})'$, $\boldsymbol{L} = (L_1, L_2, \ldots, L_7)'$. The information resulting from the use of the E option in the SAS statement, MODEL $Y = A\ \ B/\text{E}$, is displayed in Table 10.6.

On the basis of Table 10.6, we get the linear function,

$$\begin{aligned} \boldsymbol{L}'(\boldsymbol{X}'\boldsymbol{X})^{-}\boldsymbol{X}'\boldsymbol{X}\boldsymbol{\beta} &= (\mu + \alpha_{(3)} + \beta_{(3)})L_1 + (\alpha_{(1)} - \alpha_{(3)})L_2 + (\alpha_{(2)} - \alpha_{(3)})L_3 \\ &\quad + (\beta_{(1)} - \beta_{(3)})L_5 + (\beta_{(2)} - \beta_{(3)})L_6, \end{aligned} \tag{10.80}$$

which is obtained by multiplying the entries under "coefficients" by the corresponding effects and then adding up the results. We note that the number of L_i's in (10.80) is 5, which is equal to the rank of $\boldsymbol{X}$ as it should be (in this case, the rank of $\boldsymbol{X}$ is $a + b - 1 = 5$). These L_i's are arbitrary constants and can therefore be assigned any values. Hence, the coefficients of the L_i's in (10.80), namely, $\mu + \alpha_{(3)} + \beta_{(3)}$, $\alpha_{(1)} - \alpha_{(3)}$, $\alpha_{(2)} - \alpha_{(3)}$, $\beta_{(1)} - \beta_{(3)}$, $\beta_{(2)} - \beta_{(3)}$ are all estimable and should form a basis for the space of all estimable linear functions of $\boldsymbol{\beta}$ according to Lemma 10.1(b).

(b) **The E1, E2, and E3 options**

These are options that are also available in the model statement of PROC GLM. They give Type I, Type II, and Type III estimable functions, which give rise to Type I, Type II, and Type III hypotheses, respectively, for each effect in the model. In addition, SAS provides the corresponding Type I, Type II, and Type III sums of squares and F-ratios. For example, using model (10.10) and the data set from Example 10.1, the Types I, II, and III estimable functions for A and B can be derived from Tables 10.7 through 10.9, respectively.

Using Table 10.7, Type I estimable functions for A and B are obtained by multiplying the entries under A and B by the corresponding effects

TABLE 10.7
Type I Estimable Functions

Effect	A	B
μ	0	0
$\alpha_{(1)}$	L_2	0
$\alpha_{(2)}$	L_3	0
$\alpha_{(3)}$	$-L_2 - L_3$	0
$\beta_{(1)}$	$0.3143 * L_2 + 0.1143 * L_3$	L_5
$\beta_{(2)}$	$-0.0286 * L_2 - 0.4286 * L_3$	L_6
$\beta_{(3)}$	$-0.2857 * L_2 + 0.3143 * L_3$	$-L_5 - L_6$

TABLE 10.8
Type II Estimable Functions

Effect	A	B
μ	0	0
$\alpha_{(1)}$	L_2	0
$\alpha_{(2)}$	L_3	0
$\alpha_{(3)}$	$-L_2 - L_3$	0
$\beta_{(1)}$	0	L_5
$\beta_{(2)}$	0	L_6
$\beta_{(3)}$	0	$-L_5 - L_6$

TABLE 10.9
Type III Estimable Functions

Effect	A	B
μ	0	0
$\alpha_{(1)}$	L_2	0
$\alpha_{(2)}$	L_3	0
$\alpha_{(3)}$	$-L_2 - L_3$	0
$\beta_{(1)}$	0	L_5
$\beta_{(2)}$	0	L_6
$\beta_{(3)}$	0	$-L_5 - L_6$

and then adding up the results. Thus, for factor A, its Type I estimable function is of the form

$$(\alpha_{(1)} - \alpha_{(3)} + 0.3143\,\beta_{(1)} - 0.0286\,\beta_{(2)} - 0.2857\,\beta_{(3)})\,L_2 + (\alpha_{(2)} - \alpha_{(3)} + 0.1143\,\beta_{(1)} - 0.4286\,\beta_{(2)} + 0.3143\,\beta_{(3)})\,L_3$$

Note that the number of L_i's in this combination is 2, which should be equal to the number of degrees of freedom for A. Type I hypothesis for A is obtained by equating the coefficients of L_2 and L_3 to zero. Doing so, we get

$$\alpha_{(1)} - \alpha_{(3)} + 0.3143\,\beta_{(1)} - 0.0286\,\beta_{(2)} - 0.2857\,\beta_{(3)} = 0 \tag{10.81}$$

$$\alpha_{(2)} - \alpha_{(3)} + 0.1143\,\beta_{(1)} - 0.4286\,\beta_{(2)} + 0.3143\,\beta_{(3)} = 0. \tag{10.82}$$

Note that (10.81) and (10.82) can be written as

$$\alpha_{(1)} + \frac{1}{5}(3\,\beta_{(1)} + 2\,\beta_{(2)}) = \alpha_{(3)} + \frac{1}{7}(2\,\beta_{(1)} + 3\,\beta_{(2)} + 2\,\beta_{(3)}) \tag{10.83}$$

$$\alpha_{(2)} + \frac{1}{5}(2\,\beta_{(1)} + 3\,\beta_{(3)}) = \alpha_{(3)} + \frac{1}{7}(2\,\beta_{(1)} + 3\,\beta_{(2)} + 2\,\beta_{(3)}). \tag{10.84}$$

We recognize (10.83) and (10.84) as forming the Type I hypothesis for A given in (10.27) whose test statistic value is $F(\alpha \mid \mu) = 0.32$ (see Table 10.2). The Type I estimable function for B from Table 10.7 is $(\beta_{(1)} - \beta_{(3)})\,L_5 + (\beta_{(2)} - \beta_{(3)})\,L_6$, which yields the hypothesis

$$H_0 : \beta_{(1)} = \beta_{(2)} = \beta_{(3)}.$$

This is the same as the Type II hypothesis for B in (10.32) whose test statistic value is $F(\beta \mid \mu, \alpha) = 14.73$ (see Table 10.2 or Table 10.3). Note that since the SAS model here is written as $Y = A\ \ B$ with A appearing first and B second, Type I and Type II hypotheses and F-ratios for B are identical. In order to get the Type I hypothesis for B as shown in (10.28) and its corresponding F-ratio, $F(\beta \mid \mu)$, a second SAS model should be added to the SAS code in which B appears first and A second. In doing so, we get $F(\beta \mid \mu) = 12.95$.

Similarly, using Table 10.8 we get, after setting up the Type II estimable functions and equating the coefficients of the L_i's to zero, the Type II hypotheses shown in (10.31) and (10.32), respectively. The corresponding test statistics values are $F(\alpha \mid \mu, \beta) = 2.09$ and $F(\beta \mid \mu, \alpha) = 14.73$ (see Table 10.3). Table 10.9 gives the same information as Table 10.8 since for the two-way model without interaction, Type II and Type III hypotheses and F-ratios are identical.

As was pointed out earlier in Section 10.3.3, Type III analysis is preferred in general for testing hypotheses concerning all the effects in the model. This is based on the assumption that the data set under consideration contains no empty cells.

(c) **Least-squares means**

The least-squares means (LSMEANS) statement in PROC GLM computes for each effect listed in this statement estimated values of its least-squares means. This was demonstrated earlier in Example 10.1 for the case of the two-way model without interaction (see Table 10.4). As was mentioned in Example 10.1, one useful option in the LSMEANS statement is "TDIFF" which gives the t-values for the pairwise comparisons of the least-squares means for a given factor along with the corresponding p-values. If, however, it is desired to make multiple comparisons among all the least-squares means, for example, the ones for factor A, then the option "ADJUST = TUKEY" should be used in the LSMEANS statement as follows:

```
LSMEANS A / PDIFF ADJUST = TUKEY;
```

This provides a multiple comparison adjustment for the p-values for the differences of least-squares means in a manner that controls the experimentwise Type I error rate according to *Tukey's Studentized range test*. Note that "PDIFF" stands for "p-value for the difference."

10.5 A Numerical Example

An experiment was conducted on laboratory rats to study the effects of a hunger-reducing drug (factor A) and the length of time (factor B), between administration of the drug and feeding, on the amount of food ingested by the rats. Two dosage levels were applied, namely, 0.3 and 0.7 mg/kg, and three levels of time were chosen, namely, 1, 5, and 9 h. A total of 18 rats of uniform size and age were initially selected for the experiment with three rats assigned to each of the six $A \times B$ treatment combinations. At the start of the experiment, the rats were deprived of food for a certain period of time. Each rat was then inoculated with a certain dosage level of the drug and after a specific length of time, it was fed. The weight (in grams) of the food ingested by the rat was measured. However, during the course of the experiment, several rats became sick and were subsequently eliminated from the experiment. This resulted in the unbalanced data given in Table 10.10. A plot of the data points is shown in Figure 10.1. The model considered is the one in (10.36).

The error mean square for this data set is $MS_E = 0.1904$ with 7 degrees of freedom. Table 10.11 gives the Type III analysis for A, B, and $A * B$.

The corresponding hypotheses for A, B, and $A * B$ are the ones listed in (10.50), (10.51), and (10.62), respectively. We note that all three tests are highly significant.

Let us now apply formulas (10.43) and (10.44) to get the sums of squares for A and B using the method of weighted squares of means. We find that $SS_{A\omega} = 90.6558$, $SS_{B\omega} = 115.4312$, which are identical to the corresponding Type III sums of squares for A and B, respectively, in Table 10.11, as they should be.

TABLE 10.10
Weights of Food Ingested (g)

	B (time, h)		
A (Drug Dosage, mg/kg)	**1**	**5**	**9**
0.3	5.65	11.62	13.98
	5.89	12.38	
	6.14		
0.7	1.37	4.71	7.74
	1.65	5.82	8.29
	1.93		

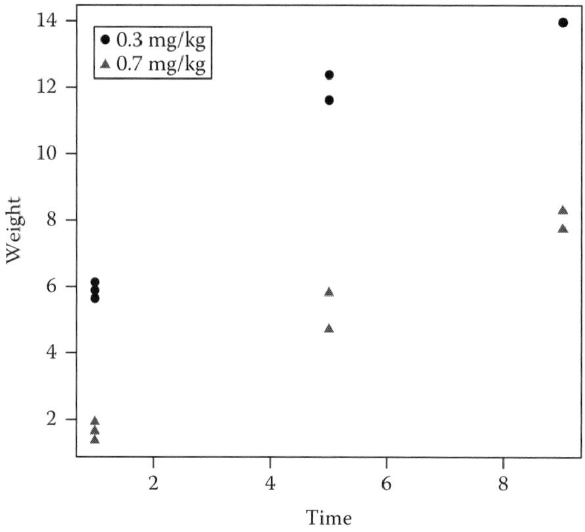

FIGURE 10.1
Table 10.10 data points.

TABLE 10.11
Type III Sums of Squares and F-Values

Source	DF	Type III *SS*	*MS*	*F*	*p*-Value
A	1	90.6558	90.6558	476.07	<0.0001
B	2	115.4312	57.7156	303.09	<0.0001
$A * B$	2	4.0017	2.0008	10.51	0.0078

Since the interaction is significant, it would be of interest to do some further analysis by testing each factor at fixed levels of the other factor. As was recommended in Section 10.3.1, since both factors are quantitative, we can consider testing the polynomial effects of each factor at fixed levels of the other factor. Taking into consideration that factor A has two levels and factor

B has three levels, the polynomial effects of A consist of just the linear effect, and those of B consist of the linear and quadratic effects.

In general, the testing of polynomial effects can be conveniently done using PROC GLM in SAS. For example, to get the sums of squares for the linear and quadratic effects of B at the fixed ith level of A ($i = 1, 2$), which we denote by $B_L(i)$, $B_Q(i)$, we can use the following SAS statements:

```
PROC GLM;
CLASS A B;
MODEL Y = A B A * B;
PROC SORT;
BY A;
PROC GLM;
BY A;
MODEL Y = B B * B;
RUN;
```

The SORT procedure sorts observations in the data set in Table 10.10 by the levels of A. The first model statement provides information for setting up Table 10.11. The second model statement is needed in order to get the sums of squares for $B_L(i)$ and $B_Q(i)$ ($i = 1, 2$). For this purpose, we only need to consider the Type I sums of squares for B and $B * B$ in the second model statement which correspond to $B_L(i)$ and $B_Q(i)$, respectively. These sums of squares are additive, and therefore statistically independent. They measure the contributions of the *orthogonal contrasts* which represent these effects (see, for example, Christensen, 1996, Section 7.12). It should be noted here that the second model statement was not preceded by the "CLASS" statement. The reason for this is that in the absence of the CLASS statement, SAS treats the model as a regression model, rather than an ANOVA model, where the model's independent variables (B and $B * B$ in this case) are treated as continuous regression variables with one degree of freedom each. The relevant Type I sums of squares are given in Table 10.12 along with the corresponding F-values. Note that these F-values were obtained by dividing each sum of squares by $MS_E = 0.1904$ for the full two-way model (the F-values obtained from using PROC SORT should not be considered since the error mean squares used in their denominators are not based on the entire data set, and are therefore different from 0.1904).

Similarly, to test the linear effect of A at the fixed jth level of B, which we denote by $A_L(j)$ ($j = 1, 2, 3$), we can use the following SAS statements:

```
PROC SORT;
BY B;
PROC GLM;
BY B;
MODEL Y = A;
RUN;
```

TABLE 10.12
Polynomial Effects of *B* at Fixed Levels of *A*

Source	DF	Type I *SS*	*F*	*p*-Value
A=0.3				
B	1	66.1865	347.6182	<0.0001
$B * B$	1	5.1088	26.8319	0.0013
A= 0.7				
B	1	49.9843	262.5226	<0.0001
$B * B$	1	0.2641	1.3871	0.2774

TABLE 10.13
Polynomial Effects of *A* at Fixed Levels of *B*

Source	DF	Type I *SS*	*F*	*p*-Value
B=1				
A	1	27.0088	141.8529	<0.0001
B=5				
A	1	45.3602	238.2363	<0.0001
B=9				
A	1	23.7208	124.5840	<0.0001

In this case, since factor A has one degree of freedom, A in the above model statement represents the linear effect of A, which is the only polynomial effect for A, at fixed levels of B. The results are given in Table 10.13.

As in Table 10.12, the error mean square, $MS_E = 0.1904$, was used to produce the F-values in Table 10.13. From Table 10.12 we can see that the linear and quadratic effects of B are significant for level 0.3 of A. However, only the linear effect of B is significant for level 0.7 of A. This can be clearly seen from examining Figure 10.2 which is obtained by plotting the estimated cell means (values of $\bar{Y}_{ij.}$) against the three levels of B. Points with the same level of A are connected. Figure 10.2 depicts the effect of the interaction and is therefore considered an *interaction plot*. It shows a quadratic trend in the mean weight under B for the low dosage, 0.3, but only a linear trend for the high dosage, 0.7. On the other hand, we note from Table 10.13 that all the linear effects of A are significant for any level of B. This is equivalent to saying that the cell means of A are significantly different for any fixed level of B.

In general, if one of the factors is qualitative, for example, factor B, and if the interaction is significant, comparisons among the cell means of B can be made using Tukey's Studentized range test at fixed levels of A, as was mentioned in Section 10.3.1. Alternatively, we can consider using PROC GLM to get an F-ratio for testing B for fixed levels of A. This can be accomplished by making use of the following SAS statements:

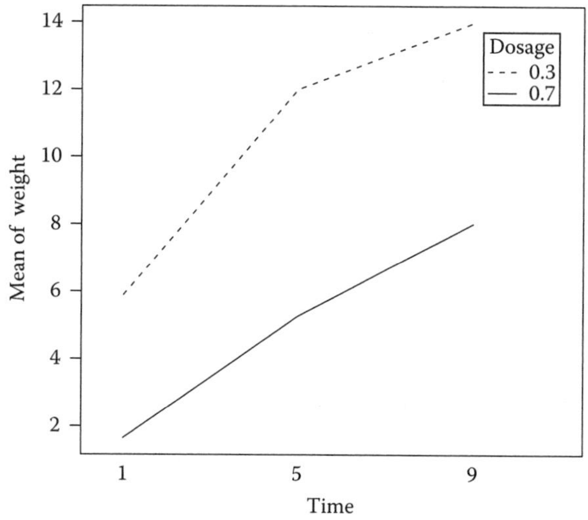

FIGURE 10.2
Interaction plot for the data in Table 10.10.

TABLE 10.14
$A * B$ Effect Sliced by A

A	DF	SS	MS	F	p-Value
0.3	2	71.2953	35.6476	187.22	<0.0001
0.7	2	50.2484	25.1242	131.95	<0.0001

```
PROC GLM;
CLASS A B;
MODEL Y = A B A * B;
LSMEANS A * B/ SLICE = A;
RUN:
```

The use of the "SLICE = A" option in the LSMEANS statement amounts to partitioning (slicing) the data according to the levels of A. For each of the "sliced" portions of the data, an F-ratio is obtained which tests equality of the least-squares means of B for the corresponding fixed level of A. The denominator of this F-ratio is MS_E, the same error mean square for the full two-way model. The SAS output from applying the above statements to the data set in Table 10.10 is shown in Table 10.14. We note that all tests are significant, which indicates that there are differences among the least-squares means of B for each level of A. Of course, in our case, B is quantitative and it would be more informative to use Table 10.12 to test its polynomial effects for each level of A.

10.6 The Method of Unweighted Means

One alternative to the method of weighted squares of means is the *method of unweighted means* (MUM), which was also introduced by Yates (1934). It provides an approximate, but computationally simple method of analysis. The MUM defines sums of squares that are analogous to those obtained for balanced data situations with one observation per treatment combination, namely, the sample mean of the corresponding cell. We shall demonstrate this method in the case of a two-way model with interaction as in (10.36). As before, we assume that the error terms in the model are independently distributed as $N(0, \sigma^2)$. In addition, we require that $n_{ij} > 0$ for all i and j, that is, the data contain no empty cells.

Using the same notation as in Section 10.3.1, let X_{ij} be defined as $X_{ij} = \bar{Y}_{ij.} = \frac{1}{n_{ij}} \sum_{k=1}^{n_{ij}} Y_{ijk}$. Let $\bar{X}_{i.} = \frac{1}{b} \sum_{j=1}^{b} X_{ij}$, $\bar{X}_{.j} = \frac{1}{a} \sum_{i=1}^{a} X_{ij}$, $\bar{X}_{..} = \frac{1}{a\,b} \sum_{i=1}^{a} \sum_{j=1}^{b} X_{ij}$. The unweighted sums of squares corresponding to factors A, B, and their interaction, $A * B$, are

$$SS_{Au} = \bar{n}_h\, b \sum_{i=1}^{a} (\bar{X}_{i.} - \bar{X}_{..})^2 \tag{10.85}$$

$$SS_{Bu} = \bar{n}_h\, a \sum_{j=1}^{b} (\bar{X}_{.j} - \bar{X}_{..})^2 \tag{10.86}$$

$$SS_{ABu} = \bar{n}_h \sum_{i=1}^{a} \sum_{j=1}^{b} (X_{ij} - \bar{X}_{i.} - \bar{X}_{.j} + \bar{X}_{..})^2, \tag{10.87}$$

where $\bar{n}_h$ is the *harmonic mean* of the cell frequencies given by

$$\bar{n}_h = a\,b \left[\sum_{i=1}^{a} \sum_{j=1}^{b} \frac{1}{n_{ij}} \right]^{-1}. \tag{10.88}$$

The rationale for the use of $\bar{n}_h$ in the above sums of squares is the following: the variance of X_{ij} is $\frac{\sigma^2}{n_{ij}}$. The average of this variance over all the ab cells is

$$\frac{\sigma^2}{a\,b} \sum_{i=1}^{a} \sum_{j=1}^{b} \frac{1}{n_{ij}} = \frac{\sigma^2}{\bar{n}_h}.$$

Thus, $\bar{n}_h$ acts like n, the common cell frequency, if the data set were balanced. Now, let $\bar{\boldsymbol{X}} = (X_{11}, X_{12}, \ldots, X_{ab})'$. Then,

$$\text{Var}(\bar{\boldsymbol{X}}) = \sigma^2 \boldsymbol{K}, \tag{10.89}$$

where $\boldsymbol{K} = \text{diag}(n_{11}^{-1}, n_{12}^{-1}, \ldots, n_{ab}^{-1})$. If the data set is balanced, then $n_{ij} = n$ for all i, j and $\boldsymbol{K} = \frac{1}{n}\boldsymbol{I}_{ab}$. The following lemma provides an added justification for the use of the harmonic mean.

Lemma 10.7 The best approximation of $\boldsymbol{K}$ in (10.89) with a diagonal matrix of the form $c\boldsymbol{I}_{ab}$ is achieved when $c = \frac{1}{\bar{n}_h}$.

Proof. By "best approximation" of $\boldsymbol{K}$ with $c\boldsymbol{I}_{ab}$ we mean finding the value of c which minimizes the Euclidean norm $\| \boldsymbol{K} - c\boldsymbol{I}_{ab} \|$, where

$$\begin{aligned} \| \boldsymbol{K} - c\,\boldsymbol{I}_{ab} \|^2 &= tr\left[(\boldsymbol{K} - c\,\boldsymbol{I}_{ab})^2\right] \\ &= \sum_{i=1}^{a}\sum_{j=1}^{b}\left(\frac{1}{n_{ij}} - c\right)^2. \end{aligned} \tag{10.90}$$

Differentiating (10.90) with respect to c and equating the derivative to zero, we get

$$-2\sum_{i=1}^{a}\sum_{j=1}^{b}\frac{1}{n_{ij}} + 2abc = 0. \tag{10.91}$$

Hence,

$$\begin{aligned} c &= \frac{1}{ab}\sum_{i=1}^{a}\sum_{j=1}^{b}\frac{1}{n_{ij}} \\ &= \frac{1}{\bar{n}_h}. \end{aligned}$$

Since the derivative of the left-hand side of (10.91) with respect to c is $2ab$, which is positive, we conclude that $\| \boldsymbol{K} - c\,\boldsymbol{I}_{ab} \|$ attains its absolute minimum when $c = \frac{1}{\bar{n}_h}$. □

The approximation of $\boldsymbol{K}$ with $\frac{1}{\bar{n}_h}\boldsymbol{I}_{ab}$ amounts to replacing the cell frequencies by their harmonic mean. Thus, if we were to pretend that $\bar{n}_h$ is a positive integer, and if in the (i, j)th cell we were to have $\bar{n}_h$ observations, all equal to $\bar{X}_{ij}$, then SS_{Au}, SS_{Bu}, and SS_{ABu} in (10.85), (10.86), and (10.87) would represent the sums of squares for A, B, and $A * B$, respectively, based on this artificial balanced data set.

10.6.1 Distributions of SS_{Au}, SS_{Bu}, and SS_{ABu}

The sums of squares in (10.85), (10.86), and (10.87) can be written as

$$SS_{Au} = \bar{n}_h \bar{X}' \left(\frac{1}{b} I_a \otimes J_b - \frac{1}{ab} J_a \otimes J_b\right) \bar{X} \tag{10.92}$$

$$SS_{Bu} = \bar{n}_h \bar{X}' \left(\frac{1}{a} J_a \otimes I_b - \frac{1}{ab} J_a \otimes J_b\right) \bar{X} \tag{10.93}$$

$$SS_{ABu} = \bar{n}_h \bar{X}' \left(I_a \otimes I_b - \frac{1}{b} I_a \otimes J_b - \frac{1}{a} J_a \otimes I_b + \frac{1}{ab} J_a \otimes J_b\right) \bar{X}, \tag{10.94}$$

where, if we recall, $\bar{X} = (X_{11}, X_{12}, \ldots, X_{ab})'$. Under the assumption of normality, independence, and equality of error variances, it can be seen that $\bar{X} \sim N(\mu, \sigma^2 K)$, where $\mu = (\mu_{11}, \mu_{12}, \ldots, \mu_{ab})'$ and K is the diagonal matrix used in (10.89).

Unlike the case of balanced data, none of the sums of squares in (10.92), (10.93), and (10.94) have scaled chi-squared distributions. For example, if we consider $\frac{1}{\sigma^2} SS_{Au}$, we find that the matrix,

$$\bar{n}_h \left(\frac{1}{b} I_a \otimes J_b - \frac{1}{ab} J_a \otimes J_b\right) K, \tag{10.95}$$

is not idempotent, which implies that $\frac{1}{\sigma^2} SS_{Au}$ does not have the chi-squared distribution (see Theorem 5.4). Furthermore, SS_{Au}, SS_{Bu}, and SS_{ABu} are not mutually independent. For, example, SS_{Au} and SS_{Bu} are not independent since

$$\left(\frac{1}{b} I_a \otimes J_b - \frac{1}{ab} J_a \otimes J_b\right) K \left(\frac{1}{a} J_a \otimes I_b - \frac{1}{ab} J_a \otimes J_b\right) \neq \mathbf{0}$$

(see Theorem 5.5). However, all three sums of squares are independent of the error sum of squares, SS_E, in (10.45) since the X_{ij}'s ($= \bar{Y}_{ij.}$) are independent of SS_E (see Example 5.5).

Let us now apply Lemma 5.1 to $\frac{1}{\sigma^2} SS_{Au}$. In doing so, we can express $\frac{1}{\sigma^2} SS_{Au}$ as a linear combination of mutually independent chi-squared variates of the form

$$\frac{1}{\sigma^2} SS_{Au} = \sum_{i=1}^{k_1} \tau_{1i} \chi^2_{\nu_{1i}}(\eta_{1i}), \tag{10.96}$$

where k_1 denotes the number of distinct nonzero eigenvalues of the matrix in (10.95), τ_{1i} is the ith of such eigenvalues with multiplicity ν_{1i}, and η_{1i} is the corresponding noncentrality parameter ($i = 1, 2, \ldots, k_1$). Similarly, for

$\frac{1}{\sigma^2} SS_{Bu}$, $\frac{1}{\sigma^2} SS_{ABu}$, we have

$$\frac{1}{\sigma^2} SS_{Bu} = \sum_{i=1}^{k_2} \tau_{2i}\, \chi^2_{\nu_{2i}}(\eta_{2i}) \tag{10.97}$$

$$\frac{1}{\sigma^2} SS_{ABu} = \sum_{i=1}^{k_3} \tau_{3i}\, \chi^2_{\nu_{3i}}(\eta_{3i}), \tag{10.98}$$

where k_j, τ_{ji}, ν_{ji}, and η_{ji} $(j = 2, 3)$ are quantities comparable to $k_1, \tau_{1i}, \nu_{1i}, \eta_{1i}$ in (10.96) and are obtained by replacing the matrix $\frac{1}{b}\boldsymbol{I}_a \otimes \boldsymbol{J}_b - \frac{1}{ab}\boldsymbol{J}_a \otimes \boldsymbol{J}_b$ with the matrices of the other two quadratic forms in (10.93) and (10.94), respectively.

Lemma 10.8

(a) $E(SS_{Au}) = \bar{n}_h\, b \sum_{i=1}^{a} (\bar{\mu}_{i.} - \bar{\mu}_{..})^2 + \sigma^2(a-1)$

(b) $E(SS_{Bu}) = \bar{n}_h\, a \sum_{j=1}^{b} (\bar{\mu}_{.j} - \bar{\mu}_{..})^2 + \sigma^2(b-1)$

(c) $E(SS_{ABu}) = \bar{n}_h \sum_{i=1}^{a} \sum_{j=1}^{b} (\mu_{ij} - \bar{\mu}_{i.} - \bar{\mu}_{.j} + \bar{\mu}_{..})^2 + \sigma^2(a-1)(b-1)$,
where $\bar{\mu}_{i.} = \frac{1}{b}\sum_{j=1}^{b} \mu_{ij}$, $\bar{\mu}_{.j} = \frac{1}{a}\sum_{i=1}^{a} \mu_{ij}$, $\bar{\mu}_{..} = \frac{1}{ab}\sum_{i=1}^{a}\sum_{j=1}^{b} \mu_{ij}$.

Proof. We shall only prove part (a) since the proofs of parts (b) and (c) will be similar.

Using Theorem 5.2, the expected value of SS_{Au} in (10.92) is given by

$$E(SS_{Au}) = \bar{n}_h\, \boldsymbol{\mu}' \left(\frac{1}{b}\boldsymbol{I}_a \otimes \boldsymbol{J}_b - \frac{1}{ab}\boldsymbol{J}_a \otimes \boldsymbol{J}_b\right) \boldsymbol{\mu} + tr\left[\bar{n}_h \left(\frac{1}{b}\boldsymbol{I}_a \otimes \boldsymbol{J}_b - \frac{1}{ab}\boldsymbol{J}_a \otimes \boldsymbol{J}_b\right)\left(\sigma^2 \boldsymbol{K}\right)\right],$$

where $\boldsymbol{\mu} = (\mu_{11}, \mu_{12}, \ldots, \mu_{ab})'$. Hence,

$$E(SS_{Au}) = \bar{n}_h\, b \sum_{i=1}^{a} (\bar{\mu}_{i.} - \bar{\mu}_{..})^2 + \sigma^2\, \bar{n}_h\, tr\left[\left(\frac{1}{b}\boldsymbol{I}_a \otimes \boldsymbol{J}_b - \frac{1}{ab}\boldsymbol{J}_a \otimes \boldsymbol{J}_b\right)\boldsymbol{K}\right]. \tag{10.99}$$

It can be verified that

$$tr\left[\left(\frac{1}{b}\boldsymbol{I}_a \otimes \boldsymbol{J}_b - \frac{1}{ab}\boldsymbol{J}_a \otimes \boldsymbol{J}_b\right)\boldsymbol{K}\right] = \frac{1}{b}\sum_{i=1}^{a}\sum_{j=1}^{b}\frac{1}{n_{ij}} - \frac{1}{ab}\sum_{i=1}^{a}\sum_{j=1}^{b}\frac{1}{n_{ij}} = \frac{1}{\bar{n}_h}(a-1).$$

By making the substitution in (10.99), we get

$$E(SS_{Au}) = \bar{n}_h\, b \sum_{i=1}^{a} (\bar{\mu}_{i.} - \bar{\mu}_{..})^2 + \sigma^2(a-1). \qquad \square$$

10.6.2 Approximate Distributions of SS_{Au}, SS_{Bu}, and SS_{ABu}

Let us again consider formulas (10.96), (10.97), and (10.98). The nonzero eigenvalues, τ_{1i} $(i = 1, 2, \ldots, k_1)$, τ_{2i} $(i = 1, 2, \ldots, k_2)$, and τ_{3i} $(i = 1, 2, \ldots, k_3)$, are all positive. This is true because, for example, the nonzero eigenvalues of the matrix in (10.95) are the same as those of

$$\bar{n}_h \boldsymbol{K}^{1/2}\left(\frac{1}{b}\boldsymbol{I}_a \otimes \boldsymbol{J}_b - \frac{1}{ab}\boldsymbol{J}_a \otimes \boldsymbol{J}_b\right)\boldsymbol{K}^{1/2},$$

which is positive semidefinite. If the n_{ij}'s are close to one another, then $\bar{n}_h\boldsymbol{K} \approx \boldsymbol{I}_{ab}$ and τ_{ji} is approximately equal to one $(i = 1, 2, \ldots, k_j;\ j = 1, 2, 3)$. In this case,

$$\frac{1}{\sigma^2} SS_{Au} \underset{approx.}{\sim} \chi^2_{a-1}(\eta_1), \tag{10.100}$$

where

$$\begin{aligned}
\eta_1 &= \frac{\bar{n}_h}{\sigma^2}\,\boldsymbol{\mu}'\left[\frac{1}{b}\boldsymbol{I}_a \otimes \boldsymbol{J}_b - \frac{1}{ab}\boldsymbol{J}_a \otimes \boldsymbol{J}_b\right]\boldsymbol{\mu} \\
&= \frac{\bar{n}_h\, b}{\sigma^2}\sum_{i=1}^{a}(\bar{\mu}_{i.} - \bar{\mu}_{..})^2.
\end{aligned}$$

(see the proof of the sufficiency part of Theorem 5.4). Similarly, we have

$$\frac{1}{\sigma^2} SS_{Bu} \underset{approx.}{\sim} \chi^2_{b-1}(\eta_2) \tag{10.101}$$

$$\frac{1}{\sigma^2} SS_{ABu} \underset{approx.}{\sim} \chi^2_{(a-1)(b-1)}(\eta_3), \tag{10.102}$$

where

$$\begin{aligned}
\eta_2 &= \frac{\bar{n}_h}{\sigma^2}\,\boldsymbol{\mu}'\left[\frac{1}{a}\boldsymbol{J}_a \otimes \boldsymbol{I}_b - \frac{1}{ab}\boldsymbol{J}_a \otimes \boldsymbol{J}_b\right]\boldsymbol{\mu} \\
&= \frac{\bar{n}_h\, a}{\sigma^2}\sum_{j=1}^{b}(\bar{\mu}_{.j} - \bar{\mu}_{..})^2,
\end{aligned}$$

and

$$\begin{aligned}
\eta_3 &= \frac{\bar{n}_h}{\sigma^2}\,\boldsymbol{\mu}'\left[\boldsymbol{I}_a \otimes \boldsymbol{I}_b - \frac{1}{b}\boldsymbol{I}_a \otimes \boldsymbol{J}_b - \frac{1}{a}\boldsymbol{J}_a \otimes \boldsymbol{I}_b + \frac{1}{ab}\boldsymbol{J}_a \otimes \boldsymbol{J}_b\right]\boldsymbol{\mu} \\
&= \frac{\bar{n}_h}{\sigma^2}\sum_{i=1}^{a}\sum_{j=1}^{b}(\mu_{ij} - \bar{\mu}_{i.} - \bar{\mu}_{.j} + \bar{\mu}_{..})^2.
\end{aligned}$$

Consequently, the F-ratio

$$\tilde{F}_A = \frac{SS_{Au}}{(a-1)MS_E} \tag{10.103}$$

tests the hypothesis that $\eta_1 = 0$, that is,

$$H_0 : \bar{\mu}_{1.} = \bar{\mu}_{2.} = \ldots = \bar{\mu}_{a.}.$$

This is the same as the hypothesis in (10.50), which equates the least-squares means for factor A. Under H_0, $\tilde{F}_A$ has approximately the F-distribution with $a-1$ and $n_{..} - ab$ degrees of freedom. Similarly, the F-ratios,

$$\tilde{F}_B = \frac{SS_{Bu}}{(b-1)MS_E}, \tag{10.104}$$

$$\tilde{F}_{AB} = \frac{SS_{ABu}}{(a-1)(b-1)MS_E}, \tag{10.105}$$

test hypotheses for B and $A * B$ that are identical to those described in Section 10.3.1.

Recall that the adequacy of the chi-squared approximations in (10.100), (10.101), and (10.102) depend on the closeness of the eigenvalues, τ_{ji}, to one ($i = 1, 2, \ldots, k_j$, $j = 1, 2, 3$), that is, when the n_{ij}'s are close to one another, as was mentioned earlier. This is equivalent to requiring that the values of

$$\lambda_A = \frac{(a-1)\sum_{i=1}^{k_1} \nu_{1i}\,\tau_{1i}^2}{\left(\sum_{i=1}^{k_1} \nu_{1i}\,\tau_{1i}\right)^2} \tag{10.106}$$

$$\lambda_B = \frac{(b-1)\sum_{i=1}^{k_2} \nu_{2i}\,\tau_{2i}^2}{\left(\sum_{i=1}^{k_2} \nu_{2i}\,\tau_{2i}\right)^2} \tag{10.107}$$

$$\lambda_{AB} = \frac{(a-1)(b-1)\sum_{i=1}^{k_3} \nu_{3i}\,\tau_{3i}^2}{\left(\sum_{i=1}^{k_3} \nu_{3i}\,\tau_{3i}\right)^2} \tag{10.108}$$

be each close to one (see Theorem 9.3). Note that, in general, $\lambda_A \geq 1$, $\lambda_B \geq 1$, $\lambda_{AB} \geq 1$. Thus, the use of the method of unweighted means (MUM) under these approximations may be inappropriate if the data set is severely unbalanced. Snedecor and Cochran (1980) stated that the MUM is appropriate only if the ratio of the largest n_{ij} to the smallest n_{ij} is less than two.

Gosslee and Lucas (1965) suggested an improvement on the chi-squared approximations by amending the degrees of freedom for the numerator mean squares in (10.103), (10.104), and (10.105) (see also Searle, 1971, pp. 365–367; Rankin, 1974). In the special case of a factorial experiment with two levels for each factor and unequal numbers of observations in the various cells, Speed and Monlezun (1979) showed that all sums of squares generated by

TABLE 10.15
Unweighted Analysis of the Data in Table 10.10

Source	DF	SS	MS	F	p-Value
A	1	90.6558	90.6558	476.1334	<0.0001
B	2	102.8635	51.4318	270.125	<0.0001
$A * B$	2	3.0838	1.5419	8.0982	0.0151
Error	7	1.3330	0.1904		

the MUM for A, B, and the interaction $A * B$ are distributed exactly as $\sigma^2 \chi_1^2$. Hence, in this case, the MUM's F-ratios follow the exact F-distribution and the hypotheses tested are the same as in a balanced data situation.

Example 10.2 Let us again consider the same data set as in Table 10.10. Applying the unweighted sums of squares formulas in (10.85), (10.86), and (10.87), we get the values listed in Table 10.15 along with their F-ratios and approximate p-values. These F-ratios were computed on the basis of formulas (10.103), (10.104), and (10.105).

We note that all tests are significant. This agrees with the results obtained from Table 10.11. The difference here is that the F-tests and corresponding p-values are approximate.

Exercises

10.1 Prove parts (a) and (b) of Theorem 10.1.

[Hint: To establish the formula for λ, show that

$$\begin{aligned}
&\boldsymbol{\beta}'\boldsymbol{X}'[\boldsymbol{X}(\boldsymbol{X}'\boldsymbol{X})^{-}\boldsymbol{X}' - \boldsymbol{X}_1(\boldsymbol{X}_1'\boldsymbol{X}_1)^{-}\boldsymbol{X}_1']\boldsymbol{X}\boldsymbol{\beta} \\
&= \boldsymbol{\beta}'[\boldsymbol{X}'\boldsymbol{X} - \boldsymbol{X}'\boldsymbol{X}_1(\boldsymbol{X}_1'\boldsymbol{X}_1)^{-}\boldsymbol{X}_1'\boldsymbol{X}]\boldsymbol{\beta} \\
&= \boldsymbol{\beta}_2'[\boldsymbol{X}_2'\boldsymbol{X}_2 - \boldsymbol{X}_2'\boldsymbol{X}_1(\boldsymbol{X}_1'\boldsymbol{X}_1)^{-}\boldsymbol{X}_1'\boldsymbol{X}_2]\boldsymbol{\beta}_2.]
\end{aligned}$$

10.2 (a) Prove formula (10.9).

(b) Show that if $\boldsymbol{X} = [\boldsymbol{X}_1 : \boldsymbol{X}_2]$ is of full column rank, then the hypothesis in (10.9) is equivalent to $H_0 : \boldsymbol{\beta}_2 = \boldsymbol{0}$.

[Hint: The matrix $[\boldsymbol{I}_n - \boldsymbol{X}_1(\boldsymbol{X}_1'\boldsymbol{X}_1)^{-}\boldsymbol{X}_1']\boldsymbol{X}_2$ is of full column rank since its rank is equal to the rank of the matrix $\boldsymbol{X}_2'[\boldsymbol{I}_n - \boldsymbol{X}_1(\boldsymbol{X}_1'\boldsymbol{X}_1)^{-}\boldsymbol{X}_1']\boldsymbol{X}_2$, which can be shown to be nonsingular.]

10.3 Consider model (10.10) with the following data set:

	B			
A	1	2	3	4
1	9 11	—	8 6	11
2	12	—	—	—
3	—	4 1 8	16	—
4	11	12 15	14 10	22

(a) Show that $\alpha_{(i)} - \alpha_{(i')}$ and $\beta_{(j)} - \beta_{(j')}$ are estimable for $i \neq i', j \neq j'$.

(b) State the hypotheses tested by $F(\alpha \mid \mu)$ and give the value of this test statistic.

(c) Give a complete Type II analysis concerning factors A and B.

(d) Find estimates of the least-squares means for the levels of factors A and B.

(e) Apply formula (10.35) to give a test statistic for testing the hypothesis, $H_0 : LSM(\alpha_{(1)}) = LSM(\alpha_{(3)})$. State your conclusion at the 5% level of significance.

10.4 Consider again the data set in Exercise 10.3.

(a) Let $\phi = \mu_{11} - 2\,\mu_{12} + \mu_{13}$, where μ_{ij} is the mean of the (i, j)th cell. Is ϕ estimable? If so, give the value of its BLUE.

(b) Test the hypothesis, $H_0 : \phi = 0$, and state your conclusion at the 5% level of significance.

(c) Obtain a 95% confidence interval on ϕ.

10.5 Consider the data set in Exercise 7.4.

(a) State the hypothesis tested by $F(\beta \mid \mu)$ and give the value of this test statistic.

(b) Test the hypothesis,

$$H_0 : \beta_{(1)} = \beta_{(2)} = \beta_{(3)},$$

at the 5% level of significance.

10.6 A certain corporation developed a new wood glue and compared its performance to the company's standard product. Ten pieces of each of eight types of wood were split and glued back together, five with the old glue and five with the new. The glue was allowed to dry, then the pieces of wood were subjected to a test to measure the amount of stress, in pounds per square inch (psi), required to break the glue bond. However, some pieces of wood broke before the glue bond broke, resulting in no measurement. The following table shows the average stress and number of observations for each glue and wood type:

	Wood Type							
Glue	A	B	C	D	E	F	G	H
New	22.65 (5)	27.43 (3)	21.06 (3)	25.98 (4)	29.73 (5)	30.80 (2)	22.62 (5)	21.03 (3)
Old	24.15 (4)	31.85 (3)	21.73 (5)	26.45 (4)	26.04 (4)	33.85 (5)	25.99 (5)	22.70 (5)

The error mean square is $MS_E = 6.93$.

(a) Use the method of weighted squares of means to test the significance of the wood type and glue factors at the 5% level of significance. State the null hypothesis and your conclusion in each case.

(b) Compare the least-squares means for the new and old glues and state your conclusion at the 5% level of significance.

(c) Compare the least-squares means for wood types E and H and state your conclusion at the 5% level of significance.

10.7 Consider again the data set in Exercise 10.6.

(a) Use the method of unweighted means to test the significance of the wood type and glue factors, and their interaction . Let $\alpha = 0.05$.

(b) Obtain a 95% confidence interval on the difference between the least-squares means for the new and old glues.

10.8 An experiment was conducted to compare responses to different doses of two drugs, denoted by A and B. Each drug was to be administered at the three doses, 5, 10, and 15 mg. A total of 18 subjects were initially selected to participate in this experiment after going through a screening procedure to eliminate any possible extraneous effects. Three subjects were assigned to each of the six drug $\times$ dose combinations. The response of interest was the reaction time of a subject to a drug dosage. However, during the course of the experiment, several subjects decided to drop out. The data from the remaining subjects are given in the following table:

	Dose (mg)		
Drug	**5**	**10**	**15**
A	15	18	22
	16	19	20
	16		
B	16	19	24
	17	22	
	19		

(a) There is a significant effect due to

  (i) The drug factor

  (ii) The dose factor

  (iii) The drug × dose interaction

  In each case, state the null hypothesis and your conclusion at the 5% level of significance.

  [Hint: Use Type III analysis.]

(b) Give the values of the Type II *F*-ratios and corresponding hypotheses for the drug and dose factors.

(c) Test the significance of the linear and quadratic effects of dose on reaction time for

  (i) Drug A

  (ii) Drug B

  State your conclusion at the 5% level of significance in each case.

(d) Compare the means of drugs A and B for each of the three doses. Let $\alpha = 0.05$.

(e) Suppose that the one observation in cell B-15 is missing. Which least-squares means for the drug and dose factors would be estimable in this case?

10.9 Consider the following data set for a two-way model with interaction:

	B		
A	**1**	**2**	**3**
1	10	9	—
	8	7	
2	13	—	8
3	—	3	14
		2	
		7	
4	12	13	14
		16	10

Use SAS to answer the following questions:

(a) Give a basis for the space of all estimable linear functions corresponding to this data set.

(b) What hypotheses are tested by $F(\alpha \mid \mu)$ and $F(\beta \mid \mu)$?

(c) What hypotheses are tested by $F(\alpha \mid \mu, \beta)$ and $F(\beta \mid \mu, \alpha)$?

(d) Is there a test for the hypothesis $H_0 : LSM(\alpha_{(1)}) = LSM(\alpha_{(2)})$?

(e) Explain why the hypothesis

$$H_0 : \frac{1}{2}(\mu_{32} + \mu_{33}) = \frac{1}{2}(\mu_{42} + \mu_{43})$$

is testable, where μ_{ij} is the mean of the (i, j)th cell. Give the value of a t-test statistic for testing H_0, then state your conclusion at the 5% level of significance.

(f) Is the no-interaction hypothesis stated in (10.62) testable in this case?

10.10 Consider a nested experiment involving factors A and B with B nested within A. Factor A has two levels and factor B has two levels for level 1 of A and three levels for level 2 of A. The model is

$$Y_{ijk} = \mu + \alpha_{(i)} + \beta_{i(j)} + \epsilon_{ij(k)},$$

where $\alpha_{(i)}$ denotes the fixed effect of level i of A and $\beta_{i(j)}$ denotes the fixed effect of level j of B nested within the ith level of A, and the $\epsilon_{ij(k)}$'s are independently distributed as $N(0, \sigma^2)$. The data set corresponding to this model is given in the table below.

(a) Show that $\beta_{i(j)} - \beta_{i(j')}$ is estimable for $i = 1, 2, j \neq j'$, and find its BLUE.

(b) Test the hypothesis, $H_0 : \beta_{1(1)} = \beta_{1(2)}$, at the 5% level of significance.

(c) Is the hypothesis

$$H_0 : \beta_{2(1)} = \beta_{2(2)} = \beta_{2(3)}$$

testable? If so, test H_0 at the 5% level of significance. If not, explain why not.

(d) Obtain simultaneous $(1-\alpha)100\%$ confidence intervals on all linear functions of the elements of $(\theta_{12}, \theta_{13})'$, where

$$\theta_{12} = \beta_{2(1)} - \beta_{2(2)}$$
$$\theta_{13} = \beta_{2(1)} - \beta_{2(3)}.$$

[Hint: Use the methodology described in Section 7.5.]

	A				
	1		2		
B	**1**	**2**	**1**	**2**	**3**
	5	8	8	6	3
		10	10	2	7
		9		1	
				3	

11

Unbalanced Random and Mixed Models

The purpose of this chapter is to provide a coverage of the analysis of unbalanced linear models which include random effects. As we may recall from Section 8.4, if all the effects in a model are randomly distributed, except for the overall mean, μ, then the model is called a *random-effects model*, or just a *random model*. If, however, some effects are fixed (besides the overall mean) and some are random (besides the experimental error), then the model is called a *mixed-effects model*, or just a *mixed model*. The determination of which factors in a given experimental situation have fixed effects and which have random effects depends on the nature of the factors and how their levels are selected in the experiment. A factor whose levels are of particular interest to the experimenter, and are therefore the only ones to be considered, is said to have fixed effects and is labeled as a *fixed factor*. If, however, the levels of the factor, which are actually used in the experiment, are selected at random from a typically large population (hypothetically infinite), P, of possible levels, then it is said to have random effects and is labeled as a *random factor*. In the latter case, the levels of the factor used in the experiment constitute a random sample chosen from P.

Such labeling of factors determines the type of analysis to be used for the associated model. More specifically, if a factor is fixed, then its effects are represented by fixed unknown parameters in the model. In this case, it would be of interest to compare the means of the levels of the factor (these levels are usually referred to as *treatments*), or possibly investigate its polynomial effects, if the factor is quantitative. On the other hand, if a factor is random, then it would be necessary to determine if a significant variation exists among its levels, that is, if the variance of the population P (from which the levels of the factor were selected) is different from zero. Thus, in this situation, inference is made with regard to the population P itself rather than the actual levels used in the experiment, as would be the case with fixed factors. For example, in a paper manufacturing process, an experimenter is interested in studying the effects of four different cooking temperatures of the pulp on the tensile strength of the paper. Five batches of pulp were prepared and three samples were taken from each batch and then cooked at one of the four temperatures. In this example, temperature is considered a fixed factor since the experimenter is interested in comparing the means of the four specific temperatures in order to perhaps determine the temperature setting that yields maximum tensile strength. The batch factor, however, should be considered as random since the batches used in the experiment represent only a sample from many other batches that could have been prepared. In this case, the experimenter would be interested in knowing if a significant variation exists among the batches, which may be quite different with regard to quality. The presence of such batch-to-batch variability should be taken into account before making any comparisons among the means of the four temperatures.

It should be noted that in a mixed model situation, the primary interest is in the estimation and testing of its fixed effects. The model's random effects play a secondary role, but are still considered important because their variances can affect any inferences made with regard to the fixed effects, as will be seen later. In the next section, methods for estimating the variances of the random effects will be discussed. These variances, which, in general, are unknown, are called *variance components*.

11.1 Estimation of Variance Components

The presence of random effects in a given model induces added variability in the response variable under consideration. It is therefore important that the model's variance components be properly estimated. Section 8.6 dealt with ANOVA (analysis of variance) estimation of variance components for balanced data. In the case of unbalanced data, estimating variance components is a more difficult task. This is due to several reasons that will be described later in this chapter. In the present section, several methods for estimating variance components, as applicable to unbalanced data, will be discussed. These include ANOVA estimation in addition to two other methods based on the principles of maximum likelihood and restricted maximum likelihood.

11.1.1 ANOVA Estimation—Henderson's Methods

We may recall from Section 8.6 that ANOVA estimates of the variance components in the balanced data case are obtained from a given ANOVA table by equating the mean squares of the random effects to their corresponding expected values, then solving the resulting equations for the estimates. The same procedure can be applied whenever the data set is unbalanced, except that we no longer have a unique ANOVA table, as is the case with balanced data. Instead, there can be several ANOVA tables depending on how the total sum of squares is partitioned (see Chapter 10). Consequently, ANOVA estimates of the variance components are not unique for a general random or mixed model in the unbalanced data case. Furthermore, they lack the nice optimal properties (such as minimum variance) that their counterparts have in a balanced data situation, as was mentioned earlier in Section 8.6.

The early development of ANOVA estimation of variance components for unbalanced data is mainly attributed to the pioneering work of Charles R. Henderson (see Henderson, 1953). He introduced three methods for estimating variance components which came to be known as *Henderson's Methods I, II,* and *III*. These methods are similar in the sense that each method uses a set of quadratic forms which are equated to their expected values in order to derive the variance components estimates. In *Method I*, the quadratic forms are analogous to those used in models for balanced data. These quadratic forms can be negative since some of them are not actually sums of squares. This method is computationally simple, but is not suitable for mixed models since it yields biased estimates of the variance components in the presence of fixed effects in the model (apart from μ). It is only suitable for random models, in which case the variance components estimates are unbiased. *Method II* was designed to correct the deficiency of Method I for mixed models. The data are first adjusted by using some estimate of the fixed effects based on the observed data. Method I is then applied to estimate the variance components from the adjusted data. This method produces unbiased estimates of the variance components. However, it cannot be applied when the mixed model contains interactions between the fixed and random effects (see Searle, 1968). If the data set is balanced, then Methods I and II are identical, and are the same as the ANOVA method for balanced data.

Due to the stated shortcomings of both Methods I and II, no further discussion of these methods will be given here. More details concerning the two methods can be found in Searle (1968), Searle (1971, Chapter 10), and Searle, Casella, and McCulloch (1992, Chapter 5).

11.1.1.1 Henderson's Method III

This method is applicable to both random and mixed models, even when interactions exist between fixed and random effects. It is therefore preferred over Methods I and II. The procedure on which *Method III* is based uses a sufficient number of reductions in sums of squares due to fitting submodels of the model under consideration. These reductions are expressed using the R-notation, as defined in Section 10.1, and are chosen so that their expected values are free from any fixed effects. Estimates of the variance components are obtained by equating these reductions to their expected values and then solving the resulting equations. Thus, Method III differs from Methods I and II in using positive R-expressions instead of quadratic forms that may not be positive.

Before demonstrating how Method III can be applied, we need to establish a certain result that is needed for the development of this method. Let us therefore consider, in general, the following model, which can be either random or mixed:

$$Y = X\beta + \epsilon, \tag{11.1}$$

where

X is a matrix or order $n \times q$

β is a vector that contains fixed and random effects

ϵ is an experimental error vector assumed to be independent of β and has a zero mean and a variance-covariance matrix, $\sigma_\epsilon^2 I_n$

Suppose that β is partitioned as $\beta = (\beta_1' : \beta_2')'$ and X is partitioned accordingly as $X = [X_1 : X_2]$, where X_i is of order $n \times q_i$ $(i = 1, 2)$. Model (11.1) can then be written as

$$Y = X_1\beta_1 + X_2\beta_2 + \epsilon. \tag{11.2}$$

From (11.2) we can have the reduced model,

$$Y = X_1\beta_1 + \epsilon. \tag{11.3}$$

The development of Method III is based on the following lemma.

Lemma 11.1 Let $R(\beta_2 \mid \beta_1) = R(\beta) - R(\beta_1)$, where $R(\beta)$ and $R(\beta_1)$ denote the regression sums of squares for the full model in (11.1) and the reduced model in (11.3), respectively, that is,

$$R(\beta) = Y'X(X'X)^- X'Y \tag{11.4}$$

$$R(\beta_1) = Y'X_1\left(X_1'X_1\right)^- X_1'Y. \tag{11.5}$$

Then,

$$E[R(\beta_2 \mid \beta_1)] = tr\left\{\left[X_2'X_2 - X_2'X_1\left(X_1'X_1\right)^- X_1'X_2\right] E\left(\beta_2\beta_2'\right)\right\} + \sigma_\epsilon^2\,[rank(X) - rank(X_1)]. \tag{11.6}$$

Proof. The mean and variance–covariance matrix of Y in (11.1) can be expressed as

$$E(Y) = XE(\beta) \tag{11.7}$$

$$\text{Var}(Y) = X\text{Var}(\beta)X' + \sigma_\epsilon^2 I_n. \tag{11.8}$$

Applying Theorem 5.2 to (11.4) in the light of (11.7) and (11.8), we get

$$\begin{aligned} E[R(\beta)] &= E(\beta')X'X(X'X)^- X'XE(\beta) \\ &\quad + tr\left\{X(X'X)^- X'\left[X\text{Var}(\beta)X' + \sigma_\epsilon^2 I_n\right]\right\} \\ &= E(\beta')X'XE(\beta) + tr[X\text{Var}(\beta)X'] + \sigma_\epsilon^2\, rank(X). \end{aligned} \tag{11.9}$$

Formula (11.9) is true because $X(X'X)^- X'X = X$ by property (c) in Section 3.7.1, and $tr[X(X'X)^- X'] = rank[X(X'X)^- X'] = rank(X)$ by property (b) in

Section 3.9 since $X(X'X)^-X'$ is idempotent with rank equal to the rank of X. Formula (11.9) can then be written as

$$\begin{aligned} E[R(\beta)] &= tr[X'XE(\beta)E(\beta') + X'X\text{Var}(\beta)] + \sigma_\epsilon^2\, rank(X) \\ &= tr[X'XE(\beta\beta')] + \sigma_\epsilon^2\, rank(X), \end{aligned} \tag{11.10}$$

since $\text{Var}(\beta) = E(\beta\beta') - E(\beta)E(\beta')$. Similarly, by applying Theorem 5.2 to (11.5), we get

$$\begin{aligned} E[R(\beta_1)] &= E(\beta')X'X_1\left(X_1'X_1\right)^- X_1'XE(\beta) \\ &\quad + tr\left\{X_1\left(X_1'X_1\right)^- X_1'\left[X\text{Var}(\beta)X' + \sigma_\epsilon^2 I_n\right]\right\} \\ &= tr\left[X'X_1\left(X_1'X_1\right)^- X_1'XE(\beta\beta')\right] + \sigma_\epsilon^2\, rank(X_1) \\ &= tr\left\{\begin{bmatrix} X_1'X_1 \\ X_2'X_1 \end{bmatrix}\left(X_1'X_1\right)^- \left[X_1'X_1 : X_1'X_2\right]E(\beta\beta')\right\} + \sigma_\epsilon^2\, rank(X_1) \\ &= tr\left\{\begin{bmatrix} X_1'X_1 & X_1'X_2 \\ X_2'X_1 & X_2'X_1\left(X_1'X_1\right)^- X_1'X_2 \end{bmatrix} E(\beta\beta')\right\} \\ &\quad + \sigma_\epsilon^2\, rank(X_1). \end{aligned} \tag{11.11}$$

From (11.10) and (11.11) we then have

$$\begin{aligned} E[R(\beta_2 \mid \beta_1)] &= E[R(\beta)] - E[(\beta_1)] \\ &= tr\left\{\left[X_2'X_2 - X_2'X_1\left(X_1'X_1\right)^- X_1'X_2\right]E\left(\beta_2\beta_2'\right)\right\} \\ &\quad + \sigma_\epsilon^2[rank(X) - rank(X_1)]. \end{aligned}$$

□

From Lemma 11.1, we note that if the elements of β_2 consist of just random effects, then $E[R(\beta_2 \mid \beta_1)]$ depends only on σ_ϵ^2 and the variance components that pertain to the random effects that make up the vector β_2. In addition, if β_1 contains fixed effects, then $E[R(\beta_2 \mid \beta_1)]$ has no terms due to those effects. This is necessary if we were to use $R(\beta_2 \mid \beta_1)$ to derive ANOVA estimates that are not dependent on the fixed effects and are therefore unbiased. Furthermore, if β_1 contains elements that are random, then $E[R(\beta_2 \mid \beta_1)]$ does not depend on $\text{Var}(\beta_1)$ or any covariances that may exist between the elements of β_1 and β_2. Also, note that the application of Lemma 11.1 requires that β_1 and β_2 make up the entire β vector for the full model.

With the help of Lemma 11.1 we can select as many R-expressions, of the form considered in this lemma, as there are variance components (excluding σ_ϵ^2) for the full model. Such expressions have no fixed effects in their expected values. Equating the selected R-expressions to their expected values and solving the resulting equations along with the equation, $E(SS_E) = [n - rank(X)]\sigma_\epsilon^2$, we get unbiased estimates of the variance components, which we refer to as Method III estimates. These estimates are not unique since the number of

TABLE 11.1
Expected Values of R-Expressions for a Random Two-Way Model

R-Expression	Expected Value
$R(\alpha, \beta, \alpha\beta \mid \mu)$	$h_{11}\,\sigma^2_{\alpha} + h_{12}\,\sigma^2_{\beta} + h_{13}\,\sigma^2_{\alpha\beta} + (ab-1)\sigma^2_{\epsilon}$
$R(\beta, \alpha\beta \mid \mu, \alpha)$	$h_{22}\,\sigma^2_{\beta} + h_{23}\,\sigma^2_{\alpha\beta} + (ab-a)\sigma^2_{\epsilon}$
$R(\alpha\beta \mid \mu, \alpha, \beta)$	$h_{33}\,\sigma^2_{\alpha\beta} + (ab-a-b+1)\sigma^2_{\epsilon}$
SS_E	$(n_{..} - ab)\sigma^2_{\epsilon}$

eligible R-expressions that can be used exceeds the number of variance components for the full model.

All Henderson's methods can give negative estimates of the variance components. Other than being unbiased, ANOVA estimators have no known optimal properties, even under normality, if the data set under consideration is unbalanced.

Let us now demonstrate the application of Lemma 11.1 and Method III by considering the following example.

Example 11.1 Consider the random two-way model,

$$Y_{ijk} = \mu + \alpha_{(i)} + \beta_{(j)} + (\alpha\beta)_{(ij)} + \epsilon_{ij(k)}, \quad i = 1, 2, \ldots, a;\ j = 1, 2, \ldots, b;$$
$$k = 1, 2, \ldots, n_{ij}.$$

The variance components for the model are therefore σ^2_{α}, σ^2_{β}, $\sigma^2_{\alpha\beta}$, and σ^2_{ϵ}. In this case, we can consider the R-expressions whose expected values are given in Table 11.1, where $n_{..}$ is the total number of observations, and $h_{11}, h_{12}, h_{13}, h_{22}, h_{23}$, and h_{33} are known constants that result from expanding the trace portion of the expected value in formula (11.6). Equating SS_E and the three R-expressions in Table 11.1 to their respective expected values and solving the resulting four equations produce Method III estimates of σ^2_{α}, σ^2_{β}, $\sigma^2_{\alpha\beta}$, and σ^2_{ϵ}.

An alternative set of R-expressions can be chosen to derive Method III estimates, as can be seen from Table 11.2, where h'_{23} is a constant possibly different from h_{23} in Table 11.1. This demonstrates the nonuniqueness of Method III estimates of the variance components.

Let us now consider the same model as before, except that $\alpha_{(i)}$ is considered fixed. Thus, the variance components are σ^2_{β}, $\sigma^2_{\alpha\beta}$, and σ^2_{ϵ}. In this case, to obtain Method III estimates of these variance components, we should only consider R-expressions that are adjusted for at least μ and α since μ and $\alpha_{(i)}$ $(i = 1, 2, \ldots, a)$ are the fixed effects in the full two-way model. We can therefore consider the R-expressions, $R(\alpha\beta \mid \mu, \alpha, \beta)$, $R(\beta, \alpha\beta \mid \mu, \alpha)$, which along with SS_E can produce Method III estimates of σ^2_{β}, $\sigma^2_{\alpha\beta}$, and σ^2_{ϵ}.

TABLE 11.2
An Alternative Set of *R*-Expressions

R-Expression	Expected Value
$R(\alpha, \beta, \alpha\beta \mid \mu)$	$h_{11}\sigma^2_{\alpha} + h_{12}\sigma^2_{\beta} + h_{13}\sigma^2_{\alpha\beta} + (ab-1)\sigma^2_{\epsilon}$
$R(\alpha, \alpha\beta \mid \mu, \beta)$	$h_{21}\sigma^2_{\alpha} + h'_{23}\sigma^2_{\alpha\beta} + (ab-b)\sigma^2_{\epsilon}$
$R(\alpha\beta \mid \mu, \alpha, \beta)$	$h_{33}\sigma^2_{\alpha\beta} + (ab-a-b+1)\sigma^2_{\epsilon}$
SS_E	$(n_{..}-ab)\sigma^2_{\epsilon}$

It should be noted that, in this example, Lemma 11.1 cannot be applied directly to *R*-expressions of the form $R(\alpha \mid \mu)$ and $R(\beta \mid \mu, \alpha)$ or $R(\beta \mid \mu)$ and $R(\alpha \mid \mu, \beta)$, which are Type I sums of squares for factors *A* and *B* (depending on how the model is written). This is so because the requirement of having $\boldsymbol{\beta}_1$ and $\boldsymbol{\beta}_2$ make up the entire $\boldsymbol{\beta}$ vector for the full model, as in Lemma 11.1, is not satisfied in any of these expressions. However, it is possible to apply Lemma 11.1 indirectly to such expressions. This is done as follows: suppose, for example, that all the effects in the full model are random. We can use the fact that

$$R(\alpha, \beta, \alpha\beta \mid \mu) = R(\alpha \mid \mu) + R(\beta \mid \mu, \alpha) + R(\alpha\beta \mid \mu, \alpha, \beta) \tag{11.12}$$

$$R(\beta, \alpha\beta \mid \mu, \alpha) = R(\beta \mid \mu, \alpha) + R(\alpha\beta \mid \mu, \alpha, \beta). \tag{11.13}$$

Solving Equations (11.12) and (11.13) for $R(\alpha \mid \mu)$ and $R(\beta \mid \mu, \alpha)$, we get

$$R(\alpha \mid \mu) = R(\alpha, \beta, \alpha\beta \mid \mu) - R(\beta, \alpha\beta \mid \mu, \alpha) \tag{11.14}$$

$$R(\beta \mid \mu, \alpha) = R(\beta, \alpha\beta \mid \mu, \alpha) - R(\alpha\beta \mid \mu, \alpha, \beta). \tag{11.15}$$

The *R*-expressions on the right-hand sides of (11.14) and (11.15) are of the type considered in Lemma 11.1. We can therefore obtain the expected values of $R(\alpha \mid \mu)$ and $R(\beta \mid \mu, \alpha)$ in terms of those of $R(\alpha\beta \mid \mu, \alpha, \beta)$, $R(\beta, \alpha\beta \mid \mu, \alpha)$, and $R(\alpha, \beta, \alpha\beta \mid \mu)$. Using Table 11.1, we get

$$\begin{aligned} E[R(\alpha \mid \mu)] &= \left[h_{11}\sigma^2_{\alpha} + h_{12}\sigma^2_{\beta} + h_{13}\sigma^2_{\alpha\beta} + (ab-1)\sigma^2_{\epsilon}\right] \\ &\quad - \left[h_{22}\sigma^2_{\beta} + h_{23}\sigma^2_{\alpha\beta} + (ab-a)\sigma^2_{\epsilon}\right] \\ &= h_{11}\sigma^2_{\alpha} + (h_{12}-h_{22})\sigma^2_{\beta} + (h_{13}-h_{23})\sigma^2_{\alpha\beta} \\ &\quad + (a-1)\sigma^2_{\epsilon} \end{aligned} \tag{11.16}$$

$$\begin{aligned} E[R(\beta \mid \mu, \alpha)] &= \left[h_{22}\sigma^2_{\beta} + h_{23}\sigma^2_{\alpha\beta} + (ab-a)\sigma^2_{\epsilon}\right] \\ &\quad - \left[h_{33}\sigma^2_{\alpha\beta} + (ab-a-b+1)\sigma^2_{\epsilon}\right] \\ &= h_{22}\sigma^2_{\beta} + (h_{23}-h_{33})\sigma^2_{\alpha\beta} + (b-1)\sigma^2_{\epsilon}. \end{aligned} \tag{11.17}$$

Equating $R(\alpha \mid \mu)$ and $R(\beta \mid \mu, \alpha)$ to their expected values and combining the resulting equations with the equations, $E[R(\alpha\beta \mid \mu, \alpha, \beta)] = R(\alpha\beta \mid \mu, \alpha, \beta)$ and $E(SS_E) = (n_{..} - ab)\sigma^2_\epsilon$, produce Method III estimates of σ^2_α, σ^2_β, $\sigma^2_{\alpha\beta}$, and σ^2_ϵ. We can similarly get other estimates by considering the expected values of $R(\beta \mid \mu)$ and $R(\alpha \mid \mu, \beta)$.

Since $R(\alpha \mid \mu)$, $R(\beta \mid \mu, \alpha)$, and $R(\alpha\beta \mid \mu, \alpha, \beta)$ are Type I sums of squares for factors A, B, and their interaction, $A * B$, if the corresponding SAS model is written as, $Y = A \;\; B \;\; A * B$, we conclude that, in general, Method III estimates of the variance components can be obtained from equating Type I sums of squares of all the random effects in the model, along with SS_E, to their expected values and then solving the resulting system of equations. Such estimates can be easily computed using PROC VARCOMP in SAS (2000) with the option "METHOD = TYPE1" included in the PROC VARCOMP statement. For example, suppose that the two-way model in Example 11.1 is random. Then, the SAS statements needed to get Method III estimates of σ^2_α, σ^2_β, $\sigma^2_{\alpha\beta}$, and σ^2_ϵ are

```
DATA;
INPUT A B Y;
CARDS;
(enter the data here)
PROC VARCOMP METHOD = TYPE1;
CLASS A B;
MODEL Y = A B A * B;
RUN;
```

If only $\alpha_{(i)}$ is fixed, then to get Method III estimates of σ^2_β, $\sigma^2_{\alpha\beta}$, σ^2_ϵ, we can use the same statements as before, except that the MODEL statement should be modified by adding the option "FIXED = 1," that is, by using the statement, MODEL Y = $A\ B\ A * B$/FIXED = 1;. In general, the MODEL option "FIXED = m" tells PROC VARCOMP that the first m effects in the MODEL statement are fixed. The remaining effects are assumed to be random. If this option is left out, then all the effects in the model are assumed random. Thus, when $\alpha_{(i)}$ is fixed in our two-way model, A should be placed first in the MODEL statement and m set equal to 1. It is therefore important that all the fixed effects, if any, be placed first in the MODEL statement.

Example 11.2 Consider the mixed two-way model,

$$Y_{ijk} = \mu + \alpha_{(i)} + \beta_{(j)} + (\alpha\beta)_{(ij)} + \epsilon_{ij(k)},$$

where $\alpha_{(i)}$ is fixed and $\beta_{(j)}$ is random. The data set used with this model is given in Table 11.3. Since $\alpha_{(i)}$ is fixed, the needed Type I R-expressions are $R(\beta \mid \mu, \alpha)$ and $R(\alpha\beta \mid \mu, \alpha, \beta)$. The corresponding expected values are given in Table 11.4. The expected values in Table 11.4 can be obtained from the output of PROC VARCOMP, which also gives directly the solution of the system

TABLE 11.3
Data Set for Example 11.2

	B (Random)			
A (Fixed)	1	2	3	4
1	20.9	25.0	23.8	70.1
	25.2	30.1	26.9	102.1
		28.9	23.6	101.2
			25.3	
2	28.9	30.9	39.8	65.3
	44.3	28.6	41.5	62.1
	33.9	31.0		
		30.9		
3	30.1	32.1	41.8	50.9
	28.9	35.9	44.3	50.8
	34.5	28.9	39.2	
		38.1		

TABLE 11.4
Expected Values of Type I *R*-Expressions

R-Expression	DF	Value of *R*-Expression	Expected Value
$R(\beta \mid \mu, \alpha)$	3	8884.2827	$25.6668\,\sigma^2_{\beta} + 8.8017\,\sigma^2_{\alpha\beta} + 3\,\sigma^2_{\epsilon}$
$R(\alpha\beta \mid \mu, \alpha, \beta)$	6	2970.3412	$16.8648\,\sigma^2_{\alpha\beta} + 6\,\sigma^2_{\epsilon}$
Error	23	908.8150	$23\,\sigma^2_{\epsilon}$

of three equations resulting from equating the values of the *R*-expressions in Table 11.4 to their expected values. We thus have the following ANOVA estimates of $\sigma^2_{\beta}, \sigma^2_{\alpha\beta}, \sigma^2_{\epsilon}$: $\hat{\sigma}^2_{\beta} = 285.9459$, $\hat{\sigma}^2_{\alpha\beta} = 162.0671$, and $\hat{\sigma}^2_{\epsilon} = 39.5137$, which are considered Method III estimates of the three variance components.

11.1.2 Maximum Likelihood Estimation

An alternative approach to variance components estimation is that of *maximum likelihood* (ML). The ML approach is based on assuming normality of the data under consideration. The likelihood function is then maximized over the parameter space under nonnegative constraints on the variance components. Hence, ML estimates must be nonnegative.

Let us consider model (11.1), which can be either random or mixed. Using the same notation as in Section 8.4, this model is expressed as

$$Y = \sum_{i=0}^{\nu-p} H_i\beta_i + \sum_{i=\nu-p+1}^{\nu+1} H_i\beta_i, \tag{11.18}$$

where

ν is the number of effects (excluding the experimental error term and the grand mean) in the model

p is a positive integer such that $p \leq \nu$, $\sum_{i=0}^{\nu-p} \boldsymbol{H}_i \boldsymbol{\beta}_i$ is the fixed portion of the model ($\boldsymbol{\beta}_i$ is a fixed unknown parameter vector for $i = 0, 1, \ldots, \nu - p$)

$\sum_{i=\nu-p+1}^{\nu+1} \boldsymbol{H}_i \boldsymbol{\beta}_i$ is its random portion

$\boldsymbol{H}_i$ is a known matrix of order $n \times c_i$ consisting of zeros and ones for $i = 1, 2, \ldots, \nu + 1$ ($\boldsymbol{H}_0 = \boldsymbol{1}_n$)

If the data set is unbalanced, then $\boldsymbol{H}_i$ is no longer expressible as a direct product of identity matrices and vector of ones, as was the case in Section 8.2. For the sake of simplicity, we shall represent the fixed portion as $\boldsymbol{X}_* \boldsymbol{g}$, where $\boldsymbol{X}_* = [\boldsymbol{H}_0 : \boldsymbol{H}_1 : \ldots : \boldsymbol{H}_{\nu-p}]$, and $\boldsymbol{g}$ is a vector containing all the fixed effects in the model. Thus, model (11.18) is written as

$$\boldsymbol{Y} = \boldsymbol{X}_* \boldsymbol{g} + \sum_{i=\nu-p+1}^{\nu+1} \boldsymbol{H}_i \boldsymbol{\beta}_i. \tag{11.19}$$

The $\boldsymbol{\beta}_i$'s are assumed to be independently distributed such that $\boldsymbol{\beta}_i \sim N(\boldsymbol{0}, \sigma_i^2 \boldsymbol{I}_{c_i})$ for $i = \nu - p + 1, \nu - p + 2, \ldots, \nu + 1$. The data vector $\boldsymbol{Y}$ is therefore normally distributed as $N(\boldsymbol{X}_* \boldsymbol{g}, \boldsymbol{\Sigma})$, where

$$\boldsymbol{\Sigma} = \sum_{i=\nu-p+1}^{\nu+1} \sigma_i^2 \boldsymbol{A}_i, \tag{11.20}$$

and $\boldsymbol{A}_i = \boldsymbol{H}_i \boldsymbol{H}_i'$ ($i = \nu - p + 1, \nu - p + 2, \ldots, \nu + 1$). On the basis of formula (4.17), the *likelihood function* associated with $\boldsymbol{Y}$ is

$$\mathcal{L} = \frac{1}{(2\pi)^{n/2}[det(\boldsymbol{\Sigma})]^{1/2}} \exp\left[-\frac{1}{2}(\boldsymbol{y} - \boldsymbol{X}_* \boldsymbol{g})' \boldsymbol{\Sigma}^{-1} (\boldsymbol{y} - \boldsymbol{X}_* \boldsymbol{g})\right]. \tag{11.21}$$

By definition, the *maximum likelihood estimates* (MLEs) of $\boldsymbol{g}$ and the variance components, $\sigma_{\nu-p+1}^2, \sigma_{\nu-p+2}^2, \ldots, \sigma_{\nu+1}^2$, are those that maximize $\mathcal{L}$ subject to the constraints, $\sigma_i^2 \geq 0$, $i = \nu - p + 1, \nu - p + 2, \ldots, \nu + 1$. Equivalently, we can consider maximizing the natural logarithm of $\mathcal{L}$, called the *log-likelihood function*, namely,

$$\log(\mathcal{L}) = -\frac{n}{2}\log(2\pi) - \frac{1}{2}\log[det(\boldsymbol{\Sigma})] - \frac{1}{2}(\boldsymbol{y} - \boldsymbol{X}_* \boldsymbol{g})' \boldsymbol{\Sigma}^{-1} (\boldsymbol{y} - \boldsymbol{X}_* \boldsymbol{g}). \tag{11.22}$$

To maximize $\log(\mathcal{L})$, we differentiate the two sides of (11.22), first with respect to $\boldsymbol{g}$, using Corollary 3.4 and Theorem 3.22, to get

$$\frac{\partial[\log(\mathcal{L})]}{\partial \boldsymbol{g}} = \boldsymbol{X}_*' \boldsymbol{\Sigma}^{-1} \boldsymbol{y} - \boldsymbol{X}_*' \boldsymbol{\Sigma}^{-1} \boldsymbol{X}_* \boldsymbol{g}. \tag{11.23}$$

Differentiating now $\log(\mathcal{L})$ with respect to σ_i^2 $(i = \nu-p+1, \nu-p+2, \ldots, \nu+1)$ and making use of the fact that,

$$\frac{\partial \log[det(\mathbf{\Sigma})]}{\partial \sigma_i^2} = tr\left(\mathbf{\Sigma}^{-1}\frac{\partial \mathbf{\Sigma}}{\partial \sigma_i^2}\right)$$
$$= tr(\mathbf{\Sigma}^{-1}\mathbf{A}_i), \quad i = \nu-p+1, \nu-p+2, \ldots, \nu+1,$$
$$\frac{\partial \mathbf{\Sigma}^{-1}}{\partial \sigma_i^2} = -\mathbf{\Sigma}^{-1}\frac{\partial \mathbf{\Sigma}}{\partial \sigma_i^2}\mathbf{\Sigma}^{-1} \text{ (see Theorem 3.24)},$$
$$= -\mathbf{\Sigma}^{-1}\mathbf{A}_i\mathbf{\Sigma}^{-1}, \quad i = \nu-p+1, \nu-p+2, \ldots, \nu+1,$$

we get

$$\frac{\partial[\log(\mathcal{L})]}{\sigma_i^2} = -\frac{1}{2}tr(\mathbf{\Sigma}^{-1}\mathbf{A}_i) + \frac{1}{2}(\mathbf{y}-\mathbf{X}_*\mathbf{g})'\mathbf{\Sigma}^{-1}\mathbf{A}_i\mathbf{\Sigma}^{-1}(\mathbf{y}-\mathbf{X}_*\mathbf{g}), \tag{11.24}$$

for $i = \nu-p+1, \nu-p+2, \ldots, \nu+1$. Equating (11.23) and (11.24) to zero and changing $\mathbf{y}$ to the random data vector, $\mathbf{Y}$, we get the equations,

$$\mathbf{X}_*'\tilde{\mathbf{\Sigma}}^{-1}\mathbf{X}_*\tilde{\mathbf{g}} = \mathbf{X}_*'\tilde{\mathbf{\Sigma}}^{-1}\mathbf{Y} \tag{11.25}$$
$$tr[\tilde{\mathbf{\Sigma}}^{-1}\mathbf{A}_i] = (\mathbf{Y}-\mathbf{X}_*\tilde{\mathbf{g}})'\tilde{\mathbf{\Sigma}}^{-1}\mathbf{A}_i\tilde{\mathbf{\Sigma}}^{-1}(\mathbf{Y}-\mathbf{X}_*\tilde{\mathbf{g}})$$
$$= \mathbf{Y}'\tilde{\mathbf{P}}\mathbf{A}_i\tilde{\mathbf{P}}\mathbf{Y}, \quad i = \nu-p+1, \nu-p+2, \ldots, \nu+1, \tag{11.26}$$

where

$$\tilde{\mathbf{P}} = \tilde{\mathbf{\Sigma}}^{-1} - \tilde{\mathbf{\Sigma}}^{-1}\mathbf{X}_*\left(\mathbf{X}_*'\tilde{\mathbf{\Sigma}}^{-1}\mathbf{X}_*\right)^{-}\mathbf{X}_*'\tilde{\mathbf{\Sigma}}^{-1} \tag{11.27}$$
$$\tilde{\mathbf{\Sigma}} = \sum_{i=\nu-p+1}^{\nu+1}\tilde{\sigma}_i^2\mathbf{A}_i, \tag{11.28}$$

and $\tilde{\mathbf{g}}$ and $\tilde{\sigma}_i^2$ are, respectively, the MLEs of $\mathbf{g}$ and σ_i^2 $(i = \nu-p+1, \nu-p+2, \ldots, \nu+1)$. Equation 11.26 follows from the fact that

$$\mathbf{Y}-\mathbf{X}_*\tilde{\mathbf{g}} = \left[\mathbf{I}_n - \mathbf{X}_*\left(\mathbf{X}_*'\tilde{\mathbf{\Sigma}}^{-1}\mathbf{X}_*\right)^{-}\mathbf{X}_*'\tilde{\mathbf{\Sigma}}^{-1}\right]\mathbf{Y}$$
$$= \tilde{\mathbf{\Sigma}}\tilde{\mathbf{P}}\mathbf{Y}.$$

Note that the left-hand side of (11.26) can be expressed as

$$tr(\tilde{\mathbf{\Sigma}}^{-1}\mathbf{A}_i) = tr(\tilde{\mathbf{\Sigma}}^{-1}\mathbf{A}_i\tilde{\mathbf{\Sigma}}^{-1}\tilde{\mathbf{\Sigma}})$$
$$= tr\left[\tilde{\mathbf{\Sigma}}^{-1}\mathbf{A}_i\tilde{\mathbf{\Sigma}}^{-1}\left(\sum_{j=\nu-p+1}^{\nu+1}\tilde{\sigma}_j^2\mathbf{A}_j\right)\right]$$
$$= \sum_{j=\nu-p+1}^{\nu+1}tr(\tilde{\mathbf{\Sigma}}^{-1}\mathbf{A}_i\tilde{\mathbf{\Sigma}}^{-1}\mathbf{A}_j)\tilde{\sigma}_j^2, \quad i = \nu-p+1, \nu-p+2, \ldots, \nu+1.$$

Formula (11.26) can then be written as

$$\sum_{j=\nu-p+1}^{\nu+1} tr(\tilde{\boldsymbol{\Sigma}}^{-1}\boldsymbol{A}_i\tilde{\boldsymbol{\Sigma}}^{-1}\boldsymbol{A}_j)\tilde{\sigma}_j^2 = \boldsymbol{Y}'\tilde{\boldsymbol{P}}\boldsymbol{A}_i\tilde{\boldsymbol{P}}\boldsymbol{Y}, \quad i = \nu-p+1, \nu-p+2, \ldots, \nu+1. \tag{11.29}$$

Equations (11.25) and (11.29) are to be solved by iteration for $\tilde{g}$ and $\tilde{\sigma}_j^2$. Only nonnegative values of $\tilde{\sigma}_j^2$ $(j = \nu-p+1, \nu-p+2, \ldots, \nu+1)$ are to be retained. A method for solving the maximum likelihood equations was developed by Hartley and Rao (1967). It should be recognized that the solution of these equations can be computationally formidable because of the need to invert a variance–covariance matrix of large order. Computational aspects of maximum likelihood estimation were discussed by several authors; see, for example, Hemmerle and Hartley (1973), Jennrich and Sampson (1976), Miller (1979), and Searle, Casella, and McCulloch (1992). Harville (1977) presented a comprehensive critique of the maximum likelihood approach to the estimation of variance components and related topics.

In SAS, PROC VARCOMP can be used to compute maximum likelihood estimates of variance components using the W-transformation developed by Hemmerle and Hartley (1973). The option "METHOD = ML" must be specified in the PROC VARCOMP statement. Furthermore, the option "FIXED = m" in the MODEL statement tells PROC VARCOMP that the first m effects in the model are fixed. The remaining effects are assumed to be random. If this option is left out, then all the effects in the model are assumed random. For example, for the mixed two-way model, $Y_{ijk} = \mu + \alpha_{(i)} + \beta_{(j)} + (\alpha\beta)_{(ij)} + \epsilon_{ij(k)}$, with $\alpha_{(i)}$ fixed and $\beta_{(j)}$ random, the statements needed to get the MLEs of σ^2_β, $\sigma^2_{\alpha\beta}$, and σ^2_ϵ are

```
DATA;
INPUT A B Y;
CARDS;
(enter the data here)
PROC VARCOMP METHOD = ML;
CLASS A B;
MODEL Y = A B A*B/FIXED=1;
RUN;
```

PROC VARCOMP, however, does not provide maximum likelihood estimates of the fixed effects in a mixed model. Furthermore, no continuous effects are allowed since the model's effects are limited to main effects, interactions, and nested effects. This restricts the use of PROC VARCOMP to just ANOVA models, but not, for example, response surface models with random effects (see Chapter 12). Because of these limitations, the more recent PROC MIXED in SAS is preferred for maximum likelihood computations.

PROC MIXED accepts a variety of mixed linear models, including those with continuous effects. It provides, among other things, maximum likelihood estimates of the variance components as well as approximate tests concerning the model's fixed effects. In addition, a variety of covariance structures can be specified for the data under consideration. The option "METHOD = ML" should be specified in the PROC MIXED statement. The MODEL statement (in PROC MIXED) includes only the fixed effects in the model. If there are none, then the MODEL statement is written as MODEL Y = ;. In this case, the model includes only random effects, in addition to μ, the overall mean. The RANDOM statement (in PROC MIXED) lists all the random effects in the model. These effects can be classification or continuous. Also, the REPEATED statement specifies the variance–covariance structure for the experimental error vector, $\boldsymbol{\epsilon}$ (see Section 11.9). If no REPEATED statement is specified, then the variance–covariance matrix for $\boldsymbol{\epsilon}$ is assumed to be $\sigma^2_{\epsilon}\boldsymbol{I}_n$.

Unlike ANOVA estimators, maximum likelihood estimators (MLEs) of variance components have interesting asymptotic properties. Under certain regularity conditions, MLEs of variance components are *consistent* (i.e., they converge in probability to the true values of the variance components as the sample size tends to infinity), *asymptotically efficient* (in the sense of attaining the *Cramér–Rao lower bound*), and *asymptotically normal*. These properties were investigated by Miller (1977). Searle (1970) and Rudan and Searle (1971) derived large-sample variances of the MLEs of the variance components. More specifically, if $\tilde{\boldsymbol{\sigma}}^2$ denotes the MLE of $\boldsymbol{\sigma}^2 = \left(\sigma^2_{\nu-p+1}, \sigma^2_{\nu-p+2}, \ldots, \sigma^2_{\nu+1}\right)'$, then, as $n \to \infty$,

$$\text{Var}(\tilde{\boldsymbol{\sigma}}^2) \to 2\boldsymbol{\Gamma}^{-1}, \tag{11.30}$$

where $\boldsymbol{\Gamma} = (\gamma_{ij})$ and γ_{ij} is given by

$$\gamma_{ij} = tr\left(\boldsymbol{\Sigma}^{-1}\boldsymbol{A}_i\boldsymbol{\Sigma}^{-1}\boldsymbol{A}_j\right), \quad i,j = \nu - p + 1, \nu - p + 2, \ldots, \nu + 1. \tag{11.31}$$

Example 11.3 Consider again the same model and data set as in Example 11.2. The following is a listing of the SAS statements needed to get the maximum likelihood estimates of $\sigma^2_{\beta}, \sigma^2_{\alpha\beta}, \sigma^2_{\epsilon}$ using PROC MIXED:

```
DATA;
INPUT A B Y;
CARDS;
(enter the data here)
PROC MIXED METHOD = ML;
CLASS A B;
MODEL Y = A;
RANDOM B A*B;
RUN;
```

On the basis of the data set in Table 11.3, the following maximum likelihood estimates were obtained: $\tilde{\sigma}^2_{\beta} = 210.65$, $\tilde{\sigma}^2_{\alpha\beta} = 123.74$, $\tilde{\sigma}^2_{\epsilon} = 39.445$.

11.1.3 Restricted Maximum Likelihood Estimation

This method is an adaptation of ML estimation of variance components brought about by maximizing only that portion of the likelihood function that is location invariant (i.e., does not depend on any fixed effects). It was first proposed for unbalanced data by Patterson and Thompson (1971). They called it *modified maximum likelihood*, but has come to be known as *restricted (or residual) maximum likelihood* (REML).

In order to understand the development of REML estimation of variance components, the following definition is needed.

Definition 11.1 An *error contrast* is a linear function, $\boldsymbol{k}'\boldsymbol{Y}$, of the data vector, $\boldsymbol{Y}$, that has zero expectation, that is, $E(\boldsymbol{k}'\boldsymbol{Y}) = 0$.

Based on this definition, if $\boldsymbol{Y}$ is represented by the model in (11.19), then $E(\boldsymbol{k}'\boldsymbol{Y}) = \boldsymbol{k}'\boldsymbol{X}_*\boldsymbol{g} = 0$ for all $\boldsymbol{g}$. Hence, $\boldsymbol{k}'\boldsymbol{Y}$ is an error contrast if and only if $\boldsymbol{k}'\boldsymbol{X}_* = \boldsymbol{0}'$, that is, $\boldsymbol{k}$ is orthogonal to the column space of $\boldsymbol{X}_*$. Since $\left[\boldsymbol{I}_n - \boldsymbol{X}_*\left(\boldsymbol{X}_*'\boldsymbol{X}_*\right)^-\boldsymbol{X}_*'\right]\boldsymbol{X}_* = \boldsymbol{0}$ and the rank of $\boldsymbol{I}_n - \boldsymbol{X}_*\left(\boldsymbol{X}_*'\boldsymbol{X}_*\right)^-\boldsymbol{X}_*'$ is equal to $n - r_*$, where $r_* = rank(\boldsymbol{X}_*)$, $\boldsymbol{k}'$ must belong to the row space of $\boldsymbol{I}_n - \boldsymbol{X}_*\left(\boldsymbol{X}_*'\boldsymbol{X}_*\right)^-\boldsymbol{X}_*'$. Thus,

$$\boldsymbol{k}' = \boldsymbol{c}'\left[\boldsymbol{I}_n - \boldsymbol{X}_*\left(\boldsymbol{X}_*'\boldsymbol{X}_*\right)^-\boldsymbol{X}_*'\right], \tag{11.32}$$

for some vector $\boldsymbol{c}$. It follows that the number of linearly independent error contrasts is equal to $n - r_*$.

Now, let us confine our attention to the vector $\boldsymbol{K}'\boldsymbol{Y}$, where $\boldsymbol{K}'$ is a matrix of order $(n - r_*) \times n$ and rank $n - r_*$ such that $\boldsymbol{K}'\boldsymbol{X}_* = \boldsymbol{0}$. Thus, the $n - r_*$ elements of $\boldsymbol{K}'\boldsymbol{Y}$ are linearly independent error contrasts that form a basis for all such error contrasts. Under the assumptions made earlier in Section 11.1.2 concerning normality of the distribution of $\boldsymbol{Y}$, $\boldsymbol{K}'\boldsymbol{Y} \sim N(\boldsymbol{0}, \boldsymbol{K}'\boldsymbol{\Sigma}\boldsymbol{K})$, since $\boldsymbol{K}'\boldsymbol{X}_*\boldsymbol{g} = \boldsymbol{0}$.

Instead of maximizing the log-likelihood function of $\boldsymbol{Y}$, as was done earlier in Section 11.1.2 for ML estimation, we can now maximize the log-likelihood function of $\boldsymbol{K}'\boldsymbol{Y}$, denoted by $\log(\mathcal{L}_r)$, which is given by

$$\log(\mathcal{L}_r) = -\frac{n - r_*}{2}\log(2\pi) - \frac{1}{2}\log[det(\boldsymbol{K}'\boldsymbol{\Sigma}\boldsymbol{K})] - \frac{1}{2}\boldsymbol{Y}'\boldsymbol{K}(\boldsymbol{K}'\boldsymbol{\Sigma}\boldsymbol{K})^{-1}\boldsymbol{K}'\boldsymbol{Y}. \tag{11.33}$$

Formula (11.33) can be derived from (11.22) by replacing n, $\boldsymbol{\Sigma}$, $\boldsymbol{Y}$, $\boldsymbol{X}_*\boldsymbol{g}$ with $n - r_*$, $\boldsymbol{K}'\boldsymbol{\Sigma}\boldsymbol{K}$, $\boldsymbol{K}'\boldsymbol{Y}$, and $\boldsymbol{K}'\boldsymbol{X}_*\boldsymbol{g} = \boldsymbol{0}$, respectively. Note that this log likelihood depends only on the variance components. By definition, the *REML estimates* of the variance components, $\sigma^2_{\nu-p+1}, \sigma^2_{\nu-p+2}, \ldots, \sigma^2_{\nu+1}$, are those that maximize $\log(\mathcal{L}_r)$ subject to the constraints $\sigma_i^2 \geq 0, i = \nu-p+1, \nu-p+2, \ldots, \nu+1$. It is interesting here to remark that $\log(\mathcal{L}_r)$ is invariant to the choice of $\boldsymbol{K}'$ as long as $\boldsymbol{K}'$ has full row rank. To show this, suppose that we were to choose another full row-rank matrix, K_1', such that $K_1'X_* = \mathbf{0}$. Then, there exists a nonsingular matrix M such that $K' = MK_1'$. Consequently, $det(K'\Sigma K) = det\left(MK_1'\Sigma K_1M'\right) = [det(M)]^2 det\left(K_1'\Sigma K_1\right)$ and

$$Y'K(K'\Sigma K)^{-1}K'Y = Y'K_1M'\left(MK_1'\Sigma K_1M'\right)^{-1}MK_1'Y$$
$$= Y'K_1\left(K_1'\Sigma K_1\right)^{-1}K_1'Y.$$

Making the substitution in (11.33), we get

$$\log(\mathcal{L}_r) = -\frac{n-r_*}{2}\log(2\pi) - \log[det(M)] - \frac{1}{2}\log\left[det(K_1'\Sigma K_1)\right]$$
$$-\frac{1}{2}Y'K_1\left(K_1'\Sigma K_1\right)^{-1}K_1'Y.$$

Since the matrix M does not depend on the variance components, the maximization of $\log(\mathcal{L}_r)$ with respect to the variance components is equivalent to the maximization of

$$-\frac{n-r_*}{2}\log(2\pi) - \frac{1}{2}\log\left[det\left(K_1'\Sigma K_1\right)\right] - \frac{1}{2}Y'K_1\left(K_1'\Sigma K_1\right)^{-1}K_1'Y,$$

which is of the same form as the right-hand side of (11.33), but with the replacement of K with K_1'. Thus, whether K or K_1' are used, the REML estimates remain unchanged.

To obtain the REML estimates of the variance components we can modify the ML equations in (11.26) by replacing Y with $K'Y$, A_i with $K'A_iK$, X_* with $K'X_* = \mathbf{0}$, and Σ with $K'\Sigma K$. We therefore get the REML equations,

$$tr\left[(K'\tilde{\Sigma}_rK)^{-1}(K'A_iK)\right] = Y'K(K'\tilde{\Sigma}_rK)^{-1}(K'A_iK)(K'\tilde{\Sigma}_rK)^{-1}K'Y,$$
$$i = \nu-p+1, \nu-p+2, \ldots, \nu+1, \quad (11.34)$$

where

$$\tilde{\Sigma}_r = \sum_{i=\nu-p+1}^{\nu+1} \tilde{\sigma}_{ri}^2 A_i, \quad (11.35)$$

and the $\tilde{\sigma}_{ri}^2$'s are the REML estimates of σ_i^2 $(i = \nu-p+1, \nu-p+2, \ldots, \nu+1)$.

Lemma 11.2

$$K(K'\Sigma K)^{-1}K' = P, \quad (11.36)$$

where

$$P = \Sigma^{-1} - \Sigma^{-1}X_*\left(X_*'\Sigma^{-1}X_*\right)^{-}X_*'\Sigma^{-1}. \quad (11.37)$$

Proof. The rows of K' span the orthogonal complement of the column space of X_* because $K'X_* = 0$ and K' is of full row rank equal to $n - r_*$. Since the rows of the matrix $I_n - X_* \left(X_*'X_*\right)^- X_*'$ are also orthogonal to the columns of X_*, we conclude that

$$I_n - X_* \left(X_*'X_*\right)^- X_*' = TK', \tag{11.38}$$

for some matrix T. Taking the transpose of both sides of (11.38), we get

$$I_n - X_* \left(X_*'X_*\right)^- X_*' = KT'. \tag{11.39}$$

Multiplying both sides of (11.39) by $(K'K)^{-1}K'$, we get $(K'K)^{-1}K' = T'$. Formula (11.39) can then be written as

$$I_n - X_* \left(X_*'X_*\right)^- X_*' = K(K'K)^{-1}K'. \tag{11.40}$$

Formula (11.40) also holds true if we were to replace X_* and K by $\Sigma^{-1/2}X_*$ and $\Sigma^{1/2}K$, respectively. This is so because $(\Sigma^{1/2}K)'(\Sigma^{-1/2}X_*) = K'X_* = 0$, and $\Sigma^{-1/2}X_*$ and $\Sigma^{1/2}K$ have the same ranks as those of X_* and K, respectively ($\Sigma^{1/2}$ is a well-defined and positive definite matrix since Σ is positive definite). From (11.40) we therefore have

$$I_n - \Sigma^{-1/2}X_* \left(X_*'\Sigma^{-1}X_*\right)^- X_*'\Sigma^{-1/2} = \Sigma^{1/2}K(K'\Sigma K)^{-1}K'\Sigma^{1/2}. \tag{11.41}$$

Formula (11.41) is equivalent to (11.36). □

Using the result of Lemma 11.2, formula (11.34) can now be expressed as

$$tr[\tilde{P}_r A_i] = Y' \tilde{P}_r A_i \tilde{P}_r Y, \quad i = \nu - p + 1, \nu - p + 2, \ldots, \nu + 1, \tag{11.42}$$

where,

$$\tilde{P}_r = \tilde{\Sigma}_r^{-1} - \tilde{\Sigma}_r^{-1} X_* \left(X_*'\tilde{\Sigma}_r^{-1} X_*\right)^- X_*'\tilde{\Sigma}_r^{-1}. \tag{11.43}$$

We note that (11.42) is of the same form as the ML equations in (11.26), except that $\tilde{\Sigma}^{-1}$ on the left-hand side of (11.26) has been replaced by $\tilde{P}_r$.

An alternative form to formula (11.42) can be derived as follows: it is easy to show that $P\Sigma P = P$. Using this fact in (11.42), we get

$$tr(\tilde{P}_r \tilde{\Sigma}_r \tilde{P}_r A_i) = Y' \tilde{P}_r A_i \tilde{P}_r Y, \quad i = \nu - p + 1, \nu - p + 2, \ldots, \nu + 1,$$

which can be written as

$$tr\left(\tilde{P}_r A_i \tilde{P}_r \sum_{j=\nu-p+1}^{\nu+1} \tilde{\sigma}_{rj}^2 A_j\right) = Y' \tilde{P}_r A_i \tilde{P}_r Y,$$
$$i = \nu - p + 1, \nu - p + 2, \ldots, \nu + 1.$$

We therefore have the equations,

$$\sum_{j=\nu-p+1}^{\nu+1} tr(\tilde{P}_r A_i \tilde{P}_r A_j)\tilde{\sigma}^2_{rj} = Y' \tilde{P}_r A_i \tilde{P}_r Y,$$

$$i = \nu - p + 1, \nu - p + 2, \ldots, \nu + 1, \quad (11.44)$$

which should be solved by using iterative procedures to obtain the REML estimates of $\sigma^2_{\nu-p+1}, \sigma^2_{\nu-p+2}, \ldots, \sigma^2_{\nu+1}$. We note that (11.44) is of the same form as (11.29) for ML estimation, except that $\tilde{P}_r$ on the left-hand side of (11.44) replaces $\tilde{\Sigma}^{-1}$ in formula (11.29). An algorithm for simplifying the computation of REML estimates was developed by Corbeil and Searle (1976a).

Let us now show that the REML estimates maximize that portion of the likelihood function that does not depend on the fixed effects. To do so, let Z_* be an $n \times r_*$ matrix of rank r_* whose columns form a basis for the column space of X_*. Let us also consider the following linear transformation of the data vector, Y:

$$Y^* = \begin{bmatrix} Z'_* \\ K' \end{bmatrix} Y$$

$$= QY.$$

Since K' is of rank $n - r_*$ and its rows span the orthogonal complement of the column space of X_*, the matrix Q must be nonsingular. It follows that the density function of Y^* is equal to the density function of Y, namely $\mathcal{L}$, divided by $\mid det(Q) \mid$. Now, let us write Y^* as

$$Y^* = \begin{bmatrix} Y_1{}^* \\ Y_2{}^* \end{bmatrix},$$

where $Y_1{}^* = Z'_* Y$ and $Y_2{}^* = K'Y$. Then, the density function of Y^* can be expressed as

$$g(y^*) = g_1\left(y_1{}^* \mid y_2{}^*\right) g_2\left(y_2^*\right),$$

where $g_2\left(y_2{}^*\right)$ is the density function of $Y_2{}^*$, which is the same as $\mathcal{L}_r$ whose natural logarithm was given earlier in (11.33), that is,

$$g_2\left(y_2{}^*\right) = \mathcal{L}_r$$

$$= \frac{1}{(2\pi)^{\frac{n-r_*}{2}} [det(K'\Sigma K)]^{1/2}} \exp\left[-\frac{1}{2} y_2{}^{*'} (K'\Sigma K)^{-1} y_2{}^*\right],$$

and $g_1\left(y_1{}^* \mid y_2{}^*\right)$ is the conditional density of $Y_1{}^*$ given $Y_2{}^* = y_2{}^*$. Hence, the likelihood function of Y can be written as

$$\mathcal{L} = \mid det(Q) \mid g_1\left(y_1{}^* \mid y_2{}^*\right) g_2\left(y_2{}^*\right). \quad (11.45)$$

Formula (11.45) shows that the REML estimates maximize only the portion of the likelihood function of Y, namely, $g_2\left(y_2{}^*\right)$, which does not depend on the fixed effects.

11.1.3.1 Properties of REML Estimators

REML estimators of variance components have properties similar to those of ML. Under certain conditions, REML estimators are asymptotically normal and asymptotically equivalent to ML estimators in the sense that the normalized differences of the corresponding estimators converge in probability to zero. These asymptotic properties were studied by Das (1979).

One interesting feature of REML estimation is that for balanced data, the solutions of the REML equations in (11.44) are identical to the ANOVA estimates (see Searle, Casella, and McCulloch, 1992, p. 253). Because of this feature, some users favor REML over ML. Comparisons between REML and ML estimators were made by Hocking and Kutner (1975), Corbeil and Searle (1976 b), and Swallow and Monahan (1984).

Furthermore, if $\tilde{\sigma}_r^2$ denotes the REML estimator of $\sigma^2 = \left(\sigma^2_{\nu-p+1}, \sigma^2_{\nu-p+2}, \ldots, \sigma^2_{\nu+1}\right)'$, then as $n \to \infty$,

$$\text{Var}\left(\tilde{\sigma}_r^2\right) \to 2\Gamma_r^{-1}, \tag{11.46}$$

where Γ_r is a matrix whose (i,j)th element is equal to

$$\gamma_{rij} = tr(\boldsymbol{PA}_i\boldsymbol{PA}_j), \quad i,j = \nu - p + 1, \nu - p + 2, \ldots, \nu + 1, \tag{11.47}$$

and $\boldsymbol{P}$ is given in (11.37). This result was reported in Searle, Casella, and McCulloch (1992, p. 253). We note that formula (11.47) is similar to formula (11.31) for ML estimation, except that $\boldsymbol{P}$ is used here in place of $\boldsymbol{\Sigma}^{-1}$ in (11.31).

Example 11.4 REML estimates can be easily obtained by using PROC MIXED in SAS. The statements needed to activate REML are the same as those for ML, except that the option "METHOD = ML" in the PROC MIXED statement should be changed to "METHOD = REML," or dropped altogether since REML is the default estimation method in PROC MIXED.

Using, for example, the model and data set for Example 11.2, along with the SAS code shown in Example 11.3 (with "METHOD = ML" changed to "METHOD = REML," or removed completely), we get the following REML estimates of σ^2_β, $\sigma^2_{\alpha\beta}$, and σ^2_ϵ: $\tilde{\sigma}^2_{r\beta} = 279.76$, $\tilde{\sigma}^2_{r\alpha\beta} = 169.60$, $\tilde{\sigma}^2_{r\epsilon} = 39.477$.

Example 11.5 A data set consisting of the weights at birth of 62 single-birth male lambs was reported in Harville and Fenech (1985). The data originated from five distinct population lines. Each lamb was the progeny of one of 23 rams, and each lamb had a different dam. The age of dam was recorded as belonging to one of three categories, numbered 1 (1–2 years), 2 (2–3 years), and 3 (over 3 years). The data are reproduced in Table 11.5, and the model considered for this experiment is given by

$$Y_{ijkl} = \mu + \alpha_{(i)} + \beta_{(j)} + \delta_{j(k)} + \epsilon_{ijk(l)}, \tag{11.48}$$

TABLE 11.5

Birth Weights (in Pounds) of Lambs

Dam Age (i)	Line (j)								
	1					**2**			
1	Sire (k)					Sire (k)			
	1	2	3	4		1	2	3	4
	6.2	13.0	9.5 10.1 11.4	10.4		—	—	12.0	11.5
2	Sire (k)					Sire (k)			
	1	2	3	4		1	2	3	4
	—	—	11.8	8.5		—	10.1	—	—
3	Sire (k)					Sire (k)			
	1	2	3	4		1	2	3	4
	—	—	12.9 13.1	—		13.5	11.0 14.0 15.5	—	10.8

Dam Age (i)	**3**					**4**		
1	Sire (k)					Sire (k)		
	1	2	3	4		1	2	3
	—	11.0	11.6	—		9.2 10.6 10.6	10.2 10.9	11.7
2	Sire (k)					Sire (k)		
	1	2	3	4		1	2	3
	9.0	10.1 11.7	—	12.0		—	—	—
3	Sire (k)					Sire (k)		
	1	2	3	4		1	2	3
	9.5 12.6	8.5 8.8 9.9 10.9 11.0 13.9	13.0	—		7.7 10.0 11.2	—	9.9

(*continued*)

TABLE 11.5 (continued)
Birth Weights (in Pounds) of Lambs

Dam Age (i)	**Line (j)**							
	5							
	Sire (k)							
	1	**2**	**3**	**4**	**5**	**6**	**7**	**8**
1	11.7 12.6	9.0	—	—	—	—	—	10.7 11.0 12.5
2	—	—	—	—	13.5	10.9	10.0 12.7	—
3	—	11.0	9.0 12.0	9.9	—	5.9	13.2 13.3	9.0 10.2

Source: Reprinted from Harville, D.A. and Fenech, A.P., *Biometrics*, 41, 137, 1985. With permission.

$i = 1,2,3$; $j = 1,2,3,4,5$; $k = 1,2,\ldots,c_j$; $l = 0,1,\ldots,n_{ijk}$ (c_j is the number of rams in the jth line, and n_{ijk} is the number of lambs that are the offspring of the kth sire in the jth population line of a dam belonging to the ith age category), where $\alpha_{(i)}$ is the effect of the ith age, $\beta_{(j)}$ is the effect of the jth line, $\delta_{j(k)}$ is the effect of the kth sire within the jth population line, and Y_{ijkl} denotes the weight at birth of the lth of those lambs that are the offspring of the kth sire in the jth population line and of a dam belonging to the ith age category.

Note that (11.48) represents a submodel of the full model for the population structure $[(i)(j:k)]:l$. The age and line effects are considered fixed, but the sire effect (nested within population line) is assumed to be random and distributed as $N\left(0, \sigma^2_{\delta(\beta)}\right)$, independently of the $\epsilon_{ijk(l)}$'s, which are assumed to be mutually independent and have the $N\left(0, \sigma^2_{\epsilon}\right)$ distribution. The SAS statements needed to obtain the ML and REML estimates of $\sigma^2_{\delta(\beta)}$ and σ^2_{ϵ} are

```
DATA;
INPUT A B C Y;
CARDS;
(enter here the data from Table 11.5)
PROC MIXED METHOD = ML;
CLASS A B C;
MODEL Y = A  B;
RANDOM C(B);
RUN;
```

TABLE 11.6
ML and REML Estimates

Variance Component	ML	REML
$\sigma^2_{\delta(\beta)}$	0	0.5171
σ^2_{ϵ}	2.9441	2.9616

Recall that for REML estimation, "METHOD = ML" can be either dropped or changed to "METHOD = REML." The corresponding ML and REML estimates of $\sigma^2_{\delta(\beta)}$ and σ^2_{ϵ} are displayed in Table 11.6.

By comparison, ANOVA's Method III estimates of $\sigma^2_{\delta(\beta)}$ and σ^2_{ϵ} are $\hat{\sigma}^2_{\delta(\beta)} = 0.7676$ and $\hat{\sigma}^2_{\epsilon} = 2.7631$.

11.2 Estimation of Estimable Linear Functions

Using model (11.19), the mean response vector is

$$E(\mathbf{Y}) = \mathbf{X}_* \mathbf{g}, \tag{11.49}$$

and the variance–covariance matrix, $\mathbf{\Sigma}$, of $\mathbf{Y}$ is given by (11.20). Let $\boldsymbol{\lambda}'\mathbf{g}$ be an estimable linear function of $\mathbf{g}$. Its best linear unbiased estimator (BLUE) is

$$\boldsymbol{\lambda}'\hat{\mathbf{g}} = \boldsymbol{\lambda}'\left(\mathbf{X}'_*\mathbf{\Sigma}^{-1}\mathbf{X}_*\right)^{-}\mathbf{X}'_*\mathbf{\Sigma}^{-1}\mathbf{Y}. \tag{11.50}$$

This is called the *generalized least-squares estimator* (GLSE) of $\boldsymbol{\lambda}'\mathbf{g}$ (see Section 6.3). It is easy to show that the variance of $\boldsymbol{\lambda}'\hat{\mathbf{g}}$ is of the form

$$\text{Var}(\boldsymbol{\lambda}'\hat{\mathbf{g}}) = \boldsymbol{\lambda}'\left(\mathbf{X}'_*\mathbf{\Sigma}^{-1}\mathbf{X}_*\right)^{-}\boldsymbol{\lambda}. \tag{11.51}$$

Formulas (11.50) and (11.51) require knowledge of $\mathbf{\Sigma}$. In general, however, $\mathbf{\Sigma}$ is unknown since it depends on the unknown variance components, $\sigma^2_{\nu-p+1}, \sigma^2_{\nu-p+2}, \ldots, \sigma^2_{\nu+1}$. It is therefore necessary to first estimate the variance components using either their ML or REML estimators. For example, if in (11.50) we were to replace the variance components with their REML estimators, $\tilde{\sigma}^2_{ri}$ $(i = \nu-p+1, \nu-p+2, \ldots, \nu+1)$, we get the so-called *estimated generalized least-squares estimator* (EGLSE) of $\boldsymbol{\lambda}'\mathbf{g}$, denoted by $\boldsymbol{\lambda}'\tilde{\mathbf{g}}_r$, which is of the form,

$$\boldsymbol{\lambda}'\tilde{\mathbf{g}}_r = \boldsymbol{\lambda}'\left(\mathbf{X}'_*\tilde{\mathbf{\Sigma}}_r^{-1}\mathbf{X}_*\right)^{-}\mathbf{X}'_*\tilde{\mathbf{\Sigma}}_r^{-1}\mathbf{Y}, \tag{11.52}$$

where $\tilde{\mathbf{\Sigma}}_r$ is given by (11.35). A comparable estimator of $\boldsymbol{\lambda}'\mathbf{g}$ can be obtained by using ML estimators of $\sigma^2_{\nu-p+1}, \sigma^2_{\nu-p+2}, \ldots, \sigma^2_{\nu+1}$ in place of their REML estimators in (11.50). Using (11.51), an approximate expression for the variance of $\boldsymbol{\lambda}'\tilde{\mathbf{g}}_r$ is

$$\text{Var}(\boldsymbol{\lambda}'\tilde{\mathbf{g}}_r) \approx \boldsymbol{\lambda}'\left(\mathbf{X}'_*\tilde{\mathbf{\Sigma}}_r^{-1}\mathbf{X}_*\right)^{-}\boldsymbol{\lambda}. \tag{11.53}$$

It should be noted that $\lambda'\tilde{g}_r$ is no longer the BLUE of $\lambda' g$ since $\tilde{\Sigma}_r$ is random and is therefore not equal to the true fixed value of Σ. Furthermore, the right-hand side of (11.53) is an estimate of $\text{Var}(\lambda'\hat{g})$ in (11.51), but is not an estimate of the true value of $\text{Var}(\lambda'\tilde{g}_r)$.

Kackar and Harville (1981) proved an interesting result concerning unbiasedness of the EGLSE of $\lambda' g$. Before giving details of the proof of this result, let $\lambda' g^*$ denote an estimator of $\lambda' g$ derived by replacing the variance components in (11.50) by their corresponding estimators, which can be of the ANOVA, ML, or REML types, or any other "suitable" estimators. By a "suitable estimator" we mean an estimator that is even and translation invariant. By definition, a statistic $S(Y)$ is an *even function* of Y if $S(-Y) = S(Y)$ for all Y. Also, $S(Y)$ is said to be *translation invariant* if $S(Y - X_* \xi) = S(Y)$ for all Y and any constant vector ξ.

Theorem 11.1 (Kackar and Harville, 1981) Let $\sigma^2 = \left(\sigma^2_{\nu-p+1}, \sigma^2_{\nu-p+2}, \ldots, \sigma^2_{\nu+1}\right)'$, and $\sigma^{*2}(Y)$ be an estimator of σ^2 whose elements are translation invariant and even functions of Y. Suppose that $\lambda' g^*$ is an estimator of $\lambda' g$ obtained by using $\sigma^{*2}(Y)$ in place of σ^2 in (11.50). Then, $\lambda' g^*$ is an unbiased estimator of $\lambda' g$ provided that the expected value of $\lambda' g^*$ is finite.

The proof of Theorem 11.1 depends on the following lemma.

Lemma 11.3 (Kackar and Harville, 1981) If Z is a random vector with a symmetric distribution around zero in the sense that Z and $-Z$ are identically distributed, and if $f(Z)$ is a random variable that is an odd function of Z, that is, $f(-Z) = -f(Z)$, then $f(Z)$ has a symmetric distribution around zero.

Proof. Let x be any real number. Then,

$$\begin{aligned} P[f(Z) \geq x] &= P[-f(Z) \leq -x] \\ &= P[f(-Z) \leq -x]. \end{aligned} \tag{11.54}$$

Since Z and $-Z$ are identically distributed, $f(Z)$ and $f(-Z)$ are also identically distributed. Hence,

$$\begin{aligned} P[f(-Z) \leq -x] &= P[f(Z) \leq -x] \\ &= P[f(-Z) \geq x]. \end{aligned} \tag{11.55}$$

From (11.54) and (11.55) we conclude that

$$\begin{aligned} P[f(Z) \geq x] &= P[f(-Z) \geq x] \\ &= P[-f(Z) \geq x]. \end{aligned}$$

This means that $f(Z)$ and $-f(Z)$ are identically distributed. □

Proof of Theorem 11.1 We need to show that $E(\lambda' g^*) = \lambda' g$. Since $\lambda' g$ is estimable, we can write $\lambda' = r' X_*$ for some vector r. We have that

$$\lambda' g^* - \lambda' g = r' X_* \left(X'_* \Sigma^{*-1} X_*\right)^- X'_* \Sigma^{*-1} Y - r' X_* g,$$

where $\boldsymbol{\Sigma}^*$ is obtained from (11.20) by replacing $\boldsymbol{\sigma}^2$ with $\boldsymbol{\sigma}^{*2}(\boldsymbol{Y})$. But,

$$\boldsymbol{X}_*\left(\boldsymbol{X}'_*\boldsymbol{\Sigma}^{*-1}\boldsymbol{X}_*\right)^-\boldsymbol{X}'_*\boldsymbol{\Sigma}^{*-1}\boldsymbol{X}_* = \boldsymbol{X}_*,$$

since it is known that

$$\boldsymbol{\Sigma}^{*-1/2}\boldsymbol{X}_*\left(\boldsymbol{X}'_*\boldsymbol{\Sigma}^{*-1}\boldsymbol{X}_*\right)^-\boldsymbol{X}'_*\boldsymbol{\Sigma}^{*-1}\boldsymbol{X}_* = \boldsymbol{\Sigma}^{*-1/2}\boldsymbol{X}_*.$$

Hence,

$$\begin{aligned}\boldsymbol{\lambda}'\boldsymbol{g}^* - \boldsymbol{\lambda}'\boldsymbol{g} &= \boldsymbol{r}'\boldsymbol{X}_*\left(\boldsymbol{X}'_*\boldsymbol{\Sigma}^{*-1}\boldsymbol{X}_*\right)^-\boldsymbol{X}'_*\boldsymbol{\Sigma}^{*-1}(\boldsymbol{Y} - \boldsymbol{X}_*\boldsymbol{g}) \\ &= \boldsymbol{r}'\boldsymbol{X}_*\left(\boldsymbol{X}'_*\boldsymbol{\Sigma}^{*-1}\boldsymbol{X}_*\right)^-\boldsymbol{X}'_*\boldsymbol{\Sigma}^{*-1}(\boldsymbol{H}_*\,\boldsymbol{\beta}_*),\end{aligned} \tag{11.56}$$

where $\boldsymbol{H}_*\,\boldsymbol{\beta}_* = \sum_{i=\nu-p+1}^{\nu+1}\boldsymbol{H}_i\boldsymbol{\beta}_i$. Now, since $\boldsymbol{\sigma}^{*2}(\boldsymbol{Y})$ is even and translation invariant, then

$$\begin{aligned}\boldsymbol{\sigma}^{*2}(\boldsymbol{Y}) &= \boldsymbol{\sigma}^{*2}(\boldsymbol{Y} - \boldsymbol{X}_*\boldsymbol{g}) \\ &= \boldsymbol{\sigma}^{*2}(\boldsymbol{H}_*\,\boldsymbol{\beta}_*) \\ &= \boldsymbol{\sigma}^{*2}(-\boldsymbol{H}_*\,\boldsymbol{\beta}_*).\end{aligned}$$

From (11.56) it follows that $\boldsymbol{\lambda}'\boldsymbol{g}^* - \boldsymbol{\lambda}'\boldsymbol{g}$ is an odd function of $\boldsymbol{\beta}_*$. But, by assumption, $\boldsymbol{\beta}_*$ is normally distributed with a zero mean vector, hence it is symmetrically distributed around zero. Consequently, by Lemma 11.3, $\boldsymbol{\lambda}'\boldsymbol{g}^* - \boldsymbol{\lambda}'\boldsymbol{g}$ has a distribution that is symmetrically distributed around zero. Hence, if $E(\boldsymbol{\lambda}'\boldsymbol{g}^*)$ is finite, then $E(\boldsymbol{\lambda}'\boldsymbol{g}^* - \boldsymbol{\lambda}'\boldsymbol{g}) = 0$, that is, $E(\boldsymbol{\lambda}'\boldsymbol{g}^*) = \boldsymbol{\lambda}'\boldsymbol{g}$. □

All three procedures for estimating variance components (Henderson's Method III, ML, and REML), as described in Section 11.1, yield even and translation invariant estimators (see Kackar and Harville, 1981). Thus, the EGLSE, $\boldsymbol{\lambda}'\tilde{\boldsymbol{g}}_r$, in (11.52) is an unbiased estimator of $\boldsymbol{\lambda}'\boldsymbol{g}$. The same is true if ML is used instead of REML.

Recall that the expression in (11.53) is just an approximation that provides a measure of precision for $\boldsymbol{\lambda}'\tilde{\boldsymbol{g}}_r$. More specifically, it is an estimate of $\boldsymbol{\lambda}'\left(\boldsymbol{X}'_*\boldsymbol{\Sigma}^{-1}\boldsymbol{X}_*\right)^-\boldsymbol{\lambda}$, which is regarded as the variance–covariance matrix of the asymptotic limiting distribution of $\boldsymbol{\lambda}'\tilde{\boldsymbol{g}}_r$ as the number of observations, n, goes to infinity. In small samples, however, the expression in (11.53) may not provide a good approximation for $\mathrm{Var}(\boldsymbol{\lambda}'\tilde{\boldsymbol{g}}_r)$. Kackar and Harville (1984) proposed an alternative, and a generally more satisfactory, approximation for $\mathrm{Var}(\boldsymbol{\lambda}'\tilde{\boldsymbol{g}}_r)$ given by

$$\mathrm{Var}(\boldsymbol{\lambda}'\tilde{\boldsymbol{g}}_r) \approx \boldsymbol{\lambda}'\left(\boldsymbol{X}'_*\boldsymbol{\Sigma}^{-1}\boldsymbol{X}_*\right)^-\boldsymbol{\lambda} + \boldsymbol{\lambda}'\boldsymbol{T}_*\boldsymbol{\lambda}, \tag{11.57}$$

where

$$T_* = \left(X_*'\Sigma^{-1}X_*\right)^- \left\{ \sum_{i=\nu-p+1}^{\nu+1} \sum_{j=\nu-p+1}^{\nu+1} \omega_{ij} \left[G_{ij} - F_i \left(X_*'\Sigma^{-1}X_*\right)^- F_j \right] \right\} \times \left(X_*'\Sigma^{-1}X_*\right)^- \tag{11.58}$$

$$G_{ij} = X_*'\Sigma^{-1}A_i\Sigma^{-1}A_j\Sigma^{-1}X_*, \quad i,j = \nu-p+1, \nu-p+2, \ldots, \nu+1 \tag{11.59}$$

$$F_i = -X_*'\Sigma^{-1}A_i\Sigma^{-1}X_*, \quad i = \nu-p+1, \nu-p+2, \ldots, \nu+1, \tag{11.60}$$

and ω_{ij} is the (i,j)th element of the asymptotic variance–covariance matrix of $\tilde{\sigma}_r^2$, the vector of REML estimators of the σ_i^2's ($i = \nu-p+1, \nu-p+2, \ldots, \nu+1$), which is given in (11.46). [If $\tilde{\sigma}^2$, the vector of ML estimators of the variance components, is used in place of $\tilde{\sigma}_r^2$ to obtain the EGLSE of $\lambda' g$, then the asymptotic variance - covariance matrix of $\tilde{\sigma}^2$ given in formula (11.30) should be used instead of (11.46)]. The expression in (11.57) was also reported in McCulloch and Searle (2001, p. 165).

An alternative expression for $\lambda' T_* \lambda$ in (11.57) is obtained as follows: let $\tau' = \lambda'\left(X_*'\Sigma^{-1}X_*\right)^- X_*'\Sigma^{-1}$. Using (11.59) and (11.60) in (11.58) we get

$$\begin{aligned}
\lambda' T_* \lambda &= \sum_{i=\nu-p+1}^{\nu+1} \sum_{j=\nu-p+1}^{\nu+1} \omega_{ij}\, \tau'[A_i\Sigma^{-1}A_j \\
&\qquad - A_i\Sigma^{-1}X_*\left(X_*'\Sigma^{-1}X_*\right)^- X_*'\Sigma^{-1}A_j]\tau \\
&= \sum_{i=\nu-p+1}^{\nu+1} \sum_{j=\nu-p+1}^{\nu+1} \omega_{ij}\, \tau' A_i P A_j \tau \\
&= tr(\Omega V),
\end{aligned} \tag{11.61}$$

where P is the matrix defined in (11.37), $\Omega = (\omega_{ij})$, and V is the matrix whose (i,j)th element is $\tau' A_i P A_j \tau$. Using (11.61) in (11.57) we get the expression

$$\text{Var}(\lambda'\tilde{g}_r) \approx \lambda'\left(X_*'\Sigma^{-1}X_*\right)^- \lambda + tr(\Omega V). \tag{11.62}$$

This approximation of $\text{Var}(\lambda'\tilde{g}_r)$ is the one derived by Kackar and Harville (1984, p. 854) and is also reported in McCulloch and Searle (2001, p. 166). Obviously, an estimate of Σ must be used in (11.62) in order to obtain an approximate computable expression for $\text{Var}(\lambda'\tilde{g}_r)$. Kackar and Harville (1984) indicated that $tr(\Omega V)$ in (11.62) is nonnegative, which implies that $\lambda'\left(X_*'\Sigma^{-1}X_*\right)^- \lambda$ represents a lower bound for $\text{Var}(\lambda'\tilde{g}_r)$, that is, $\lambda'\left(X_*'\Sigma^{-1}X_*\right)^- \lambda$ underestimates the variance of $\lambda'\tilde{g}_r$.

Kenward and Roger (1997, Section 2) noted that $\lambda'\left(X'_*\tilde{\Sigma}_r^{-1}X_*\right)^{-}\lambda$, where $\tilde{\Sigma}_r$ is given in (11.35), is a biased estimator of $\lambda'\left(X'_*\Sigma^{-1}X_*\right)^{-}\lambda$. Using a second-order Taylor's series expansion of $\lambda'\left(X'_*\tilde{\Sigma}_r^{-1}X_*\right)^{-}\lambda$ around σ^2, the vector of true variance components, it was established that

$$E\left[\lambda'\left(X'_*\tilde{\Sigma}_r^{-1}X_*\right)^{-}\lambda\right] \approx \lambda'\left(X'_*\Sigma^{-1}X_*\right)^{-}\lambda - \lambda' T_*\lambda. \qquad (11.63)$$

Hence, an approximate unbiased estimator for the right-hand side of (11.57) is given by $\lambda'\left(X'_*\tilde{\Sigma}_r^{-1}X_*\right)^{-}\lambda + 2\lambda' T_*\lambda$. An estimate of Σ can then be used in T_* in order to obtain a computable expression for this estimator.

In Sections 11.3 through 11.8, we concentrate on the study of tests and confidence intervals concerning the parameters of some unbalanced random or mixed models. Unlike balanced models, the analysis of unbalanced models can be quite complicated. This is due to the lack of a unique ANOVA table in the unbalanced case, as was stated earlier in Chapter 10. Furthermore, the sums of squares in an unbalanced ANOVA table are not, in general, independent or distributed as scaled chi-squared variates under the usual assumptions of normality, independence, and equality of variances of the individual random effects. As a result, there are no general procedures for deriving, for example, exact tests or confidence intervals concerning variance components or estimable linear functions of the fixed effects in an unbalanced mixed model situation. There are, however, certain techniques that apply to particular models. For example, the analysis of the random one-way model is discussed in Section 11.3. The random and mixed two-way models are considered in Sections 11.4 and 11.6, respectively. Sections 11.7 and 11.8 deal with the random and mixed two-fold nested models, respectively. These are not the only models for which tests and confidence intervals can be derived. However, they do provide good examples that illustrate how inference making is carried out in the case of unbalanced models with random effects. Extensions to higher-order random models are discussed in Section 11.5, and in Section 11.9, a review is given of some approximate tests for the general mixed model.

11.3 Inference Concerning the Random One-Way Model

Consider the unbalanced random one-way model,

$$Y_{ij} = \mu + \alpha_{(i)} + \epsilon_{i(j)}, \quad i = 1, 2, \ldots, k; \ j = 1, 2, \ldots, n_i, \qquad (11.64)$$

where $\alpha_{(i)}$ and $\epsilon_{i(j)}$ are independently distributed as normal variates with zero means and variances σ^2_α and σ^2_ϵ, respectively. The objective here is to derive a test concerning the hypothesis, $H_0 : \sigma^2_\alpha = 0$, and to obtain confidence intervals on σ^2_α and $\sigma^2_\alpha/\sigma^2_\epsilon$.

Taking the average of Y_{ij} over j, we get

$$\bar{Y}_{i.} = \mu + \alpha_{(i)} + \bar{\epsilon}_{i.}, \quad i = 1, 2, \ldots, k, \tag{11.65}$$

where $\bar{Y}_{i.} = \frac{1}{n_i}\sum_{j=1}^{n_i} Y_{ij}$, $\bar{\epsilon}_{i.} = \frac{1}{n_i}\sum_{j=1}^{n_i} \epsilon_{i(j)}$. Formula (11.65) can be written in vector form as

$$\bar{\boldsymbol{Y}} = \mu \boldsymbol{1}_k + \boldsymbol{\alpha} + \bar{\boldsymbol{\epsilon}}, \tag{11.66}$$

where $\bar{\boldsymbol{Y}}$, $\boldsymbol{\alpha}$, and $\bar{\boldsymbol{\epsilon}}$ are vectors whose elements are $\bar{Y}_{i.}$, $\alpha_{(i)}$, and $\bar{\epsilon}_{i.}$ ($i = 1, 2, \ldots, k$), respectively. The variance–covariance matrix of $\bar{\boldsymbol{Y}}$ is

$$\text{Var}(\bar{\boldsymbol{Y}}) = \sigma^2_\alpha \boldsymbol{I}_k + \sigma^2_\epsilon \boldsymbol{B}, \tag{11.67}$$

where

$$\boldsymbol{B} = \text{diag}\left(\frac{1}{n_1}, \frac{1}{n_2}, \ldots, \frac{1}{n_k}\right).$$

Let now $\boldsymbol{P}_1$ be a matrix of order $(k-1) \times k$ such that $\left[\frac{1}{\sqrt{k}}\boldsymbol{1}_k : \boldsymbol{P}_1'\right]'$ is an orthogonal matrix of order $k \times k$ (the choice of $\boldsymbol{P}_1$ is not unique). Hence, the rows of $\boldsymbol{P}_1$ form an orthonormal basis for the orthogonal complement of $\boldsymbol{1}_k$ in the k-dimensional Euclidean space. Let $\boldsymbol{U} = \boldsymbol{P}_1\bar{\boldsymbol{Y}}$. Then, $\boldsymbol{U}$ is normally distributed with a mean equal to $\boldsymbol{0}$ and a variance–covariance matrix given by

$$\begin{aligned} \text{Var}(\boldsymbol{U}) &= \boldsymbol{P}_1\left(\sigma^2_\alpha \boldsymbol{I}_k + \sigma^2_\epsilon \boldsymbol{B}\right)\boldsymbol{P}_1' \\ &= \sigma^2_\alpha \boldsymbol{I}_{k-1} + \sigma^2_\epsilon \boldsymbol{L}_1, \end{aligned} \tag{11.68}$$

where $\boldsymbol{L}_1 = \boldsymbol{P}_1\boldsymbol{B}\boldsymbol{P}_1'$. The *unweighted sum of squares*, SS_u, corresponding to $\alpha_{(i)}$ in model (11.64) is defined as

$$SS_u = \bar{n}_h \sum_{i=1}^{k}(\bar{Y}_{i.} - \bar{Y}^*)^2, \tag{11.69}$$

where $\bar{Y}^* = \frac{1}{k}\sum_{i=1}^{k}\bar{Y}_{i.}$ and $\bar{n}_h$ is the harmonic mean of the n_i's, namely,

$$\bar{n}_h = k\left(\sum_{i=1}^{k}\frac{1}{n_i}\right)^{-1}.$$

Using (11.69), SS_u can be written as

$$\begin{aligned} SS_u &= \bar{n}_h \bar{\boldsymbol{Y}}'\left(\boldsymbol{I}_k - \frac{1}{k}\boldsymbol{J}_k\right)\bar{\boldsymbol{Y}} \\ &= \bar{n}_h \boldsymbol{U}'\boldsymbol{U}, \end{aligned} \tag{11.70}$$

since $\boldsymbol{I}_k - \frac{1}{k}\boldsymbol{J}_k = \boldsymbol{P}_1'\boldsymbol{P}_1$, where $\boldsymbol{J}_k$ is the matrix of ones of order $k \times k$. Thus, $\boldsymbol{U}'\boldsymbol{U}$ is invariant to the choice of $\boldsymbol{P}_1$. Since $\boldsymbol{U} \sim N\left(\boldsymbol{0}, \sigma_\alpha^2 \boldsymbol{I}_{k-1} + \sigma_\epsilon^2 \boldsymbol{L}_1\right)$, the random variable,

$$X_u = \boldsymbol{U}'\left(\sigma_\alpha^2 \boldsymbol{I}_{k-1} + \sigma_\epsilon^2 \boldsymbol{L}_1\right)^{-1} \boldsymbol{U}, \tag{11.71}$$

has the chi-squared distribution with $k-1$ degrees of freedom.

Thomas and Hultquist (1978) introduced an approximation to X_u in (11.71) given by

$$\tilde{X}_u = \left(\sigma_\alpha^2 + \frac{1}{\bar{n}_h}\sigma_\epsilon^2\right)^{-1} \boldsymbol{U}'\boldsymbol{U}. \tag{11.72}$$

This approximation results from replacing $\boldsymbol{B}$ with $\frac{1}{\bar{n}_h}\boldsymbol{I}_k$, and hence $\boldsymbol{L}_1$ in (11.71) with $\frac{1}{\bar{n}_h}\boldsymbol{I}_{k-1}$. This causes $\text{Var}(\boldsymbol{U})$ in (11.68) to be approximately equal to $\left(\sigma_\alpha^2 + \frac{\sigma_\epsilon^2}{\bar{n}_h}\right)\boldsymbol{I}_{k-1}$. It follows that $\tilde{X}_u$ in (11.72) is distributed approximately as χ^2_{k-1}. From (11.70) and (11.72) we therefore have

$$\begin{aligned} SS_u &= \bar{n}_h \boldsymbol{U}'\boldsymbol{U} \\ &= \left(\bar{n}_h\sigma_\alpha^2 + \sigma_\epsilon^2\right)\tilde{X}_u \\ &\underset{approx.}{\sim} \left(\bar{n}_h\sigma_\alpha^2 + \sigma_\epsilon^2\right)\chi^2_{k-1}. \end{aligned}$$

Under $H_0: \sigma_\alpha^2 = 0$, $\frac{SS_u}{\sigma_\epsilon^2}$ is then distributed approximately as χ^2_{k-1}. Since SS_u is also independent of the error sum of squares, $SS_E = \sum_{i=1}^{k}\sum_{j=1}^{n_i}(Y_{ij} - \bar{Y}_{i.})^2$, then under H_0,

$$F_u = \frac{SS_u/(k-1)}{MS_E} \tag{11.73}$$

has approximately the F-distribution with $k-1$ and $n_. - k$ degrees of freedom, where $MS_E = SS_E/(n_. - k)$ and $n_. = \sum_{i=1}^{k} n_i$. The statistic F_u can then be used to provide an approximate F-test for H_0. Under $H_a: \sigma_\alpha^2 > 0$, $\frac{\sigma_\epsilon^2}{\bar{n}_h\sigma_\alpha^2 + \sigma_\epsilon^2} F_u$ is distributed approximately as $F_{k-1, n_.-k}$. The power of the test statistic F_u can therefore be obtained approximately for a given value of $\sigma_\alpha^2/\sigma_\epsilon^2$ on the basis of this F-distribution. The following lemma provides some justification for using $\bar{n}_h$ in setting up the approximate chi-squared distribution in (11.72).

Lemma 11.4 Let $\| \,.\, \|_2$ denote the Euclidean norm of a matrix (see Section 3.12). The best approximation of the matrix $\boldsymbol{L}_1$ in (11.68) with a diagonal matrix of the form $c_0\,\boldsymbol{I}_{k-1}$ (c_0 is a positive constant), in the sense of minimizing $\| \boldsymbol{L}_1 - c_0\,\boldsymbol{I}_{k-1} \|_2$ with respect to c_0, is achieved when $c_0 = \frac{1}{\bar{n}_h}$.

Proof. We have that

$$\begin{aligned}\| \boldsymbol{L}_1 - c_0\, \boldsymbol{I}_{k-1} \|_2^2 &= \| \boldsymbol{P}_1(\boldsymbol{B} - c_0\, \boldsymbol{I}_k)\boldsymbol{P}_1' \|_2^2 \\ &= tr\left[\boldsymbol{P}_1(\boldsymbol{B} - c_0\, \boldsymbol{I}_k)\boldsymbol{P}_1'\boldsymbol{P}_1(\boldsymbol{B} - c_0\, \boldsymbol{I}_k)\boldsymbol{P}_1'\right] \\ &= tr\left[\boldsymbol{P}_1'\boldsymbol{P}_1(\boldsymbol{B} - c_0\, \boldsymbol{I}_k)\boldsymbol{P}_1'\boldsymbol{P}_1(\boldsymbol{B} - c_0\, \boldsymbol{I}_k)\right] \\ &= tr\left[\left(\boldsymbol{I}_k - \frac{1}{k}\boldsymbol{J}_k\right)(\boldsymbol{B} - c_0\, \boldsymbol{I}_k)\left(\boldsymbol{I}_k - \frac{1}{k}\boldsymbol{J}_k\right)(\boldsymbol{B} - c_0\, \boldsymbol{I}_k)\right].\end{aligned}$$

Differentiating the right-hand side with respect to c_0 and equating the derivative to zero, we get

$$\begin{aligned}\frac{\mathrm{d}}{\mathrm{d}c_0} \| \boldsymbol{L}_1 - c_0\, \boldsymbol{I}_{k-1} \|_2^2 &= -\frac{2(k-1)}{k}\sum_{i=1}^{k}\frac{1}{n_i} + 2c_0\,(k-1) \\ &= 0.\end{aligned}$$

The solution of this equation is

$$\begin{aligned}c_0 &= \frac{1}{k}\sum_{i=1}^{k}\frac{1}{n_i} \\ &= \frac{1}{\bar{n}_h}.\end{aligned}$$

Since $\| \boldsymbol{L}_1 - c_0\, \boldsymbol{I}_{k-1} \|_2^2$ is a quadratic function of c_0 and its second derivative with respect to c_0 is positive, the solution, $c_0 = \frac{1}{\bar{n}_h}$, must be a point of absolute minimum for $\| \boldsymbol{L}_1 - c_0\, \boldsymbol{I}_{k-1} \|_2^2$. □

11.3.1 Adequacy of the Approximation

In this section, we examine the adequacy of the approximation of the distribution of $\tilde{X}_u$ in (11.72) with the chi-squared distribution. Since $\boldsymbol{U} \sim N\left(\boldsymbol{0}, \sigma_\alpha^2 \boldsymbol{I}_{k-1} + \sigma_\epsilon^2 \boldsymbol{L}_1\right)$, the exact distribution of $\tilde{X}_u$ can be determined by using Davies' (1980) algorithm (see Section 5.6). More specifically, $\tilde{X}_u$ can be written as a linear combination of independent central chi-squared random variables of the form

$$\tilde{X}_u = \sum_{i=1}^{s} \tilde{\theta}_i \chi^2_{\nu_i}, \qquad (11.74)$$

where the $\tilde{\theta}_i$'s are the distinct nonzero eigenvalues of

$$\frac{1}{\sigma_\alpha^2 + \frac{\sigma_\epsilon^2}{\bar{n}_h}}\,\text{Var}(\boldsymbol{U}) = \frac{1}{\sigma_\alpha^2 + \frac{\sigma_\epsilon^2}{\bar{n}_h}}\left(\sigma_\alpha^2 \boldsymbol{I}_{k-1} + \sigma_\epsilon^2 \boldsymbol{L}_1\right)$$
$$= \frac{1}{\sigma_\alpha^2 + \frac{\sigma_\epsilon^2}{\bar{n}_h}}\left[\left(\sigma_\alpha^2 + \frac{\sigma_\epsilon^2}{\bar{n}_h}\right)\boldsymbol{I}_{k-1} + \sigma_\epsilon^2\left(\boldsymbol{L}_1 - \frac{1}{\bar{n}_h}\boldsymbol{I}_{k-1}\right)\right]$$
$$= \boldsymbol{I}_{k-1} + \frac{\sigma_\epsilon^2}{\sigma_\alpha^2 + \frac{\sigma_\epsilon^2}{\bar{n}_h}}\left(\boldsymbol{L}_1 - \frac{1}{\bar{n}_h}\boldsymbol{I}_{k-1}\right), \tag{11.75}$$

and ν_i is the multiplicity of $\tilde{\theta}_i$ $(i = 1, 2, \ldots, s)$ such that $\sum_{i=1}^{s} \nu_i = k - 1$. Note that $\tilde{\theta}_i > 0$ for $i = 1, 2, \ldots, s$. Let $\tilde{\tau}_i$ denote the ith eigenvalue of $\left(\sigma_\alpha^2 + \frac{\sigma_\epsilon^2}{\bar{n}_h}\right)^{-1}(\boldsymbol{L}_1 - \frac{1}{\bar{n}_h}\boldsymbol{I}_{k-1})$. Then,

$$\tilde{\tau}_i = \left(\sigma_\alpha^2 + \frac{\sigma_\epsilon^2}{\bar{n}_h}\right)^{-1}\left(\tilde{\lambda}_i - \frac{1}{\bar{n}_h}\right), \quad i = 1, 2, \ldots, k-1, \tag{11.76}$$

where $\tilde{\lambda}_i$ = ith eigenvalue of $\boldsymbol{L}_1$ $(i = 1, 2, \ldots, k-1)$. Since $\boldsymbol{L}_1 = \boldsymbol{P}_1\boldsymbol{B}\boldsymbol{P}_1'$ is positive definite, $\tilde{\lambda}_i > 0$ for all i. Note that the $\tilde{\lambda}_i$'s are invariant to the choice of the matrix $\boldsymbol{P}_1$ since they are equal to the nonzero eigenvalues of $\boldsymbol{B}\boldsymbol{P}_1'\boldsymbol{P}_1 = \boldsymbol{B}(\boldsymbol{I}_k - \frac{1}{k}\boldsymbol{J}_k)$, or, equivalently, the symmetric matrix $(\boldsymbol{I}_k - \frac{1}{k}\boldsymbol{J}_k)\boldsymbol{B}(\boldsymbol{I}_k - \frac{1}{k}\boldsymbol{J}_k)$. Furthermore, from (11.75), if $\tilde{\kappa}_i$ denotes the ith eigenvalue of $\left(\sigma_\alpha^2 + \frac{\sigma_\epsilon^2}{\bar{n}_h}\right)^{-1}\text{Var}(\boldsymbol{U})$, then

$$\tilde{\kappa}_i = 1 + \sigma_\epsilon^2\,\tilde{\tau}_i, \quad i = 1, 2, \ldots, k-1. \tag{11.77}$$

Thus, the $\tilde{\theta}_i$'s in (11.74) are the distinct values of the $\tilde{\kappa}_i$'s whose average value is equal to one. This follows from Equations 11.76 and 11.77, namely,

$$\frac{1}{k-1}\sum_{i=1}^{k-1}\tilde{\kappa}_i = 1 + \frac{\sigma_\epsilon^2}{\sigma_\alpha^2 + \frac{\sigma_\epsilon^2}{\bar{n}_h}}\left(\frac{1}{k-1}\sum_{i=1}^{k-1}\tilde{\lambda}_i - \frac{1}{\bar{n}_h}\right)$$
$$= 1, \tag{11.78}$$

since

$$\frac{1}{k-1}\sum_{i=1}^{k-1}\tilde{\lambda}_i = \frac{1}{k-1}tr[\boldsymbol{B}(\boldsymbol{I}_k - \frac{1}{k}\boldsymbol{J}_k)]$$
$$= \frac{1}{k-1}\left[tr(\boldsymbol{B}) - \frac{1}{k}\boldsymbol{1}_k'\boldsymbol{B}\boldsymbol{1}_k\right]$$

$$= \frac{1}{k-1}\left[\sum_{i=1}^{k}\frac{1}{n_i} - \frac{1}{k}\sum_{i=1}^{k}\frac{1}{n_i}\right]$$
$$= \frac{1}{k}\sum_{i=1}^{k}\frac{1}{n_i}$$
$$= \frac{1}{\bar{n}_h}.$$

From (11.78) it follows that the approximation of the distribution of $\tilde{X}_u$ with χ^2_{k-1}, as proposed by Thomas and Hultquist (1978), results from formula (11.74) by replacing all the $\tilde{\theta}_i$'s by their weighted average, $\frac{1}{k-1}\sum_{i=1}^{k-1}\tilde{\kappa}_i$, which is equal to one. Thus, the closer the $\tilde{\tau}_i$'s in (11.77) are to zero, that is, the closer the $\tilde{\lambda}_i$'s are to $\frac{1}{\bar{n}_h}$ [see formula (11.76)], the better the approximation. Note that for all values of i,

$$\min_j\{\tilde{\kappa}_j\} \leq \tilde{\theta}_i \leq \max_j\{\tilde{\kappa}_j\}, \tag{11.79}$$

where from (11.76) and (11.77),

$$\min_j\{\tilde{\kappa}_j\} = 1 + \frac{\sigma^2_\epsilon}{\sigma^2_\alpha + \frac{\sigma^2_\epsilon}{\bar{n}_h}}\left[\tilde{\lambda}_{(k-1)} - \frac{1}{\bar{n}_h}\right]$$
$$= \frac{\sigma^2_\alpha + \tilde{\lambda}_{(k-1)}\sigma^2_\epsilon}{\sigma^2_\alpha + \frac{\sigma^2_\epsilon}{\bar{n}_h}}$$
$$\max_j\{\tilde{\kappa}_j\} = 1 + \frac{\sigma^2_\epsilon}{\sigma^2_\alpha + \frac{\sigma^2_\epsilon}{\bar{n}_h}}\left[\tilde{\lambda}_{(1)} - \frac{1}{\bar{n}_h}\right]$$
$$= \frac{\sigma^2_\alpha + \tilde{\lambda}_{(1)}\sigma^2_\epsilon}{\sigma^2_\alpha + \frac{\sigma^2_\epsilon}{\bar{n}_h}},$$

and $\tilde{\lambda}_{(1)}$ and $\tilde{\lambda}_{(k-1)}$ are, respectively, the largest and the smallest of the $\tilde{\lambda}_i$'s $(i = 1, 2, \ldots, k-1)$. From (11.79) and the fact that the weighted average of the $\tilde{\theta}_i$'s is one, we get

$$\frac{\sigma^2_\alpha + \tilde{\lambda}_{(k-1)}\sigma^2_\epsilon}{\sigma^2_\alpha + \frac{\sigma^2_\epsilon}{\bar{n}_h}} \leq 1 \leq \frac{\sigma^2_\alpha + \tilde{\lambda}_{(1)}\sigma^2_\epsilon}{\sigma^2_\alpha + \frac{\sigma^2_\epsilon}{\bar{n}_h}}. \tag{11.80}$$

Since $\tilde{\lambda}_{(1)} \leq \frac{1}{n_{(k)}}$ and $\tilde{\lambda}_{(k-1)} \geq \frac{1}{n_{(1)}}$, where $n_{(1)}$ and $n_{(k)}$ are, respectively, the largest and the smallest of the n_i's $(i = 1, 2, \ldots, k)$, we conclude that

$$\frac{\sigma^2_\alpha + \frac{\sigma^2_\epsilon}{n_{(1)}}}{\sigma^2_\alpha + \frac{\sigma^2_\epsilon}{\bar{n}_h}} \leq 1 \leq \frac{\sigma^2_\alpha + \frac{\sigma^2_\epsilon}{n_{(k)}}}{\sigma^2_\alpha + \frac{\sigma^2_\epsilon}{\bar{n}_h}}. \tag{11.81}$$

Thus, the closer the upper and lower bounds in (11.80) are to each other, the better the approximation of the distribution of $\tilde{X}_u$ with χ^2_{k-1}. On the basis of (11.81), this occurs when the data set is nearly balanced (in this case, $n_{(1)} \approx n_{(k)} \approx \bar{n}_h$), or when $\sigma^2_\alpha/\sigma^2_\epsilon$ is large. The same observation was made by Thomas and Hultquist (1978). The adequacy of this approximation was assessed by Khuri (2002) using graphical techniques based on the ideas presented in this section.

11.3.2 Confidence Intervals on σ^2_α and $\sigma^2_\alpha/\sigma^2_\epsilon$

An approximate confidence interval on $\sigma^2_\alpha/\sigma^2_\epsilon$ can be easily obtained by using the fact that $\frac{\sigma^2_\epsilon}{\bar{n}_h\sigma^2_\alpha+\sigma^2_\epsilon} F_u$ is approximately distributed as $F_{k-1,n_.-k}$, where F_u is given by (11.73). We thus have

$$F_{1-\frac{\alpha}{2},k-1,n_.-k} \le \left(\bar{n}_h\frac{\sigma^2_\alpha}{\sigma^2_\epsilon}+1\right)^{-1} F_u \le F_{\frac{\alpha}{2},k-1,n_.-k}$$

with a probability approximately equal to $1-\alpha$. Hence, an approximate $(1-\alpha)100\%$ confidence interval on $\sigma^2_\alpha/\sigma^2_\epsilon$ is given by

$$\frac{1}{\bar{n}_h}\left[\frac{F_u}{F_{\frac{\alpha}{2}}}-1\right] \le \frac{\sigma^2_\alpha}{\sigma^2_\epsilon} \le \frac{1}{\bar{n}_h}\left[\frac{F_u}{F_{1-\frac{\alpha}{2}}}-1\right]. \tag{11.82}$$

A confidence interval for σ^2_α can be formed by modifying the interval for σ^2_α in the balanced case. Using this approach, Thomas and Hultquist (1978) developed the following approximate $(1-\alpha)100\%$ confidence interval on σ^2_α on the basis of the so-called *Williams–Tukey formula* [see Williams (1962) and Boardman (1974)]:

$$\frac{1}{\bar{n}_h\chi^2_{\frac{\alpha}{2},k-1}}\left[SS_u-(k-1)MS_E F_{\frac{\alpha}{2},k-1,n_.-k}\right] \le \sigma^2_\alpha$$

$$\le \frac{1}{\bar{n}_h\chi^2_{1-\frac{\alpha}{2},k-1}}\left[SS_u-(k-1)MS_E F_{1-\frac{\alpha}{2},k-1,n_.-k}\right]. \tag{11.83}$$

Other approximate confidence intervals on σ^2_α are also available. A comparison of the coverage probabilities of these intervals was made by Lee and Khuri (2002) using Monte Carlo simulation. It was reported that the approximation associated with the interval in (11.83) was adequate, except in cases where $\sigma^2_\alpha/\sigma^2_\epsilon$ was small (less than 0.25) and the design was extremely unbalanced. With the exclusion of these cases, the interval in (11.83) maintained its coverage probabilities at levels close to the nominal value.

11.4 Inference Concerning the Random Two-Way Model

In this section, we discuss the analysis concerning a random two-way model with interaction of the form

$$Y_{ijk} = \mu + \alpha_{(i)} + \beta_{(j)} + (\alpha\beta)_{(ij)} + \epsilon_{ij(k)},$$
$$i = 1, 2, \ldots, a;\ j = 1, 2, \ldots, b;\ k = 1, 2, \ldots, n_{ij}, \tag{11.84}$$

where $\alpha_{(i)}$, $\beta_{(j)}$, $(\alpha\beta)_{(ij)}$, and $\epsilon_{ij(k)}$ are independently distributed as $N(0, \sigma^2_{\alpha})$, $N\left(0, \sigma^2_{\beta}\right)$, $N(0, \sigma^2_{\alpha\beta})$, and $N\left(0, \sigma^2_{\epsilon}\right)$, respectively. Note that the data set is assumed to contain no empty cells. Of interest here is the testing of hypotheses concerning σ^2_{α}, σ^2_{β}, and $\sigma^2_{\alpha\beta}$, in addition to setting up confidence intervals concerning these variance components.

11.4.1 Approximate Tests Based on the Method of Unweighted Means

We recall that the method of unweighted means (MUM) was used in Section 10.6 to provide approximate tests concerning the parameters of a fixed-effects two-way model. We now consider applying this method to the same model, but with the added assumption that the model's effects are all assumed to be randomly distributed.

Consider again the unweighted sums of squares (USSs) described in (10.85), (10.86), (10.87) corresponding to factors A, B, and their interaction $A * B$, respectively. A display of these expressions is shown again for convenience.

$$SS_{Au} = \bar{n}_h\, b \sum_{i=1}^{a} (\bar{X}_{i.} - \bar{X}_{..})^2 \tag{11.85}$$

$$SS_{Bu} = \bar{n}_h\, a \sum_{j=1}^{b} (\bar{X}_{.j} - \bar{X}_{..})^2 \tag{11.86}$$

$$SS_{ABu} = \bar{n}_h \sum_{i=1}^{a} \sum_{j=1}^{b} (X_{ij} - \bar{X}_{i.} - \bar{X}_{.j} + \bar{X}_{..})^2, \tag{11.87}$$

where $X_{ij} = \bar{Y}_{ij.}$ and $\bar{n}_h$ is the harmonic mean of the cell frequencies, namely,

$$\bar{n}_h = ab \left[\sum_{i=1}^{a} \sum_{j=1}^{b} \frac{1}{n_{ij}} \right]^{-1}. \tag{11.88}$$

Let $\bar{X} = (X_{11}, X_{12}, \ldots, X_{ab})'$ be the vector of cell means. The variance–covariance matrix of $\bar{X}$ is given by

$$\mathrm{Var}(\bar{X}) = A_1\,\sigma^2_{\alpha} + A_2\,\sigma^2_{\beta} + I_{ab}\,\sigma^2_{\alpha\beta} + \tilde{K}\,\sigma^2_{\epsilon}, \tag{11.89}$$

where $A_1 = I_a \otimes J_b$, $A_2 = J_a \otimes I_b$, and

$$\tilde{K} = \mathrm{diag}\left(\frac{1}{n_{11}}, \frac{1}{n_{12}}, \ldots, \frac{1}{n_{ab}}\right). \tag{11.90}$$

Note that the model,

$$X_{ij} = \mu + \alpha_{(i)} + \beta_{(j)} + (\alpha\beta)_{(ij)} + \bar{\epsilon}_{ij.}, \quad i = 1, 2, \ldots, a;\ j = 1, 2, \ldots, b, \tag{11.91}$$

is considered balanced with one observation, namely X_{ij}, in the (i,j)th cell ($i = 1, 2, \ldots, a$; $j = 1, 2, \ldots, b$). Hence, the rules of balanced data, as seen in Chapter 8, can apply to model (11.91). In particular, formulas (11.85), (11.86), and (11.87) are now expressible as

$$SS_{Au} = \bar{n}_h\,\bar{X}'\tilde{P}_1\bar{X} \tag{11.92}$$

$$SS_{Bu} = \bar{n}_h\,\bar{X}'\tilde{P}_2\bar{X} \tag{11.93}$$

$$SS_{ABu} = \bar{n}_h\,\bar{X}'\tilde{P}_3\bar{X}, \tag{11.94}$$

where

$$\tilde{P}_1 = \frac{1}{b}A_1 - \frac{1}{ab}A_0 \tag{11.95}$$

$$\tilde{P}_2 = \frac{1}{a}A_2 - \frac{1}{ab}A_0 \tag{11.96}$$

$$\tilde{P}_3 = A_3 - \frac{1}{b}A_1 - \frac{1}{a}A_2 + \frac{1}{ab}A_0, \tag{11.97}$$

and $A_0 = J_a \otimes J_b$, $A_3 = I_a \otimes I_b$. An application of formula (8.20) in this case gives

$$A_j\tilde{P}_i = \tilde{\kappa}_{ij}\,\tilde{P}_i, \quad i, j = 0, 1, 2, 3, \tag{11.98}$$

where $\tilde{P}_0 = \frac{1}{ab}A_0$ and $\tilde{\kappa}_{ij}$ is a scalar [see formula (8.21)]. Note that $\tilde{P}_i$ is idempotent for $i = 0, 1, 2, 3$ and $\tilde{P}_i\tilde{P}_j = \mathbf{0}$ for all $i \neq j$.

Let $\tilde{Q}$ be an orthogonal matrix that simultaneously diagonalizes both A_1 and A_2 (this matrix exists by Theorem 3.9 since $A_1A_2 = A_2A_1$). The actual construction of $\tilde{Q}$ is given in Appendix 11.A where it is shown that $\tilde{Q} = [\tilde{Q}_{10}' : \tilde{Q}_{11}' : \tilde{Q}_{12}' : \tilde{Q}_{13}']'$ so that $\tilde{Q}_{10} = \frac{1}{\sqrt{ab}}\mathbf{1}'_{ab}$ and the rows of $\tilde{Q}_{1i}$ are orthonormal and span the row space of $\tilde{P}_i$ ($i = 1, 2, 3$). Now let $\tilde{Z} = \tilde{Q}\bar{X}$, which correspondingly can be partitioned as $\left(Z_1, Z'_{\alpha}, Z'_{\beta}, Z'_{\alpha\beta}\right)'$, where

$Z_1 = \tilde{Q}_{10}\bar{X}, Z_\alpha = \tilde{Q}_{11}\bar{X}, Z_\beta = \tilde{Q}_{12}\bar{X}, Z_{\alpha\beta} = \tilde{Q}_{13}\bar{X}$. In addition, let $\tilde{U} = \tilde{Q}_1\bar{X}$, where $\tilde{Q}_1 = \left[\tilde{Q}'_{11} : \tilde{Q}'_{12} : \tilde{Q}'_{13}\right]'$. Thus, $\tilde{U} = \left(Z'_\alpha, Z'_\beta, Z'_{\alpha\beta}\right)'$, which is normally distributed with a zero mean and a variance–covariance matrix given by

$$
\begin{aligned}
\mathrm{Var}(\tilde{U}) &= \tilde{Q}_1\left[A_1\,\sigma^2_\alpha + A_2\,\sigma^2_\beta + I_{ab}\,\sigma^2_{\alpha\beta} + \tilde{K}\,\sigma^2_\epsilon\right]\tilde{Q}'_1 \\
&= \sigma^2_\alpha\,\tilde{Q}_1\left[A_1\tilde{Q}'_{11} : A_1\tilde{Q}'_{12} : A_1\tilde{Q}'_{13}\right] + \sigma^2_\beta\,\tilde{Q}_1\left[A_2\tilde{Q}'_{11} : A_2\tilde{Q}'_{12} : A_2\tilde{Q}'_{13}\right] \\
&\quad + \sigma^2_{\alpha\beta}\,I_{ab-1} + \sigma^2_\epsilon\,\tilde{Q}_1\tilde{K}\tilde{Q}'_1 \\
&= \sigma^2_\alpha\,\tilde{Q}_1\left[\tilde{\kappa}_{11}\,\tilde{Q}'_{11} : \tilde{\kappa}_{21}\,\tilde{Q}'_{12} : \tilde{\kappa}_{31}\,\tilde{Q}'_{13}\right] \\
&\quad + \sigma^2_\beta\,\tilde{Q}_1\left[\tilde{\kappa}_{12}\,\tilde{Q}'_{11} : \tilde{\kappa}_{22}\,\tilde{Q}'_{12} : \tilde{\kappa}_{32}\,\tilde{Q}'_{13}\right] \\
&\quad + \sigma^2_{\alpha\beta}\,I_{ab-1} + \sigma^2_\epsilon\,\tilde{L} \text{ (see Appendix 11.A)} \\
&= \sigma^2_\alpha\,\tilde{Q}_1\left[b\,\tilde{Q}'_{11} : \mathbf{0} : \mathbf{0}\right] + \sigma^2_\beta\,\tilde{Q}_1\left[\mathbf{0} : a\,\tilde{Q}'_{12} : \mathbf{0}\right] + \sigma^2_{\alpha\beta}\,I_{ab-1} + \sigma^2_\epsilon\,\tilde{L} \\
&= \mathrm{diag}\left(\delta_1\,I_{a-1}, \delta_2\,I_{b-1}, \delta_3\,I_{(a-1)(b-1)}\right) + \sigma^2_\epsilon\,\tilde{L}, \qquad (11.99)
\end{aligned}
$$

as can be seen from applying properties (i), (ii), and (iii) in Appendix 11.A and the fact that $\tilde{\kappa}_{11} = b$, $\tilde{\kappa}_{21} = \tilde{\kappa}_{31} = 0$, $\tilde{\kappa}_{12} = 0$, $\tilde{\kappa}_{22} = a$, $\tilde{\kappa}_{32} = 0$ [see formula (8.21)], where $\tilde{L} = \tilde{Q}_1\tilde{K}\tilde{Q}'_1$, $\delta_1 = b\,\sigma^2_\alpha + \sigma^2_{\alpha\beta}$, $\delta_2 = a\,\sigma^2_\beta + \sigma^2_{\alpha\beta}$, $\delta_3 = \sigma^2_{\alpha\beta}$. Note that $\tilde{L}$ is not diagonal unless the diagonal elements of $\tilde{K}$ are equal. This only occurs when the data set is balanced.

As in Lemma 11.4, it can be shown that the best approximation (in terms of the Euclidean norm) of $\tilde{L}$ with a diagonal matrix of the form $\tilde{c}\,I_{ab-1}$ is achieved when $\tilde{c} = \frac{1}{\bar{n}_h}$, where $\bar{n}_h$ is the harmonic mean in (11.88). Using this fact, $\mathrm{Var}(\tilde{U})$ in (11.99) is approximately equal to

$$
\mathrm{Var}(\tilde{U}) \approx \mathrm{diag}(\delta_1\,I_{a-1}, \delta_2\,I_{b-1}, \delta_3\,I_{(a-1)(b-1)}) + \frac{\sigma^2_\epsilon}{\bar{n}_h}\,I_{ab-1}. \qquad (11.100)
$$

This implies that Z_α, Z_β, $Z_{\alpha\beta}$ are approximately mutually independent and that

$$
\mathrm{Var}(Z_\alpha) \approx \delta_1 + \frac{\sigma^2_\epsilon}{\bar{n}_h} \qquad (11.101)
$$

$$
\mathrm{Var}(Z_\beta) \approx \delta_2 + \frac{\sigma^2_\epsilon}{\bar{n}_h} \qquad (11.102)
$$

$$
\mathrm{Var}(Z_{\alpha\beta}) \approx \delta_3 + \frac{\sigma^2_\epsilon}{\bar{n}_h}. \qquad (11.103)
$$

Lemma 11.5 The USSs in (11.85), (11.86), and (11.87) are equal to $\bar{n}_h\,Z'_\alpha Z_\alpha$, $\bar{n}_h\,Z'_\beta Z_\beta$, and $\bar{n}_h Z'_{\alpha\beta} Z_{\alpha\beta}$, respectively.

Proof. From Appendix 11.A we have $\tilde{Q}_{1i} = V_i \tilde{P}_i$ for some matrix V_i, $i = 0, 1, 2, 3$. Hence, $\tilde{P}_i \tilde{Q}'_{1i} = \tilde{P}_i \tilde{P}_i V'_i = \tilde{P}_i V'_i = \tilde{Q}'_{1i}$ since $\tilde{P}_i$ is idempotent ($i = 0, 1, 2, 3$). It follows that the columns of $\tilde{Q}'_{1i}$ (or the rows of $\tilde{Q}_{1i}$) are orthonormal eigenvectors of $\tilde{P}_i$ corresponding to the eigenvalue 1. Hence, we can write

$$\tilde{P}_i = \tilde{Q}'_{1i} \tilde{Q}_{1i}, \quad i = 0, 1, 2, 3. \tag{11.104}$$

It follows from (11.92), (11.93), and (11.94) that

$$\begin{aligned}
\bar{n}_h \, Z'_{\alpha} Z_{\alpha} &= \bar{n}_h \, \bar{X}' \tilde{Q}'_{11} \tilde{Q}_{11} \bar{X} \\
&= \bar{n}_h \, \bar{X}' \tilde{P}_1 \bar{X} \\
&= SS_{Au}.
\end{aligned}$$

We can similarly show that $\bar{n}_h \, Z'_{\beta} Z_{\beta} = SS_{Bu}$, $\bar{n}_h \, Z'_{\alpha\beta} Z_{\alpha\beta} = SS_{ABu}$. □

On the basis of Lemma 11.5 and formulas (11.101), (11.102), and (11.103), we conclude that

$$\frac{Z'_{\alpha} Z_{\alpha}}{\delta_1 + \frac{\sigma^2_{\epsilon}}{\bar{n}_h}} = \frac{SS_{Au}}{\bar{n}_h \, \delta_1 + \sigma^2_{\epsilon}}$$

$$\frac{Z'_{\beta} Z_{\beta}}{\delta_2 + \frac{\sigma^2_{\epsilon}}{\bar{n}_h}} = \frac{SS_{Bu}}{\bar{n}_h \, \delta_2 + \sigma^2_{\epsilon}}$$

$$\frac{Z'_{\alpha\beta} Z_{\alpha\beta}}{\delta_3 + \frac{\sigma^2_{\epsilon}}{\bar{n}_h}} = \frac{SS_{ABu}}{\bar{n}_h \, \delta_3 + \sigma^2_{\epsilon}}$$

are approximately distributed as mutually independent chi-squared variates with $a - 1$, $b - 1$, and $(a - 1)(b - 1)$ degrees of freedom, respectively. This is similar to a balanced data situation with SS_{Au}, SS_{Bu}, and SS_{ABu} acting as balanced ANOVA sums of squares and $\bar{n}_h$ being treated like n, the common cell frequency for a balanced data set. It follows that

$$F_{\alpha} = \frac{SS_{Au}/(a - 1)}{SS_{ABu}/(a - 1)(b - 1)} \tag{11.105}$$

can be used to test the hypothesis, $H_0 : \sigma^2_{\alpha} = 0$. Under H_0, F_{α} has an approximate F-distribution with $a - 1$ and $(a - 1)(b - 1)$ degrees of freedom. Similarly, to test $H_0 : \sigma^2_{\beta} = 0$, we can use

$$F_{\beta} = \frac{SS_{Bu}/(b - 1)}{SS_{ABu}/(a - 1)(b - 1)}, \tag{11.106}$$

which under H_0 has the approximate F-distribution with $b-1$ and $(a-1)(b-1)$ degrees of freedom. In addition, the interaction hypothesis, $H_0 : \sigma^2_{\alpha\beta} = 0$, can be tested by using the statistic

$$F_{\alpha\beta} = \frac{SS_{ABu}/(a-1)(b-1)}{SS_E/(n_{..} - ab)}, \tag{11.107}$$

where SS_E is the error sum of squares,

$$SS_E = \sum_{i=1}^{a}\sum_{j=1}^{b}\sum_{k=1}^{n_{ij}}(Y_{ijk} - \bar{Y}_{ij.})^2.$$

Under H_0, $F_{\alpha\beta}$ has the approximate F-distribution with $(a-1)(b-1)$ and $n_{..} - ab$ degrees of freedom.

Approximate confidence intervals on continuous functions of σ^2_α, σ^2_β, $\sigma^2_{\alpha\beta}$, and σ^2_ϵ can be easily derived by modifying the methodology presented in Section 8.7 for balanced models. Here, we treat SS_{Au}, SS_{Bu}, SS_{ABu} as mutually independent scaled chi-squared variates, which are independent of SS_E, and use $\bar{n}_h$ in place of n, the common cell frequency in a balanced data situation.

11.4.1.1 Adequacy of the Approximation

As was done earlier in Section 11.3.1, the adequacy of the approximate distributional properties concerning the USSs is examined here.

Let the right-hand side of the expression in (11.100) be denoted by $\boldsymbol{W}$. Thus,

$$\boldsymbol{W} = \text{diag}(\delta_1 \boldsymbol{I}_{a-1}, \delta_2 \boldsymbol{I}_{b-1}, \delta_3 \boldsymbol{I}_{(a-1)(b-1)}) + \frac{\sigma^2_\epsilon}{\bar{n}_h}\boldsymbol{I}_{ab-1}. \tag{11.108}$$

The properties of approximate mutual independence and chi-squaredness of SS_{Au}, SS_{Bu}, and SS_{ABu} hold exactly if and only if $\tilde{\boldsymbol{U}}'\boldsymbol{W}^{-1}\tilde{\boldsymbol{U}}$, which is equal to

$$\tilde{\boldsymbol{U}}'\boldsymbol{W}^{-1}\tilde{\boldsymbol{U}} = \frac{SS_{Au}}{\bar{n}_h\,\delta_1 + \sigma^2_\epsilon} + \frac{SS_{Bu}}{\bar{n}_h\,\delta_2 + \sigma^2_\epsilon} + \frac{SS_{ABu}}{\bar{n}_h\,\delta_3 + \sigma^2_\epsilon},$$

is distributed as a chi-squared variate (see Exercise 11.4). This amounts to requiring that $\boldsymbol{\Gamma}_w = \boldsymbol{W}^{-1}\text{Var}(\tilde{\boldsymbol{U}})$ be idempotent, that is, its nonzero eigenvalues be equal to 1 (recall Theorem 5.4). Using (11.99) and (11.108), $\boldsymbol{\Gamma}_w$ can be written as

$$\boldsymbol{\Gamma}_w = \boldsymbol{W}^{-1}\left[\boldsymbol{W} + (\tilde{\boldsymbol{L}} - \frac{1}{\bar{n}_h}\boldsymbol{I}_{ab-1})\,\sigma^2_\epsilon\right].$$

The nonzero eigenvalues of $\boldsymbol{\Gamma}_w$ are the same as those of the matrix

$$\begin{aligned}\boldsymbol{\Gamma}^*_w &= \boldsymbol{W}^{-1/2}\left[\boldsymbol{W} + (\tilde{\boldsymbol{L}} - \frac{1}{\bar{n}_h}\boldsymbol{I}_{ab-1})\,\sigma^2_\epsilon\right]\boldsymbol{W}^{-1/2} \\ &= \boldsymbol{I}_{ab-1} + \boldsymbol{\Delta}\,\sigma^2_\epsilon,\end{aligned} \tag{11.109}$$

where

$$\Delta = W^{-1/2}(\tilde{L} - \frac{1}{\bar{n}_h} I_{ab-1}) W^{-1/2}. \tag{11.110}$$

Hence, for an adequate approximation, the eigenvalues of Δ must be close to zero. We therefore require $\max_i \mid \tau_i^* \mid$ to be small, where τ_i^* is the ith eigenvalue of Δ ($i = 1, 2, \ldots, ab - 1$). The following lemma gives an upper bound on $\max_i \mid \tau_i^* \mid$. The proof of this lemma can be found in Khuri (1998, Lemma 3).

Lemma 11.6

$$\max_i \mid \tau_i^* \mid \leq \frac{\bar{n}_h}{\bar{n}_h \, \sigma^2_{\alpha\beta} + \sigma^2_{\epsilon}} \max \left\{ \left| \frac{1}{n_{\max}} - \frac{1}{\bar{n}_h} \right|, \left(\frac{1}{n_{\min}} - \frac{1}{\bar{n}_h} \right) \right\}, \tag{11.111}$$

where $n_{\min}$ and $n_{\max}$ are, respectively, the smallest and the largest of the n_{ij}'s.

The upper bound in (11.111) serves as a measure to assess the adequacy of the approximation associated with the distributions of the USSs. Small values of this upper bound indicate an adequate approximation. This occurs when $\frac{\sigma^2_{\alpha\beta}}{\sigma^2_{\epsilon}}$ is large, or when the data set is nearly balanced, since when $n_{ij} = n$, $n_{\max} = n_{\min} = \bar{n}_h = n$, and the upper bound is equal to zero. It is interesting to note that the upper bound depends on $\sigma^2_{\alpha\beta}$ and σ^2_{ϵ}, but does not depend on σ^2_{α} and σ^2_{β}.

The closeness of the upper bound in (11.111) to the actual value of $\max_i \mid \tau_i^* \mid$ was evaluated numerically by Khuri (1998). The following observations were noted concerning $\max_i \mid \tau_i^* \mid$:

(a) It is sensitive to changes in the values of $\sigma^2_{\alpha\beta}$. It decreases as $\sigma^2_{\alpha\beta}$ increases. However, it is less sensitive to changes in σ^2_{α} and σ^2_{β}.

(b) It remains fairly stable across a wide spectrum of unbalanced data situations. Its value drops significantly when the data set is nearly balanced.

(c) It is small when $\frac{\sigma^2_{\alpha\beta}}{\sigma^2_{\epsilon}}$ is large under varying degrees of data imbalance.

(d) It is fairly close to the upper bound value in (11.111), especially when the data set is nearly balanced. This upper bound is therefore an efficient measure of adequacy of the USSs as approximate balanced ANOVA sums of squares.

11.4.2 Exact Tests

We now show how to obtain exact tests concerning the variance components, σ^2_{α}, σ^2_{β}, $\sigma^2_{\alpha\beta}$. The test regarding $\sigma^2_{\alpha\beta}$ was developed by Thomsen (1975), and earlier by Spjøtvoll (1968) using a similar approach. The tests concerning σ^2_{α} and σ^2_{β} were developed by Khuri and Littell (1987).

Let us again consider $\bar{X} = (X_{11}, X_{12}, \ldots, X_{ab})'$, the vector of cell means, and the vector $\tilde{U} = \left(Z'_{\alpha}, Z'_{\beta}, Z'_{\alpha\beta}\right)'$ whose variance–covariance matrix is given in (11.99). The vectors, Z_{α}, Z_{β}, and $Z_{\alpha\beta}$ are normally distributed with zero means and variance–covariance matrices of the form

$$\text{Var}(Z_{\alpha}) = \left(b\,\sigma^2_{\alpha} + \sigma^2_{\alpha\beta}\right) I_{a-1} + \sigma^2_{\epsilon}\,\tilde{L}_1 \tag{11.112}$$

$$\text{Var}(Z_{\beta}) = \left(a\,\sigma^2_{\beta} + \sigma^2_{\alpha\beta}\right) I_{b-1} + \sigma^2_{\epsilon}\,\tilde{L}_2 \tag{11.113}$$

$$\text{Var}(Z_{\alpha\beta}) = \sigma^2_{\alpha\beta}\, I_{(a-1)(b-1)} + \sigma^2_{\epsilon}\,\tilde{L}_3, \tag{11.114}$$

where $\tilde{L}_1 = \tilde{Q}_{11}\tilde{K}\tilde{Q}'_{11}$, $\tilde{L}_2 = \tilde{Q}_{12}\tilde{K}\tilde{Q}'_{12}$, and $\tilde{L}_3 = \tilde{Q}_{13}\tilde{K}\tilde{Q}'_{13}$.

11.4.2.1 *Exact Test Concerning* $H_0 : \sigma^2_{\alpha\beta} = 0$ *(Thomsen, 1975)*

Using (11.114) we can write

$$\frac{1}{\sigma^2_{\epsilon}} Z'_{\alpha\beta}(\Delta_{\alpha\beta}\, I_{(a-1)(b-1)} + \tilde{L}_3)^{-1} Z_{\alpha\beta} \sim \chi^2_{(a-1)(b-1)},$$

where $\Delta_{\alpha\beta} = \frac{\sigma^2_{\alpha\beta}}{\sigma^2_{\epsilon}}$. It follows that

$$F(\Delta_{\alpha\beta}) = \frac{Z'_{\alpha\beta}(\Delta_{\alpha\beta}\, I_{(a-1)(b-1)} + \tilde{L}_3)^{-1} Z_{\alpha\beta}/[(a-1)(b-1)]}{SS_E/(n_{..} - ab)} \tag{11.115}$$

has the F-distribution with $(a-1)(b-1)$ and $n_{..} - ab$ degrees of freedom, where SS_E is the error sum of squares for model (11.84). Under $H_0 : \Delta_{\alpha\beta} = 0$ (i.e., $\sigma^2_{\alpha\beta} = 0$),

$$F(0) = \frac{Z'_{\alpha\beta}\tilde{L}_3^{-1} Z_{\alpha\beta}/[(a-1)(b-1)]}{SS_E/(n_{..} - ab)} \tag{11.116}$$

has the F-distribution with $(a-1)(b-1)$ and $n_{..} - ab$ degrees of freedom. Since

$$\begin{aligned} E&\left[\frac{1}{(a-1)(b-1)} Z'_{\alpha\beta}\tilde{L}_3^{-1} Z_{\alpha\beta}\right] \\ &= \frac{1}{(a-1)(b-1)}\, tr\left[\tilde{L}_3^{-1}\left(\sigma^2_{\alpha\beta}\, I_{(a-1)(b-1)} + \sigma^2_{\epsilon}\,\tilde{L}_3\right)\right] \\ &= \sigma^2_{\epsilon} + \frac{\sigma^2_{\alpha\beta}}{(a-1)(b-1)}\, tr\left(\tilde{L}_3^{-1}\right), \end{aligned}$$

we can reject $H_0 : \Delta_{\alpha\beta} = 0$ in favor of $H_a : \Delta_{\alpha\beta} > 0$ at the α-level if $F(0) \geq F_{\alpha,(a-1)(b-1),n_{..}-ab}$.

It should be noted that $F(0)$ is invariant to the choice of the matrix $\tilde{Q}_{13}$ whose rows, if we recall, are orthonormal and span the row space of $\tilde{P}_3$ [see formula (11.104)]. To show this, let Q^*_{13} be another matrix whose rows are also orthonormal and span the row space of $\tilde{P}_3$. Then, there exists a nonsingular matrix, A_{13}, such that $\tilde{Q}_{13} = A_{13}\, Q^*_{13}$. It follows that

$$\begin{aligned} Z'_{\alpha\beta}\tilde{L}_3^{-1} Z_{\alpha\beta} &= \bar{X}'\tilde{Q}'_{13}\left[\tilde{Q}_{13}\tilde{K}\tilde{Q}'_{13}\right]^{-1}\tilde{Q}_{13}\bar{X} \\ &= \bar{X}' Q^{*\prime}_{13} A'_{13}[A_{13}\, Q^*_{13}\tilde{K}\, Q^{*\prime}_{13} A'_{13}]^{-1} A_{13}\, Q^*_{13}\bar{X} \\ &= \bar{X}' Q^{*\prime}_{13}\left[Q^*_{13}\tilde{K}\, Q^{*\prime}_{13}\right]^{-1} Q^*_{13}\bar{X}. \end{aligned}$$

It is also interesting to note that $F(0)$ is identical to the Type III F-ratio,

$$F(\alpha\beta \mid \mu, \alpha, \beta) = \frac{R(\alpha\beta \mid \mu, \alpha, \beta)}{(a-1)(b-1)MS_E}, \qquad (11.117)$$

which was used in Section 10.3.1 [see formula (10.52)] to test the no-interaction hypothesis shown in (10.62) when all the effects were considered as fixed in the model. This fact was proved in Thomsen (1975). The F-test in (11.117) is known as a *Wald's test* and is described in Seely, and El-Bassiouni (1983).

Let us now consider the power of the above test. For the alternative value, $\Delta_{\alpha\beta} = \Delta^o_{\alpha\beta}$ $(\neq 0)$, $\frac{1}{\sigma^2_\epsilon} Z'_{\alpha\beta}\tilde{L}_3^{-1} Z_{\alpha\beta}$ is distributed as $\sum_{i=1}^{s^*} \kappa^*_i \chi^2_{\nu^*_i}$, where $\kappa^*_1, \kappa^*_2, \ldots, \kappa^*_{s^*}$ are the distinct nonzero eigenvalues of $\tilde{L}_3^{-1}\left(\Delta^o_{\alpha\beta}\, I_{(a-1)(b-1)} + \tilde{L}_3\right)$ with muliplicities ν^*_i $(i = 1, 2, \ldots, s^*)$ such that $\sum_{i=1}^{s^*} \nu^*_i = (a-1)(b-1)$, and the $\chi^2_{\nu^*_i}$'s are mutually independent chi-squared variates (see Lemma 5.1). It follows that under $H_a : \Delta_{\alpha\beta} = \Delta^o_{\alpha\beta}$, $F(0)$ is distributed as

$$\frac{\sum_{i=1}^{s^*} \kappa^*_i \chi^2_{\nu^*_i}/[(a-1)(b-1)]}{\chi^2_{n_{..}-ab}/(n_{..}-ab)},$$

which is basically a linear combination of F-distributed random variables. Its power function for an α-level of significance is

$$\begin{aligned} &P\left[\frac{n_{..}-ab}{(a-1)(b-1)\chi^2_{n_{..}-ab}} \sum_{i=1}^{s^*} \kappa^*_i \chi^2_{\nu^*_i} \geq F_{\alpha,(a-1)(b-1),n_{..}-ab}\right] \\ &\quad = P\left[\frac{n_{..}-ab}{(a-1)(b-1)} \sum_{i=1}^{s^*} \kappa^*_i \chi^2_{\nu^*_i} - F_{\alpha,(a-1)(b-1),n_{..}-ab}\, \chi^2_{n_{..}-ab} \geq 0\right], \qquad (11.118) \end{aligned}$$

which can be computed using Davies' (1980) algorithm since all the chi-squared variates in (11.118) are mutually independent (see Section 5.6).

Formula (11.115) can be used to obtain an exact confidence interval on $\Delta_{\alpha\beta} = \frac{\sigma^2_{\alpha\beta}}{\sigma^2_{\epsilon}}$. More specifically, we have

$$P[F_{1-\frac{\alpha}{2},(a-1)(b-1),n_{..}-ab} \le F(\Delta_{\alpha\beta}) \le F_{\frac{\alpha}{2},(a-1)(b-1),n_{..}-ab}] = 1-\alpha. \quad (11.119)$$

Since $F(\Delta_{\alpha\beta})$ is a decreasing function of $\Delta_{\alpha\beta}$, the double inequality in (11.119) can be solved to obtain an exact $(1-\alpha)100\%$ confidence interval on $\Delta_{\alpha\beta}$.

11.4.2.2 Exact Tests Concerning σ^2_{α} and σ^2_{β} (Khuri and Littell, 1987)

Let us write model (11.84) in matrix form as

$$\boldsymbol{Y} = \mu\,\boldsymbol{1}_{n_{..}} + \boldsymbol{H}_1\,\boldsymbol{\alpha} + \boldsymbol{H}_2\,\boldsymbol{\beta} + \boldsymbol{H}_3(\boldsymbol{\alpha\beta}) + \boldsymbol{\epsilon}, \quad (11.120)$$

where $\boldsymbol{H}_1 = \oplus_{i=1}^{a}\boldsymbol{1}_{n_{i.}}$, $\boldsymbol{H}_2 = \left[\oplus_{j=1}^{b}\boldsymbol{1}'_{n_{1j}} : \oplus_{j=1}^{b}\boldsymbol{1}'_{n_{2j}} : \ldots : \oplus_{j=1}^{b}\boldsymbol{1}'_{n_{aj}}\right]'$, $\boldsymbol{H}_3 = \oplus_{i=1}^{a}\oplus_{j=1}^{b}\boldsymbol{1}_{n_{ij}}$, $\boldsymbol{\alpha} = (\alpha_1, \alpha_2, \ldots, \alpha_a)'$, $\boldsymbol{\beta} = (\beta_1, \beta_2, \ldots, \beta_b)'$, and $(\boldsymbol{\alpha\beta}) = [(\alpha\beta)_{11}, (\alpha\beta)_{12}, \ldots, (\alpha\beta)_{ab}]'$. The variance–covariance matrix of $\boldsymbol{Y}$, namely $\boldsymbol{\Sigma}$, is therefore of the form

$$\boldsymbol{\Sigma} = \sigma^2_{\alpha}\,\boldsymbol{H}_1\boldsymbol{H}'_1 + \sigma^2_{\beta}\,\boldsymbol{H}_2\boldsymbol{H}'_2 + \sigma^2_{\alpha\beta}\,\boldsymbol{H}_3\boldsymbol{H}'_3 + \sigma^2_{\epsilon}\,\boldsymbol{I}_{n_{..}}. \quad (11.121)$$

Furthermore, the error sum of squares, SS_E, can be expressed as a quadratic form in $\boldsymbol{Y}$,

$$SS_E = \boldsymbol{Y}'\boldsymbol{R}\boldsymbol{Y}, \quad (11.122)$$

where $\boldsymbol{R}$ is the matrix,

$$\boldsymbol{R} = \boldsymbol{I}_{n_{..}} - \bigoplus_{i=1}^{a}\bigoplus_{j=1}^{b}\frac{1}{n_{ij}}\boldsymbol{J}_{n_{ij}}, \quad (11.123)$$

which is idempotent of rank $n_{..} - ab$. It can be verified that

$$\boldsymbol{DR} = \boldsymbol{0},\ \boldsymbol{RH}_1 = \boldsymbol{0},\ \boldsymbol{RH}_2 = \boldsymbol{0},\ \boldsymbol{RH}_3 = \boldsymbol{0}, \quad (11.124)$$

where $\boldsymbol{D} = \oplus_{i=1}^{a}\oplus_{j=1}^{b}\frac{1}{n_{ij}}\boldsymbol{1}'_{n_{ij}}$. From (11.121) and (11.124) it follows that $\boldsymbol{D\Sigma R} = \boldsymbol{0}$. Consequently, the vector of cell means, $\bar{\boldsymbol{X}}$, which can be written as $\bar{\boldsymbol{X}} = \boldsymbol{DY}$, is independent of SS_E (see Theorem 5.6). Since $\tilde{\boldsymbol{U}} = \tilde{\boldsymbol{Q}}_1\bar{\boldsymbol{X}}$, $\tilde{\boldsymbol{U}}$ is also independent of SS_E.

By the Spectral Decomposition Theorem (Theorem 3.4), the matrix $\boldsymbol{R}$ in (11.123) can be expressed as

$$\boldsymbol{R} = \boldsymbol{C\Lambda C}', \quad (11.125)$$

where

C is an orthogonal matrix

Λ is a diagonal matrix of eigenvalues of R which consist of $n_{..} - ab$ ones and ab zeros

We shall assume that

$$n_{..} > 2ab - 1. \tag{11.126}$$

Under condition (11.126), Λ and C can be partitioned as

$$\Lambda = \text{diag}(I_{a_1}, I_{a_2}, \mathbf{0}) \tag{11.127}$$

$$C = [C_1 : C_2 : C_3], \tag{11.128}$$

where $a_1 = ab - 1$, $a_2 = n_{..} - 2ab + 1$, $\mathbf{0}$ is a zero matrix of order $ab \times ab$, and C_1, C_2, C_3 are the corresponding matrices of orthonormal eigenvectors of orders $n_{..} \times a_1$, $n_{..} \times a_2$, and $n_{..} \times ab$, respectively. The integer a_2 is positive because of condition (11.126). It should be noted that the choice of C_1 depends on which a_1 columns are selected from the first $n_{..} - ab$ columns of C. The latter columns are orthonormal eigenvectors of C corresponding to the eigenvalue 1. The matrices, C_1 and C_2, are therefore not unique. Formula (11.125) can then be written as

$$R = C_1 C_1' + C_2 C_2'. \tag{11.129}$$

This results in a partitioning of SS_E in (11.122) as

$$SS_E = SS_{E1} + SS_{E2}, \tag{11.130}$$

where $SS_{E1} = Y' C_1 C_1' Y$, $SS_{E2} = Y' C_2 C_2' Y$. Note that SS_{E1} and SS_{E2} are independently distributed such that $\frac{1}{\sigma^2_\epsilon} SS_{E1} \sim \chi^2_{a_1}$, $\frac{1}{\sigma^2_\epsilon} SS_{E2} \sim \chi^2_{a_2}$.

Now, let the random vector ω be defined as

$$\omega = \tilde{U} + (\lambda_{\max} I_{a_1} - \tilde{L})^{1/2} C_1' Y, \tag{11.131}$$

where $\tilde{L}$ is the matrix used in (11.99) and $\lambda_{\max}$ is its largest eigenvalue. In (11.131), $(\lambda_{\max} I_{a_1} - \tilde{L})^{1/2}$ is a symmetric matrix with eigenvalues equal to the square roots of the eigenvalues of $\lambda_{\max} I_{a_1} - \tilde{L}$, which are nonnegative, and its eigenvectors are the same as those of $\tilde{L}$ (this results from applying the Spectral Decomposition Theorem to $\tilde{L}$). Since $\tilde{U} = \left(Z'_\alpha, Z'_\beta, Z'_{\alpha\beta}\right)'$, ω can likewise be partitioned as $\omega = \left(\omega'_\alpha, \omega'_\beta, \omega'_{\alpha\beta}\right)'$. The distributional properties of ω_α, ω_β, and $\omega_{\alpha\beta}$ are given in the next lemma.

Lemma 11.7

(i) ω_{α}, ω_{β}, and $\omega_{\alpha\beta}$ are mutually independent and normally distributed with zero means and variance–covariance matrices given by

$$\begin{aligned}\text{Var}(\boldsymbol{\omega}_{\alpha}) &= \left(b\,\sigma_{\alpha}^2 + \sigma_{\alpha\beta}^2 + \lambda_{\max}\,\sigma_{\epsilon}^2\right)\boldsymbol{I}_{a-1}\\ \text{Var}(\boldsymbol{\omega}_{\beta}) &= \left(a\,\sigma_{\beta}^2 + \sigma_{\alpha\beta}^2 + \lambda_{\max}\,\sigma_{\epsilon}^2\right)\boldsymbol{I}_{b-1}\\ \text{Var}(\boldsymbol{\omega}_{\alpha\beta}) &= \left(\sigma_{\alpha\beta}^2 + \lambda_{\max}\,\sigma_{\epsilon}^2\right)\boldsymbol{I}_{(a-1)(b-1)}.\end{aligned}$$

(ii) ω_{α}, ω_{β}, and $\omega_{\alpha\beta}$ are independent of SS_{E2}.

Proof.

(i) From (11.123) we have that $\boldsymbol{R}\boldsymbol{1}_{n_{..}} = \boldsymbol{0}$. Thus, $\left(\boldsymbol{C}_1\boldsymbol{C}_1' + \boldsymbol{C}_2\boldsymbol{C}_2'\right)\boldsymbol{1}_{n_{..}} = \boldsymbol{0}$, by virtue of (11.129). Hence, $\boldsymbol{C}_1'\boldsymbol{1}_{n_{..}} = \boldsymbol{0}$ (since $\boldsymbol{C}_1'\boldsymbol{C}_1 = \boldsymbol{I}_{a_1}$ and $\boldsymbol{C}_1'\boldsymbol{C}_2 = \boldsymbol{0}$) and $E\left(\boldsymbol{C}_1'\boldsymbol{Y}\right) = \mu\,\boldsymbol{C}_1'\boldsymbol{1}_{n_{..}} = \boldsymbol{0}$. Since $\tilde{\boldsymbol{U}}$ has a zero mean, the mean of $\boldsymbol{\omega}$ in (11.131) is then equal to $\boldsymbol{0}$. The means of $\boldsymbol{\omega}_{\alpha}$, $\boldsymbol{\omega}_{\beta}$, and $\boldsymbol{\omega}_{\alpha\beta}$ are therefore equal to zero.

Now, it is obvious that $\boldsymbol{\omega}$ is normally distributed. Furthermore, $\tilde{\boldsymbol{U}}$ is independent of $\boldsymbol{C}_1'\boldsymbol{Y}$ since $\boldsymbol{D\Sigma R} = \boldsymbol{0}$, and therefore $\boldsymbol{D\Sigma C}_1 = \boldsymbol{0}$. This implies that $\bar{\boldsymbol{X}}$, and hence $\tilde{\boldsymbol{U}}$, are independent of $\boldsymbol{C}_1'\boldsymbol{Y}$. From (11.131) we then have

$$\text{Var}(\boldsymbol{\omega}) = \text{Var}(\tilde{\boldsymbol{U}}) + (\lambda_{\max}\boldsymbol{I}_{a_1} - \tilde{\boldsymbol{L}})^{1/2}\boldsymbol{C}_1'\boldsymbol{\Sigma}\boldsymbol{C}_1(\lambda_{\max}\boldsymbol{I}_{a_1} - \tilde{\boldsymbol{L}})^{1/2}. \quad (11.132)$$

But,

$$\begin{aligned}\boldsymbol{C}_1'\boldsymbol{\Sigma}\boldsymbol{C}_1 &= \boldsymbol{C}_1'(\sigma_{\alpha}^2\,\boldsymbol{H}_1\boldsymbol{H}_1' + \sigma_{\beta}^2\,\boldsymbol{H}_2\boldsymbol{H}_2' + \sigma_{\alpha\beta}^2\,\boldsymbol{H}_3\boldsymbol{H}_3' + \sigma_{\epsilon}^2\,\boldsymbol{I}_{n_{..}})\boldsymbol{C}_1\\ &= \sigma_{\epsilon}^2\,\boldsymbol{I}_{a_1}, \qquad (11.133)\end{aligned}$$

since $\boldsymbol{C}_1'\boldsymbol{C}_1 = \boldsymbol{I}_{a_1}$, and $\boldsymbol{C}_1'\boldsymbol{H}_i = \boldsymbol{0}$, $i = 1, 2, 3$, by virtue of $\boldsymbol{R}\boldsymbol{H}_i = \boldsymbol{0}$ for $i = 1, 2, 3$ [see (11.124) and (11.129)]. From (11.99), (11.132), and (11.133), we get

$$\begin{aligned}\text{Var}(\boldsymbol{\omega}) &= \text{diag}(\delta_1\,\boldsymbol{I}_{a-1}, \delta_2\,\boldsymbol{I}_{b-1}, \delta_3\,\boldsymbol{I}_{(a-1)(b-1)}) + \sigma_{\epsilon}^2\,\tilde{\boldsymbol{L}} + \sigma_{\epsilon}^2\,(\lambda_{\max}\boldsymbol{I}_{a_1} - \tilde{\boldsymbol{L}})\\ &= \text{diag}[(\delta_1 + \lambda_{\max}\,\sigma_{\epsilon}^2)\boldsymbol{I}_{a-1}, (\delta_2 + \lambda_{\max}\,\sigma_{\epsilon}^2)\boldsymbol{I}_{b-1},\\ &\qquad \left(\delta_3 + \lambda_{\max}\,\sigma_{\epsilon}^2\right)\boldsymbol{I}_{(a-1)(b-1)}], \qquad (11.134)\end{aligned}$$

where, if we recall, $\delta_1 = b\,\sigma_{\alpha}^2 + \sigma_{\alpha\beta}^2$, $\delta_2 = a\,\sigma_{\beta}^2 + \sigma_{\alpha\beta}^2$, and $\delta_3 = \sigma_{\alpha\beta}^2$. From (11.134) we conclude that $\boldsymbol{\omega}_{\alpha}$, $\boldsymbol{\omega}_{\beta}$, and $\boldsymbol{\omega}_{\alpha\beta}$ are mutually independent and their variance–covariance matrices are as described in (i).

(ii) It is easy to see that $\tilde{U}$ is independent of $C_2'Y$ (since $D\Sigma C_2 = 0$), and hence of $SS_{E2} = Y'C_2C_2'Y$. It is also true that $C_1'Y$ is independent of $C_2'Y$, and hence of SS_{E2} (since $C_1'\Sigma C_2 = 0$ because of $C_1'H_i = 0$, $i = 1, 2, 3$, and $C_1'C_2 = 0$). This implies independence of ω from SS_{E2}. □

From Lemma 11.7 we conclude that the sums of squares, $S_\alpha = \omega'_\alpha \omega_\alpha$, $S_\beta = \omega'_\beta \omega_\beta$, $S_{\alpha\beta} = \omega'_{\alpha\beta}\omega_{\alpha\beta}$, and SS_{E2} are mutually independent and

$$S_\alpha \Big/ \left(b\,\sigma^2_\alpha + \sigma^2_{\alpha\beta} + \lambda_{\max}\,\sigma^2_\epsilon\right) \sim \chi^2_{a-1}$$
$$S_\beta \Big/ \left(a\,\sigma^2_\beta + \sigma^2_{\alpha\beta} + \lambda_{\max}\,\sigma^2_\epsilon\right) \sim \chi^2_{b-1}$$
$$S_{\alpha\beta} \Big/ \left(\sigma^2_{\alpha\beta} + \lambda_{\max}\,\sigma^2_\epsilon\right) \sim \chi^2_{(a-1)(b-1)}$$
$$SS_{E2}/\sigma^2_\epsilon \sim \chi^2_{a_2},$$

where, if we recall, $a_2 = n_{..} - 2ab + 1$. A test statistic for testing $H_0 : \sigma^2_\alpha = 0$ is therefore given by

$$F = \frac{S_\alpha/(a-1)}{S_{\alpha\beta}/[(a-1)(b-1)]}, \qquad (11.135)$$

which under H_0 has the F-distribution with $a-1$ and $(a-1)(b-1)$ degrees of freedom. The null hypothesis is rejected at the α-level if $F \geq F_{\alpha,a-1,(a-1)(b-1)}$. Similarly, the hypothesis $H_0 : \sigma^2_\beta = 0$ can be tested by

$$F = \frac{S_\beta/(b-1)}{S_{\alpha\beta}/[(a-1)(b-1)]}, \qquad (11.136)$$

which under H_0 has the F-distribution with $b-1$ and $(a-1)(b-1)$ degrees of freedom. Note that the statistic,

$$F = \frac{S_{\alpha\beta}/[(a-1)(b-1)]}{\lambda_{\max} SS_{E2}/a_2}$$

can be used to test $H_0 : \sigma^2_{\alpha\beta} = 0$, but is not recommended since it has fewer denominator degrees of freedom than the test given in (11.116).

Note that since the matrix C_1 is not chosen uniquely, the actual value of ω in (11.131), and hence the test statistic values in (11.135) and (11.136), depend on the choice of C_1. However, the distributions of these statistics (under the null and alternative hypotheses) are invariant to that choice.

If the data set is balanced, then $\tilde{K}$ in (11.90) becomes $\frac{1}{n}I_{ab}$, where n is the common cell frequency. Hence, $\tilde{L} = \tilde{Q}_1\tilde{K}\tilde{Q}_1' = \frac{1}{n}I_{ab-1}$. Consequently, $\lambda_{\max} = \frac{1}{n}$ and the vectors, ω and $\tilde{U}$ in (11.131) become identical. In addition, $n\,S_\alpha$, $n\,S_\beta$, and $n\,S_{\alpha\beta}$ reduce to the balanced ANOVA sums of squares for factors A, B, and their interaction $A * B$, respectively.

Power values concerning the test statistics in (11.135) and (11.136) can be easily derived. For example, the power of the test for σ^2_α is given by

$$P\left[\frac{S_\alpha/(a-1)}{S_{\alpha\beta}/[(a-1)(b-1)]} \geq F_{\alpha,a-1,(a-1)(b-1)} \mid \sigma^2_\alpha > 0\right]$$
$$= P\left[F_{a-1,(a-1)(b-1)} \geq \frac{\sigma^2_{\alpha\beta} + \lambda_{max}\,\sigma^2_\epsilon}{b\,\sigma^2_\alpha + \sigma^2_{\alpha\beta} + \lambda_{max}\,\sigma^2_\epsilon} F_{\alpha,a-1,(a-1)(b-1)}\right], \quad (11.137)$$

since under $H_a : \sigma^2_\alpha > 0$,

$$\frac{S_\alpha/(a-1)}{b\,\sigma^2_\alpha + \sigma^2_{\alpha\beta} + \lambda_{max}\,\sigma^2_\epsilon} \times \frac{\sigma^2_{\alpha\beta} + \lambda_{max}\,\sigma^2_\epsilon}{S_{\alpha\beta}/[(a-1)(b-1)]} \sim F_{a-1,(a-1)(b-1)}.$$

From (11.137) it can be seen that the power is a function of α, λ_{max}, which depends on the design used, and the variance ratios, $\sigma^2_\alpha/\sigma^2_\epsilon$, $\sigma^2_{\alpha\beta}/\sigma^2_\epsilon$, through the value of $\sigma^2_\alpha\Big/\left(\sigma^2_{\alpha\beta} + \lambda_{max}\,\sigma^2_\epsilon\right)$. We note that the power in (11.137) is a

(i) Monotone increasing function of $\sigma^2_\alpha/\sigma^2_\epsilon$ for a fixed value of $\sigma^2_{\alpha\beta}/\sigma^2_\epsilon$ and a given design.

(ii) Monotone decreasing function of $\sigma^2_{\alpha\beta}/\sigma^2_\epsilon$ for a fixed value of $\sigma^2_\alpha/\sigma^2_\epsilon$ and a given design.

(iii) Monotone decreasing function of λ_{max} for fixed values of $\sigma^2_\alpha/\sigma^2_\epsilon$ and $\sigma^2_{\alpha\beta}/\sigma^2_\epsilon$. Hence, small values of λ_{max} are desirable. Lemma 11.8 gives lower and upper bounds on λ_{max}.

A similar power study can be made with regard to the test for σ^2_β in (11.136).

Lemma 11.8

$$\frac{1}{\bar{n}_h} \leq \lambda_{max} \leq \frac{1}{n_{min}},$$

where n_{min} is the smallest cell frequency and $\bar{n}_h$ is the harmonic mean in (11.88).

Proof. We have that $\tilde{L} = \tilde{Q}_1\tilde{K}\tilde{Q}_1'$. Hence,

$$\tilde{L} \leq \frac{1}{n_{min}}\tilde{Q}_1\tilde{Q}_1'$$
$$= \frac{1}{n_{min}} I_{ab-1},$$

where $\tilde{L} \leq \frac{1}{n_{min}}\tilde{Q}_1\tilde{Q}_1'$ means that the matrix $\frac{1}{n_{min}}\tilde{Q}_1\tilde{Q}_1' - \tilde{L}$ is positive semidefinite. It follows that $\lambda_{max} \leq \frac{1}{n_{min}}$.

On the other hand, $\lambda_{\max}$ is greater than or equal to $tr\left(\tilde{Q}_1\tilde{K}\tilde{Q}_1'\right)\Big/(ab-1)$, which is the average of the eigenvalues of $\tilde{L}$. But,

$$\begin{aligned} tr\left(\tilde{Q}_1\tilde{K}\tilde{Q}_1'\right) &= tr\left(\tilde{Q}_1'\tilde{Q}_1\tilde{K}\right) \\ &= tr[(I_{ab} - \frac{1}{ab}J_{ab})\tilde{K}], \end{aligned}$$

since $\frac{1}{ab}J_{ab} + \tilde{Q}_1'\tilde{Q}_1 = I_{ab}$. Hence,

$$\begin{aligned} \lambda_{\max} &\geq \frac{1}{ab-1}\left[tr(\tilde{K}) - \frac{1}{ab}\mathbf{1}_{ab}'\tilde{K}\mathbf{1}_{ab}\right] \\ &= \frac{1}{ab-1}\left[\sum_{i=1}^{a}\sum_{j=1}^{b}\frac{1}{n_{ij}} - \frac{1}{ab}\sum_{i=1}^{a}\sum_{j=1}^{b}\frac{1}{n_{ij}}\right] \\ &= \frac{1}{ab}\sum_{i=1}^{a}\sum_{j=1}^{b}\frac{1}{n_{ij}} \\ &= \frac{1}{\bar{n}_h}. \end{aligned}$$

□

11.4.2.2.1 Simultaneous Confidence Intervals

Simultaneous confidence intervals on all continuous functions of σ^2_α, σ^2_β, $\sigma^2_{\alpha\beta}$, and σ^2_ϵ can be easily obtained on the basis of Lemma 11.7, just like in a balanced data situation (see Section 8.7). To accomplish this, we first need to set up individual $(1-\alpha)$ 100% confidence intervals on the following expected values:

$$\begin{aligned} E\left(\frac{S_\alpha}{a-1}\right) &= b\,\sigma^2_\alpha + \sigma^2_{\alpha\beta} + \lambda_{\max}\,\sigma^2_\epsilon \\ E\left(\frac{S_\beta}{b-1}\right) &= a\,\sigma^2_\beta + \sigma^2_{\alpha\beta} + \lambda_{\max}\,\sigma^2_\epsilon \\ E\left(\frac{S_{\alpha\beta}}{(a-1)(b-1)}\right) &= \sigma^2_{\alpha\beta} + \lambda_{\max}\,\sigma^2_\epsilon \\ E\left(\frac{SS_{E2}}{a_2}\right) &= \sigma^2_\epsilon. \end{aligned}$$

Such intervals are given by

$$\begin{aligned} C_\alpha &: \frac{S_\alpha}{\chi^2_{\alpha/2,a-1}} \leq b\,\sigma^2_\alpha + \sigma^2_{\alpha\beta} + \lambda_{\max}\,\sigma^2_\epsilon \leq \frac{S_\alpha}{\chi^2_{1-\alpha/2,a-1}} \\ C_\beta &: \frac{S_\beta}{\chi^2_{\alpha/2,b-1}} \leq a\,\sigma^2_\beta + \sigma^2_{\alpha\beta} + \lambda_{\max}\,\sigma^2_\epsilon \leq \frac{S_\beta}{\chi^2_{1-\alpha/2,b-1}} \end{aligned}$$

$$C_{\alpha\beta} : \frac{S_{\alpha\beta}}{\chi^2_{\alpha/2,(a-1)(b-1)}} \leq \sigma^2_{\alpha\beta} + \lambda_{\max}\, \sigma^2_{\epsilon} \leq \frac{S_{\alpha\beta}}{\chi^2_{1-\alpha/2,(a-1)(b-1)}}$$

$$C_{\epsilon} : \frac{SS_{E2}}{\chi_{\alpha/2,a_2}} \leq \sigma^2_{\epsilon} \leq \frac{SS_{E2}}{\chi_{1-\alpha/2,a_2}}.$$

Since $S_{\alpha}, S_{\beta}, S_{\alpha\beta}$, and SS_{E2} are mutually independent, the Cartesian product,

$$C_{\alpha\beta\epsilon} = C_{\alpha} \times C_{\beta} \times C_{\alpha\beta} \times C_{\epsilon}$$

represents a rectangular confidence region on the vector of the four expected values with a confidence coefficient equal to $1 - \alpha^* = (1 - \alpha)^4$.

Now, if $f^*\left(\sigma^2_{\alpha}, \sigma^2_{\beta}, \sigma^2_{\alpha\beta}, \sigma^2_{\epsilon}\right)$ is any continuous function of the variance components, then it can be expressed as a continuous function, f, of $E(S_{\alpha}/(a-1))$, $E(S_{\beta}/(b-1))$, $E(S_{\alpha\beta}/[(a-1)(b-1)])$, and $E(SS_{E2}/a_2)$. Then, by the method described in Section 8.7, the intervals, $[\min_{x \in C_{\alpha\beta\epsilon}} f(x), \max_{x \in C_{\alpha\beta\epsilon}} f(x)]$, for all continuous functions $f(x)$, $x \in C_{\alpha\beta\epsilon}$, provide simultaneous confidence intervals on all such functions with a joint confidence coefficient greater than or equal to $1 - \alpha^*$.

Example 11.6 Khuri and Littell (1987) reported an example that dealt with a study of the variation in fusiform rust in Southern pine tree plantations. Trees with female parents from different families (factor B) were evaluated in several test locations (factor A). The number of plots in each family $\times$ test combination ranged from 1 to 4. The proportions of symptomatic trees in each plot are reproduced in Table 11.7.

Since the data are proportions, the arcsin (square root) transformation was applied before doing the analysis. Thus, in this case, Y_{ijk} is the transformed observation from the kth plot at the ith location for the trees coming from the jth family.

Let us first do the analysis using the approximate tests based on the method of unweighted means. The values of the USSs in (11.85), (11.86), (11.87), and the error sum of squares, SS_E, in addition to the corresponding F-ratios are given in Table 11.8.

Note that the F-values for A, B, and $A * B$ were obtained by applying formulas (11.105), (11.106), and (11.107), respectively. It can be seen that the test concerning $H_0 : \sigma^2_{\alpha\beta} = 0$ is not significant, but the tests for $H_0 : \sigma^2_{\alpha} = 0$ and $H_0 : \sigma^2_{\beta} = 0$ are significant with p-values, 0.0025 and 0.0234, respectively.

In this example, $\bar{n}_h = 1.9835$, $n_{\min} = 1$, $n_{\max} = 4$. By applying formula (11.111), we get

$$\max_i \mid \tau_i^* \mid \leq \frac{1.9835}{1.9835\, \sigma^2_{\alpha\beta} + \sigma^2_{\epsilon}} \times 0.4958$$

$$\approx \frac{1}{2\, \sigma^2_{\alpha\beta} + \sigma^2_{\epsilon}}.$$

TABLE 11.7
Proportions of Symptomatic Trees

	Family Number				
Test Number	**288**	**352**	**19**	**141**	**60**
34	0.804	0.734	0.967	0.917	0.850
	0.967	0.817	0.930		
	0.970	0.833	0.889		
		0.304			
35	0.867	0.407	0.896	0.952	0.486
	0.667	0.511	0.717		0.467
	0.793	0.274			
	0.458	0.428			
40	0.409	0.411	0.919	0.408	0.275
	0.569	0.646	0.669	0.435	0.256
	0.715	0.310	0.669	0.500	
	0.487		0.450		
41	0.587	0.304	0.928	0.367	0.525
	0.538	0.428	0.855		
	0.961		0.655		
	0.300		0.800		

Source: Reprinted from Khuri, A.I. and Littell, R.C., *Biometrics*, 43, 545, 1987. With permission.

TABLE 11.8
F-Tests Based on the Method of Unweighted Means

Source	DF	*SS*	*MS*	*F*	*p*-Value
A	3	1.13967	0.37989	8.654	0.0025
B	4	0.73925	0.18481	4.210	0.0234
$A * B$	12	0.52676	0.04390	1.284	0.2738
Error	33	1.12869	0.03420		

Thus, a large value of $2\,\sigma^2_{\alpha\beta} + \sigma^2_{\epsilon}$ results in a good approximation concerning the distribution of the USSs.

Now, let us apply the exact testing procedure outlined in Sections 11.4.2.1 and 11.4.2.2. To accomplish this, the following steps are needed:

(a) The matrices $\tilde{P}_1, \tilde{P}_2$, and $\tilde{P}_3$ are obtained using the expressions in (11.95), (11.96), and (11.97), respectively, where $a = 4$, $b = 5$, $A_0 = J_4 \otimes J_5$, $A_1 = I_4 \otimes J_5$, $A_2 = J_4 \otimes I_5$, and $A_3 = I_4 \otimes I_5$.

(b) The orthonormal eigenvectors of $\tilde{P}_i$ corresponding to the eigenvalue 1 are obtained ($i = 1, 2, 3$). This can be easily done by using, for example, the "CALL EIGEN" subroutine from PROC IML in SAS (1999). These eigenvectors form the rows of $\tilde{Q}_{11}$, $\tilde{Q}_{12}$, and $\tilde{Q}_{13}$, which are of orders 3×20, 4×20, and 12×20, respectively.

(c) The vectors, $\bar{X}, Z_{\alpha}, Z_{\beta}, Z_{\alpha\beta}$, and $\tilde{U}$ are computed, where $\bar{X}$ is the 20×1 vector of cell means, $Z_{\alpha} = \tilde{Q}_{11}\bar{X}$, $Z_{\beta} = \tilde{Q}_{12}\bar{X}$, $Z_{\alpha\beta} = \tilde{Q}_{13}\bar{X}$, $\tilde{U} = \tilde{Q}_1\bar{X}$, and $\tilde{Q}_1 = \left[\tilde{Q}'_{11} : \tilde{Q}'_{12} : \tilde{Q}'_{13}\right]'$.

(d) The matrices $\tilde{L}_3 = \tilde{Q}_{13}\tilde{K}\tilde{Q}'_{13}$ and $\tilde{L} = \tilde{Q}_1\tilde{K}\tilde{Q}'_1$ are computed, where $\tilde{K}$ is given in (11.90).

(e) Formula (11.116) is applied to produce the value of $F(0)$ for testing $H_0 : \sigma^2_{\alpha\beta} = 0$. Recall that from Table 11.8, $SS_E/(n_{..} - ab) = 0.03420$.

(f) The matrix R in $SS_E = Y'RY$ is decomposed as in (11.129). In this case, C_1 and C_2 are of orders 53×19 and 53×14, respectively. This can also be carried out by using the "CALL EIGEN" subroutine as mentioned earlier. Recall that the columns of $[C_1 : C_2]$ are orthonormal eigenvectors of R corresponding to the eigenvalue 1 of R, but the choice of C_1 is not unique.

(g) The vector $\omega = \left(\omega'_{\alpha}, \omega'_{\beta}, \omega'_{\alpha\beta}\right)'$ in (11.131), and hence $S_{\alpha} = \omega'_{\alpha}\omega_{\alpha}$, $S_{\beta} = \omega'_{\beta}\omega_{\beta}$, $S_{\alpha\beta} = \omega'_{\alpha\beta}\omega_{\alpha\beta}$, are computed. Note that, in this example, $\lambda_{\max} = 1$, and upon decomposing $\tilde{L}$ as $\tilde{L} = P_{\ell}\Lambda_{\ell}P'_{\ell}$, where Λ_{ℓ} is a diagonal matrix of eigenvalues of $\tilde{L}$ and P_{ℓ} is an orthogonal matrix of corresponding eigenvectors, the matrix $(\lambda_{\max} I_{a_1} - \tilde{L})^{1/2}$ can be written as

$$\begin{aligned}(\lambda_{\max} I_{a_1} - \tilde{L})^{1/2} &= \left(I_{19} - P_{\ell}\Lambda_{\ell}P'_{\ell}\right)^{1/2} \\ &= P_{\ell}\,\mathrm{diag}\left[(1-\lambda_{\ell 1})^{1/2}, (1-\lambda_{\ell 2})^{1/2}, \ldots, (1-\lambda_{\ell 19})^{1/2}\right] P'_{\ell},\end{aligned}$$

where $\lambda_{\ell i}$ is the ith diagonal element of Λ_{ℓ} $(i = 1, 2, \ldots, 19)$.

(h) Formulas (11.116), (11.135), and (11.136) are then applied to compute the F-statistics for testing the significance of $\sigma^2_{\alpha\beta}$, σ^2_{α}, and σ^2_{β}, respectively.

On the basis of the above outline and the data set in Table 11.7, we find that $S_{\alpha} = 0.78495$ with 3 degrees of freedom, $S_{\beta} = 0.68653$ with 4 degrees of freedom, and $S_{\alpha\beta} = 0.35762$ with 12 degrees of freedom. The corresponding test statistics values concerning σ^2_{α}, σ^2_{β}, and $\sigma^2_{\alpha\beta}$ are given in Table 11.9.

TABLE 11.9
Exact Test Statistics Values

Source	DF	F	p-Value
A	3	8.780 [formula(11.135)]	0.0024
B	4	5.759 [formula (11.136)]	0.0080
$A * B$	12	0.982 [formula (11.116)]	0.4852

The test for $H_0 : \sigma^2_{\alpha\beta} = 0$ is not significant, but the tests for $H_0 : \sigma^2_{\alpha} = 0$ and $H_0 : \sigma^2_{\beta} = 0$ are significant. This agrees to a certain extent with the results from Table 11.8. Note that the F-statistic value for $A * B$, namely 0.982, is identical to the value of the Type III F-ratio in (11.117), which can be obtained directly from the main ANOVA table for a fixed-effects two-way model.

11.5 Exact Tests for Random Higher-Order Models

The exact testing procedure outlined in Section 11.4.2 can be easily extended to higher-order random models. This extension applies to any unbalanced random model provided that the data contain no empty cells and that the imbalance occurs only in the last stage of the associated design. Such a model can be written as

$$Y_\theta = \sum_{i=0}^{\nu} g_{\theta_i(\bar{\theta}_i)} + \epsilon_\theta, \tag{11.138}$$

where $\theta = \{k_1, k_2, \ldots, k_s\}$ is a complete set of subscripts that identify a typical response Y. The ith effect in the model is denoted by $g_{\theta_i(\bar{\theta}_i)}$, where $\bar{\theta}_i$ and θ_i denote the corresponding sets of rightmost and nonrightmost bracket subscripts ($i = 0, 1, \ldots, \nu$), and ϵ_θ is the experimental error term. This notation is the same as the one used in Chapter 8 for balanced data (see Section 8.2). The only difference here is that the range of subscript k_s is not constant as it depends on $k_1, k_2, \ldots, k_{s-1}$, which have constant ranges, that is,

$$k_j = \begin{cases} 1, 2, \ldots, a_j & \text{for } j = 1, 2, \ldots, s-1 \\ 1, 2, \ldots, n_\zeta & \text{for } j = s, \end{cases}$$

where $\zeta = \{k_1, k_2, \ldots, k_{s-1}\}$. For example, we can have the three-way model,

$$Y_{ijkl} = \mu + \alpha_{(i)} + \beta_{(j)} + \gamma_{(k)} + (\alpha\beta)_{(ij)} + (\alpha\gamma)_{(ik)} + (\beta\gamma)_{(jk)} + (\alpha\beta\gamma)_{(ijk)} + \epsilon_{ijk(l)},$$

where $i = 1, 2, 3$, $j = 1, 2, 3, 4$, $k = 1, 2, 3$, and $l = 1, 2, \ldots, n_{ijk}$.

In general, it is assumed that

$$N > 2 \prod_{i=1}^{s-1} a_i - 1, \tag{11.139}$$

where $N = \sum_\zeta n_\zeta$ is the total number of observations, and the summation, $\sum_\zeta$, extends over all $(s-1)$-tuples of the form $(k_1, k_2, \ldots, k_{s-1})$. Condition (11.139) is a generalization of condition (11.126). By averaging Y_θ over k_s, we get

$$\bar{Y}_\zeta = \sum_{i=0}^{\nu} g_{\theta_i(\bar{\theta}_i)} + \bar{\epsilon}_\zeta, \tag{11.140}$$

where $\bar{Y}_\zeta = \frac{1}{n_\zeta}\sum_{k_s=1}^{n_\zeta} Y_\theta$, $\bar{\epsilon}_\zeta = \frac{1}{n_\zeta}\sum_{k_s=1}^{n_\zeta} \epsilon_\theta$. Model (11.140) is considered balanced with one observation, namely $\bar{Y}_\zeta$, at the $(s-1)$-tuple, ζ.

The model in (11.138) can be written in matrix form as

$$Y = \sum_{i=0}^{\nu} H_i \beta_i + \epsilon,$$

where

H_i is a matrix of zeros and ones of order $N \times e_i$

β_i is a vector consisting of the elements of $g_{\theta_i(\bar{\theta}_i)}$ $(i = 0, 1, \ldots, \nu)$

We assume that $\beta_1, \beta_2, \ldots, \beta_\nu$, and ϵ are mutually independent such that $\beta_i \sim N\left(\mathbf{0}, \sigma_i^2 I_{e_i}\right)$ for $i = 1, 2, \ldots, \nu$, and $\epsilon \sim N\left(\mathbf{0}, \sigma_\epsilon^2 I_N\right)$.

Khuri (1990) used the above setup to derive exact F-tests concerning the hypotheses, $H_0 : \sigma_i^2 = 0$ $(i = 1, 2, \ldots, \nu)$, by combining the use of balanced models properties, as applied to model (11.140), with an extension of the methodology described in Section 11.4.2.2 for the two-way model. We shall not provide here details of this extension. The interested reader is referred to Khuri (1990) for a more thorough discussion of this topic [see also Khuri, Mathew, and Sinha (1998, Chapter 5)].

11.6 Inference Concerning the Mixed Two-Way Model (Gallo and Khuri, 1990)

Let us again consider the two-way model in (11.84) under the assumption that $\alpha_{(i)}$ is fixed, but $\beta_{(j)}$, $(\alpha\beta)_{(ij)}$, and $\epsilon_{ij(k)}$ remain independently distributed random variables such that $\beta_{(j)} \sim N\left(0, \sigma_\beta^2\right)$, $(\alpha\beta)_{(ij)} \sim N(0, \sigma_{\alpha\beta}^2)$, $\epsilon_{ij(k)} \sim N(0, \sigma_\epsilon^2)$, $i = 1, 2, \ldots, a$; $j = 1, 2, \ldots, b$; $k = 1, 2, \ldots, n_{ij}$. The matrix form of this model is given in (11.120). In this case, the variance–covariance matrix of Y is of the form

$$\Sigma = \sigma_\beta^2 H_2 H_2' + \sigma_{\alpha\beta}^2 H_3 H_3' + \sigma_\epsilon^2 I_{n_{..}}. \tag{11.141}$$

The purpose of this section is to derive exact tests concerning the variance components, σ_β^2 and $\sigma_{\alpha\beta}^2$, and estimable linear functions of the fixed effects. These tests were initially developed by Gallo and Khuri (1990) using an approach similar to the one described in Section 11.4.2.2.

11.6.1 Exact Tests Concerning σ_β^2 and $\sigma_{\alpha\beta}^2$

As before, let $\bar{X} = (X_{11}, X_{12}, \ldots, X_{ab})'$ denote the vector of cell means, where $X_{ij} = \bar{Y}_{ij.}$ $(i = 1, 2, \ldots, a; j = 1, 2, \ldots, b)$. The mean of $\bar{X}$ is

$$E(\bar{X}) = \mu \mathbf{1}_{ab} + (I_a \otimes \mathbf{1}_b)\boldsymbol{\alpha}, \tag{11.142}$$

and its variance–covariance matrix is given by

$$\text{Var}(\bar{X}) = \sigma^2_\beta A_2 + \sigma^2_{\alpha\beta} I_{ab} + \sigma^2_\epsilon \tilde{K}, \tag{11.143}$$

where A_2 and $\tilde{K}$ are the same as in (11.89). Since $E(\bar{X})$ contains the fixed parameter vector $\boldsymbol{\alpha}$, a linear transformation is first needed to eliminate $\boldsymbol{\alpha}$ in order to obtain exact tests for the variance components. This is accomplished by considering the following transformation:

$$Z^* = T_1 \bar{X}, \tag{11.144}$$

where $T_1 = I_a \otimes T_2$ and T_2 is a matrix of order $(b-1) \times b$ defined as

$$T_2 = \begin{bmatrix} \frac{1}{\sqrt{2}} & -\frac{1}{\sqrt{2}} & 0 & \cdots & 0 \\ \frac{1}{\sqrt{6}} & \frac{1}{\sqrt{6}} & -\frac{2}{\sqrt{6}} & \cdots & 0 \\ \cdot & \cdot & \cdot & \cdots & \cdot \\ \cdot & \cdot & \cdot & \cdots & \cdot \\ \cdot & \cdot & \cdot & \cdots & \cdot \\ \frac{1}{\sqrt{b(b-1)}} & \frac{1}{\sqrt{b(b-1)}} & \cdots & \cdots & -\frac{b-1}{\sqrt{b(b-1)}} \end{bmatrix}.$$

The matrix T_1 has $a(b-1)$ rows and ab columns. Its rows define orthogonal contrasts in the elements of $\bar{X}$. Hence, $E(Z^*) = \mathbf{0}$, and the variance–covariance matrix of Z^* is

$$\begin{aligned} \text{Var}(Z^*) &= T_1 \text{Var}(\bar{X}) T_1' \\ &= \sigma^2_\beta T_1 A_2 T_1' + \sigma^2_{\alpha\beta} T_1 T_1' + \sigma^2_\epsilon T_1 \tilde{K} T_1' \\ &= \sigma^2_\beta J_a \otimes I_{b-1} + \sigma^2_{\alpha\beta} I_{a(b-1)} + \sigma^2_\epsilon T_1 \tilde{K} T_1', \end{aligned} \tag{11.145}$$

since $T_2 T_2' = I_{b-1}$, and hence, $T_1 A_2 T_1' = (I_a \otimes T_2)(J_a \otimes I_b)\left(I_a \otimes T_2'\right) = J_a \otimes T_2 T_2' = J_a \otimes I_{b-1}$. In addition, $T_1 T_1' = I_a \otimes T_2 T_2' = I_a \otimes I_{b-1} = I_{a(b-1)}$. Note that the matrix $J_a \otimes I_{b-1}$ has eigenvalues a and 0 with multiplicities $b-1$ and $(a-1)(b-1)$, respectively. Since this matrix is symmetric, there exists an orthogonal matrix, P^*, of order $(ab-a) \times (ab-a)$ such that $P^*(J_a \otimes I_{b-1})P^{*'} = D^*$, where D^* is a diagonal matrix whose diagonal elements are the aforementioned eigenvalues of $J_a \otimes I_{b-1}$.

In the remainder of this section, the development of the exact tests concerning σ^2_β and $\sigma^2_{\alpha\beta}$ will be similar to the one used in Section 11.4.2.2.

Let $U^* = P^* Z^*$. Then, $E(U^*) = \mathbf{0}$, and by using (11.145), we get

$$\begin{aligned} \text{Var}(U^*) &= P^*(\sigma^2_\beta J_a \otimes I_{b-1})P^{*'} + \sigma^2_{\alpha\beta} P^* P^{*'} + \sigma^2_\epsilon P^* T_1 \tilde{K} T_1' P^{*'} \\ &= \sigma^2_\beta D^* + \sigma^2_{\alpha\beta} I_{a(b-1)} + \sigma^2_\epsilon L^*, \end{aligned}$$

where $L^* = P^* T_1 \tilde{K} T_1' P^{*'}$. Thus, $\text{Var}(U^*)$ can be written as

$$\text{Var}(U^*) = \text{diag}\left[\left(a\,\sigma^2_{\beta} + \sigma^2_{\alpha\beta}\right) I_{b-1}, \sigma^2_{\alpha\beta}\, I_{(a-1)(b-1)}\right] + \sigma^2_{\epsilon}\, L^*. \qquad (11.146)$$

Let us now consider the matrix R in (11.125) (recall that $SS_E = Y'RY$). Since R is symmetric and idempotent of rank $n_{..} - ab$, it can be decomposed as

$$R = C_* \Lambda_* C_*',$$

where C_* is an orthogonal matrix and Λ_* is a diagonal matrix whose eigenvalues are 1 of multiplicity $n_{..} - ab$ and 0 of multiplicity ab. Assuming that $n_{..}$ satisfies the condition,

$$n_{..} > 2\,ab - a, \qquad (11.147)$$

the matrix Λ_* can be partitioned as

$$\Lambda_* = \text{diag}(I_{a(b-1)}, I_{n_{..}-2ab+a}, \mathbf{0}),$$

where $\mathbf{0}$ is a zero matrix of order $ab \times ab$. Likewise, the matrix C_* is partitioned as

$$C_* = [C_{*1} : C_{*2} : C_{*3}],$$

where C_{*1}, C_{*2}, C_{*3} are matrices of orders $n_{..} \times [a(b-1)]$, $n_{..} \times (n_{..} - 2ab + a)$, and $n_{..} \times ab$, respectively. The columns of $[C_{*1} : C_{*2}]$ are orthonormal eigenvectors of R corresponding to the eigenvalue 1, whereas those of C_{*3} are orthonormal eigenvectors corresponding to the eigenvalue 0. Note that, as in Section 11.4.2.2, the choice of C_{*1} is not unique. We can then write R as

$$R = C_{*1} C_{*1}' + C_{*2} C_{*2}',$$

and, consequently, the error sum of squares is partitioned as

$$SS_E = SS^*_{E1} + SS^*_{E2}, \qquad (11.148)$$

where $SS^*_{E1} = Y' C_{*1} C_{*1}' Y$ and $SS^*_{E2} = Y' C_{*2} C_{*2}' Y$.

Let us now define the random vector, ω^*, as

$$\omega^* = U^* + \left(\lambda^*_{\max} I_{a(b-1)} - L^*\right)^{1/2} C_{*1}' Y, \qquad (11.149)$$

$\lambda^*_{\max}$ is the largest eigenvalue of L^*.

The following lemma is analogous to Lemma 11.7 and its proof is therefore similar.

Lemma 11.9 Let ω^* in (11.149) be partitioned as $\omega^* = \left(\omega^{*'}_{\beta}, \omega^{*'}_{\alpha\beta}\right)'$, where ω^*_{β} consists of the first $b-1$ elements of ω^* and $\omega^*_{\alpha\beta}$ consists of the remaining $(a-1)(b-1)$ elements. Then,

(i) ω^*_{β} and $\omega^*_{\alpha\beta}$ are independent and normally distributed with zero means and variance–covariance matrices given by

$$\text{Var}\left(\omega^*_{\beta}\right) = (a\,\sigma^2_{\beta} + \sigma^2_{\alpha\beta} + \lambda^*_{\max}\,\sigma^2_{\epsilon})\boldsymbol{I}_{b-1}$$
$$\text{Var}\left(\omega^*_{\alpha\beta}\right) = \left(\sigma^2_{\alpha\beta} + \lambda^*_{\max}\,\sigma^2_{\epsilon}\right)\boldsymbol{I}_{(a-1)(b-1)}$$

(ii) ω^*_{β} and $\omega^*_{\alpha\beta}$ are independent of SS^*_{E2} in (11.148).

From Lemma 11.9 we conclude that $S^*_{\beta} = \omega^{*'}_{\beta}\,\omega^*_{\beta}$ and $S^*_{\alpha\beta} = \omega^{*'}_{\alpha\beta}\,\omega^*_{\alpha\beta}$ are independently distributed such that

$$S^*_{\beta}\Big/\left(a\,\sigma^2_{\beta} + \sigma^2_{\alpha\beta} + \lambda^*_{\max}\,\sigma^2_{\epsilon}\right) \sim \chi^2_{b-1}$$
$$S^*_{\alpha\beta}\Big/\left(\sigma^2_{\alpha\beta} + \lambda^*_{\max}\,\sigma^2_{\epsilon}\right) \sim \chi^2_{(a-1)(b-1)}$$

Furthermore, S^*_{β} and $S^*_{\alpha\beta}$ are independent of SS^*_{E2}, which is distributed as $\sigma^2_{\epsilon}\,\chi^2_{n_{..}-2ab+a}$. Therefore, for testing $H_0 : \sigma^2_{\beta} = 0$, we can use the test statistic,

$$F = \frac{S^*_{\beta}/(b-1)}{S^*_{\alpha\beta}/[(a-1)(b-1)]},$$

which has the F-distribution with $b-1$ and $(a-1)(b-1)$ degrees of freedom under H_0. Similarly, to test $H_0 : \sigma^2_{\alpha\beta} = 0$, we can use the test statistic

$$F = \frac{S^*_{\alpha\beta}/[(a-1)(b-1)]}{\lambda^*_{\max}\,SS^*_{E2}/(n_{..} - 2ab + a)}, \qquad (11.150)$$

which has the F-distribution with $(a-1)(b-1)$ and $n_{..} - 2ab + a$ degrees of freedom under H_0.

An alternative test statistic for testing $H_0 : \sigma^2_{\alpha\beta} = 0$ is given by the Type III F-ratio, $F(\alpha\beta \mid \mu, \alpha, \beta)$ in (11.117) (see Exercise 11.7). The advantage of this test over the one in (11.150) is that it has more degrees of freedom for the denominator.

11.6.2 An Exact Test for the Fixed Effects

In this section, we give an exact test for the hypothesis $H_0 : \boldsymbol{A\alpha} = \boldsymbol{a}_0$, where $\boldsymbol{A}$ is a matrix of order $t \times a$ and rank t ($\leq a-1$) such that the t elements of $\boldsymbol{A\alpha}$ are linearly independent contrasts among the means, $\mu + \alpha_{(i)}$ ($i = 1, 2, \ldots, a$), of the fixed factor, and $\boldsymbol{a}_0$ is a constant vector.

Let us again consider the random vector, $\tilde{\boldsymbol{U}} = \tilde{\boldsymbol{Q}}_1\bar{\boldsymbol{X}}$, which was defined in Section 11.4.1, where $\tilde{\boldsymbol{Q}}_1 = \left[\tilde{\boldsymbol{Q}}'_{11} : \tilde{\boldsymbol{Q}}'_{12} : \tilde{\boldsymbol{Q}}'_{13}\right]'$. We recall that $\boldsymbol{Z}_{\alpha} = \tilde{\boldsymbol{Q}}_{11}\bar{\boldsymbol{X}}$,

$Z_{\alpha\beta} = \tilde{Q}_{13}\bar{X}$. Let $\boldsymbol{\Phi}$ be defined as $\boldsymbol{\Phi} = \left(Z'_{\alpha}, Z'_{\alpha\beta}\right)'$. Then, from (11.142), the mean of $\boldsymbol{\Phi}$ is

$$E(\boldsymbol{\Phi}) = \begin{bmatrix} \tilde{Q}_{11} \\ \tilde{Q}_{13} \end{bmatrix} [\mu \mathbf{1}_{ab} + (I_a \otimes \mathbf{1}_b)\boldsymbol{\alpha}]$$
$$= \begin{bmatrix} \tilde{Q}_{11} \\ \tilde{Q}_{13} \end{bmatrix} (I_a \otimes \mathbf{1}_b)\boldsymbol{\alpha}$$
$$= \begin{bmatrix} \tilde{Q}_{11}(I_a \otimes \mathbf{1}_b)\boldsymbol{\alpha} \\ \mathbf{0} \end{bmatrix}, \tag{11.151}$$

since $\tilde{Q}_{11}\mathbf{1}_{ab} = \mathbf{0}$, $\tilde{Q}_{13}\mathbf{1}_{ab} = \mathbf{0}$, and $\tilde{Q}_{13}(I_a \times \mathbf{1}_b) = \mathbf{0}$ by virtue of $\tilde{Q}_{13}A_1 = \mathbf{0}$, where $A_1 = I_a \otimes J_b$ (see Appendix 11.A). The variance–covariance matrix of $\boldsymbol{\Phi}$ is given by

$$\text{Var}(\boldsymbol{\Phi}) = \begin{bmatrix} \tilde{Q}_{11} \\ \tilde{Q}_{13} \end{bmatrix} \left(\sigma^2_{\beta} A_2 + \sigma^2_{\alpha\beta} I_{ab} + \sigma^2_{\epsilon} \tilde{K}\right) \left[\tilde{Q}'_{11} : \tilde{Q}'_{13}\right]$$
$$= \sigma^2_{\alpha\beta} I_{b(a-1)} + \sigma^2_{\epsilon} L^o, \tag{11.152}$$

by the fact that $\tilde{Q}_{11}A_2 = \mathbf{0}$, $\tilde{Q}_{13}A_2 = \mathbf{0}$, and $\left[\tilde{Q}'_{11} : \tilde{Q}'_{13}\right]' \left[\tilde{Q}'_{11} : \tilde{Q}'_{13}\right] = I_{b(a-1)}$ (see Appendix 11.A), where L^o is defined by

$$L^o = \begin{bmatrix} \tilde{Q}_{11} \\ \tilde{Q}_{13} \end{bmatrix} \tilde{K} \left[\tilde{Q}'_{11} : \tilde{Q}'_{13}\right]. \tag{11.153}$$

In addition, the idempotent matrix R of rank $n_{..} - ab$, which was used earlier to define the error sum of squares in (11.122), can be decomposed as in (11.129), but with different C_1 and C_2 matrices. Here, R is decomposed as

$$R = \tilde{C}_1\tilde{C}'_1 + \tilde{C}_2\tilde{C}'_2, \tag{11.154}$$

where $\tilde{C}_1$ and $\tilde{C}_2$ are of orders $n_{..} \times [b(a-1)]$, and $n_{..} \times (n_{..} - 2ab + b)$, respectively, such that the columns of $[\tilde{C}_1 : \tilde{C}_2]$ are orthonormal eigenvectors of R corresponding to the eigenvalue 1. Here, again the choice of $\tilde{C}_1$ is not unique. Note that this partitioning is possible provided that

$$n_{..} > 2ab - b \tag{11.155}$$

In order to satisfy both (11.147) and (11.155), $n_{..}$ must satisfy the condition

$$n_{..} > \max(2ab - a, 2ab - b).$$

Now, let us define the random vector $\boldsymbol{\Psi}$ as

$$\boldsymbol{\Psi} = \boldsymbol{\Phi} + \left(\lambda^o_{\max} I_{b(a-1)} - L^o\right)^{1/2} \tilde{C}'_1 Y, \tag{11.156}$$

where $\lambda^o_{\max}$ is the largest eigenvalue of $\boldsymbol{L}^o$. It is easy to verify that $E\left(\tilde{\boldsymbol{C}}_1'\boldsymbol{Y}\right) = \boldsymbol{0}$, $\boldsymbol{\Phi}$ and $\tilde{\boldsymbol{C}}_1'\boldsymbol{Y}$ are independent, $\boldsymbol{\Psi}$ is normally distributed, and that

$$\text{Var}\left(\tilde{\boldsymbol{C}}_1'\boldsymbol{Y}\right) = \sigma^2_{\epsilon}\,\boldsymbol{I}_{b(a-1)}. \tag{11.157}$$

It follows from (11.151), (11.152), (11.156), and (11.157) that $\boldsymbol{\Psi}$ has the normal distribution with mean

$$E(\boldsymbol{\Psi}) = \begin{bmatrix} \tilde{\boldsymbol{Q}}_{11}(\boldsymbol{I}_a \otimes \boldsymbol{1}_b)\boldsymbol{\alpha} \\ \boldsymbol{0} \end{bmatrix},$$

and a variance–covariance matrix given by

$$\begin{aligned} \text{Var}(\boldsymbol{\Psi}) &= \sigma^2_{\alpha\beta}\,\boldsymbol{I}_{b(a-1)} + \sigma^2_{\epsilon}\,\boldsymbol{L}^o + \left(\lambda^o_{\max}\,\boldsymbol{I}_{b(a-1)} - \boldsymbol{L}^o\right)\sigma^2_{\epsilon} \\ &= \left(\sigma^2_{\alpha\beta} + \lambda^o_{\max}\,\sigma^2_{\epsilon}\right)\boldsymbol{I}_{b(a-1)}. \end{aligned} \tag{11.158}$$

We can therefore represent the observable random vector, $\boldsymbol{\Psi}$, by the linear model

$$\boldsymbol{\Psi} = \boldsymbol{X}^*\boldsymbol{\alpha} + \boldsymbol{\epsilon}^*. \tag{11.159}$$

where $\boldsymbol{X}^*$ is the $[b(a-1)] \times a$ matrix

$$\boldsymbol{X}^* = \begin{bmatrix} \tilde{\boldsymbol{Q}}_{11}(\boldsymbol{I}_a \times \boldsymbol{1}_b) \\ \boldsymbol{0} \end{bmatrix},$$

and $\boldsymbol{\epsilon}^*$ is distributed as $N\left[\boldsymbol{0}, \left(\sigma^2_{\alpha\beta} + \lambda^o_{\max}\sigma^2_{\epsilon}\right)\boldsymbol{I}_{b(a-1)}\right]$. The model in (11.159) satisfies the usual assumptions of normality, independence, and equality of error variances.

Note that the matrix $\boldsymbol{X}^*$ is of rank $a-1$ since

$$\begin{aligned} \boldsymbol{X}^*\boldsymbol{X}^{*'} &= \begin{bmatrix} \tilde{\boldsymbol{Q}}_{11} \\ \boldsymbol{0} \end{bmatrix} \boldsymbol{A}_1 \left[\tilde{\boldsymbol{Q}}_{11}' : \boldsymbol{0}'\right] \\ &= \text{diag}(b\,\boldsymbol{I}_{a-1}, \boldsymbol{0}), \end{aligned} \tag{11.160}$$

which is of rank $a-1$ [formula (11.160) is true because $\tilde{\boldsymbol{Q}}_{11}\boldsymbol{A}_1\tilde{\boldsymbol{Q}}_{11}' = b\,\boldsymbol{I}_{a-1}$, where $\boldsymbol{A}_1 = \boldsymbol{I}_a \otimes \boldsymbol{J}_b$ (see Appendix 11.A)]. Thus, the submatrix, $\tilde{\boldsymbol{Q}}_{11}(\boldsymbol{I}_a \otimes \boldsymbol{1}_b)$, of $\boldsymbol{X}^*$ is of full row rank equal to $a-1$. Furthermore, the nonzero elements of $\boldsymbol{X}^*\boldsymbol{\alpha}$, namely the elements of $\tilde{\boldsymbol{Q}}_{11}(\boldsymbol{I}_a \otimes \boldsymbol{1}_b)\boldsymbol{\alpha}$, are linearly independent contrasts among the elements of $\boldsymbol{\alpha}$ due to the fact that $\tilde{\boldsymbol{Q}}_{11}(\boldsymbol{I}_a \otimes \boldsymbol{1}_b)\boldsymbol{1}_a = \tilde{\boldsymbol{Q}}_{11}\boldsymbol{1}_{ab} = \boldsymbol{0}$ (see Appendix 11.A). Since the number of such contrasts is $a-1$, they must form a basis for all contrasts among the elements of $\boldsymbol{\alpha}$, or equivalently, among the means $\mu + \alpha_{(i)}$ $(i = 1, 2, \ldots, a)$ of the fixed factor.

Now since the rows of A for the hypothesis $H_0 : A\alpha = a_0$ are such that the elements of $A\alpha$ are t linearly independent contrasts among the means, $\mu + \alpha_{(i)}$ $(i = 1, 2, \ldots, a)$, the rows of A must belong to the row space of X^* in model (11.159). Hence, $A\alpha$ is an estimable linear function of α under model (11.159). It follows that a test statistic for the hypothesis $H_0 : A\alpha = a_0$ is given by

$$F = \frac{(A\hat{\alpha} - a_0)'[A(X^{*'}X^*)^- A']^{-1}(A\hat{\alpha} - a_0)}{t\, MS_E^*}, \tag{11.161}$$

where

$$MS_E^* = \Psi'[I_{b(a-1)} - X^*(X^{*'}X^*)^- X^{*'}]\Psi/[(a-1)(b-1)],$$

and $\hat{\alpha} = (X^{*'}X^*)^- X^{*'}\Psi$ (see Section 7.4.2). Under H_0, F has the F-distribution with t and $(a-1)(b-1)$ degrees of freedom. In particular, if A and a_0 are chosen so that

$$A\alpha = (\alpha_{(1)} - \alpha_{(2)}, \alpha_{(1)} - \alpha_{(3)}, \ldots, \alpha_{(1)} - \alpha_{(a)})',$$

and $a_0 = 0$, then F serves as a test statistic for the hypothesis

$$H_0 : \alpha_{(1)} = \alpha_{(2)} = \ldots = \alpha_{(a)}.$$

Example 11.7 This example is taken from Gallo and Khuri (1990) and deals with the average daily gains (in pounds) of 67 steers from 9 sires (factor B) and 3 ages of dam (factor A). The data are given in Table 11.10. The age-of-dam effect, $\alpha_{(i)}$, is fixed and the sire effect, $\beta_{(j)}$, is random.

(i) Tests concerning the variance components

These tests are based on formula (11.149). Here, $a = 3, b = 9, \lambda^*_{max} = 1$, and $U^* = P^*Z^*$, where P^* is an orthogonal matrix of order 24×24 whose rows are orthonormal eigenvectors of $J_3 \otimes I_8$, and $Z^* = T_1\bar{X}, T_1 = I_3 \otimes T_2$ (T_2 is the 8×9 matrix defined in Section 11.6.1). Also, $L^* = P^*T_1\tilde{K}T_1'P^{*'}$, where $\tilde{K}$ is a diagonal matrix with diagonal elements equal to the reciprocals of the 27 cell frequencies, and C_{*1} is the matrix that consists of the first $a(b-1) = 24$ columns of the 67×67 matrix, $C_* = [C_{*1} : C_{*2} : C_{*3}]$, of orthonormal eigenvectors of R [recall that $SS_E = Y'RY$ in (11.122)].

Using formula (11.149) and the data set in Table 11.10, the vector ω^* is computed and then partitioned into ω^*_β and $\omega^*_{\alpha\beta}$ of orders 8×1 and 16×1, respectively, as in Lemma 11.9. It follows that the test statistic for testing $H_0 : \sigma^2_\beta = 0$, namely,

$$F = \frac{S^*_\beta/(b-1)}{S^*_{\alpha\beta}/[(a-1)(b-1)]},$$

has the value 0.871 with 8 and 16 degrees of freedom (p-value = 0.559). This gives no indication of a significant sire effect. Also, from (11.150), the value of the test statistic for testing $H_0 : \sigma^2_{\alpha\beta} = 0$ is $F = 1.3398$ with 16 and 16 degrees

TABLE 11.10
Average Daily Gains (in Pounds) for 67 Steers

	Age of Dam (Years)				Age of Dam (Years)		
Sire	**3**	**4**	**5-Up**	**Sire**	**3**	**4**	**5-Up**
1	2.24	2.41	2.58	6	2.30	3.00	2.25
	2.65	2.25	2.67			2.49	2.49
			2.71				2.02
			2.47				2.31
2	2.15	2.29	1.97	7	2.57	2.64	2.37
		2.26	2.14		2.37		2.22
			2.44				1.90
			2.52				2.61
			1.72				2.13
			2.75				
3	2.38	2.46	2.29	8	2.16	2.45	1.44
			2.30		2.33		1.72
			2.94		2.52		2.17
4	2.50	2.44	2.54	9	2.68	2.43	2.66
	2.44	2.15	2.74			2.36	2.46
			2.50			2.44	2.52
			2.54				2.42
5	2.65	2.52	2.79				
		2.67	2.33				
			2.67				
			2.69				

Source: Reprinted from Gallo, J. and Khuri, A.I., *Biometrics*, 46, 1087, 1990. With permission.

of freedom (p-value = 0.283). Hence, no significant interaction effect can be detected.

(ii) Testing the fixed effects

Let us now consider the hypothesis, $H_0 : \alpha_{(1)} = \alpha_{(2)} = \alpha_{(3)}$. The test for this hypothesis depends on formula (11.156). In this case, $\boldsymbol{\Phi} = \left(\boldsymbol{Z}'_{\alpha}, \boldsymbol{Z}'_{\alpha\beta}\right)'$, where $\boldsymbol{Z}_{\alpha} = \tilde{\boldsymbol{Q}}_{11}\bar{\boldsymbol{X}}$, $\boldsymbol{Z}_{\alpha\beta} = \tilde{\boldsymbol{Q}}_{13}\bar{\boldsymbol{X}}$. Furthermore, the matrix $\boldsymbol{L}^o$ is described in (11.153) and its largest eigenvalue, $\lambda^o_{\max}$, is equal to 1; the matrix $\tilde{\boldsymbol{C}}_1$ consists of the first $b(a-1) = 18$ columns of the matrix $[\tilde{\boldsymbol{C}}_1 : \tilde{\boldsymbol{C}}_2]$ of orthonormal eigenvectors of $\boldsymbol{R}$ for the eigenvalue 1 [see (11.154)]. Using these quantities in (11.156), we get the value of the vector $\boldsymbol{\Psi}$, which is then used in model (11.159) to obtain $\hat{\boldsymbol{\alpha}} = (\boldsymbol{X}^{*'}\boldsymbol{X}^*)^{-}\boldsymbol{X}^{*'}\boldsymbol{\Psi}$, where $\boldsymbol{X}^*$ is the 18×3 matrix, $\boldsymbol{X}^* = \left[\tilde{\boldsymbol{Q}}'_{11} : \boldsymbol{0}'\right]'(\boldsymbol{I}_3 \otimes \boldsymbol{1}_9)$. Finally, from (11.161) we get the test statistic value, $F = 0.3132$, with 2 and 16 degrees of freedom (p-value = 0.735). Hence, there is no significant effect due to the age of dam.

11.7 Inference Concerning the Random Two-Fold Nested Model

In this section, we address the analysis of a random two-fold nested model of the form

$$Y_{ijk} = \mu + \alpha_{(i)} + \beta_{i(j)} + \epsilon_{ij(k)},$$
$$i = 1, 2, \ldots, a;\ j = 1, 2, \ldots, b_i;\ k = 1, 2, \ldots, n_{ij}, \tag{11.162}$$

where

$\alpha_{(i)}$ and $\beta_{i(j)}$ are random effects associated with the nesting factor, A, and the nested factor, B, respectively

$\epsilon_{ij(k)}$ is a random error term

It is assumed that $\alpha_{(i)}$, $\beta_{i(j)}$, and $\epsilon_{ij(k)}$ are independently distributed as $N(0, \sigma^2_\alpha)$, $N\left(0, \sigma^2_{\beta(\alpha)}\right)$, and $N(0, \sigma^2_\epsilon)$, respectively. Of interest here is the testing of hypotheses concerning σ^2_α and $\sigma^2_{\beta(\alpha)}$.

Model (11.162) can be written in matrix form as

$$\boldsymbol{Y} = \mu \boldsymbol{1}_{n_{..}} + \left[\oplus_{i=1}^{a} \boldsymbol{1}_{n_{i.}}\right] \boldsymbol{\alpha} + \left[\oplus_{i=1}^{a} \oplus_{j=1}^{b_i} \boldsymbol{1}_{n_{ij}}\right] \boldsymbol{\beta}(\boldsymbol{\alpha}) + \boldsymbol{\epsilon}, \tag{11.163}$$

where $\boldsymbol{\alpha} = (\alpha_{(1)}, \alpha_{(2)}, \ldots, \alpha_{(a)})'$, $\boldsymbol{\beta}(\boldsymbol{\alpha}) = (\beta_{1(1)}, \beta_{1(2)}, \ldots, \beta_{1(b_1)}, \ldots, \beta_{a(1)}, \beta_{a(2)}, \ldots, \beta_{a(b_a)})'$, $n_{i.} = \sum_{j=1}^{b_i} n_{ij}$, $i = 1, 2, \ldots, a$; $n_{..} = \sum_{i=1}^{a} \sum_{j=1}^{b_i} n_{ij}$. The variance–covariance matrix, $\boldsymbol{\Sigma}$, of $\boldsymbol{Y}$ is

$$\boldsymbol{\Sigma} = \sigma^2_\alpha \left[\oplus_{i=1}^{a} \boldsymbol{J}_{n_{i.}}\right] + \sigma^2_{\beta(\alpha)} \left[\oplus_{i=1}^{a} \oplus_{j=1}^{b_i} \boldsymbol{J}_{n_{ij}}\right] + \sigma^2_\epsilon \boldsymbol{I}_{n_{..}}.$$

An exact F-test concerning the hypothesis $H_0 : \sigma^2_{\beta(\alpha)} = 0$ can be easily obtained as it is equal to the Type III F-ratio for the nested factor, namely,

$$F = \frac{R(\beta(\alpha) \mid \mu, \alpha)/(b_{.} - a)}{SS_E/(n_{..} - b_{.})}, \tag{11.164}$$

where $b_{.} = \sum_{i=1}^{a} b_i$, $R(\beta(\alpha) \mid \mu, \alpha)$ is the Type III sum of squares for factor B which is nested within factor A, and SS_E is the error sum of squares,

$$\begin{aligned} SS_E &= \sum_{i=1}^{a} \sum_{j=1}^{b_i} \sum_{k=1}^{n_{ij}} (Y_{ijk} - \bar{Y}_{ij.})^2 \\ &= \boldsymbol{Y}' \boldsymbol{\mathcal{R}} \boldsymbol{Y}, \end{aligned} \tag{11.165}$$

where $\bar{Y}_{ij.} = \frac{1}{n_{ij}} \sum_{k=1}^{n_{ij}} Y_{ijk}$, $\boldsymbol{\mathcal{R}}$ is the matrix

$$\boldsymbol{\mathcal{R}} = \boldsymbol{I}_{n_{..}} - \bigoplus_{i=1}^{a} \bigoplus_{j=1}^{b_i} \frac{1}{n_{ij}} \boldsymbol{J}_{n_{ij}}, \tag{11.166}$$

which is idempotent of rank $n_{..} - b_{.}$. It can be verified that $R(\beta(\alpha) \mid \mu, \alpha)$ is independent of SS_E, $SS_E/\sigma^2_\epsilon \sim \chi^2_{n_{..}-b_{.}}$, and under $H_0 : \sigma^2_{\beta(\alpha)} = 0$, $R(\beta(\alpha) \mid \mu, \alpha)/\sigma^2_\epsilon \sim \chi^2_{b_{.}-a}$. Hence, under H_0, the statistic in (11.164) has the F-distribution with $b_{.} - a$ and $n_{..} - b_{.}$ degrees of freedom. The test is significant at the α-level if $F \geq F_{\alpha, b_{.}-a, n_{..}-b_{.}}$. Note that this test is similar to the one used in Section 11.4.2.1 to test the interaction variance component, $\sigma^2_{\alpha\beta}$ [see formula (11.117)]. Hence, as was the case in (11.117), the F-test in (11.164) is also referred to as a Wald's test.

The test concerning $H_0 : \sigma^2_\alpha = 0$ is more involved. Tietjen (1974) used the conventional F-ratio,

$$F = \frac{\sum_{i=1}^{a} n_{i.}(\bar{Y}_{i..} - \bar{Y}_{...})^2/(a-1)}{\sum_{i=1}^{a}\sum_{j=1}^{b_i} n_{ij}\left(\bar{Y}_{ij.} - \bar{Y}_{i..}\right)^2/(b_{.} - a)}, \tag{11.167}$$

to test H_0. This test, however, does not have the exact F-distribution since neither the numerator nor the denominator in (11.167) are, in general, distributed as multiples of chi-squared variates, even under H_0. Furthermore, they are not necessarily independent. Note that if the data set is balanced, then (11.167) reduces to the usual F-test concerning σ^2_α, which has the exact F-distribution as was seen in Chapter 8. Cummings and Gaylor (1974) recommended an alternative approximate F-test that can be obtained by using a linear combination of mean squares for the error term and the nested factor in place of the denominator of the statistic in (11.167). The linear combination was chosen so that its expected value coincides with the expected value of the numerator when $\sigma^2_\alpha = 0$. Satterthwaites's approximation was then used to approximate the distribution of the denominator as a multiple of a chi-squared variate. Cummings and Gaylor (1974) also investigated the size of the resulting approximate F-test under certain unbalanced nested designs.

11.7.1 An Exact Test Concerning σ^2_α (Khuri, 1987)

The derivation of an exact F-test for $H_0 : \sigma^2_\alpha = 0$ is similar to the one used in Section 11.4.2.2 for testing the main effects' variance components for a random two-way model.

From (11.162) we get by averaging over k,

$$\bar{Y}_{ij.} = \mu + \alpha_{(i)} + \beta_{i(j)} + \bar{\epsilon}_{ij.}, \quad i = 1, 2, \ldots, a;\, j = 1, 2, \ldots, b_i, \tag{11.168}$$

where $\bar{\epsilon}_{ij.} = \frac{1}{n_{ij}}\sum_{k=1}^{n_{ij}} \epsilon_{ijk}$. We then have

$$\bar{\boldsymbol{Y}} = \mu \boldsymbol{1}_{b_{.}} + \boldsymbol{\mathcal{B}}_1 \boldsymbol{\alpha} + \boldsymbol{I}_{b_{.}} \boldsymbol{\beta}(\boldsymbol{\alpha}) + \bar{\boldsymbol{\epsilon}}, \tag{11.169}$$

where $\bar{\boldsymbol{Y}}$ and $\bar{\boldsymbol{\epsilon}}$ are vectors consisting of the $\bar{Y}_{ij.}$'s and $\bar{\epsilon}_{ij.}$'s, respectively, and $\boldsymbol{\mathcal{B}}_1 = \oplus_{i=1}^{a} \boldsymbol{1}_{b_i}$. The variance–covariance matrix of $\bar{\boldsymbol{Y}}$ is

$$\text{Var}(\bar{\boldsymbol{Y}}) = \sigma^2_\alpha \boldsymbol{\mathcal{A}}_1 + \sigma^2_{\beta(\alpha)} \boldsymbol{I}_{b_{.}} + \sigma^2_\epsilon \boldsymbol{\mathcal{K}}, \tag{11.170}$$

where $\mathcal{A}_1 = \mathcal{B}_1\mathcal{B}_1' = \oplus_{i=1}^{a} J_{b_i}$, and $\mathcal{K}$ is the diagonal matrix

$$\mathcal{K} = \text{diag}\left(\frac{1}{n_{11}}, \frac{1}{n_{12}}, \ldots, \frac{1}{n_{ab_a}}\right).$$

By the Spectral Decomposition Theorem, the matrix $\mathcal{R}$ in (11.166) can be expressed as

$$\mathcal{R} = \mathcal{C}\Upsilon\mathcal{C}', \tag{11.171}$$

where $\mathcal{C}$ is an orthogonal matrix of eigenvectors of $\mathcal{R}$ and Υ is a diagonal matrix of eigenvalues of $\mathcal{R}$. Since $\mathcal{R}$ is idempotent of rank $n_{..} - b_{.}$, Υ and $\mathcal{C}$ can be correspondingly partitioned as

$$\Upsilon = \text{diag}(\boldsymbol{I}_{b_{.}-1}, \boldsymbol{I}_{n_{..}-2b_{.}+1}, \boldsymbol{0})$$
$$\mathcal{C} = [\mathcal{C}_1 : \mathcal{C}_2 : \mathcal{C}_3],$$

where $\mathcal{C}_1$, $\mathcal{C}_2$, and $\mathcal{C}_3$ are of orders $n_{..} \times (b_{.} - 1)$, $n_{..} \times (n_{..} - 2b_{.} + 1)$, $n_{..} \times b_{.}$, respectively. The columns of $[\mathcal{C}_1 : \mathcal{C}_2]$ are orthonormal eigenvectors of $\mathcal{R}$ corresponding to the eigenvalue 1, and those of $\mathcal{C}_3$ correspond to the eigenvalue 0. This partitioning is possible under the assumption that

$$n_{..} > 2b_{.} - 1. \tag{11.172}$$

Note that the choice of $\mathcal{C}_1$ is not unique. Formulas (11.165) and (11.171) can then be written as

$$\mathcal{R} = \mathcal{C}_1\mathcal{C}_1' + \mathcal{C}_2\mathcal{C}_2' \tag{11.173}$$
$$SS_E = SS_{E1}^o + SS_{E2}^o, \tag{11.174}$$

where $SS_{E1}^o = \boldsymbol{Y}'\mathcal{C}_1\mathcal{C}_1'\boldsymbol{Y}$, $SS_{E2}^o = \boldsymbol{Y}'\mathcal{C}_2\mathcal{C}_2'\boldsymbol{Y}$.

Let us now consider the matrix $\mathcal{A}_1 = \oplus_{i=1}^{a} J_{b_i}$, which is of order $b_{.} \times b_{.}$ and rank a. Its nonzero eigenvalues are equal to $b_1, b_2, \ldots, b_a$. Then, there exists an orthogonal matrix, $\mathcal{P}$, of order $b_{.} \times b_{.}$ such that

$$\mathcal{P}\mathcal{A}_1\mathcal{P}' = \boldsymbol{\Lambda}_a, \tag{11.175}$$

where $\boldsymbol{\Lambda}_a = \text{diag}(b_1, b_2, \ldots, b_a, \boldsymbol{0})$ and $\boldsymbol{0}$ is a zero matrix of order $(b_{.} - a) \times (b_{.} - a)$. The first a rows of $\mathcal{P}$ are orthonormal eigenvectors of $\mathcal{A}_1$ corresponding to the eigenvalues $b_1, b_2, \ldots, b_a$. These eigenvectors should therefore be the same as the rows of the $a \times b_{.}$ matrix, $\mathcal{P}_1$, where

$$\mathcal{P}_1 = \bigoplus_{i=1}^{a}\left(\frac{1}{\sqrt{b_i}}\boldsymbol{1}_{b_i}'\right).$$

Let $\mathcal{P}_2$ be the $(b_. - a) \times b_.$ matrix consisting of the remaining $b_. - a$ rows of $\mathcal{P}$. Thus, $\mathcal{P} = [\mathcal{P}_1' : \mathcal{P}_2']'$. Let $\mathcal{U}_1 = \mathcal{P}\bar{Y}$. Then, from (11.169) and (11.170) we have $E(\mathcal{U}_1) = \mu\mathcal{P}\mathbf{1}_{b_.}$, and

$$\begin{aligned}\text{Var}(\mathcal{U}_1) &= \mathcal{P}\text{Var}(\bar{Y})\mathcal{P}' \\ &= \sigma^2_{\alpha}\,\boldsymbol{\Lambda}_a + \sigma^2_{\beta(\alpha)}\,\boldsymbol{I}_{b_.} + \sigma^2_{\epsilon}\,\mathcal{P}\mathcal{K}\mathcal{P}'.\end{aligned} \tag{11.176}$$

Note that the mean of $\mathcal{U}_1$ is not equal to zero, so another linear transformation is needed to get a zero mean.

There exists an orthogonal matrix, $\mathcal{Q}$, of order $b_. \times b_.$ such that the first row of $\mathcal{Q}\mathcal{P}$ is $\frac{1}{\sqrt{b_.}}\mathbf{1}'_{b_.}$ (see Appendix 11.B). Let $\mathcal{U}_2$ be defined as $\mathcal{U}_2 = \mathcal{Q}_1'\mathcal{U}_1$, where $\mathcal{Q}_1$ is such that $\mathcal{Q} = [\boldsymbol{e}_1 : \mathcal{Q}_1]'$. Here, $\boldsymbol{e}_1$ is a vector of order $b_. \times 1$ such that $\boldsymbol{e}_1 = (\boldsymbol{c}_1', \mathbf{0}')'$, where $\boldsymbol{c}_1' = \frac{1}{\sqrt{b_.}}(\sqrt{b_1}, \sqrt{b_2}, \ldots, \sqrt{b_a})$. Then, $E(\mathcal{U}_2) = \mathcal{Q}_1'E(\mathcal{U}_1) = \mu\mathcal{Q}_1'\mathcal{P}\mathbf{1}_{b_.} = \mathbf{0}$, since $\mathcal{Q}\mathcal{P}$ is orthogonal and its first row is $\frac{1}{\sqrt{b_.}}\mathbf{1}'_{b_.}$. In addition, from (11.176) we have

$$\begin{aligned}\text{Var}(\mathcal{U}_2) &= \mathcal{Q}_1'\text{Var}(\mathcal{U}_1)\mathcal{Q}_1 \\ &= \mathcal{Q}_1'\left[\sigma^2_{\alpha}\,\boldsymbol{\Lambda}_a + \sigma^2_{\beta(\alpha)}\,\boldsymbol{I}_{b_.} + \sigma^2_{\epsilon}\,\mathcal{P}\mathcal{K}\mathcal{P}'\right]\mathcal{Q}_1 \\ &= \sigma^2_{\alpha}\,\mathcal{Q}_1'\boldsymbol{\Lambda}_a\mathcal{Q}_1 + \sigma^2_{\beta(\alpha)}\,\boldsymbol{I}_{b_.-1} + \sigma^2_{\epsilon}\,\mathcal{Q}_1'\mathcal{P}\mathcal{K}\mathcal{P}'\mathcal{Q}_1.\end{aligned} \tag{11.177}$$

We now show that the matrix $\mathcal{Q}_1'\boldsymbol{\Lambda}_a\mathcal{Q}_1$ is of rank $a - 1$. For this purpose, let $\mathcal{Q}_1'$ be partitioned as $[\mathcal{Q}_{11}' : \mathcal{Q}_{12}']$, where $\mathcal{Q}_{11}'$ is $(b_. - 1) \times a$ and $\mathcal{Q}_{12}'$ is of order $(b_. - 1) \times (b_. - a)$. Then,

$$\mathcal{Q}_1'\boldsymbol{\Lambda}_a\mathcal{Q}_1 = \mathcal{Q}_{11}'\text{diag}(b_1, b_2, \ldots, b_a)\mathcal{Q}_{11}.$$

Hence, the rank of $\mathcal{Q}_1'\boldsymbol{\Lambda}_a\mathcal{Q}_1$ is the same as the rank of $\mathcal{Q}_{11}$, or the rank of $\mathcal{Q}_{11}\mathcal{Q}_{11}'$. But, from Appendix 11.B, $\boldsymbol{c}_1\boldsymbol{c}_1' + \mathcal{Q}_{11}\mathcal{Q}_{11}' = \boldsymbol{I}_a$, by the fact that the a columns of $[\boldsymbol{c}_1 : \mathcal{Q}_{11}]'$ are orthonormal. Thus, $\mathcal{Q}_{11}\mathcal{Q}_{11}' = \boldsymbol{I}_a - \boldsymbol{c}_1\boldsymbol{c}_1'$, which is idempotent of rank $a-1$. Consequently, the rank of $\mathcal{Q}_{11}\mathcal{Q}_{11}'$ is $a-1$. It follows that there exists an orthogonal matrix, $\mathcal{S}$, of order $(b_. - 1) \times (b_. - 1)$ such that

$$\mathcal{Q}_1'\boldsymbol{\Lambda}_a\mathcal{Q}_1 = \mathcal{S}\,\text{diag}(\mathcal{D}, \mathbf{0})\,\mathcal{S}',$$

where $\mathcal{D}$ is an $(a - 1) \times (a - 1)$ diagonal matrix of nonzero eigenvalues of $\mathcal{Q}_1'\boldsymbol{\Lambda}_a\mathcal{Q}_1$ and $\mathbf{0}$ is a zero matrix of order $(b_. - a) \times (b_. - a)$.

Now, let $\boldsymbol{\Phi}^* = \mathcal{S}'\mathcal{U}_2$. Then, $\boldsymbol{\Phi}^*$ has a zero mean, and by (11.177), its variance–covariance matrix is

$$\begin{aligned}\text{Var}(\boldsymbol{\Phi}^*) &= \mathcal{S}'[\sigma^2_{\alpha}\,\mathcal{Q}_1'\boldsymbol{\Lambda}_a\mathcal{Q}_1 + \sigma^2_{\beta(\alpha)}\,\boldsymbol{I}_{b_.-1} + \sigma^2_{\epsilon}\,\mathcal{Q}_1'\mathcal{P}\mathcal{K}\mathcal{P}'\mathcal{Q}_1]\mathcal{S} \\ &= \sigma^2_{\alpha}\,\text{diag}(\mathcal{D}, \mathbf{0}) + \sigma^2_{\beta(\alpha)}\,\boldsymbol{I}_{b_.-1} + \sigma^2_{\epsilon}\,\mathcal{L},\end{aligned}$$

where $\mathcal{L} = \mathcal{S}'\mathcal{Q}_1'\mathcal{P}\mathcal{K}\mathcal{P}'\mathcal{Q}_1\mathcal{S}$. Furthermore, let $\boldsymbol{\Omega}^*$ be defined as

$$\boldsymbol{\Omega}^* = \boldsymbol{\Phi}^* + (\tau_{\max}\boldsymbol{I}_{b_.-1} - \mathcal{L})^{1/2}\mathcal{C}_1'\boldsymbol{Y}, \tag{11.178}$$

where $\tau_{\max}$ is the largest eigenvalue of $\mathcal{L}$ and $\mathcal{C}_1$ is the $n_{..} \times (b_. - 1)$ matrix in (11.173). Finally, let $\boldsymbol{\Omega}^*$ be partitioned as $\boldsymbol{\Omega}^* = \left[\boldsymbol{\Omega}_\alpha^{*'} : \boldsymbol{\Omega}_\beta^{*'}\right]'$, where $\boldsymbol{\Omega}_\alpha^*$ and $\boldsymbol{\Omega}_\beta^*$ are vectors of $a - 1$ and $b_. - a$ elements, respectively. Using arguments similar to those in the proof of Lemma 11.7, the following lemma can be established [a detailed proof is given in Khuri (1987)].

Lemma 11.10 $\boldsymbol{\Omega}_\alpha^*$ and $\boldsymbol{\Omega}_\beta^*$ are independent and normally distributed random vectors with zero means and variance–covariance matrices given by

$$\text{Var}\left(\boldsymbol{\Omega}_\alpha^*\right) = \sigma_\alpha^2\,\mathcal{D} + \left(\sigma_{\beta(\alpha)}^2 + \tau_{\max}\,\sigma_\epsilon^2\right)\boldsymbol{I}_{a-1}$$

$$\text{Var}\left(\boldsymbol{\Omega}_\beta^*\right) = \left(\sigma_{\beta(\alpha)}^2 + \tau_{\max}\,\sigma_\epsilon^2\right)\boldsymbol{I}_{b_.-a}$$

Using Lemma 11.10, we can state that

$$\boldsymbol{\Omega}_\alpha^{*'}[\sigma_\alpha^2\,\mathcal{D} + (\sigma_{\beta(\alpha)}^2 + \tau_{\max}\,\sigma_\epsilon^2)\boldsymbol{I}_{a-1}]^{-1}\boldsymbol{\Omega}_\alpha^* \sim \chi_{a-1}^2$$

$$\frac{1}{\sigma_{\beta(\alpha)}^2 + \tau_{\max}\,\sigma_\epsilon^2}\,\boldsymbol{\Omega}_\beta^{*'}\boldsymbol{\Omega}_\beta^* \sim \chi_{b_.-a}^2.$$

It follows that under $H_0 : \sigma_\alpha^2 = 0$,

$$F = \frac{\boldsymbol{\Omega}_\alpha^{*'}\boldsymbol{\Omega}_\alpha^*/(a-1)}{\boldsymbol{\Omega}_\beta^{*'}\boldsymbol{\Omega}_\beta^*/(b_. - a)} \tag{11.179}$$

has the F-distribution with $a - 1$ and $b_. - a$ degrees of freedom. The test is significant at the α-level if $F \geq F_{\alpha,a-1,b_.-a}$.

Under the alternative hypothesis, $H_a : \sigma_\alpha^2 > 0$, $\boldsymbol{\Omega}_\alpha^{*'}\boldsymbol{\Omega}_\alpha^*$ is distributed as $\sum_{i=1}^{a-1}\lambda_i^*\,W_i$, where the W_i's are independently distributed such that $W_i \sim \chi_1^2$, $i = 1, 2, \ldots, a-1$, and λ_i^* is the ith eigenvalue of $\text{Var}\left(\boldsymbol{\Omega}_\alpha^*\right)$, that is, $\lambda_i^* = d_i\,\sigma_\alpha^2 + \sigma_{\beta(\alpha)}^2 + \tau_{\max}\,\sigma_\epsilon^2$, where d_i is the ith diagonal element of $\mathcal{D}$, $i = 1, 2, \ldots, a-1$. Thus, under H_a, the test statistics, F, in (11.179) can be written as

$$F = \frac{b_. - a}{(a-1)\left(\sigma_{\beta(\alpha)}^2 + \tau_{\max}\,\sigma_\epsilon^2\right)} \times \frac{\sum_{i=1}^{a-1}\lambda_i^* W_i}{\boldsymbol{\Omega}_\beta^{*'}\boldsymbol{\Omega}_\beta^* \Big/ \left(\sigma_{\beta(\alpha)}^2 + \tau_{\max}\,\sigma_\epsilon^2\right)},$$

where $\boldsymbol{\Omega}_\beta^{*'}\boldsymbol{\Omega}_\beta^* \Big/ \left(\sigma_{\beta(\alpha)}^2 + \tau_{\max}\,\sigma_\epsilon^2\right) \sim \chi_{b_.-a}^2$. Consequently, at the α-level, the power of the test in (11.179) is given by

$$P\left[F \geq F_{\alpha,a-1,b_.-a} \mid \sigma_\alpha^2 > 0\right]$$

$$= P\left[\frac{b_. - a}{(a-1)\left(\sigma_{\beta(\alpha)}^2 + \tau_{\max}\,\sigma_\epsilon^2\right)}\sum_{i=1}^{a-1}\lambda_i^*\,W_i - F_{\alpha,a-1,b_.-a}\,\chi_{b_.-a}^2 \geq 0\right].$$

This probability value can be computed by using Davies' algorithm (see Section 5.6) for given values of $\sigma^2_\alpha/\sigma^2_\epsilon$ and $\sigma^2_{\beta(\alpha)}/\sigma^2_\epsilon$.

11.8 Inference Concerning the Mixed Two-Fold Nested Model

In this section, we consider again the two-fold nested model in (11.162), where $\alpha_{(i)}$ is now fixed and $\beta_{i(j)}$ is random such that $\beta_{i(j)} \sim N\left(0, \sigma^2_{\beta(\alpha)}\right)$ independently of $\epsilon_{ij(k)}$, which is distributed as $N\left(0, \sigma^2_\epsilon\right)$. Hence, the variance–covariance matrix of $\boldsymbol{Y}$ in (11.163) is of the form

$$\boldsymbol{\Sigma} = \sigma^2_{\beta(\alpha)}\left[\oplus_{i=1}^{a} \oplus_{j=1}^{b_i} \boldsymbol{J}_{n_{ij}}\right] + \sigma^2_\epsilon \boldsymbol{I}_{n_{..}}. \tag{11.180}$$

The purpose here is to derive exact F-tests concerning $\sigma^2_{\beta(\alpha)}$ and estimable linear functions of the fixed effects. This is done in a manner similar to the one used in Section 11.6.

11.8.1 An Exact Test Concerning $\sigma^2_{\beta(\alpha)}$

From (11.169) , the mean and variance–covariance matrix of $\bar{\boldsymbol{Y}}$ are

$$E(\bar{\boldsymbol{Y}}) = \mu \boldsymbol{1}_{b_.} + \boldsymbol{\mathcal{B}}_1 \boldsymbol{\alpha} \tag{11.181}$$

$$\text{Var}(\bar{\boldsymbol{Y}}) = \sigma^2_{\beta(\alpha)} \boldsymbol{I}_{b_.} + \sigma^2_\epsilon \boldsymbol{\mathcal{K}}. \tag{11.182}$$

As before in Section 11.6.1, a linear transformation of $\bar{\boldsymbol{Y}}$ is needed in order to eliminate the dependence of the mean of $\bar{\boldsymbol{Y}}$ on $\boldsymbol{\alpha}$. For this purpose, we consider the following linear transformation:

$$\bar{\boldsymbol{Y}}^* = \boldsymbol{T}^* \bar{\boldsymbol{Y}}, \tag{11.183}$$

where $\boldsymbol{T}^* = \oplus_{i=1}^{a} \boldsymbol{T}_i^*$ and $\boldsymbol{T}_i^*$ is the matrix

$$\boldsymbol{T}_i^* = \begin{bmatrix} \frac{1}{\sqrt{2}} & -\frac{1}{\sqrt{2}} & 0 & \cdots & 0 \\ \frac{1}{\sqrt{6}} & \frac{1}{\sqrt{6}} & -\frac{2}{\sqrt{6}} & \cdots & 0 \\ \cdot & \cdot & \cdot & \cdots & \cdot \\ \cdot & \cdot & \cdot & \cdots & \cdot \\ \cdot & \cdot & \cdot & \cdots & \cdot \\ \frac{1}{\sqrt{b_i(b_i-1)}} & \frac{1}{\sqrt{b_i(b_i-1)}} & \cdots & \cdots & -\frac{b_i-1}{\sqrt{b_i(b_i-1)}} \end{bmatrix},$$

which has $b_i - 1$ rows and b_i columns, $i = 1, 2, \ldots, a$. Thus, the $\sum_{i=1}^{a}(b_i - 1) = b_. - a$ rows of $\boldsymbol{T}^*$ define orthogonal contrasts in the elements of $\bar{\boldsymbol{Y}}$. Hence,

$E(\bar{\boldsymbol{Y}}^*) = \mathbf{0}$ and from (11.182) and (11.183) we get

$$\text{Var}(\bar{\boldsymbol{Y}}^*) = \sigma^2_{\beta(\alpha)}\, \boldsymbol{I}_{b_.-a} + \sigma^2_{\epsilon}\, \boldsymbol{T}^*\boldsymbol{\mathcal{K}}\boldsymbol{T}^{*'}.$$

It follows that

$$\frac{1}{\sigma^2_{\epsilon}}\, \bar{\boldsymbol{Y}}^{*'}(\delta\, \boldsymbol{I}_{b_.-a} + \boldsymbol{T}^*\boldsymbol{\mathcal{K}}\boldsymbol{T}^{*'})^{-1}\bar{\boldsymbol{Y}}^* \sim \chi^2_{b_.-a},$$

where $\delta = \sigma^2_{\beta(\alpha)}/\sigma^2_{\epsilon}$. Since the error sum of squares , $SS_E = \boldsymbol{Y}'\boldsymbol{\mathcal{R}}\boldsymbol{Y}$, in (11.165) is independent of $\bar{\boldsymbol{Y}}$, and hence of $\bar{\boldsymbol{Y}}^*$, and $SS_E/\sigma^2_{\epsilon} \sim \chi^2_{n_{..}-b_.}$, we conclude that

$$F(\delta) = \frac{\bar{\boldsymbol{Y}}^{*'}(\delta\, \boldsymbol{I}_{b_.-a} + \boldsymbol{T}^*\boldsymbol{\mathcal{K}}\boldsymbol{T}^{*'})^{-1}\bar{\boldsymbol{Y}}^*/(b_. - a)}{SS_E/(n_{..} - b_.)} \tag{11.184}$$

has the F-distribution with $b_. - a$ and $n_{..} - b_.$ degrees of freedom. We can therefore utilize $F(\delta)$ to obtain a confidence interval on δ, and use $F(\delta_0)$ to test the hypothesis

$$H_0 : \delta \leq \delta_0 \qquad \text{against} \qquad H_a : \delta > \delta_0,$$

where δ_0 is some known quantity. Since $F(\delta)$ is a decreasing function of δ, we can reject H_0 at the α-level if $F(\delta_0) \geq F_{\alpha,b_.-a,n_{..}-b_.}$. In particular, if the data set is balanced and if $\delta_0 = 0$, then it can be shown that (11.184) reduces to the conventional F-test used in Chapter 8 to test $H_0 : \sigma^2_{\beta(\alpha)} = 0$.

11.8.2 An Exact Test for the Fixed Effects

Consider again the matrix $\boldsymbol{\mathcal{R}}$ in (11.171). Let us partition $\boldsymbol{C}$ and $\boldsymbol{\Upsilon}$ as $\boldsymbol{C} = [\boldsymbol{C}_1^* : \boldsymbol{C}_2^* : \boldsymbol{C}_3^*]$ and $\boldsymbol{\Upsilon} = \text{diag}(\boldsymbol{I}_{b_.}, \boldsymbol{I}_{n_{..}-2b_.}, \mathbf{0})$, where $\boldsymbol{C}_1^*$, $\boldsymbol{C}_2^*$, and $\boldsymbol{C}_3^*$ are of orders $n_{..} \times b_.$, $n_{..} \times (n_{..} - 2b_.)$, and $n_{..} \times b_.$, respectively, and $\mathbf{0}$ is a zero matrix of order $b_. \times b_.$. The columns of $[\boldsymbol{C}_1^* : \boldsymbol{C}_2^*]$ are orthonormal eigenvectors of $\boldsymbol{\mathcal{R}}$ corresponding to the eigenvalue 1. This partitioning is possible under the condition

$$n_{..} > 2\,b_.$$

Formula (11.171)can then be written as

$$\boldsymbol{\mathcal{R}} = \boldsymbol{C}_1^*\boldsymbol{C}_1^{*'} + \boldsymbol{C}_2^*\boldsymbol{C}_2^{*'}. \tag{11.185}$$

Let us now define the random vector $\boldsymbol{\Omega}^o$ as

$$\boldsymbol{\Omega}^o = \bar{\boldsymbol{Y}} + (\tau^o_{\max}\, \boldsymbol{I}_{b_.} - \boldsymbol{\mathcal{K}})^{1/2}\boldsymbol{C}_1^{*'}\boldsymbol{Y}, \tag{11.186}$$

where $\tau^o_{\max}$ is the largest diagonal element of $\boldsymbol{\mathcal{K}}$. It is easy to see that $\bar{\boldsymbol{Y}}$ and $\boldsymbol{C}_1^{*'}\boldsymbol{Y}$ are independent and $\boldsymbol{\Omega}^o$ has the normal distribution. Furthermore,

$$\begin{aligned}
\text{Var}(\boldsymbol{\Omega}^o) &= \text{Var}(\bar{\boldsymbol{Y}}) + \sigma^2_{\epsilon}\, (\tau^o_{\max}\, \boldsymbol{I}_{b_.} - \boldsymbol{\mathcal{K}}) \\
&= \sigma^2_{\beta(\alpha)}\, \boldsymbol{I}_{b_.} + \sigma^2_{\epsilon}\, \boldsymbol{\mathcal{K}} + \sigma^2_{\epsilon}\, (\tau^o_{\max}\, \boldsymbol{I}_{b_.} - \boldsymbol{\mathcal{K}}) \\
&= \left(\sigma^2_{\beta(\alpha)} + \tau^o_{\max}\, \sigma^2_{\epsilon}\right) \boldsymbol{I}_{b_.}.
\end{aligned} \tag{11.187}$$

The first formula in (11.187) is true because

$$\text{Var}[(\tau^o_{\max}\boldsymbol{I}_{b.} - \mathcal{K})^{1/2}\mathcal{C}_1^{*'}\boldsymbol{Y}] = (\tau^o_{\max}\boldsymbol{I}_{b.} - \mathcal{K})^{1/2}\mathcal{C}_1^{*'}\boldsymbol{\Sigma}\mathcal{C}_1^{*}(\tau^o_{\max}\boldsymbol{I}_{b.} - \mathcal{K})^{1/2},$$

where $\boldsymbol{\Sigma}$ is given in (11.180). But,

$$\begin{aligned}\mathcal{C}_1^{*'}\,\boldsymbol{\Sigma}\,\mathcal{C}_1^{*} &= \mathcal{C}_1^{*'}\left\{\sigma^2_{\beta(\alpha)}\left[\oplus_{i=1}^{a}\oplus_{j=1}^{b_i}\boldsymbol{J}_{n_{ij}}\right] + \sigma^2_{\epsilon}\,\boldsymbol{I}_{n..}\right\}\mathcal{C}_1^{*}\\ &= \sigma^2_{\epsilon}\,\mathcal{C}_1^{*'}\mathcal{C}_1^{*}\\ &= \sigma^2_{\epsilon}\,\boldsymbol{I}_{b.}.\end{aligned}$$

This follows from the fact that $\left[\oplus_{i=1}^{a}\oplus_{j=1}^{b_i}\boldsymbol{J}_{n_{ij}}\right]\mathcal{R} = \boldsymbol{0}$. Consequently,

$$\left[\oplus_{i=1}^{a}\oplus_{j=1}^{b_i}\boldsymbol{J}_{n_{ij}}\right]\left[\mathcal{C}_1^{*}\mathcal{C}_1^{*'} + \mathcal{C}_2^{*}\mathcal{C}_2^{*'}\right] = \boldsymbol{0}.$$

Multiplying both sides on the right by $\mathcal{C}_1^{*}$, we get

$$\left[\oplus_{i=1}^{a}\oplus_{j=1}^{b_i}\boldsymbol{J}_{n_{ij}}\right]\mathcal{C}_1^{*} = \boldsymbol{0},$$

since $\mathcal{C}_1^{*'}\mathcal{C}_1^{*} = \boldsymbol{I}_{b.}$ and $\mathcal{C}_2^{*'}\mathcal{C}_1^{*} = \boldsymbol{0}$.

Now, from (11.163), (11.181), and (11.186), the mean of $\boldsymbol{\Omega}^o$ is written as

$$\begin{aligned}E(\boldsymbol{\Omega}^o) &= E(\bar{\boldsymbol{Y}}) + \left(\tau^o_{\max}\boldsymbol{I}_{b.} - \mathcal{K}\right)^{1/2}\mathcal{C}_1^{*'}E(\boldsymbol{Y})\\ &= \mu\boldsymbol{1}_{b.} + \mathcal{B}_1\,\boldsymbol{\alpha} + \left(\tau^o_{\max}\boldsymbol{I}_{b.} - \mathcal{K}\right)^{1/2}\mathcal{C}_1^{*'}\{\mu\boldsymbol{1}_{n..} + [\oplus_{i=1}^{a}\boldsymbol{1}_{n_{i.}}]\,\boldsymbol{\alpha}\}\end{aligned} \tag{11.188}$$

But,

$$\mathcal{C}_1^{*'}\left[\oplus_{i=1}^{a}\boldsymbol{1}_{n_{i.}}\right] = \boldsymbol{0} \tag{11.189}$$

$$\mathcal{C}_1^{*'}\boldsymbol{1}_{n..} = \boldsymbol{0}. \tag{11.190}$$

Formula (11.189) is true because

$$\begin{aligned}\mathcal{R}\left[\oplus_{i=1}^{a}\boldsymbol{1}_{n_{i.}}\right] &= \mathcal{R}\left\{\oplus_{i=1}^{a}\left[\left(\oplus_{j=1}^{b_i}\boldsymbol{1}_{n_{ij}}\right)\boldsymbol{1}_{b_i}\right]\right\}\\ &= \oplus_{i=1}^{a}\boldsymbol{1}_{n_{i.}} - \oplus_{i=1}^{a}\left\{\left[\oplus_{j=1}^{b_i}(\boldsymbol{J}_{n_{ij}}/n_{ij})\,\boldsymbol{1}_{n_{ij}}\right]\boldsymbol{1}_{b_i}\right\}\\ &= \oplus_{i=1}^{a}\boldsymbol{1}_{n_{i.}} - \oplus_{i=1}^{a}\left[\left(\oplus_{j=1}^{b_i}\boldsymbol{1}_{n_{ij}}\right)\boldsymbol{1}_{b_i}\right]\\ &= \oplus_{i=1}^{a}\boldsymbol{1}_{n_{i.}} - \oplus_{i=1}^{a}\boldsymbol{1}_{n_{i.}}\\ &= \boldsymbol{0}.\end{aligned} \tag{11.191}$$

From (11.185) and (11.191) we conclude (11.189). Formula (11.190) follows directly from (11.189) since $\boldsymbol{1}_{n..}$ is the sum of the columns of $\oplus_{i=1}^{a}\boldsymbol{1}_{n_{i.}}$. From (11.188) we then have

$$E(\boldsymbol{\Omega}^o) = \mu\,\boldsymbol{1}_{b.} + \mathcal{B}_1\,\boldsymbol{\alpha}. \tag{11.192}$$

Using (11.187) and (11.192), we conclude that $\mathbf{\Omega}^o$ can be represented by the linear model

$$\mathbf{\Omega}^o = \mathbf{X}_0\boldsymbol{\beta} + \boldsymbol{\epsilon}^o, \tag{11.193}$$

where $\mathbf{X}_0 = [\mathbf{1}_{b_.} : \mathcal{B}_1]$, $\boldsymbol{\beta} = (\mu, \boldsymbol{\alpha}')'$, and $\boldsymbol{\epsilon}^o \sim N[\mathbf{0}, (\sigma^2_{\beta(\alpha)} + \tau^o_{\max}\,\sigma^2_{\epsilon})\boldsymbol{I}_{b_.}]$.

Now, suppose that we are interested in testing the hypothesis

$$H_0 : \mathcal{A}\,\boldsymbol{\beta} = \boldsymbol{b}_0,$$

where $\boldsymbol{b}_0$ is a known constant vector and $\mathcal{A}$ is a full row-rank matrix of rank r_0 ($\leq a$) such that $\mathcal{A}\boldsymbol{\beta}$ is estimable. Then, the corresponding test statistic is

$$F = \frac{(\mathcal{A}\,\hat{\boldsymbol{\beta}} - \boldsymbol{b}_0)'\left[\mathcal{A}\left(\boldsymbol{X}_0'\boldsymbol{X}_0\right)^{-}\mathcal{A}'\right]^{-1}(\mathcal{A}\,\hat{\boldsymbol{\beta}} - \boldsymbol{b}_0)/r_0}{\mathbf{\Omega}^{o'}\left[\boldsymbol{I}_{b_.} - \boldsymbol{X}_0\left(\boldsymbol{X}_0'\boldsymbol{X}_0\right)^{-}\boldsymbol{X}_0'\right]\mathbf{\Omega}^o/(b_. - a)}, \tag{11.194}$$

which, under H_0, has the F-distribution with r_0 and $b_. - a$ degrees of freedom, where $\hat{\boldsymbol{\beta}} = \left(\boldsymbol{X}_0'\boldsymbol{X}_0\right)^{-}\boldsymbol{X}_0'\mathbf{\Omega}^o$. The test is significant at the α-level if $F \geq F_{\alpha, r_0, b_. - a}$. In particular, we can test the hypothesis that $\alpha_{(1)} = \alpha_{(2)} = \ldots = \alpha_{(a)}$.

Example 11.8 Consider the following example where a manufacturer is interested in studying the burning rate of a rocket propellant from three production processes (factor A). Three batches of propellant were randomly selected from each of processes 1 and 2, but only two batches were selected from process 3. The batch effect is random. Several determinations of burning rate (in minutes) were made on each batch. The data set, which is given in Table 11.11, is a modification of the data given in Montgomery (2005, Problem 14-1, p. 554) where some of the data were deleted to illustrate the analysis of an unbalanced nested design.

TABLE 11.11
Burning Rate Data

Process	Batch	Burning Rate (in Minutes)
1	1	25, 30
	2	19
	3	15, 17, 14
2	1	19, 17, 14
	2	23, 24, 21
	3	35, 27, 25
3	1	20
	2	25, 33

On the basis of this data set, we find that for testing $H_0 : \delta = 0$ against $H_a : \delta > 0$, the test statistic, $F(0)$, from (11.184) has the value $F(0) = 7.53$ with 5 and 10 degrees of freedom (p-value = 0.004). There is therefore a significant variation due to batches.

For testing $H_0 : \alpha_{(1)} = \alpha_{(2)} = \alpha_{(3)}$, we find that the test statistic in (11.194), with $\boldsymbol{b}_0 = \mathbf{0}$ and

$$\mathcal{A} = \begin{bmatrix} 0 & 1 & -1 & 0 \\ 0 & 1 & 0 & -1 \end{bmatrix},$$

has the value $F = 0.288$ with 2 and 5 degrees of freedom (p-value = 0.762). Thus, no significant differences can be detected among the means of the three processes.

11.9 Inference Concerning the General Mixed Linear Model

A more general version of a mixed linear model than what has been considered thus far is written as

$$Y = \mathcal{X}\beta + \mathcal{Z}\gamma + \epsilon, \tag{11.195}$$

where

β is a fixed parameter vector
γ is a vector that contains all the random effects
$\mathcal{X}$ and $\mathcal{Z}$ are known matrices associated with the fixed and random effects, respectively
ϵ is a random error vector

It is assumed that γ and ϵ are independent and have the normal distributions such that $E(\gamma) = \mathbf{0}$, $\text{Var}(\gamma) = \boldsymbol{G}$, $E(\epsilon) = \mathbf{0}$, $\text{Var}(\epsilon) = \mathfrak{R}$. The variance–covariance matrix of $\boldsymbol{Y}$ is therefore of the form

$$\boldsymbol{\Sigma} = \mathcal{Z}\boldsymbol{G}\mathcal{Z}' + \mathfrak{R}. \tag{11.196}$$

Model (11.195) is more general than the previously considered mixed models in the sense that $\boldsymbol{G}$ does not have to be a diagonal matrix containing variance components along its diagonal, as was the case in Sections 11.4–11.8. Furthermore, $\mathfrak{R}$ does not have to be of the traditional form, $\sigma^2_\epsilon \boldsymbol{I}_n$. Such a model can therefore apply to many experimental situations that could not have been accommodated by the traditional mixed model under the standard assumptions. Examples of such situations can be found in a variety of articles and books (see, for example, Stram and Lee, 1994; Verbeke and Lesaffre, 1996; Littell, 2002; Littell et al., 1996; Verbeke, 1997; Verbeke and Molenberghs, 1997, 2000, 2003; McCulloch and Searle, 2001).

The statistical analysis for model (11.195) can be conveniently performed by using PROC MIXED in SAS (2000). It allows the specification of particular structures for $\boldsymbol{G}$ and $\boldsymbol{\Re}$ through the use of the "RANDOM" and "REPEATED" statements in PROC MIXED. The first statement is used to define the random effects in the model and the structure of the $\boldsymbol{G}$ matrix. The second statement is used to specify the matrix $\boldsymbol{\Re}$. Littell et al. (1996) provide a general discussion on the use of PROC MIXED.

The purpose of this section is to review some basic tests for the general mixed linear model using PROC MIXED.

11.9.1 Estimation and Testing of Fixed Effects

Let $\boldsymbol{H\beta}$ be a vector of estimable linear functions of $\boldsymbol{\beta}$, where $\boldsymbol{H}$ is a full row-rank matrix of rank r. Then, the *generalized least-squares estimator* (GLSE) of $\boldsymbol{H\beta}$ is

$$\boldsymbol{H}\hat{\boldsymbol{\beta}} = \boldsymbol{H}(\boldsymbol{\mathcal{X}}'\boldsymbol{\Sigma}^{-1}\boldsymbol{\mathcal{X}})^{-}\boldsymbol{\mathcal{X}}'\boldsymbol{\Sigma}^{-1}\boldsymbol{Y}, \tag{11.197}$$

whose variance–covariance matrix is of the form

$$\text{Var}(\boldsymbol{H}\hat{\boldsymbol{\beta}}) = \boldsymbol{H}(\boldsymbol{\mathcal{X}}'\boldsymbol{\Sigma}^{-1}\boldsymbol{\mathcal{X}})^{-}\boldsymbol{H}'. \tag{11.198}$$

Since $\boldsymbol{\Sigma}$ is unknown in general, it should be replaced by an estimator, $\hat{\boldsymbol{\Sigma}}$. PROC MIXED estimates $\boldsymbol{\Sigma}$ by using either maximum likelihood (ML) or restricted maximum likelihood (REML) estimates of the unknown parameters in $\boldsymbol{G}$ and $\boldsymbol{\Re}$ [see (11.196)]. For this purpose, "METHOD = ML" or "METHOD = REML" can be used as options in the PROC MIXED statement. REML is the default option. Using $\hat{\boldsymbol{\Sigma}}$ in place of $\boldsymbol{\Sigma}$ in (11.197) results in the so-called *estimated generalized least-squares estimator* (EGLSE) of $\boldsymbol{H\beta}$, which is denoted by $\boldsymbol{H}\tilde{\boldsymbol{\beta}}$. Thus,

$$\boldsymbol{H}\tilde{\boldsymbol{\beta}} = \boldsymbol{H}(\boldsymbol{\mathcal{X}}'\hat{\boldsymbol{\Sigma}}^{-1}\boldsymbol{\mathcal{X}})^{-}\boldsymbol{\mathcal{X}}'\hat{\boldsymbol{\Sigma}}^{-1}\boldsymbol{Y}. \tag{11.199}$$

Kackar and Harville (1984) showed that $\boldsymbol{H}\tilde{\boldsymbol{\beta}}$ is unbiased for $\boldsymbol{H\beta}$. Using (11.198), the variance–covariance matrix of $\boldsymbol{H}\tilde{\boldsymbol{\beta}}$ is approximately given by

$$\text{Var}(\boldsymbol{H}\tilde{\boldsymbol{\beta}}) \approx \boldsymbol{H}(\boldsymbol{\mathcal{X}}'\hat{\boldsymbol{\Sigma}}^{-1}\boldsymbol{\mathcal{X}})^{-}\boldsymbol{H}'. \tag{11.200}$$

A better approximation than the one in (11.200) was given by Kenward and Roger (1997). The latter approximation is preferred in small-sample settings.

In particular, if $\boldsymbol{H}$ consists of a single row vector, $\boldsymbol{h}'$, then a statistic for testing the hypothesis,

$$H_0 : \boldsymbol{h}'\boldsymbol{\beta} = a_0,$$

where a_0 is a known constant, is given by

$$t = \frac{h'\tilde{\beta} - a_0}{[h'(\mathcal{X}'\hat{\Sigma}^{-1}\mathcal{X})^{-}h]^{1/2}}, \quad (11.201)$$

which, under H_0, has approximately the t-distribution with η degrees of freedom. There are five methods available in PROC MIXED for estimating η. These include the method based on Satterthwaite's approximation and the one introduced by Kenward and Roger (1997), which is based on their improved procedure for approximating $\text{Var}(h'\tilde{\beta})$, as was mentioned earlier. Guerin and Stroup (2000) conducted a simulation study that demonstrated that the Kenward–Roger method substantially improved the degrees of freedom approximation (see also Littell, 2002). These two methods can be implemented in PROC MIXED by using the options, "DDFM = SATTERTH" and "DDFM = KENWARDROGER," respectively, in the MODEL statement. On the basis of (11.201), an approximate $(1-\alpha)$ 100% confidence interval on $h'\beta$ is then given by

$$h'\tilde{\beta} \pm [h'(\mathcal{X}'\hat{\Sigma}^{-1}\mathcal{X})^{-}h]^{1/2}\, t_{\alpha/2,\eta}. \quad (11.202)$$

In general , for testing the hypothesis, $H_0 : \boldsymbol{H}\beta = \boldsymbol{0}$, PROC MIXED constructs the statistic,

$$F = \frac{(\boldsymbol{H}\tilde{\beta})'[\boldsymbol{H}(\mathcal{X}'\hat{\Sigma}^{-1}\mathcal{X})^{-}\boldsymbol{H}']^{-1}(\boldsymbol{H}\tilde{\beta})}{r}, \quad (11.203)$$

where, if we recall, r is the rank of $\boldsymbol{H}$. Under H_0, F has an approximate F-distribution with r and η degrees of freedom.

It should be noted that the output from PROC MIXED includes the so-called "TESTS OF FIXED EFFECTS" table which contains hypothesis tests for the significance of each of the fixed effects specified in the MODEL statement. By default, PROC MIXED computes these tests by constructing a particular $\boldsymbol{H}$ matrix for each fixed effect, which is then used in (11.203) to compute a Type III F-ratio for the effect under consideration.

11.9.2 Tests Concerning the Random Effects

Tests of significance concerning the random effects' variance components are usually performed using the *likelihood ratio test*. The corresponding test statistic is

$$\lambda_n = \frac{\max_{H_0} \mathcal{L}_n(\theta)}{\max_{\Omega} \mathcal{L}_n(\theta)}, \quad (11.204)$$

where

$\mathcal{L}_n(\boldsymbol{\theta})$ denotes the likelihood function for model (11.195) for a sample of size n

$\boldsymbol{\theta}$ is the vector containing the unknown parameters of the model (including the variance components of the random effects)

$\max_{H_0} \mathcal{L}_n(\boldsymbol{\theta})$ denotes the maximum of $\mathcal{L}_n(\boldsymbol{\theta})$ under a null hypothesis, H_0, for a given variance component

$\max_{\Omega} \mathcal{L}_n(\boldsymbol{\theta})$ denotes the maximum of $\mathcal{L}_n(\boldsymbol{\theta})$ over the entire parameter space for $\boldsymbol{\theta}$

Usually, the null distribution of $-2\log(\lambda_n)$ is approximated with that of a chi-squared variate. The asymptotic behavior of the likelihood ratio test was discussed by Stram and Lee (1994). Using results by Self and Liang (1987), they showed that for testing a single variance component, the asymptotic null distribution of $-2\log(\lambda_n)$ is a mixture of χ_1^2 and χ_0^2 with weights equal to 0.5 (see also Verbeke, 1997, Section 3.9.1). Here, χ_0^2 denotes a discrete distribution that takes the value 0 with probability 1. Thus, if $F(x)$ denotes the asymptotic cumulative null distribution of $-2\log(\lambda_n)$, then

$$F(x) = \frac{1}{2}[F_1(x) + 1], \tag{11.205}$$

where $F_1(x)$ is the cumulative distribution function of χ_1^2. For example, the 5% critical value for such a combination is obtained by solving the equation,

$$\frac{1}{2}[F_1(x) + 1] = 0.95,$$

for x, or equivalently, $F_1(x) = 0.90$, which gives the value, $x = \chi^2_{0.10,1} = 2.71$. In general, it can be easily shown on the basis of formula (11.205) that the p-value for the likelihood ratio statistic is one half of the p-value that could have been obtained from a chi-squared distribution with one degree of freedom.

The actual value of $-2\log(\lambda_n)$ can be easily obtained from the PROC MIXED output by fitting the model twice, with and without the random effect being tested, then subtracting the corresponding values of $-2\log$ (likelihood) since by (11.204),

$$-2\log(\lambda_n) = -2\log[\max_{H_0} \mathcal{L}_n(\boldsymbol{\theta})] + 2\log[\max_{\Omega} \mathcal{L}_n(\boldsymbol{\theta})].$$

The $-2\log$ (likelihood) value can be found in the PROC MIXED output under a table entitled "FIT STATISTICS". To do so, "METHOD = ML" should be included as an option in the PROC MIXED statement.

Example 11.9 Let us again consider the data set of Example 11.7 concerning the average daily gains (in pounds) of 67 steers. The corresponding model is

TABLE 11.12
Type III Test Concerning Factor A

Effect	Numerator DF	Denominator DF	F	p-Value
A	2	19.60	0.38	0.6870

the one given in (11.84) where $\alpha_{(i)}$, the effect of level i of factor A (age of dam), is fixed ($i = 1, 2, 3$) and $\beta_{(j)}$, the effect of level j of factor B (sire), is random ($j = 1, 2, \ldots, 9$). The assumptions concerning the random effects are the same as in Section 11.6.

The F-test concerning the fixed effect of A is obtained from the output of PROC MIXED under "TYPE III TESTS OF FIXED EFFECTS" as shown in Table 11.12.

On the basis of this table, no significant differences can be detected among the three means of the age-of-dam factor. This agrees with the conclusion in Example 11.7 (part (ii)) regarding the same factor. Note that the value of the denominator degrees of freedom in Table 11.12 is based on the Kenward–Roger method.

The tests concerning the random effects of B and $A * B$ are based on the likelihood ratio log statistic, $-2\log(\lambda_n)$. Thus, for testing $H_0 : \sigma^2_{\beta} = 0$, we have the test statistic value,

$$\begin{aligned} -2\log(\lambda_n) &= -2\log\left[\max_{H_0} \mathcal{L}_n(\theta)\right] - \left\{-2\log\left[\max_{\Omega} \mathcal{L}_n(\theta)\right]\right\} \\ &= 10.2 - 9.8 \\ &= 0.4. \end{aligned} \tag{11.206}$$

The first entry on the right-hand side of (11.206) is obtained from the PROC MIXED output concerning "- 2 LOG LIKELIHOOD" under the "FIT STATISTICS" table for model (11.84), but without the effect of B. The second entry in (11.206) represents the same quantity, but for the full model with the effect of B included. In both models, "METHOD = ML" should be used in the PROC MIXED statement. The corresponding p-value for the test statistic value in (11.206) is 0.2635, which is half the p-value for a chi-squared test with one degree of freedom (=0.5270). Thus, no significant variation can be detected among sires. This also agrees with the conclusion in Example 11.7 (part (i)) regarding factor B. Similarly, by repeating the same steps, we find that the test statistic value for $H_0 : \sigma^2_{\alpha\beta} = 0$ is

$$\begin{aligned} -2\log(\lambda_n) &= 11.8 - 9.8 \\ &= 2.0. \end{aligned}$$

The corresponding p-value is 0.0786, which indicates some significant variation due to $A * B$ (the p-value for a chi-squared test with one degree of freedom is $2 \times 0.0786 = 0.1572$). We may recall that the test concerning the same hypothesis in Example 11.7 (part (i)) was not significant according to the exact test described in Section 11.6.1.

The SAS statements needed to do the aforementioned computations are given below:

```
DATA;
INPUT A B Y @@;
CARDS;
(enter here the data from Table 11.10)
PROC MIXED METHOD = ML;
CLASS A B;
MODEL Y = A/DDFM = KENWARDROGER;
RANDOM B  A*B;
PROC MIXED METHOD = ML;
CLASS A B;
MODEL Y = A;
RANDOM A*B;
PROC MIXED METHOD = ML;
CLASS A B;
MODEL Y = A;
RANDOM B;
RUN;
```

Note that the "MODEL" statement in PROC MIXED contains only the fixed effects. The "RANDOM" statement includes all the random effects. The first PROC MIXED statement is needed to get the test results concerning the fixed effects (as shown in Table 11.12) as well as the value of −2 log (likelihood) for the full two-way model. The second PROC MIXED statement is needed to get the value of −2 log (likelihood) for a model without B, and the third PROC MIXED statement gives the same quantity, but for a model without $A * B$.

Example 11.10 Consider the data set of Example 11.8 concerning the burning rate of a rocket propellant from three production processes (factor A). Batches (factor B) are nested within A. Here, A is fixed and B is random. The assumptions concerning the random effects in the two-fold nested model in (11.162) are the same as in Section 11.8.

By repeating the same steps as in Example 11.9, we find that the test for A can be obtained from the following table:

Effect	Numerator DF	Denominator DF	F	p-Value
A	2	8.61	0.43	0.6644

Hence, no significant differences can be detected among the means of the three processes. This agrees with the conclusion in Example 11.8 regarding the hypothesis, $H_0 : \alpha_{(1)} = \alpha_{(2)} = \alpha_{(3)}$.

The test for $H_0 : \sigma^2_{\beta(\alpha)} = 0$ has the test statistic value

$$\begin{aligned} -2\log(\lambda_n) &= 113.7 - 107.7 \\ &= 6.0. \end{aligned}$$

The corresponding p-value is 0.0072, which indicates a significant variation among batches. This also agrees with the conclusion in Example 11.8 regarding the nested effect of B.

The corresponding SAS statements needed for this example are:

```
DATA;
INPUT A B Y @@;
CARDS;
(enter here the data from Table 11.11)
PROC MIXED METHOD = ML;
CLASS A B;
MODEL Y = A/DDFM = KENWARDROGER;
RANDOM B(A);
PROC MIXED METHOD = ML;
CLASS A B;
MODEL Y = A;
RUN;
```

Appendix 11.A: The Construction of the Matrix $\tilde{Q}$ in Section 11.4.1

Let m_i denote the rank of $\tilde{\boldsymbol{P}}_i$ [$i = 0, 1, 2, 3$ with $m_0 = 1, m_1 = a - 1, m_2 = b - 1, m_3 = (a - 1)(b - 1)$]. Let $\tilde{Q}_{1i}$ be a full row-rank matrix of rank m_i whose rows are orthonormal and span the row space of $\tilde{\boldsymbol{P}}_i$ ($i = 0, 1, 2, 3$). Then, it is easy to see that

(i) $\tilde{Q}_{10} = \frac{1}{\sqrt{ab}}\boldsymbol{1}'_{ab}$

(ii) $\tilde{Q}_{1i}\tilde{Q}'_{1i} = \boldsymbol{I}_{m_i}, i = 0, 1, 2, 3,$ and $\tilde{Q}_{1i}\tilde{Q}'_{1j} = \boldsymbol{0}, i \neq j.$

(iii) $\boldsymbol{A}_j\tilde{Q}'_{1i} = \tilde{\kappa}_{ij}\tilde{Q}'_{1i}, \ i, j = 0, 1, 2, 3.$

Properties (ii) and (iii) follow from writing $\tilde{Q}_{1i} = \boldsymbol{V}_i\tilde{\boldsymbol{P}}_i$ for some matrix $\boldsymbol{V}_i$, $i = 0, 1, 2, 3$, and using formula (11.98) and the fact that $\tilde{\boldsymbol{P}}_i\tilde{\boldsymbol{P}}_j = \boldsymbol{0}, i \neq j$. Now, let $\tilde{Q}$ be defined as

$$\tilde{Q} = \left[\tilde{Q}'_{10} : \tilde{Q}'_{11} : \tilde{Q}'_{12} : \tilde{Q}'_{13}\right]'. \qquad (11.A.1)$$

Then, on the basis of properties (i), (ii), and (iii), the matrices $\tilde{Q}A_j\tilde{Q}'$, must be diagonal for $j = 0, 1, 2,$ and 3 since

$$\tilde{Q}A_j\tilde{Q}' = \tilde{Q}A_j\left[\tilde{Q}'_{10} : \tilde{Q}'_{11} : \tilde{Q}'_{12} : \tilde{Q}'_{13}\right]$$
$$= \tilde{Q}\,[\tilde{\kappa}_{0j}\tilde{Q}'_{10} : \tilde{\kappa}_{1j}\tilde{Q}'_{11} : \tilde{\kappa}_{2j}\tilde{Q}'_{12} : \tilde{\kappa}_{3j}\tilde{Q}'_{13}],\ j = 0, 1, 2, 3. \quad (11.A.2)$$

The right-hand side of (11.A.2) is obviously diagonal. In particular, $\tilde{Q}$ diagonalizes A_1 and A_2 simultaneously. □

Appendix 11.B: The Construction of the Matrix $\mathcal{Q}$ in Section 11.7.1

Let e_1 be the $b_. \times 1$ vector, $e_1 = (c'_1, \mathbf{0}')'$, where $c'_1 = \frac{1}{\sqrt{b_.}}(\sqrt{b_1}, \sqrt{b_2}, \ldots, \sqrt{b_a})$, and $\mathbf{0}$ is a zero vector of order $(b_. - a) \times 1$. Then, $(I_{b_.} - e_1e'_1)e_1 = \mathbf{0}$. The matrix $I_{b_.} - e_1e'_1$ is idempotent of rank $b_. - 1$ and e_1 is an eigenvector of unit length of this matrix corresponding to its zero eigenvalue. Let $\mathcal{Q} = [e_1 : \mathcal{Q}_1]'$, where $\mathcal{Q}_1$ is of order $b_. \times (b_. - 1)$ and rank $b_. - 1$ whose columns are orthonormal eigenvectors of $I_{b_.} - e_1e'_1$ corresponding to its eigenvalue one. Hence, $\mathcal{Q}$ is an orthogonal matrix and the first row of $\mathcal{Q}\mathcal{P}$ is

$$e'_1\mathcal{P} = c'_1\mathcal{P}_1$$
$$= \frac{1}{\sqrt{b_.}}\left(\sqrt{b_1}, \sqrt{b_2}, \ldots, \sqrt{b_a}\right)\bigoplus_{i=1}^{a}\left(\frac{1}{\sqrt{b_i}}\mathbf{1}'_{b_i}\right)$$
$$= \frac{1}{\sqrt{b_.}}\mathbf{1}'_{b_.}.$$

□

Exercises

11.1 Consider the data set of Example 11.5.

(a) Obtain Henderson's Method III estimates of $\sigma^2_{\delta(\beta)}$ and σ^2_{ϵ} and verify the values given at the end of Section 11.1.

(b) Give a test statistic for testing the hypothesis $H_0 : \sigma^2_{\delta(\beta)} = 0$. What distribution does this statistic have under H_0?

(c) What distribution would the test statistic in part (b) have if H_0 were false?

11.2 Five batches of raw material were randomly selected. Several samples were randomly taken from each batch and the purity of the material was determined from each sample. The following data were obtained:

Batch				
1	2	3	4	5
10.93	14.72	9.97	21.51	18.45
12.71	15.91	10.78	20.75	17.25
11.35	17.10		19.69	16.95
13.50				

The corresponding model is the one-way model in (11.64), where $\alpha_{(i)}$ denotes the effect of the ith batch ($i = 1, 2, 3, 4, 5$) such that $\alpha_{(i)} \sim N\left(0, \sigma^2_\alpha\right)$ independently of $\epsilon_{i(j)}$.

(a) Use $R(\alpha \mid \mu)$ to construct a test statistic for $H_0 : \sigma^2_\alpha = 0$.

(b) What distribution does the statistic in part (a) have if H_0 is false?

(c) Show how you can compute the power of the test in part (a) for an α-level of significance and a specific value of $\frac{\sigma^2_\alpha}{\sigma^2_\epsilon}$.

11.3 Consider again the data set in Exercise 11.2.

(a) Test the hypothesis $H_0 : \sigma^2_\alpha = 0$ using the statistic described in (11.73).

(b) Give an approximate value of the power of the test in part (a) for an α-level of significance and a specific value of $\frac{\sigma^2_\alpha}{\sigma^2_\epsilon}$.

(c) Obtain an approximate $(1 - \alpha)$ 100% confidence interval on $\frac{\sigma^2_\alpha}{\sigma^2_\epsilon}$.

(d) Obtain an approximate $(1 - \alpha)100\%$ confidence interval on σ^2_α [see the double inequality in (11.83)].

11.4 Consider Section 11.4.1.1. Show that the sums of squares, SS_{Au}, SS_{Bu}, and SS_{ABu} are mutually independent and distributed as multiples of chi-squared variates if and only if $\tilde{U}' W^{-1} \tilde{U}$ is distributed as a chi-squared variate.

11.5 Consider the data set in Table 11.3. Suppose that the effects of factors A and B are random.

(a) Test the significance of A, B, and $A*B$ at the 5% level of significance using the method of unweighted means in Section 11.4.1.

(b) Apply Lemma 11.6 to assess the adequacy of the approximate distributional properties concerning the USSs in this data situation $\left[\text{compute the upper bound in (11.111) for several values of } \frac{\sigma^2_{\alpha\beta}}{\sigma^2_\epsilon}\right]$.

(c) Use the method of unweighted means to obtain an approximate 95% confidence interval on $\sigma^2_\alpha \Big/ \left(\sigma^2_\alpha + \sigma^2_\beta + \sigma^2_{\alpha\beta} + \sigma^2_\epsilon\right)$.

11.6 Consider again the data set used in Exercise 11.5.

(a) Test the interaction hypothesis, H_0: $\sigma^2_{\alpha\beta} = 0$, using $F(0)$ in (11.116). State your conclusion at the $\alpha = 0.05$ level.

(b) Obtain a 95% confidence interval on $\Delta_{\alpha\beta} = \frac{\sigma^2_{\alpha\beta}}{\sigma^2_\epsilon}$ using the double inequality in (11.119).

(c) Find the value of an exact test statistic for testing $H_0 : \sigma^2_\alpha = 0$. State your conclusion at the $\alpha = 0.05$ level.

(d) Redo part (c) for $H_0 : \sigma^2_\beta = 0$.

(e) Obtain an expression for the power function of the test in part (c).

11.7 Consider the mixed two-way model discussed in Section 11.6. Show that $F(\alpha\beta \mid \mu, \alpha, \beta)$ can be used as an exact F-test statistic for testing $H_0 : \sigma^2_{\alpha\beta} = 0$.

11.8 Consider the data set in Table 11.3 where factor A is fixed and factor B is random.

(a) Test the hypothesis $H_0 : \sigma^2_{\alpha\beta} = 0$ in two ways:

(i) Using the test in (11.150).

(ii) Using $F(\alpha\beta \mid \mu, \alpha, \beta)$.

Give the p-value for each test.

(b) Test the hypothesis $H_0 : \sigma^2_\beta = 0$ and state your conclusion at the 5% level.

(c) Determine if there is a significant difference among the means of A. Let $\alpha = 0.05$.

(d) Obtain a 95% confidence interval on $\frac{\sigma^2_{\alpha\beta}}{\sigma^2_\epsilon}$.

11.9 Consider Section 11.7. Verify that the F-ratio in (11.164) has the F-distribution under the null hypothesis, $H_0 : \sigma^2_{\beta(\alpha)} = 0$.

11.10 Consider the F-test statistic in (11.179) concerning the hypothesis, H_0 : $\sigma^2_\alpha = 0$, for the random two-fold nested model. Show that if the data set is balanced (i.e., the n_{ij}'s are equal and the b_i's are also equal), then this statistic reduces to the ANOVA-based F-test statistic used in the balanced case to test H_0.

11.11 Consider a random two-fold nested model (see Section 11.7), where $a = 4$; $b_1 = 1$, $b_2 = 2$, $b_3 = 3$, $b_4 = 4$; $n_{11} = 4$, $n_{21} = n_{22} = 3$, $n_{31} = n_{32} = n_{33} = 2$, $n_{41} = n_{42} = n_{43} = n_{44} = 1$. The following ANOVA table is obtained:

Source	DF	SS	MS
A	$a - 1 = 3$	$R(\alpha \mid \mu)$	$R(\alpha \mid \mu)/3$
$B(A)$	$b_{.} - a = 6$	$R(\beta(\alpha) \mid \mu, \alpha)$	$R(\beta(\alpha) \mid \mu, \alpha)/6$
Error	$n_{..} - b_{.} = 10$	SS_E	MS_E

(a) Find the expected values of the mean squares in this ANOVA table.

(b) Find an approximate F-test for testing the hypothesis, $H_0 : \sigma^2_\alpha = 0$, using $R(\alpha \mid \mu)/3$ in the numerator and an appropriate linear combination of mean squares in the denominator. Use Satterthwaite's approximation to compute the denominator degrees of freedom, then state your conclusion at the approximate 5% level.

(c) Suppose that the test statistic in part (b) is written as $F = \frac{Y'Q_1Y}{Y'Q_2Y}$, where Q_1 is such that $Y'Q_1Y = R(\alpha \mid \mu)/3$ and Q_2 is the matrix associated with the linear combination of mean squares in the denominator. The actual size of the test for a nominal $\alpha = 0.05$ level is

$$P\left[\frac{Y'Q_1Y}{Y'Q_2Y} \geq F_{0.05,3,\nu} \mid \sigma^2_\alpha = 0\right],$$

where ν is the denominator degrees of freedom. This probability can be expressed as

$$P\left[Y'(Q_1 - F_{0.05,3,\nu}\, Q_2)Y \geq 0 \mid \sigma^2_\alpha = 0\right].$$

Compute the value of this probability given that $\frac{\sigma^2_{\beta(\alpha)}}{\sigma^2_\epsilon} = 4$, then compare the result with $\alpha = 0.05$. [Hint: Use Davies' algorithm (see Section 5.6)].

11.12 Consider Section 11.7.1. Show how to construct an exact $(1 - \alpha)$ 100% confidence region for $\left(\frac{\sigma^2_\alpha}{\sigma^2_\epsilon}, \frac{\sigma^2_{\beta(\alpha)}}{\sigma^2_\epsilon}\right)$. Can you use this region to obtain simultaneous confidence intervals on $\frac{\sigma^2_\alpha}{\sigma^2_\epsilon}$ and $\frac{\sigma^2_{\beta(\alpha)}}{\sigma^2_\epsilon}$ with a joint confidence coefficient greater than or equal to $1 - \alpha$?

11.13 A certain manufacturing firm purchases its raw material from three different suppliers. Four batches of the material were randomly selected from each supplier. Several samples were randomly taken from each batch and the purity of the material was determined from each sample. The resulting data are given in the following table:

	Batch			
Supplier	**1**	**2**	**3**	**4**
1	18.60	21.01	26.96	23.45
	19.50	22.72	20.68	21.86
			19.95	
2	24.57	20.80	19.55	16.05
	23.70	16.57	18.01	18.11
	20.85	21.59	21.50	20.90
3	28.65	22.58	19.03	18.25
	23.80	21.55	22.30	21.22
		19.30		

(a) Is there a significant variation due to batches? State your conclusion at the 5% level.

(b) Determine if there is a significant difference among the means of the three suppliers with regard to purity of the material. State your conclusion at the 5% level.

(c) Obtain a 95% confidence interval on $\frac{\sigma^2_{\beta(\alpha)}}{\sigma^2_\epsilon}$, where $\sigma^2_{\beta(\alpha)}$ is the variance associated with the batch effect and σ^2_ϵ is the error variance.

(d) Obtain a 95% confidence interval on the difference between the mean purities for suppliers 1 and 3. Can we conclude that these means differ at the 5% level?

11.14 Consider again Exercise 11.8.

(a) Test the hypothesis $H_0 : \sigma^2_{\alpha\beta} = 0$ using the likelihood ratio test in Section 11.9.2. Let $\alpha = 0.05$.

(b) Determine if there is a significant variation due to factor B using the likelihood ratio test. Give the corresponding p-value.

(c) Redo part (c) of Exercise 11.8 using PROC MIXED.

11.15 Consider again Exercise 11.13.

(a) Redo part (a) using the likelihood ratio test.

(b) Redo part (b) using PROC MIXED.

(c) Obtain an approximate 95% confidence interval on the difference between the mean purities for suppliers 1 and 3 using formula (11.202) (obtain first the maximum likelihood estimates of $\sigma^2_{\beta(\alpha)}$ and σ^2_ϵ).

12

Additional Topics in Linear Models

This chapter covers a variety of more recent topics in linear models. These include a study of heteroscedastic linear models, where the error variances and/or variances of some random effects are not necessarily constant, response surface models with random effects, and the analysis of linear multiresponse models.

12.1 Heteroscedastic Linear Models

Inference-making procedures concerning the fixed and random effects in a mixed linear model are often based on the assumption that the error variances are equal. However, in many experimental situations, such an assumption may not be valid. For example, in a production process, the quality of a product may be subject to some variation caused by occasional machine malfunction or human error. It would therefore be unreasonable to assume that the variance attributed to the error term in the model remains constant over the course of the experiment. Furthermore, variances attributed to some random effects in the model may also be subject to change. For example, in an industrial experiment, batches of raw material are obtained from different suppliers. The measured response is the purity of the raw material. In this case, the variation of purity among batches may be different for the different suppliers.

Interlaboratory studies performed to test, for example, a certain drug provide another example where heterogeneous variances can be expected. Measurements taken from several laboratories are made by different technicians using different equipment. It is therefore very possible that the within-laboratory variances differ substantially (see, for example, Vangel and Rukhin, 1999).

In all the above situations, lack of homogeneity of variances pertaining to the error term, or to some random effects in the model, can seriously affect the testing of the various effects in the model. A variety of models subject to such experimental conditions will be considered in this chapter.

12.2 The Random One-Way Model with Heterogeneous Error Variances

The analysis of a random one-way model with heterogeneous error variances was first studied by Cochran (1937). The analysis of the same model, but with fixed effects, is not considered in this chapter. Unlike the former model, the latter has for many years been given a great deal of attention by researchers. Welch (1947), James (1951), and Box (1954) were among the first to address the problem of heterogeneous error variances. It has long been established that the standard F-test for testing equality of the treatment means is sensitive to a failure in the assumption of equal error variances, especially if the data set is unbalanced. Several test procedures were proposed to compare the treatment means in such a situation. See, for example, Brown and Forsythe (1974c), Bishop and Dudewicz (1978), Krutchkoff (1988), Draper and Guttman (1966), and Smith and Peddada (1998).

Let us now consider the random one-way model,

$$Y_{ij} = \mu + \alpha_{(i)} + \epsilon_{i(j)}, \quad i = 1, 2, \ldots, k; \; j = 1, 2, \ldots, n_i, \tag{12.1}$$

where $\alpha_{(i)} \sim N(0, \sigma_\alpha^2)$, $\epsilon_{i(j)} \sim N(0, \sigma_i^2)$, and the $\alpha_{(i)}$'s and $\epsilon_{i(j)}$'s are mutually independent. Note that the error variances are equal within groups, that is, for a fixed i, but are unequal among groups, that is, for different values of i ($i = 1, 2, \ldots, k$). Model (12.1) can be written in vector form as

$$\boldsymbol{Y} = \mu \mathbf{1}_{n_.} + \left[\oplus_{i=1}^{k} \mathbf{1}_{n_i}\right] \boldsymbol{\alpha} + \boldsymbol{\epsilon}, \tag{12.2}$$

where

$n_. = \sum_{i=1}^{k} n_i$, $\boldsymbol{\alpha} = (\alpha_{(1)}, \alpha_{(2)}, \ldots, \alpha_{(k)})'$
$\boldsymbol{\epsilon} = (\epsilon_{1(1)}, \epsilon_{1(2)}, \ldots, \epsilon_{1(n_1)}, \ldots, \epsilon_{k(1)}, \ldots, \epsilon_{k(n_k)})'$
$\boldsymbol{Y}$ is the vector of observations

Then, $\boldsymbol{Y}$ is distributed as $N(\mu \mathbf{1}_{n_.}, \boldsymbol{\Sigma})$, where

$$\boldsymbol{\Sigma} = \sigma_\alpha^2 \oplus_{i=1}^{k} \boldsymbol{J}_{n_i} + \oplus_{i=1}^{k} \left(\sigma_i^2 \boldsymbol{I}_{n_i}\right). \tag{12.3}$$

The standard F-test statistic for testing $H_0 : \sigma_\alpha^2 = 0$ is given by

$$F = \left(\frac{n_. - k}{k - 1}\right) \frac{\boldsymbol{Y}'\boldsymbol{A}\boldsymbol{Y}}{\boldsymbol{Y}'\boldsymbol{B}\boldsymbol{Y}}, \tag{12.4}$$

where

$$A = \bigoplus_{i=1}^{k} \left(\frac{1}{n_i} J_{n_i} \right) - \frac{1}{n_.} J_{n_.}, \tag{12.5}$$

$$B = I_{n_.} - \bigoplus_{i=1}^{k} \left(\frac{1}{n_i} J_{n_i} \right). \tag{12.6}$$

If the error variances are homogeneous, that is, $\sigma_1^2 = \sigma_2^2 = \ldots = \sigma_k^2$, then under H_0, F has the F-distribution with $k-1$ and $n_. - k$ degrees of freedom. However, if the error variances are heterogeneous, then $Y'AY$ and $Y'BY$ are not independent and neither one is distributed as a multiple of a chi-squared variate (see Exercise 12.1). As a result, $F_{\alpha,k-1,n_.-k}$ cannot be used as the true α-critical value of the test statistic in (12.4). In this case, the probability, α_f, given by

$$\alpha_f = P[F \geq F_{\alpha,k-1,n_.-k} \mid \sigma_\alpha^2 = 0] \tag{12.7}$$

is not necessarily equal to the nominal level of significance, α. It is possible to get an actual value for α_f and compare it with α. This is done as follows: The right-hand side of (12.7) can be written as

$$P[F \geq F_{\alpha,k-1,n_.-k} \mid \sigma_\alpha^2 = 0] = P[Y'(A - cB)Y \geq 0 \mid \sigma_\alpha^2 = 0], \tag{12.8}$$

where $c = \frac{k-1}{n_.-k} F_{\alpha,k-1,n_.-k}$. Under $H_0 : \sigma_\alpha^2 = 0$, $Y'(A - cB)Y$ can be represented as

$$Y'(A - cB)Y = \sum_{i=1}^{s} \lambda_i \chi_{\nu_i}^2, \tag{12.9}$$

where $\lambda_1, \lambda_2, \ldots, \lambda_s$ are the distinct nonzero eigenvalues of $(A - cB)\Sigma_0$ with multiplicities $\nu_1, \nu_2, \ldots, \nu_s$, where $\Sigma_0 = \oplus_{i=1}^{k} (\sigma_i^2 I_{n_i})$, which is obtained by putting $\sigma_\alpha^2 = 0$ in (12.3), and $\chi_{\nu_1}^2, \chi_{\nu_2}^2, \ldots, \chi_{\nu_s}^2$ are mutually independent chi-squared variates (see Lemma 5.1). From (12.7) and (12.8) we then have

$$\alpha_f = P\left[\sum_{i=1}^{s} \lambda_i \chi_{\nu_i}^2 \geq 0 \right]. \tag{12.10}$$

Davies' (1980) algorithm mentioned in Section 5.6 can then be used to compute α_f for given values of $n_1, n_2, \ldots, n_k$ and $\sigma_1^2, \sigma_2^2, \ldots, \sigma_k^2$. This was carried out in Lee et al. (2007, Table 2) who reported the values of α_f for several values of k, $\sigma_1^2, \sigma_2^2, \ldots, \sigma_k^2$ and different degrees of imbalance affecting the sample sizes, $n_1, n_2, \ldots, n_k$. It was found that for $\alpha = 0.05$, for example, α_f ranged from 0.0001 to 0.8244 for $k = 3$; from 0.0003 to 0.9610 for $k = 5$, and from 0.0008 to 0.9981 for $k = 7$. This clearly shows that heterogeneity in the error variances combined with data imbalance can have a substantial effect on the level of significance of the F-test in (12.4). Lee et al. (2007) also noted that data imbalance had a larger effect on α_f than did heterogeneity in the error variances.

Even though the exact distribution of F in (12.4) is unknown under H_0, it is possible to determine its true critical value with the help of Davies' algorithm for given values of n_i and σ_i^2 $(i = 1, 2, \ldots, k)$. If F_α^* denotes the α-critical value of F, then

$$P(F \geq F_\alpha^*) = \alpha, \tag{12.11}$$

which can be written as in formula (12.10) in the form

$$P\left(\sum_{i=1}^{r} \lambda_i^* \chi_{\nu_i^*}^2 \geq 0\right) = \alpha,$$

where $\lambda_1^*, \lambda_2^*, \ldots, \lambda_r^*$ are the distinct nonzero eigenvalues of $(\boldsymbol{A} - c^*\boldsymbol{B})\boldsymbol{\Sigma}_0$, where $c^* = \frac{k-1}{n_{.}-k} F_\alpha^*$, with multiplicities $\nu_1^*, \nu_2^*, \ldots, \nu_r^*$, respectively. To compute the value of F_α^* using Davies' algorithm, we can proceed as follows: For a given α and specific values of $n_1, n_2, \ldots, n_k$; $\sigma_1^2, \sigma_2^2, \ldots, \sigma_k^2$, Davies' algorithm is used to calculate α_f from (12.10). If the ensuing value of α_f is larger, or smaller, than the nominal level, α, the value of $F_{\alpha, k-1, n_{.}-k}$ is increased, or decreased, several times until the difference between α_f and α is small enough. This results in the value of F_α^* as in (12.11). This process was illustrated in Lee et al. (2007, Table 2) who reported that the effect of heterogeneous error variances on the true α-critical value can be quite considerable.

12.2.1 An Approximate Test Concerning $H_0 : \sigma_\alpha^2 = 0$

Let us again consider model (12.1) under the same assumptions concerning the distributions of $\alpha_{(i)}$ and $\epsilon_{i(j)}$ as was stated earlier. An approximate test concerning the hypothesis, $H_0 : \sigma_\alpha^2 = 0$, was given by Jeyaratnam and Othman (1985). They proposed the following test statistic:

$$\mathcal{F} = \frac{SS_B}{SS_W}, \tag{12.12}$$

where

$$SS_B = \frac{1}{k-1} \sum_{i=1}^{k} (\bar{Y}_{i.} - \bar{Y}^*)^2, \tag{12.13}$$

$$SS_W = \frac{1}{k} \sum_{i=1}^{k} \frac{s_i^2}{n_i}, \tag{12.14}$$

and $\bar{Y}_{i.} = \frac{1}{n_i} \sum_{j=1}^{n_i} Y_{ij}$, $\bar{Y}^* = \frac{1}{k} \sum_{i=1}^{k} \bar{Y}_{i.}$, $s_i^2 = \frac{1}{n_i - 1} \sum_{j=1}^{n_i} (Y_{ij} - \bar{Y}_{i.})^2$, $i = 1, 2, \ldots, k$. The numerator of SS_B represents the unweighted sum of squares for the factor associated with $\alpha_{(i)}$ in model (12.1). It is easy to verify that

(a) SS_B and SS_W are independent.

(b) $E(SS_B) = \sigma^2_\alpha + \frac{1}{k}\sum_{i=1}^{k}\frac{\sigma_i^2}{n_i}$.

(c) $E(SS_W) = \frac{1}{k}\sum_{i=1}^{k}\frac{\sigma_i^2}{n_i}$.

Furthermore, $\frac{(n_i-1)s_i^2}{\sigma_i^2} \sim \chi^2_{n_i-1}$ for $i = 1, 2, \ldots, k$. Hence, SS_W in (12.14) is distributed as

$$SS_W = \sum_{i=1}^{k} \theta_i V_i, \tag{12.15}$$

where $\theta_i = \frac{\sigma_i^2}{k\,n_i\,(n_i-1)}$, and the V_i's are mutually independent random variables such that $V_i \sim \chi^2_{n_i-1}$ $(i = 1, 2, \ldots, k)$. In addition, by applying Lemma 5.1, we can state that $U = (k-1)SS_B$ is distributed as

$$\begin{aligned} U &= \bar{\boldsymbol{Y}}'\left(\boldsymbol{I}_k - \frac{1}{k}\boldsymbol{J}_k\right)\bar{\boldsymbol{Y}} \\ &= \sum_{i=1}^{t} \tau_i U_i, \end{aligned} \tag{12.16}$$

where $\bar{\boldsymbol{Y}} = (\bar{Y}_{1.}, \bar{Y}_{2.}, \ldots, \bar{Y}_{k.})'$, $\tau_1, \tau_2, \ldots, \tau_t$ are the distinct nonzero eigenvalues of $(\boldsymbol{I}_k - \frac{1}{k}\boldsymbol{J}_k)[\sigma^2_\alpha \boldsymbol{I}_k + \oplus_{i=1}^{k}(\sigma_i^2/n_i)]$ with multiplicities $\eta_1, \eta_2, \ldots, \eta_t$, and $U_1, U_2, \ldots, U_t$ are mutually independent random variables such that $U_i \sim \chi^2_{\eta_i}$ $(i = 1, 2, \ldots, t)$. Note that $\sigma^2_\alpha \boldsymbol{I}_k + \oplus_{i=1}^{k}\left(\frac{\sigma_i^2}{n_i}\right)$ is the variance–covariance matrix of $\bar{\boldsymbol{Y}}$.

Applying Satterthwaite's approximation (see Section 9.1) to SS_W and U in (12.15) and (12.16), respectively, we can write

$$\frac{N_1 U}{\sum_{i=1}^{t}\tau_i\eta_i} \underset{approx.}{\sim} \chi^2_{N_1}, \tag{12.17}$$

$$\frac{N_2\, SS_W}{\sum_{i=1}^{k}\theta_i\,(n_i-1)} \underset{approx.}{\sim} \chi^2_{N_2}, \tag{12.18}$$

where

$$N_1 = \frac{\left(\sum_{i=1}^{t}\tau_i\eta_i\right)^2}{\sum_{i=1}^{t}\tau_i^2\eta_i}, \tag{12.19}$$

$$N_2 = \frac{\left[\sum_{i=1}^{k} \theta_i (n_i - 1)\right]^2}{\sum_{i=1}^{k} \theta_i^2 (n_i - 1)}. \tag{12.20}$$

Note that

$$\sum_{i=1}^{t} \tau_i \eta_i = tr\left\{\left(\boldsymbol{I}_k - \frac{1}{k}\boldsymbol{J}_k\right)\left[\sigma_\alpha^2 \boldsymbol{I}_k + \bigoplus_{i=1}^{k}\left(\frac{\sigma_i^2}{n_i}\right)\right]\right\}$$

$$= \sigma_\alpha^2 (k-1) + \frac{k-1}{k}\sum_{i=1}^{k} \frac{\sigma_i^2}{n_i}, \tag{12.21}$$

$$\sum_{i=1}^{t} \tau_i^2 \eta_i = tr\left\{\left[\left(\boldsymbol{I}_k - \frac{1}{k}\boldsymbol{J}_k\right)\left(\sigma_\alpha^2 \boldsymbol{I}_k + \bigoplus_{i=1}^{k}\left(\frac{\sigma_i^2}{n_i}\right)\right)\right]^2\right\}$$

$$= \frac{1}{k^2}\left[\sum_{i=1}^{k}\left(\sigma_\alpha^2 + \frac{\sigma_i^2}{n_i}\right)\right]^2 + \frac{k-2}{k}\sum_{i=1}^{k}\left(\sigma_\alpha^2 + \frac{\sigma_i^2}{n_i}\right)^2. \tag{12.22}$$

Using formulas (12.21) and (12.22) in (12.17) and (12.19), we conclude that

$$\frac{N_1 U}{\sigma_\alpha^2 (k-1) + \frac{k-1}{k}\sum_{i=1}^{k} \frac{\sigma_i^2}{n_i}} \underset{approx.}{\sim} \chi^2_{N_1}$$

or, equivalently,

$$\frac{N_1 SS_B}{\sigma_\alpha^2 + \frac{1}{k}\sum_{i=1}^{k} \frac{\sigma_i^2}{n_i}} \underset{approx.}{\sim} \chi^2_{N_1}, \tag{12.23}$$

where

$$N_1 = \frac{\left[\sigma_\alpha^2 (k-1) + \frac{k-1}{k}\sum_{i=1}^{k} \frac{\sigma_i^2}{n_i}\right]^2}{\frac{1}{k^2}\left[\sum_{i=1}^{k}(\sigma_\alpha^2 + \frac{\sigma_i^2}{n_i})\right]^2 + \frac{k-2}{k}\sum_{i=1}^{k}\left(\sigma_\alpha^2 + \frac{\sigma_i^2}{n_i}\right)^2}. \tag{12.24}$$

Also, since $\theta_i = \frac{\sigma_i^2}{k n_i (n_i - 1)}$, $i = 1, 2, \ldots, k$, we get from (12.18),

$$\frac{N_2 SS_W}{\frac{1}{k}\sum_{i=1}^{k} \frac{\sigma_i^2}{n_i}} \underset{approx.}{\sim} \chi^2_{N_2}, \tag{12.25}$$

where from (12.20),

$$N_2 = \frac{\left[\sum_{i=1}^{k} \frac{\sigma_i^2}{n_i}\right]^2}{\sum_{i=1}^{k} \frac{\sigma_i^4}{n_i^2 (n_i - 1)}}. \tag{12.26}$$

Using properties (a) through (c), and the approximate chi-squared distributions in (12.23) and (12.25), we conclude that the statistic $\mathcal{F}$ in (12.12) can be used to test $H_0 : \sigma^2_\alpha = 0$. Under this hypothesis, $\mathcal{F}$ has the approximate F-distribution with N_1 and N_2 degrees of freedom. The test is significant at the approximate α-level if $\mathcal{F} \geq F_{\alpha,N_1,N_2}$. Since N_1 and N_2 depend on the unknown values of $\sigma^2_1, \sigma^2_2, \ldots, \sigma^2_k$, it will be necessary to replace σ^2_i by s^2_i, the sample variance for the ith group data ($i = 1, 2, \ldots, k$), in the expressions in (12.24) and (12.26) for N_1 and N_2, respectively. This gives rise to $\hat{N}_1$ and $\hat{N}_2$. Thus, under H_0, $\hat{N}_1$ and $\hat{N}_2$ are of the form

$$\hat{N}_1 = \frac{\left[(k-1)\sum_{i=1}^{k}\frac{s_i^2}{n_i}\right]^2}{\left(\sum_{i=1}^{k}\frac{s_i^2}{n_i}\right)^2 + k(k-2)\sum_{i=1}^{k}\frac{s_i^4}{n_i^2}}, \tag{12.27}$$

$$\hat{N}_2 = \frac{\left(\sum_{i=1}^{k}\frac{s_i^2}{n_i}\right)^2}{\sum_{i=1}^{k}\frac{s_i^4}{n_i^2(n_i-1)}}. \tag{12.28}$$

Simulation studies were conducted by Jeyaratnam and Othman (1985) and by Argac, Makambi, and Hartung (2001) to evaluate the performance of the test statistic in (12.12). The latter authors noted that the actual simulated significance levels of the test were somewhat close to the nominal significance level for small values of k, but the test tended to be too conservative for large k (≥ 6) and small sample sizes.

12.2.2 Point and Interval Estimation of σ^2_α

An ANOVA-type estimator of σ^2_α was given by Rao, Kaplan, and Cochran (1981), namely,

$$\hat{\sigma}^2_\alpha = \left(n_. - \frac{1}{n_.}\sum_{i=1}^{k} n_i^2\right)^{-1}\left[\sum_{i=1}^{k} n_i(\bar{Y}_{i.} - \bar{Y}_{..})^2 - \sum_{i=1}^{k}\left(1 - \frac{n_i}{n_.}\right)s_i^2\right],$$

where $\bar{Y}_{..} = \frac{1}{n_.}\sum_{i=1}^{k} n_i \bar{Y}_{i.}$. This estimator is unbiased, but can take on negative values. Another estimator that yields nonnegative estimates of σ^2_α was also given by Rao, Kaplan, and Cochran (1981) and is of the form

$$\tilde{\sigma}^2_\alpha = \frac{1}{k}\sum_{i=1}^{k} m_i^2(\bar{Y}_{i.} - \tilde{Y}_{..})^2,$$

where $m_i = \frac{n_i}{n_i+1}$, $i = 1, 2, \ldots, k$, and $\tilde{Y}_{..} = \frac{\sum_{i=1}^{k} m_i \bar{Y}_{i.}}{\sum_{i=1}^{k} m_i}$. This estimator is biased and is called the *average of the squared residuals* (ASR)-type estimator.

Confidence intervals on σ^2_α were proposed by several authors. The following confidence interval was proposed by Wimmer and Witkovsky (2003): Replacing the τ_i's in (12.16) by their weighted average,

$$\begin{aligned}\bar{\tau} &= \frac{\sum_{i=1}^{t} \tau_i \eta_i}{\sum_{i=1}^{t} \eta_i} \\ &= \frac{\sigma^2_\alpha (k-1) + \frac{k-1}{k} \sum_{i=1}^{k} \frac{\sigma_i^2}{n_i}}{k-1} \\ &= \sigma^2_\alpha + \frac{1}{k} \sum_{i=1}^{k} \frac{\sigma_i^2}{n_i},\end{aligned}$$

which results from applying formula (12.21) and the fact that $\sum_{i=1}^{t} \eta_i = rank(\boldsymbol{I}_k - \frac{1}{k}\boldsymbol{J}_k) = k-1$, we get

$$\begin{aligned}U &\underset{approx.}{\sim} \bar{\tau} \sum_{i=1}^{t} U_i \\ &= \bar{\tau} \chi^2_{k-1}.\end{aligned}$$

Thus,

$$\frac{U}{\sigma^2_\alpha + \frac{1}{k} \sum_{i=1}^{k} \frac{\sigma_i^2}{n_i}} \underset{approx.}{\sim} \chi^2_{k-1}. \tag{12.29}$$

From (12.29), an approximate $(1 - \frac{\alpha}{2})100\%$ confidence interval on σ^2_α is obtained as

$$\frac{U}{\chi^2_{\alpha/4,k-1}} - \frac{1}{k} \sum_{i=1}^{k} \frac{\sigma_i^2}{n_i} < \sigma^2_\alpha < \frac{U}{\chi^2_{1-\alpha/4,k-1}} - \frac{1}{k} \sum_{i=1}^{k} \frac{\sigma_i^2}{n_i}. \tag{12.30}$$

But, (12.30) cannot be used since it depends on the unknown values of $\sigma_1^2, \sigma_2^2, \ldots, \sigma_k^2$. Using the fact that $\frac{(n_i-1)s_i^2}{\sigma_i^2} \sim \chi^2_{n_i-1}$ independently for $i = 1, 2, \ldots, k$, the intervals,

$$\frac{(n_i-1)s_i^2}{\chi^2_{\beta/2,n_i-1}} < \sigma_i^2 < \frac{(n_i-1)s_i^2}{\chi^2_{1-\beta/2,n_i-1}}, \quad i = 1, 2, \ldots, k \tag{12.31}$$

represent simultaneous confidence intervals on $\sigma_1^2, \sigma_2^2, \ldots, \sigma_k^2$ with a joint coverage probability equal to $(1-\beta)^k$, where β is defined so that $(1-\beta)^k = 1 - \frac{\alpha}{2}$. From (12.31) we conclude that

$$\frac{1}{k} \sum_{i=1}^{k} \frac{(n_i-1)s_i^2}{n_i \chi^2_{\beta/2,n_i-1}} < \frac{1}{k} \sum_{i=1}^{k} \frac{\sigma_i^2}{n_i} < \frac{1}{k} \sum_{i=1}^{k} \frac{(n_i-1)s_i^2}{n_i \chi^2_{1-\beta/2,n_i-1}}, \tag{12.32}$$

with a probability greater than or equal to $1 - \frac{\alpha}{2}$.

Applying Bonferroni's inequality to (12.30) and (12.32), we get

$$\frac{U}{\chi^2_{\alpha/4,k-1}} - \frac{1}{k}\sum_{i=1}^{k}\frac{(n_i-1)s_i^2}{n_i\chi^2_{1-\beta/2,n_i-1}} < \sigma^2_\alpha < \frac{U}{\chi^2_{1-\alpha/4,k-1}} - \frac{1}{k}\sum_{i=1}^{k}\frac{(n_i-1)s_i^2}{n_i\chi^2_{\beta/2,n_i-1}}, \tag{12.33}$$

with an approximate coverage probability greater than or equal to $1-\alpha$. If the lower bound of this confidence interval is negative, then it can be truncated at zero. This action produces intervals with a coverage probability greater than or equal to that for the intervals in (12.33). From (12.33) we finally conclude that an approximate confidence interval on σ^2_α with a confidence coefficient greater than or equal to $1-\alpha$ is given by

$$L_1 < \sigma^2_\alpha < L_2, \tag{12.34}$$

where $L_1 = \max(0, \ell_1)$, $L_2 = \max(0, \ell_2)$, and

$$\ell_1 = \frac{U}{\chi^2_{\alpha/4,k-1}} - \frac{1}{k}\sum_{i=1}^{k}\frac{(n_i-1)s_i^2}{n_i\chi^2_{1-\beta/2,n_i-1}}$$

$$\ell_2 = \frac{U}{\chi^2_{1-\alpha/4,k-1}} - \frac{1}{k}\sum_{i=1}^{k}\frac{(n_i-1)s_i^2}{n_i\chi^2_{\beta/2,n_i-1}}.$$

12.2.3 Detecting Heterogeneity in Error Variances

The analysis concerning model (12.1) depends on whether or not the error variances are equal. Thus, there is a need to test the hypothesis

$$H_0 : \sigma_1^2 = \sigma_2^2 = \ldots = \sigma_k^2. \tag{12.35}$$

There are several procedures for testing H_0. The traditional ones, such as Bartlett's (1937) and Hartley's (1950) tests, are not recommended since they are sensitive to nonnormality of the data. By contrast, Levene's (1960) test is much more robust to nonnormality and is therefore recommended for testing H_0. Levene's test statistic has a complex form, and under H_0 has an approximate F-distribution. Large values of the test statistic are significant. A convenient way to compute the value of this statistic is to use the option, "HOVTEST = LEVENE," in the SAS statement, MEANS A, which is available in PROC GLM. Here, A denotes the factor associated with $\alpha_{(i)}$ in model (12.1). Another related and widely used homogeneity of variance test is the one by O'Brien (1978, 1979). This is a modification of Levene's test and can also be implemented using PROC GLM by changing "LEVENE" to "OBRIEN" in the statement, MEANS A/HOVTEST = LEVENE.

The following example illustrates the application of Levene's test in PROC GLM.

Example 12.1 Batches of raw material were tested for their calcium content. Several samples were taken from each of six randomly selected batches. The data are given in Table 12.1.

To test the homogeneity of variance hypothesis in (12.35) (with $k = 6$) on the basis of Levene's (1960) test, the following SAS statements can be used:

```
DATA;
INPUT BATCH Y @@;
CARDS;
(enter here the data from Table 12.1)
PROC GLM;
CLASS BATCH;
MODEL Y = BATCH;
MEANS BATCH/HOVTEST = LEVENE;
RUN;
```

From the resulting SAS output, we get Table 12.2.

Levene's test statistic has an approximate F-distribution with 5 and 30 degrees of freedom. The test statistic value is 2.39 with an approximate p-value equal to 0.0612. There is therefore some evidence of heterogeneity in the error variances among batches. We can then proceed to apply the analysis concerning σ^2_α as was described earlier in Section 12.2.1.

The value of the test statistic in (12.12) for the hypothesis $H_0 : \sigma^2_\alpha = 0$ is $\mathcal{F} = \frac{SS_B}{SS_W} = \frac{23.4543}{9.2254} = 2.54$. The estimated degrees of freedom based on

TABLE 12.1
Calcium Content Data

Batch 1	Batch 2	Batch 3	Batch 4	Batch 5	Batch 6
23.21	35.02	26.12	23.50	21.10	17.50
23.51	29.50	22.30	6.50	32.00	34.10
22.93	27.52	28.50	37.31	25.30	29.50
23.57		19.88	22.90	34.50	19.11
23.90		33.50	11.08	38.50	28.50
		32.05		40.50	38.00
		35.30		43.50	28.50
					14.50
					29.90

TABLE 12.2
Levene's Test for Homogeneity of Variances

Source	DF	SS	MS	F	p-Value
Batch	5	42748.1	8549.6	2.39	0.0612
Error	30	107221	3574.0		

formulas (12.27) and (12.28) are $\hat{N}_1 = 2.75 \approx 3$, $\hat{N}_2 = 12.30 \approx 12$, and the corresponding approximate p-value is 0.107. There is therefore an indication of a mild variability among batches with regard to calcium content. Using now formula (12.34), a confidence interval on σ^2_α with an approximate confidence coefficient greater than or equal to 0.95 is given by $0 < \sigma^2_\alpha < 189.61$. Note that the ANOVA-type and ASR-type estimates of σ^2_α (see Section 12.2.2) are $\hat{\sigma}^2_\alpha = 12.981$ and $\tilde{\sigma}^2_\alpha = 13.816$, respectively.

12.3 A Mixed Two-Fold Nested Model with Heteroscedastic Random Effects

Consider the mixed two-fold nested model,

$$Y_{ijk} = \mu + \alpha_{(i)} + \beta_{i(j)} + \epsilon_{ij(k)}, \quad i = 1, 2, \ldots, a; \; j = 1, 2, \ldots, b_i; \; k = 1, 2, \ldots, n, \tag{12.36}$$

where $\alpha_{(i)}$ is a fixed unknown parameter, $\beta_{i(j)}$ and $\epsilon_{ij(k)}$ are distributed independently such that $\beta_{i(j)} \sim N(0, \kappa_i^2)$ and $\epsilon_{ij(k)} \sim N(0, \sigma^2_\epsilon)$. We note that the variances of $\beta_{i(j)}$ are different for different i, but are constant for a given i. Furthermore, the range of subscript j depends on i, but the range of subscript k is the same for all i, j. Thus, data imbalance in this case affects only the second stage, but not the third stage of the nested design. Such a design is said to have the so-called *last-stage uniformity*.

An example to which model (12.36) can be applied concerns the testing of purity of batches of raw material obtained from different suppliers where the variation of purity among batches can be different for the different suppliers, as was mentioned earlier in Section 12.1. In another situation, the composition of soils from randomly selected plots in different agricultural areas may vary differently from one area to another.

In this section, tests are presented concerning the fixed and random effects for model (12.36). These tests were initially developed by Khuri (1992a).

Let us first define $\mathbf{Y}_{ij}$ as $\mathbf{Y}_{ij} = (Y_{ij1}, Y_{ij2}, \ldots, Y_{ijn})'$. Then,

$$E(\mathbf{Y}_{ij}) = (\mu + \alpha_{(i)})\mathbf{1}_n, \tag{12.37}$$

$$\text{Var}(\mathbf{Y}_{ij}) = \kappa_i^2 \boldsymbol{J}_n + \sigma^2_\epsilon \boldsymbol{I}_n. \tag{12.38}$$

Using the Spectral Decomposition Theorem, the matrix $\boldsymbol{J}_n$ can be represented as

$$\boldsymbol{J}_n = \boldsymbol{P}\,\text{diag}(n, \mathbf{0})\,\boldsymbol{P}',$$

where $\boldsymbol{P}$ is an orthogonal matrix of order $n \times n$ of orthonormal eigenvectors of $\boldsymbol{J}_n$. Since $\boldsymbol{J}_n\left(\frac{1}{\sqrt{n}}\boldsymbol{1}_n\right) = n\left(\frac{1}{\sqrt{n}}\boldsymbol{1}_n\right)$, $\frac{1}{\sqrt{n}}\boldsymbol{1}_n$ is a unit eigenvector of $\boldsymbol{J}_n$ with the eigenvalue n. We can therefore express $\boldsymbol{P}$ as

$$\boldsymbol{P} = \left[\frac{1}{\sqrt{n}}\boldsymbol{1}_n : \boldsymbol{Q}\right],$$

where the $n-1$ columns of $\boldsymbol{Q}$ are orthonormal eigenvectors of $\boldsymbol{J}_n$ for the eigenvalue zero. Let $\boldsymbol{Z}_{ij} = \boldsymbol{P}'\boldsymbol{Y}_{ij}$. Then,

$$\begin{aligned}
E(\boldsymbol{Z}_{ij}) &= \boldsymbol{P}'(\mu + \alpha_{(i)})\boldsymbol{1}_n \\
&= [\sqrt{n}(\mu + \alpha_{(i)}), \boldsymbol{0}']', \\
\text{Var}(\boldsymbol{Z}_{ij}) &= \boldsymbol{P}'(\kappa_i^2 \boldsymbol{J}_n + \sigma_\epsilon^2 \boldsymbol{I}_n)\boldsymbol{P} \\
&= \kappa_i^2 \,\text{diag}(n, \boldsymbol{0}) + \sigma_\epsilon^2 \boldsymbol{I}_n \\
&= \text{diag}(n\kappa_i^2 + \sigma_\epsilon^2, \sigma_\epsilon^2 \boldsymbol{I}_{n-1}).
\end{aligned}$$

Furthermore, let v_{ij} and $\boldsymbol{\omega}_{ij}$ be defined as

$$v_{ij} = \frac{1}{n}\boldsymbol{1}_n'\boldsymbol{Y}_{ij}, \tag{12.39}$$

$$\boldsymbol{\omega}_{ij} = \boldsymbol{Q}'\boldsymbol{Y}_{ij}. \tag{12.40}$$

Then, for $j = 1, 2, \ldots, b_i$ and $i = 1, 2, \ldots, a$, the v_{ij}'s and $\boldsymbol{\omega}_{ij}$'s are independently and normally distributed with means and variances given by

$$E(v_{ij}) = \mu + \alpha_{(i)}, \tag{12.41}$$

$$\text{Var}(v_{ij}) = \kappa_i^2 + \frac{\sigma_\epsilon^2}{n}, \tag{12.42}$$

$$E(\boldsymbol{\omega}_{ij}) = \boldsymbol{0}, \tag{12.43}$$

$$\text{Var}(\boldsymbol{\omega}_{ij}) = \sigma_\epsilon^2 \boldsymbol{I}_{n-1}. \tag{12.44}$$

12.3.1 Tests Concerning the Fixed Effects

The hypothesis of interest here is

$$H_0 : \alpha_{(1)} = \alpha_{(2)} = \ldots = \alpha_{(a)}. \tag{12.45}$$

The following two cases are considered:

Case 1. The data set is balanced, that is, $b_1 = b_2 = \ldots = b_a = b$.
Let $\boldsymbol{v}_j$ be defined as

$$\boldsymbol{v}_j = (v_{1j}, v_{2j}, \ldots, v_{aj})', \quad j = 1, 2, \ldots, b,$$

where v_{ij} is defined in (12.39). This vector is normally distributed, and by (12.41) and (12.42), has a mean and a variance–covariance matrix given by

$$E(v_j) = (\mu + \alpha_{(1)}, \mu + \alpha_{(2)}, \ldots, \mu + \alpha_{(a)})', \quad j = 1, 2, \ldots, b,$$
$$\mathrm{Var}(v_j) = \bigoplus_{i=1}^{a} \left(\kappa_i^2 + \frac{\sigma_\epsilon^2}{n}\right), \quad j = 1, 2, \ldots, b.$$

Let now Δ_j be defined as

$$\Delta_j = C v_j, \quad j = 1, 2, \ldots, b,$$

where C is an $(a-1) \times a$ matrix of $a-1$ orthogonal contrasts. Then,

$$E(\Delta_j) = C\alpha, \quad j = 1, 2, \ldots, b,$$

where $\alpha = (\alpha_{(1)}, \alpha_{(2)}, \ldots, \alpha_{(a)})'$ and

$$\mathrm{Var}(\Delta_j) = C \left[\bigoplus_{i=1}^{a} \left(\kappa_i^2 + \frac{\sigma_\epsilon^2}{n}\right)\right] C', \quad j = 1, 2, \ldots, b. \tag{12.46}$$

The Δ_j's are mutually independent and identically distributed as normal random vectors with zero means under H_0 and a common variance–covariance matrix given by (12.46). The hypothesis H_0 in (12.45) can then be tested by using Hotelling's T^2-statistic (see, for example, Seber, 1984, p. 63), which is given by

$$T^2 = b\bar{\Delta}' S^{-1} \bar{\Delta}, \tag{12.47}$$

where $\bar{\Delta} = \frac{1}{b}\sum_{j=1}^{b} \Delta_j$ and S is the sample variance–covariance matrix,

$$S = \frac{1}{b-1} \sum_{j=1}^{b} (\Delta_j - \bar{\Delta})(\Delta_j - \bar{\Delta})'.$$

Under H_0 and assuming that $b \geq a$,

$$F = \frac{b-a+1}{(a-1)(b-1)} T^2 \sim F_{a-1, b-a+1}. \tag{12.48}$$

The test is significant at the α-level if $F \geq F_{\alpha, a-1, b-a+1}$. Note that F is invariant to the choice of the orthogonal contrasts that make up the rows of C. This is true because the T^2-statistic in (12.47) is invariant with respect to linear transformations of the v_j's of the type described earlier (see Anderson, 1963).

Case 2. The values of b_i are not necessarily equal.
This is the more general case as it applies whether the b_i's are equal or not.

Let $\zeta_i = \bar{v}_{i.} - \bar{v}_{a.}$ $(i = 1, 2, \ldots, a-1)$, where $\bar{v}_{i.} = \frac{1}{b_i}\sum_{j=1}^{b_i} v_{ij}$ $(i = 1, 2, \ldots, a)$ and v_{ij} is defined in (12.39). From (12.42), the variance–covariance structure for the ζ_i's is given by

$$\mathrm{Var}(\zeta_i) = \frac{1}{b_i}\left(\kappa_i^2 + \frac{\sigma_\epsilon^2}{n}\right) + \frac{1}{b_a}\left(\kappa_a^2 + \frac{\sigma_\epsilon^2}{n}\right), \quad i = 1, 2, \ldots, a-1,$$

$$\mathrm{Cov}(\zeta_i, \zeta_\ell) = \mathrm{Var}(\bar{v}_{a.}) = \frac{1}{b_a}\left(\kappa_a^2 + \frac{\sigma_\epsilon^2}{n}\right), \quad i \neq \ell.$$

Let $\zeta = (\zeta_1, \zeta_2, \ldots, \zeta_{a-1})'$. Then, the variance–covariance matrix of ζ, denoted by V, is of the form

$$V = \bigoplus_{i=1}^{a-1}\left[\frac{1}{b_i}\left(\kappa_i^2 + \frac{\sigma_\epsilon^2}{n}\right)\right] + \frac{1}{b_a}\left(\kappa_a^2 + \frac{\sigma_\epsilon^2}{n}\right) J_{a-1}. \tag{12.49}$$

Since $E(\zeta) = \mathbf{0}$ under H_0, $\zeta' V^{-1}\zeta \sim \chi^2_{a-1}$. Furthermore, from (12.41) and (12.42), an unbiased estimate of $\kappa_i^2 + \frac{\sigma_\epsilon^2}{n}$ is given by t_i^2, where

$$t_i^2 = \frac{1}{b_i - 1}\sum_{j=1}^{b_i}(v_{ij} - \bar{v}_{i.})^2, \quad i = 1, 2, \ldots, a. \tag{12.50}$$

Hence, an estimate, $\hat{V}$, of V is obtained from (12.49) by replacing $\kappa_i^2 + \frac{\sigma_\epsilon^2}{n}$ by t_i^2 $(i = 1, 2, \ldots, a)$. Thus,

$$\hat{V} = \bigoplus_{i=1}^{a-1}\left(\frac{t_i^2}{b_i}\right) + \frac{t_a^2}{b_a} J_{a-1}. \tag{12.51}$$

Consequently, a test statistic for testing H_0 in (12.45) can be obtained by using $\zeta'\hat{V}^{-1}\zeta$, which under H_0 has approximately the chi-squared distribution. Large values of this test statistic are significant. James (1954) showed that the upper α-quantile of $\zeta'\hat{V}^{-1}\zeta$ is approximately given by $(\gamma_1 + \gamma_2\,\chi^2_{\alpha,a-1})\chi^2_{\alpha,a-1}$, where

$$\gamma_1 = 1 + \frac{1}{2(a-1)}\sum_{i=1}^{a}\frac{1}{b_i - 1}\left(1 - \frac{b_i}{\theta\, t_i^2}\right)^2,$$

$$\gamma_2 = \frac{3}{2(a-1)(a+1)}\sum_{i=1}^{a}\frac{1}{b_i - 1}\left(1 - \frac{b_i}{\theta\, t_i^2}\right)^2.$$

where $\theta = \sum_{i=1}^{a}\frac{b_i}{t_i^2}$ (see also, Seber, 1984, p. 446).

Note that the exact F-test described in Case 1 can be applied to Case 2 by using $v_j = (v_{1j}, v_{2j}, \ldots, v_{aj})'$, $j = 1, 2, \ldots, b_0$, where $b_0 = \min(b_1, b_2, \ldots, b_a)$ provided that $b_0 \geq a$. This amounts to discarding some of the data, which causes the test to be inefficient.

The exact and approximate tests (in Cases 1 and 2, respectively) were discussed by several authors. Ito (1969) showed that the approximate test tends to result in a slight overestimation of significance. It is, however, preferred to the exact test due to the low efficiency of the latter.

12.3.2 Tests Concerning the Random Effects

Recall that $\beta_{i(j)} \sim N(0, \kappa_i^2)$, $i = 1, 2, \ldots, a$; $j = 1, 2, \ldots, b_i$. In this section, a test is given concerning the hypothesis, $H_0 : \kappa_i^2 = 0$, $i = 1, 2, \ldots, a$. We also recall that the v_{ij}'s and ω_{ij}'s defined in (12.39) and (12.40), respectively, are independently and normally distributed with means and variances given in (12.41) through (12.44). Hence,

$$\frac{(b_i - 1)t_i^2}{\kappa_i^2 + \frac{\sigma_\epsilon^2}{n}} \sim \chi^2_{b_i - 1}, \quad i = 1, 2, \ldots, a, \tag{12.52}$$

$$\frac{\omega_{ij}'\omega_{ij}}{\sigma_\epsilon^2} \sim \chi^2_{n-1}, \quad i = 1, 2, \ldots, a; \; j = 1, 2, \ldots, b_i, \tag{12.53}$$

where t_i^2 is the sample variance in (12.50). The random variables in (12.52) and (12.53) are mutually independent for $i = 1, 2, \ldots, a$; $j = 1, 2, \ldots, b_i$. It follows that

$$\sum_{i=1}^{a} \frac{(b_i - 1)t_i^2}{\kappa_i^2 + \frac{\sigma_\epsilon^2}{n}} \sim \chi^2_{b_. - a}, \tag{12.54}$$

$$\sum_{i=1}^{a}\sum_{j=1}^{b_i} \frac{\omega_{ij}'\omega_{ij}}{\sigma_\epsilon^2} \sim \chi^2_{b_.(n-1)}, \tag{12.55}$$

where $b_. = \sum_{i=1}^{a} b_i$. Note that $n\sum_{i=1}^{a}(b_i - 1)t_i^2$ is the sum of squares for the nested effect, denoted by $SS_{B(A)}$, and that $\sum_{i=1}^{a}\sum_{j=1}^{b_i} \omega_{ij}'\omega_{ij}$ is the residual sum of squares, SS_E. This follows from the fact that

$$\begin{aligned}
n\sum_{i=1}^{a}(b_i - 1)t_i^2 &= n\sum_{i=1}^{a}\sum_{j=1}^{b_i}(v_{ij} - \bar{v}_{i.})^2 \\
&= n\sum_{i=1}^{a}\sum_{j=1}^{b_i}(\bar{Y}_{ij.} - \bar{Y}_{i..})^2 \\
&= SS_{B(A)},
\end{aligned}$$

since v_{ij} is the sample mean for the (i,j)th cell, and hence $\bar{v}_{i.} = \bar{Y}_{i..}$. Furthermore,

$$\sum_{i=1}^{a}\sum_{j=1}^{b_i} \boldsymbol{\omega}_{ij}^{'}\boldsymbol{\omega}_{ij} = \sum_{i=1}^{a}\sum_{j=1}^{b_i} \boldsymbol{Y}_{ij}^{'}\left(\boldsymbol{I}_n - \frac{1}{n}\boldsymbol{J}_n\right)\boldsymbol{Y}_{ij}$$
$$= SS_E.$$

Now, under $H_0 : \kappa_i^2 = 0,\ i = 1, 2, \ldots, a$, the statistic,

$$F = \frac{MS_{B(A)}}{MS_E}, \tag{12.56}$$

has the F-distribution with $b_. - a$ and $b_.(n-1)$ degrees of freedom, where $MS_{B(A)} = SS_{B(A)}/(b_. - a)$, $MS_E = SS_E/[b_.(n-1)]$. The test is significant at the α-level if $F \geq F_{\alpha, b_. - a, b_.(n-1)}$.

Rejecting H_0 indicates that at least one κ_i^2 is not equal to zero. In this case, it would be of interest to find out which of the κ_i^2's are different from zero. For this purpose, the hypotheses, $H_{0i} : \kappa_i^2 = 0$, can be tested individually using the ratios,

$$F_i = \frac{n\,t_i^2}{MS_E}, \quad i = 1, 2, \ldots, a \tag{12.57}$$

which under H_{0i} has the F-distribution with $b_i - 1$ and $b_.(n-1)$ degrees of freedom. If α is the level of significance for this individual test, then the *experimentwise Type I error rate* for all such tests will not exceed the value $\alpha^* = 1 - (1-\alpha)^a$. This is true because

$$P\left[\frac{n\,t_i^2}{MS_E} \leq F_{\alpha, b_i - 1, b_.(n-1)}, \text{ for all } i = 1, 2, \ldots, a\right] \geq 1 - \alpha^*. \tag{12.58}$$

Inequality (12.58) is based on a result by Kimball (1951).

Example 12.2 This example was given in Khuri (1992a). A certain firm purchases its raw material from three different suppliers. Four batches of the material were randomly selected from each supplier. Three samples were randomly taken from each batch and the purity of the raw material was determined from each sample. The data are presented in Table 12.3.

To test equality of the means of the three suppliers (that is, testing $H_0 : \alpha_{(1)} = \alpha_{(2)} = \alpha_{(3)}$), we can apply the F-test in (12.48) since $b_1 = b_2 = b_3 = 4$. This gives $F = 2.618$ with 2 and 2 degrees of freedom. The corresponding p-value is 0.276, which is not significant. On the other hand, by applying the approximate chi-squared test described in Case 2 of Section 12.3.1, we get $\boldsymbol{\zeta}'\hat{\boldsymbol{V}}^{-1}\boldsymbol{\zeta} = 10.519$, which is significant at a level > 0.06.

TABLE 12.3
Purity Data

	Batch			
Supplier	**1**	**2**	**3**	**4**
1	8.60	11.01	9.97	9.45
	13.08	10.56	10.78	10.22
	9.50	11.72	10.01	8.86
2	24.57	14.83	9.75	16.05
	23.70	15.58	12.08	19.51
	20.87	12.28	10.51	13.30
3	29.62	23.58	19.03	8.25
	23.89	21.56	17.66	9.76
	30.16	19.34	18.34	11.22

Source: Khuri, A.I., *J. Statist. Plann. Inference*, 30, 33, 1992a. With permission.

The test statistic in (12.56) concerning $H_0 : \kappa_i^2 = 0$ for $i = 1, 2, 3$ has the value $F = \frac{MS_{B(A)}}{MS_E} = 23.079$ with 9 and 24 degrees of freedom (p-value < 0.0001). This is a highly significant result. By applying the individual F-tests in (12.57), we get $F_1 = 0.348$, $F_2 = 21.989$, and $F_3 = 46.90$. Using an experimentwise Type I error rate not exceeding $\alpha^* = 0.01$, we find that F_2 and F_3 are highly significant since $F_{\alpha,3,24} = 6.004$, where $\alpha = 1-(1-\alpha^*)^{1/3} = 0.003$. We conclude that batch-to-batch purity variation within suppliers is most apparent in the material from suppliers 2 and 3.

12.4 Response Surface Models

Response surface methodology (RSM) is an area concerned with the modeling of a response variable of interest, Y, against a number of *control variables* that are believed to affect it. The development of a response model requires a careful choice of design that can provide adequate and reliable predictions concerning Y within a certain region of interest. By a *design*, it is meant the specification of the settings of the control variables to be used in a given experimental situation. Once a design is chosen and the model is fitted to the data generated by the design, regression techniques can then be implemented to assess the goodness of fit of the model and do hypothesis testing concerning the model's unknown parameters. The next stage is to use the fitted model to determine optimum operating conditions on the control variables that result in a maximum (or minimum) response within the region of experimentation. Thus, the basic components in any response surface investigation consist of (1) design selection, (2) model fitting, and (3) determination of optimum conditions.

Since the early development of RSM by Box and his co-workers in the 1950s and 1960s (see Box and Wilson, 1951; Box and Draper, 1959, 1963), it has become a very useful and effective tool for experimental research work in many diverse fields. These include, for example, chemical engineering, industrial development and process improvement, agricultural and biological research, clinical and biomedical sciences, to name just a few. For a review of RSM and its applications, see Myers, Khuri, and Carter (1989).

In a typical response surface investigation, it is quite common to fit a model of the form

$$Y = \beta_0 + \boldsymbol{f}'(\boldsymbol{x})\boldsymbol{\beta} + \epsilon, \tag{12.59}$$

where

- β_0 and the elements of $\boldsymbol{\beta}$ are fixed unknown parameters
- $\boldsymbol{x} = (x_1, x_2, \ldots, x_k)'$ is a vector of k control variables that affect the response variable, Y
- $\boldsymbol{f}(\boldsymbol{x})$ is a vector function of $\boldsymbol{x}$ such that $\beta_0 + \boldsymbol{f}'(\boldsymbol{x})\boldsymbol{\beta}$ is a polynomial in the elements of $\boldsymbol{x}$ of degree d (≥ 1)
- ϵ is a random experimental error

The control variables are fixed (that is, nonstochastic) whose values are measured on a continuous scale. Such a model is called a *polynomial model of degree d*. It basically represents an approximation of the true, but unknown, functional relationship between the response and its control variables. The most frequently used polynomial models are of degree 1 or degree 2. Thus,

$$Y = \beta_0 + \sum_{i=1}^{k} \beta_i x_i + \epsilon \tag{12.60}$$

is called a *first-degree model*, and

$$Y = \beta_0 + \sum_{i=1}^{k} \beta_i x_i + \sum_{i<j}\sum \beta_{ij} x_i x_j + \sum_{i=1}^{k} \beta_{ii} x_i^2 + \epsilon \tag{12.61}$$

is called a *full second-degree model*.

Fitting a model such as the one in (12.59) requires running a series of n experiments in each of which the settings of $x_1, x_2, \ldots, x_k$ are specified and the corresponding response variable is measured, or observed. Thus, at the uth experimental run, we have from using model (12.59),

$$Y_u = \beta_0 + \boldsymbol{f}'(\boldsymbol{x}_u)\boldsymbol{\beta} + \epsilon_u, \quad u = 1, 2, \ldots, n, \tag{12.62}$$

where

$\boldsymbol{x}_u = (x_{u1}, x_{u2}, \ldots, x_{uk})'$ and x_{ui} denotes the uth setting of x_i
Y_u is the observed response value at the uth run
ϵ_u is the corresponding experimental error ($u = 1, 2, \ldots, n$)

In a classical response surface investigation, the ϵ_u's are assumed to be mutually independent with each having a zero mean and a common variance, σ_ϵ^2. In this case, the quantity, $\beta_0 + \boldsymbol{f}'(\boldsymbol{x}_u)\boldsymbol{\beta}$ in (12.62) is the mean of Y_u and is therefore called the *mean response* at $\boldsymbol{x}_u$ and is denoted by $\eta_u = \eta(\boldsymbol{x}_u)$, where

$$\begin{aligned} \eta(\boldsymbol{x}) &= \beta_0 + \boldsymbol{f}'(\boldsymbol{x})\boldsymbol{\beta} \\ &= \boldsymbol{g}'(\boldsymbol{x})\boldsymbol{\tau} \end{aligned} \tag{12.63}$$

is the mean response at a point, $\boldsymbol{x}$, in a region of interest, where $\boldsymbol{g}'(\boldsymbol{x}) = (1, \boldsymbol{f}'(\boldsymbol{x}))$ and $\boldsymbol{\tau} = (\beta_0, \boldsymbol{\beta}')'$.

The *design matrix*, or just *design*, is an $n \times k$ matrix, $\boldsymbol{D}$, of the form

$$\boldsymbol{D} = \begin{bmatrix} x_{11} & x_{12} & \ldots & x_{1k} \\ x_{21} & x_{22} & \ldots & x_{2k} \\ \cdot & \cdot & \ldots & \cdot \\ \cdot & \cdot & \ldots & \cdot \\ \cdot & \cdot & \ldots & \cdot \\ x_{n1} & x_{n2} & \ldots & x_{nk} \end{bmatrix}. \tag{12.64}$$

Thus, the uth row of $\boldsymbol{D}$ describes the settings of $x_1, x_2, \ldots, x_k$ used in the uth experimental run ($u = 1, 2, \ldots, n$). Usually, the values of x_{ui} displayed in (12.64) ($u = 1, 2, \ldots, n$; $i = 1, 2, \ldots, k$) are coded settings of the actual levels of the control variables used in the experiment. The use of coded variables helps in simplifying the numerical calculations associated with fitting the response surface model. It also facilitates the construction of the design $\boldsymbol{D}$. In addition, coding removes the units of measurement of the control variables making them scale free (see, for example, Section 2.8 in Khuri and Cornell, 1996).

The model in (12.62) can be written in vector form as

$$\boldsymbol{Y} = \beta_0 \boldsymbol{1}_n + \boldsymbol{X}\boldsymbol{\beta} + \boldsymbol{\epsilon}, \tag{12.65}$$

where

$\boldsymbol{Y} = (Y_1, Y_2, \ldots, Y_n)'$
$\boldsymbol{\epsilon} = (\epsilon_1, \epsilon_2, \ldots, \epsilon_n)'$
$\boldsymbol{X}$ is an $n \times p$ matrix of rank p ($< n$) whose uth row consists of the elements of $\boldsymbol{f}'(\boldsymbol{x}_u)$, $u = 1, 2, \ldots, n$

Model (12.65) can also be written as

$$\boldsymbol{Y} = \tilde{\boldsymbol{X}}\boldsymbol{\tau} + \boldsymbol{\epsilon}, \tag{12.66}$$

where $\tilde{X} = [\mathbf{1}_n : X]$. Under the assumptions made earlier concerning the error term, ϵ has a zero mean vector and a variance–covariance matrix, $\sigma_\epsilon^2 I_n$. In this case, the BLUE of τ in (12.66) is given by

$$\hat{\tau} = (\tilde{X}'\tilde{X})^{-1}\tilde{X}'Y. \tag{12.67}$$

Its variance–covariance matrix is of the form

$$\text{Var}(\hat{\tau}) = \sigma_\epsilon^2\,(\tilde{X}'\tilde{X})^{-1}.$$

Using (12.67), we obtain the so-called *predicted response*, $\hat{Y}(x)$, at a point x by replacing τ in (12.63) by $\hat{\tau}$. We thus have

$$\hat{Y}(x) = g'(x)\hat{\tau}. \tag{12.68}$$

Note that $\hat{Y}(x)$ is unbiased for $\eta(x)$ and its variance is given by

$$\text{Var}[\hat{Y}(x)] = \sigma_\epsilon^2\, g'(x)(\tilde{X}'\tilde{X})^{-1}g(x). \tag{12.69}$$

This is called the *prediction variance* at x. Formulas (12.68) and (12.69) should only be used for values of x within the experimental region, that is, the region in the space of the control variables within which experimentation is carried out.

12.5 Response Surface Models with Random Block Effects

Model fitting in RSM is usually based on the assumption that the experimental runs are carried out under homogeneous conditions. However, in some experimental situations, this may not be possible or is difficult to achieve. For example, batches of raw material used in a production process may be different with regard to source, composition, or other characteristics. In another situation, experimental conditions may change over time due to certain extraneous sources of variation. In all such circumstances, the experimental runs should be carried out in groups, or *blocks*, within each of which homogeneity of conditions can be maintained. As a result, a block effect should be added to model (12.59). Unlike the control variables, which are fixed and measured on a continuous scale, the block effect is discrete and can be either fixed or random, depending on how the blocks are set up. In this chapter, we assume that the block effect is random, which is typically the case in many experimental situations.

Consider model (12.62) with the added condition that the experimental runs are arranged in b blocks of sizes $n_1, n_2, \ldots, n_b$. Thus, $n = \sum_{j=1}^{b} n_j$. Consequently, in place of model (12.62), we can now consider the model

$$Y_u = \beta_0 + f'(x_u)\beta + z_u'\gamma + f'(x_u)\Lambda z_u + \epsilon_u, \quad u = 1, 2, \ldots, n, \tag{12.70}$$

where $z_u = (z_{u1}, z_{u2}, \ldots, z_{ub})'$ and z_{uj} is a "dummy" variable that takes the value 1 if the uth run is in the jth block and is zero otherwise ($j = 1, 2, \ldots, b$; $u = 1, 2, \ldots, n$), $\boldsymbol{\gamma} = (\gamma_1, \gamma_2, \ldots, \gamma_b)'$, where γ_j denotes the effect of the jth block ($j = 1, 2, \ldots, b$). The matrix $\boldsymbol{\Lambda}$ contains interaction coefficients between the blocks and the polynomial terms in the model. Model (12.70) can be expressed in vector form as

$$\boldsymbol{Y} = \beta_0 \mathbf{1}_n + \boldsymbol{X}\boldsymbol{\beta} + \boldsymbol{Z}\boldsymbol{\gamma} + \sum_{i=1}^{p} \boldsymbol{W}_i \boldsymbol{\delta}_i + \boldsymbol{\epsilon}, \tag{12.71}$$

where $\boldsymbol{X}$ and $\boldsymbol{\beta}$ are the same as in (12.65), $\boldsymbol{Z} = \oplus_{j=1}^{b} \mathbf{1}_{n_j}$, and $\boldsymbol{W}_i$ is a matrix of order $n \times b$ whose jth column is obtained by multiplying the elements of the ith column of $\boldsymbol{X}$ with the corresponding elements of the jth column of $\boldsymbol{Z}$ ($i = 1, 2, \ldots, p$; $j = 1, 2, \ldots, b$). Furthermore, $\boldsymbol{\delta}_i$ is a vector of interaction coefficients between the blocks and the ith polynomial term ($i = 1, 2, \ldots, p$). Its elements are the same as those in the ith row of $\boldsymbol{\Lambda}$ in (12.70). Putting $\tilde{\boldsymbol{X}} = [\mathbf{1} : \boldsymbol{X}]$ and $\boldsymbol{\tau} = (\beta_0, \boldsymbol{\beta}')'$ in (12.71), we get the model

$$\boldsymbol{Y} = \tilde{\boldsymbol{X}}\boldsymbol{\tau} + \boldsymbol{Z}\boldsymbol{\gamma} + \sum_{i=1}^{p} \boldsymbol{W}_i \boldsymbol{\delta}_i + \boldsymbol{\epsilon}. \tag{12.72}$$

We assume that $\boldsymbol{\gamma}$ and $\boldsymbol{\delta}_1, \boldsymbol{\delta}_2, \ldots, \boldsymbol{\delta}_p$ are normally distributed with zero means and variance–covariance matrices, $\sigma_\gamma^2 \boldsymbol{I}_b, \sigma_1^2 \boldsymbol{I}_b, \sigma_2^2 \boldsymbol{I}_b, \ldots, \sigma_p^2 \boldsymbol{I}_b$, respectively. Furthermore, all random effects are assumed to be mutually independent of one another and of the error vector, $\boldsymbol{\epsilon}$, which has the $N(\mathbf{0}, \sigma_\epsilon^2 \boldsymbol{I}_n)$ distribution. Model (12.72) is therefore a mixed model since $\boldsymbol{\tau}$ is fixed and $\boldsymbol{\gamma}$, $\boldsymbol{\delta}_1, \boldsymbol{\delta}_2, \ldots, \boldsymbol{\delta}_p$ are random. On the basis of (12.72), the mean, $\boldsymbol{\eta}$, of $\boldsymbol{Y}$ and its variance–covariance matrix, denoted by $\boldsymbol{\Gamma}$, are given by

$$\boldsymbol{\eta} = \tilde{\boldsymbol{X}}\boldsymbol{\tau}, \tag{12.73}$$

$$\boldsymbol{\Gamma} = \sigma_\gamma^2 \boldsymbol{Z}\boldsymbol{Z}' + \sum_{i=1}^{p} \sigma_i^2 \boldsymbol{W}_i \boldsymbol{W}_i' + \sigma_\epsilon^2 \boldsymbol{I}_n. \tag{12.74}$$

We note that $\boldsymbol{\Gamma}$ depends on the design settings of $x_1, x_2, \ldots, x_k$ through $\boldsymbol{W}_1, \boldsymbol{W}_2, \ldots, \boldsymbol{W}_p$. This is quite different from the structure of $\text{Var}(\boldsymbol{Y})$, namely $\sigma_\epsilon^2 \boldsymbol{I}_n$, for a response surface model without a random block effect, as was the case in Section 12.4.

On the basis of (12.73) and (12.74), the BLUE of $\boldsymbol{\tau}$ is the *generalized least-squares estimator*,

$$\tilde{\boldsymbol{\tau}} = (\tilde{\boldsymbol{X}}'\boldsymbol{\Gamma}^{-1}\tilde{\boldsymbol{X}})^{-1}\tilde{\boldsymbol{X}}'\boldsymbol{\Gamma}^{-1}\boldsymbol{Y}, \tag{12.75}$$

and its variance–covariance matrix is

$$\text{Var}(\tilde{\boldsymbol{\tau}}) = (\tilde{\boldsymbol{X}}'\boldsymbol{\Gamma}^{-1}\tilde{\boldsymbol{X}})^{-1}. \tag{12.76}$$

In practice, $\tilde{\tau}$ cannot be computed using (12.75) since the variance components, $\sigma_\gamma^2, \sigma_1^2, \sigma_2^2, \ldots, \sigma_p^2$, and σ_ϵ^2 are unknown and should therefore be estimated. Using REML (or ML) estimates of these variance components, we get the so-called *estimated generalized least-squares estimate* (EGLSE) of τ, namely

$$\tilde{\tau}_e = (\tilde{X}'\hat{\Gamma}^{-1}\tilde{X})^{-1}\tilde{X}'\hat{\Gamma}^{-1}Y, \tag{12.77}$$

where $\hat{\Gamma}$ is obtained from Γ in (12.74) by replacing the variance components by their REML estimates, which can be computed by using PROC MIXED in SAS, as was seen earlier in Chapter 11. Note that $\tilde{\tau}_e$ is no longer the BLUE of τ, but is still unbiased by Theorem 11.1. From (12.76), the estimated variance–covariance matrix of $\tilde{\tau}_e$ is approximately equal to

$$\widehat{\text{Var}}(\tilde{\tau}_e) \approx (\tilde{X}'\hat{\Gamma}^{-1}\tilde{X})^{-1}. \tag{12.78}$$

Since the mean response, $\eta(x)$, at a point x in a region of interest, denoted by $\mathcal{R}$, is still given by (12.63), the predicted response at $x \in \mathcal{R}$, denoted here by $\hat{Y}_e(x)$, is

$$\hat{Y}_e(x) = g'(x)\tilde{\tau}_e. \tag{12.79}$$

Using (12.78), the estimated prediction variance is approximately of the form

$$\widehat{\text{Var}}\,[\hat{Y}_e(x)] \approx g'(x)(\tilde{X}'\hat{\Gamma}^{-1}\tilde{X})^{-1}g(x). \tag{12.80}$$

12.5.1 Analysis Concerning the Fixed Effects

Consider the following hypothesis concerning a linear combination of τ, the vector of unknown parameters in model (12.72):

$$H_0 : \lambda'\tau = 0, \tag{12.81}$$

where λ is known constant vector. Using formulas (12.77) and (12.78), a statistic for testing H_0 is given by

$$t = \frac{\lambda'\tilde{\tau}_e}{\left[\lambda'(\tilde{X}'\hat{\Gamma}^{-1}\tilde{X})^{-1}\lambda\right]^{1/2}}. \tag{12.82}$$

Under H_0, t has approximately the t-distribution with ν degrees of freedom, which can be obtained from using PROC MIXED in SAS, as was described in Section 11.9.1 (see formula 11.201). In particular, the Kenward and Roger (1997) method is recommended for computing ν. A special case of (12.82) is a statistic for testing the significance of the individual elements of τ. In addition, an approximate $(1 - \alpha)$ 100% confidence interval on $\lambda'\tau$ can be obtained from

$$\lambda'\tilde{\tau}_e \pm [\lambda'(\tilde{X}'\hat{\Gamma}^{-1}\tilde{X})^{-1}\lambda]^{1/2} t_{\alpha/2,\nu}. \tag{12.83}$$

12.5.2 Analysis Concerning the Random Effects

Of interest here is the testing of hypotheses concerning the variance components: σ^2_γ, for the block effect, and σ^2_i $(i = 1, 2, \ldots, p)$, the variance components for the interaction effects in model (12.72). All such tests can be carried out using Type III F-distributed ratios as shown below.

To test $H_0 : \sigma^2_\gamma = 0$, we consider the Type III F-ratio,

$$F_\gamma = \frac{R(\boldsymbol{\gamma} \mid \boldsymbol{\tau}, \boldsymbol{\delta}_1, \boldsymbol{\delta}_2, \ldots, \boldsymbol{\delta}_p)}{(b-1)MS_E}, \tag{12.84}$$

where $R(\boldsymbol{\gamma} \mid \boldsymbol{\tau}, \boldsymbol{\delta}_1, \boldsymbol{\delta}_2, \ldots, \boldsymbol{\delta}_p)$ is the Type III sum of squares for $\boldsymbol{\gamma}$ in model (12.72) and MS_E is the residual (error) mean square for the same model. Under H_0, F_γ has the F-distribution with $b-1$ and $n-b-pb$ degrees of freedom, where, if we recall, p is the number of columns of $\boldsymbol{X}$, or the number of parameters in $\boldsymbol{\beta}$, in model (12.71), and b is the number of blocks. This is true on the basis of the following facts:

(a) $R(\boldsymbol{\gamma} \mid \boldsymbol{\tau}, \boldsymbol{\delta}_1, \boldsymbol{\delta}_2, \ldots, \boldsymbol{\delta}_p)/\sigma^2_\epsilon$ is distributed as χ^2_{b-1} under H_0.

To see this, we have that

$$\frac{1}{\sigma^2_\epsilon} R(\boldsymbol{\gamma} \mid \boldsymbol{\tau}, \boldsymbol{\delta}_1, \boldsymbol{\delta}_2, \ldots, \boldsymbol{\delta}_p) = \frac{1}{\sigma^2_\epsilon} \boldsymbol{Y}'[\boldsymbol{G}(\boldsymbol{G}'\boldsymbol{G})^- \boldsymbol{G}' - \boldsymbol{G}_\gamma(\boldsymbol{G}'_\gamma \boldsymbol{G}_\gamma)^- \boldsymbol{G}'_\gamma]\boldsymbol{Y},$$

where $\boldsymbol{G} = [\tilde{\boldsymbol{X}} : \boldsymbol{Z} : \boldsymbol{W}_1 : \boldsymbol{W}_2 : \ldots : \boldsymbol{W}_p]$ and $\boldsymbol{G}_\gamma = [\tilde{\boldsymbol{X}} : \boldsymbol{W}_1 : \boldsymbol{W}_2 : \ldots : \boldsymbol{W}_p]$. Noting that the variance–covariance matrix of $\boldsymbol{Y}$ in (12.74) is $\boldsymbol{\Gamma} = \sum_{i=1}^{p} \sigma^2_i \boldsymbol{W}_i \boldsymbol{W}'_i + \sigma^2_\epsilon \boldsymbol{I}_n$ under H_0 and the fact that

$$\frac{1}{\sigma^2_\epsilon}[\boldsymbol{G}(\boldsymbol{G}'\boldsymbol{G})^- \boldsymbol{G}' - \boldsymbol{G}_\gamma(\boldsymbol{G}'_\gamma \boldsymbol{G}_\gamma)^- \boldsymbol{G}'_\gamma]\boldsymbol{\Gamma} = \boldsymbol{G}(\boldsymbol{G}'\boldsymbol{G})^- \boldsymbol{G}' - \boldsymbol{G}_\gamma(\boldsymbol{G}'_\gamma \boldsymbol{G}_\gamma)^- \boldsymbol{G}'_\gamma, \tag{12.85}$$

which is idempotent of rank $(b + pb) - (1 + pb) = b - 1$, we conclude that under H_0, $R(\boldsymbol{\gamma} \mid \boldsymbol{\tau}, \boldsymbol{\delta}_1, \boldsymbol{\delta}_2, \ldots, \boldsymbol{\delta}_p)/\sigma^2_\epsilon \sim \chi^2_{b-1}$ by Theorem 5.4. [(12.85) is true because $\boldsymbol{W}_i$ is a submatrix of both $\boldsymbol{G}$ and $\boldsymbol{G}_\gamma$, $i = 1, 2, \ldots, p$.]

(b) $SS_E/\sigma^2_\epsilon \sim \chi^2_{n-b-pb}$, where SS_E is the residual (error) sum of squares for model (12.72).

As in (a), we have that

$$\frac{1}{\sigma^2_\epsilon} SS_E = \frac{1}{\sigma^2_\epsilon} \boldsymbol{Y}'[\boldsymbol{I}_n - \boldsymbol{G}(\boldsymbol{G}'\boldsymbol{G})^- \boldsymbol{G}']\boldsymbol{Y},$$

and from (12.74), we can write

$$\frac{1}{\sigma^2_\epsilon}[\boldsymbol{I}_n - \boldsymbol{G}(\boldsymbol{G}'\boldsymbol{G})^- \boldsymbol{G}']\boldsymbol{\Gamma} = \boldsymbol{I}_n - \boldsymbol{G}(\boldsymbol{G}'\boldsymbol{G})^- \boldsymbol{G}', \tag{12.86}$$

due to the fact that $\mathbf{Z}, \mathbf{W}_1, \mathbf{W}_2, \ldots, \mathbf{W}_p$ are submatrices of $\mathbf{G}$. Since the right-hand side of (12.86) is idempotent of rank $n - b - pb$, it follows that $SS_E/\sigma^2_{\epsilon} \sim \chi^2_{n-b-pb}$.

(c) $R(\boldsymbol{\gamma} \mid \boldsymbol{\tau}, \boldsymbol{\delta}_1, \boldsymbol{\delta}_2, \ldots, \boldsymbol{\delta}_p)$ and SS_E are independent.

On the basis of (a), (b), and (c), we conclude that F_{γ} in (12.84) has the F-distribution with $b - 1$ and $n - b - pb$ degrees of freedom. The hypothesis H_0 is rejected at the α-level if $F_{\gamma} \geq F_{\alpha,b-1,n-b-pb}$.

Similarly, to test the hypothesis, $H_{0i} : \sigma^2_i = 0$, we can use the statistic,

$$F_i = \frac{R(\boldsymbol{\delta}_i \mid \boldsymbol{\tau}, \boldsymbol{\gamma}, \boldsymbol{\delta}_1, \ldots, \boldsymbol{\delta}_{i-1}, \boldsymbol{\delta}_{i+1}, \ldots, \boldsymbol{\delta}_p)}{(b-1)MS_E}, \tag{12.87}$$

which under H_{0i} has the F-distribution with $b - 1$ and $n - b - pb$ degrees of freedom. The test is significant at the α-level if $F_i \geq F_{\alpha,b-1,n-b-pb}$.

The tests concerning the interaction effects in (12.72) are important. This is true because these effects play a major role in determining the form of the prediction equation in (12.79) through the estimation of $\boldsymbol{\tau}$ as well as the structure of the prediction variance in (12.80). The predicted response, $\hat{Y}_e(\boldsymbol{x})$ and its prediction variance are directly involved in the determination of optimum conditions on $x_1, x_2, \ldots, x_k$ that maximize (or minimize) $\hat{Y}_e(\boldsymbol{x})$ over the region of experimentation. This was demonstrated in Khuri (1996a) where it was shown how to optimize $\hat{Y}_e(\boldsymbol{x})$ subject to certain constraints on the size of the prediction variance.

More recently, Khuri (2006) provided an extension of the methodology presented in this section by considering heterogeneous error variances among the blocks. Thus, the error vector, $\boldsymbol{\epsilon}$, in model (12.71) is considered here to have the normal distribution with a mean $\mathbf{0}$ and a variance–covariance matrix of the form, $\oplus_{j=1}^{b} \sigma^2_{\epsilon j} \boldsymbol{I}_{n_j}$, where $\sigma^2_{\epsilon 1}, \sigma^2_{\epsilon 2}, \ldots, \sigma^2_{\epsilon b}$ are unknown variance components that are not necessarily equal. Procedures are described for testing the fixed and random effects in this case. The test for the fixed effects is similar to what was given in Section 12.5.1, except that REML (or ML) estimates of $\sigma^2_{\epsilon 1}, \sigma^2_{\epsilon 2}, \ldots, \sigma^2_{\epsilon b}$ are to be incorporated into the variance–covariance matrix of $\boldsymbol{Y}$ (see Section 5 in Khuri, 2006). The tests concerning the random effects are different from those given in Section 12.5.2 since the Type III F-ratios are no longer valid as test statistics under heterogeneous error variances. In this case, tests are conducted by using the likelihood ratio approach. An alternative ANOVA test based on using Satterthwaite's approximation (see Chapter 9) is also given (see Section 6 in Khuri, 2006).

Example 12.3 This example was presented in Khuri (1992b). It concerns an experiment for studying the effects of two factors, curing time and temperature, on the shear strength of the bonding of galvanized steel bars with a certain adhesive. Three levels were chosen for each factor, namely, 375 °F, 400 °F, and 450 °F for temperature, and 30, 35, 40 s for time. The treatment combinations were obtained according to a 3 × 3 factorial design. Aliquots of galvanized steel were selected on each of 12 dates from the warehouse supply. The same factorial arrangement was used on all the 12 dates, except that replicates were taken at the middle set of conditions (400, 35) at the very start of the experiment (July 11), and then on August 7, September 11, and October 3. The latter three dates were chosen because each one of them represented the first time of sampling in the corresponding month. The purpose of taking such replicates was to obtain an estimate of the pure error variation to use for testing lack of fit (see, for example, Section 2.6 on *lack of fit testing* in Khuri and Cornell, 1996). The resulting date are given in Table 12.4. Note that the coded settings of temperature and time are given by

$$x_1 = \frac{\text{Temperature} - 400}{25},$$

$$x_2 = \frac{\text{Time} - 35}{5}.$$

TABLE 12.4
Design Settings and Response Values (Shear Strength in psi)

x_1	x_2	July 11	July 16	July 20	Aug. 7	Aug. 8	Aug. 14
−1	−1	1226	1075	1172	1213	1282	1142
0	−1	1898	1790	1804	1961	1940	1699
2	−1	2142	1843	2061	2184	2095	1935
−1	0	1472	1121	1506	1606	1572	1608
0	0	2010, 1882 1915, 2106	2175	2279	2450, 2355 2420, 2240	2291	2374
2	0	2352	2274	2168	2298	2147	2413
−1	1	1491	1691	1707	1882	1741	1846
0	1	2078	2513	2392	2531	2366	2392
2	1	2531	2588	2617	2609	2431	2408
x_1	x_2	**Aug. 20**	**Aug. 22**	**Sep. 11**	**Sep. 24**	**Oct. 3**	**Oct. 10**
−1	−1	1281	1305	1091	1281	1305	1207
0	−1	1833	1774	1588	1992	2011	1742
2	−1	2116	2133	1913	2213	2192	1995
−1	0	1502	1580	1343	1691	1584	1486
0	0	2471	2393	2205, 2268 2103	2142	2052, 2032 2190	2339
2	0	2430	2440	2093	2208	2201	2216
−1	1	1645	1688	1582	1692	1744	1751
0	1	2392	2413	2392	2488	2392	2390
2	1	2517	2604	2477	2601	2588	2572

Source: Khuri, A.I., *Technometrics*, 34, 26, 1992b. With permission.
Note: The original design settings of x_1 corresponding to −1, 0, 2 are 375 °F, 400 °F, and 450 °F, respectively; those for x_2 corresponding to −1, 0, 1 are 30, 35, and 40 s, respectively.

Thus, the coded design settings are $-1, 0$, and 2 for x_1 and $-1, 0$, and 1 for x_2.

A complete second-degree model in x_1 and x_2 is assumed for the mean response. Thus,

$$\eta(x) = \beta_0 + \beta_1 x_1 + \beta_2 x_2 + \beta_{12} x_1 x_2 + \beta_{11} x_1^2 + \beta_{22} x_2^2. \tag{12.88}$$

The observed response, Y_u, at the uth experimental run is given by model (12.70). In this example, the blocks are the batches of steel aliqouts with each batch being selected at random on a given date from the warehouse supply. In addition to σ_γ^2, the variance component for the block effect, and the error variance, σ_ϵ^2, there are five other variance components for the random interactions between the blocks and the polynomial terms, $x_1, x_2, x_1x_2, x_1^2, x_2^2$, in the model for Y_u. These variance components are denoted by $\sigma_1^2, \sigma_2^2, \sigma_{12}^2, \sigma_{11}^2, \sigma_{22}^2$, respectively.

The REML estimates of σ_1^2 and σ_{12}^2 from using PROC MIXED are equal to zero. This can be explained by the fact that the corresponding Type III F-ratios in (12.87) are nonsignificant with p-values equal to 0.6872 and 0.9705, respectively. The p-values for $\sigma_\gamma^2, \sigma_2^2, \sigma_{11}^2$, and σ_{22}^2 are 0.0016, 0.0701, 0.4020, and 0.1397, respectively. Thus, the batch effect, and the interaction effect associated with x_2 are significant. In addition, we have a mild interaction with x_2^2. The REML estimates of $\sigma_\gamma^2, \sigma_2^2, \sigma_{11}^2, \sigma_{22}^2$, and σ_ϵ^2 are 4124.37, 1764.00, 74.07, 1023.39, and 10709.00, respectively. Using only these estimates in (12.74), we obtain

$$\begin{aligned}\hat{\Gamma} = {} & 4124.37\boldsymbol{Z}\boldsymbol{Z}' + 1764.00\boldsymbol{W}_2\boldsymbol{W}_2' + 74.07\boldsymbol{W}_{11}\boldsymbol{W}_{11}' \\ & + 1023.39\boldsymbol{W}_{22}\boldsymbol{W}_{22}' + 10709.00\boldsymbol{I}_n,\end{aligned} \tag{12.89}$$

where $n = 118$ and $\boldsymbol{W}_2$, $\boldsymbol{W}_{11}$, and $\boldsymbol{W}_{22}$ are the $\boldsymbol{W}_i$ matrices in (12.74) corresponding to $\sigma_2^2, \sigma_{11}^2$, and σ_{22}^2, respectively.

The EGLSE estimates of the parameters in (12.88) and their corresponding estimated standard errors are computed using formulas (12.77) and (12.78). The corresponding approximate t-statistics are evaluated using (12.82). The results are given in Table 12.5.

TABLE 12.5
The EGLSE of the Parameters in Model (12.88)

Parameter	Estimate	Standard Error	t	p-Value
β_0	2187.950	26.655	82.08	<0.0001
β_1	477.770	14.256	33.51	<0.0001
β_2	255.890	17.503	14.62	<0.0001
β_{12}	−4.827	9.778	−0.49	0.6234
β_{11}	−204.670	10.500	−19.49	<0.0001
β_{22}	−45.301	22.024	−2.06	0.0562

From Table 12.5 we note that the parameters $\beta_1, \beta_2, \beta_{11}$, and β_{22} are significantly different from zero, but β_{12} is a nonsignificant parameter. Using the parameter estimates from Table 12.5, we get the following expression for the predicted response, $\hat{Y}_e(\boldsymbol{x})$:

$$\hat{Y}_e(\boldsymbol{x}) = 2187.950 + 477.770x_1 + 255.890x_2 - 4.827x_1x_2 - 204.670x_1^2 - 45.301x_2^2. \tag{12.90}$$

The equation in (12.90) can be used to determine the conditions on x_1 and x_2 that maximize the shear strength response within the region, $-1 \leq x_1 \leq 2$, $-1 \leq x_2 \leq 1$. This was demonstrated in Khuri (1996a) where the maximization of $\hat{Y}_e(\boldsymbol{x})$ was done subject to certain constraints on the size of the prediction variance.

The following is a listing of the SAS statements used to generate the various results in this example:

```
DATA;
INPUT BATCH X1 X2 Y @@;
CARDS;
(enter here the data from Table 12.4)
PROC GLM;
CLASS BATCH;
MODEL Y = X1 X2 X1 * X2 X1 * X1 X2 * X2 BATCH X1 * BATCH
X2 * BATCH X1 * X2 * BATCH X1 * X1 * BATCH X2 * X2 * BATCH;
PROC MIXED METHOD = REML;
CLASS BATCH;
MODEL Y = X1 X2 X1 * X2 X1 * X1 X2 * X2/S DDFM =
KENWARDROGER;
RANDOM BATCH X1 * BATCH X2 * BATCH X1 * X2 * BATCH
X1 * X1 * BATCH X2 * X2 * BATCH;
RUN;
```

Note that the three statements in PROC GLM are needed to obtain the Type III F-ratios for testing the significance of the random effects in the MODEL statement. The "S" option in the MODEL statement in PROC MIXED is needed to obtain the estimated generalized least-squares estimates of the parameters given in Table 12.5. The corresponding p-values are based on using the Kenward and Roger (1997) method (see Section 11.9.1).

12.6 Linear Multiresponse Models

In many experimental situations, a number of response variables can be measured for each setting of a group of control variables. For example, a certain food product may be evaluated on the basis of acceptability, nutritional value, taste, fat content, cost, and other considerations. In this case, each attribute of the product is represented by a response variable and all the relevant responses can be measured at each run of the experiment. Such an experiment is called a *multiresponse experiment*.

The analysis of data from a multiresponse experiment requires a careful recognition of the multivariate nature of the data. This means that the response variables considered in the experiment should not be treated individually or independently of one another. Instead, multivariate techniques should be used in order to combine all the relevant information from the totality of the response variables in the experiment. For example, the optimization of several response variables should take into account any interrelationships that may exist among the responses. Failure to do so can lead to meaningless results since optimal conditions for one response, optimized individually and separately from the remaining responses, may be far from optimal or even physically impractical for the other responses.

In this section, methods for the analysis of a linear multiresponse model are discussed. These include estimation and inference-making procedures concerning the unknown parameters of the model, in addition to a test for lack of fit.

12.6.1 Parameter Estimation

Suppose that m response variables, denoted by $Y_1, Y_2, \ldots, Y_m$, are measured for each setting of a group of k control variables, $x_1, x_2, \ldots, x_k$. We assume that the response variables are represented by polynomial models in $x_1, x_2, \ldots, x_k$ within a certain region, $\mathcal{R}$. If n sets of observations are taken on $Y_1, Y_2, \ldots, Y_m$, then the model for the ith response can be written as

$$\boldsymbol{Y}_i = \boldsymbol{X}_i \boldsymbol{\beta}_i + \boldsymbol{\epsilon}_i, \quad i = 1, 2, \ldots, m, \tag{12.91}$$

where

$\boldsymbol{Y}_i$ is the vector of observations on the ith response
$\boldsymbol{X}_i$ is a known matrix of order $n \times p_i$ and rank p_i
$\boldsymbol{\beta}_i$ is a vector of p_i unknown parameters
$\boldsymbol{\epsilon}_i$ is a random error vector associated with the ith response ($i = 1, 2, \ldots, m$)

We assume that

$$E(\boldsymbol{\epsilon}_i) = \boldsymbol{0}, \quad i = 1, 2, \ldots, m, \tag{12.92}$$

$$\text{Var}(\boldsymbol{\epsilon}_i) = \sigma_{ii} \boldsymbol{I}_n, \quad i = 1, 2, \ldots, m, \tag{12.93}$$

$$\text{Cov}(\boldsymbol{\epsilon}_i, \boldsymbol{\epsilon}_j) = \sigma_{ij} \boldsymbol{I}_n, \quad i, j = 1, 2, \ldots, m; \; i \neq j. \tag{12.94}$$

We note that the response variables in a given experimental run can be correlated and possibly have heteroscedastic variances. More specifically, if

$\boldsymbol{\Upsilon}$ is the matrix of random error vectors, namely,

$$\boldsymbol{\Upsilon} = [\boldsymbol{\epsilon}_1 : \boldsymbol{\epsilon}_2 : \ldots : \boldsymbol{\epsilon}_m], \tag{12.95}$$

then the rows of $\boldsymbol{\Upsilon}$ are mutually independent with each having a zero mean vector and a variance–covariance matrix, $\boldsymbol{\Sigma}_m$, whose (i,j)th element is σ_{ij} $(i,j = 1,2,\ldots,m)$. Furthermore, we assume that each row of $\boldsymbol{\Upsilon}$ has the normal distribution, $N(\mathbf{0}, \boldsymbol{\Sigma}_m)$.

The m models in (12.91) can be combined into a single model of the form

$$\boldsymbol{\mathcal{Y}} = \boldsymbol{\mathcal{X}}\boldsymbol{\Phi} + \tilde{\boldsymbol{\epsilon}}, \tag{12.96}$$

where

$\boldsymbol{\mathcal{Y}} = [\boldsymbol{Y}_1' : \boldsymbol{Y}_2' : \ldots : \boldsymbol{Y}_m']'$
$\boldsymbol{\mathcal{X}} = \oplus_{i=1}^{m} \boldsymbol{X}_i$
$\boldsymbol{\Phi} = [\boldsymbol{\beta}_1' : \boldsymbol{\beta}_2' : \ldots : \boldsymbol{\beta}_m']'$
$\tilde{\boldsymbol{\epsilon}} = [\boldsymbol{\epsilon}_1' : \boldsymbol{\epsilon}_2' : \ldots : \boldsymbol{\epsilon}_m']'$

Model (12.96) is called a *linear multiresponse model*. Under the assumptions made earlier concerning the distribution of the rows of the error matrix, $\boldsymbol{\Upsilon}$, in (12.95), the error vector, $\tilde{\boldsymbol{\epsilon}}$, in (12.96) has the normal distribution $N(\mathbf{0}, \boldsymbol{\Sigma}_m \otimes \boldsymbol{I}_n)$. Thus, the BLUE of $\boldsymbol{\Phi}$ is the generalized least-squares estimator,

$$\hat{\boldsymbol{\Phi}} = [\boldsymbol{\mathcal{X}}'(\boldsymbol{\Sigma}_m^{-1} \otimes \boldsymbol{I}_n)\boldsymbol{\mathcal{X}}]^{-1}\boldsymbol{\mathcal{X}}'(\boldsymbol{\Sigma}_m^{-1} \otimes \boldsymbol{I}_n)\boldsymbol{\mathcal{Y}}. \tag{12.97}$$

The corresponding variance–covariance matrix of $\hat{\boldsymbol{\Phi}}$ is

$$\text{Var}(\hat{\boldsymbol{\Phi}}) = [\boldsymbol{\mathcal{X}}'(\boldsymbol{\Sigma}_m^{-1} \otimes \boldsymbol{I}_n)\boldsymbol{\mathcal{X}}]^{-1}. \tag{12.98}$$

Note that equations (12.97) and (12.98) depend on $\boldsymbol{\Sigma}_m$, which is unknown. It is therefore necessary to estimate it provided that the estimate is nonsingular. One such estimate was proposed by Zellner (1962) and is given by $\hat{\boldsymbol{\Sigma}}_m = (\hat{\sigma}_{ij})$, where

$$\hat{\sigma}_{ij} = \frac{1}{n}\boldsymbol{Y}_i'[\boldsymbol{I}_n - \boldsymbol{X}_i(\boldsymbol{X}_i'\boldsymbol{X}_i)^{-1}\boldsymbol{X}_i'][\boldsymbol{I}_n - \boldsymbol{X}_j(\boldsymbol{X}_j'\boldsymbol{X}_j)^{-1}\boldsymbol{X}_j']\boldsymbol{Y}_j. \tag{12.99}$$

This estimate is computed from the residual vectors that result from fitting each response model individually by the method of ordinary least squares. Replacing $\boldsymbol{\Sigma}_m$ by $\hat{\boldsymbol{\Sigma}}_m$ in (12.97) and (12.98), we get

$$\hat{\boldsymbol{\Phi}}_e = [\boldsymbol{\mathcal{X}}'(\hat{\boldsymbol{\Sigma}}_m^{-1} \otimes \boldsymbol{I}_n)\boldsymbol{\mathcal{X}}]^{-1}\boldsymbol{\mathcal{X}}'(\hat{\boldsymbol{\Sigma}}_m^{-1} \otimes \boldsymbol{I}_n)\boldsymbol{\mathcal{Y}}. \tag{12.100}$$

This is an estimated generalized least-squares estimate (EGLSE) of $\boldsymbol{\Phi}$, and we approximately have

$$\widehat{\text{Var}}(\hat{\boldsymbol{\Phi}}_e) \approx [\boldsymbol{\mathcal{X}}'(\hat{\boldsymbol{\Sigma}}_m^{-1} \otimes \boldsymbol{I}_n)\boldsymbol{\mathcal{X}}]^{-1}. \tag{12.101}$$

The estimate $\hat{\boldsymbol{\Phi}}_e$ is no longer the BLUE of $\boldsymbol{\Phi}$. It does, however, have certain asymptotic properties as was shown in Zellner (1962).

In particular, if in models (12.91), $\boldsymbol{X}_i = \boldsymbol{X}_0$ for $i = 1, 2, \ldots, m$, then it can be easily shown that $\hat{\boldsymbol{\Phi}}$ in (12.97) reduces to

$$\hat{\boldsymbol{\Phi}} = [\boldsymbol{I}_m \otimes (\boldsymbol{X}_0'\boldsymbol{X}_0)^{-1}\boldsymbol{X}_0']\boldsymbol{\mathcal{Y}}.$$

Thus, in this case, the BLUE of $\boldsymbol{\Phi}$ does not depend on $\boldsymbol{\Sigma}_m$ and is therefore the same as the ordinary least-squares estimator obtained from fitting the m models in (12.91) individually.

The predicted response value for the ith response ($i = 1, 2, \ldots, m$) at a point, $\boldsymbol{x}$, in the region $\mathcal{R}$ is given by

$$\hat{Y}_i(\boldsymbol{x}) = \boldsymbol{f}_i'(\boldsymbol{x})\hat{\boldsymbol{\beta}}_{ie}, \quad i = 1, 2, \ldots, m, \tag{12.102}$$

where $\boldsymbol{f}_i'(\boldsymbol{x})$ is a row vector function that has the same form as a row of $\boldsymbol{X}_i$ in (12.91), but is evaluated at $\boldsymbol{x}$, $\hat{\boldsymbol{\beta}}_{ie}$ is the ith portion of $\hat{\boldsymbol{\Phi}}_e = (\hat{\boldsymbol{\beta}}_{1e}', \hat{\boldsymbol{\beta}}_{2e}', \ldots, \hat{\boldsymbol{\beta}}_{me}')'$ in (12.100). Note that $\boldsymbol{f}_i'(\boldsymbol{x})\hat{\boldsymbol{\beta}}_{ie}$ is a polynomial in the elements of $\boldsymbol{x}$ of degree d_i (≥ 1), $i = 1, 2, \ldots, m$. By combining the m equations in (12.102), we get

$$\hat{\boldsymbol{Y}}(\boldsymbol{x}) = [\oplus_{i=1}^m \boldsymbol{f}_i'(\boldsymbol{x})]\hat{\boldsymbol{\Phi}}_e, \tag{12.103}$$

where $\hat{\boldsymbol{Y}}(\boldsymbol{x}) = (\hat{Y}_1(\boldsymbol{x}), \hat{Y}_2(\boldsymbol{x}), \ldots, \hat{Y}_m(\boldsymbol{x}))'$. Using (12.101), the estimated variance–covariance matrix of $\hat{\boldsymbol{Y}}(\boldsymbol{x})$ is approximately given by

$$\widehat{\text{Var}}[\hat{\boldsymbol{Y}}(\boldsymbol{x})] \approx [\oplus_{i=1}^m \boldsymbol{f}_i'(\boldsymbol{x})]\,[\boldsymbol{\mathcal{X}}'(\hat{\boldsymbol{\Sigma}}_m^{-1} \otimes \boldsymbol{I}_n)\boldsymbol{\mathcal{X}}]^{-1}\,[\oplus_{i=1}^m \boldsymbol{f}_i(\boldsymbol{x})]. \tag{12.104}$$

12.6.2 Hypothesis Testing

We recall that the models in (12.91) were combined into a single linear multiresponse model of the form given in (12.96). Alternatively, these models can be combined using the model

$$\tilde{\boldsymbol{\mathcal{Y}}} = \tilde{\boldsymbol{\mathcal{X}}}\boldsymbol{\Psi} + \boldsymbol{\Upsilon}, \tag{12.105}$$

where

$\tilde{\boldsymbol{\mathcal{Y}}} = [\boldsymbol{Y}_1 : \boldsymbol{Y}_2 : \ldots : \boldsymbol{Y}_m]$
$\tilde{\boldsymbol{\mathcal{X}}} = [\boldsymbol{X}_1 : \boldsymbol{X}_2 : \ldots : \boldsymbol{X}_m]$
$\boldsymbol{\Psi} = \oplus_{i=1}^m \boldsymbol{\beta}_i$
$\boldsymbol{\Upsilon}$ is given in (12.95)

Consider now testing the hypothesis,

$$H_0 : \mathcal{G}\Psi = C_0, \tag{12.106}$$

where $\mathcal{G}$ and C_0 are known matrices with $\mathcal{G}$ being of full row rank, q. The rows of $\mathcal{G}$ are spanned by the rows of $\tilde{\mathcal{X}}$. Thus, $\mathcal{G}\Psi$ is estimable and, consequently, the hypothesis in (12.106) is testable. A test statistic for testing H_0 is given by $e_{\max}(S_h S_e^{-1})$, the largest eigenvalue of the matrix $S_h S_e^{-1}$ (see Roy, Gnanadesikan, and Srivastava, 1971), where

$$S_h = [\mathcal{G}(\tilde{\mathcal{X}}'\tilde{\mathcal{X}})^{-}\tilde{\mathcal{X}}'\tilde{\mathcal{Y}} - C_0]'\,[\mathcal{G}(\tilde{\mathcal{X}}'\tilde{\mathcal{X}})^{-}\mathcal{G}']^{-1}\,[\mathcal{G}(\tilde{\mathcal{X}}'\tilde{\mathcal{X}})^{-}\tilde{\mathcal{X}}'\tilde{\mathcal{Y}} - C_0], \tag{12.107}$$

$$S_e = \tilde{\mathcal{Y}}'[I_n - \tilde{\mathcal{X}}(\tilde{\mathcal{X}}'\tilde{\mathcal{X}})^{-}\tilde{\mathcal{X}}']\tilde{\mathcal{Y}}. \tag{12.108}$$

The matrices S_h and S_e are called the *matrix due to the hypothesis* and the *matrix due to the error*, respectively. The matrix S_e is positive definite with probability 1 if $n - \rho \geq m$, where ρ is the rank of $\tilde{\mathcal{X}}$ (see Roy, Gnanadesikan, and Srivastava, 1971, p. 35), and has the so-called *central Wishart distribution* with $n - \rho$ degrees of freedom. Furthermore, S_h is independent of S_e and has the *noncentral Wishart distribution* with q degrees of freedom and a noncentrality parameter matrix, $\boldsymbol{\Omega}$, given by (see, for example, Seber, 1984, p. 414)

$$\boldsymbol{\Omega} = \boldsymbol{\Sigma}_m^{-1/2}(\mathcal{G}\Psi - C_0)'\,[\mathcal{G}(\tilde{\mathcal{X}}'\tilde{\mathcal{X}})^{-}\mathcal{G}']^{-1}\,(\mathcal{G}\Psi - C_0)\boldsymbol{\Sigma}_m^{-1/2}, \tag{12.109}$$

where, if we recall, $\boldsymbol{\Sigma}_m$ is the variance–covariance matrix of the m responses corresponding to any row of $\boldsymbol{\Upsilon}$ in (12.95).

The statistic, $e_{\max}(S_h S_e^{-1})$, is called *Roy's largest root*, and the hypothesis H_0 in (12.106) is rejected at the α-level if $e_{\max}(S_h S_e^{-1})$ exceeds the α-critical value of this statistic. Such a critical value can be found in, for example, Roy, Gnanadesikan, and Srivastava (1971, Appendix B) and Seber (1984, Appendix D14).

Other test statistics for testing H_0 include *Wilks' likelihood ratio*, $det(S_e)/det(S_e + S_h)$, *Hotelling-Lawley's trace*, $tr(S_h S_e^{-1})$, and *Pillai's trace*, $tr[S_h(S_e + S_h)^{-1}]$. Small values of $det(S_e)/det(S_e + S_h)$ are significant, but large values of the remaining two test statistics are significant. These tests, along with Roy's largest root, are referred to as *multivariate tests*.

In particular, if the $\boldsymbol{\beta}_i$ vectors in (12.91) are of the same length, consisting of p elements each, then we can test two particular hypotheses, namely, the *hypothesis of concurrence* and the *hypothesis of parallelism*.

12.6.2.1 Hypothesis of Concurrence

This hypothesis is of the form

$$H_{0c} : \boldsymbol{\beta}_1 = \boldsymbol{\beta}_2 = \ldots = \boldsymbol{\beta}_m. \tag{12.110}$$

To test H_{0c}, let us first multiply the two sides of model (12.105) on the right by the $m \times (m-1)$ matrix,

$$K = \begin{bmatrix} 1 & 1 & 1 & \dots & 1 \\ -1 & 0 & 0 & \dots & 0 \\ 0 & -1 & 0 & \dots & 0 \\ . & . & . & \dots & . \\ . & . & . & \dots & . \\ . & . & . & \dots & . \\ 0 & 0 & 0 & \dots & -1 \end{bmatrix}, \qquad (12.111)$$

which is of rank $m-1$. Doing so, we get

$$\tilde{\mathcal{Y}}K = \tilde{\mathcal{X}}\Psi K + \Upsilon K. \qquad (12.112)$$

The rows of ΥK are independent and identically distributed as $N(\mathbf{0}, K'\Sigma_m K)$. The matrix ΨK has the form

$$\Psi K = \begin{bmatrix} \beta_1 & \beta_1 & \beta_1 & \dots & \beta_1 \\ -\beta_2 & \mathbf{0} & \mathbf{0} & \dots & \mathbf{0} \\ \mathbf{0} & -\beta_3 & \mathbf{0} & \dots & \mathbf{0} \\ . & . & . & \dots & . \\ . & . & . & \dots & . \\ . & . & . & \dots & . \\ \mathbf{0} & \mathbf{0} & \mathbf{0} & \dots & -\beta_m \end{bmatrix}. \qquad (12.113)$$

The hypothesis in (12.110) can now be expressed as

$$H_{0c} : G_c \Psi K = \mathbf{0}, \qquad (12.114)$$

where G_c is a matrix of order $p \times (pm)$ of the form

$$G_c = [I_p : I_p : \dots : I_p]. \qquad (12.115)$$

The hypothesis in (12.114) is testable provided that the rows of G_c are spanned by the rows of $\tilde{\mathcal{X}}$ in (12.112). In this case, we can test H_{0c} by applying Roy's largest root test statistic, $e_{\max}(S_h S_e^{-1})$, where S_h and S_e are obtained from (12.107) and (12.108), respectively, after replacing $\tilde{\mathcal{Y}}$ by $\tilde{\mathcal{Y}}K$, $\mathcal{G}$ by G_c, and C_0 by $\mathbf{0}$.

A numerical example, in which the investment models for three corporations that operate in the same branch of industry are compared, is given in Khuri (1986, Section 4).

12.6.2.2 Hypothesis of Parallelism

This hypothesis is of the form

$$H_{0p} : \beta_1^* = \beta_2^* = \dots = \beta_m^*, \qquad (12.116)$$

where $\boldsymbol{\beta}_i^*$ is equal to $\boldsymbol{\beta}_i$, except that the first element (intercept) of $\boldsymbol{\beta}_i$ has been removed. In this case, the hypothesis in (12.114) can be modified so as to produce (12.116) as follows:

$$H_{0p} : \boldsymbol{G}_p \boldsymbol{\Psi} \boldsymbol{K} = \boldsymbol{0},$$

where $\boldsymbol{G}_p$ is a matrix of order $(p-1) \times (pm)$ of the form

$$\boldsymbol{G}_p = [\boldsymbol{I}_p^* : \boldsymbol{I}_p^* : \ldots : \boldsymbol{I}_p^*],$$

where $\boldsymbol{I}_p^* = [\boldsymbol{0} : \boldsymbol{I}_{p-1}]$, which is of order $(p-1) \times p$. Thus, the testing of (12.116) can proceed as in Section 12.6.2.1 after replacing $\boldsymbol{G}_c$ by $\boldsymbol{G}_p$.

More details concerning the testing of the hypotheses H_{0c} and H_{0p}, including a discussion concerning the power of the multivariate tests, can be obtained from Khuri (1986). Furthermore, other aspects of the analysis of linear multiresponse models can be found in Chapter 7 in Khuri and Cornell (1996), and in the general review article by Khuri (1996b).

12.6.3 Testing for Lack of Fit

We may recall from Section 12.4 that a response surface model represents a polynomial approximation of the true functional relationship between a response of interest and a set of control variables. It is therefore possible that the fitted model may fail to adequately explain the behavior of the response. In this case, the model is said to suffer from *lack of fit* (LOF). This can be attributed to the omission of higher-order terms that involve the control variables in the model. For example, a first-degree model will be inadequate if in reality the response is better explained by a complete second-degree model, or perhaps a cubic model. Lack of fit can also be caused by the omission of control variables, other than those in the fitted model, which may have some effect on the response.

Testing for LOF of a single response surface model is a well-known procedure in response surface methodology (see, for example, Section 2.6 in Khuri and Cornell, 1996). A postulated model should pass the LOF test before it can be adopted and used in a given experimental situation. Such a test is therefore quite important and is considered an integral part of the repertoire of a response surface investigation.

In a multiresponse experiment, testing for LOF is more complex due to the interrelationships that may exist among the response variables. Lack of fit in one response may influence the fit of the other responses. Thus, the response models (in a multiresponse system) should not be tested individually for LOF. In this section, we show how to test for LOF of a linear multiresponse model. The development of this test is based on Khuri (1985).

Let us again consider model (12.105), assuming as before that the rows of the error matrix $\boldsymbol{\Upsilon}$ are mutually independent and distributed as $N(\boldsymbol{0}, \boldsymbol{\Sigma}_m)$. This model is said to suffer from LOF if it fails to provide an adequate representation of the true means of the m responses over the experimental region. By definition, a *multivariate lack of fit test* is a test used to check the adequacy of model (12.105). The development of this test begins with the reduction of the m responses into a single response through the creation of an arbitrary linear combination of the responses. Its corresponding model is then tested for LOF. For this purpose, let $\boldsymbol{c} = (c_1, c_2, \ldots, c_m)'$ be an arbitrary nonzero $m \times 1$ vector. Multiplying both sides of (12.105) on the right by $\boldsymbol{c}$, we get

$$\tilde{\mathcal{Y}}_c = \tilde{\mathcal{X}}\boldsymbol{\Psi}_c + \boldsymbol{\Upsilon}_c, \tag{12.117}$$

where

$\tilde{\mathcal{Y}}_c = \tilde{\mathcal{Y}}\boldsymbol{c} = \sum_{i=1}^{m} c_i \boldsymbol{Y}_i$
$\boldsymbol{\Psi}_c = \boldsymbol{\Psi}\boldsymbol{c}$
$\boldsymbol{\Upsilon}_c = \boldsymbol{\Upsilon}\boldsymbol{c}$

The vector $\tilde{\mathcal{Y}}_c$ consists of n observations on the univariate response $\tilde{y}_c = \sum_{i=1}^{m} c_i Y_i$. Note that $\boldsymbol{\Upsilon}_c$ has the normal distribution $N(\boldsymbol{0}, \sigma_c^2 \boldsymbol{I}_n)$, where $\sigma_c^2 = \boldsymbol{c}'\boldsymbol{\Sigma}_m\boldsymbol{c}$.

The multiresponse model in (12.105) is adequate if and only if the single-response models in (12.117) are adequate for all $\boldsymbol{c} \neq \boldsymbol{0}$. Equivalently, if for some $\boldsymbol{c} \neq \boldsymbol{0}$, model (12.117) is inadequate, then model (12.105) can be declared inadequate to represent the totality of the m response variables. We can then proceed to test model (12.117) for LOF for $\boldsymbol{c} \neq \boldsymbol{0}$ using the standard procedure for LOF for a single-response variable. This procedure requires the availability of replicate observations on the response variable under consideration (see Section 2.6 in Khuri and Cornell, 1996). For this purpose, we assume that the design used to fit the multiresponse model contains repeated runs on all the m responses at some points in the experimental region. This means that some rows in the matrix $\tilde{\mathcal{X}}$ in model (12.105) will be repeated. For convenience, we shall consider that the repeated runs are taken at each of the first n_0 $(1 \leq n_0 < n)$ points of the design matrix.

Since the error term in the single-response model in (12.117) has the normal distribution $N(0, \sigma_c^2 \boldsymbol{I}_n)$, then the corresponding LOF test for this model is given by (see formula (2.30) in Khuri and Cornell, 1996)

$$F_{LOF}(\boldsymbol{c}) = \frac{SS_{LOF}(\boldsymbol{c})/\nu_{LOF}}{SS_{PE}(\boldsymbol{c})/\nu_{PE}}, \tag{12.118}$$

where

$$SS_{PE}(\boldsymbol{c}) = \tilde{\mathcal{Y}}_c' \boldsymbol{M} \tilde{\mathcal{Y}}_c,$$

$$SS_{LOF}(\boldsymbol{c}) = \tilde{\mathcal{Y}}_c' [\boldsymbol{I}_n - \tilde{\mathcal{X}}(\tilde{\mathcal{X}}'\tilde{\mathcal{X}})^{-}\tilde{\mathcal{X}}' - \boldsymbol{M}] \tilde{\mathcal{Y}}_c,$$

$\boldsymbol{M}$ is the block-diagonal matrix,

$$\boldsymbol{M} = \text{diag}(\boldsymbol{M}_1, \boldsymbol{M}_2, \ldots, \boldsymbol{M}_{n_0}, \boldsymbol{0}), \tag{12.119}$$

and

$$M_i = I_{n_{0i}} - \frac{1}{n_{0i}} J_{n_{0i}}, \quad i = 1, 2, \ldots, n_0. \tag{12.120}$$

In (12.120), n_{0i} denotes the number of repeated observations on all the m responses at the ith repeat-runs site ($i = 1, 2, \ldots, n_0$). The sums of squares, $SS_{PE}(\mathbf{c})$ and $SS_{LOF}(\mathbf{c})$, are called the pure error and lack of fit sums of squares, respectively, and ν_{PE} and ν_{LOF} are their corresponding degrees of freedom given by

$$\nu_{PE} = \sum_{i=1}^{n_0} (n_{0i} - 1)$$

$$\nu_{LOF} = n - \rho - \sum_{i=1}^{n_0} (n_{0i} - 1),$$

where, if we recall, ρ is the rank of the matrix $\tilde{\mathcal{X}}$.

The lack of fit and pure error sums of squares can be written as

$$SS_{LOF}(\mathbf{c}) = \mathbf{c}' \mathbf{H}_1 \mathbf{c}, \tag{12.121}$$

$$SS_{PE}(\mathbf{c}) = \mathbf{c}' \mathbf{H}_2 \mathbf{c}, \tag{12.122}$$

where

$$\mathbf{H}_1 = \tilde{\mathcal{Y}}' [\mathbf{I}_n - \tilde{\mathcal{X}}(\tilde{\mathcal{X}}'\tilde{\mathcal{X}})^- \tilde{\mathcal{X}}' - \mathbf{M}] \tilde{\mathcal{Y}}, \tag{12.123}$$

$$\mathbf{H}_2 = \tilde{\mathcal{Y}}' \mathbf{M} \tilde{\mathcal{Y}}. \tag{12.124}$$

Thus, $F_{LOF}(\mathbf{c})$ in (12.118) can be expressed as

$$F_{LOF}(\mathbf{c}) = \frac{\nu_{PE}}{\nu_{LOF}} \frac{\mathbf{c}' \mathbf{H}_1 \mathbf{c}}{\mathbf{c}' \mathbf{H}_2 \mathbf{c}}. \tag{12.125}$$

A large value of $F_{LOF}(\mathbf{c})$ casts doubt on the adequacy of model (12.117) for $\mathbf{c} \neq \mathbf{0}$. Since model (12.105) is inadequate if and only if at least one of the models in (12.117) is inadequate for some $\mathbf{c} \neq \mathbf{0}$, we conclude that model (12.105) has a significant LOF if

$$\max_{\mathbf{c} \neq \mathbf{0}} \left[\frac{\mathbf{c}' \mathbf{H}_1 \mathbf{c}}{\mathbf{c}' \mathbf{H}_2 \mathbf{c}} \right]$$

exceeds a certain critical value. But, by using Theorems 3.10 and 3.11, it is easy to show that

$$\max_{\mathbf{c} \neq \mathbf{0}} \left[\frac{\mathbf{c}' \mathbf{H}_1 \mathbf{c}}{\mathbf{c}' \mathbf{H}_2 \mathbf{c}} \right] = e_{\max}(\mathbf{H}_2^{-1} \mathbf{H}_1), \tag{12.126}$$

where $e_{\max}(H_2^{-1}H_1)$ is the largest eigenvalue of $H_2^{-1}H_1$, which is the same as Roy's largest root test statistic. If λ_α denotes the upper $100\alpha\%$ point of the distribution of $e_{\max}(H_2^{-1}H_1)$ when model (12.105) is the true model, then a significant LOF for model (12.105) can be detected at the α level if

$$e_{\max}(H_2^{-1}H_1) \geq \lambda_\alpha. \tag{12.127}$$

This test is referred to as a *multivariate lack of fit test*. Note that the matrix H_2 is positive definite with probability 1 if $\nu_{PE} \geq m$ (see Roy, Gnanadesikan, and Srivastava, 1971, p. 35). Tables for the critical value λ_α are available in Roy, Gnanadesikan, and Srivastava (1971) and Morrison (1976); those for two and three responses are in Foster and Rees (1957) and Foster (1957), respectively.

12.6.3.1 Responses Contributing to LOF

In the event the multivariate LOF test is significant, it would be of interest to determine which of the m responses, or combinations thereof, are contributing to LOF. A significant value of $e_{\max}(H_2^{-1}H_1)$ indicates that there exists at least one linear combination of the responses, given by some nonzero value of c, for which model (12.117) is inadequate. In particular, the vector c^* which maximizes $c'H_1c/c'H_2c$ is such a vector. We thus have

$$\max_{c\neq 0} \frac{c'H_1c}{c'H_2c} = e_{\max}(H_2^{-1}H_1)$$
$$= \frac{c^{*'}H_1c^*}{c^{*'}H_2c^*}. \tag{12.128}$$

From (12.128) we conclude that

$$c^{*'}\,[H_1 - e_{\max}(H_2^{-1}H_1)H_2]\,c^* = 0,$$

or equivalently,

$$c^{*'}\,H_2^{1/2}\left[e_{\max}(H_2^{-1}H_1)I_m - H_2^{-1/2}H_1H_2^{-1/2}\right]H_2^{1/2}\,c^* = 0. \tag{12.129}$$

Since the matrix

$$e_{\max}(H_2^{-1}H_1)I_m - H_2^{-1/2}H_1H_2^{-1/2}$$

is positive semidefinite, the equality in (12.129) is true if and only if

$$\left[e_{\max}(H_2^{-1}H_1)I_m - H_2^{-1/2}H_1H_2^{-1/2}\right]H_2^{1/2}c^* = \mathbf{0}.$$

Multiplying both sides on the left by $-H_2^{1/2}$, we get

$$[H_1 - e_{max}(H_2^{-1}H_1)H_2]\, c^* = 0,$$

or equivalently,

$$H_2^{-1}H_1c^* = e_{max}(H_2^{-1}H_1)\, c^*. \qquad (12.130)$$

Formula (12.130) indicates that c^* is an eigenvector of $H_2^{-1}H_1$ corresponding to its largest eigenvalue, $e_{max}(H_2^{-1}H_1)$. Using c^*, we get the following linear combination of the responses representing the univariate response $\tilde{\mathcal{Y}}_{c^*}$,

$$\tilde{\mathcal{Y}}_{c^*} = \sum_{i=1}^{m} c_i^* Y_i, \qquad (12.131)$$

where c_i^* is the ith element of c^* ($i = 1, 2, \ldots, m$). The responses that correspond to nonzero values of c_i^* in (12.131) are believed to contribute to LOF. Since the response variables may be measured in different units, they must be standardized. We can therefore rewrite (12.131) as

$$\tilde{\mathcal{Y}}_{c^*} = \sum_{i=1}^{m} d_i^* Z_i, \qquad (12.132)$$

where $Z_i = \frac{1}{\|Y_i\|} Y_i$, $d_i^* = c_i^* \parallel Y_i \parallel$, and $\parallel Y_i \parallel$ is the Euclidean norm of Y_i ($i = 1, 2, \ldots, m$). Large values of $\mid d_i^* \mid$ correspond to responses that are influential contributors to LOF of model (12.105). In addition, we can consider subsets of the m responses and determine if the responses in each subset contribute significantly to LOF. For this purpose, let $\mathcal{S} = \{Y_{i_1}, Y_{i_2}, \ldots, Y_{i_s}\}$ denote a nonempty subset of the m responses ($1 \leq i_1 < i_2 < \ldots < i_s \leq m$). We can then compute $e_{max}(H_2^{-1}H_1)$ based on only the elements of $\mathcal{S}$. Let such a value be denoted by $e_{max}(H_2^{-1}H_1)_{\mathcal{S}}$. This is the maximum of $c'H_1c/c'H_2c$ constrained by putting $c_i = 0$ for all the responses that do not belong to $\mathcal{S}$. Thus,

$$e_{max}(H_2^{-1}H_1)_{\mathcal{S}} \leq e_{max}(H_2^{-1}H_1), \qquad (12.133)$$

for all nonempty subsets of the m responses. A subset $\mathcal{S}$ suffices to produce a significant LOF if

$$e_{max}(H_2^{-1}H_1)_{\mathcal{S}} \geq \lambda_\alpha, \qquad (12.134)$$

where λ_α is the α-critical value of $e_{max}(H_2^{-1}H_1)$. We can see from (12.133) and (12.134) that whenever (12.134) holds for a nonempty subset $\mathcal{S}$, the inequality in (12.127) must also hold. It follows that if E and $E_{\mathcal{S}}$ are the events

$$E = \{e_{max}(H_2^{-1}H_1) \geq \lambda_\alpha\},$$
$$E_{\mathcal{S}} = \{e_{max}(H_2^{-1}H_1)_{\mathcal{S}} \geq \lambda_\alpha\},$$

then

$$P\left(\bigcup_{\mathcal{S}} E_{\mathcal{S}}\right) \leq P(E) = \alpha. \tag{12.135}$$

The inequality in (12.135) implies that when all nonempty subsets of the m responses are examined for LOF, the probability of falsely detecting a significant subset cannot exceed the value α. Thus, the use of the same critical value, λ_α, in these simultaneous tests helps control the experimentwise Type I error rate at a level not exceeding α.

In summary, a multiresponse model should be tested for LOF before it can be used in the analysis of a multiresponse experiment. If LOF is detected, then a subsequent investigation should be made to determine which responses contribute significantly to LOF. This can be achieved by an examination of the elements of the eigenvector of $\boldsymbol{H}_2^{-1}\boldsymbol{H}_1$ corresponding to its largest eigenvalue, followed by an inspection of the values of $e_{max}(\boldsymbol{H}_2^{-1}\boldsymbol{H}_1)_{\mathcal{S}}$ for all nonempty subsets of the responses. The information gained from this investigation can then be used to upgrade the models for only those responses deemed to be influential contributors to LOF. This is contingent on the ability of the design (the one used for model 12.105) to support the fit of the upgraded models. If this is not possible, then the design should be augmented with additional experimental runs. Such model upgrades can be helpful in eliminating LOF, hence improving model adequacy for the entire multiresponse system.

Example 12.4 This example is given in Khuri (1985). It concerns an experiment in food science in which three researchers (Richert, Morr, and Cooney, 1974) investigated the effects of five control variables, namely, heating temperature (x_1), pH level (x_2), redox potential (x_3), sodium oxalate (x_4), and sodium lauryl sulfate (x_5) on the foaming properties of whey protein concentrates (WPC), which are of considerable interest to the food industry. Measurements were made on the following three responses: Y_1 = whipping time (the total elapsed time required to produce peaks of foam formed during whipping of a liquid sample), Y_2 = maximum overrun [determined by weighing 5-oz. paper cups of foam and unwhipped liquid sample and calculating the expression, 100 (weight of liquid − weight of foam)/weight of foam], and Y_3 = percent soluble protein. The design used consisted of a one-half fraction of a 2^5 factorial design in addition to five pairs of symmetric (with respect to the origin) axial points at the same distance from the origin, and five center point replications. Such a design is called a *central composite design* (see Khuri and Cornell, 1996, Section 4.5.3), and is suitable for fitting second-degree models. The original and coded levels of the five control variables are given in Table 12.6.

A listing of the points of the central composite design (in coded form) and the corresponding multiresponse data (values of Y_1, Y_2, and Y_3) is given in Table 12.7.

TABLE 12.6
The Original and Coded Levels of the Control Variables

Variable		Coded Levels				
		−2	−1	0	1	2
Heating temperature (°C)	x_1	65.0	70.0	75.0	80.0	85.0
pH	x_2	4.0	5.0	6.0	7.0	8.0
Redox potential (volt)	x_3	−0.025	0.075	0.175	0.275	0.375
Sodium oxalate (molar)	x_4	0.0	0.0125	0.025	0.0375	0.05
Sodium lauryl sulfate (%)	x_5	0.0	0.05	0.10	0.15	0.20

Source: Khuri, A.I., *Technometrics*, 27, 213, 1985. With permission.

TABLE 12.7
Central Composite Design Points and the Multiresponse Data

x_1	x_2	x_3	x_4	x_5	Y_1 (min)	Y_2 (%)	Y_3 (%)
0	0	0	0	0	3.5	1179	104
0	0	0	0	0	3.5	1183	107
0	0	0	0	0	4.0	1120	104
0	0	0	0	0	3.5	1180	101
0	0	0	0	0	3.0	1195	103
−1	−1	−1	−1	1	4.75	1082	81.4
1	−1	−1	−1	−1	4.0	824	69.6
−1	1	−1	−1	−1	5.0	953	105
1	1	−1	−1	1	9.5	759	81.2
−1	−1	1	−1	−1	4.0	1163	80.8
1	−1	1	−1	1	5.0	839	76.3
−1	1	1	−1	1	3.0	1343	103
1	1	1	−1	−1	7.0	736	76.9
−1	−1	−1	1	−1	5.25	1027	87.2
1	−1	−1	1	1	5.0	836	74.0
−1	1	−1	1	1	3.0	1272	98.5
1	1	−1	1	−1	6.5	825	94.1
−1	−1	1	1	1	3.25	1363	95.9
1	−1	1	1	−1	5.0	855	76.8
−1	1	1	1	−1	2.75	1284	100
1	1	1	1	1	5.0	851	104
−2	0	0	0	0	3.75	1283	100
2	0	0	0	0	11.0	651	50.5
0	−2	0	0	0	4.5	1217	71.2
0	2	0	0	0	4.0	982	101
0	0	−2	0	0	5.0	884	85.8
0	0	2	0	0	3.75	1147	103
0	0	0	−2	0	3.75	1081	104
0	0	0	2	0	4.75	1036	89.4
0	0	0	0	−2	4.0	1213	105
0	0	0	0	2	3.5	1103	113

Source: Khuri, A.I., *Technometrics*, 27, 213, 1985. With permission.

The fitted model for each of the three responses is a full quadratic model in $x_1, \ldots, x_5$ of the form:

$$Y = \beta_0 + \sum_{i=1}^{5} \beta_i x_i + \sum_{i<j}\sum \beta_{ij} x_i x_j + \sum_{i=1}^{5} \beta_{ii} x_i^2 + \epsilon.$$

In this case, the matrix $\tilde{\boldsymbol{\mathcal{X}}} = [\boldsymbol{X}_1 : \boldsymbol{X}_2 : \boldsymbol{X}_3]$ in model (12.105) is of order 31×63 and rank $\rho = 21$ since $\boldsymbol{X}_1 = \boldsymbol{X}_2 = \boldsymbol{X}_3$. The matrix $\boldsymbol{M}$ in (12.119) is equal to $\boldsymbol{M} = \text{diag}(\boldsymbol{M}_1, \boldsymbol{0})$, where $\boldsymbol{M}_1 = \boldsymbol{I}_5 - \frac{1}{5}\boldsymbol{J}_5$ since the center point is the only replicated design point. Hence, the pure error and lack of fit degrees of freedom are $\nu_{PE} = 4$ and $\nu_{LOF} = 6$. The corresponding matrices, $\boldsymbol{H}_2$ and $\boldsymbol{H}_1$, can be computed using formulas (12.124) and (12.123), respectively.

The value of Roy's largest test statistic in (12.127) is $e_{\max}(\boldsymbol{H}_2^{-1}\boldsymbol{H}_1) = 245.518$. The corresponding critical value at the 10% level is $\lambda_{0.10} = 85.21$. Hence, a significant lack of fit can be detected at the 10% level.

Let us now assess the contribution of the three responses to lack of fit as was discussed in Section 12.6.3.1. The eigenvector, $\boldsymbol{c}^*$, of $\boldsymbol{H}_2^{-1}\boldsymbol{H}_1$ corresponding to its largest eigenvalue, 245.518, is $\boldsymbol{c}^* = (3.2659, 0.0385, -0.0904)'$. The Euclidean norms of the response data vectors are $\| \boldsymbol{Y}_1 \| = 27.60$, $\| \boldsymbol{Y}_2 \| = 5929.27$, $\| \boldsymbol{Y}_3 \| = 517.49$. Thus, the linear combination of standardized responses in (12.132) is written as

$$\tilde{\mathcal{Y}}_{c^*} = 90.139 Z_1 + 228.277 Z_2 - 46.781 Z_3.$$

The right-hand side is proportional to

$$0.395 Z_1 + Z_2 - 0.205 Z_3. \tag{12.136}$$

Using the size of the absolute values of the coefficients of Z_1, Z_2, and Z_3 in (12.136), we conclude that the response Y_2 is most influential with respect to lack of fit, followed by the response Y_1. The next step is to consider values of $e_{\max}(\boldsymbol{H}_2^{-1}\boldsymbol{H}_1)_{\mathcal{S}}$ for all nonempty subsets, $\mathcal{S}$, of Y_1, Y_2, and Y_3. These values are given in Table 12.8.

From Table 12.8 we can see that in addition to the subset of all three responses, the only other significant subset at the $\alpha = 0.10$ level is $\{Y_1, Y_2\}$. This is consistent with our earlier finding about Y_1 and Y_2.

On the basis of the above analysis, we can state that the responses Y_1 and Y_2 together produce a significant lack of fit. Hence, any future upgrade of the models should target these two responses in order to improve the adequacy of the multiresponse system.

Note that the value of $e_{\max}(\boldsymbol{H}_2^{-1}\boldsymbol{H}_1)_{\mathcal{S}}$ for each individual response in Table 12.8 is the ratio of the lack of fit sum of squares to the pure error sum of squares resulting from the analysis of each of the three individually-fitted response models. If each such ratio is multiplied by $\nu_{PE}/\nu_{LOF} = 2/3$, we obtain the value of the F-statistic that would result from applying the univariate lack of fit test to each of the three responses.

TABLE 12.8
Values of $e_{max}(H_2^{-1}H_1)_{\mathcal{S}}$ for All Nonempty Subsets of the Responses

Subset, $\mathcal{S}$	$e_{max}(H_2^{-1}H_1)_{\mathcal{S}}$	Critical Value ($\lambda_{0.10}$)
Y_1, Y_2, Y_3	245.518*	85.21
Y_1, Y_2	214.307*	85.21
Y_1, Y_3	45.532	85.21
Y_2, Y_3	32.107	85.21
Y_1	14.936	85.21
Y_2	19.573	85.21
Y_3	28.495	85.21

Source: Khuri, A.I., *Technometrics*, 27, 213, 1985. With permission.
* Significant at the 10% level.

Exercises

12.1 Consider Section 12.2. Show that if the error variances, namely, $\sigma_1^2, \sigma_2^2, \ldots, \sigma_k^2$, are heterogeneous, then $\boldsymbol{Y}'\boldsymbol{A}\boldsymbol{Y}$ and $\boldsymbol{Y}'\boldsymbol{B}\boldsymbol{Y}$ are not independent and not distributed as scaled chi-squared variates, where $\boldsymbol{A}$ and $\boldsymbol{B}$ are given by (12.5) and (12.6), respectively. The normality assumptions concerning the distributions of $\alpha_{(i)}$ and $\epsilon_{i(j)}$ in model (12.1) are considered valid.

12.2 Consider model (12.1), where $k = 5$, $n_1 = 14$, $n_2 = 21$, $n_3 = 6$, $n_4 = 4$, $n_5 = 5$, $\sigma_1^2 = 0.10$, $\sigma_2^2 = 0.10$, $\sigma_3^2 = 10.00$, $\sigma_4^2 = 0.10$, $\sigma_5^2 = 1.00$. Use formula (12.10) to compute the probability,

$$\alpha_f = P[F \geq F_{0.05,4,45} \mid \sigma_\alpha^2 = 0],$$

where F is given by (12.4).

12.3 Consider the confidence interval on σ_α^2 given in (12.34). Suppose that $k = 10$, $\alpha = 0.05$, $n_1 = 2$, $n_2 = 2$, $n_3 = 10$, $n_4 = 20$, $n_5 = 30$, $n_6 = 40$, $n_7 = 50$, $n_8 = 60$, $n_9 = 70$, $n_{10} = 80$; $\sigma_1^2 = 0.001$, $\sigma_2^2 = 0.01$, $\sigma_3^2 = 0.10$, $\sigma_4^2 = 1.0$, $\sigma_5^2 = 5.0$, $\sigma_6^2 = 10.0$, $\sigma_7^2 = 20.0$, $\sigma_8^2 = 30.0$, $\sigma_9^2 = 50.0$, $\sigma_{10}^2 = 100.0$.

Use computer simulation to estimate the coverage probability for this confidence interval given that $\sigma_\alpha^2 = 0.50$ [without loss of generality, assign μ the value 0 in model (12.1)].

12.4 Consider the data set given below from a completely randomized design associated with model (12.1) with $k = 6$.

(a) Test the homogeneity of variances hypothesis in (12.35) (with $k = 6$). Let $\alpha = 0.05$.

(b) Test the hypothesis $H_0 : \sigma^2_\alpha = 0$ at the 5% level using the test statistic in (12.12).

(c) Use the interval in (12.34) to obtain a confidence interval on σ^2_α with an approximate confidence coefficient greater than or equal to 0.95.

Sample 1	**Sample 2**	**Sample 3**	**Sample 4**	**Sample 5**	**Sample 6**
44	65	50	45	40	32
35	55	40	18	66	66
38		59	75	49	55
59		40	41	68	35
29		62	12	76	50
		68		84	78
		72		86	52
					22
					68

12.5 The scores on a certain achievement test were compared for three specific metropolitan areas in a certain state. Four schools were randomly selected in each area and the average scores for three classes in each school were obtained. The results are shown in the following table:

Metropolitan Area	**School**			
	1	**2**	**3**	**4**
1	80.6	82.9	79.5	90.5
	85.2	87.1	78.1	85.8
	79.1	81.5	75.3	95.1
2	80.1	65.1	77.2	80.9
	70.3	59.7	75.3	82.3
	65.7	50.3	68.7	70.5
3	91.8	78.8	84.4	90.3
	82.5	61.3	81.3	66.5
	80.8	65.5	76.6	55.1

(a) Test equality of the mean scores for the three metropolitan areas. Let $\alpha = 0.05$.

(b) Test the hypothesis $H_0 : \kappa_i^2 = 0,\ i = 1, 2, 3$, where κ_i^2 denotes the variability associated with schools in metropolitan area $i\ (i = 1, 2, 3)$. Let $\alpha = 0.05$.

(c) If the test in part (b) is significant, determine which metropolitan areas have significant variations in the test scores among their schools. Let $\alpha^* = 0.05$, where α^* is the experimentwise Type I error rate.

12.6 An experiment was conducted to compare the effects of four fertilizer treatments, denoted by FT1, FT2, FT3, and FT4, on the yield of a variety of corn. Each treatment was assigned to three farms selected at random in some agricultural area. The corn yields (in kg) obtained from three plots chosen in each farm were recorded (all chosen plots had equal areas and similar soil compositions). The data are given in the following table:

Treatment	**Farm**		
	1	**2**	**3**
FT1	43.5	59.7	49.9
	65.5	52.8	54.3
	47.0	63.5	54.1
FT2	122.9	74.2	48.8
	118.5	77.9	60.4
	104.4	61.4	52.6
FT3	148.1	117.9	95.2
	119.5	107.8	88.3
	150.8	96.7	91.7
FT4	81.5	64.2	49.3
	91.7	65.4	57.2
	76.3	60.5	51.9

(a) Test equality of the means of the four fertilizer treatments. Let $\alpha = 0.05$.

(b) Test the hypothesis $H_0 : \kappa_i^2 = 0,\ i = 1, 2, 3, 4$, where κ_i^2 is the variability associated with farms within treatment i ($i = 1, 2, 3, 4$). Let $\alpha = 0.05$.

(c) If the test in part (b) is significant, determine which of the κ_i^2's are different from zero. Let $\alpha^* = 0.05$, where α^* is the experimentwise Type I error rate.

12.7 Two types of fertilizers were applied to experimental plots to assess their effects on the yield of peanuts measured in pounds per plot. The amount (in lb) of each fertilizer applied to a plot was determined by the coordinate settings of a central composite design. The data in the following table consist of the original and coded settings (denoted by x_1 and x_2) of the two fertilizers and the corresponding yield values:

Fertilizer 1	Fertilizer 2	x_1	x_2	Yield
60	20	-1	-1	7.59
110	20	1	-1	13.31
60	30	-1	1	14.55
110	30	1	1	17.48
49.64	25	$-\sqrt{2}$	0	8.66
120.36	25	$\sqrt{2}$	0	14.97
85	17.93	0	$-\sqrt{2}$	8.10
85	32.07	0	$\sqrt{2}$	16.49
85	25	0	0	15.71

(a) Fit a second-degree model to the peanut yields in the coded variables, $x_1 = (\text{Fertilizer 1} - 85)/25$, $x_2 = (\text{Fertilizer 2} - 25)/5$.

(b) Give an estimate of the error variance, σ^2_ϵ, using MS_E, the error (residual) mean square from the corresponding ANOVA table.

(c) Use the estimate of σ^2_ϵ in part (b) to obtain estimates of the variances of the least-squares estimates of the parameters in the fitted model in part (a).

(d) Find the predicted response values at the points, $x_1 = -0.5$, $x_2 = 0.5$; $x_1 = 0$, $x_2 = -0.5$. What are the estimated prediction variance values at these locations?

12.8 Verify the entries in Table 12.5 by using PROC MIXED.

12.9 An experiment was performed to determine the effects of storage conditions on the quality of freshly harvested "Delicious" apples. The apples were harvested on the same date from four different orchards, which were selected at random from a certain agricultural district. Two control variables were considered, namely, X_1 = number of weeks in storage after harvest and X_2 = storage temperature (°C) after harvest. The quality of apples was measured by Y = amount of extractable juice (mL/100 g fluid weight). The same 4×4 factorial design was used to run the experiment in all four orchards. The design settings in the original levels of the control variables and the corresponding response values are given in the table below.

Suppose that the following second-degree model in x_1 and x_2 is fitted to the responses data:

$$\eta(x) = \beta_0 + \beta_1 x_1 + \beta_2 x_2 + \beta_{12} x_1 x_2 + \beta_{11} x_1^2 + \beta_{22} x_2^2,$$

X_1	X_2	Y			
		Orchard 1	**Orchard 2**	**Orchard 3**	**Orchard 4**
1	0	75.3	74.1	74.5	75.2
2	0	72.9	74.9	75.4	70.0
3	0	74.8	74.5	73.2	76.1
4	0	74.0	73.4	72.4	76.0
1	4.4	75.1	73.5	74.1	76.0
2	4.4	73.4	73.3	75.4	72.6
3	4.4	71.9	74.0	74.5	75.4
4	4.4	73.0	73.8	74.1	76.4
1	12.8	73.1	74.0	72.2	72.1
2	12.8	72.0	71.6	72.5	70.4
3	12.8	71.0	72.5	71.0	70.0
4	12.8	70.2	72.0	70.2	69.6
1	20.0	74.6	72.5	72.4	70.1
2	20.0	71.1	73.1	71.3	68.5
3	20.0	69.8	68.3	69.0	71.3
4	20.0	68.1	69.0	69.9	69.0

where x_1 and x_2 are the coded variables,

$$x_1 = \frac{X_1 - 2.5}{1.12},$$
$$x_2 = \frac{X_2 - 9.3}{7.70}.$$

(a) Obtain the estimated generalized least-squares estimates of the model's parameters. Then, test the significance of these parameters at the 5% level (use PROC MIXED).

(b) Test the significance of the variance components, σ_γ^2, σ_1^2, σ_2^2, σ_{12}^2, σ_{11}^2, and σ_{22}^2 at the 5% level, where σ_γ^2 is the variance component associated with the orchard effect, and the remaining variance components correspond to interactions between the orchards and the polynomial terms x_1, x_2, x_1x_2, x_1^2, and x_2^2, respectively.

(c) Obtain an expression for the predicted response, $\hat{Y}_e(\boldsymbol{x})$.

(d) Use part (c) to find the maximum value of $\hat{Y}_e(\boldsymbol{x})$ over the region, $-1.34 \leq x_1 \leq 1.34$, $-1.21 \leq x_2 \leq 1.39$.

12.10 Consider the multiresponse data given in the table below.

x_1	x_2	x_3	x_4	Y_1	Y_2	Y_3
0	0	0	0	61.21	44.77	24.01
0	0	0	0	62.36	46.70	26.30
0	0	0	0	62.90	44.99	27.50
0	0	0	0	61.55	47.00	30.11
0	0	0	0	63.61	48.09	30.98
−1	−1	−1	−1	58.89	57.95	34.40
1	−1	−1	−1	79.30	54.16	41.60
−1	1	−1	−1	46.71	52.79	25.51
1	1	−1	−1	72.60	35.09	25.29
−1	−1	1	−1	46.29	58.81	27.31
1	−1	1	−1	65.61	53.81	33.10
−1	1	1	−1	34.11	57.81	20.01
1	1	1	−1	58.21	47.10	28.66
−1	−1	−1	1	70.62	41.12	29.89
1	−1	−1	1	85.59	30.41	24.12
−1	1	−1	1	62.06	36.66	24.90
1	1	−1	1	76.11	34.85	29.01
−1	−1	1	1	61.02	64.18	38.56
1	−1	1	1	75.36	44.30	34.29
−1	1	1	1	51.96	65.34	36.12
1	1	1	1	65.61	37.92	27.25

The fitted models are of the form

$$Y_1 = \beta_{10} + \sum_{i=1}^{4} \beta_{1i} x_i + \beta_{114} x_1 x_4 + \epsilon_1,$$

$$Y_2 = \beta_{20} + \beta_{21} x_1 + \beta_{23} x_3 + \beta_{24} x_4 + \beta_{234} x_3 x_4 + \epsilon_2,$$

$$Y_3 = \beta_{30} + \beta_{31} x_1 + \beta_{32} x_2 + \beta_{34} x_4 + \beta_{314} x_1 x_4 + \beta_{324} x_2 x_4 + \epsilon_3.$$

(a) Test for lack of fit of the multiresponse model at the 10% level. [Note: Here, $\nu_{PE} = 4$, $\nu_{LOF} = n - \rho - \nu_{PE} = 9$, where ρ is the rank of $\tilde{\mathcal{X}}$, which is equal to 8. Using Foster's (1957) tables, the 10% critical value for Roy's largest root test statistic is $\lambda_{0.10} = 127.205$.]

(b) Assess the contributions of the three responses to lack of fit, if any.

13

Generalized Linear Models

The purpose of this chapter is to provide an introduction to the subject of generalized linear models. This is a relatively new area in the field of statistical modeling that was originally developed to establish a unified framework for the modeling of discrete as well as continuous data. The topics covered in this chapter include an introduction to the exponential family of distributions, estimation of parameters for a generalized linear model using the method of maximum likelihood, measures of goodness of fit, hypothesis testing, confidence intervals, and gamma-distributed responses.

13.1 Introduction

The models discussed in the previous chapters were of the linear type representing a response variable, Y, which can be measured on a continuous scale. Its distribution was, for the most part, assumed to be normal. There are, however, many experimental situations where linearity of the postulated model and/or normality of the response Y may not be quite valid. For example, Y may be a *discrete random variable,* that is, it has values that can be counted, or put in a one-to-one correspondence with the set of positive integers. A *binary response* having two possible outcomes, labeled as success or failure, is one example of a discrete random variable. In this case, one is usually interested in modeling the probability of success. In another situation, the response may have values in the form of counts (number of defects in a given lot, or the number of traffic accidents at a busy intersection in a given time period). Other examples of response variables that are not normally distributed can be found in biomedical applications, clinical trials, and quality engineering, to name just a few.

Generalized linear models (GLMs) were introduced as an extension of the class of linear models. Under the framework of GLMs, discrete as well as continuous response variables can be accommodated, and the usual assumptions of normality and constant variance are not necessarily made on the response under consideration. Thus, GLMs provide a unified representation for a large class of models for discrete and continuous response variables. They have therefore proved to be very effective in several areas of application. For example, in biological assays, reliability and quality engineering, survival analysis, and a variety of applied biomedical fields, the use of GLMs has become quite prevalent.

Generalized linear models were first introduced by Nelder and Wedderburn (1972). A classic book on the topic is the one by McCullagh and Nelder (1989). In addition, the more recent books by Lindsey (1997), McCulloch and Searle (2001), Dobson (2008), and Myers, Montgomery, and Vining (2002) provide added insights into the application and usefulness of GLMs.

13.2 The Exponential Family

There are three components that define GLMs, they are

(a) The elements of a response vector, $\boldsymbol{Y}$, are distributed independently according to a certain probability distribution considered to belong to the *exponential family* with a probability mass function (or a density function for a continuous distribution) given by

$$f(y,\theta,\phi) = \exp\left[\frac{\theta y - b(\theta)}{a(\phi)} + c(y,\phi)\right], \tag{13.1}$$

where

$a(.)$, $b(.)$, and $c(.)$ are known functions

θ is called the *canonical parameter*, which is a function of the mean, μ, of the distribution

ϕ is a *dispersion parameter* (see, for example, McCullagh and Nelder, 1989, p. 28).

(b) A linear model of the form

$$\eta(\boldsymbol{x}) = \boldsymbol{f}'(\boldsymbol{x})\boldsymbol{\beta}, \tag{13.2}$$

which relates the so-called *linear predictor*, η, to a set of k control variables, $x_1, x_2, \ldots, x_k$, where $\boldsymbol{x} = (x_1, x_2, \ldots, x_k)'$ and $\boldsymbol{f}(\boldsymbol{x})$ is a known q-component vector function of $\boldsymbol{x}$, and $\boldsymbol{\beta}$ is a vector of q unknown parameters.

(c) A *link function* $g(.)$ which relates $\eta(\boldsymbol{x})$ in (13.2) to the mean response at $\boldsymbol{x}$, denoted by $\mu(\boldsymbol{x})$, so that

$$\eta(\boldsymbol{x}) = g[\mu(\boldsymbol{x})], \tag{13.3}$$

where $g(.)$ is a monotone differentiable function whose inverse function is $h(.)$, that is,

$$\mu(\boldsymbol{x}) = h[\eta(\boldsymbol{x})]. \tag{13.4}$$

Thus, in a generalized linear model, a particular transformation of the mean response, namely $g[\mu(\boldsymbol{x})]$, is represented as a linear model in terms of $x_1, x_2, \ldots, x_k$. Since $h(.)$ can be a nonlinear function, formula (13.4) indicates that the mean response $\mu(\boldsymbol{x})$ is, in general, represented by a nonlinear model.

In particular, if $g(.)$ is the identity function and the response Y has the normal distribution, we obtain the special class of linear models. Also, if $g(\mu) = \theta$, where θ is the canonical parameter in (13.1), then $g(.)$ is called the *canonical link*.

The most commonly used link functions are

(i) The logit function,

$$\eta(\boldsymbol{x}) = \log\left[\frac{\mu(\boldsymbol{x})}{1 - \mu(\boldsymbol{x})}\right].$$

(ii) The probit function,

$$\eta(\boldsymbol{x}) = F^{-1}[\mu(\boldsymbol{x})],$$

where $F^{-1}(.)$ is the inverse of the cumulative distribution function of the standard normal distribution.

(iii) The logarithmic function,

$$\eta(\boldsymbol{x}) = \log[\mu(\boldsymbol{x})].$$

(iv) The inverse polynomial function,

$$\eta(\boldsymbol{x}) = \frac{1}{\mu(\boldsymbol{x})}.$$

This is also called the *reciprocal link*.

A listing of other link functions can be found in McCullagh and Nelder (1989, Table 2.1).

Several distributions can be classified as belonging to the exponential family. Here are some examples.

(a) The normal distribution

This is a well-known member of the exponential family. The density function for a normal random variable Y with mean μ and variance σ^2 is

$$f(y, \mu, \sigma^2) = \frac{1}{\sqrt{2\pi\sigma^2}} \exp\left[-\frac{1}{2\sigma^2}(y - \mu)^2\right],$$

which can be written as

$$f(y, \mu, \sigma^2) = \exp\left\{\frac{\mu y - \frac{\mu^2}{2}}{\sigma^2} - \frac{1}{2}\left[\frac{y^2}{\sigma^2} + \log(2\pi\sigma^2)\right]\right\}.$$

Comparing this with (13.1), we find that $\theta = \mu$, $b(\theta) = \frac{\mu^2}{2}$, $a(\phi) = \phi = \sigma^2$, and

$$c(y, \phi) = -\frac{1}{2}\left[\frac{y^2}{\sigma^2} + \log(2\pi\sigma^2)\right].$$

Note that the canonical parameter θ is equal to μ, which is a location parameter, and the dispersion parameter is σ^2 (the canonical link function here is the identity).

(b) The Poisson distribution

This is a discrete distribution concerning a random variable Y that takes the values $0, 1, 2, \ldots$ according to the probability mass function,

$$f(y, \lambda) = \frac{\exp(-\lambda)\,\lambda^y}{y!}, \quad y = 0, 1, 2, \ldots,$$

where λ is the mean, μ, of Y. This probability mass function can be expressed as

$$f(y, \lambda) = \exp[y \log\lambda - \lambda - \log(y!)]. \tag{13.5}$$

By comparing (13.5) with (13.1) we find that $\theta = \log\lambda$, $b(\theta) = \lambda = \exp(\theta)$, $a(\phi) = 1$, and $c(y, \phi) = -\log(y!)$. Hence, the canonical parameter is $\log\lambda$ and the dispersion parameter is $\phi = 1$. The corresponding canonical link function is

$$\begin{aligned} g(\mu) &= \theta \\ &= \log\lambda \\ &= \log\mu, \end{aligned}$$

which is the logarithmic link function mentioned earlier.

The Poisson distribution is used to model count data such as the number of occurrences of some event in a particular time period, or in a given space such as the number of defects in a production process.

(c) The binary distribution

Consider a series of independent trials in each of which we have two possible outcomes, success or failure. The probability of success on a single trial is denoted by p. Let Y be a random variable that takes the value 1 if success is attained on a given trial, otherwise, it takes the value 0. In this case, the mean, μ, of Y is equal to p, and the corresponding probability mass function is

$$f(y, p) = p^y (1-p)^{1-y}, \quad y = 0, 1.$$

This can be expressed as

$$f(y, p) = \exp\{y [\log p - \log(1-p)] + \log(1-p)\}, \quad y = 0, 1, \qquad (13.6)$$

which has the same form as (13.1) with $\theta = \log p - \log(1-p) = \log\left(\frac{p}{1-p}\right)$, $b(\theta) = -\log(1-p) = \log[1 + \exp(\theta)]$, $a(\phi) = \phi = 1$, $c(y, \phi) = 0$. The corresponding canonical link function is

$$\begin{aligned} g(\mu) &= \theta \\ &= \log\left(\frac{p}{1-p}\right) \\ &= \log\left(\frac{\mu}{1-\mu}\right), \end{aligned} \qquad (13.7)$$

which is the logit function mentioned earlier. It is also called the *logistic link function*. Using the linear predictor in (13.2) at a point $\boldsymbol{x} = (x_1, x_2, \ldots, x_k)'$, namely $\eta(\boldsymbol{x})$, it is possible to use (13.7) to express the probability of success, $p = \mu$, as a function of $\boldsymbol{x}$, namely $p(\boldsymbol{x})$, of the form

$$p(\boldsymbol{x}) = \frac{\exp[\boldsymbol{f}'(\boldsymbol{x})\boldsymbol{\beta}]}{1 + \exp[\boldsymbol{f}'(\boldsymbol{x})\boldsymbol{\beta}]}. \qquad (13.8)$$

This is a nonlinear model for $p(\boldsymbol{x})$ called the *logistic regression model*. It is used with data that are observed in the form of proportions.

Other possible link functions for the binary distribution include the *probit function*, $F^{-1}(p)$, that was mentioned earlier, and the *complementary log-log function*, $\log[-\log(1-p)]$.

A closely related distribution to the binary distribution is the binomial distribution. In the latter case, if Y denotes the number of successes in the series of n independent trials mentioned earlier, where the probability of success, p, on a single trial is the same in all trials, then Y is called a *binomial random variable*. Its probability mass function is

$$f(y, p) = \binom{n}{y} p^y (1-p)^{n-y}, \quad y = 0, 1, 2, \ldots, n.$$

This distribution belongs to the exponential family with $\theta = \log\left(\frac{p}{1-p}\right)$, $b(\theta) = n\log(1 + \exp(\theta))$, $a(\phi) = \phi = 1$, $c(y, \phi) = \log\binom{n}{y}$. Its mean and variance are given by $\mu = np$ and $\sigma^2 = np(1-p)$, and the canonical link function is $g(\mu) = \theta = \log\left(\frac{p}{1-p}\right) = \log\left(\frac{\mu}{n-\mu}\right)$. Note that the binary distribution is a special case of this distribution with $n = 1$.

13.2.1 Likelihood Function

In formula (13.1), $f(y, \theta, \phi)$ is considered a function of y for fixed θ and ϕ, where y is a value of the random variable Y. By definition, the *likelihood function* associated with Y is a function, $\mathcal{L}(\theta, \phi, Y)$, of θ and ϕ such that for an observed value, y, of Y,

$$\mathcal{L}(\theta, \phi, y) = f(y, \theta, \phi). \tag{13.9}$$

The placement of the arguments θ and ϕ first is to emphasize that $\mathcal{L}$ is basically a function of θ and ϕ. Since this function is determined by the outcome of Y, $\mathcal{L}(\theta, \phi, Y)$ is therefore a random variable. The *log-likelihood function* is the logarithm of $\mathcal{L}(\theta, \phi, Y)$ and is denoted by $\ell(\theta, \phi, Y)$. Thus,

$$\ell(\theta, \phi, Y) = \log \mathcal{L}(\theta, \phi, Y). \tag{13.10}$$

Under certain conditions on the likelihood function (which are true for the exponential family), it can be shown that (see, for example, Bickel and Doksum, 1977, p. 139)

$$E\left[\frac{\partial \ell(\theta, \phi, Y)}{\partial \theta}\right] = 0, \tag{13.11}$$

$$E\left[\frac{\partial^2 \ell(\theta, \phi, Y)}{\partial \theta^2}\right] + E\left\{\left[\frac{\partial \ell(\theta, \phi, Y)}{\partial \theta}\right]^2\right\} = 0. \tag{13.12}$$

Using the expression in (13.1), we have

$$\ell(\theta, \phi, Y) = \frac{\theta\, Y - b(\theta)}{a(\theta)} + c(Y, \phi). \tag{13.13}$$

Differentiating both sides of (13.13) with respect to θ once then twice, we get

$$\frac{\partial \ell(\theta, \phi, Y)}{\partial \theta} = \frac{Y - b'(\theta)}{a(\phi)}, \tag{13.14}$$

$$\frac{\partial^2 \ell(\theta, \phi, Y)}{\partial \theta^2} = -\frac{b''(\theta)}{a(\phi)}, \tag{13.15}$$

where $b'(\theta)$ and $b''(\theta)$ denote the first and second derivatives of $b(\theta)$, respectively. From (13.11) and (13.14) we then have

$$\mu = b'(\theta), \tag{13.16}$$

where $\mu = E(Y)$. Furthermore, from (13.12), (13.14) through (13.16), we obtain

$$-\frac{b''(\theta)}{a(\phi)} + \frac{\mathrm{Var}(Y)}{a^2(\phi)} = 0. \tag{13.17}$$

Hence, the variance of Y is given by

$$\text{Var}(Y) = a(\phi)\, b''(\theta). \tag{13.18}$$

Note that Var(Y) is the product of $a(\phi)$, which depends only on the dispersion parameter ϕ, and $b''(\theta)$, which depends only on the canonical parameter θ, and hence on the mean μ of the distribution of Y. The latter quantity is called the *variance function* and is denoted by $V(\mu)$.

To verify the above results, we can apply formulas (13.16) and (13.18) to the normal, Poisson, and binary distributions. In the normal case, $\theta = \mu$, $b(\theta) = \frac{1}{2}\mu^2$, $a(\phi) = \phi = \sigma^2$. Hence, $b(\theta) = \frac{1}{2}\theta^2$ and $b'(\theta) = \mu$. Also, $a(\phi)b''(\theta)$ gives the variance σ^2. For the Poisson distribution, $\theta = \log\lambda$, $b(\theta) = \exp(\theta)$, $a(\phi) = 1$. Hence, $b'(\theta) = \exp(\theta) = \lambda$, which is the mean of this distribution. In addition, $a(\phi)b''(\theta) = \exp(\theta) = \lambda$, which is the variance of the Poisson distribution. As for the binary distribution, we have that $\theta = \log[p/(1-p)]$, $b(\theta) = \log[1+\exp(\theta)]$, $a(\phi) = 1$. In this case, $b'(\theta) = \exp(\theta)/[1+\exp(\theta)] = p$, which is the mean. Also, $a(\phi)b''(\theta) = \exp(\theta)/[1+\exp(\theta)]^2 = p(1-p)$, which is equal to the variance of the binary distribution.

13.3 Estimation of Parameters

In this section, we show how to estimate the parameter vector, $\boldsymbol{\beta}$, in the model for the linear predictor, η, in (13.2). The method of maximum likelihood is used for this purpose.

Suppose that we have n independent random variables, $Y_1, Y_2, \ldots, Y_n$, each of which has the probability distribution described in (13.1) such that the mean of Y_i is μ_i and the corresponding canonical parameter is θ_i ($i = 1, 2, \ldots, n$). The Y_i's constitute a sample of n observations on some response Y. Let $\eta_i = g(\mu_i)$, where g is an appropriate link function such that by (13.2), $\eta_i = \eta(\boldsymbol{x}_i) = \boldsymbol{f}'(\boldsymbol{x}_i)\boldsymbol{\beta}$; $\boldsymbol{x}_i$ is the value of $\boldsymbol{x}$ at which $Y = Y_i$ ($i = 1, 2, \ldots, n$). From (13.1) we then have

$$f(y_i, \theta_i, \phi) = \exp\left[\frac{\theta_i y_i - b(\theta_i)}{a(\phi)} + c(y_i, \phi)\right], \quad i = 1, 2, \ldots, n, \tag{13.19}$$

where the dispersion parameter ϕ is considered to have a fixed value (that is, it does not change over the values of y_i). The corresponding log-likelihood function associated with $Y_1, Y_2, \ldots, Y_n$ is therefore of the form

$$\ell(\boldsymbol{\theta}, \phi, \boldsymbol{Y}) = \sum_{i=1}^{n}\left[\frac{\theta_i Y_i - b(\theta_i)}{a(\phi)} + c(Y_i, \phi)\right], \tag{13.20}$$

where $\boldsymbol{\theta} = (\theta_1, \theta_2, \ldots, \theta_n)'$, $\boldsymbol{Y} = (Y_1, Y_2, \ldots, Y_n)'$. Since the linear predictor η depends on $\boldsymbol{\beta}$ and the mean response is a function of η by (13.4), the means $\mu_1, \mu_2, \ldots, \mu_n$ are therefore functions of $\boldsymbol{\beta}$. The maximum likelihood estimates of the elements of $\boldsymbol{\beta}$ are obtained by solving the equations

$$\frac{\partial \ell}{\partial \beta_j} = 0, \quad j = 1, 2, \ldots, q, \tag{13.21}$$

where β_j is the jth element of $\boldsymbol{\beta}$ and $\frac{\partial \ell}{\partial \beta_j}$ is the partial derivative of $\ell(\boldsymbol{\theta}, \phi, \boldsymbol{Y})$ in (13.20) with respect to β_j ($j = 1, 2, \ldots, q$). From (13.20) we have

$$\frac{\partial \ell}{\partial \beta_j} = \sum_{i=1}^{n} \frac{\partial \ell_i}{\partial \beta_j}, \quad j = 1, 2, \ldots, q, \tag{13.22}$$

where

$$\ell_i = \frac{\theta_i Y_i - b(\theta_i)}{a(\phi)} + c(Y_i, \phi), \quad i = 1, 2, \ldots, n. \tag{13.23}$$

But,

$$\begin{aligned} \frac{\partial \ell_i}{\partial \beta_j} &= \frac{\partial \ell_i}{\partial \theta_i} \frac{\partial \theta_i}{\partial \mu_i} \frac{\partial \mu_i}{\partial \eta_i} \frac{\partial \eta_i}{\partial \beta_j}, \\ \frac{\partial \ell_i}{\partial \theta_i} &= \frac{Y_i - b'(\theta_i)}{a(\phi)} \\ &= \frac{Y_i - \mu_i}{a(\phi)}, \end{aligned}$$

since $\mu_i = b'(\theta_i)$ by (13.16), $i = 1, 2, \ldots, n$, and

$$\begin{aligned} \frac{\partial \theta_i}{\partial \mu_i} &= \left[\frac{\partial \mu_i}{\partial \theta_i}\right]^{-1} \\ &= \frac{1}{b''(\theta_i)}, \\ \frac{\partial \mu_i}{\partial \eta_i} &= \left[\frac{\partial \eta_i}{\partial \mu_i}\right]^{-1} \\ &= \frac{1}{g'(\mu_i)}, \\ \frac{\partial \eta_i}{\partial \beta_j} &= f_j(\boldsymbol{x}_i), \end{aligned}$$

where, if we recall, $\eta_i = f'(x_i)\beta$ and $f_j(x_i)$ is the jth element of $f'(x_i)$, $i = 1,2,\ldots,n; j = 1,2,\ldots,q$. We conclude that

$$\frac{\partial \ell_i}{\partial \beta_j} = \frac{Y_i - \mu_i}{a(\phi)} \frac{1}{b''(\theta_i)} \frac{1}{g'(\mu_i)} f_j(x_i)$$

$$= \frac{Y_i - \mu_i}{\text{Var}(Y_i)} \frac{f_j(x_i)}{g'(\mu_i)}, \quad i = 1,2,\ldots,n,$$

since $\text{Var}(Y_i) = a(\phi)\, b''(\theta_i)$. From (13.22) we then have

$$\frac{\partial \ell}{\partial \beta_j} = \sum_{i=1}^{n} \frac{Y_i - \mu_i}{\text{Var}(Y_i)} \frac{f_j(x_i)}{g'(\mu_i)}, \quad j = 1,2,\ldots,q,$$

which can be written as

$$\frac{\partial \ell}{\partial \beta_j} = \sum_{i=1}^{n} \frac{Y_i - \mu_i}{\omega_i} g'(\mu_i) f_j(x_i), \quad j = 1,2,\ldots,q, \tag{13.24}$$

where

$$\omega_i = \text{Var}(Y_i)\ [g'(\mu_i)]^2, \quad i = 1,2,\ldots,n. \tag{13.25}$$

Thus, the maximum likelihood equations in (13.21) can be written as

$$\sum_{i=1}^{n} \frac{Y_i - \mu_i}{\omega_i} g'(\mu_i) f_j(x_i) = 0, \quad j = 1,2,\ldots,q. \tag{13.26}$$

The maximum likelihood (ML) estimate of β, denoted by $\hat{\beta}$, is obtained as the solution of equations (13.26). Note that these equations are nonlinear in $\beta_1, \beta_2, \ldots, \beta_q$ since $\mu_1, \mu_2, \ldots, \mu_n$ are, in general, nonlinear functions of $\beta_1, \beta_2, \ldots, \beta_q$. Therefore, equations (13.26) may not have a closed-form solution. However, the equations can be solved iteratively by using *Fisher's method of scoring*, which is based on the *Newton–Raphson method* (see Section 8.8 in Khuri, 2003). This is done as follows.

Let $\beta^{(0)}$ be an initial estimate of β, and let $\beta^{(m)}$ be the estimate of β at the mth iteration ($m \geq 1$). Then,

$$\beta^{(m+1)} = \beta^{(m)} - \left\{ E\left[H_\ell(\beta)\right]\Big|_{\beta=\beta^{(m)}} \right\}^{-1} \frac{\partial \ell}{\partial \beta}\Big|_{\beta=\beta^{(m)}}, \quad m = 0,1,\ldots, \tag{13.27}$$

where $\frac{\partial \ell}{\partial \beta}\Big|_{\beta=\beta^{(m)}}$ is the vector whose jth element is $\frac{\partial \ell}{\partial \beta_j}$ ($j = 1,2,\ldots,q$), evaluated using $\beta^{(m)}$ in place of β, and $H_\ell(\beta)$ is the *Hessian matrix* of ℓ given by

$$H_\ell(\beta) = \frac{\partial}{\partial \beta}\left[\frac{\partial \ell}{\partial \beta'}\right]. \tag{13.28}$$

Note that $\boldsymbol{H}_\ell$ is a symmetric $q \times q$ matrix whose (j,k)th element is $\frac{\partial^2 \ell}{\partial \beta_j \partial \beta_k}$, and the expected value in (13.27) is taken with respect to the given distribution, then evaluated using $\boldsymbol{\beta}^{(m)}$ in place of $\boldsymbol{\beta}$ (see formula (8.77) in Khuri, 2003). Making use of (13.24), $\frac{\partial \ell}{\partial \boldsymbol{\beta}}$ can be written as

$$\frac{\partial \ell}{\partial \boldsymbol{\beta}} = \boldsymbol{X}'\boldsymbol{W}^{-1}\boldsymbol{D}(\boldsymbol{Y} - \boldsymbol{\mu}), \tag{13.29}$$

where $\boldsymbol{X}$ is an $n \times q$ matrix whose ith row is $\boldsymbol{f}'(\boldsymbol{x}_i)$ [this is called the *model matrix* for the linear predictor in (13.2)], $\boldsymbol{W} = \oplus_{i=1}^{n} \omega_i$, $\boldsymbol{D} = \oplus_{i=1}^{n}[g'(\mu_i)]$, $\boldsymbol{Y} = (Y_1, Y_2, \ldots, Y_n)'$, and $\boldsymbol{\mu} = (\mu_1, \mu_2, \ldots, \mu_n)'$. Furthermore, using (13.24), the (j,k)th element of the Hessian matrix $\boldsymbol{H}_\ell(\boldsymbol{\beta})$ is

$$\frac{\partial^2 \ell}{\partial \beta_j \partial \beta_k} = \sum_{i=1}^{n}(Y_i - \mu_i)\frac{\partial}{\partial \beta_k}\left[\frac{g'(\mu_i)}{\omega_i} f_j(\boldsymbol{x}_i)\right] + \sum_{i=1}^{n}\frac{g'(\mu_i)}{\omega_i} f_j(\boldsymbol{x}_i)\frac{\partial}{\partial \beta_k}(Y_i - \mu_i).$$

But,

$$\begin{aligned}\frac{\partial}{\partial \beta_k}(Y_i - \mu_i) &= -\frac{\partial \mu_i}{\partial \eta_i}\frac{\partial \eta_i}{\partial \beta_k}\\ &= -\frac{1}{g'(\mu_i)} f_k(\boldsymbol{x}_i),\end{aligned}$$

since, if we recall, $\frac{\partial \mu_i}{\partial \eta_i} = \frac{1}{g'(\mu_i)}$ and $\frac{\partial \eta_i}{\partial \beta_k} = f_k(\boldsymbol{x}_i)$. Hence,

$$\frac{\partial^2 \ell}{\partial \beta_j \partial \beta_k} = \sum_{i=1}^{n}(Y_i - \mu_i)\frac{\partial}{\partial \beta_k}\left[\frac{g'(\mu_i)}{\omega_i} f_j(\boldsymbol{x}_i)\right] - \sum_{i=1}^{n}\frac{f_j(\boldsymbol{x}_i)}{\omega_i} f_k(\boldsymbol{x}_i). \tag{13.30}$$

Taking the expected values of both sides of (13.30), we get

$$E\left[\frac{\partial^2 \ell}{\partial \beta_j \partial \beta_k}\right] = -\sum_{i=1}^{n}\frac{1}{\omega_i} f_j(\boldsymbol{x}_i) f_k(\boldsymbol{x}_i).$$

Consequently, the expected value of $\boldsymbol{H}_\ell(\boldsymbol{\beta})$ is of the form

$$E[\boldsymbol{H}_\ell(\boldsymbol{\beta})] = -\boldsymbol{X}'\boldsymbol{W}^{-1}\boldsymbol{X}. \tag{13.31}$$

Using (13.29) and (13.31), formula (13.27) is then expressed as

$$\begin{aligned}\boldsymbol{\beta}^{(m+1)} &= \boldsymbol{\beta}^{(m)} + [(\boldsymbol{X}'\boldsymbol{W}^{-1}\boldsymbol{X})^{-1}\boldsymbol{X}'\boldsymbol{W}^{-1}\boldsymbol{D}(\boldsymbol{Y} - \boldsymbol{\mu})]_{\boldsymbol{\beta}=\boldsymbol{\beta}^{(m)}}\\ &= [(\boldsymbol{X}'\boldsymbol{W}^{-1}\boldsymbol{X})^{-1}\boldsymbol{X}'\boldsymbol{W}^{-1}]_{\boldsymbol{\beta}=\boldsymbol{\beta}^{(m)}}\left\{\boldsymbol{X}\boldsymbol{\beta}^{(m)} + [\boldsymbol{D}(\boldsymbol{Y} - \boldsymbol{\mu})]_{\boldsymbol{\beta}=\boldsymbol{\beta}^{(m)}}\right\},\\ &\qquad m = 0, 1, \ldots\end{aligned} \tag{13.32}$$

The expression on the right-hand side of (13.32) has an interesting interpretation as we now show:

Suppose that a first-order Taylor's series approximation of $g(Y)$ is taken in a neighborhood of μ, that is, $g(Y) \approx Z$, where

$$\begin{aligned} Z &= g(\mu) + (Y - \mu)\, g'(\mu) \\ &= \eta + (Y - \mu)\, g'(\mu) \\ &= f'(x)\beta + (Y - \mu)\, g'(\mu). \end{aligned} \tag{13.33}$$

Thus, the mean of Z is η and its variance is $\text{Var}(Y)\,[g'(\mu)]^2$. Evaluating (13.33) at $x_1, x_2, \ldots, x_n$, we get

$$Z_i = f'(x_i)\beta + (Y_i - \mu_i)\, g'(\mu_i), \quad i = 1, 2, \ldots, n. \tag{13.34}$$

Equations (13.34) can be written in matrix form as

$$\boldsymbol{Z} = \boldsymbol{X}\beta + \boldsymbol{D}(\boldsymbol{Y} - \mu), \tag{13.35}$$

where $\boldsymbol{Z} = (Z_1, Z_2, \ldots, Z_n)'$. The variance–covariance matrix of $\boldsymbol{Z}$ is

$$\begin{aligned} \text{Var}(\boldsymbol{Z}) &= \boldsymbol{D}\left\{\bigoplus_{i=1}^{n}[\text{Var}(Y_i)]\right\}\boldsymbol{D} \\ &= \bigoplus_{i=1}^{n}[g'(\mu_i)]^2\ \text{Var}(Y_i) \\ &= \boldsymbol{W}. \end{aligned} \tag{13.36}$$

The value of $\boldsymbol{Z}$ at the mth iteration is denoted by $\boldsymbol{Z}^{(m)}$. From (13.32), (13.35), and (13.36) we conclude that $\beta^{(m+1)}$ in (13.32) has the same form as a generalized (or weighted) least-squares estimator of β in a linear model whose response vector is $\boldsymbol{Z}^{(m)}$, rather than $\boldsymbol{Y}$, with a variance–covariance matrix given by $\boldsymbol{W}^{(m)}$, the value of $\boldsymbol{W}$ at the mth iteration. Several iterations of formula (13.32) can then be carried out until some convergence is achieved in the resulting values. We can therefore state that the solution of the maximum likelihood equations in (13.26) is obtained by performing an iterative weighted least-squares procedure using a linearized form of the link function applied to Y. The data vector $\boldsymbol{y}$ can be used as a first estimate of μ from which we get a first estimate of η, namely η^0, whose ith element is $g(y_i)$ (y_i is the ith element of $\boldsymbol{y}$, $i = 1, 2, \ldots, n$). From this we obtain initial values for $g'(\mu_i)$, $V(\mu_i)$, the variance function at μ_i ($i = 1, 2, \ldots, n$), and $\boldsymbol{Z}$, the latter is chosen as η^0. These are sufficient to get the iterative procedure started (see Exercise 13.7).

13.3.1 Estimation of the Mean Response

An estimate of the linear predictor, $\eta(x)$, in (13.2) is given by

$$\hat{\eta}(x) = f'(x)\hat{\beta}, \tag{13.37}$$

where $\hat{\beta}$ is the ML estimate of β obtained as a result of the iterative process as described earlier. Using formula (13.4), an estimate of the mean response, $\mu(x)$, at a given point x is obtained from the following expression:

$$\begin{aligned}\hat{\mu}(x) &= h[\hat{\eta}(x)] \\ &= h[f'(x)\hat{\beta}],\end{aligned} \quad (13.38)$$

where $h(.)$ is the inverse of the link function $g(.)$. This estimate is also referred to as the *predicted response* at x.

13.3.2 Asymptotic Distribution of $\hat{\beta}$

The precision associated with the ML estimation of β can be assessed asymptotically by using the so-called *Fisher's information matrix* for β, which is denoted by $I(\beta)$ and is defined as

$$I(\beta) = -\mathrm{E}[H_{\ell}(\beta)], \quad (13.39)$$

where E $[H_{\ell}(\beta)]$ is the expected value of the Hessian matrix in (13.28). Using formula (13.31), $I(\beta)$ can be expressed as

$$I(\beta) = X'W^{-1}X. \quad (13.40)$$

It is known that as $n \to \infty$, where n is the sample size, the ML estimator of β is asymptotically normally distributed with mean β and a variance–covariance matrix given by $[I(\beta)]^{-1}$. We can then write

$$\hat{\beta} \approx \mathrm{AN}[\beta, (X'W^{-1}X)^{-1}], \quad (13.41)$$

where "AN" denotes asymptotic normality. This result can be found in, for example, McCulloch and Searle (2001, p. 306) (see also Searle, Casella, and McCulloch, 1992, p. 473). Thus, for a given sample size, the variance–covariance matrix of $\hat{\beta}$ is approximately equal to

$$\mathrm{Var}(\hat{\beta}) \approx (X'W^{-1}X)^{-1}. \quad (13.42)$$

Using (13.37) and (13.42), the variance of $\hat{\eta}(x)$ is approximately equal to

$$\mathrm{Var}[\hat{\eta}(x)] \approx f'(x)(X'W^{-1}X)^{-1}f(x). \quad (13.43)$$

Now, in order to obtain an approximate expression for the variance of $\hat{\mu}(x)$ in (13.38), we first obtain a first-order Taylor's series approximation of $h[\hat{\eta}(x)]$ in a neighborhood of $\eta(x)$ of the form

$$h[\hat{\eta}(x)] \approx h[\eta(x)] + [\hat{\eta}(x) - \eta(x)]h'[\eta(x)], \quad (13.44)$$

where $h^{'}[\eta(x)]$ is the derivative of $h(.)$ with respect to η. Using formulas (13.38), (13.43), and (13.44), we obtain

$$\begin{aligned}\text{Var}[\hat{\mu}(x)] &\approx \{h^{'}[\eta(x)]\}^2\,\text{Var}[\hat{\eta}(x)] \\ &= \{h^{'}[\eta(x)]\}^2 f'(x)(X'W^{-1}X)^{-1}f(x).\end{aligned} \tag{13.45}$$

The variance of $\hat{\mu}(x)$ is called the *prediction variance* at x, and the right-hand side of (13.45) provides an approximate expression for this variance.

13.3.3 Computation of $\hat{\beta}$ in SAS

The actual computation of $\hat{\beta}$ on the basis of the iterative process described earlier in Section 13.3 can be conveniently done using PROC GENMOD in SAS. A number of link functions (logit, probit, log, complementary log-log) and probability distributions (normal, binomial, Poisson, gamma) can be specified in the MODEL statement in PROC GENMOD. The default initial parameter values are weighted least-squares estimates based on using y, the data vector, as an initial estimate of μ, as was seen in Section 13.3. For example, to fit a generalized linear model for a Poisson-distributed response, Y, using a logarithmic link function and a linear predictor of the form, $\eta = \beta_0 + \beta_1 x_1 + \beta_2 x_2 + \beta_{12} x_1 x_2 + \beta_{11} x_1^2 + \beta_{22} x_2^2$, the following SAS statements are needed:

```
DATA;
INPUT X1 X2 Y;
CARDS;
(data are entered here)
PROC GENMOD;
MODEL Y = X1 X2 X1 * X2 X1 * X1 X2 * X2/DIST=POISSON LINK=LOG;
RUN;
```

In case of binomial data, the response Y in the MODEL statement is replaced by "S/N," where N is the number of trials in a given experimental run and S is the number of successes. In addition, S and N must be specified in the INPUT statement in place of Y. For the corresponding "DIST" and "LINK", we can use "DIST=BINOMIAL" and "LINK=LOGIT," respectively. Note that "LOG" and "LOGIT" are the default link functions for the Poisson and binomial distributions, respectively, since they are the canonical links for their respective distributions. Among other things, the SAS output provides ML estimates of the elements of the parameter vector β for the linear predictor η and their corresponding standard errors, which are the square roots of the diagonal elements of the matrix $(X'W^{-1}X)^{-1}$ given in formula (13.42), where W is replaced by an estimate. This is demonstrated in the next example.

Example 13.1 An experiment was conducted to determine the effects of the control variables, x_1 = burner setting, x_2 = amount of vegetable oil (in table

TABLE 13.1
Coded Design Settings and Response Values

x_1	x_2	x_3	Y (Number of Inedible Kernels)
1	1	0	20
−1	0	1	21
1	0	1	42
1	−1	0	36
0	1	1	11
−1	0	−1	120
0	−1	1	33
−1	1	0	36
0	−1	−1	32
−1	−1	0	38
1	0	−1	20
0	0	0	17
0	1	−1	49

spoons), x_3 = popping time (in seconds) on the number, Y, of inedible kernels of popcorn. Each run of the experiment used 1/4 cup of unpopped popcorns. A description of this experiment is given in Vining and Khuri (1991). The actual levels of x_1, x_2, and x_3 used in the experiment are 5, 6, 7 for x_1; 2, 3, 4 for x_2; and 75, 90, 105 for x_3. The three levels for each variable are coded as $-1, 0, 1$, respectively. The design settings (in coded form) and corresponding response values are given in Table 13.1.

Note that the design used is of the *Box-Behnken* type (see Khuri and Cornell, 1996, Section 4.5.2, for a general description of the Box-Behnken design). This design is suitable for fitting a second-degree response surface model in x_1, x_2, and x_3 of the form

$$\eta = \beta_0 + \beta_1 x_1 + \beta_2 x_2 + \beta_3 x_3 + \beta_{12} x_1 x_2 + \beta_{13} x_1 x_3 + \beta_{23} x_2 x_3 + \beta_{11} x_1^2 + \beta_{22} x_2^2 + \beta_{33} x_3^2. \quad (13.46)$$

Assuming a Poisson distribution for Y and a log link, the following SAS code was used to obtain the ML estimates of the parameters in model (13.46):

```
DATA;
INPUT X1 X2 X3 Y;
CARDS;
(enter here the data from Table 13.1)
PROC GENMOD;
MODEL Y = X1 X2 X3 X1*X2 X1*X3 X2*X3 X1*X1 X2*X2 X3*X3
/DIST=POISSON LINK=LOG;
RUN;
```

The resulting parameter ML estimates and corresponding standard errors are given in Table 13.2.

TABLE 13.2
ML Estimates and Standard Errors

Parameter	Estimate	Standard Error
β_0	2.8332	0.2425
β_1	−0.2066	0.0617
β_2	−0.1585	0.0660
β_3	−0.3160	0.0634
β_{12}	−0.1401	0.0897
β_{13}	0.6246	0.0894
β_{23}	−0.3709	0.0969
β_{11}	0.4589	0.1446
β_{22}	0.1501	0.1445
β_{33}	0.3470	0.1454

Since the inverse, $h(.)$, of the link function is the exponential function, we get from (13.38) and Table 13.2 the following expression for the predicted response $\hat{\mu}(x)$:

$$\hat{\mu}(x) = \exp(2.8332 - 0.2066x_1 - 0.1585x_2 - 0.3160x_3 - 0.1401x_1x_2 + 0.6246x_1x_3 - 0.3709x_2x_3 + 0.4589x_1^2 + 0.1501x_2^2 + 0.3470x_3^2). \quad (13.47)$$

Model (13.47) can be utilized to determine the optimal settings of x_1, x_2, and x_3 that result in the minimization of the number of inedible kernels within the experimental region (in the coded space), $x_1^2 + x_2^2 + x_3^2 \leq 2$. This was demonstrated in Paul and Khuri (2000).

13.4 Goodness of Fit

In this section, two measures are presented for assessing the goodness of fit of a given generalized linear model. These measures are the *deviance* and *Pearson's chi-square statistic*.

13.4.1 The Deviance

The fitting of a model amounts to deriving estimates of its parameters that can be used to provide information about the mean response at various locations inside the region of experimentation, which we denote by $\mathcal{R}$. If the number of observations used to fit the model is equal to n, and if these observations are attained at distinct locations inside $\mathcal{R}$, then the maximum number of parameters that can be estimated in the model is equal to n. (If at some locations in $\mathcal{R}$, replicate observations on the response Y are obtained, then the maximum number of parameters that can be estimated in the model is equal to the number of distinct locations used to collect all the response data. Such locations, or points, make up the associated response surface design.) A model having as many parameters as there are points in the corresponding response surface design is called a *saturated* (or full) *model*. Such a model is not informative since it does not summarize the response data. However, the likelihood function for the saturated model is larger than any other likelihood function (for the same data with the same distribution and link function), if the latter function is based on a model with fewer parameters than the saturated model. Let $\mathcal{L}_{\max}(\phi, Y)$ denote the likelihood function for the saturated model, and let $\mathcal{L}(\phi, \hat{\beta})$ denote the likelihood function, maximized over β, for a given generalized linear model with q parameters ($q < n$), where $\hat{\beta}$ is the ML estimate of β. Then, for a given data vector, y, $\mathcal{L}_{\max}(\phi, y) > \mathcal{L}(\phi, \hat{\beta})$. Thus, the likelihood ratio,

$$\Lambda = \frac{\mathcal{L}(\phi, \hat{\beta})}{\mathcal{L}_{\max}(\phi, y)}, \tag{13.48}$$

provides a measure of goodness of fit for the given (or assumed) model. A small value of Λ (close to 0) indicates that the assumed model does not provide a good fit to the data. Alternatively, we can consider the quantity,

$$-2\log\Lambda = 2\,[\log\mathcal{L}_{\max}(\phi, y) - \log\mathcal{L}(\phi, \hat{\beta})], \tag{13.49}$$

as a measure of goodness of fit for the assumed model. In this case, a large value of $-2\log\Lambda$ is an indication of a bad fit. If we denote the estimates of the canonical parameters under the assumed and the saturated models by $\hat{\theta}$ and $\tilde{\theta}$, respectively, then by formulas (13.20) and (13.49),

$$\begin{aligned}-2\log\Lambda &= 2\sum_{i=1}^{n}\frac{(\tilde{\theta}_i - \hat{\theta}_i)y_i - b(\tilde{\theta}_i) + b(\hat{\theta}_i)}{a(\phi)}\\ &= \frac{\mathcal{D}(\hat{\beta}, y)}{a(\phi)},\end{aligned} \tag{13.50}$$

where

$$\mathcal{D}(\hat{\beta}, y) = 2\sum_{i=1}^{n}[(\tilde{\theta}_i - \hat{\theta}_i)y_i - b(\tilde{\theta}_i) + b(\hat{\theta}_i)] \tag{13.51}$$

is called the *deviance* for the assumed model. When $a(\phi) = \phi$, the expression in (13.50) equals

$$\begin{aligned}\mathcal{D}^*(\hat{\beta}, y) &= \frac{\mathcal{D}(\hat{\beta}, y)}{\phi}\\ &= -2\log\Lambda,\end{aligned} \tag{13.52}$$

which is called the *scaled deviance*. A small value of $\mathcal{D}^*$ is desirable for a good fit.

It is known that $\mathcal{D}^* = -2\log\Lambda$ is asymptotically distributed as χ^2_{n-q} for large n, where q is the number of parameters in the linear predictor $\eta(\boldsymbol{x})$ in (13.2). Since the expected value of χ^2_{n-q} is $n-q$, then a value of $\mathcal{D}^*/(n-q)$ much larger than 1 gives an indication of a bad fit for the assumed model. This is of course contingent on the assumption that the asymptotic approximation with the chi-squared distribution is satisfactory for small samples.

Examples of deviance expressions are given below using the normal, Poisson, and binomial distributions that were mentioned earlier.

(a) The normal distribution

Consider the log-likelihood function for a given response vector, $\boldsymbol{Y} = (Y_1, Y_2, \ldots, Y_n)'$, such that $Y_1, Y_2, \ldots, Y_n$ are mutually independent and $Y_i \sim N(\mu_i, \sigma^2)$, $i = 1, 2, \ldots, n$. Then, for a realized value, $\boldsymbol{y}$, of $\boldsymbol{Y}$, this function is

$$-\frac{n}{2}\log(2\pi\sigma^2) - \frac{1}{2\sigma^2}\sum_{i=1}^{n}(y_i - \mu_i)^2, \tag{13.53}$$

where the dispersion parameter, ϕ, is equal to σ^2. For the saturated model, μ_i is estimated by y_i. Hence,

$$\log \mathcal{L}_{\max}(\sigma^2, \boldsymbol{y}) = -\frac{n}{2}\log(2\pi\sigma^2).$$

For a given model with q parameters ($q < n$), μ_i is estimated by $\hat{\mu}_i$, its ML estimate. Using $\hat{\mu}_i$ in place of μ_i in (13.53), we get

$$-\frac{n}{2}\log(2\pi\sigma^2) - \frac{1}{2\sigma^2}\sum_{i=1}^{n}(y_i - \hat{\mu}_i)^2.$$

From (13.49) we then have

$$-2\log\Lambda = \frac{1}{\sigma^2}\sum_{i=1}^{n}(y_i - \hat{\mu}_i)^2.$$

Hence, the deviance $\mathcal{D}$ is equal to $\sum_{i=1}^{n}(y_i - \hat{\mu}_i)^2$ and the scaled deviance is $\mathcal{D}^* = \mathcal{D}/\sigma^2$. In this case, $\mathcal{D}$ is just the residual sum of squares.

(b) The Poisson distribution

For the data vector, $\boldsymbol{Y} = (Y_1, Y_2, \ldots, Y_n)'$, where the Y_i's are mutually independent and have the Poisson distributions with parameters λ_i ($i = 1, 2, \ldots, n$), the log-likelihood function is

$$\sum_{i=1}^{n}[Y_i \log\lambda_i - \lambda_i - \log(Y_i!)],$$

where $\phi = 1$. For the saturated model and a realized value, $\boldsymbol{y}$, of $\boldsymbol{Y}$, λ_i is estimated by y_i, but for a less-than-saturated model, λ_i is estimated by $\hat{\lambda}_i$, its ML estimate. Hence, the deviance and scaled deviance are equal to

$$-2\log\Lambda = 2\sum_{i=1}^{n}\left[y_i \log\left(\frac{y_i}{\hat{\lambda}_i}\right) - \left(y_i - \hat{\lambda}_i\right)\right].$$

(c) The binomial distribution

In this case, $\boldsymbol{Y} = (Y_1, Y_2, \ldots, Y_n)'$ whose elements are mutually independent such that $Y_i \sim$ binomial with parameters n_i and p_i. The log-likelihood function for a realized value, $\boldsymbol{y}$, of $\boldsymbol{Y}$ is

$$\sum_{i=1}^{n}\left[y_i \log\left(\frac{p_i}{1-p_i}\right) + n_i \log(1-p_i) + \log\binom{n_i}{y_i}\right],$$

where $\phi = 1$. For the saturated model, the mean of Y_i, namely $\mu_i = n_i p_i$, is estimated by y_i, where y_i is the ith element of $\boldsymbol{y}$ $(i = 1, 2, \ldots, n)$. Hence, p_i is estimated by y_i/n_i. Consequently,

$$\log \mathcal{L}_{\max}(\phi, \boldsymbol{y}) = \sum_{i=1}^{n} y_i \log\left(\frac{y_i}{n_i - y_i}\right) + n_i \log\left(1 - \frac{y_i}{n_i}\right) + \log\binom{n_i}{y_i}.$$

For a less-than-saturated model, p_i is estimated by $\frac{\hat{\mu}_i}{n_i}$, where $\hat{\mu}_i$ is the ML estimate of μ_i $(i = 1, 2, \ldots, n)$. Thus,

$$\log \mathcal{L}(\phi, \hat{\boldsymbol{\beta}}) = \sum_{i=1}^{n} y_i \log\left(\frac{\hat{\mu}_i}{n_i - \hat{\mu}_i}\right) + n_i \log\left(1 - \frac{\hat{\mu}_i}{n_i}\right) + \log\binom{n_i}{y_i}.$$

It follows that

$$\begin{aligned}-2\log\Lambda = 2\sum_{i=1}^{n}\Bigg\{ & y_i\left[\log\left(\frac{y_i}{n_i - y_i}\right) - \log\left(\frac{\hat{\mu}_i}{n_i - \hat{\mu}_i}\right)\right] \\ & + n_i\left[\log\left(1 - \frac{y_i}{n_i}\right) - \log\left(1 - \frac{\hat{\mu}_i}{n_i}\right)\right]\Bigg\}. \qquad (13.54)\end{aligned}$$

Both the deviance and scaled deviance are equal to the right-hand side of (13.54).

13.4.2 Pearson's Chi-Square Statistic

Another measure of goodness of fit for a data vector $\boldsymbol{Y} = (Y_1, Y_2, \ldots, Y_n)'$ is

$$\begin{aligned}\chi_s^2 &= \sum_{i=1}^{n}\frac{(Y_i - \hat{\mu}_i)^2}{\widehat{\text{Var}}(Y_i)} \\ &= \frac{1}{a(\phi)}\sum_{i=1}^{n}\frac{(Y_i - \hat{\mu}_i)^2}{V(\hat{\mu}_i)},\end{aligned}$$

where, if we recall from formula (13.18), $\text{Var}(Y_i) = a(\phi)b''(\theta_i) = a(\phi)V(\mu_i)$, $V(\mu_i)$ is the variance function for the ith mean, and $\hat{\mu}_i$ is the ML estimate of μ_i $(i = 1, 2, \ldots, n)$. For a realized value, $\boldsymbol{y}$, of $\boldsymbol{Y}$ and when $a(\phi) = \phi$, χ_s^2 is written as

$$\chi_s^2 = \frac{1}{\phi}\sum_{i=1}^{n}\frac{(y_i - \hat{\mu}_i)^2}{V(\hat{\mu}_i)}. \tag{13.55}$$

The quantity in (13.55) is called the *scaled Pearson's chi-square*, and the expression

$$\chi^2 = \sum_{i=1}^{n}\frac{(y_i - \hat{\mu}_i)^2}{V(\hat{\mu}_i)} \tag{13.56}$$

is called *Pearson's chi-square statistic*. For large n, χ_s^2 is asymptotically distributed as χ_{n-q}^2. A large value of χ^2 is an indication of a bad fit.

In situations where the dispersion parameter ϕ is not known, an estimate can be used in the expressions for the scaled deviance and the scaled Pearson's chi-square, as is the case when using PROC GENMOD. These scaled values should then be used instead of their corresponding unscaled values in assessing the model's goodness of fit.

13.4.3 Residuals

Residuals are used to assess the fit of the model at individual points in a region $\mathcal{R}$ where the data values are obtained. For a given data vector, $\boldsymbol{y} = (y_1, y_2, \ldots, y_n)'$, the ith *raw residual* is defined as $y_i - \hat{\mu}_i$, where $\hat{\mu}_i$ is the ML estimate of μ_i. Since in a generalized linear model situation, $\text{Var}(Y_i)$ is not constant, the use of raw residuals is not appropriate. For this reason, two other types of residuals are considered in parts (a) and (b) below.

(a) Pearson's residuals

These are given by

$$r_{i,p} = \frac{y_i - \hat{\mu}_i}{\sqrt{V(\hat{\mu}_i)}}, \quad i = 1, 2, \ldots, n. \tag{13.57}$$

Note that $\sum_{i=1}^{n} r_{i,p}^2$ is equal to χ^2 in (13.56). When $a(\phi) = \phi$, the quantity $r_{i,p}/\sqrt{\phi}$ is called the ith *scaled Pearson's chi-square residual* $(i = 1, 2, \ldots, n)$. In this case, $\frac{1}{\phi}\sum_{i=1}^{n} r_{i,p}^2 = \chi_s^2$, which is the scaled Pearson's chi-square in (13.55).

(b) Deviance residuals

These are defined as

$$r_{i,d} = [\text{sign}(y_i - \hat{\mu}_i)]\sqrt{d_i}, \quad i = 1, 2, \ldots, n, \tag{13.58}$$

where $d_i \geq 0$ so that $\sum_{i=1}^{n} r_{i,d}^2 = \mathcal{D}(\hat{\beta}, y)$, the deviance for the fitted model as given in (13.51). Thus, $\sqrt{d_i}$ represents the square root of the contribution of the ith observation to the deviance $\mathcal{D}$ and $\text{sign}(y_i - \hat{\mu}_i)$ is the sign of the ith raw residual ($i = 1, 2, \ldots, n$). When $a(\phi) = \phi$, dividing $r_{i,d}$ by $\sqrt{\phi}$ yields the ith *scaled deviance residual* ($i = 1, 2, \ldots, n$). In this case, $\frac{1}{\phi} \sum_{i=1}^{n} r_{i,d}^2 = \mathcal{D}^*$ is the scaled deviance in (13.52).

(c) Studentized residuals

In classical linear models, residuals are usually standardized so that they become scale free and have the same precision. This makes it more convenient to compare residuals at various locations in the region of experimentation, $\mathcal{R}$. If $\boldsymbol{Y} = \boldsymbol{X\beta} + \boldsymbol{\epsilon}$ is a given linear model where it is assumed that the error vector $\boldsymbol{\epsilon}$ has a zero mean and a variance-covariance matrix given by $\sigma^2 \boldsymbol{I}_n$, then the ith raw residual, denoted here by e_i, is standardized by dividing it by $\sqrt{\text{Var}(e_i)} = \sqrt{\sigma^2(1 - h_{ii})}$, where h_{ii} is the ith diagonal element of the so-called *hat matrix*, denoted by $\boldsymbol{H}$ and is given by $\boldsymbol{H} = \boldsymbol{X}(\boldsymbol{X}'\boldsymbol{X})^{-1}\boldsymbol{X}'$ ($i = 1, 2, \ldots, n$). In this case, $e_i/\sqrt{\sigma^2(1 - h_{ii})}$ has a zero mean and a variance equal to 1. Since σ^2 is unknown, it can be replaced by MS_E, the error mean square. Doing so leads to the *Studentized* ith *residual*, namely, $e_i/\sqrt{MS_E(1 - h_{ii})}$, $i = 1, 2, \ldots, n$. The Studentized residuals are scale free and are very useful in checking model adequacy and the assumptions concerning the error distribution. This can be accomplished by using various plots of the Stundentized residuals against the corresponding predicted response values and against the control variables in the fitted model. Other types of residual plots can also be used (see, for example, Atkinson, 1985; Draper and Smith, 1998).

For a generalized linear model, the $\boldsymbol{X}$ matrix for the linear model mentioned earlier is replaced by $\boldsymbol{W}^{-1/2}\boldsymbol{X}$, where $\boldsymbol{W} = \oplus_{i=1}^{n} \omega_i$ and ω_i is defined in (13.25), $i = 1, 2, \ldots, n$, and $\boldsymbol{X}$ is the model matrix for the linear predictor. Consequently, the "hat" matrix $\boldsymbol{H}$ is replaced by

$$\mathcal{H} = \boldsymbol{W}^{-1/2}\boldsymbol{X}(\boldsymbol{X}'\boldsymbol{W}^{-1}\boldsymbol{X})^{-1}\boldsymbol{X}'\boldsymbol{W}^{-1/2}. \tag{13.59}$$

Let $\tilde{h}_{ii}$ denote the ith diagonal element of $\mathcal{H}$ ($i = 1, 2, \ldots, n$). Then, the ith *Studentized Pearson's residual* is defined as $r_{i,p}/\sqrt{\phi\,(1 - \tilde{h}_{ii})}$, where $r_{i,p}$ is given in (13.57). Hence,

$$\frac{r_{i,p}}{\sqrt{\phi\,(1 - \tilde{h}_{ii})}} = \frac{y_i - \hat{\mu}_i}{\sqrt{\phi\, V(\hat{\mu}_i)(1 - \tilde{h}_{ii})}}, \quad i = 1, 2, \ldots, n, \tag{13.60}$$

where ϕ is considered known, otherwise, an estimate of ϕ can be used. Note that this Studentized residual has a unit asymptotic variance.

As for the deviance residual, its Studentized version is defined as

$$\frac{r_{i,d}}{\sqrt{\phi\,(1-\tilde{h}_{ii})}} = \frac{\text{sign}(y_i - \hat{\mu}_i)\sqrt{d_i}}{\sqrt{\phi\,(1-\tilde{h}_{ii})}}, \quad i = 1, 2, \ldots, n, \tag{13.61}$$

where $r_{i,d}$ is given in (13.58). This Studentized residual has also a unit asymptotic variance.

The Studentized residuals in (13.60) and (13.61) can be plotted against $\hat{\eta}_i$ [see formula (13.37)] or against $\hat{\mu}_i$ $(i = 1, 2, \ldots, n)$, as well as against each of the control variables in the linear predictor model. These plots are analogous to the common residual plots used in a classical linear model and have similar interpretations.

Example 13.2 Let us again consider Example 13.1. Using the SAS code given earlier in that example, we get the following information concerning the deviance and Pearson's chi-square statistic for model (13.46):

Criteria for Assessing Goodness of Fit

Criterion	DF	Value	Value/DF
Deviance	3	1.2471	0.4157
Scaled deviance	3	1.2471	0.4157
Pearson's chi-square	3	1.2506	0.4169
Scaled Pearson's chi-square	3	1.2506	0.4169

We note that the values of the scaled deviance and the scaled Pearson's chi-square statistic are small relative to the degrees of freedom. This indicates that the model fits the data well. We also note that the deviance and Pearson's chi-square values are identical to their scaled counterparts. This follows from the fact that the dispersion parameter ϕ is equal to 1 for the Poisson distribution.

In order to get information concerning Pearson's residuals [as in formula (13.57)] and the deviance residuals [as in formula (13.58)], the option "OBSTATS" should be added to the MODEL statement in PROC GENMOD. This option also provides values of the predicted response, that is, values of $\hat{\mu}(x_i)$ for $i = 1, 2, \ldots, n$, where $n = 13$ in this example, in addition to several other items. As for the information concerning the Studentized values of the deviance and Pearson's residuals [as in (13.61) and (13.60), respectively], the option "RESIDUALS" should also be added to the MODEL statement. Thus, this statement can now be rewritten as

$$\text{MODEL } Y = X_1\ X_2\ X_3\ X_1 * X_2\ X_1 * X_3\ X_2 * X_3\ X_1 * X_1\ X_2 * X_2\ X_3 * X_3$$
$$/\text{DIST} = \text{POISSON LINK} = \text{LOG OBSTATS RESIDUALS;}$$

The corresponding SAS output is given in Table 13.3.

From Table 13.3 we note the close agreement between Y and the predicted response values. This is reflected in the small values of the deviance and

TABLE 13.3
Observation Statistics

Y	Pred	Resraw	Resdev	Reschi	StResdev	StReschi
20	18.8589	1.14107	0.26017	0.26276	0.41869	0.42285
21	18.2663	2.73372	0.62460	0.63963	1.09076	1.11702
42	42.1402	−0.14018	−0.02161	−0.02159	−0.04553	−0.04551
36	34.2672	1.73284	0.29357	0.29602	0.61776	0.62290
11	12.0009	−1.00089	−0.29308	−0.28892	−0.42365	−0.41763
120	119.8598	0.14018	0.01280	0.01280	0.04550	0.04551
33	34.5927	−1.59266	−0.27291	−0.27079	−0.54276	−0.53855
36	37.7328	−1.73284	−0.28430	−0.28210	−0.62776	−0.62290
32	30.9991	1.00089	0.17881	0.17977	0.41541	0.41763
38	39.1411	−1.14107	−0.18328	−0.18239	−0.42493	−0.42285
20	22.7337	−2.73372	−0.58546	−0.57335	−1.14061	−1.11702
17	17.0000	0.00000	0.00000	0.00000	0.00000	0.00000
49	47.4073	1.59266	0.23004	0.23131	0.53557	0.53855

Note: Pred, predicted response; Resraw, raw residual; Resdev, deviance residual; Reschi, Pearson's chi-square residual; StResdev, Studentized deviance residual; StReschi, Studentized Pearson's chi-square residual.

Pearson's chi-square residuals. The residual plots are shown in Figures 13.1 through 13.6. Figure 13.1 gives a plot of Y_i against the predicted response $\hat{\mu}(x_i)$ $(i = 1, 2, \ldots, 13)$, which clearly shows the closeness of the corresponding

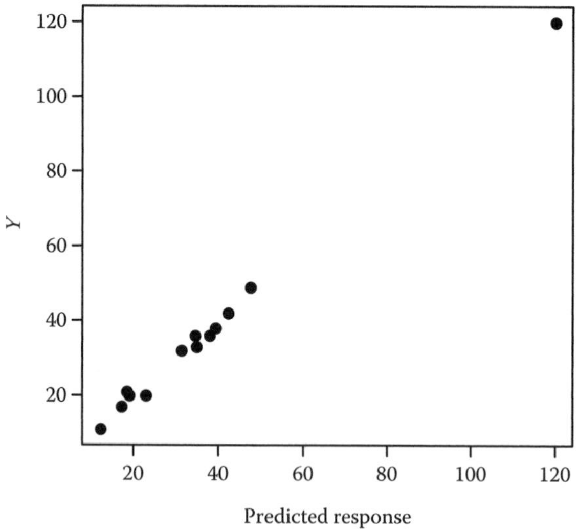

FIGURE 13.1
Plot of the values of Y (number of inedible kernels) against the predicted response values.

values. Figures 13.2 and 13.3 give plots of the Studentized deviance and Pearson's chi-square residuals, respectively, against the predicted response values. Both plots show no systematic changes in the residuals with respect to

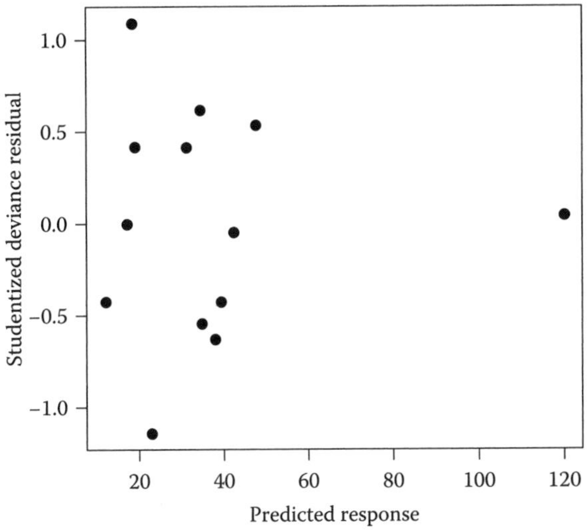

FIGURE 13.2
Plot of the Studentized deviance residuals against the predicted response values.

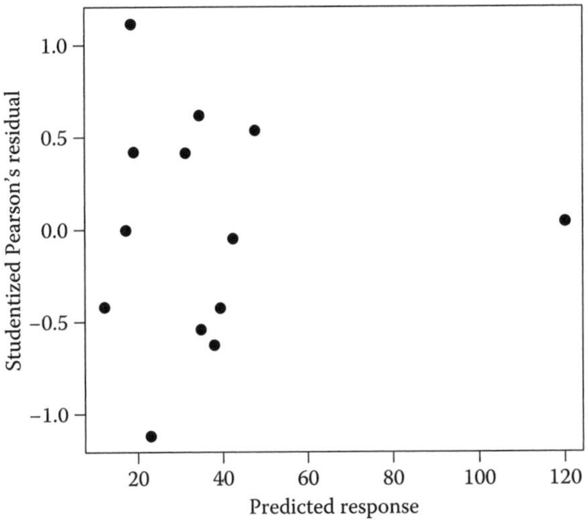

FIGURE 13.3
Plot of the Studentized Pearson's residuals against the the predicted response values.

the predicted response. Note that the remoteness of the point corresponding to $x_1 = -1$, $x_2 = 0$, $x_3 = -1$, and $Y = 120$ is understandable since with low temperature and short popping time, a large number of inedible kernels is rather expected. Figures 13.4 through 13.6 give plots of the Studentized deviance residuals against x_1, x_2, and x_3, respectively. Here again, the plots

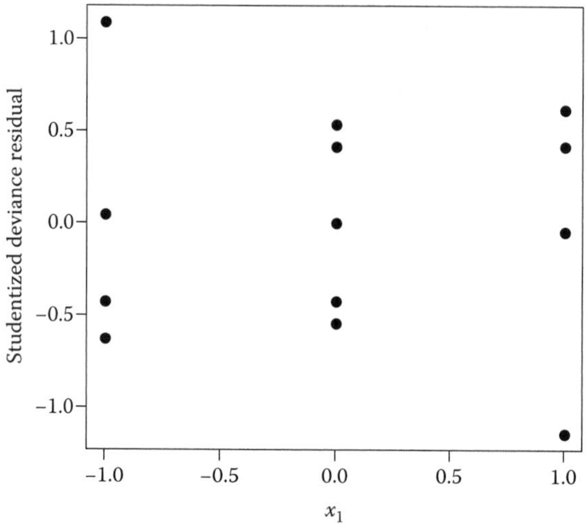

FIGURE 13.4
Plot of the Studentized deviance residuals against x_1.

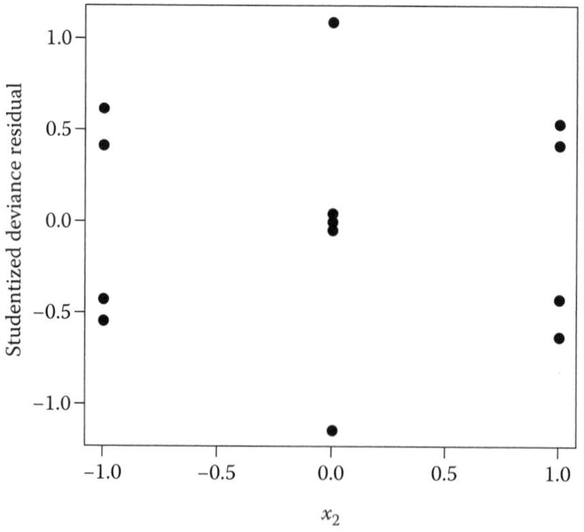

FIGURE 13.5
Plot of the Studentized deviance residuals against x_2.

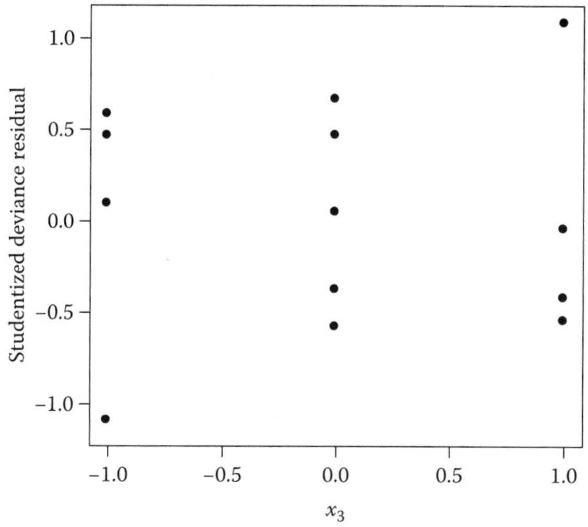

FIGURE 13.6
Plot of the Studentized deviance residuals against x_3.

reveal no systematic trends, which indicate no problems with the assumptions concerning the fitted model. Since the Studentized deviance residuals are very close to their corresponding Studentized Pearson's chi-square residuals, plots of the latter against x_1, x_2, and x_3 were omitted.

13.5 Hypothesis Testing

Hypothesis testing in the case of generalized linear models can be carried out using two types of inference, namely, the *Wald inference* and the *likelihood ratio inference*.

13.5.1 Wald Inference

This type of inference is based on the asymptotic normality of the ML estimator, $\hat{\boldsymbol{\beta}}$, of $\boldsymbol{\beta}$ as was seen in Section 13.3.2 [see (13.41)]. Consider, for example, testing the hypothesis

$$H_0 : \boldsymbol{A\beta} = \boldsymbol{b}, \tag{13.62}$$

where

$\boldsymbol{A}$ is a known matrix of order $s \times q$ and rank s ($\leq q$)
$\boldsymbol{b}$ is a known vector

Then, for large n, $A\hat{\beta}$ is asymptotically normal with mean $A\beta$ and a variance–covariance matrix $A(X'W^{-1}X)^{-1}A'$. It follows that under H_0,

$$(A\hat{\beta} - b)'[A(X'\hat{W}^{-1}X)^{-1}A']^{-1}(A\hat{\beta} - b) \qquad (13.63)$$

has an approximate chi-squared distribution with s degrees of freedom, where $\hat{W}$ is an estimate of W. The statistic given in (13.63) is called *Wald's test statistic*. The test is significant at the approximate α-level if this statistic is greater than or equal to $\chi^2_{\alpha,s}$. In particular, to test the significance of the individual elements of β (that is, testing $H_0 : \beta_i = 0$, where β_i is the ith element of β, $i = 1, 2, \ldots, q$), we can consider the random variable,

$$\frac{\hat{\beta}_i}{\sqrt{d_{ii}}}, \quad i = 1, 2, \ldots, q, \qquad (13.64)$$

which is distributed asymptotically as $N(0, 1)$ under $H_0 : \beta_i = 0$, where d_{ii} is the ith diagonal element of $(X'W^{-1}X)^{-1}$. Hence, under H_0,

$$\frac{\hat{\beta}_i^2}{\hat{d}_{ii}}, \quad i = 1, 2, \ldots, q, \qquad (13.65)$$

has an approximate chi-squared distribution with one degree of freedom, where $\hat{d}_{ii}$ is the ith diagonal element of $(X'\hat{W}^{-1}X)^{-1}$.

13.5.2 Likelihood Ratio Inference

Suppose that the vector β is partitioned as $\beta = (\beta_1' : \beta_2')'$, where β_1 and β_2 have q_1 and q_2 elements, respectively. Consider the hypothesis, $H_0 : \beta_1 = 0$. The ML estimate, $\hat{\beta}$, is partitioned accordingly as $(\hat{\beta}_1' : \hat{\beta}_2')'$. We may recall from Section 13.4.1 that $\mathcal{L}(\phi, \hat{\beta})$ is the likelihood function maximized over β. Let $\tilde{\beta}_2$ denote the ML estimate of β_2 under the restriction that $\beta_1 = 0$. Then, $\mathcal{L}(\phi, \tilde{\beta}_2) < \mathcal{L}(\phi, \hat{\beta})$, where $\mathcal{L}(\phi, \tilde{\beta}_2)$ is the likelihood function maximized over β_2 while β_1 is set equal to zero. If Λ_1 is the likelihood ratio (LR),

$$\Lambda_1 = \frac{\mathcal{L}(\phi, \tilde{\beta}_2)}{\mathcal{L}(\phi, \hat{\beta})},$$

then the hypothesis $H_0 : \beta_1 = 0$ can be tested using the statistic

$$-2 \log \Lambda_1 = 2[\log \mathcal{L}(\phi, \hat{\beta}) - \log \mathcal{L}(\phi, \tilde{\beta}_2)], \qquad (13.66)$$

which represents the difference between the scaled deviance for the full model (that contains all of β) and the scaled deviance for the reduced model (that contains only β_2). Under H_0, this statistic has approximately the chi-squared distribution whose number of degrees of freedom is equal to the difference between the degrees of freedom for the two scaled deviances, that is, $(n-q_2)-(n-q)=q-q_2=q_1$, where q_i is the number of elements of $\boldsymbol{\beta}_i$ $(i=1,2)$. The hypothesis H_0 can then be rejected at the approximate α-level if $-2\log\Lambda_1 \geq \chi^2_{\alpha,q_1}$. Note that the same hypothesis can also be tested by Wald's test.

The LR procedure can be used to test the significance of the individual elements of $\boldsymbol{\beta}$, just like Wald's test. For large samples, both tests give similar results, but can be different for small samples. In general, the LR test is preferred over Wald's test because the former's asymptotic distribution provides a better approximation than the latter in case of small and moderate-sized samples (see McCulloch and Searle, 2001, Section 5.5). However, Wald's test has a computational advantage over the LR test since it requires less computing time. This is true because the LR test requires fitting a submodel for each parameter tested.

Testing of the individual elements of $\boldsymbol{\beta}$ is available in PROC GENMOD. For example, to apply the LR test, "TYPE3" is selected as an option in the MODEL statement. A Type 3 analysis is similar to Type III sum of squares used in PROC GLM, except that likelihood ratios are used instead of sums of squares (see SAS, 1997, Chapter 10). As a result, we get a table entitled "LR Statistics for Type 3 Analysis," which gives the LR statistics and the corresponding p-values (from the chi-squared approximation) for each parameter in the model. Alternatively, a Wald's test statistic can be obtained for each parameter along with the corresponding p-value (from the chi-squared approximation) by adding the options "TYPE3" and "WALD" to the MODEL statement. PROC GENMOD also gives a table entitled "Analysis of Parameter Estimates," which provides a listing of the parameters in the model, their ML estimates and standard errors, chi-squared values [based on the test statistic given in (13.65)], and the corresponding p-values. These tests are identical to those obtained under Wald's inference.

13.6 Confidence Intervals

Large-sample confidence intervals on the individual elements of $\boldsymbol{\beta}=(\beta_1, \beta_2,\ldots,\beta_q)'$ can be constructed using either the likelihood ratio inference or the Wald inference.

13.6.1 Wald's Confidence Intervals

Using the information given in Section 13.5, we have that for large n,

$$\frac{\hat{\beta}_i-\beta_i}{\sqrt{\hat{d}_{ii}}}, \quad i=1,2,\ldots,q,$$

is distributed approximately as $N(0,1)$, where $\hat{\beta}_i$ is the ML estimate of β_i and $\hat{d}_{ii}$ is the ith diagonal element of $(\mathbf{X}'\hat{\mathbf{W}}^{-1}\mathbf{X})^{-1}$. It follows that an approximate $(1-\alpha)100\%$ confidence interval on β_i is given by

$$\hat{\beta}_i \pm \sqrt{\hat{d}_{ii}}\, z_{\alpha/2}, \quad i = 1, 2, \ldots, q. \tag{13.67}$$

Such intervals can be obtained from PROC GENMOD by adding the option "WALDCI" to the MODEL statement. The confidence coefficient can be selected with the "ALPHA=" option. The default value is 0.95.

Wald's confidence intervals can also be constructed for the mean response, $\mu(\mathbf{x}) = h[\eta(\mathbf{x})]$, at the points of the design under consideration, where $\eta(\mathbf{x}) = \mathbf{f}'(\mathbf{x})\boldsymbol{\beta}$ and $h(.)$ is the inverse of the link function $g(.)$ [see formulas (13.2) and (13.4)]. An approximate $(1-\alpha)100\%$ confidence interval on $\eta(\mathbf{x})$ is first computed, namely,

$$\mathbf{f}'(\mathbf{x})\hat{\boldsymbol{\beta}} \pm [\mathbf{f}'(\mathbf{x})(\mathbf{X}'\hat{\mathbf{W}}^{-1}\mathbf{X})^{-1}\mathbf{f}(\mathbf{x})]^{1/2}\, z_{\alpha/2}.$$

Then, an approximate $(1-\alpha)100\%$ confidence interval on $\mu(\mathbf{x})$ is given by

$$h\left\{\mathbf{f}'(\mathbf{x})\hat{\boldsymbol{\beta}} \pm [\mathbf{f}'(\mathbf{x})(\mathbf{X}'\hat{\mathbf{W}}^{-1}\mathbf{X})^{-1}\mathbf{f}(\mathbf{x})]^{1/2}\, z_{\alpha/2}\right\}. \tag{13.68}$$

The end points of this interval can be found in the output from PROC GENMOD in a table entitled "Observation Statistics," and are referred to as "LOWER" and "UPPER." For this purpose, the option "OBSTATS" must be added to the MODEL statement. The printed values of "LOWER" and "UPPER" correspond to each observation in the data set under consideration. The confidence coefficient is specified with the "ALPHA=" option in the same MODEL statement.

Another way to construct a Wald's confidence interval on $\mu(\mathbf{x})$ is to make use of formula (13.45) which gives an approximate expression for $\text{Var}[\hat{\mu}(\mathbf{x})]$. Using this expression, we obtain the following approximate $(1-\alpha)100\%$ confidence interval on $\mu(\mathbf{x}) = h[\eta(\mathbf{x})] = h[\mathbf{f}'(\mathbf{x})\boldsymbol{\beta}]$:

$$h[\mathbf{f}'(\mathbf{x})\hat{\boldsymbol{\beta}}] \pm \{[h'(\mathbf{f}'(\mathbf{x})\hat{\boldsymbol{\beta}})]^2\mathbf{f}'(\mathbf{x})(\mathbf{X}'\hat{\mathbf{W}}^{-1}\mathbf{X})^{-1}\mathbf{f}(\mathbf{x})\}^{1/2}\, z_{\alpha/2}, \tag{13.69}$$

where $h'(.)$ is the derivative of $h[\eta(\mathbf{x})]$ with respect to η.

13.6.2 Likelihood Ratio-Based Confidence Intervals

Confidence intervals can be constructed for each element of $\boldsymbol{\beta} = (\beta_1, \beta_2, \ldots, \beta_q)'$ using the likelihood ratio inference in Section 13.5.2. Suppose, for example, we consider β_i $(i = 1, 2, \ldots, q)$. Let $\mathcal{L}(\phi, \beta_i, \hat{\boldsymbol{\beta}}_{(i)})$ denote the likelihood function maximized with respect to all the elements of $\boldsymbol{\beta}$, except for β_i, which is held fixed, that is,

$$\mathcal{L}(\phi, \beta_i, \hat{\boldsymbol{\beta}}_{(i)}) = \max_{\boldsymbol{\beta}_{(i)}} \mathcal{L}(\phi, \beta_i, \boldsymbol{\beta}_{(i)}), \quad i = 1, 2, \ldots, q. \tag{13.70}$$

Here, $\boldsymbol{\beta}_{(i)}$ denotes the vector $\boldsymbol{\beta}$ with β_i removed, and $\hat{\boldsymbol{\beta}}_{(i)}$ is the ML estimate of $\boldsymbol{\beta}_{(i)}$ under the restriction that β_i is fixed. Thus, $\mathcal{L}(\phi, \beta_i, \hat{\boldsymbol{\beta}}_{(i)})$ is a partially maximized likelihood function and is called the *profile likelihood* function for β_i $(i = 1, 2, \ldots, q)$. Then, as in Section 13.5.2,

$$2\,[\log \mathcal{L}(\phi, \hat{\boldsymbol{\beta}}) - \log \mathcal{L}(\phi, \beta_i, \hat{\boldsymbol{\beta}}_{(i)})]$$

has an asymptotic chi-squared distribution with one degree of freedom, where $\hat{\boldsymbol{\beta}}$ is the usual ML estimate of $\boldsymbol{\beta}$. Consequently, an approximate $(1-\alpha)100\%$ confidence set on β_i is defined by the values of β_i that satisfy the inequality

$$2\,[\log \mathcal{L}(\phi, \hat{\boldsymbol{\beta}}) - \log \mathcal{L}(\phi, \beta_i, \hat{\boldsymbol{\beta}}_{(i)})] \le \chi^2_{\alpha,1}. \tag{13.71}$$

To obtain a confidence interval on β_i using this method, (13.71) must be solved numerically for all values of β_i that satisfy the preceding inequality. The resulting interval is called a *profile likelihood confidence interval*. This is done in PROC GENMOD by adding the option "LRCI" to the MODEL statement. The confidence coefficient can be chosen with the "ALPHA=" option. The default value is 0.95. It should be noted here that the computation needed to derive these intervals involves an iterative procedure and can therefore be more time consuming than in the case of the Wald intervals. However, the latter intervals are not thought to be as accurate as the likelihood ratio intervals, especially for small sample sizes.

Example 13.3 In order to demonstrate the use of PROC GENMOD in deriving tests and confidence intervals concerning the elements of $\boldsymbol{\beta}$, let us consider once more the same data set and model as in Example 13.1. Table 13.4 gives the parameter ML estimates and corresponding chi-squared and p-values [based on formula (13.65)]. These estimates were previously given in Table 13.2

TABLE 13.4
Analysis of Parameter Estimates

Parameter	DF	Estimate	Chi-Squared	p-Value
β_0	1	2.8332	136.46	<0.0001
β_1	1	−0.2066	11.20	0.0008
β_2	1	−0.1585	5.77	0.0163
β_3	1	−0.3160	24.84	<0.0001
β_{12}	1	−0.1401	2.44	0.1183
β_{13}	1	0.6246	48.85	<0.0001
β_{23}	1	−0.3709	14.65	0.0001
β_{11}	1	0.4589	10.07	0.0015
β_{22}	1	0.1501	1.08	0.2987
β_{33}	1	0.3470	5.70	0.0170

TABLE 13.5
Wald's Statistics for Type 3 Analysis

Source	DF	Chi-Squared	p-Value
x_1	1	11.20	0.0008
x_2	1	5.77	0.0163
x_3	1	24.84	<0.0001
x_1x_2	1	2.44	0.1183
x_1x_3	1	48.85	<0.0001
x_2x_3	1	14.65	0.0001
x_1^2	1	10.07	0.0015
x_2^2	1	1.08	0.2987
x_3^2	1	5.70	0.0170

TABLE 13.6
Likelihood Ratio Statistics for Type 3 Analysis

Source	DF	Chi-Squared	p-Value
x_1	1	11.32	0.0008
x_2	1	5.83	0.0157
x_3	1	25.78	<0.0001
x_1x_2	1	2.46	0.1167
x_1x_3	1	58.08	<0.0001
x_2x_3	1	15.53	<0.0001
x_1^2	1	11.53	0.0007
x_2^2	1	1.13	0.2873
x_3^2	1	6.29	0.0121

and are repeated here for convenience. We note that all parameters are significantly different from zero, except for β_{12} and β_{22}. Tables 13.5 and 13.6 show the results of Wald's and LR tests, respectively, concerning the model's parameters, except for the intercept β_0. Note that the chi-squared values in Table 13.5 concerning $\beta_1, \beta_2, \beta_3, \beta_{12}, \beta_{13}, \beta_{23}, \beta_{11}, \beta_{22}$, and β_{33} are identical to the corresponding values in Table 13.4. We also note that the results of the Wald and LR tests are very similar concerning the significance of all the effects in the model. Table 13.7 gives approximate 95% confidence intervals on all the parameters on the basis of the Wald and LR inferences. Finally, approximate 95% confidence intervals on the mean responses, $\mu(x_i)$, $i = 1, 2, \ldots, 13$, that correspond to the 13 observations in the data set, are presented in Table 13.8. Note that the observed (Y) and predicted response values in this table are repeated here (from Table 13.3) for convenience.

TABLE 13.7
95% Confidence Intervals on the Parameters

Parameter	Wald's Confidence Limits	Likelihood Ratio Confidence Limits
β_0	(2.3579, 3.3086)	(2.3170, 3.2737)
β_1	(−0.3276, −0.0856)	(−0.3284, −0.0861)
β_2	(−0.2878, −0.0292)	(−0.2887, −0.0298)
β_3	(−0.4403, −0.1917)	(−0.4418, −0.1929)
β_{12}	(−0.3160, 0.0357)	(−0.3177, 0.0348)
β_{13}	(0.4495, 0.7998)	(0.4544, 0.8057)
β_{23}	(−0.5608, −0.1809)	(−0.5650, −0.1842)
β_{11}	(0.1754, 0.7423)	(0.1870, 0.7569)
β_{22}	(−0.1330, 0.4333)	(−0.1214, 0.4479)
β_{33}	(0.0620, 0.6319)	(0.0732, 0.6461)

TABLE 13.8
Wald's 95% Confidence Intervals on the Mean Responses

Y	Predicted	Lower Limit	Upper Limit
20	18.8589	13.2418	26.8589
21	18.2663	12.5422	26.6028
42	42.1402	32.3054	54.9689
36	34.2672	25.5233	46.0065
11	12.0009	7.9761	18.0567
120	119.8598	100.9405	142.3252
33	34.5927	25.9349	46.1405
36	37.7328	28.3905	50.1494
32	30.9991	22.5608	42.5935
38	39.1411	29.5042	51.9255
20	22.7337	15.9756	32.3507
17	17.0000	10.5682	27.3461
49	47.4073	36.6610	61.3038

A listing of the SAS statements needed to get all the results in Tables 13.2 through 13.8 is given below

```
DATA;
INPUT X1 X2 X3 Y;
CARDS;
(enter here the data from Table 13.1)
PROC GENMOD;
MODEL Y = X1 X2 X3 X1 * X2 X1 * X3 X2 * X3 X1 * X1 X2 * X2 X3 * X3
/DIST=POISSON LINK=LOG OBSTATS RESIDUALS TYPE3 WALD
WALDCI;
PROC GENMOD;
```

```
MODEL Y = X1 X2 X3 X1 * X2 X1 * X3 X2 * X3 X1 * X1 X2 * X2 X3 * X3
    /DIST=POISSON LINK=LOG TYPE3 LRCI;
RUN;
```

The first GENMOD statement generated the results in Tables 13.2 through 13.5 and 13.8. The options, "TYPE3" and "WALD," should appear together in the MODEL statement in order to get Wald's chi-squared results. The option, "WALDCI," was used to obtain Wald's approximate 95% (default value) confidence intervals on the model's parameters (as in Table 13.7). Table 13.6 was obtained as a result of using the option "TYPE3" in the second MODEL statement, but without including the "WALD" option. To get approximate 95% (default value) likelihood ratio confidence intervals on the model's parameters (as in Table 13.7), the option "LRCI" should be added to the second MODEL statement.

13.7 Gamma-Distributed Response

The gamma distribution is a member in the family of continuous distributions. Its density function is of the form

$$f(y, \alpha, \beta) = \frac{y^{\alpha-1}\exp(-y/\beta)}{\Gamma(\alpha)\,\beta^{\alpha}}, \quad y \geq 0, \tag{13.72}$$

where α and β are positive constants. A random variable, Y, having this distribution is said to be *gamma distributed* with parameters α and β. This fact is denoted by writing $Y \sim G(\alpha, \beta)$. The density function in (13.72) can also be written as

$$f(y, \alpha, \beta) = \exp\left[-\frac{y}{\beta} + (\alpha - 1)\log y - \alpha\log\beta - \log\Gamma(\alpha)\right]. \tag{13.73}$$

Comparing (13.73) with (13.1) we find that

$$\begin{aligned}
\theta &= -\frac{1}{\alpha\,\beta}, \\
a(\phi) &= \frac{1}{\alpha}, \\
b(\theta) &= \log\alpha + \log\beta \\
&= -\log(-\theta), \\
c(y, \phi) &= \alpha\log\alpha - \log\Gamma(\alpha) + (\alpha - 1)\log y.
\end{aligned}$$

We conclude that the gamma distribution belongs to the exponential family. Using formulas (13.16) and (13.18), we find that the mean and variance of this distribution are

$$\begin{aligned} E(Y) &= \mu \\ &= b'(\theta) \\ &= -\frac{1}{\theta} \\ &= \alpha\,\beta, \end{aligned} \tag{13.74}$$

$$\begin{aligned} \text{Var}(Y) &= a(\phi)\,b''(\theta) \\ &= \frac{1}{\alpha\,\theta^2} \\ &= \alpha\,\beta^2. \end{aligned} \tag{13.75}$$

From (13.74) and (13.75) we note that the ratio,

$$\begin{aligned} \frac{\text{Var}(Y)}{\mu^2} &= \frac{1}{\alpha} \\ &= a(\phi), \end{aligned}$$

is constant. This ratio, which is denoted by ρ^2, is the square of the coefficient of variation, ρ. It follows that the gamma distribution has a constant coefficient of variation.

The canonical link for the gamma distribution is given by

$$\begin{aligned} \eta &= g(\mu) \\ &= \frac{1}{\mu}. \end{aligned} \tag{13.76}$$

This is the *reciprocal link*. Hence, the inverse link function is

$$\begin{aligned} \mu &= h(\eta) \\ &= \frac{1}{\eta}. \end{aligned}$$

If the linear predictor, η, is expressed as a linear model as in (13.2), then the mean response at a point $\boldsymbol{x}$ is given by

$$\mu(\boldsymbol{x}) = \frac{1}{\boldsymbol{f}'(\boldsymbol{x})\boldsymbol{\beta}}.$$

The predicted response $\hat{\mu}(\boldsymbol{x})$ is then of the form

$$\hat{\mu}(\boldsymbol{x}) = \frac{1}{\boldsymbol{f}'(\boldsymbol{x})\hat{\boldsymbol{\beta}}}, \tag{13.77}$$

where $\hat{\boldsymbol{\beta}}$ is the ML estimate of $\boldsymbol{\beta}$. Note that since $\mu(\boldsymbol{x})$ is positive, it would be desirable for $\hat{\mu}(\boldsymbol{x})$ to be also positive. But, $\boldsymbol{f}'(\boldsymbol{x})\hat{\boldsymbol{\beta}}$ may be negative for some values of $\boldsymbol{x}$, which is undesirable. Hence, care should be exercised in order to avoid negative values of $\hat{\mu}(\boldsymbol{x})$ when using the canonical link function. For this reason, the log link may be considered in place of the reciprocal link since it does not result in negative values of $\hat{\mu}(\boldsymbol{x})$.

13.7.1 Deviance for the Gamma Distribution

Consider a response vector, $\boldsymbol{Y} = (Y_1, Y_2, \ldots, Y_n)'$, such that $Y_1, Y_2, \ldots, Y_n$ are mutually independent and $Y_i \sim G(\alpha, \beta_i), i = 1, 2, \ldots, n$. Here, α is considered to be a constant that does not depend on i. Let $y_1, y_2, \ldots, y_n$ be realized values of $Y_1, Y_2, \ldots, Y_n$, respectively. Then, by formula (13.51), the deviance for the gamma distribution, on the basis of the sample $y_1, y_2, \ldots, y_n$, is given by

$$\mathcal{D} = 2\sum_{i=1}^{n}[(\tilde{\theta}_i - \hat{\theta}_i)\, y_i - b(\tilde{\theta}_i) + b(\hat{\theta}_i)], \tag{13.78}$$

where, if we recall, $\hat{\theta}_i$ and $\tilde{\theta}_i$ are the estimates of θ_i under the assumed and saturated models, respectively, $i = 1, 2, \ldots, n$. For the saturated model, $\tilde{\theta}_i = -\frac{1}{y_i}$ since $\theta = -\frac{1}{\mu}$, and for the assumed model, $\hat{\theta}_i = -\frac{1}{\hat{\mu}_i}$, where $\hat{\mu}_i$ is the ML estimate of μ_i $(i = 1, 2, \ldots, n)$. From (13.78) we then get

$$\begin{aligned}\mathcal{D} &= 2\sum_{i=1}^{n}\left[\left(-\frac{1}{y_i} + \frac{1}{\hat{\mu}_i}\right) y_i + \log\left(\frac{1}{y_i}\right) - \log\left(\frac{1}{\hat{\mu}_i}\right)\right] \\ &= 2\sum_{i=1}^{n}\left[-\log\left(\frac{y_i}{\hat{\mu}_i}\right) + \frac{y_i - \hat{\mu}_i}{\hat{\mu}_i}\right].\end{aligned}$$

13.7.2 Variance–Covariance Matrix of $\hat{\beta}$

Using formula (13.42), the variance–covariance matrix of $\hat{\beta}$, the ML estimator of β for the linear predictor, is approximately equal to

$$\text{Var}(\hat{\beta}) \approx (\boldsymbol{X}'\boldsymbol{W}^{-1}\boldsymbol{X})^{-1}, \tag{13.79}$$

where $\boldsymbol{X}$ is the model matrix for the linear predictor and $\boldsymbol{W} = \oplus_{i=1}^{n} [g'(\mu_i)]^2\, \text{Var}(Y_i)$ [see formula (13.36)]. The link function, $g(.)$, for the gamma distribution can be either the reciprocal link or the log link. Note that by (13.75),

$$\begin{aligned}\text{Var}(Y_i) &= \frac{1}{\alpha}\, b''(\theta_i) \\ &= \rho^2\, V(\mu_i), \quad i = 1, 2, \ldots, n,\end{aligned}$$

where

ρ is the coefficient of variation

$V(\mu_i)$ is the variance function, which, if we recall, is equal to $b''(\theta_i)$, $i = 1, 2, \ldots, n$

Making the substitution in (13.79), we get

$$\text{Var}(\hat{\beta}) \approx \rho^2 \left\{\boldsymbol{X}'\left(\bigoplus_{i=1}^{n} \frac{1}{[g'(\mu_i)]^2\, V(\mu_i)}\right)\boldsymbol{X}\right\}^{-1}.$$

In general, ρ^2 is unknown and should be estimated. Since

$$\rho^2 = \frac{\text{Var}(Y_i)}{\mu_i^2}, \quad i = 1, 2, \ldots, n,$$

an estimator of ρ^2 can be obtained as follows:

$$\begin{aligned}\rho^2 &= \frac{\text{E}(Y_i - \mu_i)^2}{\mu_i^2} \\ &= \text{E}\left(\frac{Y_i - \mu_i}{\mu_i}\right)^2, \quad i = 1, 2, \ldots, n.\end{aligned}$$

Hence, a consistent estimator of ρ^2 is given by

$$\hat{\rho}^2 = \frac{1}{n}\sum_{i=1}^{n}\left(\frac{Y_i - \hat{\mu}_i}{\hat{\mu}_i}\right)^2,$$

where $\hat{\mu}_i = h[f'(x_i)\hat{\beta}]$ and $h(.)$ is the inverse of the link function ($i = 1, 2, \ldots, n$). Thus, an estimate of $\text{Var}(\hat{\beta})$ is approximately given by

$$\widehat{\text{Var}}(\hat{\beta}) \approx \hat{\rho}^2 \left\{ X' \left(\bigoplus_{i=1}^{n} \frac{1}{[g'(\hat{\mu}_i)]^2 \, V(\hat{\mu}_i)} \right) X \right\}^{-1}.$$

Example 13.4 The gamma distribution has applications in a wide variety of situations. For example, it is used to represent the distributions of lifetimes (or failure times) in industrial experiments, reliability and survival data, the resistivity of test wafers (see Myers, Montgomery, and Vining, 2002, Section 5.8), and daily rainfall data (see McCullagh and Nelder, 1989, Section 8.4.3). Such distributions are known to have a heavy right tail where the variance is proportional to the square of the mean (that is, have constant coefficients of variation).

Consider, as an example, an experiment conducted to study the effect of exterior temperature, x, on the life, Y, of a certain type of batteries to be used in an electronic device. The plate materials for the batteries are the same. Three batteries were tested at each of three levels of temperature, namely, 20, 70, 120°F. These levels were coded as $-1, 0, 1$. The corresponding values of Y (in h) are given in Table 13.9.

A generalized linear model was fitted using a reciprocal link and a gamma distribution for the response. The model for the linear predictor is of the second degree of the form

$$\eta(x) = \beta_0 + \beta_1 x + \beta_{11} x^2. \tag{13.80}$$

TABLE 13.9
Lifetimes Data (in h)

Coded Temperature (°F)[a]	Y
−1	140.2, 165.9, 158.8
0	140.4, 127.3, 129.7
1	74.8, 68.1, 79.6

[a] The coded levels, −1, 0, and 1 correspond to 20, 70, and 120 °F, respectively.

The following SAS statements were used

```
DATA;
INPUT X Y;
CARDS;
(enter here the data from Table 13.9)
PROC GENMOD;
MODEL Y = X X * X/DIST = GAMMA LINK = POWER (-1) OBSTATS
RESIDUALS TYPE3 WALD;
PROC GENMOD;
MODEL Y = X X * X/DIST = GAMMA LINK = POWER (-1) TYPE3;
RUN;
```

Note that the reciprocal link is denoted in SAS as "LINK = POWER (-1)".

From the corresponding SAS output we find that the deviance and Pearson's chi-square statistic values are 0.0328 and 0.0323, respectively, with 6 degrees of freedom. Their scaled values divided by the degrees of freedom are 1.5009 and 1.4808, respectively. Table 13.10 gives information concerning the analysis of parameter estimates and the LR tests for model (13.80). Values of the deviance and Pearson's chi-square residuals are presented in Table 13.11. All this information indicates that the model fits the data well and that the model parameters are all significantly different from zero.

TABLE 13.10
Analysis of Model (13.80) Using the Reciprocal Link

Parameter	DF	Estimate	Chi-Squared	p-Value
Analysis of parameter estimates				
β_0	1	0.0075	824.09	<0.0001
β_1	1	0.0035	182.28	<0.0001
β_{11}	1	0.0024	42.73	<0.0001
Source	**DF**	**Chi-Squared**	**p-Value**	
Likelihood ratio statistics for Type 3 analysis				
x	1	29.21	<0.0001	
$x * x$	1	15.60	<0.0001	

TABLE 13.11
Observation Statistics for Model (13.80) Using the Reciprocal Link

Y	x	Pred	Resdev	Reschi	StResdev	StReschi
140.2	−1	154.967	−0.098	−0.095	−1.999	−1.934
158.8	−1	154.967	0.025	0.025	0.498	0.502
165.9	−1	154.967	0.069	0.071	1.400	1.432
127.3	0	132.467	−0.040	−0.039	−0.802	−0.792
129.7	0	132.467	−0.021	−0.021	−0.427	−0.424
140.4	0	132.467	0.059	0.060	1.192	1.216
68.1	1	74.167	−0.084	−0.082	−1.708	−1.660
74.8	1	74.167	0.009	0.009	0.173	0.173
79.6	1	74.167	0.072	0.073	1.452	1.487

TABLE 13.12
Analysis of Model (13.80) Using the Log Link

Parameter	DF	Estimate	Chi-Squared	p-Value
Analysis of parameter estimates				
β_0	1	4.8863	19676.10	<0.0001
β_1	1	−0.3684	223.75	<0.0001
β_{11}	1	−0.2116	24.59	<0.0001

Source	DF	Chi-Squared	p-Value
Likelihood ratio statistics for Type 3 analysis			
x	1	29.21	<0.0001
$x * x$	1	12.01	0.0005

Using now the log link instead of the reciprocal link (the log link is often used with the gamma distribution), we find that the results concerning the deviance, Person's chi-square statistic, and their scaled values are identical to those obtained under the reciprocal link. The analysis of parameter estimates and the LR tests are displayed in Table 13.12. We note that the conclusions concerning the significance of the model parameters are consistent with the reciprocal link case, but the actual parameter estimates are different. The results concerning the deviance and Pearson's chi-square residuals are identical to those shown in Table 13.11.

Exercises

13.1 Consider the likelihood function for the Poisson distribution. Verify that formulas (13.11) and (13.12) are satisfied for this distribution.

13.2 Establish the validity of formulas (13.11) and (13.12).

[Hint: Use the following relationship:

$$\frac{\partial^2 \ell(\theta, \phi, Y)}{\partial \theta^2} = \left[\left(\frac{\partial^2 \mathcal{L}(\theta, \phi, Y)}{\partial \theta^2}\right) \Big/ \mathcal{L}(\theta, \phi, Y)\right] - \left[\frac{\partial \ell(\theta, \phi, Y)}{\partial \theta}\right]^2 \Bigg].$$

13.3 Let $\tau(Y)$ be a transformation of a random variable Y whose mean is μ. Assume that $\tau(.)$ is differentiable in a neighborhood of μ such that $\tau'(\mu) \neq 0$, where $\tau'(\mu)$ denotes the derivative of $\tau(\mu)$ with respect to μ.

(a) Show that if $\mathrm{Var}[\tau(Y)] = c$, where c is a constant, then

$$\mathrm{Var}(Y) \approx \frac{c}{[\tau'(\mu)]^2}.$$

(b) Deduce that if $\log(Y)$ has a constant variance, then $\mathrm{Var}(Y) \approx c\,\mu^2$.

(c) If Y has the gamma distribution, what can you say about $\mathrm{Var}[\log(Y)]$?

13.4 The negative binomial distribution has the probability mass function

$$f(y, p) = \binom{r + y - 1}{r - 1} p^r (1 - p)^y, \quad y = 0, 1, \ldots,$$

where r is a known positive integer and $0 < p < 1$.

(a) Show that this distribution belongs to the exponential family.

(b) Apply formulas (13.16) and (13.18) to show that if Y has this distribution, then

$$\mathrm{E}(Y) = \frac{r(1 - p)}{p},$$

$$\mathrm{Var}(Y) = \frac{r(1 - p)}{p^2}.$$

13.5 Consider again the negative binomial distribution in Exercise 13.4.

(a) Find the variance function for a given observation.

(b) Give an expression for the deviance on the basis of a sample of size n from this distribution.

(c) Give an expression for Pearson's chi-square statistic for a sample of size n.

13.6 Consider the linear predictor in (13.2).

(a) Use the Wald inference to obtain an approximate $(1 - \alpha)100\%$ confidence region on $\boldsymbol{\beta}$.

(b) Let $\lambda(\boldsymbol{\beta})$ be a scalar function of $\boldsymbol{\beta}$ which is assumed to be continuous. Show how you can obtain an approximate confidence interval on $\lambda(\boldsymbol{\beta})$ with an approximate confidence coefficient greater than or equal to $1 - \alpha$.

[Hint: Consider the maximum and minimum of $\lambda(\boldsymbol{\beta})$ over the confidence region in part (a)].

13.7 On the basis of formula (13.32), we can write

$$\boldsymbol{\beta}^{(m+1)} = [(\mathbf{X}'\mathbf{W}^{-1}\mathbf{X})^{-1}\mathbf{X}'\mathbf{W}^{-1}]_{\boldsymbol{\beta}=\boldsymbol{\beta}^{(m)}}\,\mathbf{Z}^{(m)}, \quad m = 0, 1, \ldots,$$

where

$\mathbf{Z}^{(m)}$ is the value of $\mathbf{Z}$ at the mth iteration

$\mathbf{Z}$ is given in (13.35)

Hence, for $m = 0$,

$$\boldsymbol{\beta}^{(1)} = [(\mathbf{X}'\mathbf{W}^{-1}\mathbf{X})^{-1}\mathbf{X}'\mathbf{W}^{-1}]_0\,\mathbf{Z}^{(0)},$$

where $[(\mathbf{X}'\mathbf{W}^{-1}\mathbf{X})^{-1}\mathbf{X}'\mathbf{W}^{-1}]_0$ is the value of $(\mathbf{X}'\mathbf{W}^{-1}\mathbf{X})^{-1}\mathbf{X}'\mathbf{W}^{-1}$ when $\boldsymbol{\beta} = \boldsymbol{\beta}^0$, that is, when $\boldsymbol{\eta} = \boldsymbol{\eta}^0 = \mathbf{X}\boldsymbol{\beta}^0$, or when $\boldsymbol{\mu} = \boldsymbol{\mu}^0$, where $\boldsymbol{\mu}^0$ is the value of $\boldsymbol{\mu}$ corresponding to $\boldsymbol{\eta}^0$. Choose $\boldsymbol{\mu}^0 = \boldsymbol{y}$, the data vector, and $\mathbf{Z}^0$ to be the vector whose ith element is $g(y_i)$, where y_i is the ith element of $\boldsymbol{y}$ $(i = 1, 2, \ldots, n)$. This results in a value of $\boldsymbol{\beta}^{(1)}$, the first iterated value of $\boldsymbol{\beta}$. Subsequent iterated values of $\boldsymbol{\beta}$ can now be obtained by applying formula (13.32).

Apply this procedure in case of Example 13.1 to obtain the first five iterated values of the parameter estimates for model (13.46).

13.8 Consider the same model, response distribution, link function, and data set as in Example 13.1.

(a) Obtain an expression for the profile likelihood function for β_1.

(b) Use (13.71) directly to obtain an approximate 95% profile likelihood confidence interval on β_1 without using the "LRCI" model option in PROC GENMOD.

13.9 A biomedical study was conducted to study the effects of two agents, whose levels are denoted by X_1 and X_2, on the number, Y, of cells that exhibit differentiation after exposure to the two agents. A 4×4 factorial experiment was carried out using the levels 0, 2, 8, and 80 for X_1 and 0, 6, 16, and 80 for X_2. At each combination of X_1 and X_2, 100 cells were examined and the number, y, of cells differentiating was recorded. The data are given in the following table.

X_1	X_2	y
0	0	5
0	6	8
0	16	9
0	80	19
2	0	10
2	6	17
2	16	25
2	80	35
8	0	15
8	6	33
8	16	34
8	80	63
80	0	51
80	6	84
80	16	88
80	80	95

Assume a Poisson distribution on Y and consider using a log link.

(a) Obtain the ML estimates for the model

$$\eta = \beta_0 + \beta_1 x_1 + \beta_2 x_2 + \beta_{12} x_1 x_2 + \beta_{11} x_1^2 + \beta_{22} x_2^2,$$

where x_1 and x_2 denote the coded values, $x_i = (X_i - 40)/40, i = 1, 2$.

(b) Find the deviance and Pearson's chi-square statistic values for this model. What can you say about the fit of the model?

(c) Obtain values of the deviance and Pearson's chi-square residuals.

(d) Test the significance of the model parameters. Let $\alpha = 0.05$.

(e) Obtain the 95% Wald and likelihood ratio confidence intervals on the model parameters.

13.10 In a cancer research experiment, a logistic regression model was utilized in a cytotoxicity study to investigate the dose-response curve for a combination of two agents. Their respective concentration levels are denoted by x_1 and x_2. Cell cytotoxicity for the combination of x_1 and x_2 was evaluated by counting the number of viable and dead cells. Of interest was the modeling of the probability, p, of dead cells as a function of x_1 and x_2. The following data were obtained as a result of running a 4×4 factorial experiment:

x_1	x_2	**Total Number of Cells, N**	**Number of Dead Cells, S**
0	0	95	18
0	3	86	22
0	15	92	55
0	100	86	67
3	0	90	14
3	3	93	18
3	15	91	42
3	100	94	80
15	0	89	16
15	3	82	11
15	15	88	37
15	100	84	63
35	0	90	17
35	3	91	35
35	15	95	57
35	100	88	75

The fitted linear predictor is

$$\eta(\mathbf{x}) = \beta_0 + \beta_1 x_1 + \beta_2 x_2 + \beta_{11} x_1^2 + \beta_{22} x_2^2,$$

where $\mathbf{x} = (x_1, x_2)'$.

(a) Obtain the ML estimates of the model parameters, then provide a representation for $\hat{p}(\mathbf{x})$, the predicted value of p at $\mathbf{x}$.

(b) Find the deviance and Pearson's chi-square statistic values. Comment on the goodness of fit.

(c) Obtain values of the Studentized deviance and Pearson's residuals.

(d) Assess the significance of the model parameters. Let $\alpha = 0.05$.

(e) Give the predicted values, $\hat{p}(\mathbf{x}_i)$, at the 16 design points, then obtain Wald's 95% confidence intervals on $p(\mathbf{x}_i)$, $i = 1, 2, \ldots, 16$.

Bibliography

Airy, G. B. (1861). *On the Algebraical and Numerical Theory of Errors of Observations and the Combinations of Observations*. MacMillan, London, U.K.

Aitken, A. C. (1935). On least squares and linear combinations of observations, *Proc. Roy. Soc. Edinb.*, 55, 42–48.

Aitken, A. C. (1950). On the statistical independence of quadratic forms in normal variates, *Biometrika*, 37, 93–96.

Aitken, A. C. (1958). *Determinants and Matrices*, 9th ed. Oliver and Boyd, London, U.K.

Aldrich, J. (1997). R. A. Fisher and the making of maximum likelihood 1912–1922, *Statist. Sci.*, 12, 162–176.

Ali, M. M. and Silver, J. L. (1985). Tests for equality between sets of coefficients in two linear regressions under heteroscedasticity, *J. Am. Stat. Assoc.*, 80, 730–735.

Ames, M. H. and Webster, J. T. (1991). On estimating approximate degrees of freedom, *Am. Stat.*, 45, 45–50.

Anderson, M. R. (1971). A characterization of the multivariate normal distribution, *Ann. Math. Stat.*, 42, 824–827.

Anderson, T. W. (1963). A test for equality of means when covariance matrices are unequal, *Ann. Math. Stat.*, 34, 671–672.

Anderson, R. L. and Bancroft, T. A. (1952). *Statistical Theory in Research*. McGraw-Hill, New York.

Anderson, T. W. and Gupta, S. D. (1963). Some inequalities on characteristic roots of matrices, *Biometrika*, 50, 522–524.

Anderson, T. W., Olkin, I., and Underhill, L. G. (1987). Generation of random orthogonal matrices, *SIAM J. Sci. Stat. Comput.*, 8, 625–629.

Angellier, H., Choisnard, L., Molina-Boisseau, S., Ozil, P., and Dufresne, A. (2004). Optimization of the preparation of aqueous suspensions of waxy maize starch nanocrystals using a response surface methodology, *Biomacromolecules*, 5, 1545–1551.

Argac, D., Makambi, K. H., and Hartung, J. (2001). A note on testing the nullity of the between group variance in the one-way random effects model under variance heterogeneity, *J. Appl. Stat.*, 28, 215–222.

Arnold, B. C. and Shavelle, R. M. (1998). Joint confidence sets for the mean and variance of a normal distribution, *Am. Stat.*, 52, 133–140.

Arnold, S. F. (1981). *The Theory of Linear Models and Multivariate Analysis*. Wiley, New York.

Atkinson, A. C. (1985). *Plots, Transformations, and Regression*. Oxford University Press, Oxford.

Bargmann, R. E. and Nel, D. G. (1974). On the matrix differentiation of the characteristic roots of matrices, *S. Afr. Stat. J.*, 8, 135–144.

Bartlett, M. S. (1937). Properties of sufficiency and statistical tests, *Proc. Roy. Soc. Ser. A*, 160, 268–282.

Basilevsky, A. (1983). *Applied Matrix Algebra in the Statistical Sciences*. North-Holland, New York.

Bellman, R. (1997). *Introduction to Matrix Analysis*, 2nd ed. SIAM, Philadelphia, PA.

Bhat, B. R. (1962). On the distribution of certain quadratic forms in normal variates, *J. Roy. Stat. Soc. Ser. B*, 24, 148–151.

Bickel, P. J. and Doksum, K. A. (1977). *Mathematical Statistics*. Holden-Day, San Francisco, CA.

Birnbaum, Z. W. (1942). An inequality for Mill's ratio, *Ann. Math. Stat.*, 13, 245–246.

Bishop, T. A. and Dudewicz, E. J. (1978). Exact analysis of variance with unequal variances: Test procedures and tables, *Technometrics*, 20, 419–430.

Boardman, T. J. (1974). Confidence intervals for variance components—A comparative Monte Carlo study, *Biometrics*, 30, 251–262.

Bose, R. C. (1944). The fundamental problem of linear estimation, *Proc. 31th Indian Sci. Congress*, Part III, 2–3.

Box, G. E. P. (1954). Some theorems on quadratic forms applied in the study of analysis of variance problems, I. Effect of inequality of variances in the one-way classification, *Ann. Math. Stat.*, 25, 290–302.

Box, G. E. P. and Draper, N. R. (1959). A basis for the selection of a response surface design, *J. Am. Stat. Assoc.*, 54, 622–654.

Box, G. E. P. and Draper, N. R. (1963). The choice of a second order rotatable design, *Biometrika*, 50, 335–352.

Box, G. E. P. and Wilson, K. B. (1951). On the experimental attainment of optimum conditions (with discussion), *J. Roy. Stat. Soc. Ser. B*, 13, 1–45.

Broemeling, L. D. (1969). Confidence regions for variance ratios of random models, *J. Am. Stat. Assoc.*, 64, 660–664.

Brown, W. C. (1988). *A Second Course in Linear Algebra*. Wiley, New York.

Brown, M. B. and Forsythe, A. B. (1974a). The ANOVA and multiple comparisons for data with heterogeneous variances, *Biometrics*, 30, 719–724.

Brown, M. B. and Forsythe, A. B. (1974b). Robust tests for the equality of variances, *J. Am. Stat. Assoc.*, 69, 364–367.

Brown, M. B. and Forsythe, A. B. (1974c). The small sample behavior of some statistics which test the equality of several means, *Technometrics*, 16, 129–132.

Brownlee, K. A. (1965). *Statistical Theory and Methodology*, 2nd ed. Wiley, New York.

Burdick, R. K. and Graybill, F. A. (1984). Confidence intervals on linear combinations of variance components in the unbalanced one-way classification, *Technometrics*, 26, 131–136.

Burdick, R. K. and Graybill, F. A. (1992). *Confidence Intervals on Variance Components*. Dekker, New York.

Burdick, R. K. and Larsen, G. A. (1997). Confidence intervals on measures of variability in R & R studies, *J. Qual. Technol.*, 29, 261–273.

Cain, M. (1994). The moment-generating function of the minimum of bivariate normal random variables, *Am. Stat.*, 48, 124–125.

Carlson, R. and Carlson, J. E. (2005). Canonical analysis of response surfaces: A valuable tool for process development, *Org. Process Res. Dev.*, 9, 321–330.

Carter, Jr., W. H., Chinchilli, V. M., Campbell, E. D., and Wampler, G. L. (1984). Confidence interval about the response at the stationary point of a response surface, with an application to preclinical cancer therapy, *Biometrics*, 40, 1125–1130.

Casella, G. and Berger, R. L. (2002). *Statistical Inference*, 2nd ed. Duxbury, Pacific Grove, CA.

Chauvenet, W. (1863). *A Manual of Spherical and Practical Astronomy, 2: Theory and Use of Astronomical Instruments*. Lippincott, Philadelphia, PA.

Christensen, R. (1996). *Analysis of Variance, Design and Regression*. Chapman & Hall/CRC, Boca Raton, FL.

Cochran, W. G. (1934). The distribution of quadratic forms in a normal system, with applications to the analysis of covariance, *Proc. Camb. Phil. Soc.*, 30, 178–191.

Cochran, W. G. (1937). Problems arising in the analysis of a series of similar experiments, *J. Roy. Stat. Soc.*, 4 (Suppl.), 102–118.

Cochran, W. G. (1951). Testing a linear relation among variances, *Biometrics*, 7, 17–32.

Conover, W. J., Johnson, M. E., and Johnson, M. M. (1981). A comparative study of tests for homogeneity of variances, with applications to the outer continental shelf bidding data, *Technometrics*, 23, 351–361.

Corbeil, R. R. and Searle, S. R. (1976a). Restricted maximum likelihood (REML) estimation of variance components in the mixed model, *Technometrics*, 18, 31–38.

Corbeil, R. R. and Searle, S. R. (1976b). A comparison of variance component estimators, *Biometrics*, 32, 779–791.

Craig, A. T. (1943). Note on the independence of certain quadratic forms, *Ann. Math. Stat.*, 14, 195–197.

Cramér, H. (1946). *Mathematical Methods of Statistics*. Princeton University Press, Princeton, NJ.

Crump, S. L. (1946). The estimation of variance components in analysis of variance, *Biometrics Bull.*, 2, 7–11.

Crump, S. L. (1947). The estimation of components of variance in multiple classification, PhD thesis, Iowa State University, Ames, IA.

Crump, S. L. (1951). The present status of variance components analysis, *Biometrics*, 7, 1–16.

Cummings, W. B. and Gaylor, D. W. (1974). Variance component testing in unbalanced nested designs, *J. Am. Stat. Assoc.*, 69, 765–771.

Daniels, H. E. (1954). Saddlepoint approximations in statistics, *Ann. Math. Stat.*, 25, 631–650.

Das, K. (1979). Asymptotic optimality of restricted maximum likelihood estimates for the mixed model, *Calcutta Stat. Assoc. Bull.*, 28, 125–142.

Davenport, J. M. (1975). Two methods of estimating the degrees of freedom of an approximate F, *Biometrika*, 62, 682–684.

Davenport, J. M. and Webster, J. T. (1972). Type-I error and power of a test involving a Satterthwaite's approximate F-statistic, *Technometrics*, 14, 555–569.

Davenport, J. M. and Webster, J. T. (1973). A comparison of some approximate F-tests, *Technometrics*, 15, 779–789.

Davies, R. B. (1973). Numerical inversion of a chracteristic function, *Biometrika*, 60, 415–417.

Davies, R. B. (1980). The distribution of a linear combination of χ^2 random variables, *Appl. Stat.*, 29, 323–333.

Dey, A., Hande, S., and Tiku, M. L. (1994). Statistical proofs of some matrix results, *Linear Multilinear Algebra*, 38, 109–116.

Dobson, A. J. (2008). *An Introduction to Generalized Linear Models*, 3rd ed. Chapman & Hall/CRC, Boca Raton, FL.

Draper, N. R. and Guttman, I. (1966). Unequal group variances in the fixed-effects one-way analysis of variance: A Bayesian sidelight, *Biometrika*, 53, 27–35.

Draper, N. R. and Smith, H. (1998). *Applied Regression Analysis*, 3rd ed. Wiley, New York.

Driscoll, M. F. and Gundberg, Jr., W. R. (1986). A history of the development of Craig's theorem, *Am. Stat.*, 40, 65–70.

Driscoll, M. F. and Krasnicka, B. (1995). An accessible proof of Craig's theorem in the general case, *Am. Stat.*, 49, 59–62.

Dufour, J. M. (1986). Bias of S^2 in linear regressions with dependent errors, *Am. Stat.*, 40, 284–285.

Dwyer, P. S. (1967). Some applications of matrix derivatives in multivariate analysis, *J. Am. Stat. Assoc.*, 62, 607–625.

Eaton, M. L. (1970). Gauss–Markov estimation for multivariate linear models: A coordinate-free approach, *Ann. Math. Stat.*, 41, 528–538.

Edwards, A. W. F. (1974). The history of likelihood, *Int. Stat. Rev.*, 42, 9–15.

Eisenhart, C. (1947). The assumptions underlying the analysis of variance, *Biometrics*, 3, 1–21.

Evans, M., Hastings, N., and Peacock, B. (2000). *Statistical Distributions*, 3rd ed. Wiley, New York.

Eves, H. (1969). *An Introduction to the History of Mathematics*, 3rd ed. Holt, Rinehart, and Winston, New York.

Farebrother, R. W. (1997). A. C. Aitken and the consolidation of matrix theory, *Linear Algebra Appl.*, 43, 3–12.

Feller, W. (1957). *An Introduction to Probability Theory and Its Application*, Vol. I, 2nd ed. Wiley, New York.

Fisher, R. A. (1915). Frequency distribution of the values of the correlation coefficient in samples from an indefinitely large population, *Biometrika*, 10, 507–521.

Fisher, R. A. (1918). The correlation between relatives on the supposition of Mendelian inheritance, *Trans. Roy. Soc. Edinb.*, 52, 399–433.

Fisher, R. A. (1922). On the mathematical foundation of theoretical statistics, *Phil. Trans. Roy. Soc. A*, 222, 308–358.

Fisher, R. A. (1925). Theory of statistical estimation, *Proc. Camb. Phil. Soc.*, 22, 700–725.

Fisher, L. (1973). An alternative approach to Cochran's theorem, *Am. Stat.*, 27, 109.

Foster, F. G. (1957). Upper percentage points of the generalized beta distribution II, *Biometrika*, 44, 441–453.

Foster, F. G. and Rees, D. H. (1957). Upper percentage points of the generalized beta distribution I, *Biometrika*, 44, 237–247.

Fuchs, C. and Sampson, A. R. (1987). Simultaneous confidence intervals for the general linear model, *Biometrics*, 43, 457–469.

Fulks, W. (1978). *Advanced Calculus*, 3rd ed. Wiley, New York.

Gallo, J. and Khuri, A. I. (1990). Exact tests for the random and fixed effects in an unbalanced mixed two-way cross-classification model, *Biometrics*, 46, 1087–1095.

Gantmacher, F. R. (1959). *The Theory of Matrices*, Vol. I. Chelsea, New York.

Gaylor, D. W. and Hopper, F. N. (1969). Estimating the degrees of freedom for linear combinations of mean squares by Satterthwaite's formula, *Technometrics*, 11, 691–706.

Geary, R. C. (1936). The distribution of Student's ratio for non-normal samples, *J. Roy. Stat. Soc. Suppl.*, 3, 178–184.

Geisser, S. (1956). A note on the normal distribution, *Ann. Math. Stat.*, 27, 858–859.

Ghosh, M. (1996). Wishart distribution via induction, *Am. Stat.*, 50, 243–246.

Ghurye, S. G. and Olkin, I. (1962). A characterization of the multivariate normal distribution, *Ann. Math. Stat.*, 33, 533–541.

Golub, G. H. and Van Loan, C. F. (1983). *Matrix Computations*. The Johns Hopkins University Press, Baltimore, MD.

Good, I. J. (1963). On the independence of quadratic expressions, *J. Roy. Stat. Soc. Ser. B*, 25, 377–382.

Good, I. J. (1969). Conditions for a quadratic form to have a chi-squared distribution, *Biometrika*, 56, 215–216.

Gosslee, D. G. and Lucas, H. L. (1965). Analysis of variance of disproportionate data when interaction is present, *Biometrics*, 21, 115–133.

Graham, A. (1981). *Kronecker Products and Matrix Calculus: With Applications*. Wiley, New York.

Grattan–Guiness, I. (1994). A new type of question: On the pre-history of linear and non-linear programming, 1770–1940. In: *The History of Modern Mathematics*, Vol. III. Academic Press, Boston, MA, pp. 43–89.

Graybill, F. A. (1976). *Theory and Application of the Linear Model*. Duxbury, North Scituate, MA.

Graybill, F. A. (1983). *Matrices with Applications in Statistics*, 2nd ed. Wadsworth, Belmont, CA.

Guerin, L. and Stroup, W. W. (2000). A simulation study to evaluate PROC MIXED analysis of repeated measures data. In: *Proceedings of the 12th Annual Conference on Applied Statistics in Agriculture*. Kansas State University, Manhattan, KS, pp. 170–203.

Gurland, J. (1955). Distribution of definite and of indefinite quadratic forms, *Ann. Math. Stat.*, 26, 122–127.

Haberman, S. J. (1975). How much do Gauss–Markov and least squares estimates differ? A coordinate-free approach, *Ann. Stat.*, 3, 982–990.

Hartley, H. O. (1950). The maximum F-ratio as a short-cut test for heterogeneity of variance, *Biometrika*, 37, 308–312.

Hartley, H. O. and Rao, J. N. K. (1967). Maximum likelihood estimation for the mixed analysis of variance model, *Biometrika*, 54, 93–108.

Harville, D. A. (1977). Maximum-likelihood approaches to variance component estimation and to related problems, *J. Amer. Stat. Assoc.*, 72, 320–338.

Harville, D. A. (1997). *Matrix Algebra from a Statistician's Perspective*. Springer, New York.

Harville, D. A. and Fenech, A. P. (1985). Confidence intervals for a variance ratio, or for heritability, in an unbalanced mixed linear model, *Biometrics*, 41, 137–152.

Heiberger, R. M., Velleman, P. F., and Ypelaar, M. A. (1983). Generating test data with independently controllable features for multivariate general linear forms, *J. Am. Stat. Assoc.*, 78, 585–595.

Hemmerle, W. J. and Hartley, H. O. (1973). Computing maximum likelihood estimates for the mixed A.O.V. model using the W-transformation, *Technometrics*, 15, 819–831.

Henderson, C. R. (1953). Estimation of variance and covariance components, *Biometrics*, 9, 226–252.

Henderson, H. V. and Searle, S. R. (1981). The vec-permutation matrix, the vec operator and Kronecker products: A review, *Linear and Multilinear Algebra*, 9, 271–288.

Henderson, H. V., Pukelsheim, F., and Searle, S. R. (1983). On the history of the Kronecker product, *Linear and Multilinear Algebra*, 14, 113–120.

Herr, D. G. (1980). On the history of the use of geometry in the general linear model, *Am. Stat.*, 34, 43–47.

Herr, D. G. (1986). On the history of ANOVA in unbalanced, factorial designs: The first 30 years, *Am. Stat.*, 40, 265–270.

Herstein, I. N. (1964). *Topics in Algebra*. Blaisdell, Waltham, MA.

Hocking, R. R. and Kutner, M. H. (1975). Some analytical and numerical comparisons of estimators for the mixed A.O.V. model, *Biometrics*, 31, 19–28.

Hogg, R. V. and Craig, A. T. (1978). *Introduction to Mathematical Statistics*, 4th ed. Macmillan, New York.

Hudson, J. D. and Krutchkoff, R. G. (1968). A Monte-Carlo investigation of the size and power of tests employing Satterthwaite's synthetic mean squares, *Biometrika*, 55, 431–433.

Imhof, J. P. (1961). Computing the distribution of quadratic forms in normal variables, *Biometrika*, 48, 419–426.

Ito, K. (1969). On the effect of heteroscedasticity and nonnormality upon some multivariate test procedures. In: *Multivariate Analysis II*, P. R. Krishnaiah (ed.). Academic Press, New York, pp. 87–120.

Jain, M. C. (2001). *Vector Spaces and Matrices in Physics*. CRC, Boca Raton, FL.

James, G. S. (1951). The comparison of several groups of observations when the ratios of the population variances are unknown, *Biometrika*, 38, 324–329.

James, G. S. (1954). Tests of linear hypotheses in univariate and multivariate analysis when the ratios of the population variances are unknown, *Biometrika*, 41, 19–43.

Jennrich, R. I. and Sampson, P. F. (1976). Newton–Raphson and related algorithms for maximum likelihood variance component estimation, *Technometrics*, 18, 11–17.

Jeske, D. R. (1994). Illustrating the Gauss–Markov theorem, *Am. Stat.*, 48, 237.

Jeyaratnam, S. and Othman, A. R. (1985). Test of hypothesis in one-way random effects model with unequal error variances, *J. Stat. Comput. Simul.*, 21, 51–57.

Johnson, N. L. and Kotz, S. (1970). *Continuous Univariate Distributions-2*. Wiley, New York.

Johnson, N. L. and Kotz, S. (1972). *Distributions in Statistics: Continuous Multivariate Distributions*. Wiley, New York.

Johnson, N. L. and Leone, F. C. (1964). *Statistics and Experimental Design*, Vol. II. Wiley, New York.

Kackar, R. N. and Harville, D. A. (1981). Unbiasedness of two-stage estimation and prediction procedures for mixed linear models, *Commun. Stat. Theor. Meth.*, 10, 1249–1261.

Kackar, R. N. and Harville, D. A. (1984). Approximations for standard errors of estimators of fixed and random effects in mixed linear models, *J. Am. Stat. Assoc.*, 79, 853–862.

Kagan, A. M., Linnik, Y. V., and Rao, C. R. (1973). *Characterization Problems in Mathematical Statistics*. Wiley, New York.

Kaiser, L. D. and Bowden, D. C. (1983). Simultaneous confidence intervals for all linear contrasts of means with heterogeneous variances, *Commun. Stat. Theor. Meth.*, 12, 73–88.

Kawada, Y. (1950). Independence of quadratic forms of normally correlated variables, *Ann. Math. Stat.*, 21, 614–615.

Kempthorne, O. (1952). *The Design and Analysis of Experiments*. Wiley, New York.

Kendall, M. G. and Stuart, A. (1963). *The Advanced Theory of Statistics*, Vol. 1, 2nd ed. Hafner, New York.

Kenward, M. G. and Roger, J. H. (1997). Small sample inference for fixed effects from restricted maximum likelihood, *Biometrics*, 53, 983–997.

Khuri, A. I. (1981). Simultaneous confidence intervals for functions of variance components in random models, *J. Am. Stat. Assoc.*, 76, 878–885.

Khuri, A. I. (1982). Direct products: A powerful tool for the analysis of balanced data, *Commun. Stat. Theor. Meth.*, 11, 2903–2920.

Khuri, A. I. (1985). A test for lack of fit of a linear multiresponse model, *Technometrics*, 27, 213–218.

Khuri, A. I. (1986). Exact tests for the comparison of correlated response models with an unknown dispersion matrix, *Technometrics*, 28, 347–357.

Khuri, A. I. (1987). An exact test for the nesting effect's variance component in an unbalanced random two-fold nested model, *Stat. Prob. Lett.*, 5, 305–311.

Khuri, A. I. (1990). Exact tests for random models with unequal cell frequencies in the last stage, *J. Stat. Plan. Infer.*, 24, 177–193.

Khuri, A. I. (1992a). Tests concerning a nested mixed model with heteroscedastic random effects, *J. Stat. Plan. Infer.*, 30, 33–44.

Khuri, A. I. (1992b). Response surface models with random block effects, *Technometrics*, 34, 26–37.

Khuri, A. I. (1993). A note on Scheffé's confidence intervals, *Am. Stat.*, 47, 176–178.

Khuri, A. I. (1994). The probability of a negative linear combination of independent mean squares, *Biometrical J.*, 36, 899–910.

Khuri, A. I. (1995a). A test to detect inadequacy of Satterthwaite's approximation in balanced mixed models, *Statistics*, 27, 45–54.

Khuri, A. I. (1995b). A measure to evaluate the closeness of Satterthwaite's approximation, *Biometrical J.*, 37, 547–563.

Khuri, A. I. (1996a). Response surface models with mixed effects, *J. Qual. Technol.*, 28, 177–186.

Khuri, A. I. (1996b). Multiresponse surface methodology. In: *Handbook of Statistics*, Vol. 13, S. Ghosh and C. R. Rao (eds.). Elsevier Science B. V., Amsterdam, pp. 377–406.

Khuri, A. I. (1998). On unweighted sums of squares in unbalanced analysis of variance, *J. Stat. Plan. Infer.*, 74, 135–147.

Khuri, A. I. (1999). A necessary condition for a quadratic form to have a chi-squared distribution: an accessible proof, *Int. J. Math. Educ. Sci. Technol.*, 30, 335–339.

Khuri, A. I. (2002). Graphical evaluation of the adequacy of the method of unweighted means, *J. Appl. Stat.*, 29, 1107–1119.

Khuri, A. I. (2003). *Advanced Calculus with Applications in Statistics*, 2nd ed. Wiley, New York.

Khuri, A. I. (2006). Mixed response surface models with heterogeneous within-block error variances, *Technometrics*, 48, 206–218.

Khuri, A. I. and Cornell, J. A. (1996). *Response Surfaces*, 2nd ed. Dekker, New York.

Khuri, A. I. and Good, I. J. (1989). The parameterization of orthogonal matrices: A review mainly for statisticians, *S. Afr. Stat. J.*, 23, 231–250.

Khuri, A. I. and Littell, R. C. (1987). Exact tests for the main effects variance components in an unbalanced random two-way model, *Biometrics*, 43, 545–560.

Khuri, A. I. and Sahai, H. (1985). Variance components analysis: A selective literature survey, *Int. Stat. Rev.*, 53, 279–300.

Khuri, A. I., Mathew, T., and Sinha, B. K. (1998). *Statistical Tests for Mixed Linear Models*. Wiley, New York.

Kimball, A. W. (1951). On dependent tests of significance in the analysis of variance, *Ann. Math. Stat.*, 22, 600–602.

Koerts, J. and Abrahamse, A. P. J. (1969). *On the Theory and Application of the General Linear Model*. Rotterdam University Press, Rotterdam.

Kowalski, C. J. (1973). Non-normal bivariate distributions with normal marginals, *Am. Stat.*, 27, 103–105.

Kramer, C. Y. (1956). Extension of multiple range tests to group means with unequal numbers of replications, *Biometrics*, 12, 307–310.

Kruskal, W. H. (1961). The coordinate-free approach to Gauss–Markov estimation and its application to missing and extra observations. In: *4th Berkeley Symposium on Mathematical Statistics and Probability*, Vol. 1. Statistical Laboratory of the University of California, Berkeley, CA, pp. 435–451.

Kruskal, W. H. (1968). When are Gauss–Markov and least-squares estimators identical? A coordinate-free approach, *Ann. Math. Stat.*, 39, 70–75.

Kruskal, W. H. (1975). The geometry of generalized inverses, *J. Roy. Stat. Soc. Ser. B*, 37, 272–283.

Krutchkoff, R. G. (1988). One-way fixed effects analysis of variance when the error variances may be unequal, *J. Stat. Comput. Simul.*, 30, 259–271.

Kwon, J. H., Bélanger, J. M. R., and Paré, J. R. J. (2003). Optimization of microwave-assisted extraction (MAP) for ginseng components by response surface methodology, *J. Agric. Food Chem.*, 51, 1807–1810.

Laha, R. G. (1956). On the stochastic independence of two second-degree polynomial statistics in normally distributed variates, *Ann. Math. Stat.*, 27, 790–796.

Lancaster, P. (1969). *Theory of Matrices*. Academic Press, New York.

Lang, S. (1987). *Linear Algebra*, 3rd ed. Springer-Verlag, New York.

Lee, J. and Khuri, A. I. (2002). Comparison of confidence intervals on the among-group variance component for the unbalanced one-way random model, *Commun. Stat. Simul.*, 31, 35–47.

Lee, J., Khuri, A. I., Kim, K. W., and Lee, S. (2007). On the size of the *F*-test for the one-way random model with heterogeneous error variances, *J. Stat. Comput. Simul.*, 77, 443–455.

Levene, H. (1960). Robust tests for equality of variances. In: *Contributions to Probability and Statistics*, I. Olkin (ed.). Stanford University Press, Palo Alto, CA, pp. 278–292.

Lieberman, O. (1994). Saddlepoint approximation for the distribution of a ratio of quadratic forms in normal variables, *J. Am. Stat. Assoc.*, 89, 924–928.

Lindgren, B. W. (1976). *Statistical Theory*, 3rd ed. Macmillan, New York.

Lindsey, J. K. (1997). *Applying Generalized Linear Models*. Springer, New York.

Littell, R. C. (2002). Analysis of unbalanced mixed model data: A case study comparison of ANOVA versus REML/GLS, *J. Agric. Biol. Environ. Statist.*, 7, 472–490.

Littell, R. C., Milliken, G. A., Stroup, W. W., and Wolfinger, R. D. (1996). *SAS System for Mixed Models*. SAS Institute, Inc., Cary, NC.

Lowerre, J. M. (1983). An integral of the bivariate normal and an application, *Am. Stat.*, 37, 235–236.

Lugannani, R. and Rice, S. O. (1984). Distribution of the ratio of quadratic forms in normal variables—Numerical methods, *SIAM J. Sci. Stat. Comput.*, 5, 476–488.

Lukacs, E. (1942). A characterization of the normal distribution, *Ann. Math. Stat.*, 13, 91–93.

Lütkepohl, H. (1996). *Handbook of Matrices*. Wiley, New York.

Magnus, J. R. and Neudecker, H. (1988). *Matrix Differential Calculus with Applications in Statistics and Econometrics*. Wiley, New York.

Marcus, M. and Minc, H. (1964). *A Survey of Matrix Theory and Matrix Inequalities*. Dover, New York.

Marcus, M. and Minc, H. (1965). *Introduction to Linear Algebra*. Dover, New York.

Marsaglia, G. and Styan, G. P. H. (1974). Equalities and inequalities for ranks of matrices, *Linear and Multilinear Algebra*, 2, 269–292.

Matérn, B. (1949). Independence of non-negative quadratic forms in normally correlated variables, *Ann. Math. Stat.*, 20, 119–120.

May, W. G. (1970). *Linear Algebra*. Scott, Foresman and Company, Glenview, IL.

McCullagh, P. and Nelder, J. A. (1989). *Generalized Linear Models*, 2nd ed. Chapman & Hall, London, U.K.

McCulloch, C. E. and Searle, S. R. (1995). On an identity derived from unbiasedness in linear models, *Am. Stat.*, 49, 39–42.

McCulloch, C. E. and Searle, S. R. (2001). *Generalized, Linear, and Mixed Models*. Wiley, New York.

Melnick, E. L. and Tenenbein, A. (1982). Misspecifications of the normal distribution, *Am. Stat.*, 36, 372–373.

Miller, J. J. (1977). Asymptotic properties of maximum likelihood estimates in the mixed model of the analysis of variance, *Ann. Stat.*, 5, 746–762.

Miller, J. J. (1979). Maximum likelihood estimation of variance components—A Monte Carlo study, *J. Stat. Comput. Simul*, 8, 175–190.

Milliken, G. A. and Albohali, M. (1984). On necessary and sufficient conditions for ordinary least squares estimators to be best linear unbiased estimators, *Am. Stat.*, 38, 298–299.

Milliken, G. A. and Johnson, D. E. (1984). *Analysis of Messy Data*. Lifetime Learning Publications, Belmont, CA.

Montgomery, D. C. (2005). *Design and Analysis of Experiments*, 6th ed. Wiley, New York.

Mood, A. M., Grayhill, F. A., and Boes, D. C. (1973). *Introduction to the Theory of Statistics*, 3rd ed. McGraw-Hill, New York.

Moore, E. H. (1920). On the reciprocal of the general algebraic matrix, *Bull., Am. Math. Soc.*, 26, 394–395.

Morrison, D. F. (1976). *Multivariate Statistical Methods*, 2nd ed. McGraw-Hill, New York.

Muirhead, R. J. (1982). *Aspects of Multivariate Statistical Theory*. Wiley, New York.

Myers, R. H. and Howe, R. B. (1971). On alternative approximate *F* tests for hypotheses involving variance components, *Biometrika*, 58, 393–396.

Myers, R. H., Khuri, A. I., and Carter, W. H. (1989). Response surface methodology: 1966–1988, *Technometrics*, 31, 137–157.

Myers, R. H., Montgomery, D. C., and Vining, G. G. (2002). *Generalized Linear Models*. Wiley, New York.

Nagarsenker, P. B. (1984). On Bartlett's test for homogeneity of variances, *Biometrika*, 71, 405–407.

Nel, D. G. (1980). On matrix differentiation in statistics, *S. Afr. Stat. J.*, 14, 137–193.

Nelder, J. A. and Wedderburn, R. W. M. (1972). Generalized linear models, *J. Roy. Stat. Soc. Ser. A*, 135, 370–384.

Neudecker, H. (1969). Some theorems on matrix differentiation with special reference to Kronecker products, *J. Am. Stat. Assoc.*, 64, 953–963.

Newcomb, R. W. (1960). On the simultaneous diagonalization of two semidefinite matrices, *Quart. Appl. Math.*, 19, 144–146.

Neyman, J. (1934). On the two different aspects of the representative method: The method of stratified sampling and the method of purposive selection, *J. Roy. Stat. Soc.*, 97, 558–625.

Nolan, D. and Speed, T. (2000). *Stat Labs. Mathematical Statistics through Applications*. Springer, Berlin.

Norden, R. H. (1972). A survey of maximum likelihood estimation, *Int. Stat. Rev.*, 40, 329–354.

O'Brien, R. G. (1978). Robust techniques for testing heterogeneity of variance effects in factorial designs, *Psychometrika*, 43, 327–344.

O'Brien, R. G. (1979). A general ANOVA method for robust tests of additive models for variances, *J. Am. Stat. Assoc.*, 74, 877–880.

Ogawa, J. (1950). On the independence of quadratic forms in a non-central normal system, *Osaka Math. J.*, 2, 151–159.

Olkin, I. (1990). Interface between statistics and linear algebra. In: *Matrix Theory and Applications*, Vol. 40, C. R. Johnson (ed.). *Am. Math. Soc.*, Providence, RI, pp. 233–256.

Ott, R. L. and Longnecker, M. T. (2004). *A First Course in Statistical Methods.* Brooks/Cole, Belmont, CA.

Patterson, H. D. and Thompson, R. (1971). Recovery of inter-block information when block sizes are unequal, *Biometrika*, 58, 545–554.

Paul, S. and Khuri, A. I. (2000). Modified ridge analyses under nonstandard conditions, *Commun. Stat. Theor. Meth.*, 29, 2181–2200.

Penrose, R. A. (1955). A generalized inverse for matrices, *Proc. Camb. Phil. Soc.*, 51, 406–413.

Piepho, H. P. and Emrich, K. (2005). Simultaneous confidence intervals for two estimable functions and their ratio under a linear model, *Am. Stat.*, 59, 292–300.

Pierce, D. A. and Dykstra, R. L. (1969). Independence and the normal distribution, *Am. Stat.*, 23, 39.

Plackett, R. L. (1972). Studies in the history of probability and statistics. XXIX The discovery of the method of least squares, *Biometrika*, 59, 239–251.

Price, G. B. (1947). Some identities in the theory of determinants, *Am. Math. Month.*, 54, 75–90.

Rankin, N. O. (1974). The harmonic mean method for one-way and two-way analyses of variance, *Biometrika*, 61, 117–122.

Rao, C. R. (1952). *Advanced Statistical Methods in Biometric Research.* Wiley, New York.

Rao, C. R. (1962). A note on a generalized inverse of a matrix with applications to problems in mathematical statistics, *J. Roy. Stat. Soc. Ser. B*, 24, 152–158.

Rao, C. R. (1966). Generalized inverse for matrices and its applications in mathematical statistics. In: *Research Papers in Statistics: Festschrift for J. Neyman*, F. N. David (ed.). Wiley, London, pp. 263–279.

Rao, C. R. (1973a). *Linear Statistical Inference and its Applications*, 2nd ed. Wiley New York.

Rao, C. R. (1973b). Unified theory of least squares, *Commun. Stat. Theor. Meth.*, 1, 1–8.

Rao, P. S. R. S., Kaplan, J., and Cochran, W. G. (1981). Estimators for the one-way random effects model with unequal error variances, *J. Am. Stat. Assoc.*, 76, 89–97.

Reid, J. G. and Driscoll, M. F. (1988). An accessible proof of Craig's theorem in the noncentral case, *Am. Stat.*, 42, 139–142.

Richert, S. H., Morr, C. V., and Cooney, C. M. (1974). Effect of heat and other factors upon foaming properties of whey protein concentrates, *J. Food Sci.*, 39, 42–48.

Robinson, D. L. (1987). Estimation and use of variance components, *The Statistician*, 36, 3–14.

Rogers, G. S. (1980). *Matrix Derivatives*. Dekker, New York.

Roy, S. N., Gnanadesikan, R., and Srivastava, J. N. (1971). *Analysis and Design of Certain Quantitative Multiresponse Experiments*. Pergamon Press, Oxford.

Rudan, J. W. and Searle, S. R. (1971). Large sample variances of maximum likelihood estimators of variance components in the three-way nested classification, random model, with unbalanced data, *Biometrics*, 27, 1087–1091.

Sahai, H. and Anderson, R. L. (1973). Confidence regions for variance ratios of random models for balanced data, *J. Am. Stat. Assoc.*, 68, 951–952.

SAS Institute, Inc. (1997). *SAS/STAT Software: Changes and Enhancements through Release 6.12*. Author, Cary, NC.

SAS Institute, Inc. (1999). *SAS/IML User's Guide*, Version 8. Author, Cary, NC.

SAS Institute, Inc. (2000). *Online Doc*, Version 8. Author, Cary, NC.

Satterthwaite, F. E. (1941). Synthesis of variance, *Psychometrika*, 6, 309–316.

Satterthwaite, F. E. (1946). An approximate distribution of estimates of variance components, *Biometrics Bull.*, 2, 110–114.

Schey, H. M. (1985). A geometric description of orthogonal contrasts in one-way analysis of variance, *Am. Stat.*, 39, 104–106.

Seal, H. L. (1967). Studies in the history of probability and statistics. XV The historical development of the Gauss linear model, *Biometrika*, 54, 1–24.

Searle, S. R. (1968). Another look at Henderson's methods of estimating variance components, *Biometrics*, 24, 749–778.

Searle, S. R. (1970). Large sample variances of maximum likelihood estimators of variance components using unbalanced data, *Biometrics*, 26, 505–524.

Searle, S. R. (1971). *Linear Models*. Wiley, New York.

Searle, S. R. (1982). *Matrix Algebra Useful for Statistics*. Wiley, New York.

Searle, S. R. (1987). *Linear Models for Unbalanced Data*. Wiley, New York.

Searle, S. R. (1994). Analysis of variance computing package output for unbalanced data from fixed-effects models with nested factors, *Am. Stat.*, 48, 148–153.

Searle, S. R. (1995). An overview of variance component estimation, *Metrika*, 42, 215–230.

Searle, S. R. (1999). The infusion of matrices into statistics. Technical Report BU-1444-M, Department of Biometrics and Statistical Science, Cornell University, Ithaca, NY.

Searle, S. R., Speed, F. M., and Milliken, G. A. (1980). Population marginal means in the linear model: An alternative to least squares means, *Am. Stat.*, 34, 216–221.

Searle, S. R., Speed, F. M., and Henderson, H. V. (1981). Some computational and model equivalences in analyses of variance of unequal-subclass-numbers data, *Am. Stat.*, 35, 16–33.

Searle, S. R., Casella, G., and McCulloch, C. E. (1992). *Variance Components*. Wiley, New York.

Seber, G. A. F. (1984). *Multivariate Observations*. Wiley, New York.

Seely, J. (1977). Estimability and linear hypotheses, *Am. Stat.*, 31, 121–123.

Seely, J. F. and El-Bassiouni, Y. (1983). Applying Wald's variance component test, *Ann. Stat.*, 11, 197–201.

Self, S. G. and Liang, K. Y. (1987). Asymptotic properties of maximum likelihood estimators and likelihood ratio tests under nonstandard conditions, *J. Am. Stat. Assoc.*, 82, 605–610.

Shanbhag, D. N. (1966). On the independence of quadratic forms, *J. Roy. Stat. Soc. Ser. B*, 28, 582–583.

Šidák, Z. (1967). Rectangular confidence regions for the means of multivariate normal distributions, *J. Am. Stat. Assoc.*, 62, 626–633.

Simonnard, M. (1966). *Linear Programming*. Prentice-Hall, Englewood Cliffs, NJ.

Smith, D. E. (1958). *History of Mathematics*, Vol. 1. Dover, New York.

Smith, P. J. and Choi, S. C. (1982). Simple tests to compare two dependent regression lines, *Technometrics*, 24, 123–126.

Smith, T. and Peddada, S. D. (1998). Analysis of fixed effects linear models under heteroscedastic errors, *Stat. Prob. Lett.*, 37, 399–408.

Snedecor, G. W. (1934). *Analysis of Variance and Covariance*. Collegiate Press, Ames, IA.

Snedecor, G. W. and Cochran, W. G. (1980). *Statistical Methods*, 7th ed. Iowa State University Press, Ames, IA.

Speed, F. M. and Hocking, R. R. (1976). The use of the $R()$-notation with unbalanced data, *Am. Stat.*, 30, 30–33.

Speed, F. M. and Monlezun, C. J. (1979). Exact F tests for the method of unweighted means in a 2^k experiment, *Am. Stat.*, 33, 15–18.

Speed, F. M., Hocking, R. R., and Hackney, O. P. (1978). Method of analysis of linear models with unbalanced data, *J. Am. Stat. Assoc.*, 73, 105–112.

Spjøtvoll, E. (1968). Confidence intervals and tests for variance ratios in unbalanced variance components models, *Rev. Int. Stat. Inst.*, 36, 37–42.

Spjøtvoll, E. (1972). Joint confidence intervals for all linear functions of means in the one-way layout with unknown group variances, *Biometrika*, 59, 683–685.

Stablein, D. M., Carter, Jr., W. H., and Wampler, G. L. (1983). Confidence regions for constrained optima in response-surface experiments, *Biometrics*, 39, 759–763.

Stigler, S. M. (1981). Gauss and the invention of least squares, *Ann. Stat.*, 9, 465–474.

Stigler, S. M. (1984). Kruskal's proof of the joint distribution of $\bar{X}$ and s^2, *Am. Stat.*, 38, 134–135.

Stigler, S. M. (1986). *The History of Statistics*. The Balknap Press of Harvard University Press, Cambridge, MA.

Stram, D. O. and Lee, J. W. (1994). Variance components testing in the longitudinal mixed effects model, *Biometrics*, 50, 1171–1177.

Styan, G. P. H. (1970). Notes on the distribution of quadratic forms in singular normal variables, *Biometrika*, 57, 567–572.

Swallow, W. H. and Monahan, J. F. (1984). Monte-Carlo comparison of ANOVA, MIVQUE, REML, and ML estimators of variance components, *Technometrics*, 26, 47–57.

Tamhane, A. C. (1979). A comparison of procedures for multiple comparisons of means with unequal variances, *J. Am. Stat. Assoc.*, 74, 471–480.

Thibaudeau, Y. and Styan, G. P. H. (1985). Bounds for Chakrabarti's measure of imbalance in experimental design. In: *Proceedings of the First International Tampere Seminar on Linear Statistical Models and Their Applications*, T. Pukkila and S. Puntanen (eds.). University of Tampere, Tampere, Finland, pp. 323–347.

Thomas, J. D. and Hultquist, R. A. (1978). Interval estimation for the unbalanced case of the one-way random effects model, *Ann. Stat.*, 6, 582–587.

Thomsen, I. (1975). Testing hypotheses in unbalanced variance components models for two-way layouts, *Ann. Stat.*, 3, 257–265.

Tietjen, G. L. (1974). Exact and approximate tests for unbalanced random effects designs, *Biometrics*, 30, 573–581.

Tong, Y. L. (1990). *The Multivariate Normal Distribution*. Springer-Verlag, New York.

Trenkler, G. (2004). An extension of Lagrange's identity to matrices, *Int. J. Math. Educ. Sci. Technol.*, 35, 245–315.

Turnbull, H. W. and Aitken, A. C. (1932). *An Introduction to the Theory of Canonical Matrices*. Blackie & Sons, London, U.K.

Vangel, M. G. and Rukhin, A. L. (1999). Maximum likelihood analysis for heteroscedastic one-way random effects ANOVA in interlaboratory studies, *Biometrics*, 55, 129–136.

Verbeke, G. (1997). Linear mixed models for longitudinal data. In: *Linear Mixed Models in Practice*, G. Verbeke and G. Molenberghs (eds.). Springer, New York, pp. 63–153.

Verbeke, G. and Lesaffre, E. (1996). A linear mixed-effects model with heterogeneity in the random-effects population, *J. Am. Stat. Assoc.*, 91, 217–221.

Verbeke, G. and Molenberghs, G. (eds.) (1997). *Linear Mixed Models in Practice*. Springer, New York.

Verbeke, G. and Molenberghs, G. (2000). *Linear Mixed Models for Longitudinal Data*. Springer, New York.

Verbeke, G. and Molenberghs, G. (2003). The use of score tests for inference on variance components, *Biometrics*, 59, 254–262.

Vining, G. G. and Khuri, A. I. (1991). A modified ridge analysis for experiments with attribute data, Technical Report, Department of Statistics, University of Florida, Gainesville, FL.

Wald, A. (1940). A note on the analysis of variance with unequal cell frequencies, *Ann. Math. Stat.*, 11, 96–100.

Wald, A. (1941). On the analysis of variance in case of multiple classifications with unequal cell frequencies, *Ann. Math. Stat.*, 12, 346–350.

Walpole, R. E. and Myers, R. H. (1985). *Probability and Statistics for Engineers and Scientists*, 3rd ed. Macmillan, New York.

Wasan, M. T. (1970). *Parameter Estimation*. McGraw-Hill, New York.

Welch, B. L. (1947). The generalization of Student's problem when several different population variances are involved, *Biometrika*, 34, 28–35.

Wichura, M. J. (2006). *The Coordinate-Free Approach to Linear Models*. Cambridge University Press, New York.

Wicksell, S. D. (1930). Remarks on regression, *Ann. Math. Stat.*, 1, 3–13.

Williams, J. S. (1962). A confidence interval for variance components, *Biometrika*, 49, 278–281.

Wimmer, G. and Witkovsky, V. (2003). Between group variance component interval estimation for the unbalanced heteroscedastic one-way random effects model, *J. Stat. Comput. Simul.*, 73, 333–346.

Wolkowicz, H. (1994). Solution to Problem 93–17, *SIAM Rev.*, 36, 658.

Wolkowicz, H. and Styan, G. P. H. (1980). Bounds for eigenvalues using traces, *Linear Algebra Appl.*, 29, 471–506.

Yates, F. (1934). The analysis of multiple classifications with unequal numbers in the different classes, *J. Am. Stat. Assoc.*, 29, 51–66.

Zellner, A. (1962). An efficient method of estimating seemingly unrelated regressions and tests for aggregation bias, *J. Am. Stat. Assoc.*, 57, 348–368.

Zemanian, A. H. (1987). *Generalized Integral Transformations*. Dover, New York.

Zyskind, G. (1962). On structure, relation, Σ, and expectation of mean squares, *Sankhyā Ser. A*, 24, 115–148.

Index

M

N